Complex Multiplication and Lifting Problems

Mathematical
Surveys
and
Monographs

Volume 195

Complex Multiplication and Lifting Problems

Ching-Li Chai
Brian Conrad
Frans Oort

American Mathematical Society
Providence, Rhode Island

2010 *Mathematics Subject Classification.* Primary 11G15, 14K02, 14L05, 14K15, 14D15.

For additional information and updates on this book, visit
www.ams.org/bookpages/surv-195

Library of Congress Cataloging-in-Publication Data

Chai, Ching-Li, author.
Complex multiplication and lifting problems / Ching-Li Chai, Brian Conrad, Frans Oort.
pages cm — (Mathematical surveys and monographs ; volume 195)
Includes bibliographical references and index.
ISBN 978-1-4704-1014-8 (alk. paper)
1. Multiplication, Complex. 2. Abelian varieties. I. Conrad, Brian, 1970– author. II. Oort, Frans, 1935– author. III. Title.

QA564.C44 2014
516.3′53—dc23

2013036892

∞ The paper used in this book is acid-free and falls within the guidelines
established to ensure permanence and durability.
Visit the AMS home page at `http://www.ams.org/`

10 9 8 7 6 5 4 3 2 1 19 18 17 16 15 14

This book is dedicated to

John Tate

for what he taught us, and for inspiring us

Contents

Preface

During the *Workshop on Abelian Varieties* in Amsterdam in May 2006, the three authors of this book formulated two refined versions of a problem concerning lifting into characteristic 0 for abelian varieties over a finite field. These problems address the phenomenon of *CM lifting*: the lift into characteristic 0 is required to be a CM abelian variety (in the sense defined in 1.3.8.1). The precise formulations appear at the end of Chapter 1 (see 1.8.5), as problems (I) and (IN).

Abelian surface counterexamples to (IN) were found at that time; see 2.3.1–2.3.3, and see 4.1.2 for a more thorough analysis. To our surprise, the same counterexamples (typical among *toy models* as defined in 4.1.3) play a crucial role in the general solution to problems (I) and (IN). This book is the story of our adventure guided by CM lifting problems.

Ching-Li Chai thanks Hsiao-Ling for her love and support during all these years. He also thanks Utrecht University for hospitality during many visits, including the May 2006 Spring School on Abelian Varieties which concluded with the workshop in Amsterdam. Support by NSF grants DMS 0400482, DMS 0901163, and DMS120027 is gratefully acknowledged.

Brian Conrad thanks the many participants in the "CM seminar" at the University of Michigan for their enthusiasm on the topic of complex multiplication, as well as Columbia University for its hospitality during a sabbatical visit, and gratefully acknowledges support by NSF grants DMS 0093542, DMS 0917686, and DMS 1100784.

Frans Oort thanks the University of Pennsylvania for hospitality and stimulating environment during several visits.

We are also grateful to Burcu Baran, Bas Edixhoven, Ofer Gabber, Johan de Jong, Bill Messing, Ben Moonen, James Parson, René Schoof, and Jonathan Wise for insightful and memorable discussions.

Introduction

I restricted myself to characteristic zero: for a short time, the quantum jump to $p \neq 0$ was beyond the range ... but it did not take me too long to make this jump.

— Oscar Zariski

The arithmetic of abelian varieties with complex multiplication over a number field is fascinating. However this will not be our focus. We study the theory of complex multiplication in mixed characteristic.

Abelian varieties over finite fields. In 1940 Deuring showed that an elliptic curve over a finite field can have an endomorphism algebra of rank 4 [**33**, §2.10]. For an elliptic curve in characteristic zero with an endomorphism algebra of rank 2 (rather than rank 1, as in the "generic" case), the j-invariant is called a *singular j-invariant.* For this reason elliptic curves with even more endomorphisms, in positive characteristic, are called *supersingular.*[1]

Mumford observed as a consequence of results of Deuring that for any elliptic curves E_1 and E_2 over a finite field κ of characteristic $p > 0$ and any prime $\ell \neq p$, the natural map

$$\mathbb{Z}_\ell \otimes_{\mathbb{Z}} \mathrm{Hom}(E_1, E_2) \longrightarrow \mathrm{Hom}_{\mathbb{Z}_\ell[\mathrm{Gal}(\overline{\kappa}/\kappa)]}(T_\ell(E_1), T_\ell(E_2))$$

(where on the left side we consider only homomorphisms "defined over κ") is an isomorphism [**118**, §1]. The interested reader might find it an instructive exercise to reconstruct this (unpublished) proof by Mumford. Tate proved in [**118**] that the analogous result holds for all abelian varieties over a finite field and he also incorporated the case $\ell = p$ by using p-divisible groups. He generalized this result into his influential conjecture [**117**]:

> *An ℓ-adic cohomology class[2] that is fixed under the Galois group should be a $\mathbb{Q}_\ell$-linear combination of fundamental classes of algebraic cycles when the ground field is finitely generated over its prime field.*

Honda and Tate gave a classification of isogeny classes of simple abelian varieties A over a finite field κ (see [**50**] and [**121**]), and Tate refined this by describing

[1] Of course, a supersingular elliptic curve isn't singular. A purist perhaps would like to say "an elliptic curve with supersingular j-value". However we will adopt the generally used terminology "supersingular elliptic curve" instead.

[2] The prime number ℓ is assumed to be invertible in the base field.

the structure of the endomorphism algebra $\mathrm{End}^0(A)$ (working in the isogeny category over κ) in terms of the Weil q-integer of A, with $q = \#\kappa$; see [**121**, Thm. 1]. It follows from Tate's work (see 1.6.2.5) that an abelian variety A over a finite field κ admits sufficiently many complex multiplications in the sense that its endomorphism algebra $\mathrm{End}^0(A)$ contains a CM subalgebra[3] L of rank $2\dim(A)$. We will call such an abelian variety (in any characteristic) a CM *abelian variety* and the embedding $L \to \mathrm{End}^0(A)$ a CM *structure* on A.

Grothendieck showed that over any algebraically closed field K, an abelian variety that admits sufficiently many complex multiplications is isogenous to an abelian variety defined over a finite extension of the prime field [**89**]. This was previously known in characteristic zero (by Shimura and Taniyama), and in that case there is a number field $K' \subset K$ such that the abelian variety can be defined over K' (in the sense of 1.7.1). However in positive characteristic such abelian varieties can fail to be defined over a finite subfield of K; examples exist in every dimension > 1 (see Example 1.7.1.2).

Abelian varieties in mixed characteristic. In characteristic zero, an abelian variety A gives a representation of the endomorphism algebra $D = \mathrm{End}^0(A)$ on the Lie algebra $\mathrm{Lie}(A)$ of A. If A has complex multiplication by a CM algebra L of degree $2\dim(A)$ then the isomorphism class of the representation of L on $\mathrm{Lie}(A)$ is called the CM *type* of the CM structure $L \hookrightarrow \mathrm{End}^0(A)$ on A (see Lemma 1.5.2 and Definition 1.5.2.1).

As we noted above, every abelian variety over a finite field is a CM abelian variety. Thus, it is natural to ask whether every abelian variety over a finite field can be "CM lifted" to characteristic zero (in various senses that are made precise in 1.8.5). One of the obstacles[4] in this question is that in characteristic zero there is the notion of CM type that is invariant under isogenies, whereas in positive characteristic whatever can be defined in an analogous way is *not* invariant under isogenies. For this reason we will use the terminology "CM type" only in characteristic zero.

For instance, the action of the endomorphism ring $R = \mathrm{End}(A_0)$ of an abelian variety A_0 on the Lie algebra of A_0 in characteristic $p > 0$ defines a representation of R/pR on $\mathrm{Lie}(A_0)$. Given an isogeny $f : A_0 \to B_0$ we get an identification $\mathrm{End}^0(A_0) = \mathrm{End}^0(B_0)$ of endomorphism algebras, but even if $\mathrm{End}(A_0) = \mathrm{End}(B_0)$ under this identification, the representations of this common endomorphism ring on $\mathrm{Lie}(A_0)$ and $\mathrm{Lie}(B_0)$ may well be non-isomorphic since $\mathrm{Lie}(f)$ may not be an isomorphism. Moreover, if we have a lifting A of A_0 over a local domain of characteristic 0, in general the inclusion $\mathrm{End}(A) \subset \mathrm{End}(A_0)$ is not an equality. If the inclusion $\mathrm{End}^0(A) \subset \mathrm{End}^0(A_0)$ is an equality then the character of the representation of $\mathrm{End}(A_0)$ on $\mathrm{Lie}(A_0)$ is the reduction of the character of the representation of $\mathrm{End}(A)$ on $\mathrm{Lie}(A)$. This relation can be viewed as an obstruction to the existence of CM lifting with the full ring of integers of a CM algebra operating on the lift; see 4.1.2, especially 4.1.2.3–4.1.2.4, for an illustration.

In the case when $\mathrm{End}(A_0)$ contains the ring of integers $\mathcal{O}_L$ of a CM algebra $L \subset \mathrm{End}^0(A_0)$ with $[L : \mathbb{Q}] = 2\dim(A_0)$, the representation of $\mathcal{O}_L/p\mathcal{O}_L$ on $\mathrm{Lie}(A_0)$ turns out to be quite useful, despite the fact that it is not an isogeny invariant. Its class in a suitable K-group will be called the *Lie type* of $(A_0, \mathcal{O}_L \hookrightarrow \mathrm{End}(A_0))$.

---

[3]A CM algebra is a finite product of CM fields; see Definition 1.3.3.1.

[4]surely also part of the attraction

The above discrepancy between the theories in characteristic zero and characteristic $p > 0$ is the basic phenomenon underlying this entire book. Before discussing its content, we recall the following theorem of Honda and Tate ([**50**, §2, Thm. 1] and [**121**, Thm. 2]).

> *For an abelian variety A_0 over a finite field κ there is a finite extension κ' of κ and an isogeny $(A_0)_{\kappa'} \to B_0$ such that B_0 admits a* CM *lifting over a local domain of characteristic zero with residue field κ'.*

This result has been used in the study of Shimura varieties, for settings where the ground field is an algebraic closure of $\mathbb{F}_p$ and isogeny classes (of structured abelian varieties) are the objects of interest; see [**135**]. Our starting point comes from the following questions which focus on controlling ground field extensions and isogenies.

> *For an abelian variety A_0 over a finite field κ, to ensure the existence of a* CM *lifting over a local domain with characteristic zero and residue field κ' of finite degree over κ,*
> (a) *may we choose $\kappa' = \kappa$?*
> (b) *is an isogeny $(A_0)_{\kappa'} \to B_0$ necessary?*

These questions are formulated in various precise forms in 1.8.

An isogeny is necessary. Question (b) was answered in 1992 (see [**93**]) as follows.

> *There exist (many) abelian varieties over $\overline{\mathbb{F}}_p$ that do not admit any* CM *lifting to characteristic zero.*

The main point of [**93**] is that a CM liftable abelian variety over $\overline{\mathbb{F}}_p$ can be defined over a small finite field. This idea is further pursued in Chapter 3, where the *size*, or more accurately the *minima*[5] of the size, of all possible *fields of definition* of the p-divisible group of a given abelian variety over $\overline{\mathbb{F}}_p$ is turned into an obstruction for the existence of a CM lifting to characteristic 0. This is used to show (in 3.8.3) that in "most" isogeny classes of non-ordinary abelian varieties of dimension $\geqslant 2$ over finite fields there is a member that has no CM lift to characteristic 0. (In dimension 1 a CM lift to characteristic 0 always exists, over the valuation ring of the minimal possible p-adic field, by Deuring Lifting Theorem; see 1.7.4.6.) We also provide effectively computable examples of abelian varieties over explicit finite fields such that there is no CM lift to characteristic 0.

A field extension might be necessary—depending on what you ask. Bearing in mind the necessity to modify a given abelian variety over a finite field to guarantee the existence of a CM lifting, we rephrase question (a) in a more precise version (a)′ below.

> (a)′ *Given an abelian variety A_0 over a finite field κ of characteristic p, is it necessary to extend scalars to a strictly larger finite field $\kappa' \supset \kappa$ (depending on A_0) to ensure the existence of a κ'-rational isogeny $(A_0)_{\kappa'} \to B_0$ such that B_0 admits a* CM *lifting over a characteristic* 0 *local domain R with residue field κ'?*

It turns out there are two quite different answers to question (a)′, depending on whether one requires the local domain R of characteristic 0 to be normal. The subtle distinction between using normal or general local domains for the lifting

[5]The size of a finite field κ_1 is *smaller* than the size of a finite field κ_2 if κ_1 is isomorphic to a subfield of κ_2, or equivalently if $\#\kappa_1 \mid \#\kappa_2$. Among the sizes of a family of finite fields there may not be a unique minimal element.

went unnoticed for a long time. Once this distinction came in focus, answers to the resulting questions became available.

If we ask for a CM lifting over a *normal* domain up to isogeny, in general a base field extension *before* modification by an isogeny is necessary. This is explained in 2.1.2, where we formulate the "residual reflex obstruction", the idea for which goes as follows. Over an algebraically closed field K of characteristic zero, we know that a simple CM abelian variety B with K-valued CM type Φ (for the action of a CM field L) is defined over a number field in K containing the reflex field $E(\Phi)$ of Φ. Suppose that for every K-valued CM type Φ of L, the residue field of $E(\Phi)$ at any prime above p is not contained in the finite field κ with which we began in question (a). In such cases, for every CM structure $L \to \mathrm{End}^0(A_0)$ on A_0 and any abelian variety B_0 over κ which is κ-isogenous to A_0, there is no L-linear CM lifting of B_0 over a *normal* local domain R of characteristic zero with residue field κ.[6] In 2.3.1–2.3.3 we give such an example, a supersingular abelian surface A_0 over $\mathbb{F}_{p^2}$ with $\mathrm{End}(A_0) = \mathbb{Z}[\zeta_5]$ for any $p \equiv \pm 2 \pmod 5$. A much broader class of examples is given in 2.3.5, consisting of absolutely simple abelian varieties (with arbitrarily large dimension) over $\mathbb{F}_p$ for infinitely many p.

Note that passing to the normalization of a complete local noetherian domain generally enlarges the residue field. Hence, if we drop the condition that the mixed characteristic local domain R be normal then the obstruction in the preceding consideration dissolves. And in fact we were put on the right track by mathematics itself. The phenomenon is best illustrated in the example in 4.1.2, which is the same as the example in 2.3.1 already mentioned: an abelian surface C_0 over $\mathbb{F}_{p^2}$ with CM order $\mathbb{Z}[\zeta_5]$ that, even up to isogeny, is not CM liftable to a *normal* local domain of characteristic zero. On the other hand, this abelian surface C_0 is CM liftable to an abelian scheme C over a mixed characteristic non-normal local domain of characteristic zero, though the maximal subring of $\mathbb{Z}[\zeta_5]$ whose action lifts to C is non-Dedekind locally at p; see 4.1.2.[7] This example is easy to construct, and the proof of the existence of a CM lifting, possibly after applying an $\mathbb{F}_{p^2}$-rational isogeny, is not difficult either.

In Chapter 4 we show that the general question of existence of a CM lifting after an appropriate isogeny can be reduced to the same question for (a mild generalization of) the example in 4.1.2, enabling us to prove:

> *every abelian variety* A_0 *over a finite field* κ *admits an isogeny* $A_0 \to B_0$ *over* κ *such that* B_0 *admits a* CM *lifting to a mixed characteristic local domain with residue field* κ.

There are refined lifting problems, such as specifying at the beginning which CM structure on A_0 is to be lifted, or even what its CM type should be on a geometric fiber in characteristic 0. These matters will also be addressed.

[6]The source of obstructions is that the *base field* κ might be too small to contain at least one characteristic p residue field of the reflex field $E(\Phi)$ for at least one CM type Φ on L. Thus, the field of definition of the generic fiber of the hypothetical lift may be too big. Likewise, an obstruction for question (b) is that the field of definition of the p-divisible group $A_0[p^\infty]$ may be too big (in a sense that is made precise in 3.8.3 and illustrated in 3.8.4–3.8.5).

[7]No modification by isogeny is necessary in this example, but the universal deformation for C_0 with its $\mathbb{Z}[\zeta_5]$-action is a *non-algebraizable* formal abelian scheme over $W(\mathbb{F}_{p^2})$.

Our basic method is to "localize" various CM lifting problems to the corresponding problems for p-divisible groups. Although global properties of abelian varieties are often lost in this localization process, the non-rigid nature of p-divisible groups can be an advantage. In Chapter 3 the size of fields of definition of a p-divisible group in characteristic p appears as an obstruction to the existence of CM lifting. The reduction steps in Chapter 4 rely on a classification and descent of CM p-divisible groups in characteristic p with the help of their Lie types (see 4.2.2, 4.4.2). In addition, the "Serre tensor construction" is applied to p-divisible groups, both in characteristic p and in mixed characteristic $(0, p)$; see 1.7.4 and 4.3.1 for this general construction.

Survey of the contents. In **Chapter 1** we start with a survey of general facts about CM abelian varieties and their endomorphism algebras. In particular, we discuss the deformation theory of abelian varieties and p-divisible groups, and we review results in Honda-Tate theory that describe isogeny classes and endomorphism algebras of abelian varieties over a finite field in terms of Weil integers. We conclude by formulating various CM lifting questions in 1.8. These are studied in the following chapters. We will see that the questions can be answered with some precision.

In **Chapter 2** we formulate and study the "residual reflex condition". Using this condition we construct several examples of abelian varieties over finite fields κ such that, even after applying a κ-isogeny, there is no CM lifting to a *normal local domain* with characteristic zero and residue field of finite degree over κ; see 2.3. It is remarkable that many such examples exist, but we do not know whether we have characterized all possible examples; see 2.3.7.

We then study algebraic Hecke characters and review part of the theory of complex multiplication due to Shimura and Taniyama. Using the relationship between algebraic Hecke characters for a CM field L and CM abelian varieties with CM by L (the precise statement of which we review and prove), we use global methods to show that the residual reflex condition is the *only* obstruction to the existence of CM lifting up to isogeny over a *normal* local domain of characteristic zero. We also give another proof by local methods (such as p-adic Hodge theory).

In **Chapter 3** we take up methods described in [**93**]. In that paper classical CM theory in characteristic zero was used. Here we use p-divisible groups instead of abelian varieties and show that the size of fields of definition of a p-divisible group in characteristic p is a non-trivial obstruction to the existence of a CM lifting. In 3.3 we study the notion of isogeny for p-divisible groups over a base scheme (including its relation with duality). We show, in one case of the CM lifting problem left open in [**93**, Question C], that an isogeny is necessary. Our methods also provide effectively computed examples. Some facts about CM p-divisible groups explained in 3.7 are used in 3.8 to get an upper bound of a field of definition for the closed fiber of a CM p-divisible group.

In Appendix 3.9, we use the construction (in 3.7) of a p-divisible group with any given p-adic CM type over the reflex field to produce a semisimple abelian crystalline p-adic representation of the local Galois group such that its restriction to the inertia group is "algebraic" with algebraic part that we may prescribe arbitrarily in accordance with some necessary conditions (see 3.9.4 and 3.9.8).

In **Chapter 4** we show CM liftability after an isogeny over the finite ground field (lifting over a characteristic zero local domain that need not be normal). That is,

> *every* CM *structure* $(A_0, L \to \mathrm{End}^0(A_0))$ *over a finite field* κ *has an isogeny over* κ *to a* CM *structure* $(B_0, L \to \mathrm{End}^0(B_0))$ *that admits a* CM *lifting;*

(see 4.1.1). This statement is immediately reduced to the case when L is a CM field (not just a CM algebra) and the whole ring $\mathcal{O}_L$ of integers of L operates on A_0, which we assume.

Our motivation comes from the proof in 4.1.2 (using an algebraization argument at the end of 4.1.3) that the counterexample in 2.3.1 to CM lifting over a *normal* local domain satisfies this property. In general, after an easy reduction to the isotypic case, we apply the Serre-Tate deformation theorem to localize the problem at p-adic places v of the maximal totally real subfield L^+ of a CM field $L \subseteq \mathrm{End}^0(A_0)$ of degree $2\dim(A_0)$. This reduces the existence of a CM lifting for the abelian variety A_0 to a corresponding problem for the CM p-divisible group $A_0[v^\infty]$ attached to v.[8]

We formulate several properties of v with respect to the CM field L; any one of them ensures the existence of a CM lifting of $A_0[v^\infty]_{\overline{\kappa}}$ after applying a κ-isogeny to $A_0[v^\infty]$ (see 4.1.6, 4.1.7, and 4.5.7). These properties involve the ramification and residue fields of L and L^+ relative to v. If v violates all of these properties then we call it *bad* (with respect to L/L^+ and κ). Let $L_v := L \otimes_{L^+} L^+_v$. After applying a preliminary κ-isogeny to arrange that $\mathcal{O}_L \subset \mathrm{End}(A_0)$, for v that are not bad we apply an $\mathcal{O}_L$-linear κ-isogeny to arrange that the Lie type of the $\mathcal{O}_{L,v}$-factor of $\mathrm{Lie}(A_0)$ (i.e., its class in a certain K-group of $(\mathcal{O}_{L,v}/(p)) \otimes \kappa$-modules) is "self-dual". Under the self-duality condition (defined in 4.4.3) we produce an $\mathcal{O}_{L,v}$-linear CM lifting of $A_0[v^\infty]_{\overline{\kappa}}$ by specializing a suitable $\mathcal{O}_{L,v}$-linear CM v-divisible group in mixed characteristic; see 4.4.6. We use an argument with deformation rings to eliminate the intervention of $\overline{\kappa}$: if every p-adic place v of L^+ is not bad then there exists a κ-isogeny $A_0 \to B_0$ such that $\mathcal{O}_L \subset \mathrm{End}(B_0)$ and the pair $(B_0, \mathcal{O}_L \hookrightarrow \mathrm{End}(B_0))$ admits a lift to characteristic 0 without increasing κ.

If some p-adic place v of the totally real field L^+ is bad then the above argument does not work because in that case no member of the $\mathcal{O}_{L,v}$-linear κ-isogeny class of the p-divisible group $A_0[v^\infty]$ has a self-dual Lie type. Instead we change $A_0[v^\infty]$ by a suitable $\mathcal{O}_{L,v}$-linear κ-isogeny so that its Lie type becomes as symmetric as possible, a condition whose precise formulation is called "striped". Such a p-divisible group is shown to be isomorphic to the Serre tensor construction applied to a special class of 2-dimensional p-divisible groups of height 4 that are similar to the ones arising from the abelian surface counterexamples in 2.3.1; we call these *toy models* (see 4.1.3, especially 4.1.3.2).

These "toy models" are sufficiently special that we can analyze their CM lifting properties directly; see 4.2.10 and 4.5.15(iii). After this key step we deduce the existence of a CM lifting of $A_0[v^\infty]_{\overline{\kappa}}$ from corresponding statements for (the p-divisible group version of) toy models. In the final step, once again we use deformation theory to produce an abelian variety B_0 isogenous to (the original) A_0 over κ and a CM lifting of B_0 over a possibly *non-normal* 1-dimensional complete local noetherian domain of characteristic 0 with residue field κ. Although $\mathcal{O}_L$ acts

[8]See 1.4.5.3 for the statement of the Serre–Tate deformation theorem, and 2.2.3 and 4.6.3.1 for a precise statement of the algebraization criterion that is used in this localization step.

on the closed fiber, we can only ensure that a subring of $\mathcal{O}_L$ of finite index[9] acts on the lifted abelian scheme (see 4.6.4).

Appendix A. In Appendix A.1 we provide a self-contained development of the proof of the p-part of Tate's isogeny theorem for abelian varieties over finite fields of characteristic p, as well as a proof of Tate's formula for the local invariants at p-adic places for endomorphism algebras of simple abelian varieties over such fields. (An exposition of these results is also given in [**79**]; our treatment uses less input from non-commutative algebra.) Appendices A.2 and A.3 provide purely algebraic proofs of the Main Theorem of Complex Multiplication for abelian varieties, as well as a converse result, both of which are used in essential ways in Chapter 2. In Appendix A.4 we use Shimura's method to show that an algebraic Hecke character with a given algebraic part can be constructed over the field of moduli of the algebraic part, with control over places of bad reduction.

In the special case of the reflex norm of a CM type (L, Φ), combining this construction of algebraic Hecke characters with the converse to the Main Theorem of CM in A.3 proves that over the associated field of moduli $M \subset \overline{\mathbb{Q}}$ (a subfield of the Hilbert class field of the reflex field $E(L, \Phi)$) there exists a CM abelian variety A with CM type (L, Φ) such that A has good reduction at all p-adic places of M; see A.4.6.5. Since M is the smallest possible field of definition given (L, Φ), this existence result is optimal in terms of its field of definition. Typically $M \neq E(L, \Phi)$, and this is regarded as a "class group obstruction" to finding A with its CM structure by L over $E(L, \Phi)$, a well-known phenomenon in the classical CM theory of elliptic curves.

(In the "local" setting of CM p-divisible groups over p-adic integer rings there are no class group problems and one gets a better result: in 3.7 we *use* the preceding global construction over the field of moduli to prove that for any p-adic CM type (F, Φ) and the associated p-adic reflex field $E \subset \overline{\mathbb{Q}}_p$ over $\mathbb{Q}_p$ there exists a CM p-divisible group over $\mathcal{O}_E$ with p-adic CM type (F, Φ).)

Appendix B. In Appendices B.1 and B.2, we give two versions of a more direct (but more complicated) proof of the existence of CM liftings for a higher-dimensional generalization of the toy model.[10] The first version uses Raynaud's theory of group schemes of type $(p, \ldots, p)$. The second version uses recent developments in p-adic Hodge theory. We hope that material described there will be useful in the future. In Appendix B.3 we compare several Dieudonné theories over a perfect base field of characteristic $p > 0$. In Appendix B.4 we give a formula for the Dieudonné module of the closed fiber of a finite flat commutative group scheme, constructed using integral p-adic Hodge theory; this formula is used in B.2.

[9] This subring of finite index can be taken to be $\mathbb{Z} + p\mathcal{O}_L$.

[10] In the original proof of our main CM lifting result in 4.1.1, the case of a bad place $v|p$ of L^+ was reduced through the Serre tensor construction to this existence result. Both B.1 and B.2 are logically independent of results in Chapter 4. Readers who cannot wait to see a proof of the existence of a CM lifting (without modification by any isogeny) for such a higher-dimensional toy model may proceed directly to B.1 or B.2, after consulting 4.2 for the definition of the Lie type of an $\mathcal{O}$-linear p-divisible group and related notation.

References

(1) ABELIAN VARIETIES. In Mumford's book [**82**] the theory of abelian varieties is developed over an algebraically closed base field, and we need the theory over a general field; references addressing this extra generality are Milne's article [**76**] (which rests on [**82**]) and the forthcoming book [**45**]. Since [**45**] is not yet in final form we do not refer to it in the main text, but the reader should keep in mind that many results for which we refer to [**82**] and [**76**] are also treated in [**45**]. We refer the reader to [**83**, Ch. 6, §1–§2] for a self-contained development of the elementary properties of abelian schemes, which we freely use. (For example, the group law is necessarily commutative and is determined by the identity section, as in the theory over a field.)

(2) SEMISIMPLE ALGEBRAS. We assume familiarity with the classical theory of finite-dimensional semisimple algebras over fields (including the theory of their splitting fields and maximal commutative subfields). A suitable reference for this material is [**53**, §4.1–4.6]; another reference is [**11**]. In 1.2.2–1.2.4 we review some of the facts that we need from that theory.

(3) DESCENT THEORY AND FORMAL SCHEMES. In many places, we need to use the techniques of descent theory and Grothendieck topologies (especially the fppf topology, though in some situations we use the fpqc topology to perform descent from a completion). This is required for arguments with group schemes, even over a field, such as in considerations with short exact sequences. For accounts of descent theory, we refer the reader to [**10**, §6.1–6.2], and to [**39**, Part 1] for a more exhaustive discussion. These techniques are discussed in a manner well-suited to group schemes in [**98**] and [**30**, Exp. IV–VI_{A}].

Our arguments with deformation theory rest on the theory of formal schemes, especially Grothendieck's formal GAGA and algebraization theorems. A succinct overview of these matters is given in [**39**, Part 4], and the original references [**34**, I, §10; III_1, §4–§5] are also highly recommended.

(4) DIEUDONNÉ THEORY AND p-DIVISIBLE GROUPS. To handle p-torsion phenomena in characteristic $p > 0$ we use Dieudonné theory and p-divisible groups. Brief surveys of some basic definitions and properties in this direction are given in 1.4, 3.1.2–3.1.6, and B.3.5.1–B.3.5.5. We refer the reader to [**119**], [**71**] and [**110**, §6] for more systematic discussions of basic facts concerning p-divisible groups, and to [**29**] and [**41**, Ch. II–III] for self-contained developments of (contravariant) Dieudonné theory, with applications to p-divisible groups. Contravariant Dieudonné theory is used in Chapters 1–4.

Covariant Dieudonné theory is used in Appendix B.1 because the alternative proof there of the main result of Chapter 4 uses a covariant version of p-adic Hodge theory. A brief summary of covariant Dieudonné can be found in B.3.5.6–B.3.6.7. We recommend [**136**] for Cartier theory; an older standard reference is [**69**].

A very useful technique within the deformation theory of p-divisible groups is Grothendieck–Messing theory, which is developed from scratch in [**75**]. Although we do not provide an introduction to this topic, we hope that our applications of it may inspire an interested reader who is not familiar with this technique to learn more about it.

Notation and terminology

- Numerical labeling of text items and displayed expressions.
  - We use "$x.y.z$", "$x.y.z.w$", etc. for text items (sub-subsections, results, remarks, definitions, etc.), arranged lexicographically without repetition.
  - Any labeling of displayed expressions (equations, commutative diagrams, etc.) is indicated with parentheses, so "see ($x.y.z$)" means that one should look at the zth displayed expression in subsection $x.y$. This convention avoids confusion with the use of "$x.y.z$" to label a text item.
  - Any label for a text item is *uniquely assigned*, so even though "see $x.y.z$" does not indicate if it is a sub-subsection or theorem (or lemma, etc.), there is no ambiguity for finding it in this book.
- Convention on notation.
  - p denotes a prime number.
  - CM fields are usually denoted by L.
  - K often stands for an arbitrary field, κ is usually used to denote either a residue field or a finite field of characteristic p.
  - $V^\vee$ denotes the dual of a finite-dimensional vector space V over a field.
  - k denotes a perfect field, often of characteristic $p > 0$. In 4.2–4.6, k is an algebraically closed field of characteristic p.
  - K_0 is the fraction field of $W(k)$, where k is a perfect field of characteristic $p > 0$ and $W(k)$ is the ring of p-adic Witt vectors with entries in k.
  - Abelian varieties are usually written as A, B, or C, and p-divisible groups are often denoted as G or as X or Y.
  - The p-divisible group attached to an abelian variety or an abelian scheme A is denoted by $A[p^\infty]$; its subgroup scheme of p^n-torsion points is $A[p^n]$.
- Fields and their extensions.
  - For a field K, we write $\overline{K}$ to denote an algebraic closure and K_s to denote a separable closure.
  - An extension of fields K'/K is *primary* if K is separably algebraically closed in K' (i.e., the algebraic closure of K in K' is purely inseparable over K).
  - For a number field L we write $\mathcal{O}_L$ to denote its ring of integers. Similar notation is used for non-archimedean local fields.
  - If q is a power of a prime p, $\mathbb{F}_q$ denotes a finite field with size q (sometimes understood to be the unique subfield of order q in a fixed algebraically closed field of characteristic p). If κ and κ' are abstract finite fields with respective sizes $q = p^n$ and $q' = p^{n'}$ for integers $n, n' \geqslant 1$ then $\kappa \cap \kappa'$ denotes the unique subfield of either κ or κ' with size $p^{\gcd(n,n')}$; the context will always make clear if this is being considered as a subfield of either κ or κ'. Likewise, $\kappa\kappa'$ denotes $\kappa \otimes_{\kappa\cap\kappa'} \kappa'$, a common extension of κ and κ' with size $p^{\mathrm{lcm}(n,n')}$.
- Base change.
  - If $T \to S$ is a map of schemes and S' is an S-scheme, then $T_{S'}$ denotes the S'-scheme $T \times_S S'$ if S is understood from context.
  - When $S = \mathrm{Spec}(R)$ and $S' = \mathrm{Spec}(R')$ are affine, we may write $T_{R'}$ to denote $T \otimes_R R' := T \times_{\mathrm{Spec}(R)} \mathrm{Spec}(R')$ when R is understood from context.

- Abelian varieties and homomorphisms between them.
  - The dual of an abelian variety A is denoted A^t.
  - For an abelian variety A over a field K and a prime ℓ not divisible by $\mathrm{char}(K)$, upon choosing a separable closure K_s of K (often understood from context) the *ℓ-adic Tate module* $T_\ell(A)$ denotes $\varprojlim A[\ell^n](K_s)$ and $V_\ell(A)$ denotes $\mathbb{Q}_\ell \otimes_{\mathbb{Z}_\ell} T_\ell(A)$.
  - For any abelian varieties A and B over a field K, $\mathrm{Hom}(A, B)$ denotes the group of homomorphisms $A \to B$ over K, and $\mathrm{Hom}^0(A, B)$ denotes $\mathbb{Q} \otimes_{\mathbb{Z}} \mathrm{Hom}(A, B)$.
    (Since $\mathrm{Hom}(A, B) \to \mathrm{Hom}(A_{\overline{K}}, B_{\overline{K}})$ is injective, $\mathrm{Hom}(A, B)$ is a finite free $\mathbb{Z}$-module since the same holds over $\overline{K}$ by [**82**, §19, Thm. 3].)
  - When $B = A$ we write $\mathrm{End}(A)$ and $\mathrm{End}^0(A)$ respectively, and call $\mathrm{End}^0(A)$ the *endomorphism algebra* of A (over K). The endomorphism algebra $\mathrm{End}^0(A)$ is an invariant which only depends on A up to isogeny over K, in contrast with the endomorphism ring $\mathrm{End}(A)$.
  - We write $A \sim B$ to denote that abelian varieties A and B over K are K-isogenous.
  - To avoid any possible confusion with notation found in the literature, we emphasize that what we call $\mathrm{Hom}(A, B)$ and $\mathrm{Hom}^0(A, B)$ are sometimes denoted by others as $\mathrm{Hom}_K(A, B)$ and $\mathrm{Hom}_K^0(A, B)$.[11]
- Adeles and local fields.
  - We write $\mathbb{A}_L$ to denote the adele ring of a number field L, $\mathbb{A}_{L,f}$ to denote the factor ring of finite adeles, and $\mathbb{A}$ and $\mathbb{A}_f$ in the case $L = \mathbb{Q}$.
  - If v is a place of a number field L then L_v denotes the completion of L with respect to v; $\mathcal{O}_{L,v}$ denotes the valuation ring $\mathcal{O}_{L_v}$ of L_v in case v is non-archimedean, with residue field κ_v whose size is denoted q_v.
  - For a place w of $\mathbb{Q}$ we define $L_w := \mathbb{Q}_w \otimes_{\mathbb{Q}} L = \prod_{v|w} L_v$, and in case w is the ℓ-adic place for a prime ℓ we define $\mathcal{O}_{L,\ell} := \mathbb{Z}_\ell \otimes_{\mathbb{Z}} \mathcal{O}_L = \prod_{v|\ell} \mathcal{O}_{L,v}$.
- Class field theory and reciprocity laws.
  - The Artin maps of local and global class field theory are taken with the arithmetic normalization, which is to say that local uniformizers are carried to arithmetic Frobenius elements.[12]
  - $\mathrm{rec}_L : \mathbb{A}_L^\times / L^\times \to \mathrm{Gal}(L^{\mathrm{ab}}/L)$ denotes the arithmetically normalized global reciprocity map for a number field L.
  - The composition of $\mathbb{A}_L^\times \twoheadrightarrow \mathbb{A}_L^\times / L^\times$ with rec_L is denoted r_L.
  - For a non-archimedean local field F we write $r_F : F^\times \to \mathrm{Gal}(F^{\mathrm{ab}}/F)$ to denote the arithmetically normalized local reciprocity map.
- Frobenius and Verschiebung.
  - For a commutative group scheme N over an $\mathbb{F}_p$-scheme S, $N^{(p)}$ denotes the base change of N by the absolute Frobenius endomorphism of S. The relative Frobenius homomorphism is denoted $\mathrm{Fr}_{N/S} : N \to N^{(p)}$, and the

[11] with the notation $\mathrm{Hom}(A, B)$ and $\mathrm{Hom}^0(A, B)$ then reserved to mean the analogues for $A_{\overline{K}}$ and $B_{\overline{K}}$ over $\overline{K}$, or equivalently for A_{K_s} and B_{K_s} over K_s (see Lemma 1.2.1.2).

[12] Recall that for a non-archimedean local field F with residue field of size q, an element of $\mathrm{Gal}(F_s/F)$ is called an arithmetic (resp. geometric) Frobenius element if its effect on the residue field of F_s is the automorphism $x \mapsto x^q$ (resp. $x \mapsto x^{1/q}$); this automorphism of the residue field is likewise called the arithmetic (resp. geometric) Frobenius automorphism. We choose the arithmetic normalization of class field theory so that uniformizers correspond to Frobenius endomorphisms of abelian varieties in the Main Theorem of Complex Multiplication.

Verschiebung homomorphism for S-flat N of finite presentation denoted $\mathrm{Ver}_{N/S}\colon N^{(p)} \to N$ see [**30**, VII$_{\mathrm{A}}$, 4.2–4.3]). If S is understood from context then we may denote these as Fr_N and Ver_N respectively.
For $n \geqslant 1$, the p^n-fold relative Frobenius and Verschiebung homomorphisms $N \to N^{(p^n)}$ and $N^{(p^n)} \to N$ are respectively denoted $\mathrm{Fr}_{N/S,p^n}$ and $\mathrm{Ver}_{N/S,p^n}$.

- For a perfect field k with $\mathrm{char}(k) = p > 0$ and the unique lift $\sigma : W(k) \to W(k)$ of the Frobenius automorphism $y \mapsto y^p$ of k, a *Dieudonné module* over k is a $W(k)$-module M equipped with additive endomorphisms $\mathcal{F} : M \to M$ and $\mathcal{V} : M \to M$ such that $\mathcal{F} \circ \mathcal{V} = [p]_M = \mathcal{V} \circ \mathcal{F}$, $\mathcal{F}(cm) = \sigma(c)\,\mathcal{F}(m)$, and $c\,\mathcal{V}(m) = \mathcal{V}(\sigma(c)\,m)$ for all $c \in W(k)$ and $m \in M$; these are the left modules over the *Dieudonné ring* D_k (see 1.4.3.1).
- The semilinear operators $\mathcal{F}$ and $\mathcal{V}$ on a Dieudonné module M correspond to respective $W(k)$-linear maps $M^{(p)} \to M$ and $M \to M^{(p)}$, where $M^{(p)} := W(k) \otimes_{\sigma, W(k)} M$.

CHAPTER 1

Algebraic theory of complex multiplication

The theory of complex multiplication... is not only the most beautiful part of mathematics but also of all science.

—David Hilbert

1.1. Introduction

1.1.1. Lifting questions. A natural question early in the theory of abelian varieties is whether every abelian variety in positive characteristic admits a lift to characteristic 0. That is, given an abelian variety A_0 over a field κ with $\text{char}(\kappa) > 0$, does there exist a local domain R of characteristic zero with residue field κ and an abelian scheme A over R whose special fiber A_κ is isomorphic to A_0? We may also wish to demand that a specified polarization of A_0 or subring of the endomorphism algebra of A_0 (or both) also lifts to A. (The functor $A \rightsquigarrow A_\kappa$ from abelian R-schemes to abelian varieties over κ is faithful, by consideration of finite étale torsion levels; see the beginning of 1.4.4.)

Suppose there is an affirmative solution A to such a lifting problem over some local domain R as above. Let's see that we can arrange for a solution to be found over a local *noetherian* domain (that is even complete). This rests on a direct limit technique (that is very useful throughout algebraic geometry), as follows. Observe that for the directed system of noetherian local subrings R_i with local inclusions $R_i \hookrightarrow R$, we have $R = \varinjlim R_i$. In [**34**, IV_3, §8–§12; IV_4, §17] there is an exhaustive development of the technique of descent through direct limits. The principle is that if $\{D_i\}$ is a directed system of rings with limit D, and if we are given a "finitely presented" algebro-geometric situation over D (a diagram of finitely many D-schemes of finite presentation, equipped with with finitely many D-morphisms among them and perhaps some finitely presented quasi-coherent sheaves on them, some of which may be D-flat, etc.) then the entire structure descends to D_i for sufficiently large i. Moreover, if we increase i enough then we can also descend "reasonable" properties (such as flatness for morphisms or sheaves, and properness, surjectivity, smoothness, and having geometrically connected fibers for morphisms), any two descents become isomorphic after increasing i some more, and so on.

The results of this direct limit formalism are intuitively plausible, but their proofs can be rather non-obvious to the uninitiated (e.g., descending the properties of flatness and surjectivity). We will often use this limit formalism without much explanation, and we hope that the plausibility of such results is sufficient for a non-expert reader to follow the ideas. Everything we need is completed proved in the cited sections of [**34**]. As a basic example, since the condition of being an abelian scheme amounts to a group scheme diagram for a smooth proper R-scheme having geometrically connected fibers, the abelian scheme A over R descends to an abelian scheme over R_{i_0} for some sufficiently large i_0.

The residue field κ_{i_0} of R_{i_0} is merely a subfield of κ. By [**34**, 0_{III}, 10.3.1], there is a faithfully flat local extension $R_{i_0} \to R'$ with R' *noetherian* and having residue field κ over κ_{i_0}. By faithful flatness, every minimal prime of $\widehat{R'}$ has residue characteristic 0, so we can replace $\widehat{R'}$ with its quotient by such a prime to obtain a solution over a *complete* local noetherian domain with residue field κ.

Typically our liftings will be equipped with additional structure such as a polarization, and so the existence of an affirmative solution for our lifting problem (for a given A_0) often amounts to an appropriate deformation ring $\mathcal{R}$ for A_0 (over a Cohen ring for κ) admitting a generic point in characteristic 0; the coordinate ring of the corresponding irreducible component of $\mathrm{Spec}(\mathcal{R})$ is such an R. If $\kappa \to \kappa'$ is an extension of fields and $W \to W'$ is the associated extension of Cohen rings then often there is a natural isomorphism $\mathcal{R}' \simeq W'\widehat{\otimes}_W \mathcal{R}$ relating the corresponding deformation rings for A_0 and $(A_0)_{\kappa'}$ (see 1.4.4.5, 1.4.4.13, and 1.4.4.14). Thus, if $\mathcal{R}'$ has a generic point of characteristic 0 then so does $\mathcal{R}$. Hence, to prove an affirmative answer to lifting questions as above it is usually enough to consider algebraically closed κ. For example, the general lifting problem for polarized abelian varieties (allowing polarizations for which the associated symmetric isogeny $A_0 \to A_0^t$ is not separable) was solved affirmatively by Norman-Oort [**85**, Cor. 3.2] when $\kappa = \overline{\kappa}$, and the general case follows by deformation theory (via 1.4.4.14 with $\mathcal{O} = \mathbb{Z}$).

1.1.2. Refinements. When a lifting problem as above has an affirmative solution, it is natural to ask if the (complete local noetherian) base ring R for the lifting can be chosen to satisfy nice ring-theoretic properties, such as being normal or a discrete valuation ring. Slicing methods allow one to find an R with $\dim(R) = 1$ (see 2.1.1 for this argument), but normalization generally increases the residue field. Hence, asking that the complete local noetherian domain R be normal or a discrete valuation ring with a specified residue field κ is a non-trivial condition unless κ is algebraically closed.

We are interested in versions of the lifting problem for finite κ when we lift not only the abelian variety but also a large commutative subring of its endomorphism algebra. To avoid counterexamples it is sometimes necessary to weaken the lifting problem by permitting the initial abelian variety A_0 to be replaced with another in the same isogeny class over κ. In 1.8 we will precisely formulate several such lifting problems involving complex multiplication, and the main result of our work is a rather satisfactory solution to these lifting problems.

1.1.3. Purpose of this chapter. Much of the literature on complex multiplication involves either (i) working over an algebraically closed ground field, (ii) making unspecified finite extensions of the ground field, or (iii) restricting attention to simple abelian varieties. To avoid any uncertainty about the degree of generality in which various foundational results in the theory are valid, as well as to provide a convenient reference for subsequent considerations, in this chapter we provide an extensive review of the algebraic theory of complex multiplication over a general base field. This includes special features of the theory over finite fields and over fields of characteristic 0, and for some important proofs we refer to the original literature (e.g., papers of Tate). Some arithmetic aspects (such as reflex fields and the Main Theorem of Complex Multiplication) are discussed in Chapter 2, and Appendix A provides proofs of the Main Theorem of Complex Multiplication and some results of Tate over finite fields.

1.2. Simplicity, isotypicity, and endomorphism algebras

1.2.1. Simple abelian varieties. An abelian variety A over a field K is *simple* (over K) if it is non-zero and contains no non-zero proper abelian subvarieties. Simplicity is not generally preserved under extension of the base field; see Example 1.6.3 for some two-dimensional examples over finite fields and over $\mathbb{Q}$. An abelian variety A over K is *absolutely simple* (over K) if $A_{\overline{K}}$ is simple.

1.2.1.1. Lemma. *If A is absolutely simple over a field K then for any field extension K'/K, the abelian variety $A_{K'}$ over K' is simple.*

PROOF. This is an application of direct limit and specialization arguments, as we now explain. Assume for some K'/K that there is a non-zero proper abelian subvariety $B' \subset A_{K'}$. By replacing K' with an algebraic closure we may arrange that K' and then especially K is algebraically closed. (The algebraically closed property for K' is unimportant, but it is crucial that we have it for K.) By expressing K' as a direct limit of finitely generated K-subalgebras, there is a finitely generated K-subalgebra $R \subset K'$ such that $B' = B_{K'}$ for an abelian scheme $B \to \text{Spec}(R)$ that is a closed R-subgroup of A_R.

The constant positive dimension of the fibers of $B \to \text{Spec}(R)$ is strictly less than $\dim(A)$, as we may check using the K'-fiber $B' \subset A_{K'}$. Since K is algebraically closed we can choose a K-point x of $\text{Spec}(R)$. The fiber B_x is a non-zero proper abelian subvariety of A, contrary to the simplicity of A over K. □

For a pair of abelian varieties A and B over a field K, $\text{Hom}^0(A_{K'}, B_{K'})$ can be strictly larger than $\text{Hom}^0(A, B)$ for some separable algebraic extension K'/K. For example, if E is an elliptic curve over $\mathbb{Q}$ then considerations with the tangent line over $\mathbb{Q}$ force $\text{End}^0(E) = \mathbb{Q}$, but it can happen that $\text{End}^0(E_L) = L$ for an imaginary quadratic field L (e.g., $E : y^2 = x^3 - x$ and $L = \mathbb{Q}(\sqrt{-1})$).

Scalar extension from number fields to $\mathbb{C}$ or from an imperfect field to its perfect closure are useful techniques in the study of abelian varieties, so there is natural interest in considering ground field extensions that are not separable algebraic (e.g., non-algebraic or purely inseparable). It is an important fact that allowing such general extensions of the base field does not lead to more homomorphisms:

1.2.1.2. Lemma (Chow). *Let K'/K be an extension of fields that is primary (i.e., K is separably algebraically closed in K'). For abelian varieties A and B over K, the natural map* $\text{Hom}(A, B) \to \text{Hom}(A_{K'}, B_{K'})$ *is bijective.*

PROOF. See [**23**, Thm. 3.19] for a proof using faithfully flat descent (which is reviewed at the beginning of [**23**, §3]). An alternative proof is to show that the locally finite type Hom-scheme $\underline{\text{Hom}}(A, B)$ over K is étale. □

We shall be interested in certain commutative rings acting faithfully on abelian varieties, so we need non-trivial information about the structure of endomorphism algebras of abelian varieties. The study of such rings rests on the following fundamental result.

1.2.1.3. Theorem (Poincaré reducibility). *Let A be an abelian variety over a field K. For any abelian subvariety $B \subset A$, there is is abelian subvariety $B' \subset A$ such that the multiplication map $B \times B' \to A$ is an isogeny.*

In particular, if $A \neq 0$ then there exist pairwise non-isogenous simple abelian varieties $C_1, \ldots, C_s$ over K such that A is isogenous to $\prod C_i^{e_i}$ for some $e_i \geqslant 1$.

PROOF. When K is algebraically closed this result is proved in [**82**, §19, Thm. 1]. The same method works for perfect K, as explained in [**76**, Prop. 12.1]. (Perfectness is implicit in the property that the underlying reduced scheme of a finite type K-group is a K-subgroup. For a counterexample over any imperfect field, see [**25**, Ex. A.3.8].) The general case can be pulled down from the perfect closure via Lemma 1.2.1.2; see the proof of [**23**, Cor. 3.20] for the argument. □

1.2.1.4. Corollary. *For a non-zero abelian variety A over a field K and a primary extension of fields K'/K, every abelian subvariety B' of $A_{K'}$ has the form $B_{K'}$ for a unique abelian subvariety $B \subset A$.*

PROOF. By the Poincaré reducibility theorem, abelian subvarieties of A are precisely the images of maps $A \to A$, and similarly for $A_{K'}$. Since scalar extension commutes with the formation of images, the assertion is reduced to the bijectivity of $\mathrm{End}(A) \to \mathrm{End}(A_{K'})$, which follows from Lemma 1.2.1.2. □

Since any map between simple abelian varieties over K is either 0 or an isogeny, by general categorical arguments the collection of C_i's (up to isogeny) in the Poincaré reducibility theorem is unique up to rearrangement, and the multiplicities e_i are also uniquely determined.

1.2.1.5. Definition. The C_i's in the Poincaré reducibility theorem (considered up to isogeny) are the *simple factors* of A.

By the uniqueness of the simple factors up to isogeny, we deduce:

1.2.1.6. Corollary. *Let A be a non-zero abelian variety over a field, with simple factors $C_1, \ldots, C_s$. The non-zero abelian subvarieties of A are generated by the images of maps $C_i \to A$ from the simple factors.*

1.2.2. Central simple algebras. Using notation from the Poincaré reducibility theorem, for a non-zero abelian variety A we have

$$\mathrm{End}^0(A) \simeq \prod \mathrm{Mat}_{e_i}(\mathrm{End}^0(C_i))$$

where $\{C_i\}$ is the set of simple factors of A and the e_i's are the corresponding multiplicities. Each $\mathrm{End}^0(C_i)$ is a division algebra, by simplicity of the C_i's. Thus, to understand the structure of endomorphism algebras of abelian varieties we need to understand matrix algebras over division algebras, especially those of finite dimension over $\mathbb{Q}$. We therefore next review some general facts about such rings.

Although we have used K to denote the ground field for abelian varieties above, in what follows we will use K to denote the ground field for central simple algebras; the two are certainly not to be confused, since for abelian varieties in positive characteristic the endomorphism algebras are over fields of characteristic 0.

1.2.2.1. Definition. A *central simple algebra* over a field K is a non-zero associative K-algebra of finite dimension such that K is the center and the underlying ring is simple (i.e., has no non-trivial two-sided ideals).

A *central division algebra* over K is a central simple algebra over K whose underlying ring is a division algebra.

Among the most basic examples of central simple algebras over a field K are the matrix algebras $\mathrm{Mat}_n(K)$ for $n \geqslant 1$. The most general case is given by:

1.2.2.2. Proposition (Wedderburn's Theorem). *Every central simple algebra D over a field K is isomorphic to $\mathrm{Mat}_n(\Delta) = \mathrm{End}_\Delta(\Delta^{\oplus n})$ for some $n \geqslant 1$ and some central division algebra Δ over K (where $\Delta^{\oplus n}$ is a left Δ-module). Moreover, n is uniquely determined by D, and Δ is uniquely determined up to K-isomorphism.*

Proof. This is a special case of a general structure theorem for simple rings; see [**53**, Thm. 4.2] and [**53**, §4.4, Lemma 2]. □

In addition to matrix algebras, another way to make new central simple algebras from old ones is to use tensor products:

1.2.2.3. Lemma. *If D and D' are central simple algebras over a field K, then so is $D \otimes_K D'$. For any extension field K'/K, $D_{K'} := K' \otimes_K D$ is a central simple K'-algebra.*

Proof. The first part is [**53**, §4.6, Cor. 3]; the second is [**53**, §4.6, Cor. 1, 2]. □

1.2.3. Splitting fields. It is a general fact that for any central division algebra Δ over a field K, Δ_{K_s} is a matrix algebra over K_s (so $[\Delta : K]$ is a square). In other words, Δ is split by a finite separable extension of K. There is a refined structure theory concerning splitting fields and maximal commutative subfields of central simple algebras over fields; [**53**, §4.1–4.6] gives a self-contained development of this material. An important result in this direction is:

1.2.3.1. Proposition. *Let D be a central simple algebra over a field F, with $[D : F] = n^2$. An extension field F'/F with degree n embeds as an F-subalgebra of D if and only if F' splits D (i.e., $D_{F'} \simeq \mathrm{Mat}_n(F')$). Moreover, if D is a division algebra then every maximal commutative subfield of D has degree n over F.*

Proof. The first assertion is a special case of [**53**, Thm. 4.12]. Now assume that D is a division algebra and consider a maximal commutative subfield F'. In such cases F' splits D (by [**53**, §4.6, Cor. to Thm. 4.8]), so $n|[F' : F]$ by [**53**, Thm. 4.12]. To establish the reverse divisibility it suffices to show that for *any* central simple algebra D of dimension n^2 over F, every commutative subfield of D has F-degree at most n. If A is any simple F-subalgebra of D and its centralizer in D is denoted $Z_D(A)$ then $n^2 = [A : F][Z_D(A) : F]$ by [**53**, §4.6, Thm. 4.11]. Thus, if A is also commutative (so A is contained in $Z_D(A)$) then $[A : F] \leqslant n$. □

The second assertion in Proposition 1.2.3.1 does not generalize to central simple algebras; e.g., perhaps $D = \mathrm{Mat}_n(F)$ with F having no degree-n extension fields.

In general, for a splitting field F'/F of a central simple F-algebra D, the choice of isomorphism $D_{F'} \simeq \mathrm{Mat}_n(F')$ is ambiguous up to composition with the action of $\mathrm{Aut}_{F'}(\mathrm{Mat}_n(F'))$, so it is useful to determine this automorphism group. The subgroup of *inner automorphisms* is $\mathrm{GL}_n(F')/F'^{\times}$, arising from conjugation against elements of $\mathrm{Mat}_n(F')^{\times} = \mathrm{GL}_n(F')$. In general, the inner automorphisms are the only ones:

1.2.3.2. Theorem (Skolem–Noether). *For a central simple algebra D over a field F, the inclusion $D^{\times}/F^{\times} \hookrightarrow \mathrm{Aut}_F(D)$ carrying $u \in D^{\times}$ to $(d \mapsto udu^{-1})$ is an equality. That is, all automorphisms are inner.*

PROOF. This is [**53**, §4.6, Cor. to Thm. 4.9]. □

We finish our discussion of central simple algebras by using the Skolem–Noether theorem to build the K-linear *reduced trace* map $\mathrm{Trd}_{D/K} : D \to K$ for a central simple algebra D over a field K.

1.2.3.3. Construction. Let D be a central simple algebra over an arbitrary field K. It splits over a separable closure K_s, which is to say that there is a K_s-algebra isomorphism $f : D_{K_s} \simeq \mathrm{Mat}_n(K_s)$ onto the $n \times n$ matrix algebra for some $n \geqslant 1$. By the Skolem-Noether theorem, all automorphisms of a matrix algebra are given by conjugation by an invertible matrix. Hence, f is well-defined up to composition with an inner automorphism.

The matrix trace map $\mathrm{Tr} : \mathrm{Mat}_n(K_s) \to K_s$ is invariant under inner automorphisms and is equivariant for the natural action of $\mathrm{Gal}(K_s/K)$, so the composition of the matrix trace with f is a K_s-linear map $D_{K_s} \to K_s$ that is independent of f and $\mathrm{Gal}(K_s/K)$-equivariant. Thus, this descends to a K-linear map $\mathrm{Trd}_{D/K} : D \to K$ that is defined to be the *reduced trace*. In other words, the reduced trace map is a twisted form of the usual matrix trace, just as D is a twisted form of a matrix algebra. (For $d \in D$, the K-linear left multiplication map $x \mapsto d \cdot x$ on D has trace $\sqrt{[D:K]}\,\mathrm{Trd}_{D/K}(x)$, as we can see by scalar extension to K_s and a direct computation for matrix algebras. The elimination of the coefficient $\sqrt{[D:K]}$ is the reason for the word "reduced".)

1.2.4. Brauer groups. For applications to abelian varieties it is important to classify division algebras of finite dimension over $\mathbb{Q}$ (such as the endomorphism algebra of a simple abelian variety over a field). If Δ is such a ring then its center Z is a number field and Δ is a central division algebra over Z. More generally, the set of isomorphism classes of central division algebras over an arbitrary field has an interesting abelian group structure. This comes out of the following definition.

1.2.4.1. Definition. Central simple algebras D and D' over a field K are *similar* if there exist $n, n' \geqslant 1$ such that the central simple K-algebras $D \otimes_K \mathrm{Mat}_n(K) = \mathrm{Mat}_n(D)$ and $D' \otimes_K \mathrm{Mat}_{n'}(K) = \mathrm{Mat}_{n'}(D')$ are K-isomorphic.

The *Brauer group* $\mathrm{Br}(K)$ is the set of similarity classes of central simple algebras over K, and $[D]$ denotes the similarity class of D. For classes $[D]$ and $[D']$, define

$$[D][D'] := [D \otimes_K D'].$$

This composition law on $\mathrm{Br}(K)$ is well-defined and makes it into an abelian group with inversion given by $[D]^{-1} = [D^{\mathrm{opp}}]$, where D^{opp} is the "opposite algebra". By Proposition 1.2.2.2, each element in $\mathrm{Br}(K)$ is represented (up to isomorphism) by a unique central division algebra over K. In this sense, $\mathrm{Br}(K)$ is an abelian group structure on the set of isomorphism classes of such division algebras.

1.2.4.2. Example. The computation of the Brauer group of a number field involves computing the Brauer groups of local fields, so we now clear up any possible confusion concerning sign conventions in the description of Brauer groups for non-archimedean local fields. Upon choosing a separable closure K_s of an arbitrary field K, there are two natural procedures to define a functorial group isomorphism $\mathrm{Br}(K) \simeq \mathrm{H}^2(K_s/K, K_s^\times)$: a conceptual method via non-abelian cohomology as in [**107**, Ch. X, §5] and an explicit method via crossed-product algebras. By [**107**, Ch. X, §5, Exer. 2], these procedures are negatives of each other. We use the conceptual method of non-abelian cohomology, but we do not need to make that method explicit here and so we refer the interested reader to [**107**] for the details.

Let K be a non-archimedean local field with residue field κ and let K^{un} denote its maximal unramified subextension in K_s (with $\overline{\kappa}$ the residue field of K^{un}). It is known from local class field theory that the natural map $\mathrm{H}^2(K^{\mathrm{un}}/K, K^{\mathrm{un}\times}) \to \mathrm{H}^2(K_s/K, K_s^\times)$ is an isomorphism, and the normalized valuation mapping $K^{\mathrm{un}\times} \to \mathbb{Z}$ induces an isomorphism

$$\begin{aligned} \mathrm{H}^2(K^{\mathrm{un}}/K, K^{\mathrm{un}\times}) \simeq \mathrm{H}^2(K^{\mathrm{un}}/K, \mathbb{Z}) &\overset{\delta}{\simeq} \mathrm{H}^1(\mathrm{Gal}(K^{\mathrm{un}}/K), \mathbb{Q}/\mathbb{Z}) \\ &= \mathrm{H}^1(\mathrm{Gal}(\overline{\kappa}/\kappa), \mathbb{Q}/\mathbb{Z}). \end{aligned}$$

There now arises the question of choice of topological generator for $\mathrm{Gal}(\overline{\kappa}/\kappa)$: arithmetic or geometric Frobenius? We choose to work with arithmetic Frobenius. (In [**103**, §1.1] and [**107**, Ch. XIII, §3] the arithmetic Frobenius generator is also used.)

Via evaluation on the chosen topological generator, our conventions lead to a composite isomorphism

$$\mathrm{inv}_K : \mathrm{Br}(K) \simeq \mathbb{Q}/\mathbb{Z}$$

for non-archimedean local fields K. If one uses the geometric Frobenius convention, then by also adopting the crossed-product algebra method to define the isomorphism

$$\mathrm{Br}(K) \simeq \mathrm{H}^2(K_s/K, K_s^\times)$$

one would get the *same* composite isomorphism inv_K since the two sign differences cancel out in the composite. (Beware that in [**103**] and [**107**] the Brauer group of a general field K is *defined* to be $\mathrm{H}^2(K_s/K, K_s^\times)$, and so the issue of choosing between non-abelian cohomology or crossed-product algebras does not arise in the foundational aspects of the theory. However, this issue implicitly arises in the relationship of Brauer groups and central simple algebras, such as in [**103**, Appendix to §1] where the details are omitted.)

Since $\mathrm{Br}(\mathbb{R})$ is cyclic of order 2 and $\mathrm{Br}(\mathbb{C})$ is trivial, for archimedean local fields K there is a unique injective homomorphism $\mathrm{inv}_K : \mathrm{Br}(K) \hookrightarrow \mathbb{Q}/\mathbb{Z}$.

By [**103**, §1.1, Thm. 3], for a finite extension K'/K of non-archimedean local fields, composition with the natural map $r^K_{K'} : \mathrm{Br}(K) \to \mathrm{Br}(K')$ satisfies

$$\mathrm{inv}_{K'} \circ r^K_{K'} = [K' : K] \cdot \mathrm{inv}_K. \tag{1.2.4.1}$$

By [**107**, Ch. XIII, §3, Cor. 3], $\mathrm{inv}_K(\Delta)$ has order $\sqrt{[\Delta : K]}$ for any central division algebra Δ over K. These assertions are trivially verified to hold for archimedean local fields K as well.

1.2.4.3. Theorem. *Let L be a global field. There is an exact sequence*

$$0 \longrightarrow \mathrm{Br}(L) \longrightarrow \bigoplus_v \mathrm{Br}(L_v) \xrightarrow{\sum \mathrm{inv}_{L_v}} \mathbb{Q}/\mathbb{Z} \longrightarrow 0$$

where the direct sum is taken over all places of L and the first map is defined via extension of scalars.

PROOF. This is [**120**, §9.7, §11.2]. □

For a global field L and central division algebra Δ over L, $\mathrm{inv}_v(\Delta)$ denotes $\mathrm{inv}_{L_v}(\Delta_{L_v})$. Theorem 1.2.4.3 says that a central division algebra Δ over a global field L is uniquely determined up to isomorphism by its invariants $\mathrm{inv}_v(\Delta)$, and that these may be arbitrarily assigned subject to the conditions $\mathrm{inv}_v(\Delta) = 0$ for all but finitely many v and $\sum \mathrm{inv}_v(\Delta) = 0$. Moreover, the order of $[\Delta]$ in $\mathrm{Br}(L)$ is the least common "denominator" of the local invariants $\mathrm{inv}_v(\Delta) \in \mathbb{Q}/\mathbb{Z}$.

If K is any field then for a class $c \in \mathrm{Br}(K)$ its *period* is its order and its *index* is $\sqrt{[\Delta : K]}$ with Δ the unique central division algebra over K representing the class c. It is a classical fact that the period divides that index and that these integers have the same prime factors (see [**107**, X.5], especially Lemma 1 and Exercise 3), but in general equality does not hold. For example, there are function fields of complex 3-folds for which some order-2 elements in the Brauer group cannot be represented by a quaternion algebra; examples are given in [**61**, §4], and there are examples with less interesting fields as first discovered by Brauer. We have noted above that over local fields there is equality of period and index (the archimedean case being trivial). The following deep result is an analogue over global fields.

1.2.4.4. Theorem. *For a central division algebra Δ over a global field L, the order of $[\Delta]$ in $\mathrm{Br}(L)$ is $\sqrt{[\Delta : L]}$.*

As a special (and very important) case, elements of order 2 in $\mathrm{Br}(L)$ are precisely the Brauer classes of quaternion division algebras for a global field L; as noted above, this fails for more general fields. Since Theorem 1.2.4.4 does not seem to be explicitly stated in any of the standard modern references on class field theory (though there is an allusion to it at the end of [**4**, Ch. X, §2]), and the structure theory of endomorphism algebras of abelian varieties rests on it, here is a proof.

PROOF. Let Δ have degree n^2 over L and let d be the order of $[\Delta]$ in $\mathrm{Br}(L)$, so $d|n$. Note that d is the least common multiple of the local orders d_v of $[\Delta_{L_v}] \in \mathrm{Br}(L_v)$ for each place v of L, with $d_v = 1$ for complex v, $d_v|2$ for real v, and $d_v = 1$ for all but finitely many v. Using these formal properties of the d_v's, we may call upon the full power of global class field theory via Theorem 6 in [**4**, Ch. X] to infer the existence of a cyclic extension L'/L of degree d such that $[L'_{v'} : L_v]$ is a multiple of d_v for every place v of L (here, v' is any place on L' over v, and the constraint on the local degree is only non-trivial when $d_v > 1$). In the special case $d = 2$ (the only case we will require) one only needs weak approximation and Krasner's Lemma rather than class field theory: take L' to split a separable quadratic polynomial over L that closely approximates ones that define quadratic separable extensions of L_v for each v such that $d_v = 2$.

By (1.2.4.1), restriction maps on local Brauer groups induce multiplication by the local degree on the local invariants, so $\Delta_{L'}$ is locally split at all places of L'. Thus, by the injectivity of the map from the global Brauer group into the direct sum of the local ones (for L') we conclude that the Galois extension L'/L *of degree* d splits Δ. (The existence of cyclic splitting fields for all Brauer classes is proved for number fields in [**120**] and is proved for all global fields in [**128**], but neither reference seems to control the degree of the global cyclic extension.) It is a general fact for Brauer groups of arbitrary fields [**107**, Ch. X, §5, Lemma 1] that every Brauer class split by a Galois extension of degree r is represented by a central simple algebra with degree r^2. Applying this fact from algebra in our situation, $[\Delta] = [D]$ for a central simple algebra D of degree d^2 over L. But each Brauer class is represented by a unique central division algebra, and so D must be L-isomorphic to a matrix algebra over Δ. Since $[D : L] = d^2$ and $[\Delta : L] = n^2$ with $d|n$, this forces $d = n$ as desired. □

1.2.5. Homomorphisms and isotypicity. The study of maps between abelian varieties over a field rests on the following useful injectivity result.

1.2.5.1. Proposition. *Let A and B be abelian varieties over a field K. For any prime ℓ (allowing $\ell = \mathrm{char}(K)$), the natural map*

$$\mathbb{Z}_\ell \otimes_{\mathbb{Z}} \mathrm{Hom}(A, B) \to \mathrm{Hom}(A[\ell^\infty], B[\ell^\infty])$$

is injective, where the target is the $\mathbb{Z}_\ell$-module of maps of ℓ-divisible groups over K (i.e., compatible systems of K-group maps $A[\ell^n] \to B[\ell^n]$ for all $n \geqslant 1$).

PROOF. Without loss of generality, K is algebraically closed (and hence perfect). When $\ell \neq \mathrm{char}(K)$ the assertion is a reformulation of the well-known analogous injectivity with ℓ-adic Tate modules (and such injectivity in turn underlies the proof of $\mathbb{Z}$-module finiteness of $\mathrm{Hom}(A, B)$). The proof in terms of Tate modules is given in [**82**, §19, Thm. 3] for $\ell \neq \mathrm{char}(K)$, and when phrased in terms of ℓ-divisible groups it works even when $\ell = p = \mathrm{char}(K) > 0$. For the convenience of the reader, we now provide the argument for $\ell = p$ in such terms. We will use that the torsion-free $\mathbb{Z}$-module $\mathrm{Hom}(A, B)$ is finitely generated, and our argument works for any ℓ (especially $\ell = \mathrm{char}(K)$).

Choose a $\mathbb{Z}$-basis $\{f_1, \ldots, f_n\}$ of $\mathrm{Hom}(A, B)$. For $c_1, \ldots, c_n \in \mathbb{Z}_\ell$ it suffices to show that if $\sum c_i f_i$ kills $A[\ell]$ then $\ell | c_i$ for all i. Indeed, if we can prove this then consider the case when $\sum c_i f_i$ kills $A[\ell^\infty]$. Certainly $c_i = \ell c_i'$ for some $c_i' \in \mathbb{Z}_\ell$, and $(\sum c_i' f_i) \cdot \ell$ kills $A[\ell^n]$ for all $n > 0$. But the map $A[\ell^n] \to A[\ell^{n-1}]$ induced by ℓ-multiplication is faithfully flat since it is the pullback along $A[\ell^{n-1}] \hookrightarrow A$ of the faithfully flat map $\ell : A \to A$, so $\sum c_i' f_i$ kills $A[\ell^{n-1}]$ for all $n > 0$. In other words, the kernel of the map in the Proposition would be ℓ-divisible, yet this kernel is a finitely generated $\mathbb{Z}_\ell$-module, so it would vanish as desired.

Now consider $c_1, \ldots, c_n \in \mathbb{Z}_\ell$ such that $\sum c_i f_i$ kills $A[\ell]$. For the purpose of proving $c_i \in \ell\mathbb{Z}_\ell$ for all i, it is harmless to add to each c_i any element of $\ell\mathbb{Z}_\ell$. Hence, we may and do assume $c_i \in \mathbb{Z}$ for all i, so $\sum c_i f_i : A \to B$ makes sense and kills $A[\ell]$. Since $\ell : A \to A$ is a faithfully flat homomorphism with kernel $A[\ell]$, by fppf descent theory any K-group scheme homomorphism $A \to G$ that kills $A[\ell]$ factors through $\ell : A \to A$ (see [**30**, IV, 5.1.7.1] and [**98**]). Thus, $\sum c_i f_i = \ell \cdot h$ for some $h \in \mathrm{Hom}(A, B)$. Writing $h = \sum m_i f_i$ with $m_i \in \mathbb{Z}$, we get $\sum c_i \otimes f_i = \ell \cdot \sum 1 \otimes m_i f_i$ in $\mathbb{Z}_\ell \otimes_{\mathbb{Z}} \mathrm{Hom}(A, B)$. This implies $c_i = \ell m_i$ for all i, so we are done. □

A weakening of simplicity that is sometimes convenient is:

1.2.5.2. Definition. An abelian variety A over a field K is *isotypic* if it is isogenous to C^e for a simple abelian variety C over K with $e \geqslant 1$; that is, up to isogeny, A has a unique simple factor. For a simple factor C of an abelian variety A over K, the *C-isotypic part* of A is the isotypic subvariety of A generated by the images of all maps $C \to A$. An *isotypic part* of A is a C-isotypic part for some such C.

Clearly $\mathrm{End}^0(A)$ is a semisimple $\mathbb{Q}$-algebra. It is simple if and only if A is isotypic, and it is a division algebra if and only if A is simple.

By the Poincaré reducibility theorem, every non-zero abelian variety A over a field K is naturally isogenous to the product of its distinct isotypic parts, and these distinct parts admit no non-zero maps between them. Hence, if $\{B_i\}$ is the set of isotypic parts of A then $\mathrm{End}^0(A) = \prod \mathrm{End}^0(B_i)$ with each $\mathrm{End}^0(B_i)$ a simple algebra of finite dimension over $\mathbb{Q}$. Explicitly, if C_i is the unique simple factor of B_i then a choice of isogeny $B_i \to C_i^{e_i}$ defines an isomorphism from $\mathrm{End}^0(B_i)$ onto the matrix algebra $\mathrm{Mat}_{e_i}(\mathrm{End}^0(C_i))$ over the division algebra $\mathrm{End}^0(C_i)$. Beware that the composite "diagonal" ring map $\mathrm{End}^0(C_i) \to \mathrm{Mat}_{e_i}(\mathrm{End}^0(C_i)) \simeq \mathrm{End}^0(B_i)$ is canonical only when $\mathrm{End}^0(C_i)$ is commutative.

In general isotypicity is not preserved by extension of the ground field. To make examples illustrating this possibility, as well as other examples in the theory of abelian varieties, we need the operation of *Weil restriction of scalars*. For a field K and finite K-algebra K', the Weil restriction functor $\mathrm{Res}_{K'/K}$ from quasi-projective K'-schemes to separated (even quasi-projective) K-schemes of finite type is characterized by the functorial identity $\mathrm{Res}_{K'/K}(X')(A) = X'(K' \otimes_K A)$ for K-algebras A. Informally, Weil restriction is an algebraic analogue of viewing a complex manifold as a real manifold with twice the dimension. In particular, if K'/K is an extension of fields then $\mathrm{Res}_{K'/K}(X')$ is K'-smooth and equidimensional when X' is K-smooth and equidimensional, with

$$\dim(\mathrm{Res}_{K'/K}(X')) = [K' : K] \cdot \dim(X').$$

We refer the reader to [**10**, §7.6] for a self-contained development of the construction and properties of Weil restriction (replacing K with more general rings), and to [**25**, A.5] for a discussion of further properties (especially of interest for group schemes). In general the formation of Weil restriction naturally commutes with any extension of the base field, and for K' equal to the product ring K^n we have that $\mathrm{Res}_{K'/K}$ carries a disjoint union $\coprod_{i=1}^n S_i$ of quasi-projective K-schemes (viewed as a K'-scheme) to the product $\prod S_i$. Thus, the natural isomorphism

$$\mathrm{Res}_{K'/K}(X')_{K_s} \simeq \mathrm{Res}_{(K' \otimes_K K_s)/K_s}(X'_{K' \otimes_K K_s})$$

implies that if K' is a field separable over K then $\mathrm{Res}_{K'/K}(A')$ is an abelian variety over K of dimension $[K' : K]\dim(A')$ for any abelian variety A' over K' (since $K' \otimes_K K_s \simeq K_s^{[K':K]}$). If K'/K is a field extension of finite degree that is not separable then $\mathrm{Res}_{K'/K}(X')$ is never proper when X' is smooth and proper of positive dimension [**25**, Ex. A.5.6].

1.2.6. Example. Consider a separable quadratic extension of fields K'/K and a simple abelian variety A' over K'. Let $\sigma \in \mathrm{Gal}(K'/K)$ be the non-trivial element, so $K' \otimes_K K' \simeq K' \times K'$ via $x \otimes y \mapsto (xy, \sigma(x)y)$. Thus, the Weil restriction $A := \mathrm{Res}_{K'/K}(A')$ satisfies $A_{K'} \simeq A' \times \sigma^*(A')$, so $A_{K'}$ is not isotypic if and only if A' is not isogenous to its σ-twist. Hence, for $K = \mathbb{R}$ examples of non-isotypic

$A_{K'}$ are obtained by taking A' to be an elliptic curve over $\mathbb{C}$ with analytic model $\mathbb{C}/(\mathbb{Z} \oplus \mathbb{Z}\tau)$ for $\tau \in \mathbb{C} - \mathbb{R}$ such that $1, \tau, \overline{\tau}, \tau\overline{\tau}$ are $\mathbb{Q}$-linearly independent. (In Example 1.6.4 we give examples with $K = \mathbb{Q}$.)

In cases when $A_{K'}$ is non-isotypic, A is necessarily simple. Indeed, if A is not simple then a simple factor of A would be a K-descent of a member of the isogeny class of A', contradicting that A' and $\sigma^*(A')$ are not isogenous. Thus, we have exhibited examples in characteristic 0 for which isotypicity is lost after a ground field extension.

The failure of isotypicity to be preserved after a ground field extension does not occur over finite fields:

1.2.6.1. Proposition. *If A is an isotypic abelian variety over a finite field K then $A_{K'}$ is isotypic for any extension field K'/K.*

PROOF. By the Poincaré reducibility theorem, it is equivalent to show that $\mathrm{End}^0(A_{K'})$ is a simple $\mathbb{Q}$-algebra, so by Lemma 1.2.1.2 we may replace K' with the algebraic closure of K in K'. That is, we can assume that K'/K is algebraic. Writing $K' = \varinjlim K'_i$ with $\{K'_i\}$ denoting the directed system of subfields of finite degree over K, we have $\mathrm{End}(A_{K'}) = \varinjlim \mathrm{End}(A_{K'_i})$. But $\mathrm{End}(A_{K'})$ is finitely generated as a $\mathbb{Z}$-module, so for large enough i we have $\mathrm{End}^0(A_{K'}) = \mathrm{End}^0(A_{K'_i})$. We may therefore replace K' with K'_i for sufficiently large i to reduce to the case when K'/K is of finite degree. Let $q = \#K$.

The key point is to show that for any abelian variety B' over K' and any $g \in \mathrm{Gal}(K'/K)$, B' and $g^*(B')$ are isogenous. Since $\mathrm{Gal}(K'/K)$ is generated by the q-Frobenius σ_q, it suffices to show that B' and $B'^{(q)} := \sigma_q^*(B')$ are isogenous. The purely inseparable relative q-Frobenius morphism $B' \to B'^{(q)}$ (arising from the absolute q-Frobenius map $B' \to B'$ over the q-Frobenius of $\mathrm{Spec}(K')$) is such an isogeny. Hence, the Weil restriction $\mathrm{Res}_{K'/K}(B')$ satisfies $\mathrm{Res}_{K'/K}(B')_{K'} \simeq \prod_g g^*(B') \sim B'^{[K':K]}$.

Take B' to be a simple factor of $A_{K'}$ (up to isogeny), so $\mathrm{Res}_{K'/K}(B')$ is an isogeny factor of $\mathrm{Res}_{K'/K}(A_{K'}) \sim A^{[K':K]}$. By the simplicity of A and the Poincaré reducibility theorem, it follows that $\mathrm{Res}_{K'/K}(B')$ is isogenous to a power of A. Extending scalars, $\mathrm{Res}_{K'/K}(B')_{K'}$ is therefore isogenous to a power of $A_{K'}$. But $\mathrm{Res}_{K'/K}(B')_{K'} \sim B'^{[K':K]}$, so non-trivial powers of $A_{K'}$ and B' are isogenous. By the simplicity of B' and Poincaré reducibility, this forces B' to be the only simple factor of $A_{K'}$ (up to isogeny), so $A_{K'}$ is isotypic. □

1.3. Complex multiplication

1.3.1. Commutative subrings of endomorphism algebras. The following fact motivates the study of complex multiplication in the sense that we shall consider.

1.3.1.1. Theorem. *Let A be an abelian variety over a field K with $g := \dim(A) > 0$, and let $P \subset \mathrm{End}^0(A)$ be a commutative semisimple $\mathbb{Q}$-subalgebra. Then $[P : \mathbb{Q}] \leqslant 2g$, and if equality holds then P is its own centralizer in $\mathrm{End}^0(A)$. If equality holds* *and moreover P is a field of degree $2g$ over $\mathbb{Q}$ then A is isotypic and P is a maximal commutative subfield of* $\mathrm{End}^0(A)$.

PROOF. Consider the decomposition $P = \prod L_i$ into a product of fields. Using the primitive idempotents of P, we get a corresponding decomposition $\prod A_i$ of A in the isogeny category of abelian varieties over K, with each $A_i \neq 0$ and each L_i a commutative subfield of $\mathrm{End}^0(A_i)$ compatibly with the inclusion $\prod \mathrm{End}^0(A_i) \subset \mathrm{End}^0(A)$ and the equality $\prod L_i = P$. Since $\dim(A) = \sum \dim(A_i)$, to prove that $[P : \mathbb{Q}] \leqslant 2g$ it suffices to treat the A_i's separately, which is to say that we may and do assume that $P = L$ is a field.

Since $D = \mathrm{End}^0(A)$ is of finite rank over $\mathbb{Q}$, clearly $[L : \mathbb{Q}]$ is finite. Choose a prime ℓ different from $\mathrm{char}(K)$. Recall that $V_\ell(A)$ denotes $\mathbb{Q}_\ell \otimes_{\mathbb{Z}_\ell} T_\ell(A)$ for $T_\ell(A) := \varprojlim A[\ell^n](K_s)$. The injectivity of the natural map

$$L_\ell := \mathbb{Q}_\ell \otimes_{\mathbb{Q}} L \hookrightarrow \mathrm{End}_{\mathbb{Q}_\ell}(V_\ell(A))$$

(see Proposition 1.2.5.1) implies that L_ℓ acts faithfully on the $\mathbb{Q}_\ell$-vector space $V_\ell(A)$ of rank $2g$. But $L_\ell = \prod_{w|\ell} L_w$, where w runs over all ℓ-adic places of L, so each corresponding factor module $V_\ell(A)_w$ over L_w is non-zero as a vector space over L_w. Hence,

$$2g = \dim_{\mathbb{Q}_\ell} V_\ell(A) = \sum_{w|\ell} \dim_{\mathbb{Q}_\ell} V_\ell(A)_w \geqslant \sum_{w|\ell} [L_w : \mathbb{Q}_\ell] = [L : \mathbb{Q}]$$

with equality if and only if $V_\ell(A)$ is free of rank 1 over L_ℓ.

Assume that equality holds, so $V_\ell(A)$ is free of rank 1 over L_ℓ. If A is not isotypic then by passing to an isogenous abelian variety we may arrange that $A = B \times B'$ with B and B' non-zero abelian varieties such that $\mathrm{Hom}(B, B') = 0 = \mathrm{Hom}(B', B)$. Hence, $\mathrm{End}^0(A) = \mathrm{End}^0(B) \times \mathrm{End}^0(B')$ and so L embeds into $\mathrm{End}^0(B)$. But $2\dim(B) < 2\dim(A) = [L : \mathbb{Q}]$, so we have a contradiction (since $B \neq 0$).

It remains to prove, without assuming P is a field, that if $[P : \mathbb{Q}] = 2g$ then P is its own centralizer in $\mathrm{End}^0(A)$. (In case P is a field, so A is isotypic and hence $\mathrm{End}^0(A)$ is simple, such a centralizer property would imply that P is a maximal commutative subfield of $\mathrm{End}^0(A)$, as desired.) Consider once again the ring decomposition $P = \prod L_i$ and the corresponding isogeny decomposition $\prod A_i$ of A as at the beginning of this proof. We have $[L_i : \mathbb{Q}] \leqslant 2\dim(A_i)$ for all i, and these inequalities add up to an equality when summed over all i, so in fact $[L_i : \mathbb{Q}] = 2\dim(A_i)$ for all i. The preceding analysis shows that each $V_\ell(A_i)$ is free of rank 1 over $L_{i,\ell} := \mathbb{Q}_\ell \otimes_{\mathbb{Q}} L_i$, and so likewise $V_\ell(A)$ is free of rank 1 over P_ℓ. Hence, $\mathrm{End}_{P_\ell}(V_\ell(A)) = P_\ell$, so if $Z(P)$ denotes the centralizer of P in $\mathrm{End}^0(A)$ then the P_ℓ-algebra map

$$Z(P)_\ell = \mathbb{Q}_\ell \otimes_{\mathbb{Q}} Z(P) \to \mathrm{End}_{\mathbb{Q}_\ell}(V_\ell(A))$$

is injective (Proposition 1.2.5.1) and lands inside $\mathrm{End}_{P_\ell}(V_\ell(A)) = P_\ell$. In other words, the inclusion $P \subset Z(P)$ of $\mathbb{Q}$-algebras becomes an equality after scalar extension to $\mathbb{Q}_\ell$, so $P = Z(P)$ as desired. □

The preceding theorem justifies the interest in the following concept.

1.3.1.2. Definition. An abelian variety A of dimension $g > 0$ over a field K admits *sufficiently many complex multiplications* (over K) if there exists a commutative semisimple $\mathbb{Q}$-subalgebra P in $\mathrm{End}^0(A)$ with rank $2g$ over $\mathbb{Q}$.

The reason for the terminology in Definition 1.3.1.2 is due to certain examples with $K = \mathbb{C}$ and P a number field such that the analytic uniformization of $A(\mathbb{C})$ expresses the P-action in terms of multiplication of complex numbers; see Example 1.5.3. The classical theory of complex multiplication focused on the case of Definition 1.3.1.2 in which P is a field, but it is useful to allow P to be a product of several fields (i.e., a commutative semisimple $\mathbb{Q}$-algebra). For example, by Theorem 1.3.1.1 this is necessary if we wish to consider the theory of complex multiplication with A that is not isotypic, or more generally if we want Definition 1.3.1.2 to be preserved under the formation of products. The theory of Shimura varieties provides further reasons not to require P to be a field.

Note that we do not consider A to admit sufficiently many complex multiplications merely if it does so after an extension of the base field K.

1.3.2. Example. The elliptic curve $y^2 = x^3 - x$ admits sufficiently many complex multiplications over $\mathbb{Q}(\sqrt{-1})$ but not over $\mathbb{Q}$. More generally, $\mathrm{End}^0(E) = \mathbb{Q}$ for every elliptic curve E over $\mathbb{Q}$ (since the tangent line at the origin is too small to support a $\mathbb{Q}$-linear action by an imaginary quadratic field), so in our terminology an elliptic curve over $\mathbb{Q}$ does not admit sufficiently many complex multiplications.

1.3.2.1. Proposition. *Let A be a non-zero abelian variety over a field K. The following are equivalent.*

(1) *The abelian variety A admits sufficiently many complex multiplications.*
(2) *Each isotypic part of A admits sufficiently many complex multiplications.*
(3) *Each simple factor of A admits sufficiently many complex multiplications.*

See Definition 1.2.5.2 for the terminology used in (2).

PROOF. Let $\{B_i\}$ be the set of isotypic parts of A, so $\mathrm{End}^0(B_i) \simeq \mathrm{Mat}_{e_i}(\mathrm{End}^0(C_i))$ where C_i is the unique simple factor of B_i and $e_i \geqslant 1$ is its multiplicity as such. Since $\mathrm{End}^0(A) = \prod \mathrm{End}^0(B_i)$, (2) implies (1). It is clear that (3) implies (2).

Conversely, assume that $\mathrm{End}^0(A)$ contains a $\mathbb{Q}$-algebra P satisfying $[P : \mathbb{Q}] = 2\dim(A)$. There is a unique decomposition $P = \prod L_i$ with fields $L_1, \ldots, L_s$, and $\sum [L_i : \mathbb{Q}] = 2\dim(A)$. We saw in the proof of Theorem 1.3.1.1 that by replacing A with an isogenous abelian variety we may arrange that $A = \prod A_i$ with each A_i a non-zero abelian variety having $L_i \subset \mathrm{End}^0(A_i)$ compatibly with the embedding $\prod \mathrm{End}^0(A_i) \subset \mathrm{End}^0(A)$ and the equality $\prod L_i = P$. Thus, $[L_i : \mathbb{Q}] \leqslant 2\dim(A_i)$ for all i (by Theorem 1.3.1.1), and adding this up over all i yields an equality, so each A_i admits sufficiently many complex multiplications using L_i. Since each simple factor of A is a simple factor of some A_i, to prove (3) we are therefore reduced to the case when $P = L$ is a field.

Applying Theorem 1.3.1.1 once again, L is its own centralizer in $\mathrm{End}^0(A)$ and A is isotypic, say with unique simple factor C appearing with multiplicity e. In particular, $\mathrm{End}^0(A) = \mathrm{Mat}_e(D)$ for the division algebra $D = \mathrm{End}^0(C)$ of finite rank over $\mathbb{Q}$. If Z denotes the center of D then D is a central division algebra over Z, and L contains Z since L is its own centralizer in $\mathrm{End}^0(A) = \mathrm{Mat}_e(D)$. Letting $d = \dim(C)$, $\mathrm{Mat}_e(D)$ contains the maximal commutative subfield L of degree $2g/[Z : \mathbb{Q}] = (2d/[Z : \mathbb{Q}])e$ over Z.

As we noted in the proof of Proposition 1.2.3.1 (parts of which are carried out for central simple algebras that may not be division algebras), the Z-degree of

$\mathrm{Mat}_e(D)$ is the product of the Z-degrees of L and the centralizer of L in $\mathrm{Mat}_e(D)$. But L is its own centralizer, so

$$e^2[D : Z] = \dim_Z \mathrm{Mat}_e(D) = [L : Z]^2 = e^2(2d/[Z : \mathbb{Q}])^2.$$

We conclude that $2d/[Z : \mathbb{Q}] = \sqrt{[D : Z]}$, so (by Proposition 1.2.3.1) $2d/[Z : \mathbb{Q}]$ is the common Z-degree of all maximal commutative subfields of the central division algebra $D = \mathrm{End}^0(C)$ over Z, or equivalently $2d$ is the $\mathbb{Q}$-degree of all such fields. But $2d = 2\dim(C)$, so choosing any maximal commutative subfield of D shows that C admits sufficiently many complex multiplications. □

1.3.3. CM algebras and CM abelian varieties. The following three conditions on a number field L are equivalent:

(1) L has no real embeddings but is quadratic over a totally real subfield,
(2) for every embedding $j : L \to \mathbb{C}$, the subfield $j(L) \subset \mathbb{C}$ is stable under complex conjugation and the involution $x \mapsto j^{-1}(\overline{j(x)})$ in $\mathrm{Aut}(L)$ is non-trivial and independent of j,
(3) there is a non-trivial involution $\tau \in \mathrm{Aut}(L)$ such that for every embedding $j : L \to \mathbb{C}$ we have $j(\tau(x)) = \overline{j(x)}$ for all $x \in L$.

The proof of the equivalence is easy. When these conditions hold, τ in (3) is unique and its fixed field is the maximal totally real subfield $L^+ \subset L$ (over which L is quadratic). The case $L^+ = \mathbb{Q}$ corresponds to the case when L is an imaginary quadratic field.

1.3.3.1. Definition. A CM *field* is a number field L satisfying the equivalent conditions (1), (2), and (3) above. A CM *algebra* is a product $L_1 \times \cdots \times L_s$ of finitely many CM fields (with $s \geqslant 1$).

The reason for this terminology is due to the following important result (along with Example 1.5.3).

1.3.4. Theorem (Tate). *Let A be an abelian variety of dimension $g > 0$ over a field K. Suppose A admits sufficiently many complex multiplications. Then there exists a* CM *algebra $P \subset \mathrm{End}^0(A)$ with $[P : \mathbb{Q}] = 2\dim(A)$. In case A is isotypic we can take P to be a* CM *field.*

The proof of this theorem (which ends with the proof of Lemma 1.3.7.1) will require some effort, especially since we consider an arbitrary base field K. Before we start the proof, it is instructive to consider an example.

1.3.4.1. Example. Consider $A = E^2$ with an elliptic curve E over $K = \mathbb{C}$ such that $L := \mathrm{End}^0(E)$ is an imaginary quadratic field. The endomorphism algebra $\mathrm{End}^0(A) = \mathrm{Mat}_2(L)$ is simple and contains as its maximal commutative subfields all quadratic extensions of L. Those extensions which are biquadratic over $\mathbb{Q}$ are CM fields, and the rest are not CM fields. Hence, in the setup of Theorem 1.3.4, even when A is isotypic and $\mathrm{char}(K) = 0$ there can be maximal commutative semisimple subalgebras of $\mathrm{End}^0(A)$ that are not CM algebras. However, if $\mathrm{char}(K) = 0$ and A is simple (over K) then $\mathrm{End}^0(A)$ is a CM field; see Proposition 1.3.6.4.

1.3.5. We will begin the proof of Theorem 1.3.4 now, but at a certain point we will need to use deeper input concerning the fine structure of endomorphism algebras of simple abelian varieties over general fields. At that point we will digress to review the required structure theory, and then we will complete the argument.

By Proposition 1.3.2.1, every simple factor of A admits sufficiently many complex multiplications. Thus, to prove the existence of the CM subalgebra P in Theorem 1.3.4 it suffices to treat the case when A is simple. Note that in the simple case such a CM subalgebra is automatically a field, since the endomorphism algebra is a division algebra. Let us first show that the result in the simple case implies that in the general isotypic case we can find P as a CM field. For isotypic A, by passing to an isogenous abelian variety we can arrange that $A = A'^m$ for a simple abelian variety A' over K and some $m \geqslant 1$. Thus, if $g' = \dim A'$ then $g = mg'$ and $\mathrm{End}^0(A')$ contains a CM field P' of degree $2g'$ over $\mathbb{Q}$. But $\mathrm{End}^0(A) \simeq \mathrm{Mat}_m(\mathrm{End}^0(A'))$ and this contains $\mathrm{Mat}_m(P')$. To find a CM field $P \subset \mathrm{End}^0(A)$ of degree $2g = 2g'm$ over $\mathbb{Q}$ it therefore suffices to construct a degree-m extension field P of P' such that P is a CM field.

Let P'^+ be the maximal totally real subfield of P', so for any totally real field P^+ of finite degree over P'^+ the ring $P = P^+ \otimes_{P'^+} P'$ is a field quadratic over P^+ and it is totally complex, so it is a CM field and clearly $[P : \mathbb{Q}] = [P : P'][P' : \mathbb{Q}] = 2g'[P^+ : P'^+]$. Hence, to find the required CM field P in the isotypic case it suffices to construct a degree-m totally real extension of P'^+. To do this, first recall the following basic fact from number theory [**15**, §6]:

1.3.5.1. Theorem (weak approximation). *For any number field L and finite set S of places of L, the map $L \to \prod_{v \in S} L_v$ has dense image.*

PROOF. This is [**15**, §6]. □

Applying this to P'^+, we can construct a monic polynomial f of degree m in $P'^+[u]$ that is very close to a totally split monic polynomial of degree m at each real place and is very close to an irreducible (e.g., Eisenstein) polynomial at a single non-archimedean place. It follows that f is totally split at each real place of P'^+ and is irreducible over P'^+, so the ring $P^+ = P'^+[u]/(f)$ is a totally real field of degree m over P'^+ as required.

1.3.5.2. We may and do assume for the remainder of the argument that A is simple. In this case $D = \mathrm{End}^0(A)$ is a central division algebra over a number field Z, so the commutative semisimple $\mathbb{Q}$-subalgebra $P \subset D$ is a field, and the proof of Proposition 1.3.2.1 shows that the common $\mathbb{Q}$-degree of all maximal commutative subfields of D is $2g$. Hence, our problem is to construct a maximal commutative subfield of D that is a CM field.

Let $\mathrm{Trd}_{D/\mathbb{Q}} = \mathrm{Tr}_{Z/\mathbb{Q}} \circ \mathrm{Trd}_{D/Z}$, where $\mathrm{Trd}_{D/Z}$ is the reduced trace. An abelian variety over any field admits a polarization, so choose a polarization of A over K. Let $x \mapsto x^*$ denote the associated Rosati involution on D (so $(xy)^* = y^*x^*$ and $x^{**} = x$).

1.3.5.3. Lemma. *The quadratic form $x \mapsto \mathrm{Trd}_{D/\mathbb{Q}}(xx^*)$ on D is positive-definite.*

PROOF. For any central simple algebra D over any field K whatsoever, let $n = \sqrt{[D : K]}$ and define the variant $\mathrm{Trm}_{D/K} : D \to K$ of the reduced trace to be the map that sends each $y \in D$ to the trace of the K-linear map $m_y : D \to D$ defined by $d \mapsto yd$. We have $\mathrm{Trm}_{D/K} = n \cdot \mathrm{Trd}_{D/K}$, as may be checked by extending scalars to K_s and directly computing with elementary matrices (see 1.2.3.3). Hence, in the setting of interest with $D = \mathrm{End}^0(A)$ and $K = Z$ we see that it is equivalent to prove positive-definiteness for the quadratic form $x \mapsto \mathrm{Trm}_{D/\mathbb{Q}}(xx^*)$, where $\mathrm{Trm}_{D/\mathbb{Q}} = \mathrm{Tr}_{Z/\mathbb{Q}} \circ \mathrm{Trm}_{D/Z}$. The positive-definiteness for $\mathrm{Trm}_{D/\mathbb{Q}}$ can be verified by replacing D with $\mathrm{End}^0_{\overline{K}}(A_{\overline{K}})$, to which [**82**, §21, Thm. 1] applies. □

Lemma 1.3.5.3 says that $x \mapsto x^*$ is a *positive involution* of D (relative to the linear form $\mathrm{Trd}_{D/\mathbb{Q}}$). The existence of such an involution severely constrains the possibilities for D. First we record the consequences for the center Z.

1.3.5.4. Lemma. *The center Z of $D = \mathrm{End}^0(A)$ is either totally real or a* CM *field, and in the latter case its canonical complex conjugation is induced by the Rosati involution defined by any polarization of A over K.*

PROOF. Fix a polarization and consider the associated Rosati involution $x \mapsto x^*$ on the center Z of D. Clearly Z is stable under this involution. The positive-definite $\mathrm{Trd}_{D/\mathbb{Q}}(xx^*)$ on D restricts to $\sqrt{[D:Z]} \cdot \mathrm{Tr}_{Z/\mathbb{Q}}(xx^*)$ on Z, so $\mathrm{Tr}_{Z/\mathbb{Q}}(xx^*)$ is positive-definite on Z. If $x^* = x$ for all $x \in Z$ then the rational quadratic form $\mathrm{Tr}_{Z/\mathbb{Q}}(x^2)$ is positive-definite on Z, so by extending scalars to $\mathbb{R}$ we see that $\mathrm{Tr}_{(\mathbb{R}\otimes_{\mathbb{Q}}Z)/\mathbb{R}}(x^2)$ is positive-definite. This forces the finite étale $\mathbb{R}$-algebra $\mathbb{R} \otimes_{\mathbb{Q}} Z$ to have no complex factors. Hence, Z is a totally real field in such cases.

It remains to show that if the involution $x \mapsto x^*$ is non-trivial on Z for some choice of polarization then Z is a CM field (so the preceding argument would imply that the Rosati involution arising from *any* polarization of A is non-trivial on Z) and its intrinsic complex conjugation is equal to this involution on Z. Let Z^+ be the subfield of fixed points in Z for this involution, so $[Z : Z^+] = 2$ and $2\,\mathrm{Tr}_{Z^+/\mathbb{Q}}$ is the restriction to Z^+ of $\mathrm{Tr}_{Z/\mathbb{Q}}$. Hence, $\mathrm{Tr}_{Z^+/\mathbb{Q}}(x^2)$ is positive-definite on Z^+, so Z^+ is totally real. We aim to prove that Z has no real places, so we assume otherwise and seek a contradiction.

Let v be a real place of Z. Since the involution $x \mapsto x^*$ is non-trivial on Z and the field $Z_v \simeq \mathbb{R}$ has no non-trivial field automorphisms, the real place v on Z is not fixed by the involution $x \mapsto x^*$. Thus, the real place v^* obtained from v under the involution is a real place of Z distinct from v, and so the positive-definiteness of $\mathrm{Tr}_{Z/\mathbb{Q}}(xx^*)$ implies (after scalar extension to $\mathbb{R}$) the positive-definiteness of $\mathrm{Tr}_{(Z_v \times Z_{v^*})/\mathbb{R}}(xx^*)$, where $x \mapsto x^*$ on $Z_v \times Z_{v^*} = \mathbb{R} \times \mathbb{R}$ is the involution that swaps the factors. In other words, this is the quadratic form $(c, c') \mapsto 2cc'$, which by inspection is not positive-definite. □

1.3.6. Albert's classification. To go further with the proof of Theorem 1.3.4, we need to review properties of endomorphism algebras of simple abelian varieties over arbitrary fields.

1.3.6.1. Definition. An *Albert algebra* is a pair consisting of a division algebra D of finite dimension over $\mathbb{Q}$ and a positive involution $x \mapsto x^*$ on D.

For any Albert algebra D and any algebraically closed field K, there exists a simple abelian variety A over K such that $\mathrm{End}^0(A)$ is $\mathbb{Q}$-isomorphic to D (with the given involution on D arising from a polarization on A); see [**1**], [**2**], [**3**], [**112**, §4.1, Thm. 5], and [**46**, Thm. 13]. For a survey and further references on this topic, see [**92**]. We will not need this result.

Instead, we are interested in the non-trivial constraints on the Albert algebras that arise from polarized simple abelian varieties A over an arbitrary field K when $\mathrm{char}(K)$ and $\dim A$ are fixed. Before listing these constraints, it is convenient to record Albert's classification of general Albert algebras (omitting a description of the possibilities for the involution).

1.3.6.2. Theorem (Albert). *Let $(D, (\cdot)^*)$ be an Albert algebra. For any place v of the center Z, let v^* denote the pullback of v along $x \mapsto x^*$. Exactly one of the following occurs:*

Type I*: $D = Z$ is a totally real field.*

Type II*: D is a central quaternion division algebra over a totally real field Z such that D splits at each real place of Z.*

Type III*: D is a central quaternion division algebra over a totally real field Z such that D is non-split at each real place of Z.*

Type IV*: D is a central division algebra over a* CM *field Z such that for all finite places v of Z, $\mathrm{inv}_v(D) + \mathrm{inv}_{v^*}(D) = 0$ in $\mathbb{Q}/\mathbb{Z}$ and moreover D splits at such a v if $v = v^*$.*

PROOF. See [**82**, §21, Thm. 2] (which also records the possibilities for the involution). □

1.3.6.3. Let A be a simple abelian variety over a field K, $D = \mathrm{End}^0(A)$, and Z the center of D. Let Z^+ be the maximal totally real subfield of Z, so either $Z = Z^+$ or Z is a totally complex quadratic extension of Z^+. The invariants $e = [Z : \mathbb{Q}]$, $e_0 = [Z^+ : \mathbb{Q}]$, $d = \sqrt{[D : Z]}$, and $g = \dim(A)$ satisfy some divisibility restrictions:

- whenever $\mathrm{char}(K) = 0$, the integer $ed^2 = [D : \mathbb{Q}]$ divides $2g$ (proof: there is a subfield $K_0 \subseteq K$ finitely generated over $\mathbb{Q}$ such that A descends to an abelian variety A_0 over K_0 and the D-action on A in the isogeny category over K descends to an action on A_0 in the isogeny category over K_0, so upon choosing an embedding $K_0 \hookrightarrow \mathbb{C}$ we get a $\mathbb{Q}$-linear action by the division algebra D on the $2g$-dimensional homology $\mathrm{H}_1(A_0(\mathbb{C}), \mathbb{Q})$),
- the action of D on $V_\ell(A)$ with $\ell \neq \mathrm{char}(K)$ implies (via Cor. to Thm. 4 of [**82**, §19], whose proof is valid over any base field) that $ed | 2g$,
- the structure of symmetric elements in

$$\mathbb{Q} \otimes_{\mathbb{Z}} \mathrm{Hom}(A, A^t) \simeq \mathbb{Q} \otimes_{\mathbb{Z}} \mathrm{Pic}(A)/\mathrm{Pic}^0(A)$$

  (via [**82**, §20, Cor. to Thm. 3], whose proof is valid over any base field) yields that $[L : \mathbb{Q}] | g$ for every commutative subfield $L \subset D$ whose elements are invariant under the involution.
- for Type II in any characteristic we have $2e | g$ (which coincides with the general divisibility $ed^2 | 2g$ when $\mathrm{char}(K) = 0$ since $d = 2$ for Type II). To prove it uniformly across all characteristics, first note that for Type II we have

$$\mathbb{R} \otimes_{\mathbb{Q}} D = (\mathbb{R} \otimes_{\mathbb{Q}} Z) \otimes_Z D = \prod_{v | \infty} Z_v \otimes_Z D \simeq \mathrm{Mat}_2(Z_v)^e.$$

Moreover, by [**82**, §21, Thm. 2] it can be arranged that under this composite isomorphism the positive involution on D goes over to transpose on each factor $\mathrm{Mat}_2(Z_v) = \mathrm{Mat}_2(\mathbb{R})$. Thus, for D of Type II the fixed part of the involution on D has $\mathbb{Q}$-dimension $2e$ and hence Z-degree 2. By centrality of Z in the division algebra D, the condition $x^* = x$ for x in D of Type II therefore defines a necessarily commutative quadratic extension Z' of Z inside D, so g is divisible by $[Z' : \mathbb{Q}] = 2e$ as desired.

The preceding results are summarized in the following table, taken from the end of [**82**, §21]. (As we have just seen, the hypothesis there that K is algebraically closed is not necessary.) The invariants of $D = \mathrm{End}^0(A)$ are given in the first three columns. In the last two columns we give some necessary divisibility restrictions on these invariants.

Type	e	d	$\mathrm{char}(K) = 0$	$\mathrm{char}(K) > 0$
I	$e = e_0$	1	$e \mid g$	$e \mid g$
II	$e = e_0$	2	$2e \mid g$	$2e \mid g$
III	$e = e_0$	2	$2e \mid g$	$e \mid g$
IV	$e = 2e_0$	d	$e_0 d^2 \mid g$	$e_0 d \mid g$

We refer the reader to [**82**, §21], and to [**92**] for further information on these invariants. Using the above table, we can prove the following additional facts when the simple A admits sufficiently many complex multiplications.

1.3.6.4. Proposition. *Let A be a simple abelian variety of dimension $g > 0$ over a field K, and assume that A admits sufficiently many complex multiplications. Let $D = \mathrm{End}^0(A)$.*

(1) *If* $\mathrm{char}(K) = 0$ *then D is of Type* IV *with $d = 1$ and $e = 2g$ (so D is a* CM *field, by Theorem* 1.3.6.2*).*
(2) *If* $\mathrm{char}(K) > 0$ *then D is of Type* III *or Type* IV.

PROOF. By simplicity, D is a division algebra. Its center Z is a commutative field.

First suppose $\mathrm{char}(K) = 0$. Let $P \subset D$ be a commutative semisimple $\mathbb{Q}$-subalgebra with $[P : \mathbb{Q}] = 2g$. Since D is a division algebra, P is a field. The above table (or the discussion preceding it) says that the degree $[D : \mathbb{Q}] = ed^2$ divides $[P : \mathbb{Q}] = 2g$, so the inclusion $P \subset D$ is an equality. Thus, D is commutative (i.e., $d = 1$), so $D = Z$ is a commutative field and hence $e := [Z : \mathbb{Q}] = 2g$ by the complex multiplication hypothesis. The table shows that in characteristic 0 we have $e|g$ for Types I, II, and III, so D is of Type IV.

Suppose $\mathrm{char}(K) > 0$. In view of the divisibility relations in the table in positive characteristic, D is not of Type I since in such cases D is a commutative field whose $\mathbb{Q}$-degree divides $\dim(A)$, contradicting the existence of sufficiently many complex multiplications. For Type II we have $2e|g$, yet the integer $2e = 2[Z : \mathbb{Q}]$ is the $\mathbb{Q}$-degree of a maximal commutative subfield of the central quaternion division algebra D over Z, so there are no such subfields with $\mathbb{Q}$-degree $2g$. Since a commutative semisimple $\mathbb{Q}$-subalgebra of D is a field (as D is a division algebra), Type II is not possible if the simple A has sufficiently many complex multiplications. □

1.3.7. Returning to the proof of Theorem 1.3.4, recall that we reduced the proof to the case of simple A. Proposition 1.3.6.4(1) settles the case of characteristic 0, and Proposition 1.3.6.4(2) gives that $D = \mathrm{End}^0(A)$ is an Albert algebra of Type III or IV when $\mathrm{char}(K) > 0$. If D is of Type III then the center Z is totally real and d is even, whereas if D is of Type IV then Z is CM. Thus, we can apply the following general lemma to conclude the proof.

1.3.7.1. Lemma (Tate). *Let D be a central division algebra of degree d^2 over a number field Z that is totally real or* CM. *If Z is totally real then assume that d is even. There exists a maximal commutative subfield $L \subset D$ that is a* CM *field.*

The parity condition on d is necessary when Z is totally real, since $d = [L : Z]$ by maximality of L in D.

PROOF. By Proposition 1.2.3.1, any degree-d extension of Z that splits D is a maximal commutative subfield of D. Hence, we just need to find a degree-d extension L of Z that is a CM field and splits D. Let Σ be a finite non-empty set of finite places of Z containing the finite places at which D is non-split. By the structure of Brauer groups of local fields, for any $v \in \Sigma$ the central simple Z_v-algebra

$$D_v := Z_v \otimes_Z D$$

of rank d^2 over Z_v is split by any extension of Z_v of degree d.

First assume that Z is totally real, so d is even. By weak approximation (Theorem 1.3.5.1), there is a monic polynomial f over Z of degree $d/2$ that is close to a monic irreducible polynomial of degree $d/2$ over Z_v for all $v \in \Sigma$ (and in particular f is irreducible over all such Z_v, and hence over Z since Σ is non-empty). We can also arrange that for each real place v of Z the polynomial f viewed over $Z_v \simeq \mathbb{R}$ is close to a totally split monic polynomial of degree $d/2$ and hence is totally split over Z_v. Thus, $Z' := Z[u]/(f)$ is a totally real extension field of Z with degree $d/2$. By the same method, we can construct a quadratic extension L/Z' that is unramified quadratic over each place v' over a place in Σ and is also totally complex (by using approximations to irreducible quadratics over $\mathbb{R}$ at the real places of Z'). This L is a CM field and it is designed so that $Z_v \otimes_Z L$ is a degree-d field extension of Z_v for all $v \in \Sigma$. Hence, D_L is split at all places of L (the archimedean ones being obvious), so D_L is split.

Assume next that Z is a CM field. Let $Z^+ \subset Z$ be the maximal totally real subfield. By the same weak approximation procedure as above (replacing $d/2$ with d), we can construct a degree d totally real extension Z'^+/Z^+ such that for each place v_0 of Z^+ beneath a place $v \in \Sigma$, the extension Z'^+/Z^+ has a unique place v_0' over v_0 and is totally ramified (resp. unramified) at v_0' when Z/Z^+ is unramified (resp. ramified) at v. Hence, $(Z'^+)_{v_0'}$ and Z_v are linearly disjoint over $(Z^+)_{v_0}$. We conclude that Z'^+ and Z are linearly disjoint over Z', so $L := Z'^+ \otimes_{Z^+} Z$ is a field and each $v \in \Sigma$ has a unique place w over it in L. Clearly $[L_w : Z_v] = d$ for all such w, so L splits D. By construction, L is visibly CM. We have proved Lemma 1.3.7.1. This also finishes the proof of Theorem 1.3.4. $\square$

1.3.7.2. Corollary. *An isotypic abelian variety A with sufficiently many complex multiplications remains isotypic after any extension of the base field.*

PROOF. By Theorem 1.3.4, the endomorphism algebra $\mathrm{End}^0(A)$ contains a commutative field with $\mathbb{Q}$-degree $2\dim(A)$. This property is preserved after any ground field extension (even though the endomorphism algebra may get larger), so by the final part of Theorem 1.3.1.1 isotypicity is preserved as well. □

1.3.8. CM abelian varieties. It turns out to be convenient to view the CM algebra P in Theorem 1.3.4 as an abstract ring in its own right, and to thereby regard the embedding $P \hookrightarrow \mathrm{End}^0(A)$ as additional structure on A. This is encoded in the following concept.

1.3.8.1. Definition. Let A be an abelian variety over a field K, and assume that A has sufficiently many complex multiplications. Let $j : P \hookrightarrow \mathrm{End}^0(A)$ be an embedding of a CM algebra P with $[P : \mathbb{Q}] = 2\dim(A)$. Such a pair (A, j) is called a CM *abelian variety* (with complex multiplication by P).

Note that in this definition we are requiring P to be embedded in the endomorphism algebra of A over K (and not merely in the endomorphism algebra after an extension of K). For example, according to this definition, no elliptic curve over $\mathbb{Q}$ admits a CM structure (even if such a structure exists after an extension of the base field).

As an application of Theorem 1.3.4, we establish the following result concerning the possibilities for Z of Type III in Proposition 1.3.6.4(2). This will not be used later.

1.3.8.2. Proposition. *Let A, K, and D be as in Proposition* 1.3.6.4(2) *with* $p = \mathrm{char}(K) > 0$, *and let Z be the center of D,* $g = \dim(A)$, $d = \sqrt{[D : Z]}$, *and* $e = [Z : \mathbb{Q}]$. *We have $ed = 2g$, and if D is of Type* III *(so $d = 2$) then either $Z = \mathbb{Q}$ or $Z = \mathbb{Q}(\sqrt{p})$.*

Note that in this proposition, K is an arbitrary field with $\mathrm{char}(K) > 0$; K is not assumed to be finite.

PROOF. We always have $ed|2g$, but $ed = \sqrt{[D : \mathbb{Q}]}$ and D contains a field P of $\mathbb{Q}$-degree $2g$, so $2g|ed$. Thus, $ed = 2g$. Now we can assume A is of Type III, so the field Z is totally real.

Since A is of finite type over K and D is finite-dimensional over $\mathbb{Q}$, by direct limit considerations we can descend to the case when K is finitely generated over $\mathbb{F}_p$. Let S be a separated integral $\mathbb{F}_p$-scheme of finite type whose function field is K. Since A is an abelian variety over the direct limit K of the coordinate rings of the non-empty affine open subschemes of S, by replacing S with a sufficiently small non-empty affine open subscheme we can arrange that A is the generic fiber of an abelian scheme $\mathscr{A} \to S$. Since S is connected, the fibers of the map $\mathscr{A} \to S$ all have the same dimension, and this common dimension is g (as we may compute using the generic fiber A).

The $\mathbb{Z}$-module $\mathrm{End}(A)$ is finitely generated, and each endomorphism of A extends uniquely to a U-endomorphism of $\mathscr{A}_U$ for some non-empty open U in S (with U perhaps depending on the chosen endomorphism). Using a finite set of endomorphisms that spans $\mathrm{End}(A)$ allows us to shrink S so that all elements of $\mathrm{End}(A)$ extend to S-endomorphisms of $\mathscr{A}$, or in other words $\mathrm{End}(A) = \mathrm{End}(\mathscr{A})$. We therefore have a specialization map $D = \mathrm{End}^0(A) \to \mathrm{End}^0(\mathscr{A}_s)$ for every $s \in S$.

Fix a prime $\ell \neq p$. Since S is connected and $\mathscr{A}[\ell^n]$ is finite étale over S, an S-endomorphism of $\mathscr{A}[\ell^n]$ is uniquely determined by its effect on a single geometric fiber over S. But maps between abelian varieties are uniquely determined by their effect on ℓ-adic Tate modules when ℓ is a unit in the base field, so we conclude (via consideration of ℓ-power torsion) that the specialization map $D \to \mathrm{End}^0(\mathscr{A}_s)$ is injective for all $s \in S$. We can therefore speak of an element of $\mathrm{End}^0(\mathscr{A}_s)$ "lifting" over K in the sense that it is the image of a unique element of $D = \mathrm{End}^0(A)$ under the specialization mapping at s. This will be of interest when s is a closed point and we consider the q_s-Frobenius endomorphism of $\mathscr{A}_s$ over the finite residue field $\kappa(s)$ at s (with $q_s = \#\kappa(s)$).

By Theorem 1.3.4, we can choose a CM field $L \subset D$ with $[L : \mathbb{Q}] = 2g$. In particular, for each $s \in S$ the field L embeds into $\mathrm{End}^0(\mathscr{A}_s)$ with $[L : \mathbb{Q}] = 2g = 2\dim(\mathscr{A}_s)$, so each $\mathscr{A}_s$ is isotypic. By Theorem 1.3.1.1, L is its own centralizer in $\mathrm{End}^0(\mathscr{A}_s)$. Take s to be a closed point of S, and let q_s denote the size of the finite residue field $\kappa(s)$ at s. The q_s-Frobenius endomorphism $\varphi_s \in \mathrm{End}^0(\mathscr{A}_s)$ is central, so it centralizes L and hence must lie in the image of L. In particular, φ_s lifts to an element of $\mathrm{End}^0(A) = D$ that is necessarily central (as we may compute after applying the injective specialization map $D \hookrightarrow \mathrm{End}^0(\mathscr{A}_s)$). That is, $\varphi_s \in Z \subset D$ for all closed points $s \in S$.

Let Z' be the subfield of Z generated over $\mathbb{Q}$ by the lifts of the endomorphisms φ_s as s varies through all closed points of S. Each $\mathbb{Q}[\varphi_s]$ is a totally real field since Z is totally real. By Weil's Riemann Hypothesis for abelian varieties over finite fields (see the discussion following Definition 1.6.1.2), under any embedding $\iota : \mathbb{Q}[\varphi_s] \hookrightarrow \mathbb{C}$ we have each $\iota(\varphi_s)\overline{\iota(\varphi_s)} = q_s$ for $q_s = \#\kappa(s) \in p^{\mathbb{Z}}$, so the real number $\iota(\varphi_s)$ is a power of $\sqrt{p}$. Hence, the subfield $\mathbb{Q}[\varphi_s] \subset Z$ is either $\mathbb{Q}$ or $\mathbb{Q}(\sqrt{p})$, so the subfield $Z' \subset Z$ is either $\mathbb{Q}$ or $\mathbb{Q}(\sqrt{p})$.

Let η be a geometric generic point of S, and let Γ be the associated absolute Galois group for the function field of S. Because each $\mathscr{A}[\ell^n]$ is finite étale over S, the representation of Γ on $V_\ell(A)$ factors through the quotient $\pi_1(S, \eta)$. The Chebotarev Density Theorem for $\pi_1(S, \eta)$ [**97**, App. B.9] says that the Frobenius elements at the closed points of S generate a dense subgroup of the quotient $\pi_1(S, \eta)$ of Γ. Thus, the image of $\mathbb{Q}_\ell[\Gamma]$ in $\mathrm{End}_{\mathbb{Q}_\ell}(V_\ell(A))$ is equal to the subalgebra $Z'_\ell := \mathbb{Q}_\ell \otimes_{\mathbb{Q}} Z'$ generated by the endomorphisms φ_s. We therefore have an injective map

$$\mathbb{Q}_\ell \otimes_{\mathbb{Q}} D \hookrightarrow \mathrm{End}_{\mathbb{Q}_\ell[\Gamma]}(V_\ell(A)) = \mathrm{End}_{Z'_\ell}(V_\ell(A)).$$

By Zarhin's theorem [**134**] (see [**80**, XII, §2] for a proof valid for all p, especially allowing $p = 2$) this injection is an isomorphism, so we conclude that Z_ℓ is central in $\mathrm{End}_{Z'_\ell}(V_\ell(A))$. But the center of this latter matrix algebra is Z'_ℓ, so the inclusion $Z'_\ell \subset Z_\ell$ is an equality. Hence, the inclusion $Z' \subset Z$ is an equality as well. Since Z' is either $\mathbb{Q}$ or $\mathbb{Q}(\sqrt{p})$, we are done. □

1.4. Dieudonné theory, p-divisible groups, and deformations

To solve problems involving lifts from characteristic p to characteristic 0, we need a technique for handling p-torsion phenomena in characteristic $p > 0$. The two main tools for this purpose in what we shall do are Dieudonné theory and p-divisible groups. For the convenience of the reader we review the basic facts in this direction, and for additional details we refer to [**119**], [**110**, §6], and [**75**] for

p-divisible groups, and [**41**, Ch. II–III] for (contravariant) Dieudonné theory with applications to p-divisible groups.

1.4.1. Exactness. We shall frequently use exact sequence arguments with abelian varieties and finite group schemes over fields, as well as with their relative analogues over more general base schemes. It is assumed that the reader has some familiarity with these notions, but we now provide a review of this material.

1.4.1.1. Definition. Let S be a scheme, and let $\mathscr{T}$ be a Grothendieck topology on the category of S-schemes. (For our purposes, only the étale, fppf, and fpqc topologies will arise.) A diagram $1 \to G' \to G \xrightarrow{f} G'' \to 1$ of S-group schemes is *short exact* for the topology $\mathscr{T}$ if $G' \to G$ is an isomorphism onto $\ker(f)$ and the map f is a $\mathscr{T}$-covering.

By [**30**, Exp. IV, 5.1.7.1], in such cases G'' represents the quotient sheaf G/G' for the chosen Grothendieck topology. By [**31**, Exp. V, Thm. 4.1(iv), Rem. 5.1], if G is a quasi-projective group scheme over a noetherian ring R and if G' is a finite flat R-subgroup of G then the fppf quotient sheaf G/G' is represented by a quasi-projective R-group (also denoted G/G'), and the resulting map of group schemes $G \to G/G'$ is an fppf G'-torsor (so G/G' is R-flat if G is).

1.4.1.2. The *Cartier dual* N^D of a commutative finite locally free group scheme N over a base scheme S is the commutative finite locally free group scheme which represents the fppf sheaf functor $\mathscr{H}om(N, \mathbb{G}_{\mathrm{m}}) : S' \rightsquigarrow \mathrm{Hom}_{S'\text{-gp}}(N_{S'}, \mathbb{G}_{\mathrm{m}})$ on the category of S-schemes. The structure sheaf $\mathscr{O}_{N^D}$ of N^D is canonically isomorphic to the $\mathscr{O}_S$-linear dual of the structure sheaf $\mathscr{O}_N$ of N, and the co-multiplication (respectively multiplication) map for $\mathscr{O}_{N^D}$ is the $\mathscr{O}_S$-linear dual of the multiplication (respectively co-multiplication) map for $\mathscr{O}_N$.

The functor $N \rightsquigarrow N^D$ on the category of commutative finite locally free group schemes over S swaps closed immersions and quotient maps, preserves exactness, and is an involution in the sense that there is a natural isomorphism $f_N : N \simeq (N^D)^D$ satisfying $(f_N)^D = f_{N^D}^{-1}$. See [**87**, Prop. 2.9] for further details.

As an application, if the S-homomorphism $j : G' \hookrightarrow G$ is a closed immersion between finite locally free commutative group schemes then we can use Cartier duality to give a direct proof that the the fppf quotient sheaf G/G' is represented by a finite locally free S-group (without needing to appeal to general existence results for quotients by G'-actions on quasi-projective S-schemes). The key point is that the Cartier dual map $j^D : G^D \to G'^D$ between finite flat S-schemes is faithfully flat, as this holds on fibers over S (since injective maps between Hopf algebras over a field are always faithfully flat [**126**, 14.1]). Such flatness implies that $H := \ker(j^D)$ is a finite locally free commutative S-group, so H^D makes sense and the dual map $q : G \simeq (G^D)^D \to H^D$ is faithfully flat. It is clear that $G' \subset \ker(q)$, and this inclusion between finite locally free S-schemes is an isomorphism by comparison of fibral degrees, so H^D represents G/G'.

The following result is useful for constructing commutative group schemes $G \to S$ that are finite and fppf (equivalently, finite and locally free over S).

1.4.1.3. Proposition. *Let S be a scheme, and let G' and G'' be finitely presented separated S-group schemes with G' affine and flat over S. For any exact sequence*

$$1 \to G' \to G \to G'' \to 1$$

of group sheaves for the fppf topology on the category of S-schemes, G is represented by a finitely presented S-group that is flat and affine over G''. Moreover, G' and G'' are S-finite if and only if G is S-finite.

See [**87**, 17.4] for a generalization (using the fpqc topology).

PROOF. For any G''-scheme T viewed as an S-scheme, let $g'' \in G''(T)$ correspond to the given S-morphism $T \to G''$. Consider the set $E_{g''}(T)$ that is the preimage under $G(T) \to G''(T)$ of g''. This is a sheaf of sets on the category of G''-schemes equipped with the fppf topology, and as such it is a left G'-torsor (strictly speaking, a left torsor for the G''-group $G'_{G''}$) due to the given exact sequence. In particular, the fppf sheaves of sets $E_{g''}$ and $G'_{G''}$ over G'' are isomorphic fppf-locally over G''.

Since G' is fppf affine over S and fppf descent is effective for affine morphisms, it follows that $E_{g''}$ as an fppf sheaf of sets over G'' is represented by an affine fppf G''-scheme (which is therefore affine fppf over S when G'' is). It is elementary to check that this affine G''-scheme viewed as an S-scheme has its functor of points naturally identified with G (since for any S-scheme T and $g \in G(T)$, visibly $g \in E_{g''}(T)$ for the point $g'' \in G''(T)$ arising from g), so G is represented by an S-group.

Separatedness of G'' over S and exactness imply that G' is closed in G. Moreover, $G \to G''$ is a left $G'_{G''}$-torsor for the fppf topology over G'', so it is finite when G' is S-finite. Thus, if G' and G'' are S-finite then G is S-finite. Conversely, if G is S-finite then its closed subscheme G' is S-finite, so the quotient G/G' exists as an S-finite scheme. But G'' represents this quotient, so G'' is S-finite too. □

1.4.1.4. Remark. If $1 \to G' \to G \to G'' \to 1$ is an exact sequence of separated fppf S-groups with G' and G'' abelian schemes then G is an abelian scheme. Indeed, since $G \to G''$ is an fppf torsor for the G''-group $G'_{G''}$ that is smooth and proper with geometrically connected fibers, $G \to G''$ is smooth and proper with geometrically connected fibers. The map $G'' \to S$ is also smooth and proper with geometrically connected fibers, so $G \to S$ is as well. Hence, G is an abelian scheme.

It is also true that if G is an abelian scheme and G' is a closed S-subgroup of G that is also an abelian scheme then the fppf quotient sheaf G/G' is represented by an abelian scheme. We will give an elementary proof of this over fields in Lemma 1.7.4.4 using the Poincaré reducibility theorem (which is only available over fields). In general the proof requires a detour through algebraic spaces.

1.4.2. Duality for abelian schemes. In [**83**, §6.1], duality is developed for projective abelian schemes, building on the case of an algebraically closed ground field. Projectivity is imposed primarily due to the projectivity hypotheses in Grothendieck's work on Hilbert schemes. The projective case is sufficient for our needs because any abelian scheme over a discrete valuation ring is projective (this follows from Lemma 2.1.1.1, to which the interested reader may now turn). For both technical and aesthetic reasons, it is convenient to avoid the projectivity hypothesis. We now sketch Grothendieck's results on duality in the projective case, as well as Artin's improvements that eliminated the projectivity assumption.

1.4.2.1. Let $A \to S$ be an abelian scheme, and let $\mathrm{Pic}_{A/S}$ be the functor assigning to any S-scheme T the group of isomorphism classes of pairs $(\mathscr{L}, i)$ consisting of an invertible sheaf $\mathscr{L}$ on A_T and a trivialization $i : e_T^*(\mathscr{L}) \simeq \mathcal{O}_T$ along the identity section e_T of $A_T \to T$. This is an fppf group sheaf on the category of S-schemes, and its restriction to the category of S'-schemes (for an S-scheme S') is $\mathrm{Pic}_{A_{S'}/S'}$.

Let $\mathrm{Pic}^0_{A/S} \subset \mathrm{Pic}_{A/S}$ be the subfunctor classifying pairs $(\mathscr{L}, i)$ that lie in the identity component of the Picard scheme on geometric fibers. By [**7**, Exp. XIII, Thm. 4.7(i)] (see [**39**, §9.6] for the projective case), the inclusion $j : \mathrm{Pic}^0_{A/S} \to \mathrm{Pic}_{A/S}$ is an open subfunctor; i.e., it is relatively representable by open immersions. This means that for any S-scheme T and $(\mathscr{L}, i) \in \mathrm{Pic}_{A/S}(T)$, $\mathrm{Pic}^0_{A/S} \times_{\mathrm{Pic}_{A/S}} T$ as a functor on T-schemes is represented by an open subscheme $U \subset T$; explicitly, there is an open subscheme $U \subset T$ such that a T-scheme T' lies over U if and only if the T'-pullback of $(\mathscr{L}, i)$ lies in Pic^0 on geometric fibers over T'.

By Grothendieck's work on Picard schemes (see [**39**, Part 5]), if $A \to S$ is projective Zariski-locally on S then $\mathrm{Pic}_{A/S}$ is represented by a locally finitely presented and separated S-scheme and the open subscheme A^t representing $\mathrm{Pic}^0_{A/S}$ is quasi-projective Zariski-locally on S and finitely presented. For noetherian S, functorial criteria show that A^t is proper and smooth (see [**83**, §6.1]), hence an abelian scheme; the case of general S (with A projective Zariski-locally on S) then follows by descent to the noetherian case.

To drop the projectivity hypothesis, one has to use algebraic spaces. Informally, an algebraic space over S is an fppf sheaf on the category of S-schemes that is "well-approximated" by a representable functor (relative to the étale topology), so concepts from algebraic geometry such as smoothness, properness, and connectedness can be defined and behave as expected; see [**60**]. By Artin's work on relative Picard functors as algebraic spaces (see [**5**, Thm. 7.3]), $\mathrm{Pic}_{A/S}$ is always a separated algebraic space locally of finite presentation, and by [**7**, Exp. XIII, Thm. 4.7(iii)] the open algebraic subspace $\mathrm{Pic}^0_{A/S}$ is finitely presented over S.

The functorial arguments that prove smoothness and properness for $\mathrm{Pic}^0_{A/S}$ when A is projective work without projectivity because the same criteria are available for algebraic spaces. Thus, $\mathrm{Pic}^0_{A/S}$ is smooth and proper over S in the sense of algebraic spaces. Consequently, by a theorem of Raynaud (see [**38**, Thm. 1.9]), $\mathrm{Pic}^0_{A/S}$ is represented by an S-scheme A^t; this must be an abelian scheme, called the *dual abelian scheme*. Its formation commutes with any base change on S, and it is contravariant in A in an evident manner.

1.4.2.2. Over $A \times A^t$ there is a Poincaré bundle $\mathscr{P}_{A/S}$ provided by the universal property of A^t, exactly as in the theory of duality for abelian varieties over a field. In particular, $\mathscr{P}_{A/S}$ is canonically trivialized along $e \times \mathrm{id}_{A^t}$. Let $e' \in A^t(S)$ be the identity, so for any S-scheme T the point $e'_T \in A^t(T)$ corresponds to $\mathcal{O}_{A_T}$ equipped with the canonical trivialization of $e_T^*(\mathcal{O}_{A_T})$. Thus, setting $T = A$ gives that $\mathscr{P}_{A/S}$ is also canonically trivialized along $\mathrm{id}_A \times e'$. Hence, the pullback of $\mathscr{P}_{A/S}$ along the flip $A^t \times A \simeq A \times A^t$ defines a canonical S-morphism $\iota_{A/S} : A \to A^{tt}$. This morphism carries the identity to the identity, so it is a homomorphism. By applying the duality theory over fields to the fibers of A over S, it follows that $\iota_{A/S}$ is an isomorphism; in other words, the pullback of $\mathscr{P}_{A/S}$ along the flip $A^t \times A \simeq A \times A^t$ is uniquely isomorphic to $\mathscr{P}_{A^t/S}$ respecting trivializations along the identity sections of both factors. Such uniqueness implies that $\iota^t_{A/S}$ is inverse to $\iota_{A^t/S}$.

A homomorphism $f : A \to A^t$ is *symmetric* when the map

$$f^t \circ \iota_{A/S} : A \simeq A^{tt} \to A^t$$

is equal to f. Writing $f^\dagger := f^t \circ \iota_{A/S}$, the equality $\iota^t_{A/S} = \iota^{-1}_{A^t/S}$ and the functoriality of $\iota_{A/S}$ in A (applied with respect to f) implies $f^{\dagger\dagger} = f$, so if we abuse notation by writing f^t rather than $f^\dagger$ then $(f^t)^t = f$. We say f is *symmetric* when $f^t = f$ (or more accurately, $f^\dagger = f$). This property holds if it does so on fibers over S, because homomorphisms $f, f' : A \rightrightarrows B$ between abelian schemes coincide if $f_s = f'_s$ for all $s \in S$. Indeed, for noetherian S such rigidity is [**83**, Cor. 6.2], and the general case reduces to this because equality on all fibers descends through direct limits (since it says that the finitely presented ideal of $(f, f')^{-1}(\Delta_{A/S})$ in $\mathcal{O}_A$ is nilpotent).

A *polarization* of A is a homomorphism $f : A \to A^t$ that is a polarization on geometric fibers. Any such f is necessarily symmetric. The properties of polarizations are developed in [**83**, §6.2] for projective abelian schemes, but the only purpose of imposing projectivity at the outset (even though it is a consequence of the definition, due to [**34**, IV_3, 9.6.4]) is to ensure the existence of the dual abelian scheme, so such an assumption may be eliminated.

1.4.2.3. Definition. A homomorphism $\varphi : A \to B$ between abelian schemes over a scheme S is an *isogeny* when it is surjective with finite fibers. (Equivalently, the homomorphims φ_s are isogenies in the sense of abelian varieties for each $s \in S$.)

Since quasi-finite proper morphisms are finite by [**34**, IV_4, 18.12.4] (or by [**34**, IV_3, 8.11.1] with finite presentation hypothesis, which suffices for us), any isogeny between abelian schemes is a finite morphism. Moreover, by the fibral flatness criterion [**34**, IV_3, 11.3.11], such maps are flat. Hence, if φ as above is an isogeny then it is finite locally free (and surjective), so the closed subgroup $\ker(\varphi)$ is a finite locally free commutative S-group scheme. Thus, B represents the fppf quotient sheaf $A/\ker(\varphi)$. For example, setting $\varphi = [n]_A$ for $n \geqslant 1$ gives $A/A[n] \simeq A$.

Turning this around, suppose we are given the abelian scheme A and a closed S-subgroup $N \subset A$ that is finite locally free over S. Consider the fppf quotient sheaf A/N. We claim that this quotient is (represented by) an abelian scheme, so the map $A \to A/N$ with kernel N is an isogeny. It suffices to work Zariski-locally on S, so we may assume that $N \to S$ has all fibers with the same order $n \geqslant 1$. We then have $N \subset A[n]$, due to the following result (proved in [**123**, §1]):

1.4.2.4. Theorem (Deligne)**.** *Let S be a scheme and let H be a commutative S-group scheme for which the structural morphism $H \to S$ is finite and locally free. If the fibers H_s have rank n for all $s \in S$ then H is killed by n.*

The quotient sheaf A/N is an fppf torsor over $A/A[n] \simeq A$ with fppf covering group $A[n]/N$ that is finite (and hence *affine*) over S. It then follows from effective fppf descent for affine morphisms that the quotient A/N is represented by a scheme finite over $A/A[n] = A$, and the map $A \to A/N$ is an fppf $A[n]/N$-torsor, so the S-proper S-smooth A is finite locally free over A/N (as $A[n]/N$ is finite locally free over S). Hence, A/N is proper and smooth since A is, and likewise its fibers over S are geometrically connected. Thus, A/N is an abelian scheme as desired.

1.4.2.5. Theorem. *Let $\varphi : A \to B$ be an isogeny between abelian schemes over a scheme S, and let $N = \ker(\varphi)$. Duality applied to the exact sequence*

$$0 \to N \longrightarrow A \xrightarrow{\varphi} B \to 0$$

functorially yields an exact sequence

$$0 \to N^D \longrightarrow B^t \xrightarrow{\varphi^t} A^t \to 0.$$

That is, the map φ^t is an isogeny whose kernel is canonically isomorphic to N^D.

Moreover, double duality for abelian schemes and for finite locally free commutative group schemes are compatible up to a sign: if we identify φ and φ^{tt} via $\iota_{A/S}$ and $\iota_{B/S}$ then the natural isomorphism $(N^D)^D \simeq \ker((\varphi^t)^t) = \ker(\varphi^{tt}) \simeq \ker(\varphi) = N$ is the negative *of the canonical isomorphism provided by Cartier duality.*

We refer the reader to [**86**, Thm. 1.1, Cor. 1.3] for a proof based on arguments that relativize the ones over an algebraically closed field in [**82**]. (An alternative approach, at least for the first part, is [**87**, Thm. 19.1], resting on the link between dual abelian schemes and Ext-sheaves given in [**87**, Thm. 18.1].) The special case $\varphi = [n]_A : A \to A$ implies that naturally $A[n]^D = A^t[n]$ for every $n \geqslant 1$ because $[n]_A^t = [n]_{A^t}$ (by [**87**, 18.3]); this identification respects multiplicative change in n.

1.4.3. Constructions and definitions. Let us now focus on constructions specific to the theory of finite commutative group schemes over a perfect field k of characteristic $p > 0$. Let $W = W(k)$ be the ring of Witt vectors of k; e.g., if k is finite of size $q = p^r$ then W is the ring of integers in an unramified extension of $\mathbb{Q}_p$ of degree r. Let σ be the unique automorphism of W that reduces to the map $x \mapsto x^p$ on the residue field k.

1.4.3.1. Definition. The *Dieudonné ring* D_k over k is $W[\mathcal{F}, \mathcal{V}]$, where $\mathcal{F}$ and $\mathcal{V}$ are indeterminates subject to the relations

(1) $\mathcal{F}\mathcal{V} = \mathcal{V}\mathcal{F} = p$,
(2) $\mathcal{F}c = \sigma(c)\mathcal{F}$ and $c\mathcal{V} = \mathcal{V}\sigma(c)$ for all $c \in W$.

Explicitly, elements of D_k have unique expressions as finite sums

$$a_0 + \sum_{j>0} a_j \mathcal{F}^j + \sum_{j>0} b_j \mathcal{V}^j$$

with coefficients in W (so the center of D_k is clearly $\mathbb{Z}_p[\mathcal{F}^r, \mathcal{V}^r]$ if k has finite size p^r and it is $\mathbb{Z}_p$ otherwise; i.e., if k is infinite).

Some of the main conclusions in classical Dieudonné theory, as developed from scratch in [**41**, Ch. I–III], are summarized in the following theorem.

1.4.3.2. Theorem. *There is an additive anti-equivalence of categories $G \rightsquigarrow \mathrm{M}^*(G)$ from the category of finite commutative k-group schemes of p-power order to the category of left D_k-modules of finite W-length. Moreover, the following hold.*

(1) *A group scheme G has order $p^{\ell_W(\mathrm{M}^*(G))}$, where $\ell_W(\cdot)$ denotes W-length.*
(2) *If $k \to k'$ is an extension of perfect fields with associated extension $W \to W'$ of Witt rings (e.g., the absolute Frobenius automorphism of k) then the functor $W' \otimes_W (\cdot)$ on Dieudonné modules is naturally identified with the base-change*

functor on finite commutative group schemes. In particular, $\mathrm{M}^*(G^{(p)}) \simeq \sigma^*(\mathrm{M}^*(G))$ *as W-modules.*

(3) *Let* $\mathrm{Fr}_{G/k} : G \to G^{(p)}$ *be the relative Frobenius morphism. The σ-semilinear action on* $\mathrm{M}^*(G)$ *induced by* $\mathrm{M}^*(\mathrm{Fr}_{G/k})$ *through the isomorphism* $\mathrm{M}^*(G^{(p)}) \simeq \sigma^*(\mathrm{M}^*(G))$ *equals the action of $\mathcal{F}$, and G is connected if and only if $\mathcal{F}$ is nilpotent on* $\mathrm{M}^*(G)$.

(4) *There is a natural k-linear isomorphism* $\mathrm{M}^*(G)/\mathcal{F}\mathrm{M}^*(G) \simeq \mathrm{Lie}(G)^\vee$ *respecting extension of the perfect base field.*

(5) *For the Cartier dual G^D, naturally* $\mathrm{M}^*(G^D) \simeq \mathrm{Hom}_W(\mathrm{M}^*(G), K/W)$ *with* $K = W[1/p]$, *using the operators* $\mathcal{F}(\ell) : m \mapsto \sigma(\ell(\mathcal{V}(m)))$ *and* $\mathcal{V}(\ell) : m \mapsto \sigma^{-1}(\ell(\mathcal{F}(m)))$ *on K/W-valued linear forms ℓ.* □

For an abelian scheme $A \to S$ with fibers of constant dimension $g \geqslant 1$ and its finite commutative p^n-torsion subgroup scheme $A[p^n]$ with order $(p^n)^{2g}$, the directed system $A[p^\infty] := (A[p^n])_{n \geqslant 1}$ satisfies the following definition (with $h = 2g$).

1.4.3.3. Definition. A *p-divisible group of height $h \geqslant 0$* over a scheme S is a directed system $G = (G_n)_{n \geqslant 1}$ of commutative S-groups G_n such that: G_n is killed by p^n, each $G_n \to S$ is finite and locally free, $[p]\colon G_{n+1} \to G_n$ is faithfully flat for every $n \geqslant 1$, $G_1 \to S$ has constant degree p^h, and G_n is identified with $G_{n+1}[p^n]$ for all $n \geqslant 1$.

The (Serre) *dual p-divisible group* G^t is the directed system (G_n^D) of Cartier dual group schemes G_n^D with the transition maps $G_n^D \to G_{n+1}^D$ that are Cartier dual to the quotient maps $[p] : G_{n+1} \to G_n$.

As an illustration, if $A \to S$ is an abelian scheme with fibers of dimension $g \geqslant 1$ then the isomorphisms $A[n]^D \simeq A^t[n]$ respecting multiplicative change in n (as noted immediately below Theorem 1.4.2.5) yield a canonical isomorphism between the Serre dual $A[p^\infty]^t$ and the p-divisible group $A^t[p^\infty]$ of the dual abelian scheme A^t (see [**86**, Prop. 1.8] or [**87**, Thm. 19.1]).

1.4.3.4. Remark. In view of the sign discrepancy for comparisons of double duality in Theorem 1.4.2.5, if $\varphi : A \to A^t$ is an S-homomorphism and

$$f : A[p^\infty] \to A^t[p^\infty] \simeq A[p^\infty]^t$$

is the associated homomorphism between p-divisible groups then the dual homomorphism $\varphi^t : A \to A^t$ (strictly speaking, $\varphi^t \circ \iota_{A/S}$ via double duality for abelian schemes) has as its associated homomorphism $A[p^\infty] \to A[p^\infty]^t$ the *negative*[1] of f^t (using double duality for p-divisible groups).

It follows that if φ is symmetric with respect to double duality for abelian schemes then f is skew-symmetric with respect to double duality for p-divisible groups. The converse is also true: we can see immediately via skew-symmetry of f that φ and φ^t induce the same homomorphism between p-divisible groups, and to conclude that $\varphi = \varphi^t$ it suffices to check on fibers due to the rigidity of abelian schemes (as in 1.4.2.2). On fibers we can apply the faithfulness of passage to p-divisible groups over fields via 1.2.5.1 with $\ell = p$.

[1] A related sign issue in the double duality for commutative finite group schemes over perfect fields is discussed in a footnote at the end of B.3.5.5.

1.4.3.5. Example. If $G = (G_n)$ is a p-divisible group over S (with height h) and $H = (H_n)$ is a p-divisible subgroup of G (with height $h' \leqslant h$) in the sense that H_n is a closed S-subgroup of G_n compatibly in n, then $G/H := (G_n/H_n)$ is also a p-divisible group over S. Indeed, a computation with fppf abelian sheaves shows that the complex

$$0 \to G_1/H_1 \to G_{n+1}/H_{n+1} \overset{[p]}{\to} G_n/H_n$$

is left exact in the sense of fppf abelian sheaves and hence the induced map

$$(G_{n+1}/H_{n+1})/(G_1/H_1) \to G_n/H_n$$

between finite locally free commutative S-groups is a closed immersion (as is any proper monomorphism), so it is an isomorphism for order reasons. This shows that the map $[p] : G_{n+1}/H_{n+1} \to G_n/H_n$ is faithfully flat with kernel G_1/H_1 of order $p^{h-h'}$, so induction on n implies that $(G_n/H_n)[p^m]$ is faithfully flat of order $p^{m(h-h')}$ for any $m \leqslant n$. In particular, the closed immersion $G_n/H_n \to (G_{n+1}/H_{n+1})[p^n]$ is an isomorphism for order reasons, so (G_n/H_n) is a p-divisible group.

In the preceding example, clearly the natural map $q : G \to G/H$ has functorial kernel H and has the mapping property of a quotient: any homomorphism of p-divisible groups $G \to G''$ that kills H uniquely factors through q. Hence, it is appropriate to define a *short exact sequence* of p-divisible groups to be a complex

$$0 \to G' \to G \to G'' \to 0$$

such that G' is a p-divisible subgroup of G and the induced map $G/G' \to G''$ is an isomorphism, or equivalently the induced complex of finite locally free commutative S-groups

$$0 \to G'_n \to G_n \to G''_n \to 0$$

is short exact for all $n \geqslant 1$.

For example, if $0 \to A' \to A \to A'' \to 0$ is a short exact sequence abelian schemes (in the sense of fppf abelian sheaves on the category of S-schemes), or equivalently $A \to A''$ is faithfully flat with kernel A', or equivalently it is short exact on geometric fibers over every point of S, then a computation with the snake lemma for fppf abelian sheaves shows that the induced complex

$$0 \to A'[p^\infty] \to A[p^\infty] \to A''[p^\infty] \to 0$$

is short exact. Also, by the definition of the Serre dual p-divisible group in terms of Cartier duality at finite levels, the Serre dual of a short exact sequence of p-divisible groups is short exact.

1.4.3.6. Example. An important example of a short exact sequence of p-divisible groups is the *connected-étale sequence* for a p-divisible group over a complete local noetherian ring R with residue characteristic $p > 0$. To define this, first recall that for any finite flat commutative R-group scheme H, the connected component H^0 of the identity section is an open and closed R-subgroup (in particular, it inherits R-flatness from H, so it is a finite flat R-group) and the associated finite flat quotient $H^{\text{ét}} := H/H^0$ is finite étale; these properties can be seen via the special fiber. The short exact sequence

$$0 \to H^0 \to H \to H^{\text{ét}} \to 0$$

of R-group schemes is the *connected-étale sequence* for H.

A p-divisible group $G = (G_n)$ over R is *connected* if each G_n is connected; equivalently, every G_n has infinitesimal special fiber. By a snake lemma argument with fppf abelian sheaves and the connected-étale sequence for finite flat commutative R-group schemes, if $G = (G_n)$ is a p-divisible group over R then $G^0 := (G_n^0)$ is a p-divisible group (called the *connected component* of G) and $G^{\text{ét}} := (G_n^{\text{ét}})$ is a p-divisible group (called the *étale part* of G). We call

$$0 \to G^0 \to G \to G^{\text{ét}} \to 0$$

the *connected-étale sequence* for G.

Somewhat deeper lies the fact (see [**119**, §2.2] and [**75**, II, 3.3.18, 4.5]) that if G is a connected p-divisible group over R then $\mathcal{O}(G) := \varprojlim \mathcal{O}(G_n)$ is a formal power series ring in finitely many variables over R such that the induced formal R-group structure makes $[p]^* : \mathcal{O}(G) \to \mathcal{O}(G)$ finite flat, and moreover there is an equivalence

$$G \rightsquigarrow \widehat{G} := \operatorname{Spf}(\mathcal{O}(G))$$

from the category of connected p-divisible groups over R to the category of commutative formal Lie groups Γ over R for which $[p]_\Gamma$ is an isogeny. The quasi-inverse functor is $\Gamma \rightsquigarrow (\Gamma[p^n])$. This equivalence defines the (relative) *dimension* and *Lie algebra* for a connected p-divisible group over R, via analogous notions for formal Lie groups over R. By [**119**, §2.3, Prop. 3], $\dim(G) + \dim(G^t)$ is the height of G.

For later purposes, here is how this construction works in the important example of the p-divisible group $G = A[p^\infty]$ arising from an abelian R-scheme A. What is $\widehat{G}$? Let $\mathcal{C}_R$ denote the category of artinian local R-algebras that are module-finite over R (and hence killed by some power of the maximal ideal of R). Every point of A valued in such an algebra and supported at the identity of the special fiber is a point of the commutative formal Lie group $\widehat{A} := \operatorname{Spf}(\mathcal{O}^\wedge_{A,0})$. A computation with formal group laws shows that all such points have p-power torsion, due to R having residue characteristic p. (This calculation will be given in a self-contained manner in the proof of Proposition 1.4.4.3.) Thus, for any $C \in \mathcal{C}_R$, the formal Lie group $\widehat{A}$ has each of its C-points supported in some $A[p^n]^0$. It follows that $\widehat{A}$ and $\widehat{G}$ pro-represent the same functor on $\mathcal{C}_R$, so the natural map $\widehat{G} \to \widehat{A}$ is an isomorphism. In particular, the p-divisible group of A has the same (relative) dimension and Lie algebra as A does.

1.4.3.7. Now consider p-divisible groups over a perfect field k of characteristic $p > 0$. For any p-divisible group $G = (G_n)_{n \geqslant 1}$ over k with height $h \geqslant 1$ we let $\mathrm{M}^*(G)$ denote the D_k-module $\varprojlim \mathrm{M}^*(G_n)$. By the same style of arguments used to work out the $\mathbb{Z}_\ell$-module structure of Tate modules of abelian varieties in characteristic $\neq \ell$ (resting on knowledge of the orders of the ℓ-power torsion subgroups), we use W-length to replace counting to infer that $\mathrm{M}^*(G)$ is a free W-module of rank h and

$$\mathrm{M}^*(G)/p^n\mathrm{M}^*(G) \to \mathrm{M}^*(G_n)$$

is an isomorphism for all $n \geqslant 1$. The p-divisible group G is connected if and only if $\mathcal{F}$ is topologically nilpotent on $\mathrm{M}^*(G)$ (since this is equivalent to the nilpotence of $\mathcal{F}$ on each $\mathrm{M}^*(G_n)$).

The Dieudonné module functor defines an anti-equivalence between the category of p-divisible groups over k (using the evident notion of morphism) and the category of left D_k-modules that are finite and free as W-modules; the W-rank of $\mathrm{M}^*(G)$ is equal to the height of G.

The notion of *isogeny* for p-divisible groups over a general scheme will be discussed in 3.3.3–3.3.5, but for our present purposes we only need the case when the base is a perfect field. This special case is easier to develop, and it is also convenient to have it available (on geometric fibers) when considering the relative case. Thus, we now define and briefly study this concept over perfect fields.

1.4.3.8. Definition. A homomorphism $f : X \to Y$ between p-divisible groups over a perfect field K is an *isogeny* if $\ker f$ is a finite K-group and the heights of X and Y coincide.

We first explain what this means in more concrete terms when $\mathrm{char}(K) \neq p$ by using p-adic Tate modules, and then the interesting case of perfect K of characteristic p will proceed similarly by using Dieudonné modules.

Assume $\mathrm{char}(K) \neq p$, so all p-divisible groups over K are étale (in the sense that each p^n-torsion subgroup is étale over K). Via the formation of p-adic Tate modules, the category of p-divisible groups over K is equivalent to the category of continuous linear representations of $\mathrm{Gal}(K_s/K)$ on finite free $\mathbb{Z}_p$-modules. It follows that f is an isogeny if and only if the induced map between p-adic Tate modules becomes an isomorphism after inverting p. Thus, a homomorphism $f : X \to Y$ between p-divisible groups is an isogeny if and only if there is a homomorphism $f' : Y \to X$ such that $f' \circ f = [p^n]_X$ and $f \circ f' = [p^n]_Y$ for some $n \geqslant 0$, and such an f is a quotient modulo the finite kernel $\ker(f)$ in the sense that any homomorphism of p-divisible groups $X \to X''$ over K that kills $\ker(f)$ factors uniquely through f.

By forming a quotient Tate module, we likewise see that for any finite K-subgroup $G \subset X$ there is an isogeny of p-divisible groups $f : X \to Y$ over K with $\ker(f) = G$, and f is unique up to unique isomorphism in an evident sense. We call Y the *quotient* of X modulo G and denote it as X/G. (It is not entirely trivial to describe $Y[p^n]$ in terms of X and G, and this will make the analogous construction over a general base scheme less straightforward.)

Now assume $\mathrm{char}(K) = p$ (and K is perfect). By arguing with Dieudonné modules in the role of p-adic Tate modules, it is elementary to check that a homomorphism $f : X \to Y$ between p-divisible groups over K is an isogeny if and only if the induced map $\mathrm{M}^*(f)$ between Dieudonné modules becomes an isomorphism after inverting p. Consequently, we again obtain that f is an isogeny if and only if there is a homomorphism $f' : Y \to X$ such that $f' \circ f = [p^n]_X$ and $f \circ f' = [p^n]_Y$ for some $n \geqslant 0$, and that f has the expected universal mapping property for homomorphisms from X that kill the finite kernel of f. Likewise, for any finite K-subgroup $G \subset X$ the induced map of (contravariant!) Dieudonné modules $\mathrm{M}^*(X) \to \mathrm{M}^*(G)$ is surjective (since $G \subset X[p^n]$ for large n), so the kernel of this surjection is W-finite free of the same rank as $\mathrm{M}^*(X)$. The corresponding p-divisible group is denoted X/G because the evident map $X \to X/G$ is an isogeny with kernel G.

As with abelian varieties, the *isogeny category* of p-divisible groups over a perfect field k of characteristic $p > 0$ is defined either by formally inverting isogenies or more concretely by using as the Hom-sets $\mathrm{Hom}^0(X, Y) = \mathrm{Hom}(X, Y)[1/p]$. For an abelian variety A of dimension $g > 0$ over k, the D_k-module $\mathrm{M}^*(A[p^\infty])$ is finite free of rank $2g$ over W, so it is an analogue of the ℓ-adic Tate module for $\ell \neq \mathrm{char}(k)$ even though it is contravariant in A. The D_k-module structure is the analogue of the Galois action on ℓ-adic Tate modules, though the action by $\mathcal{F}$ and $\mathcal{V}$ is highly non-trivial even when $k = \bar{k}$ (whereas the Galois action on Tate modules is trivial for such k). As an analogue of the Poincaré reducibility theorem for abelian varieties, the isogeny category of p-divisible groups over k is semisimple when $k = \overline{k}$; see Theorem 3.1.3. (This semisimplicity fails more generally, even over finite fields, as we see in the étale case: Galois groups can have non-semisimple representations on finite-dimensional $\mathbb{Q}_p$-vector spaces.)

To illustrate the use of Dieudonné modules as a replacement for Tate modules in the case $\ell = p$, we have the following result that will be important in our later study of a notion of "complex multiplication" for p-divisible groups.

1.4.3.9. Proposition. *Let G be a p-divisible group of height $h > 0$ over a field κ of characteristic p, and let k be a perfect extension of κ.*

1. *If F is a commutative semisimple $\mathbb{Q}_p$-subalgebra of $\mathrm{End}^0(G) := \mathbb{Q}_p \otimes_{\mathbb{Z}_p} \mathrm{End}(G)$ then $[F : \mathbb{Q}_p] \leqslant h$, with equality if and only if $\mathrm{M}^*(G_k)[1/p]$ is free of rank 1 as a $W(k) \otimes_{\mathbb{Z}_p} F$-module.*
2. *When equality holds, F is its own centralizer in $\mathrm{End}^0(G)$. If moreover the maximal $\mathbb{Z}_p$-order $\mathcal{O}_F$ in F lies in $\mathrm{End}(G)$ then $\mathrm{M}^*(G_k)$ is free of rank 1 as a $W(k) \otimes_{\mathbb{Z}_p} \mathcal{O}_F$-module.*

In 3.1.8 we will show that $\mathrm{End}(G)$ is finitely generated as a $\mathbb{Z}_p$-module, but this fact is not needed here.

PROOF. We may and do replace κ with k, so κ is perfect. (In particular, we may use the notion of isogeny as in 1.4.3.8.) Letting $K_0 = W(\kappa)[1/p]$, we view $\mathrm{M}^*(G)[1/p]$ as a finite module over the semisimple ring $K_0 \otimes_{\mathbb{Q}_p} F$. The second condition in (2) is immediate from the freeness in (1) (as $W(\kappa) \otimes_{\mathbb{Z}_p} \mathcal{O}_F$ is a finite product of discrete valuation rings that are $W(\kappa)$-finite, and $\mathrm{M}^*(G)$ is finite free as a $W(\kappa)$-module), so it is harmless to pass to an F-linearly isogenous p-divisible group. Thus, we may decompose G according to the idempotents of F to reduce to the case when F is a p-adic field. Let k_0 be its finite residue field, $F_0 = W(k_0)[1/p]$, κ' a compositum of k_0 with κ over $\mathbb{F}_p$, and $K_0' = W(\kappa')[1/p]$. Consider the decomposition

$$K_0 \otimes_{\mathbb{Q}_p} F = (K_0 \otimes_{\mathbb{Q}_p} F_0) \otimes_{F_0} F \simeq \prod_{j : k_0 \to \kappa'} (K_0' \otimes_{j, F_0} F)$$

where j varies through the embeddings over $k_0 \cap \kappa$ ($\subset k_0$). This is a finite product of copies of totally ramified finite extensions of K_0', and the factor fields are permuted transitively by the natural F-linear action of the Galois group $\mathrm{Gal}(k_0/(k_0 \cap \kappa))$. Note that this Galois group is generated by a power of the absolute Frobenius.

We conclude that any $K_0 \otimes_{\mathbb{Q}_p} F$-module M canonically decomposes in a compatible F-linear way as $\prod M_j$ for vector spaces M_j over the factors fields. Hence, if M is equipped with an injective F-linear endomorphism $\mathcal{F}$ that is semilinear over the absolute Frobenius of K_0' then $\mathcal{F}$ must be an F-linear automorphism that transitively permutes the M_j's via F-linear isomorphisms. In particular, if such an M is non-zero then each M_j is a non-zero vector space over the factor field indexed by j, so M as a $K_0 \otimes_{\mathbb{Q}_p} F$-module would be free of some positive rank ρ and hence of K_0-dimension $[F : \mathbb{Q}_p]\rho$.

Now set $M = \mathrm{M}^*(G)[1/p]$, whose K_0-dimension is h. This is equipped with the required Frobenius-semilinear injective endomorphism (that is moreover F-linear by functoriality), so it is free of some rank $\rho \geqslant 1$ over $K_0 \otimes_{\mathbb{Q}_p} F$ and hence $h = [F : \mathbb{Q}_p]\rho$. It follows that $[F : \mathbb{Q}_p] \leqslant h$, with equality if and only if $\rho = 1$. This proves (1).

Finally, assuming we are in this rank-1 case, it remains to prove that F is its own centralizer in $\mathrm{End}^0(G)$. To compute the centralizer of F, first observe that the Dieudonné module functor (on the isogeny category) is valued in the category of K_0-vector spaces, so every element $f \in \mathrm{End}^0(G)$ that commutes with the F-action induces a $K_0 \otimes_{\mathbb{Q}_p} F$-linear endomorphism of $\mathrm{M}^*(G)[1/p]$. We know that $\mathrm{M}^*(G)[1/p]$ is free of rank 1 over $K_0 \otimes_{\mathbb{Q}_p} F$, so $\mathrm{M}^*(f)$ acts as multiplication by some $c \in K_0 \otimes_{\mathbb{Q}_p} F$. Since $\mathrm{M}^*(f)$ also commutes with the action of $\mathcal{F}$ that is semilinear over the absolute Frobenius σ of K_0, we have $(\sigma \otimes 1)(c) = c$. This forces $c \in F$, as desired. □

1.4.4. Deformation theory. Let R be a local ring with residue field κ. The functor $A \rightsquigarrow A_\kappa$ from abelian schemes over R to abelian varieties over κ is faithful. This follows from two facts: the collection of finite étale subgroup schemes $A[N]$ for N not divisible by $\mathrm{char}(\kappa)$ is schematically dense in A (due to the fiberwise denseness and [**34**, IV_3, 11.10.9]), and passage to the special fiber is faithful on finite étale R-schemes. When considering deformation problems for abelian varieties equipped with endomorphisms or a polarization (viewed as a special kind of isogeny), this faithfulness result is implicitly used without comment.

1.4.4.1. Remark. For abelian R-schemes A and B, the injective reduction map

$$\mathrm{Hom}^0(A, B) := \mathbb{Q} \otimes_{\mathbb{Z}} \mathrm{Hom}(A, B) \to \mathrm{Hom}^0(A_\kappa, B_\kappa)$$

gives meaning to the intersection $\mathrm{Hom}^0(A, B) \bigcap \mathrm{Hom}(A_\kappa, B_\kappa)$. This intersection contains $\mathrm{Hom}(A, B)$ but can be strictly larger. To make an example, let R be a discrete valuation ring with fraction field K of characteristic 0 and residue field κ of characteristic $p > 0$, and let E be an elliptic curve over R such that $E[p]_K$ is constant and E_κ is ordinary. There are $p+1$ cyclic subgroups of $E[p]_K$ with order p, so via R-flat closure there are $p+1$ closed R-flat subgroups $C \subset E$ of order p. For any local extension of discrete valuation rings $R \hookrightarrow R'$, the R'-subgroups $\{C_{R'}\}$ of $E_{R'}$ exhaust the $p+1$ possibilities over R'. Due to the connected-étale sequence over $\widehat{R}$, it follows that exactly one such $C \subset E$ is connected, so the p others are étale and hence have reduction equal to the same (unique) étale κ-subgroup of $E_\kappa[p]$.

If $C, C' \subset E[p]$ are distinct étale subgroups of order p then the kernels of the isogenies $f : E \to E/C$ and $f' : E \to E/C'$ are distinct over K but the same over κ. Since the reductions of f and f' have the same kernel, in the isogeny category of elliptic curves over R the element $f \circ (f')^{-1} \in \mathrm{Hom}^0(E/C', E/C)$ has reduction that is a morphism of elliptic curves (and even an isomorphism, with inverse given by the reduction of $f' \circ f^{-1}$). If $f \circ {f'}^{-1}$ were a morphism of elliptic curves over R then it would have to be an isomorphism (since its reduction is an isomorphism), and so C' would be in the orbit of C under $\mathrm{Aut}(E) = \mathrm{Aut}(E_K)$. These orbits have size at most $\#\mathrm{Aut}(E_K)/2 \leqslant 3$, so we can find C' and C not in the same $\mathrm{Aut}(E)$-orbit whenever $j(E_K) \neq 0, 1728$ or $p > 3$.

The analogous faithfulness result for p-divisible groups is more subtle when $\mathrm{char}(\kappa) = p$, since it is false without a noetherian condition:

1.4.4.2. Example. Let $R = \mathbb{Z}_p[\zeta_{p^\infty}]$ be the (non-noetherian) valuation ring of the p-power cyclotomic extension of $\mathbb{Q}_p$, and let $\{\zeta_{p^n}\}$ a compatible system of primitive p-power roots of unity in R. The R-homomorphism between p-divisible groups $\mathbb{Q}_p/\mathbb{Z}_p \to \mu_{p^\infty}$ defined by $p^{-n} \mapsto \zeta_{p^n}$ is an isomorphism between the generic fibers and induces the zero map between the special fibers.

Under a noetherian hypothesis, the preceding pathology cannot occur:

1.4.4.3. Proposition. *Let $(R, \mathfrak{m})$ be a noetherian local ring with residue field κ of characteristic $p > 0$. The functor $G \rightsquigarrow G_\kappa$ from p-divisible groups over R to p-divisible groups over κ is faithful.*

PROOF. The problem is to prove that if $f : G' \to G$ is a homomorphism between p-divisible groups over R and $f_\kappa = 0$ then $f = 0$. For each $n \geqslant 1$, the induced map $f_n : G'[p^n] \to G[p^n]$ between finite flat R-group schemes is described by a matrix over R upon choosing R-bases of the coordinate rings (as finite free R-modules). Hence, by the Krull intersection theorem it suffices to prove the vanishing result over $R/\mathfrak{m}^N$ for all $N \geqslant 1$, so we may and do assume that R is an artinian local ring. By the functoriality of the connected-étale sequence, it suffices to treat the following separate cases: G' and G are both connected, G' and G are both étale, or G' is étale and G is connected.

The case when both are étale is obvious. When G' is étale and G is connected then we claim that $\mathrm{Hom}(G', G) = 0$. By faithfully flat base change to the (artinian local) strict henselization R^{sh} we may assume G' is constant, so it is a power of $\mathbb{Q}_p/\mathbb{Z}_p$. Hence, we can assume $G' = \mathbb{Q}_p/\mathbb{Z}_p$, so $\mathrm{Hom}(G', G) = \varprojlim G[p^n](R)$ (inverse limit via p-power maps). The equivalence between connected p-divisible groups and formal Lie groups over R on which multiplication by p is an isogeny (see Example 1.4.3.6) identifies this inverse limit with the p-adic Tate module of $\widehat{G}(R)$, where $\widehat{G}$ is the formal Lie group associated to G. Hence, the desired vanishing is reduced to proving that $\widehat{G}(R)$ has no non-zero infinitely p-divisible elements. In fact, we claim that $[p^N]$ kills $\widehat{G}(R)$ for sufficiently large N.

Upon choosing formal parameters for $\widehat{G}$, we may identify the set $\widehat{G}(R)$ with the set of ordered d-tuples in $\mathfrak{m}$, where $d = \dim G$. If $g \in \widehat{G}(R)$ has coordinates in an ideal I of R then $[p](g)$ has coordinates in (pI, I^2) since $[p]$ has linear part given by p-multiplication on the coordinates. Hence, if we define the sequence of ideals $J_0 = \mathfrak{m}$ and $J_{n+1} = (pJ_n, J_n^2)$ then we just need $J_N = 0$ for sufficiently large N. More generally, for any ring whatsoever and any ideal J_0, an elementary induction argument shows that $J_n \subset (p, J_0)^n$. The nilpotence of $\mathfrak{m}$ then does the job.

Finally, we address the most interesting case, which is connected G' and G. In this case we switch to the perspective of formal Lie groups and aim to prove that for commutative formal Lie groups Γ and Γ' over R such that $[p]_{\Gamma'}$ is an isogeny, $\mathrm{Hom}(\Gamma', \Gamma) \to \mathrm{Hom}(\Gamma'_\kappa, \Gamma_\kappa)$ is injective. Consider $f \in \mathrm{Hom}(\Gamma, \Gamma')$ that vanishes modulo an ideal $I \subseteq \mathfrak{m}$. Choose formal coordinates $\{x'_i\}$ and $\{x_j\}$ for Γ' and Γ respectively, so the coefficients of all monomials in $f^*(x_j)$ lie in I. Hence, the formal power series $(f \circ [p]_{\Gamma'})^*(x_j) = [p]_{\Gamma'}^*(f^*(x_j))$ has all coefficients of all monomials lying in (pI, I^2). Iterating, if $f \in \text{Hom}(\Gamma', \Gamma)$ vanishes over κ then $f \circ [p^n]_{\Gamma'}$ vanishes modulo the ideal J_n, where $J_0 = \mathfrak{m}$ and $J_{n+1} = (pJ_n, J_n^2)$. We have already seen that $J_N = 0$ for sufficiently large N, so $f \circ [p^N]_{\Gamma'} = 0$ for large N. By hypothesis the isogenous endomorphism $[p^N]_{\Gamma'} = [p]_{\Gamma'}^N$ of Γ' induces an injective endomorphism of the coordinate ring, so $f = 0$. □

An important fact in the study of lifting problems for abelian varieties from characteristic p to characteristic 0 is that infinitesimal lifting for such an abelian variety is entirely controlled by that of its p-divisible group. This will be made precise in Theorem 1.4.5.3 (and Example 1.4.5.4 will address algebraization aspects in the limit). We now focus on the existence and structure of deformation rings for abelian varieties and p-divisible groups as well as the behavior of these deformation rings relative to extension of the residue field.

1.4.4.4. Definition. Let A_0 be an abelian variety of dimension g over a field κ. For a complete local noetherian ring R with residue field κ, a *deformation* of A_0 over R is a pair (A, i) consisting of an abelian scheme A over R and an isomorphism $i : A_\kappa \simeq A_0$ over κ.

There is an evident notion of *isomorphism* between two deformations of A_0 over R, and such deformations have no non-trivial automorphisms. Likewise, if A_0 is equipped with a polarization $\phi_0 : A_0 \to A_0^t$ or an injective homomorphism $\alpha_0 : \mathcal{O} \to \text{End}(A_0)$ from a specified $\mathbb{Z}$-finite associative ring $\mathcal{O}$ (or both!), we define in an evident way the notion of *deformation* for A_0 equipped with this additional structure. In the case of polarizations, any R-homomorphism $\phi : A \to A^t$ lifting ϕ_0 is necessarily a polarization. Indeed, the symmetry of ϕ is inherited from ϕ_0 (due to faithfulness of passage to the special fiber for abelian schemes over a local ring), and the ampleness on A of the pullback $(1, \phi)^*(\mathscr{P}_A)$ of the Poincaré bundle $\mathscr{P}_A$ is inherited from the ampleness on A_0 of its restriction $(1, \phi_0)^*(\mathscr{P}_{A_0})$ due to [**34**, IV_3, 9.6.4].

Fix a complete local noetherian ring Λ with residue field κ (e.g., a Cohen ring for κ), and let $\mathcal{C}_\Lambda$ be the category of artinian local Λ-algebras R with local structure map $\Lambda \to R$ and residue field κ. The *deformation functor*

$$\text{Def}_\Lambda(A_0) : \mathcal{C}_\Lambda \to \text{Set}$$

assigns to every R in $\mathcal{C}_\Lambda$ the set of isomorphism classes of deformations of A_0 over R. Likewise, if A_0 is equipped with a polarization $\phi_0 : A_0 \to A_0^t$ and endomorphism structure $\alpha_0 : \mathcal{O} \hookrightarrow \text{End}(A_0)$ (for a $\mathbb{Z}$-finite associative ring $\mathcal{O}$) then we define the deformation functor $\text{Def}_\Lambda(A_0, \phi_0, \alpha_0)$ similarly. This is a subfunctor of $\text{Def}_\Lambda(A_0)$.

A covariant functor $F : \mathcal{C}_\Lambda \to \text{Set}$ is *pro-representable* if there is a complete local noetherian Λ-algebra $\mathcal{R}$ with local structure map $\Lambda \to \mathcal{R}$ and residue field κ such that $F \simeq \text{Hom}_\Lambda(\mathcal{R}, \cdot)$ (using local Λ-algebra homomorphisms). A *formal deformation ring* for A_0 (if one exists) is an $\mathcal{R}$ that pro-represents $\text{Def}_\Lambda(A_0)$. The reason we say "formal" is that over such an $\mathcal{R}$ there is merely a universal *formal* abelian scheme (which is however universal modulo $\mathfrak{m}_\mathcal{R}^n$ among abelian scheme deformations of A_0 over objects in $\mathcal{C}_\Lambda$ whose maximal ideal has vanishing nth power, for each $n \geqslant 1$).

When we include a polarization as part of the deformation problem, if this enhanced problem admits a pro-representing ring $\mathcal{R}$ then by Grothendieck's formal

GAGA algebraization theorems [**34**, III$_1$, 5.4.1, 5.4.5] there is a universal deformation (i.e., a polarized abelian scheme deformation over $\mathcal{R}$ that represents the deformation functor on the category of all complete local noetherian Λ-algebras with residue field κ).

1.4.4.5. Theorem. *The deformation functor* $\mathrm{Def}_\Lambda(A_0)$ *is pro-representable and formally smooth, with tangent space canonically isomorphic to* $\mathrm{Lie}(A_0^t) \otimes_\kappa \mathrm{Lie}(A_0)$ *as a* κ*-vector space. In particular, this formal deformation ring is a formal power series ring over* Λ *in* g^2 *variables.*

Each deformation functor $\mathrm{Def}_\Lambda(A_0, \phi_0, \alpha_0)$ *and* $\mathrm{Def}_\Lambda(A_0, \alpha_0)$ *is pro-represented by a quotient of the formal deformation ring for* A_0.

PROOF. The case of $\mathrm{Def}_\Lambda(A_0)$ is due to Grothendieck and is explained in detail in [**88**, Thm. 2.2.1] (where the description of the tangent space to the deformation functor is given in the proof). To show that a deformation functor F of the form $\mathrm{Def}_\Lambda(A_0, \phi_0, \alpha_0)$ or $\mathrm{Def}_\Lambda(A_0, \alpha_0)$ is pro-represented by a quotient of the formal deformation ring $(\mathcal{R}, \mathfrak{m})$ for A_0, for each integer $n \geqslant 1$ we consider the full subcategory $\mathcal{C}_{\Lambda,n}$ of objects $R \in \mathcal{C}_\Lambda$ whose maximal ideal has vanishing nth power. The restriction $\mathrm{Def}_\Lambda(A_0)|_{\mathcal{C}_{\Lambda,n}}$ is represented by $\mathcal{R}/\mathfrak{m}^n$. For each n, suppose there is an ideal $I_n \subset \mathcal{R}/\mathfrak{m}^n$ such that $(\mathcal{R}/\mathfrak{m}^n)/I_n$ represents the subfunctor $F|_{\mathcal{C}_{\Lambda,n}}$ of $\mathrm{Def}_\Lambda(A_0)|_{\mathcal{C}_{\Lambda,n}}$. Since $\mathcal{C}_{\Lambda,n}$ is a full subcategory of $\mathcal{C}_{\Lambda,n+1}$, by universality we see that I_{n+1} has image I_n under $\mathcal{R}/\mathfrak{m}^{n+1} \to \mathcal{R}/\mathfrak{m}^n$. Hence, there is a unique ideal $I \subset \mathcal{R}$ such that $I_n = (I + \mathfrak{m}^n)/\mathfrak{m}^n$ for all $n \geqslant 1$, so $\mathcal{R}/I$ is the desired quotient.

We are reduced to the following general problem for abelian schemes (applied to the universal deformation of A_0 over $\mathcal{R}/\mathfrak{m}^n$ for every $n \geqslant 1$ and the structure (ϕ_0, α_0) on its reduction A_0 modulo the nilpotent ideal $\mathfrak{m}/\mathfrak{m}^n$). Let A and B be abelian schemes over a noetherian scheme S, and let $\mathscr{I} \subset \mathcal{O}_S$ be a nilpotent coherent ideal sheaf defining a closed subscheme $S_0 \subset S$. For a homomorphism $f_0 : A_0 \to B_0$ over S_0, the condition on an S-scheme T that $(f_0)_{T_0}$ lifts (necessarily uniquely!) to a T-homomorphism $A_T \to B_T$ is represented by a closed subscheme of S (visibly containing S_0). We will prove this by using Hom-schemes.

Consider the functor $\underline{\mathrm{Hom}}(A, B) : T \rightsquigarrow \mathrm{Hom}_{T\text{-gp}}(A_T, B_T)$ on S-schemes. We shall prove this is represented by an S-scheme locally of finite type (avoiding projectivity hypotheses on A and B). Grothendieck's construction of Hom-schemes from Hilbert schemes (via graph arguments) for schemes that are proper, flat, and finitely presented over the base requires projectivity because this hypothesis is needed to ensure representability of Hilbert functors. But Artin showed (see [**5**, Cor. 6.2]) that the Hilbert functor of a proper, flat, and finitely presented S-scheme X is an algebraic space that is separated and locally of finite type over S. Consequently, the same holds for Hom-functors between such schemes, and so also for the subfunctors that impose compatibility with group laws.

We conclude that $H := \underline{\mathrm{Hom}}(A, B)$ is an algebraic space that is separated and locally of finite type over S. For all $s \in S$ the fibers H_s are étale (by the functorial criterion), and an algebraic space that is separated and locally quasi-finite over a noetherian scheme is a scheme [**60**, II, 6.16]. Thus, H is represented by a separated and locally finite type S-scheme that we denote also as H.

The given f_0 defines a section to $H_0 := H \times_S S_0 \to S_0$. We claim that the closed subscheme $Z_0 \hookrightarrow H_0$ underlying this section is stable under generization. Suppose not, so there exists a discrete valuation ring R and an element $h_0 \in H_0(R)$ whose generic point lands outside Z_0 and whose closed point lands inside Z_0. Making a base change by the resulting map $\mathrm{Spec}(R) \to S_0$ yields a pair of R-homomorphisms $(A_0)_R \rightrightarrows (B_0)_R$ (one coming from h_0, and the other from f_0) that agree on the closed fibers but are distinct on generic fibers, contradicting faithfulness of passage to the special fiber for abelian schemes over a local ring.

By stability under generization, the closed subscheme Z_0 in H_0 is topologically open, so the open subscheme $U \subset H$ with the underlying space of Z_0 is a union of connected components of H. The structure morphism $U \to S$ is a homeomorphism, so it is of finite type, not just locally of finite type. The map $H \to S$ satisfies the valuative criterion for properness since an abelian scheme over a discrete valuation ring is the Néron model of its generic fiber [**10**, 1.2/8], so the open and closed U in H also satisfies the valuative criterion over S. This proves that the finite type map $U \to S$ is proper, yet it has étale fibers of degree 1, so it is a closed immersion (defined by a nilpotent ideal). The closed subscheme $U \hookrightarrow S$ represents the lifting condition on f_0. □

1.4.4.6. Remark. A globalization of the formal smoothness of the infinitesimal deformation theory of an abelian variety is Grothendieck's result that if R is a ring containing an ideal J satisfying $J^2 = 0$ then every abelian scheme A_0 over $R_0 := R/J$ lifts to an abelian scheme over R. We sketch the proof, building on the key case of an artinian local base that follows from the formal smoothness of the infinitesimal deformation theory (and is a key step in the *proof* of the formal smoothness in [**88**, Thm. 2.2.1]).

By direct limit arguments we may and do assume R is noetherian. The obstruction to lifting A_0 to a smooth proper R-scheme A is a certain class $\xi \in \mathrm{H}^2(A_0, (\Omega^1_{A_0/R_0})^\vee \otimes_{R_0} J)$. The formation of ξ is compatible with base change on R (relative to base change morphisms for the cohomology of quasi-coherent sheaves), so by Zariski localization and completion we see that the vanishing of ξ is reduced to the case when R is a complete local noetherian ring. By the Theorem on Formal Functions [**34**, III_1, 4.2.1], the vanishing of ξ is reduced to the settled case when R is an artinian local ring.

Now return to a general noetherian R, and fix a smooth proper R-scheme A lifting A_0. By smoothness we may choose a lift $e \in A(R)$ of the identity section $e_0 \in A_0(R_0)$. We claim that the subtraction morphism $\mu_0 : A_0 \times A_0 \to A_0$ defined by $(x, y) \mapsto x - y$ uniquely lifts to an R-morphism $\mu : A \times A \to A$ carrying (e, e) to e. Once such a μ exists, it is the subtraction for a unique group law due to rigidity arguments explained in [**83**, Ch. 6, §3]. In particular, over an arbitrary ring R (without noetherian hypotheses) μ is unique if it exists. Hence, by Zariski localization it suffices to prove the existence of μ when R is a local noetherian ring, and by fpqc descent with respect to $R \to \widehat{R}$ we may assume R is complete. Formal GAGA for morphisms [**83**, III_1, 5.4.1] then reduces the existence problem to the case of artinian local R, so by length induction we may assume J is killed by the maximal ideal of R. This case is settled in [**83**, Ch. 6, §3, Prop. 6.15].

The deformation theory of p-divisible groups ends up with results similar to the case of abelian varieties but proceeds by another path. To describe this, let κ be a field of characteristic $p > 0$, Λ a complete local noetherian ring with residue field κ, and X_0 a p-divisible group of height $h \geqslant 0$ and dimension $d \geqslant 0$ over κ (so

$\dim(X_0^t) = h - d$ by [**119**, §2.3, Prop. 3]). A *deformation* of X_0 over a complete local noetherian ring R with residue field κ is a pair $(\mathcal{X}, i)$ consisting of a p-divisible group $\mathcal{X}$ over R and an isomorphism $i : \mathcal{X}_\kappa \simeq X_0$.

There is an evident notion of *morphism* between such pairs. By Proposition 1.4.4.3, deformations of X_0 have no non-trivial automorphisms (lifting the identity on X_0). Hence, it is reasonable to study the functor $\mathrm{Def}_\Lambda(X_0)$ assigning to any $R \in \mathcal{C}_\Lambda$ the set of isomorphism classes of deformations of X_0 over R. In contrast with the deformation theory of non-zero abelian varieties, for which the universal formal deformation is never algebraizable beyond the case of elliptic curves, the universal formal deformation of a p-divisible group is also a universal deformation relative to all complete local noetherian Λ-algebras with residue field κ since p-divisible groups are built from torsion-levels that are *finite* flat over the base.

1.4.4.7. Theorem. *The functor* $\mathrm{Def}_\Lambda(X_0)$ *is pro-represented by a power series ring over* Λ *in* $d(h-d)$ *variables. The tangent space* t_{X_0} *to this functor is canonically isomorphic to* $\mathrm{Lie}(X_0^t) \otimes_\kappa \mathrm{Lie}(X_0)$ *as a* κ*-vector space.*

The pro-representability for connected X_0 over perfect κ is established in [**124**] by using formal group laws to verify Schlessinger's criteria; the perfectness is required to carry out a Dieudonné module computation establishing that $\dim_\kappa(t_{X_0}) < \infty$ (equal to $d(h-d)$). This approach does not prove formal smoothness. Over algebraically closed fields the pro-representability for general X_0 is deduced formally from the connected case in the proof of [**16**, Thm. 4.4 (2)]. The general case over any κ may be deduced from Schlessinger's criteria and [**51**, 4.4]; the latter ingredient is proved via the cotangent complex (also see [**51**, 4.8] for perfect κ). For the convenience of the reader, here is a proof for general κ that avoids the machinery of the cotangent complex.

PROOF. First, we address the formal smoothness. The case of connected X_0 is a special case of the unobstructedness of lifting commutative formal Lie groups, which can be proved over any ring via Cartier theory; see [**136**, Thm. 4.46]. For disconnected X_0 consider a deformation X of X_0 over an artinian local Λ-algebra R with residue field κ. There is a unique (up to unique isomorphism) étale p-divisible group E over R that lifts $X_0^{\text{ét}}$ over κ, so $X^{\text{ét}}$ is uniquely isomorphic to E as deformations of $X_0^{\text{ét}}$.

Since X^0 is a deformation of the identity component of X_0, we see that the construction of such X comes in two steps: (i) deform X^0 to a (necessarily connected) p-divisible group over R (this step is unobstructed, by the settled connected case), and (ii) construct extensions over R of E by the chosen deformation of X_0^0. Such extensions in the sense of fppf abelian sheaves on the category $\mathscr{C}_R$ of finite R-algebras (equipped with the fppf topology) arise from p-divisible groups, due to:

1.4.4.8. Lemma. *Let* R *be an local artinian ring with residue characteristic* $p > 0$*, and choose a connected* p*-divisible group* G *over* R *and an étale* p*-divisible group* E *over* R*. For any short exact sequence of fppf abelian sheaves*

$$0 \to G \to Y \to E \to 0 \tag{1.4.4.1}$$

on the category $\mathscr{C}_R$ *of finite* R*-algebras (with the fppf topology),* Y *is a* p*-divisible group (so the given short exact sequence is the connected-étale sequence of* Y*).*

PROOF. Since G and E are p-power torsion sheaves, the sheaf Y is the union of its subsheaves $Y[p^n]$ for $n \geqslant 1$. The snake lemma gives that $[p] : Y \to Y$ is an epimorphism and provides short exact sequences of abelian sheaves

$$0 \to G[p^n] \to Y[p^n] \to E[p^n] \to 0$$

for all $n \geqslant 1$. The outer terms are represented by finite flat R-group schemes, so by Proposition 1.4.1.3 the middle term must also be represented by a finite flat R-group scheme. Thus, Y is a p-divisible group. □

Over any base scheme S, the study of extensions of an étale p-divisible group E by a given p-divisible group G over S can be reduced to the special case $E = \mathbb{Q}_p/\mathbb{Z}_p$ at the cost of replacing G with a p-divisible group denoted $E^\vee \otimes G$, as follows.

Let $E^\vee \otimes G$ be the direct limit (over $n \to \infty$) of the tensor products

$$E[p^n]^\vee \otimes_{\mathbb{Z}/(p^n)} G[p^n]$$

of $G[p^n]$ against the $\mathbb{Z}/(p^n)$-linear dual $E[p^n]^\vee$ of the étale sheaf $E[p^n]$, using the evident transition maps. This is easily seen to be a p-divisible group. Note that if S is the spectrum of a complete local noetherian ring with residue characteristic p and if G is connected then $E^\vee \otimes G$ is connected with dimension $\mathrm{height}(E) \cdot \dim(G)$.

There is a general categorical equivalence (compatible with base change) from the category of extensions of E by G to the category of extensions of the constant p-divisible group $\mathbb{Q}_p/\mathbb{Z}_p$ by $E^\vee \otimes G$. In one direction, for an extension (1.4.4.1) apply $E^\vee \otimes (\cdot)$ and then pull back the short exact sequence

$$0 \to E^\vee \otimes G \to E^\vee \otimes Y \to E^\vee \otimes E \to 0$$

along $\mathbb{Q}_p/\mathbb{Z}_p \hookrightarrow E^\vee \otimes E$ corresponding to the identity map in $E[p^n]^\vee \otimes E[p^n] = \mathscr{E}nd(E[p^n])$ for $n \geqslant 1$. In the other direction, given an extension of $\mathbb{Q}_p/\mathbb{Z}_p$ by $E^\vee \otimes G$, we apply $E \otimes (\cdot)$ to the given exact sequence and push out along the evaluation map $E \otimes E^\vee \otimes G \to G$. In an evident way, these are quasi-inverse constructions.

To summarize, the formal smoothness of $\mathrm{Def}_\Lambda(X_0)$ for disconnected X_0 is reduced to cases with $X_0^{\text{ét}} = \mathbb{Q}_p/\mathbb{Z}_p$. Since the deformation theory is formally smooth in the connected case, the formal smoothness of $\mathrm{Def}_\Lambda(X_0)$ is reduced to the following assertion.

1.4.4.9. Lemma. *Let $(R, \mathfrak{m})$ be artinian local with residue field κ of characteristic $p > 0$, J a non-zero ideal in R such that $\mathfrak{m}J = 0$, and $R_0 := R/J$. For a connected p-divisible group G over R with reduction G_0 over R_0,*

$$\mathrm{Ext}^1_R(\mathbb{Q}_p/\mathbb{Z}_p, G) \to \mathrm{Ext}^1_{R_0}(\mathbb{Q}_p/\mathbb{Z}_p, G_0)$$

is surjective.

PROOF. As above, let $\mathscr{C}_R$ denote the category of finite R-algebras, equipped with the fppf topology, and define $\mathscr{C}_{R_0}$ similarly; the Ext-groups are computed in the categories of abelian fppf sheaves on the respective sites $\mathscr{C}_R$ and $\mathscr{C}_{R_0}$. By writing $\mathbb{Q}/\mathbb{Z}$ as the direct sum of its p-primary part $\mathbb{Q}_p/\mathbb{Z}_p$ and its prime-to-p part M,

$$\mathrm{Ext}^1_R(\mathbb{Q}/\mathbb{Z}, G) = \mathrm{Ext}^1_R(\mathbb{Q}_p/\mathbb{Z}_p, G) \oplus \mathrm{Ext}^1_R(M, G).$$

The final Ext-term vanishes: any extension E of M by G is a torsion sheaf, so the decomposition of E as a direct sum of its p-primary part and prime-to-p part splits the extension structure. Thus, our problem is proving the surjectivity of

$$\mathrm{Ext}^1_R(\mathbb{Q}/\mathbb{Z}, G) \to \mathrm{Ext}^1_{R_0}(\mathbb{Q}/\mathbb{Z}, G_0).$$

Consider the evident commutative diagram of long exact sequences

$$\begin{array}{ccccccccc} 0 & \longrightarrow & G(R) & \longrightarrow & \mathrm{Ext}^1_R(\mathbb{Q}/\mathbb{Z}, G) & \longrightarrow & \mathrm{Ext}^1_R(\mathbb{Q}, G) & \longrightarrow & \mathrm{H}^1(R, G) \\ & & \downarrow & & \downarrow f & & \downarrow f' & & \downarrow f'' \\ 0 & \longrightarrow & G_0(R_0) & \longrightarrow & \mathrm{Ext}^1_{R_0}(\mathbb{Q}/\mathbb{Z}, G_0) & \longrightarrow & \mathrm{Ext}^1_{R_0}(\mathbb{Q}, G_0) & \longrightarrow & \mathrm{H}^1(R_0, G_0) \end{array}$$

We want f to be surjective. The functor G on $\mathscr{C}_R$ is pro-represented by the coordinate ring $\mathcal{O}(G)$ of the associated formal Lie group, so it is formally smooth. Hence, the left vertical map is surjective, so by the 5-lemma it suffices to show that f' is surjective and f'' is injective.

We first show that f'' is injective, for which it suffices to show that any G-torsor fppf sheaf of sets X on $\mathscr{C}_R$ is formally smooth. There is a local finite flat cover R' of R such that $X|_{\mathscr{C}_{R'}}$ is pro-represented by $\mathcal{O}(G)_{R'}$, so X is pro-represented by an R-descent A of $\mathcal{O}(G)_{R'}$; this descent is easily checked to be a complete noetherian local R-algebra (and its functor on $\mathscr{C}_R$ is computed using local ring homomorphisms). Clearly A is R-flat and the scalar extension $\kappa' \otimes_\kappa A_\kappa$ over the residue field κ' of R' is a formal power series ring over κ'. It follows that $k \otimes_\kappa A_\kappa$ is regular for any finite extension k of κ (it suffices to consider k containing κ', as a noetherian ring with a regular faithfully flat extension is regular [**73**, Thm. 23.7]), so A_κ is "geometrically regular" over κ. Thus, A_κ is formally smooth over κ relative to its max-adic topology [**73**, Thm. 28.7], so R-flatness ensures that A is formally smooth over R relative to its max-adic topology by [**34**, 0_{IV}, 19.7.1].

It remains to show that f' is surjective, so choose a short exact sequence

$$0 \to G_0 \to E_0 \to \mathbb{Q} \to 0$$

representing a class $\xi_0 \in \mathrm{Ext}^1_{R_0}(\mathbb{Q}, G_0)$. The vanishing of $\mathrm{H}^1(S_0, G_0)$ for all finite R_0-algebras S_0 implies that we obtain a short exact sequence on S_0-points for any S_0, and constant Zariski sheaves on $\mathscr{C}_R$ are sheaves for the finite flat topology, so pushforward along $j : \mathrm{Spec}(R_0) \to \mathrm{Spec}(R)$ gives an exact sequence

$$0 \to j_*(G_0) \to j_*(E_0) \to \mathbb{Q} \to 0.$$

Since $\mathfrak{m}J = 0$ and G is formally smooth, we have a short exact sequence

$$0 \to \mathscr{K} \to G \to j_*(G_0) \to 0$$

on $\mathscr{C}_R$ where $\mathscr{K}(S) = \mathrm{Lie}(G_\kappa) \otimes_\kappa JS$ for any finite R-algebra S. Thus, $p : G \to G$ factors uniquely through a map $h : j_*(G_0) \to G$ since $p\mathcal{K} = 0$, and the pushout of $j_*(E_0)$ along h is an extension of $\mathbb{Q}$ by G whose pullback over R_0 is $p\xi_0$. Since p acts invertibly on $\mathbb{Q}$, we conclude that f' is surjective. ◻

We have settled the formal smoothness in Theorem 1.4.4.7. Before we address the other assertions, we make a practical observation:

1.4.4.10. Remark. The infinitesimal liftability of p-divisible groups, Zariski locally on the base, is a hypothesis which underlies the Grothendieck–Messing crystalline Dieudonné theory [**75**, Ch. IV, V] that classifies the deformations of p-divisible groups. It is a deep theorem of Grothendieck–Illusie that this liftability property holds without Zariski-localization over any affine base on which p is nilpotent (see [**51**, 4.4]). To apply Grothendieck–Messing theory over artinian local base rings (which is all we really need in this book) it suffices to use the formal smoothness established above.

Continuing with the proof of Theorem 1.4.4.7, we address Schlessinger's criteria for pro-representability of $F = \mathrm{Def}_\Lambda(X_0)$ (and compute its tangent space). Consider a pair of maps $R_1, R_2 \rightrightarrows R_0$ in $\mathcal{C}_\Lambda$ with $R_1 \to R_0$ surjective. Since deformations of p-divisible groups over artinian local rings have no non-trivial automorphisms, the bijectivity of

$$F(R_1 \times_{R_0} R_2) \to F(R_1) \times_{F(R_0)} F(R_2)$$

is immediate from part (2) of the following general result (which will be very useful in our later work with algebraization of formal CM abelian schemes; see Theorem 2.2.3).

1.4.4.11. Proposition (Ferrand). *Let $p_1 : R_1 \to R_0$ and $p_2 : R_2 \to R_0$ be maps of rings with p_1 surjective. Let R denote the fiber product ring $R_1 \times_{R_0} R_2$.*

1. *If M is a flat R-module and M_j denotes the flat R_j-module $M \otimes_R R_j$ then the natural map $M \to M_1 \times_{M_0} M_2$ is an isomorphism. Conversely, if M_j is an R_j-module and there are given isomorphisms $R_0 \otimes_{R_1} M_1 \simeq M_0 \simeq R_0 \otimes_{R_2} M_2$ then for the R-module $M = M_1 \times_{M_0} M_2$ the natural maps $R_j \otimes_R M \to M_j$ are isomorphisms, and M is R-flat when each M_j is R_j-flat.*
2. *Let R'_j be a finitely generated R_j-algebra and suppose there are given isomorphisms of R_0-algebras $R_0 \otimes_{R_1} R'_1 \simeq R'_0 \simeq R_0 \otimes_{R_2} R'_2$. The R-algebra $R' := R'_1 \times_{R'_0} R'_2$ is finitely generated, and if each R'_j is flat and finitely presented over R_j then R' is flat and finitely presented over R.*
3. *Assume that p_2 is surjective or that all elements of $\ker(p_1)$ are nilpotent. The functor*

$$X \rightsquigarrow (X_{R_1}, X_{R_2}, (X_{R_1})_{R_0} \simeq (X_{R_2})_{R_0})$$

from the category of flat R-schemes to the category of triples (X_1, X_2, f) consisting of flat schemes X_j over R_j and an R_0-isomorphism $f : (X_1)_{R_0} \simeq (X_2)_{R_0}$ is an equivalence, and X is finite type (respectively flat and finitely presented) over R if and only if each X_j is finite type (respectively flat and finitely presented) over R_j.

*An R-map $f : X \to Y$ between flat finitely presented R-schemes satisfies property **P** if and only if the pullback maps f_{R_1} and f_{R_2} satisfy **P**, where **P** is any of the properties: separated, proper, finite, flat, smooth, étale, isomorphism, geometric fibers of pure dimension d, connected geometric fibers.*

Generalizations of parts (2) and (3) are given in [**101**, Appendix A] (and references therein). For applications to Schlessinger's criteria, part (3) is used with

$\ker(p_1)$ nilpotent. In other situations (such as gluing along closed subschemes, which we need in the proof of Theorem 2.2.3) part (3) is used with surjective p_2.

PROOF. Part (1) is [**40**, Thm. 2.2(iv)] (upon noting that by the *proof* of [**40**, Thm. 2.2(iii)], the kernel vanishes when M is R-flat). We prove part (2) by using a limit argument suggested by D. Rydh. First assume that each R'_j is finitely generated over R_j. Consider the directed system $\{R'_\alpha\}$ of finitely generated R-subalgebras of R', and let $R'_{j,\alpha} = R'_\alpha \otimes_R R_j$, so $\varinjlim R'_{j,\alpha} = R'_j$ for each j (since $R' \otimes_R R_j \to R'_j$ is an isomorphism by part (1) applied to $M = R'$). Since each R'_j is finitely generated over R_j, it follows that for sufficiently large α_0 the maps $R'_{j,\alpha_0} \to R'_j$ are surjective for all j. In other words, for the cokernel R-module $M = \operatorname{coker}(R'_{\alpha_0} \to R')$, both $M \otimes_R R_1$ and $M \otimes_R R_2$ vanish. Hence, $M = 0$ by [**40**, Thm. 2.2(ii)]. This says that the inclusion $R'_{\alpha_0} \subset R'$ is an equality, so R' is finitely generated over R.

Now assume that each R'_j is flat and finitely presented over R_j, so R' is flat and finitely generated over R. Form a presentation

$$0 \to I \to R[t_1, \ldots, t_n] \to R' \to 0.$$

By the R-flatness of R', I is R-flat and this sequence remains exact after applying $R_j \otimes_R (\cdot)$. Thus, the finite presentation of R'_j over R_j implies that the ideal $I_j = R_j \otimes_R I$ is finitely generated as a module over $P_j = R_j[t_1, \ldots, t_n]$ for each j. By part (1) applied to the R-flat I, we have $I = I_1 \times_{I_0} I_2$. Letting $P = R[t_1, \ldots, t_n]$, clearly $P = P_1 \times_{P_0} P_2$ for surjections $P_1, P_2 \rightrightarrows P_0$, and $I_j = P_j \otimes_P I$. Since $I = I_1 \times_{I_0} I_2$, a variation on the limit argument in the proof of (2) (now applied to modules over R rather than algebras over R) shows that the P-module $I = I_1 \times_{I_0} I_2$ is finitely generated since each I_j is finitely generated over P_j.

To prove (3), first we prove the equivalences of categories. Assume p_2 is surjective. The key point in this case is that for closed immersions of schemes $Z \hookrightarrow Y$ and $Z \hookrightarrow Y'$ the associated pushout of ringed spaces $Y \coprod_Z Y'$ (topological space gluing and fiber product of structure sheaves) is again a scheme, identified in the evident manner with the spectrum of a ring-theoretic fiber product when Y, Y', and Z are affine. The proof of this assertion is elementary and left to the reader; it is made easier by first showing that the ringed space gluing has the expected universal property among ringed spaces and is compatible with topological localization, and then proving it is isomorphic in the expected way to the desired affine scheme when Y, Y', and Z are affine. (See [**21**, §2] for further discussion of this argument, and [**40**, §7], [**62**, Thm. 38], [**81**, §3], [**101**, Appendix A] for more general existence results for pushouts.)

By part (1), the formation of the pushout $Y \coprod_Z Y'$ is compatible with flat base change over the pushout, so if X is a flat R-scheme then it is the pushout of its closed subschemes X_{R_1} and X_{R_2} along their common closed subscheme $X_{R_0} = (X_{R_1})_{R_0} = (X_{R_2})_{R_0}$. The equivalence of categories in (3) therefore follows from (1) and the existence of the general "gluing" of schemes along a closed subscheme and its compatibility with Zariski-localization (so we may carry out computations in the affine setting).

Suppose instead that the elements of $\ker(p_1)$ are nilpotent. Now the relevant pushout we must construct is relative to a closed immersion of schemes $j : Z \hookrightarrow Y$ that is topologically an equality (i.e., all sections of the defining quasi-coherent ideal sheaf are locally nilpotent) and an affine map $f' : Z \to Y'$. For every affine open

$U' \subset Y'$, the preimage affine open $f'^{-1}(U')$ in Z underlies an open subscheme of Y that is also affine (since affineness is determined by the underlying reduced scheme [**24**, A.2]). We define the pushout ringed space $Y \coprod_Z Y'$ to have the same topological space as Y' and structure sheaf $f'_*(\mathscr{O}_Y) \times_{f'_*(\mathscr{O}_Z)} \mathscr{O}_{Y'}$ (via the identification of $\mathscr{O}_Y$ as a sheaf on the topological space of Z). By working Zariski-locally on Y' this is easily checked to be a scheme, coinciding in the evident manner with the spectrum of the expected fiber product ring when Y', Y, and Z are affine. Thus, the equivalence of categories in (3) follows via (1) as in the case when p_2 is surjective.

Finally, we analyze the behavior of various properties $\mathbf{P}$ of R-scheme morphisms $h : X \to Y$ between two flat R-schemes of finite presentation when either p_2 is surjective or $\ker(p_1)$ consists of nilpotent elements. Let $Y_j = Y_{R_j}$ and $X_j = X_{R_j}$. Note that if $\ker(p_1)$ consists of nilpotent elements then the map $\mathrm{Spec}(R) \to \mathrm{Spec}(R_2)$ is a closed immersion defined by an ideal of nilpotent elements, so the same holds for $X \to X_2$ and $Y \to Y_2$. The two cases (surjective p_2, or $\ker(p_1)$ consisting of nilpotent elements) now will be treated simultaneously.

Clearly the map induced by h between fibers over a point in $\mathrm{Spec}(R)$ is identified with the analogous fiber map for the one of the pullbacks $h_j : X_j \to Y_j$ over R_j with $j \in \{1, 2\}$. Hence, since X and Y are R-flat of finite presentation, the fibral flatness criterion [**34**, IV_3, 11.3.11] implies that h is flat if and only if h_1 and h_2 are flat. By the fibral smoothness, étaleness, and isomorphism criteria [**34**, IV_4, 17.8.2, 17.9.5], the cases when $\mathbf{P}$ is "smooth", "étale", or "isomorphism" are also settled. The cases of geometric fibers being connected or of pure dimension d are obvious.

Since h is finite if and only if it is quasi-finite and proper [**34**, IV_3, 8.11.1], and the finite type h is quasi-finite if and only if h_1 and h_2 are quasi-finite, the case when $\mathbf{P}$ is "finite" is reduced the case when $\mathbf{P}$ is "proper". When h_1 and h_2 are universally closed it is clear that h is universally closed, so it remains to treat the case of separatedness; i.e., closedness of the diagonal $\Delta_{X/Y}$. Topologically this map is visibly a gluing of Δ_{X_1/Y_1} and Δ_{X_2/Y_2} when p_2 is surjective, and it coincides topologically with Δ_{X_2/Y_2} when $\ker(p_1)$ consists of nilpotent elements. Thus, in both cases separatedness of h_1 and h_2 implies separatedness of h. □

We have established enough compatibility for $F = \mathrm{Def}_\Lambda(X_0)$ with respect to fiber products in $\mathcal{C}_\Lambda$ to equip the tangent space t_{X_0} with a natural κ-vector space structure. Thus, to complete the verification of Schlessinger's criteria we need to address the finite-dimensionality of t_{X_0}.

Equip the square-zero kernel of $\kappa[\epsilon] \twoheadrightarrow \kappa$ with trivial divided powers, so by Grothendieck–Messing theory [**75**, Ch. IV, V] we can classify the deformations X of X_0 over $\kappa[\epsilon]$ in terms of the subbundle $\mathrm{Lie}(X^t)^\vee$ of the Lie algebra of the universal vector extension $\mathrm{E}(X)$ of X; here, $\mathrm{E}(X)$ is universal among extensions of X by the vector group $\mathrm{Lie}(X^t)^\vee$. (This application of Grothendieck–Messing theory does not rest on the caveats as in Remark 1.4.4.10 since $\kappa \to \kappa[\epsilon]$ has a section.) Via the choice of divided powers, the Lie algebra $\mathrm{Lie}(\mathrm{E}(X))$ is *canonically* isomorphic to the Lie algebra

$$\mathrm{Lie}(\mathrm{E}((X_0)_{\kappa[\epsilon]})) = \mathrm{Lie}(\mathrm{E}(X_0)) \otimes_\kappa \kappa[\epsilon]$$

associated to the constant deformation (the "origin" of the tangent space), and there are canonical exact sequences

$$0 \to \mathrm{Lie}(X^t)^\vee \to \mathrm{Lie}(\mathrm{E}(X)) \to \mathrm{Lie}(X) \to 0$$

over $\kappa[\epsilon]$ and

$$0 \to \mathrm{Lie}(X_0^t)^\vee \to \mathrm{Lie}(\mathrm{E}(X_0)) \to \mathrm{Lie}(X_0) \to 0$$

over κ.

Grothendieck–Messing theory identifies t_{X_0} with the set of $\kappa[\epsilon]$-subbundles of $\mathrm{Lie}(\mathrm{E}(X_0)) \otimes_\kappa \kappa[\epsilon]$ which lift $\mathrm{Lie}(X_0^t)^\vee$ modulo ϵ. Such subbundles are parameterized by

$$\begin{aligned} \mathrm{Hom}(\mathrm{Lie}(X_0^t)^\vee, \mathrm{Lie}(\mathrm{E}(X_0)))/\,\mathrm{End}(\mathrm{Lie}(X_0^t)^\vee) &= \mathrm{Hom}(\mathrm{Lie}(X_0^t)^\vee, \mathrm{Lie}(X_0)) \\ &= \mathrm{Lie}(X_0^t) \otimes_\kappa \mathrm{Lie}(X_0) \end{aligned}$$

by assigning to any representative κ-linear map $L : \mathrm{Lie}(X_0^t)^\vee \to \mathrm{Lie}(\mathrm{E}(X_0))$ the $\kappa[\epsilon]$-subbundle that is the image of the map $\mathrm{Lie}(X_0^t)^\vee \otimes_\kappa \kappa[\epsilon] \to \mathrm{Lie}(\mathrm{E}(X_0)) \otimes_\kappa \kappa[\epsilon]$ defined by $x + y\epsilon \mapsto x + (y + L(x))\epsilon$. Our computation of t_{X_0} respects the κ-linear structure on t_{X_0}, so this completes the proof of Theorem 1.4.4.7. □

As an extension of Theorem 1.4.4.7, in Theorem 1.4.5.5 we will establish a "p-divisible group" version of the second part of Theorem 1.4.4.5. We finish the present discussion of deformation theory by addressing the useful topic of how deformation rings behave with respect to "change of coefficients". This is relevant when trying to reduce certain structural questions about deformation rings (e.g., is there a characteristic-0 point?) to the case of an algebraically closed residue field.

1.4.4.12. Example. Consider a local map $\Lambda \to \Lambda'$ between complete local noetherian rings with respective residue fields κ and κ', and let X be a proper κ-scheme such that any proper flat deformations of X over artin local Λ-algebras with residue field κ admit no non-trivial automorphisms, and similarly for $X_{\kappa'}$ using Λ'-algebras. The deformation functor $\mathrm{Def}_\Lambda(X)$ assigns to any artin local Λ-algebra R with residue field κ the set of isomorphism classes of proper flat deformations of X over R, and we define $\mathrm{Def}_{\Lambda'}(X_{\kappa'})$ similarly.

It is a theorem of Schlessinger that $\mathrm{Def}_\Lambda(X)$ is pro-represented by a complete local noetherian Λ-algebra $\mathcal{R}$ with residue field κ, and likewise $\mathrm{Def}_{\Lambda'}(X_{\kappa'})$ is pro-represented by a complete local noetherian Λ'-algebra $\mathcal{R}'$ with residue field κ'. Note that $\Lambda'\widehat{\otimes}_\Lambda\mathcal{R}$ is a complete local noetherian Λ'-algebra with residue field κ'. By the local flatness criterion [**73**, 22.4], base change along $\mathcal{R} \to \Lambda'\widehat{\otimes}_\Lambda\mathcal{R}$ carries flat formal $\mathcal{R}$-schemes of finite type to flat formal $\Lambda'\widehat{\otimes}_\Lambda\mathcal{R}$-schemes of finite type.

Let $\mathfrak{X}$ and $\mathfrak{X}'$ be the universal formal deformations of X and $X_{\kappa'}$ over $\mathcal{R}$ and $\mathcal{R}'$ respectively. Clearly

$$\mathfrak{X} \times_{\mathrm{Spf}(\mathcal{R})} \mathrm{Spf}(\Lambda'\widehat{\otimes}_\Lambda\mathcal{R}) \to \mathrm{Spf}(\Lambda'\widehat{\otimes}_\Lambda\mathcal{R})$$

is a proper flat deformation of $X_{\kappa'}$, so it arises as the pullback of $\mathfrak{X}'$ along a unique local Λ'-algebra map

$$\mathcal{R}' \to \Lambda'\widehat{\otimes}_\Lambda\mathcal{R}. \tag{1.4.4.2}$$

Is this map an isomorphism?

To demonstrate the usefulness of an affirmative answer, suppose κ and κ' are perfect of characteristic $p > 0$, and $\Lambda = W(\kappa)$ and $\Lambda' = W(\kappa')$. (The case $\kappa' = \overline{\kappa}$ will be of most interest.) Consider the problem of whether or not X admits a proper flat formal lift over a complete local noetherian domain with residue field κ and characteristic 0. Suppose we can establish that $X_{\kappa'}$ admits a proper flat formal lift over a complete local noetherian domain D' with residue field κ' and characteristic 0. This lift corresponds to a local $W(\kappa')$-algebra homomorphism

$\mathcal{R}' \to D'$, so if (1.4.4.2) is an isomorphism then we obtain a local $W(\kappa)$-algebra homomorphism $\mathcal{R} \to D'$. Hence, $\mathcal{R}[1/p] \neq 0$, so by [**54**, 7.1.9] (which interprets the closed points of $\mathrm{Spec}(\mathcal{R}[1/p])$ in terms of rigid-analytic geometry) there is a local $W(\kappa)$-homomorphism from $\mathcal{R}$ into the valuation ring of a field F of finite degree over $W(\kappa)[1/p]$. The image of $\mathcal{R}$ in $\mathcal{O}_F$ is a (possibly non-normal) local domain that is finite flat over $W(\kappa)$ and has residue field κ. (See 2.1.1 for another instance of this slicing argument in deformation rings.)

To summarize, if (1.4.4.2) is an isomorphism then the problem of finding a formal lift of X to characteristic 0 with residue field κ is reduced to the same problem for $X_{\kappa'}$ with residue field κ'. In other words, the formal flat lifting problem with a *fixed* residue field is unaffected by making a preliminary scalar extension to κ'! This is useful when $\kappa' = \overline{\kappa}$ because it is often easier to make constructions without increasing the residue field when the residue field is algebraically closed.

The problem of compatibility of the deformation ring with respect to change of the coefficients (i.e., the isomorphism problem for analogues of (1.4.4.2)) arises in applications ranging from moduli problems in algebraic geometry to Galois deformation theory and beyond. In the special case that Λ' is the finite étale extension of Λ corresponding to a finite separable extension κ' of κ, an affirmative answer to the isomorphism problem for the pair (Λ, Λ') can be established within the axiomatic deformation theory framework presented by Rim in [**100**, 1.15–1.19], which considers more general Λ' for which κ' is just finitely generated over κ. However, to include the case $\kappa' = \overline{\kappa}$ all finiteness hypotheses on κ'/κ must be avoided. There is an axiomatic approach to the "change of coefficients" problem (in the spirit of Schlessinger's criteria), independent of the methods of SGA7, but for the present purposes it is simpler to give direct proofs in the cases we need. Later we will need the abstract criterion, at which point we will state and prove it. (See Proposition 1.4.5.6.)

For the formal deformation rings of abelian varieties and p-divisible groups (without extra structure), the isomorphism property for the analogue of (1.4.4.2) is easy to verify:

1.4.4.13. Proposition. *Let $\Lambda \to \Lambda'$ be a local map between complete local noetherian rings, inducing an extension $\kappa \to \kappa'$ of residue fields. Let G_0 be an abelian variety or p-divisible group over κ, and $G_0' = (G_0)_{\kappa'}$. In the case of p-divisible groups, assume* $\mathrm{char}(\kappa) = p$.

For the deformation rings $\mathcal{R}$ and $\mathcal{R}'$ pro-representing $\mathrm{Def}_\Lambda(G_0)$ *and* $\mathrm{Def}_{\Lambda'}(G_0')$ *respectively, the natural map*

$$\mathcal{R}' \to \Lambda' \widehat{\otimes}_\Lambda \mathcal{R}$$

is an isomorphism.

PROOF. In both cases, the map in question is between formal power series rings over Λ'. Hence, it suffices to check that the induced map between relative tangent spaces over κ' is an isomorphism. In each case, the tangent map is identified with the natural map

$$\kappa' \otimes_\kappa (\mathrm{Lie}(G_0^t) \otimes_\kappa \mathrm{Lie}(G_0)) \to \mathrm{Lie}((G_0')^t) \otimes_{\kappa'} \mathrm{Lie}(G_0') \tag{1.4.4.3}$$

that is an isomorphism. Indeed, in the case of abelian varieties this compatibility follows from the functoriality in the ground field for the general identification of the tangent space of the deformation functor of a proper κ-scheme Y_0 with $\mathrm{H}^1(Y_0, (\Omega^1_{Y_0/\kappa})^\vee)$. In the case of p-divisible groups it follows from the compatibility of Grothendieck–Messing theory with respect to base change. □

Deformation rings for geometric objects equipped with extra structure (e.g., polarized abelian varieties with endomorphism structure) are not as easy to describe as in the formally smooth setting used in the preceding proof. In such cases we can sometimes use a global moduli scheme to establish an isomorphism result for the "change of coefficients" map:

1.4.4.14. Proposition. *Let Λ be a complete local noetherian ring with residue field κ, and let (A_0, ϕ_0, α_0) be a polarized abelian variety of dimension $g > 0$ with endomorphism structure $\alpha_0 : \mathcal{O} \hookrightarrow \mathrm{End}(A_0)$ for a $\mathbb{Z}$-finite associative ring $\mathcal{O}$. The analogue of* (1.4.4.2) *for (A_0, ϕ_0, α_0) is an isomorphism for any Λ'.*

The same "change of coefficients" isomorphism holds for the formal deformation ring of any pair (A_0, α_0).

PROOF. First we treat the polarized cases by relating the deformation problem to global moduli schemes via auxiliary "level structure", and then we modify the method to apply in the absence of polarizations. The introduction of auxiliary level structure can be carried out without increasing κ by using non-constant "finite étale level structure" arising inside A_0, as follows.

Fix an integer $n \geqslant 3$ not divisible by $\mathrm{char}(\kappa)$. The finite étale group scheme $A_0[n]$ over κ uniquely lifts to a finite étale group scheme $\mathcal{G}$ over Λ. Consider the functor F on the category of Λ-schemes that assigns to any Λ-scheme S the set of isomorphism classes of quadruples (A, ϕ, α, τ) where (A, ϕ) is a polarized abelian scheme over S of relative dimension g with ϕ of constant square degree $\deg(\phi_0)$, $\alpha : \mathcal{O} \to \mathrm{End}(A)$ is a homomorphism, and $\tau : A[n] \simeq \mathcal{G}_S$ is an S-group isomorphism ("level structure"). These quadruples have no nontrivial automorphisms since $n \geqslant 3$ and $n \in \Lambda^\times$.

By standard moduli space arguments (using Hilbert schemes), the functor F is represented by a Λ-scheme $\mathcal{M}$ locally of finite type. The triple (A_0, ϕ_0, α_0) and the canonical isomorphism $\tau_0 : \mathcal{G}_\kappa \simeq A_0[n]$ define a point $\xi \in \mathcal{M}(\kappa)$, and the formal deformation ring for (A_0, ϕ_0, α_0) is naturally isomorphic to the completed local ring $\mathcal{O}^\wedge_{\mathcal{M},\xi}$ at ξ (since the finite étale "level structure" τ_0 uniquely lifts through any infinitesimal deformation).

Since $\mathcal{M}_{\Lambda'}$ represents the restriction of F to the category of Λ'-schemes, it is straightforward to verify that the analogue of (1.4.4.2) for the present situation is the inverse of the natural isomorphism

$$\Lambda' \widehat{\otimes}_\Lambda \mathcal{O}^\wedge_{\mathcal{M},\xi} \simeq \mathcal{O}^\wedge_{\mathcal{M}_{\Lambda'},\xi_{\kappa'}},$$

so it is an isomorphism. (This style of argument applies whenever we can relate the infinitesimal deformation problem to the formal structure on a global moduli scheme over Λ.)

Now we treat the case of formal deformation rings for (A_0, α_0) and (A_0', α_0') relative to some $\Lambda \to \Lambda'$. The absence of a polarization eliminates the option to use global moduli schemes for abelian schemes. Instead, we work with formal Hom-schemes attached to formal abelian schemes. (The same procedure can be used to handle the polarized case above.) Let $\mathfrak{A} \to \mathrm{Spf}(\mathcal{R})$ be the universal deformation of A_0, so for $\mathcal{R}' := \Lambda'\widehat{\otimes}_\Lambda \mathcal{R}$ it follows from Proposition 1.4.4.13 that the universal deformation of A_0' is the base change

$$\mathfrak{A}' = \mathfrak{A} \times_{\mathrm{Spf}(\mathcal{R})} \mathrm{Spf}(\mathcal{R}') \to \mathrm{Spf}(\mathcal{R}').$$

For any formal abelian schemes $\mathfrak{B}$ and $\mathfrak{C}$ over a complete local noetherian ring R, applying the arguments with Hom-functors from the proof of Theorem 1.4.4.5 over the artinian quotients of R proves that the Hom-functor $\underline{\mathrm{Hom}}(\mathfrak{B}, \mathfrak{C}) : \mathfrak{Y} \rightsquigarrow \mathrm{Hom}_{\mathfrak{Y}\text{-gp}}(\mathfrak{B}_{\mathfrak{Y}}, \mathfrak{C}_{\mathfrak{Y}})$ on the category of formal (adic) R-schemes $\mathfrak{Y}$ locally of finite type is represented by a separated and locally finite type formal R-scheme.

Thus, we get a separated and locally finite type formal $\mathcal{R}$-scheme $\underline{\mathrm{End}}(\mathfrak{A})$ classifying endomorphisms of $\mathfrak{A}$, and its formation commutes with local base change to any complete local noetherian ring (such as $\mathcal{R}'$). The definition of an action on (a base change of) $\mathfrak{A}$ by the $\mathbb{Z}$-finite ring $\mathcal{O}$ underlying α_0 amounts to giving several points of $\underline{\mathrm{End}}(\mathfrak{A})$ satisfying finitely many relations. These relations correspond to a formal closed subscheme in a fiber power of $\mathrm{End}(\mathfrak{A})$, so we get an adic formal moduli scheme $\mathfrak{M}$ locally of finite type over $\mathcal{R}$ that classifies $\mathcal{O}$-actions on $\mathfrak{A}$. Moreover, the formation of $\mathfrak{M}$ commutes with base change along any local homomorphism from $\mathcal{R}$ to another complete local noetherian ring.

The given α_0 corresponds to a rational point $\xi \in \mathfrak{M}(\kappa)$ in the special fiber, and the deformation ring pro-representing $\mathrm{Def}_\Lambda(A_0, \alpha_0)$ is the completed local ring $\mathcal{O}^\wedge_{\mathfrak{M},\xi}$. Similarly, $\mathfrak{M}' := \mathfrak{M}_{\mathcal{R}'}$ contains the κ'-point ξ' in its special fiber that corresponds to α_0' and arises by base change from ξ, so there is a natural isomorphism

$$\Lambda'\widehat{\otimes}_\Lambda \mathcal{O}^\wedge_{\mathfrak{M},\xi} \simeq \mathcal{O}^\wedge_{\mathfrak{M}',\xi'}.$$

The inverse of this isomorphism is the "change of coefficients" map that we wanted to prove is an isomorphism. □

1.4.5. Hodge–Tate decomposition and Serre–Tate lifts. We finish our summary of the theory of p-divisible groups by recording (for later reference) two fundamental theorems. The first is a deep result of Tate.

1.4.5.1. Theorem (Tate). *Let R be a complete discrete valuation ring with perfect residue field of characteristic $p > 0$ and fraction field K of characteristic* 0. *For any p-divisible groups G and G' over R, the natural injective map* $\mathrm{Hom}(G, G') \to \mathrm{Hom}(G_K, G'_K)$ *is bijective. Moreover, if $\mathbb{C}_K$ denotes the completion of an algebraic closure $\overline{K}$ of K then there is a canonical $\mathbb{C}_K$-linear* $\mathrm{Gal}(\overline{K}/K)$*-equivariant isomorphism*

$$\mathbb{C}_K \otimes_{\mathbb{Q}_p} V_p(G_{\overline{K}}) \simeq (\mathbb{C}_K(1) \otimes_K \mathrm{Lie}(G)_K) \oplus (\mathbb{C}_K \otimes_K \mathrm{Lie}(G^t)^\vee_K). \tag{1.4.5.1}$$

Proof. The full faithfulness of $G \rightsquigarrow G_K$ is [**119**, 4.2], and the construction of the isomorphism (1.4.5.1) occupies most of [**119**]. □

A useful application of the full faithfulness in Tate's theorem is "completed unramified descent" for p-divisible groups:

1.4.5.2. Corollary. *Let R be as in Theorem* 1.4.5.1 *and let K' be the completion of the maximal unramified extension of K inside $\overline{K}$. Let G' be a p-divisible group over the valuation ring R' of K', and assume that the generic fiber $G'_{K'}$ is equipped with a descent to a p-divisible group X over K.*

There is a pair (G, α) consisting of a p-divisible group G over R and an isomorphism $\alpha : G_K \simeq X$ over K. Moreover, there is a unique isomorphism $G_{R'} \simeq G'$ recovering the given identification of generic fibers $X \otimes_K K' \simeq G' \otimes_{R'} K'$.

A special case of this corollary is that any "unramified twist" of the p-adic Tate module of the generic fiber of a p-divisible group over R is also the p-adic Tate module of the generic fiber of a p-divisible group over R (since both p-adic Galois lattices have the same inertial restriction).

PROOF. The discrete valuation ring R' is the completion of the strict henselization R^{sh} of R inside $\overline{K}$, so $R^{\mathrm{sh}} \to R'$ is a local inclusion with relative ramification degree 1 and induces an isomorphism on residue fields. Thus, for any affine R'-scheme Y', descent of $Y'_{K'}$ to $K^{\mathrm{sh}} = \mathrm{Frac}(R^{\mathrm{sh}})$ is equivalent to descent of Y' to R^{sh} [**10**, 6.2, Prop. D.4]. Applying this to each $G'[p^n]$, we see that G' uniquely descends to a p-divisible group $\mathscr{G}$ over R^{sh} compatibly with the descent $X_{K^{\mathrm{sh}}}$ of $G'_{K'}$.

For each $\sigma \in \mathrm{Gal}(K^{\mathrm{sh}}/K)$ and the associated continuous K-automorphism σ' of the completion K', the canonical isomorphism $\sigma^*(X_{K^{\mathrm{sh}}}) \simeq X_{K^{\mathrm{sh}}}$ induces an isomorphism $\sigma'^*(G'_{K'}) \simeq G'_{K'}$. By Tate's full faithfulness result (applied over R'), this extends to an isomorphism $\sigma'^*(G') \simeq G'$ of p-divisible groups over R'. The uniqueness of the descent from R' to R^{sh} implies that this latter isomorphism uniquely descends to an R^{sh}-isomorphism $\alpha_\sigma : \sigma^*(\mathscr{G}) \simeq \mathscr{G}$ extending the canonical isomorphism $\sigma^*(X_{K^{\mathrm{sh}}}) \simeq X_{K^{\mathrm{sh}}}$.

The 1-cocycle condition $\alpha_\tau \circ \tau^*(\alpha_\sigma) = \alpha_{\tau\sigma}$ over R^{sh} is inherited from the generic fiber. Thus, since R^{sh} is a direct limit of R-subalgebras that are Galois local finite étale over R, on each *finite* p^n-torsion level the R^{sh}-isomorphisms α_σ amount to an étale descent datum relative to $R \to R^{\mathrm{sh}}$ [**10**, 6.2/B]. The resulting effective descent of $\mathscr{G}$ to a p-divisible group G over R is equipped with a canonical K-isomorphism $\alpha : G_K \simeq X$. By Tate's full faithfulness theorem applied over R', the K'-isomorphism

$$(G_{R'}) \otimes_{R'} K' = G_K \otimes_K K' \simeq X_{K'} \simeq G' \otimes_{R'} K'$$

uniquely extends to an R'-isomorphism $G_{R'} \simeq G'$. □

The link between the deformation theories of abelian varieties and p-divisible groups in characteristic p is provided by the Serre-Tate deformation theorem:

1.4.5.3. Theorem (Serre–Tate). *Let R be a ring in which a prime p is nilpotent, and let I be an ideal in R such that $I^n = 0$ for some $n \geqslant 1$. Define $R_0 = R/I$, and for an abelian scheme A and p-divisible group G over R let A_0 and G_0 denote their respective reductions modulo I.*

For any abelian scheme A over R, let $\epsilon_A : A[p^\infty]_0 \simeq A_0[p^\infty]$ denote the canonical isomorphism. The functor $A \rightsquigarrow (A_0, A[p^\infty], \epsilon_A)$ from the category of abelian schemes over R to the category of triples $(A_0, G, \epsilon : A[p^\infty]_0 \simeq G_0)$ is an equivalence.

See [**57**, 1.2.1] for a proof of this result. □

The most important application of the Serre-Tate deformation theorem is that for an abelian variety A_0 over a field of characteristic $p > 0$, the infinitesimal deformation theory of A_0 coincides with that of its p-divisible group. Likewise, if we fix a subring $\mathscr{O} \subset \mathrm{End}(A_0)$ or a polarization of A_0 (or both) then via the injection $\mathrm{End}(A_0) \subset \mathrm{End}(A_0[p^\infty])$ (Proposition 1.2.5.1) and the identification of

$A_0^t[p^\infty]$ with the Cartier dual of $A_0[p^\infty]$, the infinitesimal deformation theory of A_0 equipped with this extra structure is the same as that of its p-divisible group equipped with the analogous induced extra structure.

Consider a homomorphism $\phi_0 : A_0 \to A_0^t$ and the associated homomorphism

$$f_0 : A_0[p^\infty] \to A_0^t[p^\infty] \simeq A_0[p^\infty]^t.$$

Clearly ϕ_0 is an isogeny if and only if f_0 is an isogeny, and we saw in 1.4.3.4 that ϕ_0 is symmetric with respect to double duality of abelian varieties if and only if f_0 is skew-symmetric (i.e., $f_0^t = -f_0$) with respect to the canonical isomorphism of a p-divisible group with its Serre double dual.

This leads us to define a *quasi-polarization* of a p-divisible group G over a complete local noetherian ring to be a skew-symmetric homomorphism $f : G \to G^t$ that induces an isogeny between the special fibers. (We will see in 3.3.8 that it is equivalent to say the skew-symmetric f is an isogeny in the sense of 3.3.5, so we can thereby define the notion of quasi-polarization over any base scheme.) The ampleness aspect of a polarization cannot be encoded in terms of p-divisible groups, but quasi-polarizations are nonetheless a helpful concept when using p-divisible groups to study abelian varieties and their deformations, as the following example illustrates.

1.4.5.4. Example. As a special case of the Serre-Tate deformation theorem, if R is a complete local noetherian ring of residue characteristic $p > 0$ and A_0 is an abelian variety over the residue field, then a deformation of $A_0[p^\infty]$ to a p-divisible group G over R corresponds to a deformation of A_0 to a formal abelian scheme $\mathfrak{A}$ over R. If A_0 is equipped with a CM structure and we demand that this structure lifts to $\mathfrak{A}$ (via the injection $\mathrm{End}(\mathfrak{A}) \hookrightarrow \mathrm{End}(A_0)$) then $\mathfrak{A}$ can fail to be algebraic (i.e., it may not be the formal completion of a proper R-scheme). Explicit CM examples of this type are given in 4.1.2; also see the discussion immediately following the statement of Theorem 2.2.3.

To ensure algebraicity of $\mathfrak{A}$, we need to encode the deformation of a polarization. More specifically, choose a polarization $\phi_0 : A_0 \to A_0^t$ and suppose that the corresponding map $f : A_0[p^\infty] \to A_0^t[p^\infty] = A_0[p^\infty]^t$ lifts to R (as can happen in at most one way, by Proposition 1.4.4.3). Let $\phi : \mathfrak{A} \to \mathfrak{A}^t$ be the corresponding unique homomorphism that lifts ϕ_0 (in accordance with the Serre–Tate deformation theorem). If $\mathfrak{P}$ denotes the formal Poincaré bundle on $\mathfrak{A} \times \mathfrak{A}^t$ (which lifts the Poincaré bundle $\mathscr{P}_0$ on $A_0 \times A_0^t$) then $(1, \phi)^*\mathfrak{P}$ is a line bundle on $\mathfrak{A}$ lifting the line bundle $(1, \phi_0)^*\mathscr{P}_0$ on A_0 that is ample (due to ϕ_0 being a polarization). Hence, by Grothendieck's algebraization theorems [**34**, III_1, 5.4.1, 5.4.5], in such cases $\mathfrak{A}$ is algebraic, so it arises from a unique abelian R-scheme A deforming A_0.

There is a special case in which the liftability of all polarizations comes "for free": the *Serre–Tate canonical lifting* of an *ordinary* abelian variety A_0 over a perfect field k of characteristic $p > 0$. To explain these concepts, we first note that by the perfectness of k, the connected-étale sequence of every p-divisible group X over k is split (exactly as for finite commutative k-groups), so X is (uniquely) the product of an étale p-divisible group and a connected p-divisible group; see [**87**, I.2]. Letting $g = \dim(A_0)$, since an étale p-divisible group over k has connected Serre dual (as we may check over $\overline{k}$) and $A_0[p^\infty] = A_0[p^\infty]^0 \times A_0[p^\infty]^{\text{ét}}$ is of height $2g$ yet isogenous to its Serre dual $A_0[p^\infty]^t \simeq A_0^t[p^\infty]$ (as A_0 is isogenous to A_0^t), we see that the *p-rank* of $A_0[p^\infty]$ (i.e., height of $A_0[p^\infty]^{\text{ét}}$) is at most g. We say that A_0

is *ordinary* when $A_0[p^\infty]$ has the maximal possible p-rank, namely g; equivalently, $A_0[p^\infty]^0$ has height g.

Now assume that A_0 is ordinary. Since

$$A_0^t[p^\infty] \simeq A_0[p^\infty]^t = (A_0[p^\infty]^0)^t \times (A_0[p^\infty]^{\text{ét}})^t$$

with A_0^t isogenous to A_0, for height reasons it follows that the dual of $A_0[p^\infty]^0$ is étale. In other words, if A_0 is ordinary then canonically $A_0[p^\infty] \simeq X_0'^t \times X_0$ for étale p-divisible groups X_0 and X_0' that are functorial in $A_0[p^\infty]$. For any complete local noetherian ring R with residue field k, X_0 and X_0' uniquely lift to respective étale p-divisible groups X and X' over R, so the deformation $X'^t \times X$ of $X_0'^t \times X_0$ corresponds to a canonical formal deformation $\mathfrak{A}$ of A_0 over R. We claim that the formal abelian scheme $\mathfrak{A}$ is algebraic; its algebraization is called the Serre–Tate *canonical lifting*.

Choose a polarization $\phi_0 : A_0 \to A_0^t$. The skew-symmetry of the associated quasi-polarization

$$X_0'^t \times X_0 = A_0[p^\infty] \to A_0^t[p^\infty] \simeq A_0[p^\infty]^t = X_0^t \times X_0',$$

forces it to have the form $-f_0^t \times f_0$ for a homomorphism $f_0 : X_0 \to X_0'$. There is a unique lifting $f : X \to X'$ of f_0 since X_0 and X_0' are étale, so $-f^t \times f$ lifts $-f_0^t \times f_0$. In other words, the map induced by ϕ_0 between p-divisible groups lifts (necessarily uniquely, by 1.4.4.3) to a homomorphism $\mathfrak{A}[p^\infty] \to \mathfrak{A}^t[p^\infty]$, as suffices for the algebraicity of $\mathfrak{A}$.

Quasi-polarizations yield a p-divisible group analogue of Proposition 1.4.4.14:

1.4.5.5. Theorem. *Let Λ be a complete local noetherian ring with residue field κ of characteristic $p > 0$, let X_0 be a p-divisible group over κ, and let $\alpha_0 : \mathcal{O} \hookrightarrow$* $\mathrm{End}(X_0)$ *be an injective homomorphism from an associative finite flat $\mathbb{Z}_p$-algebra. Let $\phi_0 : X_0 \to X_0^t$ be a quasi-polarization of X_0.*

(1) *The functors* $\mathrm{Def}_\Lambda(X_0, \alpha_0)$ *and* $\mathrm{Def}_\Lambda(X_0, \phi_0, \alpha_0)$ *on $\mathcal{C}_\Lambda$ are pro-represented by quotients of the deformation ring for* $\mathrm{Def}_\Lambda(X_0)$.
(2) *Let $\Lambda \to \Lambda'$ be a local map between complete local noetherian rings, with $\kappa \to \kappa'$ the induced map between residue fields. Let $(X_0', \phi_0', \alpha_0') = (X_0, \phi_0, \alpha_0)_{\kappa'}$, and let $\mathcal{R}$ and $\mathcal{R}'$ be the respective rings pro-representing* $\mathrm{Def}_\Lambda(X_0, \phi_0, \alpha_0)$ *and* $\mathrm{Def}_{\Lambda'}(X_0', \phi_0', \alpha_0')$. *The natural map*

(1.4.5.2) $$\mathcal{R}' \to \Lambda' \widehat{\otimes}_\Lambda \mathcal{R}$$

analogous to (1.4.4.2) *is an isomorphism. The same holds for the deformation rings of (X_0, α_0) and (X_0', α_0'), as well as for (X_0, ϕ_0) and (X_0', ϕ_0').*

PROOF. Since α_0 is encoded in terms of finitely many endomorphisms of X_0, and ϕ_0 is a homomorphism $X_0 \to X_0^t$, for the proof of (1) it suffices to establish the following general claim (applied to the universal deformation of X_0 and its dual). Let X and Y be p-divisible groups over a complete local noetherian ring $(\mathcal{R}, \mathfrak{m})$ with residue field κ of characteristic p, and let $f_0 : X_0 \to Y_0$ be a homomorphism between the special fibers. We claim there exists an ideal $I \subset \mathcal{R}$ such that for any local map $\mathcal{R} \to R$ to an artinian local ring with residue field κ, a lift $X_R \to Y_R$ of f_0 exists if and only if I has vanishing image in R.

Consider the set-valued functor F on $\mathcal{C}_{\mathcal{R}}$ that carries an object R to the set of all R-homomorphisms $X_R \to Y_R$ lifting f_0 (i.e., $F(R)$ is empty if there is no such lift and $F(R)$ has a single element when a lift exists). The problem is to show that F is pro-represented by a quotient of $\mathcal{R}$. By the functorial aspect of Proposition 1.4.4.11, it is straightforward to check that for a pair of maps $R_1, R_2 \rightrightarrows R_0$ in $\mathcal{C}_{\mathcal{R}}$ at least one of which is surjective, the natural map

$$F(R_1 \times_{R_0} R_2) \to F(R_1) \times_{F(R_0)} F(R_2)$$

is bijective. Moreover, $F(\kappa[\epsilon])$ consists of a single element (the constant deformation of f_0 as a homomorphism from $X \otimes_{\mathcal{R}} \kappa[\epsilon] = X_0 \otimes_\kappa \kappa[\epsilon]$ to $Y_0 \otimes_\kappa \kappa[\epsilon]$), so it vanishes as a κ-vector space. Thus, by Schlessinger's criteria, F is pro-represented by a complete local noetherian $\mathcal{R}$-algebra with residue field κ and has vanishing relative tangent space (over $\mathcal{R}$), so the ring pro-representing F is a quotient of $\mathcal{R}$. This completes the proof of (1).

In view of the proof of (1), to prove (2) it suffices to show that for $(\mathcal{R}, I)$ as above and any local homomorphism $\mathcal{R} \to \mathcal{R}'$ to a complete local noetherian ring with residue field $\kappa' \supset \kappa$, $I\mathcal{R}'$ is the analogous ideal inside $\mathcal{R}'$ relative to $X_{\mathcal{R}'}$, $Y_{\mathcal{R}'}$, and $(f_0)_{\kappa'}$. This seems difficult to verify directly, so we digress to prove an abstract isomorphism criterion for maps such as (1.4.5.2) and then apply it to establish (2).

To formulate an abstract isomorphism criterion for the "change of coefficients" map for deformation rings, we need to assume that the functor is defined on a larger class of rings than the artinian ones. For a complete local noetherian ring Λ with residue field κ, define Inf_Λ to be the category of pairs (Λ', R') consisting of a complete local noetherian Λ-algebra Λ' and a local Λ'-algebra $(R', \mathfrak{m})$ such that

(i) $\Lambda' \to R'$ is local and induces an isomorphism on residue fields,
(ii) $\mathfrak{m}^n = 0$ for some $n \geqslant 1$.

A *morphism* $(\Lambda'_1, R'_1) \to (\Lambda'_2, R'_2)$ consists of a local Λ-algebra map $f : \Lambda'_1 \to \Lambda'_2$ and a local homomorphism $R'_1 \to R'_2$ over f. In an evident way, Inf_Λ contains $\mathcal{C}_{\Lambda'}$ as a full subcategory for any Λ'. Also, if $n \geqslant 1$ and $\{(\Lambda', R'_i)\}$ is a directed system in Inf_Λ such that $\mathfrak{m}^n_{R'_i} = 0$ for all i then $R' := \varinjlim R'_i$ equipped with its evident Λ'-algebra structure is an object in Inf_Λ whose maximal ideal has vanishing nth power. As an important special case, for any (Λ', R') in Inf_Λ with $\mathfrak{m}^n_{R'} = 0$, the directed system $\{R'_i\}$ of artinian local Λ'-subalgebras of R' provides a directed system $\{(\Lambda', R'_i)\}$ in Inf_Λ with all maximal ideals having vanishing nth power and $\varinjlim R'_i = R'$.

Consider a covariant set-valued functor F on Inf_Λ and any directed system $\{(\Lambda', R'_i)\}$ as above. There is a natural map

$$\varinjlim F(\Lambda', R'_i) \to F(\Lambda', \varinjlim R'_i).$$

If this is always bijective then we say that F *commutes with direct limits*. If we only consider such directed systems with a fixed Λ' (such as Λ) then we say F *commutes with direct limits over* Λ'. In each of these definitions it suffices to consider direct limits with R'_i that are artinian. For example, if $(\mathcal{R}, \mathfrak{m})$ is a complete local noetherian Λ-algebra with residue field κ and F is defined to be the functor $(\Lambda', R') \rightsquigarrow \mathrm{Hom}_\Lambda(\mathcal{R}, R')$ (using local Λ-algebra maps) then F commutes with direct limits because $\mathcal{R}/\mathfrak{m}^n$ is artinian with finite Λ-length for every n.

Choose Λ' with residue field κ' and assume the restriction $F|_{\mathcal{C}_{\Lambda'}}$ is pro-represented by a complete local noetherian Λ'-algebra $(\mathcal{R}', \mathfrak{m}')$ with residue field κ'. Also assume $F|_{\mathcal{C}_\Lambda}$ is pro-represented by a complete local noetherian Λ-algebra $(\mathcal{R}, \mathfrak{m})$

with residue field κ. For each $n \geqslant 1$ there is a universal element $\xi_n \in F(\mathcal{R}/\mathfrak{m}^n)$, so for any $m \geqslant 1$ the induced element in $F((\Lambda'/\mathfrak{m}_{\Lambda'}^m) \otimes_\Lambda (\mathcal{R}/\mathfrak{m}^n))$ is classified by a map of local Λ'-algebras

$$\mathcal{R}' \to (\Lambda'/\mathfrak{m}_{\Lambda'}^m) \otimes_\Lambda (\mathcal{R}/\mathfrak{m}^n)$$

that is compatible with change in m and n (since the ξ_n's are compatible with change in n). Passing to the inverse limit defines a map of complete local noetherian Λ'-algebras

$$\mathcal{R}' \to \Lambda' \widehat{\otimes}_\Lambda \mathcal{R} \tag{1.4.5.3}$$

that recovers (1.4.4.2) in the setting of Example 1.4.4.12.

We seek abstract conditions on F which ensure that the map (1.4.5.3) is an isomorphism. Such an isomorphism property for all Λ' (assuming $F|_{\mathcal{C}_{\Lambda'}}$ is pro-represented by a complete local noetherian Λ'-algebra for all Λ') says exactly that $F = \mathrm{Hom}_\Lambda(\mathcal{R}, \cdot)$, since we have seen the necessity of commutation with direct limits in such cases and every object (Λ', R') in Inf_Λ is the direct limit of its artinian local Λ'-subalgebras.

1.4.5.6. Proposition. *Let Λ' be a complete local noetherian Λ-algebra with residue field κ', and let F be a covariant set-valued functor on Inf_Λ such that $F|_{\mathcal{C}_\Lambda}$ and $F|_{\mathcal{C}_{\Lambda'}}$ are pro-represented by complete local noetherian rings $(\mathcal{R}, \mathfrak{m})$ and $(\mathcal{R}', \mathfrak{m}')$ with residue fields κ and κ' respectively. The map (1.4.5.3) is an isomorphism if and only if the following conditions hold:*

(i) *F commutes with direct limits over Λ,*
(ii) *for any $(\Lambda', R') \in \mathcal{C}_{\Lambda'}$ and the local Λ-subalgebra $R = R' \times_{\kappa'} \kappa \subset R'$ with residue field κ, the natural map $F(\Lambda, R) \to F(\Lambda', R')$ is bijective.*

This result is an abstract version of an argument of Faltings in the setting of Galois deformations; see [**129**, pp. 457-8]. Note that R in (ii) is not noetherian when $[\kappa' : \kappa]$ is not finite and $R' \neq \kappa'$.

PROOF. The necessity of (i) has been explained, and the necessity of (ii) is obvious. To prove sufficiency, we first make a general construction that has nothing to do with (i) or (ii).

For any (Λ', R') in Inf_Λ, each local Λ-algebra map $f : \mathcal{R} \to R'$ factors through a local Λ-algebra map $f_n : \mathcal{R}/\mathfrak{m}^n \to R'$ for some $n \geqslant 1$. The map

$$F(f_n) : \mathrm{Hom}_\Lambda(\mathcal{R}, \mathcal{R}/\mathfrak{m}^n) = F(\mathcal{R}/\mathfrak{m}^n) \to F(R')$$

produces an element of $F(R')$ that is independent of n and functorial in (Λ', R'), so it defines a natural transformation of functors $\mathrm{Hom}_\Lambda(\mathcal{R}, \cdot) \to F$ on Inf_Λ.

Our problem is precisely to prove that this is an isomorphism on $\mathcal{C}_{\Lambda'}$. By (ii), it suffices to work on the category of pairs (Λ, R). Now using (i), we are done. □

The abstract criteria in Proposition 1.4.5.6 will now be used to establish part (2) of Theorem 1.4.5.5 (taking Λ in the abstract criteria to be the universal deformation ring of X_0 as in Theorem 1.4.5.5). For a pair of p-divisible groups X and Y over Λ and a homomorphism $f_0 : X_0 \to Y_0$ between the special fibers, define the covariant set-valued functor F on Inf_Λ to carry (Λ', R') to the set of deformations of $(f_0)_{\kappa'}$ to R' (where κ' is the residue field of Λ'). The set $F(\Lambda', R')$ is empty when there is no lift and it consists of a single element when there exists a lift and R' is artinian (as the uniqueness of such a lift for artinian R' follows from Proposition 1.4.4.3). In fact $F(\Lambda', R')$ always consists of a single element when it is non-empty, even when R' is not noetherian. The key point is that $\mathfrak{m}^n = 0$ for some $n \geqslant 1$, so we can apply:

1.4.5.7. Lemma. *Let R be a ring in which a prime p is nilpotent and let $J \subset R$ be an ideal such that $J^n = 0$ with $n \geqslant 1$. Let $f : X \to Y$ be a homomorphism between p-divisible groups over R. If f vanishes modulo J then $f = 0$.*

PROOF. The case $n = 1$ is trivial, and by induction on n we may assume $n = 2$. Our problem is comparing two R-homomorphisms with the same reduction modulo J. Equipping J with trivial divided powers, this problem will be solved used Grothendieck–Messing theory. Consider the Lie algebras of the universal vector extensions $\mathrm{E}(X)$ and $\mathrm{E}(Y)$, equipped with their respective Hodge subbundles $\mathrm{Lie}(X^t)^\vee$ and $\mathrm{Lie}(Y^t)^\vee$. For any R-homomorphism $u : X \to Y$, Grothendieck–Messing theory shows that the map $\mathrm{Lie}(\mathrm{E}(u))$ respecting Hodge subbundles uniquely determines u and only depends on $u_0 := u \bmod J$. Hence, if $u_0 = 0$ then $u = 0$. □

We now apply Proposition 1.4.5.6 to the functor F defined above whose value on any $(\Lambda', R') \in \mathrm{Inf}_\Lambda$ is empty or a singleton. The established part (1) of Theorem 1.4.5.5 (with varying Λ) implies the pro-representability hypothesis in Proposition 1.4.5.6. Thus, we just have to verify conditions (i) and (ii) in Proposition 1.4.5.6. Condition (ii) is immediate via Proposition 1.4.4.11. To establish (i) for $(\Lambda', R') = \varinjlim(\Lambda', R'_i)$, we just have to show that if $F(\Lambda', R')$ is non-empty then so is $F(\Lambda', R'_i)$ for some i. We may rename Λ' as Λ, R' as R, and R'_i as R_i for simplicity of notation. Our problem is to show that if f_0 lifts to an R-homomorphism $f : X_R \to Y_R$ then f descends to an R_i-homomorphism $X_{R_i} \to Y_{R_i}$ for some large i. We shall induct on the integer $n \geqslant 1$ such that the maximal ideal $\mathfrak{m}$ of R and maximal ideal $\mathfrak{m}_i$ of every R_i have vanishing nth power, the case $n = 1$ being trivial.

Let $\overline{R} = R/\mathfrak{m}^{n-1}$ and $\overline{R}_i = R_i/\mathfrak{m}_i^{n-1}$, so $\varinjlim \overline{R}_i = \overline{R}$. We may assume $n \geqslant 2$ and (by induction) that the lift $\overline{f} := f_{\overline{R}}$ of f_0 descends to a (necessarily unique) lift $\overline{f}_{i_0} : X_{\overline{R}_{i_0}} \to Y_{\overline{R}_{i_0}}$ of f_0. For $i \geqslant i_0$ let $\overline{f}_i = \overline{f}_{i_0} \otimes_{\overline{R}_{i_0}} \overline{R}_i$. For $i \geqslant i_0$, if there is a lift $f_i : X_{R_i} \to Y_{R_i}$ of $\overline{f}_i$ then $f_i \otimes_{R_i} R$ is a lift of f_0 and thus coincides with f (by Lemma 1.4.5.7). Hence, it is necessary and sufficient to find $i \geqslant i_0$ so that $\overline{f}_i$ lifts over R_i.

By Grothendieck–Messing theory (see [**75**, IV, 2.5; V, 1.6]), for every $i \geqslant i_0$ there is a canonical map

$$L_i : \mathrm{Lie}(\mathrm{E}(X_{R_i})) \to \mathrm{Lie}(\mathrm{E}(Y_{R_i}))$$

that only depends on $\overline{f}_i$ and has reduction $\mathrm{Lie}(\mathrm{E}(\overline{f}_i))$ modulo the square-zero ideal $\mathfrak{m}_i^{n-1} \subset R_i$, and moreover L_i respects the Hodge subbundles if and only if $\overline{f}_i$ lifts to an R_i-homomorphism $X_{R_i} \to Y_{R_i}$. Thus, it is necessary and sufficient to prove that L_i respects the Hodge subbundles for large i. Compatibility with base change ensures that $L_{i'} = L_i \otimes_{R_i} R_{i'}$ whenever $i' \geqslant i \geqslant i_0$ and that $L_i \otimes_{R_i} R = \mathrm{Lie}(\mathrm{E}(f))$. But this latter map respects the Hodge subbundles since it arises from an R-homomorphism f lifting $\overline{f}$. Hence, by standard limit arguments (and the compatibility of the Hodge subbundles with respect to base change) it follows that L_i respects the Hodge subbundles for sufficiently large i. This completes the proof of Theorem 1.4.5.5. □

1.5. CM types

Let A be an isotypic abelian variety of dimension $g > 0$ over a field K such that A admits sufficiently many complex multiplications. By Theorem 1.3.4, we may and do choose a CM field $L \subset \mathrm{End}^0(A)$ with degree $2g$. (Conversely, by Theorem 1.3.1.1, the existence of such an L forces A to be isotypic.) It turns out that the L-linear isogeny class of A is encoded in terms of a rather simple discrete invariant when $\mathrm{char}(K) = 0$ and K is algebraically closed. We wish to review the basic features of this invariant, called the CM type, and to discuss some useful replacements for it in positive characteristic.

The order $\mathscr{O} = L \cap \mathrm{End}(A)$ in L acts on A over K, and is called the CM *order*. It acts K-linearly on the tangent space $T = \mathrm{Lie}(A)$ at the origin, so if $\mathrm{char}(K) = 0$ then $L = \mathscr{O} \otimes_{\mathbb{Z}} \mathbb{Q}$ acts K-linearly on T, whereas if $\mathrm{char}(K) = p > 0$ then $\mathscr{O}/(p)$ acts K-linearly on T. In particular, if $\mathrm{char}(K) = 0$ then T is an $L \otimes_{\mathbb{Q}} K$-module whose isomorphism class is an invariant of the L-linear isogeny class of A over K; nothing of the sort is true when $\mathrm{char}(K) = p > 0$.

1.5.1. Characteristic 0. We now focus on the case $\mathrm{char}(K) = 0$. Let K'/K be an algebraically closed extension. Since $L \otimes_{\mathbb{Q}} K \simeq \prod K_i$ for finite (separable) extensions K_i/K, any $L \otimes_{\mathbb{Q}} K$-module M canonically decomposes as $\prod M_i$ for a K_i-vector space M_i. Thus, if $\dim_K M$ is finite then the isomorphism class of M is determined by the numbers $\dim_{K_i} M_i$, which in turn are determined by the isomorphism class of the $L \otimes_{\mathbb{Q}} K'$-module $M \otimes_K K'$.

The K'-algebra $L \otimes_{\mathbb{Q}} K'$ has a very simple form: it is $\prod_{\varphi} K'_{\varphi}$ where φ ranges through all field embeddings $L \to K'$ and K'_{φ} denotes K' viewed as an L-algebra via φ. Hence, any $L \otimes_{\mathbb{Q}} K'$-module M' decomposes into a corresponding product of eigenspaces M'_{φ} over K' on which L acts through φ. We conclude that for an $L \otimes_{\mathbb{Q}} K$-module M with finite K-dimension, the isomorphism class of M is determined by the numbers $\dim_{K'}(M \otimes_K K')_{\varphi}$ as φ varies through $\mathrm{Hom}(L, K')$.

On the set $\mathrm{Hom}(L, K') = \mathrm{Hom}(L, \overline{\mathbb{Q}})$ (with $\overline{\mathbb{Q}}$ the algebraic closure of $\mathbb{Q}$ in K') there is a natural involution defined by precomposition with the intrinsic complex conjugation ι of the CM field L (i.e., the non-trivial automorphism of L over its maximal totally real subfield L^+). This decomposes the set $\mathrm{Hom}(L, K')$ of size $2g$ into g "conjugate pairs" of embeddings. In the special case $K' = \mathbb{C}$ we can also compute the involution on $\mathrm{Hom}(L, K')$ by using composition with complex conjugation on $K' = \mathbb{C}$.

An especially interesting example is the $L \otimes_{\mathbb{Q}} K'$-module $M = T \otimes_K K'$ with $T = \mathrm{Lie}(A)$ for a CM abelian variety A over K with complex multiplication by L. There is a non-trivial constraint on the eigenspaces $(T \otimes_K K')_{\varphi}$ for the L-action on $T \otimes_K K'$ (with φ varying through the embeddings $\varphi : L \to K'$):

1.5.2. Lemma. *When* $\mathrm{char}(K) = 0$, *each eigenspace* $(T \otimes_K K')_{\varphi}$ *is at most* 1-*dimensional over* K'. *If* Φ *denotes the set of* g *distinct embeddings* $\varphi : L \to K'$ *for which there is a* φ*-eigenline in* $T \otimes_K K'$ *then* Φ *contains no "conjugate pairs". That is, we have a disjoint union decomposition* $\mathrm{Hom}(L, K') = \Phi \coprod (\Phi \circ \iota)$.

PROOF. By considerations with direct limits (as in the proof of Proposition 1.2.6.1), we may and do first arrange that K is finitely generated over $\mathbb{Q}$. The choice of algebraically closed extension K'/K does not matter, so we can replace

K' with $\overline{K}$. We may then reduce to the case $K = K' = \mathbb{C}$, in which case a proof is given via the complex-analytic uniformization in [**82**, §22]. □

The preceding considerations motivate the following concepts.

1.5.2.1. Definition. (i) Let L be a CM field of degree $2g$ over $\mathbb{Q}$ and K an algebraically closed field of characteristic 0. An K-valued CM *type* for L is a subset $\Phi \subset \mathrm{Hom}(L, K)$ of representatives for the g orbits of the action by the complex conjugation ι of L. That is, Φ consists of g distinct elements such that $\varphi \circ \iota \notin \Phi$ for all $\varphi \in \Phi$, or equivalently $\mathrm{Hom}(L, K) = \Phi \coprod (\Phi \circ \iota)$. To emphasize the role of L, we often refer to the pair (L, Φ) as a CM *type*.

(ii) Let $L = L_1 \times \cdots \times L_s$ be a CM algebra, where each L_i is a CM field. A CM *type* for L is a subset $\coprod \Phi_i \subset \coprod \mathrm{Hom}(L_i, K) = \mathrm{Hom}(L, K)$ where (L_i, Φ_i) is a CM type for each i.

If K is a field of characteristic 0 and K'/K is an algebraically closed extension, then the tangent space to a CM abelian variety A over K with complex multiplication by a CM algebra L determines a K'-valued CM type Φ for L (apply Lemma 1.5.2 to the isogeny factors of A determined by the primitive idempotents of L). This is an invariant of the L-linear isogeny class of A over K.

1.5.2.2. Remark. In general, a CM type takes values in the algebraic closure $\overline{\mathbb{Q}}$ of $\mathbb{Q}$ in K', so if we first choose this algebraic closure as an abstract field and then take K' to be equipped with a specified embedding of this $\overline{\mathbb{Q}}$ then we can regard the CM type as being independent of K'; this is sometimes useful for passing between different choices of K' (such as $\mathbb{C}$ and $\overline{\mathbb{Q}}_p$).

1.5.3. Example. Let L be a CM field and Φ a $\mathbb{C}$-valued CM type on L. Let $(\mathbb{R} \otimes_{\mathbb{Q}} L)_\Phi$ denote $\mathbb{R} \otimes_{\mathbb{Q}} L = \prod_{v|\infty} L_v$ endowed with the complex structure defined via the isomorphism $L_v \simeq \mathbb{C}$ using the unique $\varphi_v \in \Phi$ pulling back the standard absolute value of $\mathbb{C}$ to the place v of L for each $v|\infty$. In other words, $(\mathbb{R} \otimes_{\mathbb{Q}} L)_\Phi = \prod_{\varphi \in \Phi} \mathbb{C}_\varphi$ where $\mathbb{C}_\varphi$ denotes $\mathbb{C}$ equipped with the L-action via $\varphi : L \to \mathbb{C}$. The ring of integers $\mathcal{O}_L$ is a lattice in $\mathbb{R} \otimes_{\mathbb{Q}} L = \mathbb{R} \otimes_{\mathbb{Z}} \mathcal{O}_L$ in the natural way, so $(\mathbb{R} \otimes_{\mathbb{Q}} L)_\Phi / \mathcal{O}_L$ is a complex torus of dimension $[L : \mathbb{Q}]/2$.

In the complex-analytic theory [**82**, §22] it is proved (using that L is a CM field) that this complex torus admits a Riemann form (with respect to which the action of each $c \in L$ has adjoint given by the complex conjugate $\overline{c} \in L$), and hence is an abelian variety. Let A_Φ be the corresponding abelian variety over $\mathbb{C}$. By construction (and GAGA), we get an action by $\mathcal{O}_L$ on A_Φ and hence an embedding $L \hookrightarrow \mathrm{End}^0(A_\Phi)$ as a subfield of $\mathbb{Q}$-degree $[L : \mathbb{Q}] = 2\dim(A_\Phi)$. This makes A_Φ into a CM abelian variety over $\mathbb{C}$ with complex multiplication by L. The action by any $c \in \mathcal{O}_L \subset \mathrm{End}(A_\Phi)$ on

$$\mathrm{Lie}(A_\Phi) = \mathrm{Lie}(A_\Phi^{\mathrm{an}}) = (\mathbb{R} \otimes_{\mathbb{Q}} L)_\Phi = \prod_{\varphi \in \Phi} \mathbb{C}_\varphi$$

is the map $(a_\varphi) \mapsto (\varphi(c) a_\varphi)$ involving multiplication in $\mathbb{C}$. Thus, A_Φ equipped with the embedding $L \hookrightarrow \mathrm{End}^0(A_\Phi)$ gives rise to the CM type Φ on L.

The CM abelian varieties A_Φ are generally not simple; see Remark 1.5.4.2 for further discussion of the simplicity aspect. It is shown in the classical theory [**82**, §22, First Ex., Thm.] that as we vary Φ through all CM types on L, the A_Φ vary (without repetition) through all L-linear isogeny classes of (necessarily isotypic) CM abelian varieties over $\mathbb{C}$ with complex multiplication by L.

1.5.3.1. Definition. Let A be an isotypic abelian variety of dimension $g > 0$ over a field K, and let L be a CM field of degree $2g$ equipped with an embedding $j : L \hookrightarrow \mathrm{End}^0(A)$. The *dual* CM *structure* on the dual abelian variety A^t is the embedding $L \hookrightarrow \mathrm{End}^0(A^t)$ defined by $x \mapsto j(\overline{x})^t$, where $x \mapsto \overline{x}$ is complex conjugation on L.

It is easy to check that this definition respects double duality (i.e., if A^{tt} is equipped with the CM structure dual to the one on A^t then the canonical isomorphism $A \simeq A^{tt}$ is L-linear). The reason for the appearance of complex conjugation on L in the definition of the dual CM structure is that when K is algebraically closed of characteristic 0 it gives A^t the *same* (K-valued) CM type as A.

To verify this equality of CM types we may reduce to the case when $K = \mathbb{C}$ and then use the exhaustive construction in the complex-analytic theory as in Example 1.5.3. Alternatively, still working over $\mathbb{C}$, consider the functorial isomorphism $\mathrm{Lie}(A^t) \simeq \mathrm{H}^1(A, \mathcal{O}_A)$ and the functorial Hodge decomposition

$$\mathbb{C} \otimes_{\mathbb{Q}} \mathrm{H}_1(A(\mathbb{C}), \mathbb{Q}) \simeq \mathrm{H}^1(A(\mathbb{C}), \mathbb{C})^{\vee} \simeq \mathrm{Lie}(A) \oplus \mathrm{H}^1(A, \mathcal{O}_A)^{\vee}.$$

Since $\mathrm{H}_1(A(\mathbb{C}), \mathbb{Q})$ is 1-dimensional as an L-vector space, when A^t is equipped with the dual action $j(x)^t$ (without the intervention of complex conjugation) then its CM type is $\mathrm{Hom}(L, \mathbb{C}) - \Phi = \overline{\Phi}$.

1.5.4. Descent to a number field. For a CM abelian variety over an algebraically closed field K of characteristic 0, we may make the CM type essentially be independent of K by replacing K with $\overline{\mathbb{Q}}$ (see Remark 1.5.2.2). This enables us to use the complex-analytic theory to prove the following purely algebraic result.

1.5.4.1. Proposition. *Let K be an algebraically closed field of characteristic* 0*. Let L be a* CM *field, and consider a* CM *abelian variety A over K with complex multiplication via $j : L \hookrightarrow \mathrm{End}^0(A)$. The L-linear isogeny class of A is uniquely determined by the K-valued* CM *type Φ on L associated to (A, j), and every* CM *type on L arises in this way from some (A, j) over K.*

The hypothesis $K = \overline{K}$ cannot be weakened. For example, if K is a number field containing a Galois closure of L over $\mathbb{Q}$ (so all $\overline{K}$-valued CM types on L are K-valued) then any quadratic twist of A (equipped with the evident L-linear structure) has the same CM type as (A, j) but is generally not K-isogenous to A.

PROOF. In view of Lemma 1.2.1.2, by expressing K as a direct limit of algebraically closed subfields of finite transcendence degree over $\mathbb{Q}$ we can reduce to the case when K has finite transcendence degree over $\mathbb{Q}$. To show that the CM type determines the L-linear isogeny class it suffices (again by Lemma 1.2.1.2) to treat the case $K = \mathbb{C}$. This case was addressed in Example 1.5.3 via the complex-analytic theory, where it was also seen that every CM type Φ on L does arise when $K = \mathbb{C}$.

It remains to show that every CM type Φ on L arises when $K = \overline{\mathbb{Q}}$. Consider the CM abelian variety A_Φ over $\mathbb{C}$ with complex multiplication by L and CM type Φ as in Example 1.5.3. Recall that $\mathcal{O}_L = L \cap \mathrm{End}(A_\Phi)$ inside $\mathrm{End}^0(A_\Phi)$. By expressing $\mathbb{C}$ as a direct limit of its finitely generated $\overline{\mathbb{Q}}$-subalgebras, there is such a subalgebra R for which A with its $\mathcal{O}_L$-action over $\mathbb{C}$ descends to an abelian scheme $\mathscr{A}$ over R equipped with an $\mathcal{O}_L$-action.

By localization of R, we can arrange that the tangent space $\mathrm{Lie}(\mathscr{A})$ is finite and free as an R-module, and by increasing R to contain the integer ring of the Galois closure of L in $\mathbb{C}$ we can arrange that the $\mathcal{O}_L$-action on $\mathrm{Lie}(\mathscr{A})$ decomposes as $\prod_{\varphi\in\Phi} R_\varphi$ with R_φ equal to R having action by $c \in \mathcal{O}_L$ through multiplication by $\varphi(c) \in \overline{\mathbb{Q}} \subset R$. For any maximal ideal $\mathfrak{m}$ of R, the natural map of $\overline{\mathbb{Q}}$-algebras $\overline{\mathbb{Q}} \to R/\mathfrak{m}$ is an isomorphism. Thus, passing to the fiber of $\mathscr{A}$ at a closed point of $\mathrm{Spec}(R)$ gives a pair (A, j) over $\overline{\mathbb{Q}}$ with CM type Φ. □

This proposition has an important consequence for descending the field of definition of a CM abelian variety in characteristic 0, as we will see in Theorem 1.7.2.1.

1.5.4.2. Remark. By Theorem 1.3.1.1, for any (A, L) as in Proposition 1.5.4.1, A has a unique simple factor C (in the sense of Definition 1.2.1.5). By Proposition 1.3.2.1, C is a CM abelian variety with complex multiplication by the CM field $L' := \mathrm{End}^0(C)$ (see Proposition 1.3.6.4(1)). Since L' is canonically identified with the center of $\mathrm{End}^0(A)$, it naturally embeds into L. Hence, there is a K-valued CM type Ψ on L' arising from C, and the pair (L', Ψ) is determined by (L, Φ) since A with its complex multiplication by L is determined up to L-linear isogeny by Φ (and $L' = L$ if and only if $A = C$, which is to say that A is simple). It is therefore natural to seek an intrinsic recipe to directly construct (L', Ψ) from (L, Φ), and in particular to characterize in terms of Φ whether or not A is simple.

The criterion is this: among the CM fields in L from which Φ is obtained by full preimage under restriction, (L', Ψ) is the unique such pair with $[L' : \mathbb{Q}]$ minimal and Φ the full preimage of Ψ. Indeed, since the CM type is $\overline{\mathbb{Q}}$-valued (Remark 1.5.2.2) and the base field K is algebraically closed, it suffices to treat the case $K = \mathbb{C}$. In this case the desired recipe is established in the complex-analytic theory (see [**82**, §22, Rem. (1)]).

1.5.5. Positive characteristic. Assume $\mathrm{char}(K) > 0$, and let A be an abelian variety over K of dimension $g > 0$ admitting an action by an order $\mathcal{O}$ in a CM field L of degree $2g$ over $\mathbb{Q}$. There is no action by $L = \mathbb{Q} \otimes_{\mathbb{Z}} \mathcal{O}$ on the tangent space $T := \mathrm{Lie}(A)$ of A at the origin since L is a $\mathbb{Q}$-algebra and T is a K-vector space. Thus, there is not a good notion of CM type on L associated to the embedding of L into $\mathrm{End}^0(A)$. More specifically, for the CM order $\mathcal{O} := L \cap \mathrm{End}(A)$ in L, T has a K-linear action by $\mathcal{O}/(p)$ and there is generally no constraint on this action akin to the eigenspace decomposition considered in characteristic 0 (as in Lemma 1.5.2). The lack of such a constraint occurs for a couple of reasons, as we now explain.

1.5.5.1. Example. If p divides the discriminant of $\mathcal{O}$ over $\mathbb{Z}$ or $p\mathcal{O}_L$ is not prime in $\mathcal{O}_L$ then $\mathcal{O}/p\mathcal{O}$ fails to be a field. In such cases, the K-linear $\mathcal{O}/(p)$-action on T admits no notion of eigenspace decomposition that closely resembles the situation in characteristic 0.

1.5.5.2. Example. Suppose that $\mathcal{O}$ has discriminant not divisible by p (so $\mathcal{O}_{(p)} = \mathcal{O}_{L,(p)}$) and that p is totally inert in L. In such cases $\kappa := \mathcal{O}/(p)$ is a finite field of degree $2g$ over $\mathbb{F}_p$ and $\mathrm{Aut}(L/\mathbb{Q})$ injects into $\mathrm{Gal}(\kappa/\mathbb{F}_p)$, so the canonical complex conjugation on L induces a non-trivial involution on κ.

For an algebraically closed extension K'/K we can consider the eigenspace decomposition of $T \otimes_K K'$ over $\kappa \otimes_{\mathbb{F}_p} K' = \prod_\varphi K'_\varphi$ where φ ranges over the $2g$ distinct embeddings of κ into K'. This could fail to resemble the CM types that arise in characteristic 0 because (as we shall see in later examples, such as in Remark 2.3.4) there may be conjugate pairs occurring among the φ for which $T \otimes_K K'$ has a non-zero φ-eigenspace with respect to its K'-linear κ-action.

In such cases, the composite action

$$\mathcal{O} \to \mathrm{End}(A) \to \mathrm{End}(A)/(p) \to \mathrm{End}_K(T)$$

does not "look like the reduction of a CM type", and so this provides an obstruction for A equipped with its $\mathcal{O}$-action to lift to characteristic 0. There is no dimension obstruction to such lifting: each φ-eigenspace in $T \otimes_K K' = \mathrm{Lie}(A_{K'})$ has K'-dimension at most 1. To prove this, first note that the Dieudonné module $D := \mathrm{M}^*(A_{K'}[p^\infty])$ is free of rank 1 over

$$\mathcal{O} \otimes_{\mathbb{Z}} W(K') = \mathcal{O}_{(p)} \otimes_{\mathbb{Z}_{(p)}} W(K') = \mathcal{O}_L \otimes_{\mathbb{Z}} W(K')$$

by Proposition 1.4.3.9(2) (or Proposition 1.2.5.1 and $W(K')$-rank considerations), so D/pD is free of rank 1 over $\kappa \otimes_{\mathbb{F}_p} K'$. The formal group $\widehat{A}_{K'}$ is the identity component of the p-divisible group $A_{K'}[p^\infty]$ (Example 1.4.3.6), and its tangent space coincides with that of $A_{K'}$. Hence, by 1.4.3.2(4),

$$T \otimes_K K' \simeq \mathrm{Lie}(A_{K'}[p^\infty]) \simeq (D/\mathcal{F}(D))^\vee,$$

where $\mathcal{F} : D \to D$ denotes the semilinear Frobenius endomorphism. By naturality, this composite isomorphism is $\kappa \otimes_{\mathbb{F}_p} K'$-linear, so $T \otimes_K K'$ is monogenic over $\kappa \otimes_{\mathbb{F}_p} K'$ since the K'-linear dual of a monogenic $\kappa \otimes_{\mathbb{F}_p} K'$-module is monogenic (as $\kappa \otimes_{\mathbb{F}_p} K'$ is a finite étale K'-algebra). Each φ-eigenspace of $T \otimes_K K'$ is therefore monogenic over K', which is to say is of dimension at most 1 over K'.

To go beyond Example 1.5.5.2, an obstruction to the existence of a CM lift over a *normal* local domain of characteristic 0 will be formulated precisely later (see 2.1.5 and 4.1.2). This will be used to exhibit examples (e.g., in 4.1.2) of abelian varieties over finite fields for which there is no such lift. Such examples are interesting due to Corollary 1.6.2.5 below, according to which *every* abelian variety over a finite field admits sufficiently many complex multiplications.

Although the Lie algebra fails to be an isogeny invariant for the study of CM abelian varieties in positive characteristic (and $\mathrm{End}^0(A)$ does not act on the tangent space when $\mathrm{char}(K) > 0$), there is an alternative linear object attached to a CM abelian variety A that serves as a good substitute when $\mathrm{char}(K) = p > 0$: the p-divisible group $A[p^\infty]$, or its (contravariant) Dieudonné module $\mathrm{M}^*(A[p^\infty])$ when K is perfect.

Letting $B = A$ in Proposition 1.2.5.1, we see that $\mathbb{Z}_p \otimes_{\mathbb{Z}} \mathrm{End}(A)$ acts faithfully on $A[p^\infty]$. Hence, $\mathbb{Q}_p \otimes_{\mathbb{Q}} \mathrm{End}^0(A)$ acts faithfully on $A[p^\infty]$ in the isogeny category of p-divisible groups over K. In particular, if K is perfect (e.g., finite) and A is an isotypic CM abelian variety over K with complex multiplication by the CM field L (see Theorem 1.3.4) then $L_p := \mathbb{Q}_p \otimes_{\mathbb{Q}} L$ acts faithfully and linearly on the vector space $\mathrm{M}^*(A[p^\infty])[1/p]$ of rank $2g$ over the absolutely unramified p-adic field $W(K)[1/p]$.

This $W(K)[1/p]$-linear faithful L_p-action for perfect K with $\mathrm{char}(K) = p$ is an analogue of a classical construction when $\mathrm{char}(K) = 0$: the action of L on the algebraic de Rham cohomology $\mathrm{H}^1_{\mathrm{dR}}(A/K)$ (a filtered K-vector space of dimension $2g$). It will be useful in later considerations (e.g., the proof of Theorem 2.2.3) with lifting problems from positive characteristic to characteristic 0. (Note that when $\mathrm{char}(K) = 0$, $\mathrm{H}^1_{\mathrm{dR}}(A/K)$ provides essentially the same information as the CM type arising from the L-action on $\mathrm{Lie}(A) = \mathrm{H}^0(A, \Omega^1_{A/K})^\vee$, in view of the Hodge filtration on $\mathrm{H}^1_{\mathrm{dR}}(A/K)$; cf. Definition 1.5.3.1 and the subsequent discussion there.)

1.6. Abelian varieties over finite fields

In this section we work over a finite field κ with $\mathrm{char}(\kappa) = p$.

1.6.1. Tate's theorem and Weil numbers. A fundamental fact in the theory of abelian varieties over finite fields is:

1.6.1.1. Theorem (Tate's isogeny theorem). *For abelian varieties A and B over a finite field κ, the natural injective map*

$$\mathbb{Z}_\ell \otimes_{\mathbb{Z}} \mathrm{Hom}(A, B) \to \mathrm{Hom}(A[\ell^\infty], B[\ell^\infty])$$

is bijective for every prime ℓ (including $\ell = \mathrm{char}(\kappa)$).

PROOF. By passing to $A \times B$, it suffices to treat the case $A = B$, as we shall now consider. The case $\ell \neq \mathrm{char}(\kappa)$ is the main result in [**118**]; see [**82**, App. I, Thm. 1] for a proof as well. Unfortunately, Tate did not publish his proof for the case $\ell = p$ (though his argument was published in [**79**]). See Appendix A.1 for a proof. □

Tate's proof of his isogeny theorem is closely tied up with his analysis of the general structure of endomorphism algebras of abelian varieties over finite fields. The essential case, and the one on which we will now focus, is a simple abelian variety A over a finite field κ. In this case $D := \mathrm{End}^0(A)$ is a division algebra of finite dimension over $\mathbb{Q}$. If $q = \#\kappa$ then the q-Frobenius endomorphism

$$\pi = \pi_A : A \longrightarrow A$$

is central in D since the q-Frobenius is functorial on the category of κ-schemes. Hence, the number field $\mathbb{Q}[\pi] = \mathbb{Q}(\pi)$ is contained in the center of D.

Even without simplicity or isotypicity hypotheses on A, Tate proved (see [**82**, App. I, Thm. 3(a)]) that the commutative $\mathbb{Q}$-algebra $\mathbb{Q}[\pi]$ is the center of $\mathrm{End}^0(A)$ for any abelian variety A over κ.

1.6.1.2. Definition. Let $q = p^n$ for a positive integer n and prime number p. Let F be a field of characteristic 0.

(i) A *Weil q-integer* in F (or a *Weil q-integer of weight* 1, to be precise) is an algebraic integer $z \in F$ whose $\mathbb{Q}$-conjugates in $\mathbb{C}$ have absolute value $q^{1/2}$.[2]

(ii) Let w be an integer. A *Weil q-number* of weight w is an algebraic number z such that $\mathrm{ord}_v(z) = 0$ for all finite places v of $\mathbb{Q}(z)$ prime to q and $|\tau(y)| = q^{w/2}$ for all injective ring homomorphisms $\tau\colon \mathbb{Q}(z) \to \mathbb{C}$.

Note that a Weil q-integer as defined above is precisely a Weil q-number of weight 1 such that $\mathrm{ord}_v(z) \geqslant 0$ for all p-adic places v of $\mathbb{Q}(z)$. The interest in Definition 1.6.1.2 is that Weil proved (see [**95**, §3]) that for any non-zero abelian

[2] What is called a Weil q-integer here is often called a "Weil q-number" or "Weil q-number of weight 1" in the literature.

variety A over κ and any $\ell \neq p := \mathrm{char}(\kappa)$, the $\mathbb{Q}_\ell$-linear q-Frobenius action on $V_\ell(A)$ has characteristic polynomial $f_{A,q} \in \mathbb{Z}[T]$ that is independent of ℓ and has all roots in $\mathbb{C}$ equal to Weil q-integers.

In Tate's work, he also proved (see [**82**, App. I, Thm. 3(e)]) that A is isotypic if and only if the common characteristic polynomial $f_{A,q} \in \mathbb{Z}[T]$ of the q-Frobenius action on the Tate modules has a single monic irreducible factor over $\mathbb{Q}$, in which case this irreducible factor is obviously the minimal polynomial f_π over $\mathbb{Q}$ for the q-Frobenius endomorphism $\pi \in \mathrm{End}^0(A)$ (since π is central). The polynomial f_π only depends on A through its isogeny class (due to the functoriality of q-Frobenius on κ-schemes), and by Weil's Riemann Hypothesis its $\mathrm{Gal}(\overline{\mathbb{Q}}/\mathbb{Q})$-conjugacy class of roots in $\mathbb{C}$ consists of Weil q-integers, where $\overline{\mathbb{Q}}$ is the algebraic closure of $\mathbb{Q}$ in $\mathbb{C}$.

1.6.2. The Honda-Tate theorem. Fix an abstract algebraic closure $\overline{\mathbb{Q}}$ of $\mathbb{Q}$ and let Weil(q) denote the set of Weil q-integers in $\overline{\mathbb{Q}}$. Elements of Weil(q) are *equivalent* when they lie in the same $\mathrm{Gal}(\overline{\mathbb{Q}}/\mathbb{Q})$-orbit; i.e., they have the same minimal polynomial over $\mathbb{Q}$. The following remarkable result relates Weil q-integers to isogeny classes of simple abelian varieties over a finite field of size q.

1.6.2.1. Theorem (Honda-Tate). *Let κ be a finite field of size q. The assignment $A \mapsto \pi_A$ defines a bijection from the set of isogeny classes of simple abelian varieties over κ to the set of* $\mathrm{Gal}(\overline{\mathbb{Q}}/\mathbb{Q})$*-conjugacy classes of Weil q-integers.*

PROOF. We refer the reader to [**50**], [**121**], and [**95**] for a discussion of the proof of the Honda-Tate theorem. The proof of injectivity in Theorem 1.6.2.1 rests on the work of Tate related to Theorem 1.6.1.1. The proof of surjectivity uses abelian varieties in characteristic 0 (in fact, it uses descents to number fields of CM abelian varieties over $\mathbb{C}$; see Theorem 1.7.2.1). We are not aware of a proof of surjectivity that avoids abelian varieties in characteristic 0. □

The following consequence of the (proof of the) Honda–Tate theorem describes the possibilities for the division algebra $D = \mathrm{End}^0(A)$ in terms of whether the center Z is $\mathbb{Q}$ or $\mathbb{Q}(\sqrt{p})$ (the totally real cases) or a CM field.

1.6.2.2. Corollary. *Let A be a simple abelian variety over a finite field κ of size q and characteristic p. Define $D = \mathrm{End}^0(A)$, so $Z := \mathbb{Q}(\pi)$ is its center. Let $\pi \in D$ be the q-Frobenius endomorphism. Exactly one of the following occurs.*

(1) *We have $\pi^2 = q = p^n$ with n even. This is precisely the case $Z = \mathbb{Q}$, and occurs exactly when D is a central quaternion division algebra over $\mathbb{Q}$, in which case it is the unique quaternion division algebra over $\mathbb{Q}$ ramified at exactly $\{p, \infty\}$.*

  Each of the isogeny classes of simple abelian varieties with $\pi \in \{\pm p^{n/2}\}$ consists of supersingular elliptic curves E over κ for which all endomorphisms of $E_{\overline{\kappa}}$ are defined over κ (equivalently, the geometric endomorphism algebra $\mathrm{End}^0(E_{\overline{\kappa}})$ *coincides with* $\mathrm{End}^0(E)$*).*

(2) *We have $\pi^2 = q = p^n$ with n odd. This is precisely the case $Z = \mathbb{Q}(\sqrt{p})$, and occurs if and only if D is the unique central quaternion division algebra over Z ramified at exactly the two infinite places of Z.*

  The corresponding isogeny class of simple abelian varieties is represented by the 2-dimensional Weil restriction $\mathrm{Res}_{\kappa'/\kappa}(E')$ *where κ'/κ is a quadratic extension and E' is a supersingular elliptic curve over κ' whose geometric endomorphism algebra is defined over κ'.*

(3) *The field Z is a* CM *field. In such cases, D is the central division algebra over Z that is split at all places of Z away from p and for each p-adic place v of Z has local invariant* $\text{inv}_v(D) = (\text{ord}_v(\pi)/\text{ord}_v(q))[Z_v : \mathbb{Q}_p] \bmod \mathbb{Z} \in \mathbb{Q}/\mathbb{Z}$.

The members of the corresponding isogeny class of simple abelian varieties over κ have dimension $g = (1/2)[Z : \mathbb{Q}] \cdot \sqrt{[D : Z]}$.

PROOF. Since π is a Weil q-integer, for any embedding $j : Z \to \mathbb{C}$ the subfield $j(Z)$ in $\mathbb{C}$ is stable under complex conjugation and the effect of this involution on Z is given by the intrinsic formula $\pi \mapsto q/\pi$ that is independent of j. Thus, Z is either totally real or a CM field and the totally real cases are precisely when $\pi^2 = q = p^n$, with $Z = \mathbb{Q}$ for even n and $Z = \mathbb{Q}(\sqrt{p})$ for odd n.

In all cases, D is split at the finite places of Z away from p. Indeed, Tate's isogeny theorem away from p implies that for a prime $\ell \neq p$ the map

$$\mathbb{Q}_\ell \otimes_{\mathbb{Q}} D \to \text{End}_{Z_\ell}(V_\ell(A)) \tag{1.6.2.1}$$

is an isomorphism (using that the Z-action on $V_\ell(A)$ encodes the action of the Galois group $\text{Gal}(\overline{\kappa}/\kappa)$). The right side of (1.6.2.1) is visibly a product of matrix algebras over factor fields of $Z_\ell = \prod_{v|\ell} Z_v$, so D splits at all ℓ-adic places of Z. Writing $d^2 = [D : Z]$, the Z_v-algebra isomorphism $D_v \simeq \text{End}_{Z_v}(V_v(A))$ for $v|\ell$ implies that $V_v(A)$ has Z_v-dimension d for all such v, so $V_\ell(A)$ is free of rank d over Z_ℓ. Hence, $[Z_\ell : \mathbb{Q}_\ell]d = 2g$, so $g = (1/2)[Z : \mathbb{Q}]\sqrt{[D : Z]}$. This is the asserted dimension formula in case (3), and the proof also applies in cases (1) and (2) (as will be used below). The formula for $\text{inv}_v(D)$ in case (3) with $v|p$ is proved in A.1.3, resting on preliminary work in A.1.1 and A.1.2, and that proof is applicable regardless of whether Z is CM or totally real. This completes the proof of case (3), and in cases (1) and (2) (so $\pi^2 = q$) it establishes the formula $\text{inv}_v(D) = [Z_v : \mathbb{Q}_p]/2 \bmod \mathbb{Z}$ for p-adic places v of Z.

Consider case (1) (equivalently, $\pi = \pm p^{n/2}$ with n even), so D is a central division algebra over $Z = \mathbb{Q}$ split away from p and ∞ with $\text{inv}_p(D) = 1/2 \bmod \mathbb{Z}$. This forces $\text{inv}_\infty(D) = 1/2 \bmod \mathbb{Z}$, so D is a quaternion division algebra over $\mathbb{Q}$. In particular, the dimension formula yields $g = 1$, so A is an elliptic curve. In view of the other possibilities for Z, these are the only cases for which D is a central quaternion division algebra over $\mathbb{Q}$. The elliptic curves E that arise in such cases must be supersingular (since it is classical that $\text{End}^0(E_{\overline{\kappa}})$ is commutative in the ordinary case). Moreover, since it is classical that $\text{End}^0(E_{\overline{\kappa}})$ is a quaternion division algebra in the supersingular case, it follows for $\mathbb{Q}$-dimension reasons that the injection $D = \text{End}^0(E) \to \text{End}^0(E_{\overline{\kappa}})$ is an equality. In other words, all endomorphisms of $E_{\overline{\kappa}}$ are defined over κ. This settles case (1).

Finally, consider case (2). Since the numbers $\pm p^{n/2}$ with odd n are Galois conjugate over $\mathbb{Q}$, there is exactly one isogeny class that arises in this case. For the unique p-adic place v of $Z = \mathbb{Q}(\sqrt{p})$, the formula for $\text{inv}_v(D)$ vanishes. Hence, D splits away from the two real places of Z, so its order in $\text{Br}(Z)$ divides 2 and the dimension formula says $g = \sqrt{[D : Z]}$. Thus, either $D = Z$ and A is an elliptic curve or D is the unique central quaternion division algebra over Z split away from the real places and A is an abelian surface. The first case cannot happen, since otherwise the quadratic field Z would provide a CM structure on the elliptic curve, contradicting Proposition 1.3.6.4(2) since $D = Z$ is a real quadratic field.

For the quadratic extension κ' of κ, the abelian surface $A_{\kappa'}$ is isotypic (by 1.2.6.1). But its q^2-Frobenius is $\pi_{\kappa'}^2 = q$, which by the settled case (1) over κ' is the Weil q^2-integer associated to some supersingular elliptic curve over κ'. Hence, $A_{\kappa'}$ cannot also be κ'-simple, so it is isogenous to $E' \times E'$ for an elliptic curve E' over κ' that must be in the isogeny class which occurs in case (1) over κ'. Choosing such an isogeny decomposition provides a non-zero homomorphism $A_{\kappa'} \to E'$ over κ'. By the universal property of Weil restriction, we thereby get a non-zero κ-homomorphism $A \to \mathrm{Res}_{\kappa'/\kappa}(E')$. Since A is κ-simple, by dimension reasons this must be an isogeny. The existence of A is guaranteed by the Honda–Tate classification, so it follows that for E' as in case (1) over κ', the abelian surface $\mathrm{Res}_{\kappa'/\kappa}(E')$ is necessarily κ-simple and lies in the unique isogeny class occurring in case (2) over κ. □

1.6.2.3. Remark. In the terminology of Theorem 1.3.6.2, the three cases in Corollary 1.6.2.2 correspond to A that are respectively of Type III with $e = 1$, Type III with $e = 2$, and Type IV. We saw in the proof of 1.6.2.2 that the formula for $\mathrm{inv}_v(D)$ for p-adic places v of Z in case (3) works in cases (1) and (2), and that the dimension formula in case (3) works in cases (1) and (2) (though these facts in cases (1) and (2) are also clear by inspection).

The common $\mathbb{Q}$-degree $[Z : \mathbb{Q}]\sqrt{[D : Z]}$ of maximal commutative subfields of the division algebra D is $2g$ in each case of Corollary 1.6.2.2, so *simple* abelian varieties over finite fields always have sufficiently many complex multiplications.

1.6.2.4. Example. Observe that in part (3) of Corollary 1.6.2.2, for any elliptic curve case that arises necessarily the CM field $Z = \mathbb{Q}[\pi]$ is imaginary quadratic and $D = Z$. Writing $q = p^r$, the elliptic curves that arise in this way are as follows, depending on the behavior of p in the imaginary quadratic field $Z = \mathbb{Q}[\pi]$.

There are several possibilities for the splitting behavior of p in Z: (i) p splits in Z with π generating the rth power of one of the two primes of Z over p, (ii) p is inert in Z with r even and $\pi = p^{r/2}\zeta$ for an imaginary quadratic root of unity $\zeta \neq \pm 1$ such that p is inert in $\mathbb{Q}(\zeta)$, or (iii) p is ramified in Z and π generates the rth power of the unique prime of Z over p. Cases (ii) and (iii) are exactly the supersingular cases, and since $D = Z$ in theses cases, the geometric endomorphism algebra is not entirely defined over κ. Hence, part (1) of Corollary 1.6.2.2 gives *all* supersingular elliptic curves over κ (up to isogeny) whose geometric endomorphism algebra is defined over κ.

By passing to products and using Theorem 1.3.4, we obtain the following result.

1.6.2.5. Corollary (Tate). *Every abelian variety A over a finite field admits sufficiently many complex multiplications. If A is isotypic then it admits a structure of* CM *abelian variety with complex multiplication by a* CM *field.* □

1.6.3. Example. As an application of Corollary 1.6.2.2, here are examples of simple abelian surfaces A over prime fields of any characteristic $p \not\equiv 1 \pmod{12}$ such that A is not absolutely simple. Let κ be a finite field of size p^2, with p a prime such that $p \not\equiv 1 \pmod 4$ (resp. $p \not\equiv 1 \pmod 3$). Choose ζ such that $\zeta^2 + 1 = 0$ (resp. $\zeta^2 + \zeta + 1 = 0$), so $Z := \mathbb{Q}(\zeta)$ is an imaginary quadratic field of class number

1 in which p is not split. Let $\pi = \pm p\zeta$ when p is inert in Z, and let π generate the unique prime over p in Z when p is ramified in Z, so π is a Weil p^2-integer and $Z = \mathbb{Q}(\pi)$. (Note that $\pi \neq \pm 2\sqrt{-1} = (1 \pm \sqrt{-1})^2$ since 2 is ramified in $\mathbb{Q}(\sqrt{-1})$.)

By Corollary 1.6.2.2(3) (also see Example 1.6.2.4), a simple abelian variety E_0 over κ with p^2-Frobenius equal to π must have endomorphism algebra Z and dimension 1. The elliptic curve E_0 is supersingular because p is not split in Z. The isogeny class of E_0 contains no member that is the scalar extension of an elliptic curve over $\mathbb{F}_p$, as otherwise π would have a square root $\pi_0 \in Z$, which is visibly absurd by inspection since $\pi \neq \pm 2\sqrt{-1}$.

The abelian surface $A_0 := \mathrm{Res}_{\kappa/\mathbb{F}_p}(E_0)$ satisfies $(A_0)_\kappa \simeq E_0 \times E_0^{(p)}$, so $(A_0)_\kappa$ is not simple. But A_0 is simple, as otherwise there would be a non-zero homomorphism $E_0' \to A_0$ from an elliptic curve E_0' over $\mathbb{F}_p$ and hence (by the universal property of Weil restriction) a non-zero homomorphism $(E_0')_\kappa \to E_0$, contrary to what we just saw concerning the isogeny class of E_0. (Note that $(A_0)_\kappa$ is isotypic, since $E_0^{(p)}$ is isogenous to E_0 via the relative Frobenius morphism $E_0 \to E_0^{(p)}$.)

Taking $K/\mathbb{Q}$ to be a quadratic field in which p is inert, we can lift E_0 over $\mathscr{O}_{K,(p)}$ to get an elliptic curve E over K having good reduction E_0 at $p\mathscr{O}_K$. Then $A := \mathrm{Res}_{K/\mathbb{Q}}(E)$ is an abelian surface over $\mathbb{Q}$ having good reduction $\mathrm{Res}_{\kappa/\mathbb{F}_p}(E_0)$ at p that is simple over $\mathbb{F}_p$, so (via consideration of Néron models over $\mathbb{Z}_{(p)}$) A is simple over $\mathbb{Q}$. However, $A_K \simeq E \times E'$ where E' is the twist $\sigma^*(E)$ by the non-trivial automorphism σ of K over $\mathbb{Q}$, so A_K is not simple.

1.6.4. Example. Pushing the end of Example 1.6.3 further over $\mathbb{Q}$, we now prove that if $\pi = \pm p\zeta$ and $p \equiv -1 \pmod 4$ with $\zeta^2 + 1 = 0$ (resp. $p \equiv -1 \pmod 3$ with $\zeta^2 + \zeta + 1 = 0$) then E and E' over K are not isogenous (so A_K is *not* isotypic, in contrast with its reduction $(A_0)_\kappa$). Suppose that there were an isogeny $\psi : E \to E'$, and choose it with minimal degree. In particular, ψ is not divisible by $[p]_E$. We claim that $\mathrm{ord}_p(\deg \psi)$ is odd (and in particular, is positive). Suppose otherwise, so $\deg \psi = mp^{2n}$ with $n \geqslant 0$ and $p \nmid m$. Consider the reduction $\psi_0 : E_0 \to E_0^{(p)}$ of ψ, also an isogeny with degree mp^{2n}. In particular, $\ker(\psi_0)$ is a finite subgroup scheme of E_0 with order mp^{2n}, so its p-part has order p^{2n}. But E_0 is supersingular, so it has a unique subgroup scheme of each p-power order. Hence, the p-part of $\ker(\psi_0)$ is $E_0[p^n]$, so $\psi_0 = \psi_0' \circ [p^n]_{E_0}$ with $\psi_0' : E_0 \to E_0^{(p)}$ of degree m.

Consider the composite isogeny

$$E_0 \stackrel{\psi_0'}{\to} E_0^{(p)} \to E_0^{(p^2)} = E_0$$

using the Frobenius isogeny of $E_0^{(p)}$. This is an endomorphism of E_0 with degree pm. Since $\mathrm{End}(E_0)$ is an order in $\mathbb{Z}[\zeta]$ on which the degree is computed as the norm to $\mathbb{Z}$, we get an element of $\mathbb{Z}[\zeta]$ whose norm in $\mathbb{Z}$ is divisible exactly once by p. That is impossible since p is prime in $\mathbb{Z}[\zeta]$, and so completes the verification that $\deg \psi$ has p-part p^j for some odd j.

We conclude that the finite K-subgroup $N := \ker(\psi) \subset E$ has non-trivial p-part, and this p-part has cyclic geometric fiber (as otherwise it would contain $E[p]$, contradicting that we arranged ψ to not be divisible by $[p]_E$). By cyclicity, $N[p]$ is a K-subgroup of E with order p. Consider its scheme-theoretic closure G in the Néron model of E at $p\mathscr{O}_{K,(p)}$. This is a finite flat group scheme over $\mathscr{O}_{K,(p)}$ of order p, and its special fiber G_κ is an order-p subgroup scheme of the supersingular elliptic curve E, so $G_\kappa \simeq \alpha_p$ as κ-groups (since G_κ is local-local of order p, so $\mathrm{M}^*(G_\kappa) = \kappa$ with $\mathcal{F} = \mathcal{V} = 0$). But $\mathcal{O}_{K,(p)}$ is an absolutely unramified discrete valuation ring, so there are no finite flat group schemes over $\mathcal{O}_{K,(p)}$ with special fiber α_p (by the classification results in [**123**]). This contradiction shows that E and E' are not isogenous (so A_K is not isotypic), as claimed. In fact, we have proved something stronger: if K_p denotes the p-adic completion of K then A_{K_p} is not isotypic.

1.6.5. CM lifting after a field extension and isogeny. The proof of the surjectivity aspect of the Honda-Tate theorem requires constructing abelian varieties having prescribed properties over finite fields. The idea is to relate simple abelian varieties over finite fields to simple factors of reductions of CM abelian varieties over number fields, at least after some finite extension on the initial finite field. (See [**50**] or [**121**, Lemme 3] for details.)

One can ask (as Honda implicitly did at the end of [**50**, §2]) to do better by arranging simplicity to hold for the reduction of a CM abelian variety over a number field (thereby eliminating the need to pass to a simple factor). Building on earlier work of Honda, such an improved lifting theorem was proved by Tate [**121**, Thm. 2] (and is really the starting point for the many lifting questions about CM abelian varieties that we consider in this book):

1.6.5.1. Theorem (Honda, Tate). *For any isotypic abelian variety A over a finite field κ, there is a finite extension κ'/κ such that $A_{\kappa'}$ is isogenous to the reduction of a* CM *abelian variety with good reduction over a p-adic field with residue field κ'.*

PROOF. By Corollary 1.6.2.5, there is a CM field $L \subset \mathrm{End}^0(A)$ with $[L : \mathbb{Q}] = 2\dim(A)$. The field L is its own centralizer in $\mathrm{End}^0(A)$, so it contains an element π which acts by the q-Frobenius endomorphism on A, where $q = \#\kappa$. Let $g = \dim(A)$. Since A is κ-isotypic, Tate's work on isogenies among abelian varieties over finite fields [**118**] gives two results for A: (i) the common characteristic polynomial over $\mathbb{Q}$ for the action of π on the Tate modules of A is a power of an irreducible polynomial f_π over $\mathbb{Q}$ (necessarily the minimal polynomial of π over $\mathbb{Q}$), and (ii) A is κ-isogenous to any g-dimensional isotypic abelian variety over κ whose q-Frobenius is a zero of f_π. Moreover, these properties persist after replacing κ with any finite extension κ' (and replacing π with $\pi^{[\kappa':\kappa]}$), due to Proposition 1.2.6.1.

By [**121**, §3, Thm. 2] (which is stated in the simple case but holds in the isotypic case by the same proof), there exists a number field $K \subset \overline{\mathbb{Q}}_p$, a g-dimensional abelian variety B over K with good reduction at the induced p-adic place v of K, an embedding of finite fields $\kappa \hookrightarrow \kappa_v$, and an action of $\mathcal{O}_L$ on B such that the reduction B_0 at v has q_v-Frobenius in $\mathcal{O}_L$ given by the action of $\pi_v = \pi^{[\kappa_v:\kappa]} \in \mathcal{O}_L$. (Here, $q_v = \#\kappa_v$.) Since B_0 admits a CM structure over κ_v by a *field* (namely, L), it is κ_v-isotypic. Thus, since $\dim(B_0) = \dim(B)$ and A_{κ_v} satisfies $\mathrm{Fr}_{A_{\kappa_v},q_v} = \mathrm{Fr}_{A,q}^{[\kappa_v:\kappa]} = \pi_v$, it follows from the results (i) and (ii) in [**118**] recalled above that there exists a κ_v-isogeny $\phi : B_0 \to A_{\kappa_v}$. □

1.6.5.2. Remark. The κ_v-isogeny $\phi : B_0 \to A_{\kappa_v}$ at the end of the proof of Theorem 1.6.5.1 might not be L-linear, though it is $\mathbb{Q}(\pi_v)$-linear since it is compatible with q_v-Frobenius endomorphisms. We can exploit the $\mathbb{Q}(\pi_v)$-linearity of ϕ to find an L-linear κ_v-isogeny $B_0 \to A_{\kappa_v}$ as follows.

Since the q_v-Frobenius generates the center of the endomorphism algebra of any abelian variety over κ_v, the Skolem-Noether theorem ensures that any two $\mathbb{Q}(\pi_v)$-embeddings of L into the central simple $\mathbb{Q}(\pi_v)$-algebra $\mathrm{End}^0(A_{\kappa_v})$ are related through conjugation by a unit. Hence, there is an isogeny $u \in \mathrm{End}(A_{\kappa_v})$ such that $u \circ \phi$ is L-linear. By renaming this as ϕ, we may arrange that ϕ is L-linear. Thus, in Theorem 1.6.5.1 we may choose the CM lift so that the action of a *specified* degree-$2g$ CM field $L \subset \mathrm{End}^0(A)$ also lifts.

It is natural to ask for a strengthening of Theorem 1.6.5.1 in which the isogeny is applied prior to making a residue field extension (to acquire a CM lifting). As we will record near the end of 1.8, such a stronger form is true and follows from one of the main results proved later in this book.

1.7. A theorem of Grothendieck and a construction of Serre

1.7.1. Isogenies and fields of definition. Let A be an abelian variety over a field K and let $K_1 \subset K$ be a subfield. We say that A *is defined over* K_1 if there exists an abelian variety A_1 over K_1 and an isomorphism $f : A \simeq (A_1)_K$. We use similar terminology for a map $h : A \to B$ between abelian varieties over K (i.e., there exists a map $h_1 : A_1 \to B_1$ between abelian varieties over K_1 such that $(h_1)_K$ is identified with h).

For example, suppose K/K_1 is a primary extension of fields (i.e., K_1 is separably algebraically closed in K) and consider abelian varieties A and A' over K such that there are isomorphisms $f : A \simeq (A_1)_K$ and $f' : A' \simeq (A'_1)_K$ for abelian varieties A_1 and A'_1 over K_1. By Lemma 1.2.1.2, the pairs (A_1, f) and (A'_1, f') are unique up to unique isomorphism and every map $A \to A'$ as abelian varieties over K is defined over K_1 in the sense that it uniquely descends to a map $A_1 \to A'_1$ as abelian varieties over K_1. Likewise, by Corollary 1.2.1.4, all abelian subvarieties of A are defined over K_1 (and even uniquely arise from abelian subvarieties of A_1). For general extensions K/K_1 such K_1-descents may not exist, and when (A_1, f) does exist it is not necessarily unique (up to isomorphism).

1.7.1.1. Example. Assume $\mathrm{char}(K) = 0$ and let K'/K be an algebraically closed extension (a basic example of interest being $K' = \mathbb{C}$). We claim that each member of the isogeny class of $A_{K'}$ is defined over the algebraic closure $\overline{K}$ of K in K' (and hence over a finite extension of K inside K'). To prove this, observe that the kernel of any isogeny $\psi : A_{K'} \to B$ over K' is contained in some torsion subgroup $A[n]_{K'}$, and $A[n]$ becomes constant over $\overline{K}$ (since $A[n]$ is K-étale, as $\mathrm{char}(K) = 0$). Hence, we can descend $\ker(\psi)$ to a constant finite subgroup of $A_{\overline{K}}$, and the quotient of $A_{\overline{K}}$ by this gives a descent of (B, ψ) to $\overline{K} \subset K'$.

1.7.1.2. Example. When $\mathrm{char}(K) = p > 0$, the naive analogue of Example 1.7.1.1 fails. An interesting counterexample is $A = E^2$ for a supersingular elliptic curve E over a field K of characteristic $p > 0$. The kernel H of the Frobenius isogeny $E \to E^{(p)}$ is a local-local K-group of order p, and it is the unique infinitesimal subgroup of E with order p (as any commutative infinitesimal K-group of order p has vanishing Frobenius morphism).

The only local-local finite commutative group scheme of order p over K is α_p. For perfect K this is easily proved by a computation with Dieudonné modules (as we noted in Example 1.6.4). In Proposition 3.1.10 we will prove that this property in general descends from the perfect closure (set $r = 1$ there).

Fix a choice of K-subgroup inclusion $\alpha_p \hookrightarrow E$ over K, so we get a canonical copy of α_p^2 in $A = E^2$ (as the kernel of the Frobenius isogeny $A \to A^{(p)}$). Over an arbitrary field of characteristic $p > 0$ the Frobenius and Verschiebung morphisms of α_p^2 vanish, so over any such field the non-trivial proper subgroups of α_p^2 are naturally parameterized by lines in the 2-dimensional tangent space $\mathrm{Lie}(\alpha_p^2)$; this parameterization is given by the tangent line of the subgroup (see [**23**, Thm. 3.18] for more details). In particular, the non-trivial proper K-subgroups of α_p^2 are parameterized by $\mathbb{P}^1(K)$, and if K'/K is an extension field then the non-trivial proper K'-subgroups of $(\alpha_p)_{K'}^2$ are parameterized by $\mathbb{P}^1(K')$ (with the subset $\mathbb{P}^1(K)$ consisting of the tangent lines to the K'-subgroups defined over K).

We conclude that if K'/K is a non-trivial extension field then there are K'-subgroups $G' \subset A' := A_{K'}$ of order p that are contained in $(\alpha_p^2)_{K'}$ and do not arise from a K-subgroup of A. In contrast with what we saw in Example 1.7.1.1 for isogeny classes over algebraically closed fields of characteristic 0, we claim that if K is separably closed (or more generally if K is separably closed in K', with G' not defined over K inside $A' = A_{K'}$) then the isogenous quotient A'/G' of $A' = A_{K'}$ cannot be defined over K as an abstract abelian variety!

Indeed, if there were an isomorphism $A'/G' \simeq B_{K'}$ for an abelian variety B over K then the resulting isogeny

$$A_{K'} = A' \twoheadrightarrow A'/G' \simeq B_{K'}$$

descends to an isogeny $A \to B$ over K by Lemma 1.2.1.2 (since K'/K is primary). The kernel of this latter isogeny is a K-subgroup of A that descends $G' \subset A'$, contrary to how G' was chosen. Thus, no such B exists.

One lesson we learn from Example 1.7.1.1 and Example 1.7.1.2 is that fields of definition for abelian varieties in positive characteristic are rather more subtle than in characteristic 0, even when working over algebraically closed base fields.

1.7.2. Grothendieck's theorem. To fully appreciate the significance of Example 1.7.1.2, we turn our attention to a striking result of Grothendieck concerning the field of definition of an abelian variety with sufficiently many complex multiplications in positive characteristic. Before stating Grothendieck's result, we record the analogue in characteristic 0 that is a source of inspiration.

1.7.2.1. Theorem (Shimura–Taniyama). *Every non-zero abelian variety A with sufficiently many complex multiplications over an algebraically closed field K of characteristic* 0 *is defined* (*along with its entire endomorphism algebra*) *over a number field inside K.*

PROOF. By Example 1.7.1.1, without loss of generality we may replace A with an isogenous abelian variety. Thus, by Proposition 1.3.2.1 we can pass to the isotypic (and even simple) case, and so by Theorem 1.3.4 the abelian variety A over K admits complex multiplication by a CM field L. Let Φ be the resulting CM type on L. Letting $\overline{\mathbb{Q}}$ denote the algebraic closure of $\mathbb{Q}$ in K, we may view Φ as a $\overline{\mathbb{Q}}$-valued CM type on L.

By Proposition 1.5.4.1 (applied over the algebraically closed base field $\overline{\mathbb{Q}}$) there is a CM abelian variety B over $\overline{\mathbb{Q}}$ with complex multiplication by L and CM type Φ (viewed as valued in $\overline{\mathbb{Q}}$). The abelian variety B_K over K admits complex multiplication by L with associated CM type Φ (viewed as valued in K), so by applying Proposition 1.5.4.1 over K we see that B_K is L-linearly isogenous to A over K (as these two abelian varieties over K are endowed with complex multiplication by L yielding the same CM type Φ on L). Fix such an isogeny $f : B_K \to A$. By Example 1.7.1.1 (applied to $K/\overline{\mathbb{Q}}$), the finite kernel of f descends to a finite subgroup of B. The quotient of B by this descent of $\ker(f)$ is a descent A_0 of $A = B_K/\ker(f)$ to an abelian variety over $\overline{\mathbb{Q}}$.

By Lemma 1.2.1.2, $\mathrm{End}(A_0) \to \mathrm{End}(A)$ is an isomorphism. Thus, we may assume the base field K is an algebraic closure $\overline{\mathbb{Q}}$ of $\mathbb{Q}$. Express $\overline{\mathbb{Q}}$ as a direct limit of number fields to see that the $\overline{\mathbb{Q}}$-group A descends to an abelian variety over a number field inside $\overline{\mathbb{Q}}$. The same direct limit argument as used at the start of the proof of Proposition 1.2.6.1 then shows that we can choose the descent of A to a number field so that all elements of $\mathrm{End}(A)$ descend as well. □

Theorem 1.7.2.1 can be formulated with a general ground field K of characteristic 0, but the nature of the descent becomes a bit more subtle. Namely, if A is an abelian variety over a field K of characteristic 0 and if A admits sufficiently many complex multiplications, then there is a finite extension K'/K such that $A_{K'}$ descends (along with its entire endomorphism algebra) to an abelian variety over a number field contained in K'. In this formulation it is crucial to introduce the finite extension K'/K, even if we just wish to descend the abelian variety (and not any specific endomorphisms). This is illustrated by quadratic twists of elliptic curves:

1.7.2.2. Example. Consider a CM elliptic curve over $\mathbb{C}$ and extend scalars to $K = \mathbb{C}(t)$. Let E be the quadratic twist of this scalar extension by a quadratic extension K'/K, so E is a CM elliptic curve over K whose ℓ-adic representation for $\mathrm{Gal}(K_s/K)$ is non-trivial. No member of the isogeny class of E over K can be defined over $\mathbb{C}$ (let alone over $\overline{\mathbb{Q}}$), as all members of the isogeny class have non-trivial action by $\mathrm{Gal}(K_s/K)$ for their ℓ-adic representations. Of course, if we pass up to $\overline{K}$ then the effect of quadratic twisting goes away and there is no obstruction to descent to $\overline{\mathbb{Q}}$.

Over number fields, CM abelian varieties extend to abelian schemes over the entire ring of integers at the cost of a finite extension of the ground field. This is an application of the semi-stable reduction theorem for abelian varieties (see [**109**, Thm. 6]), and we record it here for later reference:

1.7.2.3. Theorem. *Every* CM *abelian variety over a number field has potentially good reduction at all places.*

Since every abelian variety over an algebraic closure of $\mathbb{F}_p$ descends to a finite field and hence has sufficiently many complex multiplications (by Corollary 1.6.2.5), a first guess for an analogue of Theorem 1.7.2.1 in positive characteristic is that CM abelian varieties over algebraically closed fields K with positive characteristic can be descended to the algebraic closure of the prime field inside K (or equivalently, to a finite subfield of K).

Example 1.7.1.2 shows that this guess is false, since A'/G' as built there admits a CM structure since supersingular elliptic curves over K always admit a CM structure (as they descend to elliptic curves over the algebraic closure of $\mathbb{F}_p$ inside K). Allowing isogenies does not eliminate the need for a finite extension when the ground field K is not algebraically closed:

1.7.2.4. Example. For a finite field κ, Example 1.7.2.2 adapts to work over $K = \kappa(t)$ by beginning with an elliptic curve over κ (for which complex multiplication by an imaginary quadratic field exists in general; see Corollary 1.6.2.5). One can do likewise over $\kappa(t)$ with κ an algebraic closure of $\mathbb{F}_p$.

Motivated by the above examples, Grothendieck proved a reasonable analogue of Theorem 1.7.2.1 in positive characteristic:

1.7.2.5. Theorem (Grothendieck). *Let A be an abelian variety over a field K with* $\mathrm{char}(K) = p > 0$, *and assume A admits sufficiently many complex multiplications. Then there exists a finite extension K' of K, a finite subfield $\kappa \subset K'$, and an abelian variety B over κ such that the scalar extensions $A \times_{\mathrm{Spec}(K)} \mathrm{Spec}(K')$ and $B \times_{\mathrm{Spec}(\kappa)} \mathrm{Spec}(K')$ over K' are isogenous.* □

For an exposition of Grothendieck's proof, see [**89**]. The essential difficulty in the proof in contrast with characteristic 0 is that the isogeny cannot be avoided, even when $K = \overline{K}$ (due to Example 1.7.1.2). The proof of Theorem 1.7.2.5 is immediately reduced to the case when K is finitely generated over $\mathbb{F}_p$. Grothendieck used the theory of potentially good reduction to find the required K'/K and made a descent from K' to a finite subfield via a Chow trace (in the sense of [**23**, §6]).

1.7.3. There is a refinement of Grothendieck's theorem, due to C-F. Yu, that clarifies the role of the isogeny and proceeds in a simpler way by using moduli spaces of abelian varieties. This refinement is given in 1.7.5. We will not need that result, but the main ingredient in its proof is a technique to modify the endomorphism ring that will be very useful later, so we now explain that technique.

As motivation, consider an abelian variety A of dimension $g > 0$ over a field K such that A admits sufficiently many complex multiplications, and let $P \subset \mathrm{End}^0(A)$ be a commutative semisimple $\mathbb{Q}$-subalgebra with $[P : \mathbb{Q}] = 2g$. The intersection $\mathcal{O} := P \cap \mathrm{End}(A)$ is an order in P that may not be maximal (i.e., it may not equal $\mathcal{O}_P := \prod \mathcal{O}_{L_i}$, where $\prod L_i$ is the decomposition of P into a finite product of number fields). It is natural to ask if there is an isogenous abelian variety for which the non-maximality problem goes away. The following discussion addresses this issue.

1.7.3.1. Example. Consider the preceding setup with $K = \mathbb{C}$. In this case we have an analytic uniformization $A^{\mathrm{an}} = V/\Lambda$ in which V is a $\mathbb{C}$-vector space equipped with a $\mathbb{C}$-linear action by P and Λ is a lattice in V stable under the order $\mathcal{O}$. Then $\Lambda' := \mathcal{O}_P \cdot \Lambda$ is an $\mathcal{O}_P$-stable lattice in V and V/Λ' is an isogenous quotient of A^{an} on which $\mathcal{O}_P$ naturally acts. This algebraizes to an isogenous quotient A' of A such that under the identification $\mathrm{End}^0(A') = \mathrm{End}^0(A)$ we have $P \cap \mathrm{End}(A') = \mathcal{O}_P$.

We need an algebraic variant of the analytic construction in Example 1.7.3.1. Observe that $\mathcal{O}_P \cdot \Lambda$ is the image of the natural map $\mathcal{O}_P \otimes_{\mathcal{O}} \Lambda \to V$. Inspired by this, we are led to ask if these is a way to enlarge an endomorphism ring via a "tensor product" against a finite-index extension of coefficient rings. There is a construction of this sort due to Serre [**104**], applicable over any base scheme, though it turns out to not be applicable to the above situation because $\mathcal{O}_P$ is not a projective $\mathcal{O}$-module when $\mathcal{O} \neq \mathcal{O}_P$. We wish to adapt Serre's construction to the above situation over a field, so we digress to explain Serre's procedure.

1.7.4. Serre's tensor construction. Consider a scheme S, a commutative ring $\mathcal{O}$, and an $\mathcal{O}$-module scheme A over S. (The main example of such an A to keep in mind is an abelian scheme, but there are other interesting examples, such the n-torsion subgroups of such abelian schemes for $n \geqslant 1$.) Let M be a projective $\mathcal{O}$-module of finite rank.

The projectivity of M ensures that the functor $T \rightsquigarrow M \otimes_{\mathcal{O}} A(T)$ on S-schemes is represented by an S-scheme, denoted $M \otimes_{\mathcal{O}} A$, and that $M \otimes_{\mathcal{O}} A$ inherits many nice properties from A such as flatness, smoothness, properness, good behavior with respect to analytification over $\mathbb{C}$, etc. The interested reader can see [**22**, §7] for details (where non-commutative $\mathcal{O}$ are also considered), but the idea of the construction of $M \otimes_{\mathcal{O}} A$ is simple, as follows.

If $\mathcal{O}^r \xrightarrow{\varphi} \mathcal{O}^s \to M \to 0$ is a presentation then we want to define $M \otimes_{\mathcal{O}} A$ to be the cokernel of the S-group map $A^r \to A^s$ induced by the matrix of φ. Without the projectivity assumption on the $\mathcal{O}$-module M, over a general base scheme S such a quotient may not exist. However, since M is locally free of finite rank as an $\mathcal{O}$-module (by the projectivity hypothesis) we can instead begin with a presentation of the dual module $M^\vee$ and then dualize to get a left-exact sequence $0 \to M \to \mathcal{O}^s \to \mathcal{O}^r$ with suitable local splitting properties to enable us to construct $M \otimes_{\mathcal{O}} A$ as a scheme-theoretic kernel. More explicitly, by projectivity there is an integer $r \geqslant 1$ and $\mathcal{O}$-module M' such that $\mathcal{O}^r \simeq M \oplus M'$, so there is an $\mathcal{O}$-linear idempotent endomorphism e of $\mathcal{O}^r$ such that $M = \ker(e)$. The kernel of the associated $\mathcal{O}$-linear endomorphism of A^r represents $M \otimes_{\mathcal{O}} A$.

1.7.4.1. Example. Let L be a CM field, and let (A, i) and (A', i') be CM abelian varieties over a field K, where i and i' respectively define complex multiplication by L. Assume that via these embeddings, $\mathcal{O}_L$ lies in the endomorphism rings of the abelian varieties. Finally, assume that there exists a non-zero $\mathcal{O}_L$-linear map $A' \to A$. (By Proposition 1.5.4.1, when K is algebraically closed of characteristic 0 it is equivalent to assume that the associated CM types Φ' and Φ on L coincide.)

We claim that $M := \mathrm{Hom}((A', i'), (A, i))$ is an invertible $\mathcal{O}_L$-module whose formation is unaffected by extension of the ground field and that if $\mathrm{char}(K) = 0$ then the evaluation map

$$M \otimes_{\mathcal{O}_L} A' \to A$$

is an isomorphism. (Hence, over an algebraically closed field of characteristic 0 the Serre tensor construction defines a natural *transitive* action of the finite group $\mathrm{Pic}(\mathcal{O}_L)$ on the set of isomorphism classes of CM abelian varieties with a fixed CM type (L, Φ) and CM order $\mathcal{O}_L$. The argument below will show that the action is simply transitive.)

The non-zero L-vector space $M_{\mathbb{Q}} = \mathrm{Hom}^0((A', i'), (A, i))$ has dimension exactly 1 since for $\ell \neq \mathrm{char}(K)$ the natural map

$$\mathbb{Q}_\ell \otimes_{\mathbb{Q}} M_{\mathbb{Q}} \to \mathrm{Hom}_{L_\ell}(V_\ell(A'), V_\ell(A))$$

is injective and the target is free of rank 1 over L_ℓ. Hence, the finitely generated torsion-free $\mathcal{O}_L$-module M is invertible. Also, the injective map

$$\mathrm{Hom}((A', i'), (A, i)) \to \mathrm{Hom}((A'_{K_s}, i'), (A_{K_s}, i))$$

between finitely generated $\mathcal{O}_L$-modules has image with finite index since it becomes an equality after applying $\mathbb{Q} \otimes_{\mathbb{Z}} (\cdot)$ (for L-dimension reasons); let n be the index. It follows that all L-linear K_s-homomorphisms $f : A'_{K_s} \to A_{K_s}$ are $\mathrm{Gal}(K_s/K)$-invariant because nf is defined over K. This shows that the formation of M is unaffected by ground field extension to K_s, and hence by any ground field extension (due to Lemma 1.2.1.2).

Now we assume that $\mathrm{char}(K) = 0$ and seek to prove that $M \otimes_{\mathcal{O}_L} A' \to A$ is an isomorphism. We may assume K is finitely generated, and then that $K = \mathbb{C}$. The $\mathcal{O}_L$-modules $\mathrm{H}_1(A(\mathbb{C}), \mathbb{Z})$ and $\mathrm{H}_1(A'(\mathbb{C}), \mathbb{Z})$ are each invertible (due to being $\mathbb{Z}$-flat of rank $[L : \mathbb{Q}]$). By Example 1.5.3, we get $\mathcal{O}_L$-linear isomorphisms $A(\mathbb{C}) = (\mathbb{R} \otimes_{\mathbb{Q}} L)_\Phi/\mathfrak{a}$ and $A'(\mathbb{C}) = (\mathbb{R} \otimes_{\mathbb{Q}} L)_\Phi/\mathfrak{a}'$ for non-zero ideals $\mathfrak{a}, \mathfrak{a}' \subset \mathcal{O}_L$. Hence, elements of M are precisely multiplication on $(\mathbb{R} \otimes_{\mathbb{Q}} L)_\Phi$ by those $c \in L$ such that $c\mathfrak{a}' \subseteq \mathfrak{a}$. We conclude that $M = \mathrm{Hom}_{\mathcal{O}_L}(\mathfrak{a}', \mathfrak{a}) = \mathfrak{a}\mathfrak{a}'^{-1}$, with $M \otimes_{\mathcal{O}_L} A' \to A$ given by the evident evaluation pairing on $\mathbb{C}$-points. This is an isomorphism because the induced map on homology lattices is the natural map $M \otimes_{\mathcal{O}_L} \mathfrak{a}' \to \mathfrak{a}$ that is clearly an isomorphism.

The isomorphism property for the map $M \otimes_{\mathcal{O}_L} A' \to A$ in Example 1.7.4.1 fails away from characteristic 0, even for elliptic curves over finite fields. For example, if L is imaginary quadratic with class number 1 then the relative Frobenius isogeny provides counterexamples (using elliptic curves whose j-invariant is not in the prime field). More explicitly, for $p \equiv -1 \bmod 4$ and $\kappa := \mathbb{Z}[i]/(p) \simeq \mathbb{F}_{p^2}$ with $p > 3$, consider the elliptic curves $E_\pm = E = \{y^2 = x^3 - x\}$ over κ with CM by $\mathcal{O}_L$ via the actions $[i](x, y) = (-x, \pm iy)$. These are not $\mathcal{O}_L$-linearly isomorphic (since $\mathrm{Aut}(E) = \mu_4 \subset \mathcal{O}_L^\times$, as $p > 3$) but the Frobenius isogeny $E \to E^{(p)}$ is an $\mathcal{O}_L$-linear isogeny $E_+ \to E_-$. Thus, the module M of $\mathcal{O}_L$-linear homomorphisms from E_+ to E_- is non-zero but the $\mathcal{O}_L$-linear map $M \otimes_{\mathcal{O}_L} E_+ \to E_-$ cannot be an isomorphism (since $M \simeq \mathcal{O}_L$ as $\mathcal{O}_L$-modules).

1.7.4.2. Example. Let (A, i, L) be as in Example 1.7.4.1 over a field K, so $\mathcal{O}_L \subset \mathrm{End}(A)$. For an invertible $\mathcal{O}_L$-module M, another such abelian variety is given by $M \otimes_{\mathcal{O}_L} A$. Any $m \in M$ defines an $\mathcal{O}_L$-linear map $e_m : A \to M \otimes_{\mathcal{O}_L} A$ via $x \mapsto m \otimes x$. For $\ell \neq \mathrm{char}(K)$, the map $T_\ell(e_m)$ induced by e_m on ℓ-adic Tate modules is the map $T_\ell(A) \to M \otimes_{\mathcal{O}_L} T_\ell(A)$ given by $v \mapsto m \otimes v$, so $e_m \neq 0$ when $m \neq 0$. In particular, the module $\mathrm{Hom}_{\mathcal{O}_L}(A, M \otimes_{\mathcal{O}_L} A)$ of $\mathcal{O}_L$-linear homomorphisms is non-zero and therefore invertible (by Example 1.7.4.1).

The natural map of invertible $\mathcal{O}_L$-modules

$$e_{A,M} : M \to \mathrm{Hom}_{\mathcal{O}_L}(A, M \otimes_{\mathcal{O}_L} A)$$

is injective, hence of finite index. We shall now show that this map is an isomorphism. It suffices to check the result after applying $\mathbb{Z}_\ell \otimes_{\mathbb{Z}} (\cdot)$ for *every* prime ℓ (allowing $\ell = \mathrm{char}(K)$). This scalar extension is the first map in the diagram

$$M_\ell \to \mathbb{Z}_\ell \otimes_{\mathbb{Z}} \mathrm{Hom}_{\mathcal{O}_L}(A, M \otimes_{\mathcal{O}_L} A) \to \mathrm{Hom}_{\mathcal{O}_{L,\ell}}(A[\ell^\infty], M_\ell \otimes_{\mathcal{O}_{L,\ell}} A[\ell^\infty])$$

whose second map is injective (Proposition 1.2.5.1) and composition is the canonical homomorphism that is injective. It therefore suffices to show that the composite map is an isomorphism. But M_ℓ is free of rank 1 as an $\mathcal{O}_{L,\ell}$-module, so it is equivalent to show that the natural map

$$\mathcal{O}_{L,\ell} \to \mathrm{End}_{\mathcal{O}_{L,\ell}}(A[\ell^\infty])$$

is an isomorphism for all ℓ.

The case $\ell \neq \mathrm{char}(K)$ is trivial, as then the ℓ-adic Tate module $T_\ell(A)$ is free of rank 1 over $\mathcal{O}_{L,\ell}$ (since $V_\ell(A)$ is free of rank 1 over L_ℓ, due to faithfulness and $\mathbb{Q}_\ell$-dimension reasons). Now assume $\mathrm{char}(K) = p > 0$ and $\ell = p$. Decomposing $A[p^\infty]$ according to the primitive idempotents of $\mathcal{O}_{L,p}$, for each p-adic place v of L the v-factor has height at least $[L_v : \mathbb{Q}_p]$ (by Proposition 1.4.3.9) and hence height exactly $[L_v : \mathbb{Q}_p]$. Thus, it suffices to prove more generally that if X is a p-divisible group over K of height $h > 0$ and F is a p-adic field of degree h over $\mathbb{Q}_p$ such that $\mathcal{O}_F \subset \mathrm{End}(X)$ then $\mathcal{O}_F$ is its own centralizer in $\mathrm{End}(X)$.

We may and do assume that K is algebraically closed, so the Dieudonné module $\mathrm{M}^*(X)$ makes sense and is free of rank 1 as a $W(K) \otimes_{\mathbb{Z}_p} \mathcal{O}_F$-module (Proposition 1.4.3.9). Thus, any $\mathcal{O}_F$-linear endomorphism f of X gives rise to a $W(K) \otimes_{\mathbb{Z}_p} \mathcal{O}_F$-linear endomorphism of $\mathrm{M}^*(X)$, so $\mathrm{M}^*(f)$ must be multiplication by some $c \in W(K) \otimes_{\mathbb{Z}_p} \mathcal{O}_F$. But $\mathrm{M}^*(f)$ commutes with the $\mathcal{F}$ operator on $\mathrm{M}^*(X)$, so c is invariant under the absolute Frobenius automorphism σ of $W(K)$. This forces $c \in W(K)^{\sigma=1} \otimes_{\mathbb{Z}_p} \mathcal{O}_F = \mathcal{O}_F$, so f is an $\mathcal{O}_F$-multiplier, as desired.

1.7.4.3. Remark. The $\mathcal{O}$-linear projectivity hypothesis on M in the construction of $M \otimes_{\mathcal{O}} A$ cannot be dropped, even when the base is the spectrum of a discrete valuation ring. For example, if R is a p-adic discrete valuation ring and E is an elliptic curve over R with endomorphism ring $\mathcal{O} = \mathbb{Z}[p\sqrt{-p}]$ (as can be easily constructed using classical CM theory for elliptic curves), then for the non-projective $\mathcal{O}$-module $M = \mathbb{Z}[\sqrt{-p}]$ the fppf sheafification of the functor $T \rightsquigarrow M \otimes_{\mathcal{O}} E(T)$ on R-schemes is not representable. (The idea is as follows. First one proves that a representing object, if one exists, must be an elliptic curve $\mathscr{E}$. By presenting M over $\mathcal{O}$ using two generators and two relations, we get a quotient homomorphism $E \times E \to \mathscr{E}$ whose kernel must be an R-flat divisor in $E \times E$. Studying its defining equation in the formal group of $E \times E$ leads to a contradiction.)

For any homomorphism $f : A \to A'$ between abelian varieties over a field, the image $f(A)$ is an abelian subvariety over A' and the map $A \to f(A)$ is flat. Such good properties for $f(A)$ generally fail for homomorphisms between abelian schemes over a more general base, but their availability over fields enables us to push through the initial cokernel idea for the Serre tensor construction over fields. In this way we can avoid dualizing M and hence make the construction work with weaker hypotheses on M than projectivity. We shall now give a version of this for abelian varieties (and in 4.3.1 there is a version for p-divisible groups), but first we require a general lemma:

1.7.4.4. Lemma. *For an abelian variety B over a field K and an abelian subvariety B', the quotient sheaf B/B' for the fppf topology on the category of K-schemes is represented by an abelian variety.*

PROOF. Recall from the Poincaré reducibility theorem over K (see Theorem 1.2.1.3) that there is an abelian subvariety $B'' \subset B$ over K that is an isogeny complement to B' in the sense that the natural map $f : B' \times B'' \to B$ is an isogeny. Since f is a finite flat surjection, so $B \simeq (B' \times B'')/\ker(f)$ as fppf abelian sheaves, we likewise have $B/B' = B''/(B' \cap B'')$ as fppf sheaves. Thus, the problem for B/B' is the same as for the quotient of B'' by the finite K-subgroup scheme $B' \cap B''$. Hence, it suffices to show that B''/G is (represented by) an abelian variety for any finite K-subgroup $G \subset B''$.

Rather than appealing to existence results for quotients by the free action of a finite group scheme on a quasi-projective scheme, here is a more direct argument via fppf descent theory and a special property of abelian varieties: for any $n \geqslant 1$, the map $[n]_{B''} : B'' \to B''$ is a finite flat surjection, so $B''/B''[n] \simeq B''$ as fppf abelian sheaves. To exploit this, note that the K-group G is killed by its order n. (Indeed, we may assume $\mathrm{char}(K) = p > 0$ and use the connected-étale sequence for G and kernels of relative Frobenius morphisms to reduce to the case when $\mathrm{Fr}_{G/K}$ vanishes. In such cases p kills G since $[p]_G = \mathrm{Ver}_{G^{(p)}/K} \circ \mathrm{Fr}_{G/K}$.) Thus, $G \subset B''[n]$, so as fppf abelian sheaves B''/G is a $B''[n]/G$-torsor over $B''/B''[n] \simeq B''$.

Since $B''[n]/G$ is represented by a finite K-scheme, by effective descent for finite morphisms we see that the quotient sheaf B''/G is therefore represented by a finite flat B''-scheme over which B'' is a finite flat cover (even a G-torsor). This implies that B''/G is proper, smooth, and connected, so it is an abelian variety, as desired. □

Here is the promised generalized Serre tensor construction over fields (allowing non-projective modules).

1.7.4.5. Proposition. *Let A be an abelian variety or finite commutative group scheme over a field K. Let $\mathcal{O} \to \mathrm{End}(A)$ be a homomorphism from a commutative ring. For any finitely generated $\mathcal{O}$-module M, the functor $T \rightsquigarrow M \otimes_{\mathcal{O}} A(T)$ on K-schemes has fppf sheafification that is respectively represented by an abelian variety or finite commutative group scheme $M \otimes_{\mathcal{O}} A$.*

Suppose A is an abelian variety. For an injective map $M \to N$ between torsion-free $\mathcal{O}$-modules with finite cokernel, the induced map $M \otimes_{\mathcal{O}} A \to N \otimes_{\mathcal{O}} A$ is an isogeny. In particular, if $\mathcal{O}'$ is a $\mathbb{Z}$-flat $\mathcal{O}$-algebra that is finitely generated as an $\mathcal{O}$-module and $\mathcal{O}_{\mathbb{Q}} \to \mathcal{O}'_{\mathbb{Q}}$ is an isomorphism then the natural map of abelian varieties $A \to A' := \mathcal{O}' \otimes_{\mathcal{O}} A$ is an isogeny and the identification $\mathrm{End}^0(A) = \mathrm{End}^0(A')$ carries $\mathcal{O}' \subset \mathrm{End}^0(A)$ into $\mathrm{End}(A')$.

The notation $\mathcal{O}' \otimes_{\mathcal{O}} A$ should not be confused with the standard notation for affine base change of schemes. Also, if M is killed by a non-zero element of $\mathcal{O}$ then it is killed by a non-zero integer and hence the abelian variety $M \otimes_{\mathcal{O}} A$ vanishes.

PROOF. Choose a finite presentation of $\mathcal{O}$-modules

$$\mathcal{O}^r \xrightarrow{\varphi} \mathcal{O}^s \longrightarrow M \to 0.$$

The map φ is given by an $s \times r$ matrix over $\mathcal{O}$, so it defines an analogous map $[\varphi] : A^r \to A^s$ between K-groups. Since we are working over a field, if A is an abelian variety then the map $[\varphi]$ has image that is an abelian subvariety of A^s onto which $[\varphi]$ is faithfully flat. If instead A is a finite commutative K-group then there is a finite flat quotient map $A^r \to A^r/\ker[\varphi]$. The induced map $A^r/\ker[\varphi] \to A^s$

between finite K-groups has trivial kernel, so it is a closed immersion. We denote this closed K-subgroup as $[\varphi](A^r)$, so (as with abelian varieties) the map $[\varphi]$ is faithfully flat onto a closed K-subgroup $[\varphi](A^r) \subset A^s$.

Using Lemma 1.7.4.4 in the abelian variety case and the more elementary theory of quotients for finite commutative K-group schemes in the finite case, the quotient $M \otimes_{\mathcal{O}} A := A^s/[\varphi](A^r)$ as an abelian variety or finite K-group scheme represents the cokernel of $[\varphi]$ in the sense of fppf abelian sheaves on the category of K-schemes. It follows (via the right-exactness of algebraic tensor products) that the K-group scheme $M \otimes_{\mathcal{O}} A$ represents the fppf sheafification of $T \rightsquigarrow M \otimes_{\mathcal{O}} A(T)$.

Now assume that A is an abelian variety. Let $M \to N$ be an injective map between torsion-free $\mathcal{O}$-modules with finite cokernel. There is a map $N \to M$ such that both composites $M \to M$ and $N \to N$ are multiplication by a common non-zero integer n. Hence, we get maps in both directions between $M \otimes_{\mathcal{O}} A$ and $N \otimes_{\mathcal{O}} A$ whose composites are each equal to multiplication by n, so both maps between $M \otimes_{\mathcal{O}} A$ and $N \otimes_{\mathcal{O}} A$ are isogenies.

The assertions concerning $\mathcal{O}'$ follow by considering the functor $T \rightsquigarrow \mathcal{O}' \otimes_{\mathcal{O}} A(T)$ and the abelian variety over K representing it. □

As an application of the Serre tensor construction with non-projective modules when the base is a field, we prove a precise form of the "lifting" part of the Deuring Lifting Theorem:

1.7.4.6. Theorem (Deuring). *Let E_0 be an elliptic curve over $\mathbb{F}_q$. For any $f_0 \in \mathrm{End}(E_0)$ generating an imaginary quadratic field $L \subset \mathrm{End}^0(E_0)$ and p-adic place $\mathfrak{p}$ of $\mathcal{O}_L$, let R be the valuation ring of the compositum $W(\mathbb{F}_q)[1/p] \cdot L_{\mathfrak{p}}$ over $\mathbb{Q}_p$.*

There exists a CM elliptic curve E over R equipped with an endomorphism f such that (E, f) has special fiber isomorphic to (E_0, f_0).

By 1.6.2.5, for any E_0 over $\mathbb{F}_q$ there is an imaginary quadratic field L inside $\mathrm{End}^0(E_0)$. The CM structure forces R to have residue field $\mathbb{F}_q$, as we shall see in the proof below.

PROOF. If E_0 is ordinary then $L_{\mathfrak{p}} = \mathbb{Q}_p$ and we can choose E to be the Serre–Tate canonical lift over $W(\mathbb{F}_q)$, to which all endomorphisms of E_0 uniquely lift; see 1.4.5.4. Suppose instead that E_0 is supersingular. It suffices to show that for any imaginary quadratic field $L \subset \mathrm{End}^0(E_0)$ and $\mathcal{O} := \mathrm{End}(E_0) \bigcap \mathcal{O}_L$, we can lift (E_0, α_0) over R where $\alpha_0 : \mathcal{O} \hookrightarrow \mathrm{End}(E_0)$ is the natural inclusion.

Consider the canonical $\mathcal{O}$-linear isogeny

$$h : E_0 \to E_0' := \mathcal{O}_L \otimes_{\mathcal{O}} E_0$$

(see 1.7.4.5). The key point is to show that p does not divide the degree of h (so h induces an isomorphism on p-divisible groups). The degree of h is the order of the finite $\mathbb{F}_q$-group $\ker(h)$, so if this kernel is étale then its order must be relatively prime to p because a supersingular elliptic curve has infinitesimal p-torsion. Thus, we may assume $\ker(h)$ is not étale, so the infinitesimal identity component of $\ker(h)$ is nontrivial and therefore the relative Frobenius morphism of $\ker(h)$ has nontrivial kernel. The $\mathcal{O}$-linear relative Frobenius morphism $\mathrm{Fr}_{E_0/\mathbb{F}_q} : E_0 \to E_0^{(p)}$ for the elliptic curve E_0 has kernel of order p, so this latter kernel must lie inside $\ker(h)$.

We conclude that if $\ker(h)$ is not étale then h factors through $\mathrm{Fr}_{E_0/\mathbb{F}_q}$, so there is a commutative diagram

$$\begin{array}{ccc} E_0 & \xrightarrow{h} & \mathcal{O}_L \otimes_{\mathcal{O}} E_0 \\ {\scriptstyle \mathrm{Fr}_{E_0/\mathbb{F}_q}}\big\downarrow & \nearrow & \big\downarrow{\scriptstyle 1\otimes \mathrm{Fr}_{E_0/\mathbb{F}_q}} \\ E_0^{(p)} & \xrightarrow[h^{(p)}]{} & \mathcal{O}_L \otimes_{\mathcal{O}} E_0^{(p)} \end{array}$$

of $\mathcal{O}$-linear isogenies. Since $h^{(p)}$ is the initial $\mathcal{O}$-linear map from $E_0^{(p)}$ to an $\mathcal{O}_L$-linear module scheme over $\mathbb{F}_q$, the right vertical map must be an isomorphism. This latter map is the relative Frobenius isogeny for the elliptic curve $\mathcal{O}_L \otimes_{\mathcal{O}} E_0$, as that also makes the outside square commute (and commutativity uniquely determines the right vertical map in terms of the other maps on the outside of the diagram). The Frobenius isogeny for an elliptic curve over $\mathbb{F}_q$ is not an isomorphism, so we have a contradiction. Thus, $\ker(h)$ is étale and hence has order not divisible by p.

We conclude that h induces an isomorphism on p-divisible groups, so by the Serre–Tate deformation theorem it follows that the formal deformation theory of E_0 is the same as that of E_0'. Any formal deformation of E_0 over a complete local noetherian $W(\mathbb{F}_q)$-algebra is a scheme, since the inverse ideal sheaf of the identity section provides a canonical algebraization, so by using formal GAGA for morphisms [**34**, III_1, 5.4.1] to keep track of the CM structure we may replace E_0 with E_0' to arrange that $\mathcal{O}_L \subset \mathrm{End}(E_0)$.

The $\mathcal{O}_L$-action on $\mathrm{Lie}(E_0)$ selects a prime over p in $\mathcal{O}_L$ and embeds its residue field into $\mathbb{F}_q$. Pre-composing the $\mathcal{O}_L$-action on E_0 with the involution of L if necessary, we can arrange that $\mathcal{O}_L$ acts on $\mathrm{Lie}(E_0)$ through an embedding $\mathcal{O}_L/\mathfrak{p} \hookrightarrow \mathbb{F}_q$, so the formal group corresponding to $E_0[p^\infty]$ is a formal $\mathcal{O}_{L,\mathfrak{p}}$-module over $\mathbb{F}_q$ of dimension 1. Both Lubin–Tate theory and the deformation theory of 1-dimensional formal modules [**49**, 22.4.4] ensure that this lifts to a formal $\mathcal{O}_{L,\mathfrak{p}}$-module over R, so we obtain the desired $\mathcal{O}_L$-linear formal lift of E_0 over R. □

1.7.5. Variant on Grothendieck's theorem. C-F. Yu's variant on Theorem 1.7.2.5 asserts that we can *first* apply an isogeny and *then* pass to a finite extension on K (with no further isogeny involved) to get to a situation that descends to a finite field. This goes as follows. Consider the setup in Theorem 1.7.2.5. By Proposition 1.3.2.1, the simple factors have sufficiently many complex multiplications, so we may focus on the case of simple abelian varieties A. Choose a polarization, so the division algebra $D = \mathrm{End}^0(A)$ is endowed with a positive involution. By [**133**, 2.2], there is a maximal commutative subfield $L \subset D$ that is stable under the involution, so L is either totally real or CM. We claim that L is a CM field, or in other words L is not totally real.

To prove this property of L, first note that by Proposition 1.3.6.4 (in positive characteristic) the division algebra D is either of Type III or Type IV (in the sense of Theorem 1.3.6.2). Since L contains the center Z of D, for Type IV we get the CM property for L from the fact that Z is CM in such cases. For Type III, the key point is that Z is totally real and D is non-split at all real places of Z. We know that D_L is split over L since L is a maximal commutative subfield of D, so L is not totally real. Hence, once again L is a CM field.

Applying Proposition 1.7.4.5, we can pass to an isogenous abelian variety so that $\mathcal{O}_L \subset \mathrm{End}(A)$. In this special case, we may conclude via the following theorem.

1.7.5.1. Theorem (Yu). *Let K be a field with positive characteristic, and A a* CM *abelian variety over K with* CM *structure provided by a* CM *field $L \subset \mathrm{End}^0(A)$. If $\mathcal{O}_L \subset \mathrm{End}(A)$ then there is a finite extension K'/K such that $A_{K'}$ equipped with its $\mathcal{O}_L$-action descends to a finite field contained in K'.*

This result is [**133**, Thm. 1.3]; it will not be used in what follows. (Note that it suffices just to descend the abelian variety $A_{K'}$ to a finite subfield of K', as then a further finite extension on K' will enable us to descend the abelian variety along with its $\mathcal{O}_L$-action, by Lemma 1.2.1.2.)

1.8. CM lifting questions

1.8.1. Basic definitions and examples. Let κ be a field of characteristic $p > 0$, and consider an abelian variety A_0 over κ. By Corollary 1.6.2.5, if κ is finite and A_0 is isotypic then we may endow it with a structure of CM abelian variety having complex multiplication by a CM field. Inspired in part by Theorem 1.6.5.1, we wish to pose several questions related to the problem of lifting A_0 to characteristic 0 in the presence of CM structures. First we make a general definition unrelated to complex multiplication.

1.8.1.1. Definition. A *lifting* of A_0 to characteristic 0 is a triple (R, A, ϕ) consisting of a domain R of characteristic 0, an abelian scheme A over R, a surjective map $R \twoheadrightarrow \kappa$, and an isomorphism $\phi : A_\kappa \simeq A_0$ of abelian varieties over κ.

We may replace R with its localization at the maximal ideal $\ker(R \twoheadrightarrow \kappa)$ so that R is local with residue field κ. For $K := \mathrm{Frac}(R)$, if A_K admits sufficiently many complex multiplications then we say A is a CM *lift* of A_0 to characteristic 0. The injective map $\mathrm{End}(A) \to \mathrm{End}(A_K)$ has torsion-free cokernel:

1.8.2. Lemma. *For abelian schemes A, B over an integral scheme S with generic point η, the injective map $\mathrm{Hom}(A, B) \to \mathrm{Hom}(A_\eta, B_\eta)$ has torsion-free cokernel.*

PROOF. Consider $f : A_\eta \to B_\eta$ such that $n \cdot f$ extends to an S-group map $h : A \to B$ for a non-zero integer n. The restriction $h : A[n] \to B[n]$ between finite flat S-groups vanishes because such vanishing holds on the generic fiber over the integral S. Since $[n] : A \to A$ is an fppf covering with kernel $A[n]$, it follows that h factors through this map over S, which is to say $h = n \cdot \widetilde{f}$ for some S-group map $\widetilde{f} : A \to B$. Hence, the map $\widetilde{f}_\eta - f \in \mathrm{Hom}(A_\eta, B_\eta)$ is killed by n, so $\widetilde{f}_\eta = f$. □

The injective map in Lemma 1.8.2 can fail to be surjective:

1.8.3. Example. Let p be a prime with $p \equiv 3 \pmod 4$, so p is prime in $\mathbb{Z}[i]$ (with $i^2 = -1$). Let $R = \mathbb{Z}_{(p)} + p\mathbb{Z}_{(p)}[i]$, so $[\mathbb{Z}_{(p)}[i] : R] = p$ and $\mathrm{Frac}(R) = \mathbb{Q}(i)$. Let E be the elliptic curve $y^2 = x^3 - x$ over R, so the generic fiber $E_{\mathbb{Q}(i)}$ has endomorphism ring $\mathbb{Z}[i]$ via the action $[i](x, y) = (-x, -iy)$. Because $[i]^*(dx/y) = i \cdot dx/y$, $\mathbb{Z}[i]$ acts on $\mathrm{Lie}(E_{\mathbb{Q}(i)})$ through scaling via the canonical inclusion $\mathbb{Z}[i] \hookrightarrow \mathbb{Q}(i)$.

We claim that $\mathrm{End}(E) = \mathbb{Z}$ (so $\mathrm{End}^0(E) := \mathbb{Q} \otimes_{\mathbb{Z}} \mathrm{End}(E) = \mathbb{Q}$, even though the generic fiber $E_{\mathbb{Q}(i)}$ has endomorphism algebra $\mathbb{Q}(i)$). Indeed, if not then $\mathrm{End}(E)$ is an order in $\mathbb{Z}[i] = \mathrm{End}(E_{\mathbb{Q}(i)})$, so $\mathrm{End}(E) = \mathbb{Z}[i]$ by Lemma 1.8.2. In particular, the action by i on $E_{\mathbb{Q}(i)}$ would extend to an action on E, so the resulting multiplier action by i on the tangent line $\mathrm{Lie}(E_{\mathbb{Q}(i)}) = \mathrm{Lie}(E) \otimes_R \mathbb{Q}(i)$ would preserve the R-submodule $\mathrm{Lie}(E)$. But $\mathrm{Lie}(E)$ is a free R-module of rank 1 since R is local and E is R-smooth, so the i-action on this R-module is multiplication by some element $r \in R$. By working over $\mathbb{Q}(i)$ we have seen that we get the multiplier i, so necessarily $r = i$. Since $i \notin R$ due to the definition of R, we have a contradiction.

In Example 1.8.3, the base ring R is not normal. This is essential, since in the normal case there is no obstruction to extending maps between abelian schemes:

1.8.4. Lemma. *For a normal domain R with fraction field K, the functor $A \rightsquigarrow A_K$ from abelian schemes over R to abelian varieties over K is fully faithful.*

PROOF. This is a special case of a general lemma of Faltings [**36**, §2, Lemma 1] concerning homomorphisms between semi-abelian schemes over a normal scheme (the proof of which simplifies considerably in the case of abelian schemes). □

1.8.4.1. For normal R, Lemma 1.8.4 provides a specialization map

$$\mathrm{End}^0(A_K) = \mathrm{End}^0(A) := \mathbb{Q} \otimes_{\mathbb{Z}} \mathrm{End}(A) \to \mathrm{End}^0(A_\kappa)$$

between endomorphism algebras, and likewise for endomorphism rings. This makes normality a natural property to impose on R when studying questions about CM lifts. If R is not normal then $\mathrm{End}^0(A_K)$ may be larger than $\mathrm{End}^0(A)$, so it is not evident how to compare endomorphism algebras of the K-fiber and κ-fiber.

Hence, for general R we just work with the specialization map of endomorphism algebras $\mathrm{End}^0(A) \to \mathrm{End}^0(A_\kappa)$. This map can fail to be surjective. An elementary example is an elliptic curve over $\mathbb{Z}_{(p)}$ for a prime p (since elliptic curves over finite fields always admit complex multiplication, by Corollary 1.6.2.5, whereas elliptic curves over $\mathbb{Q}$ have endomorphism algebra $\mathbb{Q}$). In contrast with Lemma 1.8.2, the specialization map of endomorphism rings $\mathrm{End}(A) \to \mathrm{End}(A_\kappa)$ can have cokernel that is not torsion-free, even when R is normal. (In Chapter 4 we will see many natural examples of this phenomenon in our study of CM lifting problems, when we consider lifting questions for specific orders in CM fields; e.g., see 4.1.2.)

1.8.4.2. Remark. In the setting of Lemma 1.8.4, if $\lambda : A \to B$ is a homomorphism between abelian schemes over R and λ_K is an isogeny then λ is an isogeny (i.e., λ is fiberwise surjective with finite kernel; see §3.3 for a general discussion of isogenies for abelian schemes). To prove this, choose a K-homomorphism $\mu_K : B_K \to A_K$ such that $\mu_K \circ \lambda_K$ is multiplication by a non-zero integer n. The homomorphism $\mu : B \to A$ extending μ_K therefore satisfies $\mu \circ \lambda = [n]_A$, so λ has a fiberwise finite kernel and hence is an isogeny by fibral dimension considerations.

1.8.5. CM lifting problems. To formulate the lifting questions that we study in subsequent chapters, let $\mathbb{F}_q$ be a finite field of size q and let B be an abelian variety of dimension $g > 0$ over $\mathbb{F}_q$. Assume B is isotypic over $\mathbb{F}_q$ (necessary and sufficient for B to admit a structure of CM abelian variety with complex multiplication by a CM field, by Theorem 1.3.1.1 and Corollary 1.6.2.5). Let B_κ denote the scalar extension of B over a finite extension field $\kappa/\mathbb{F}_q$. Consider the following five assertions concerning the existence of a CM lifting of B or B_κ to characteristic 0.

- (CML) CM *lifting*: there is a local domain R with characteristic 0 and residue field $\mathbb{F}_q$, an abelian scheme A over R with relative dimension g equipped with a CM field $L \subset \mathrm{End}^0(A) := \mathbb{Q} \otimes_{\mathbb{Z}} \mathrm{End}(A)$ satisfying $[L : \mathbb{Q}] = 2g$, and an isomorphism $\phi : A_{\mathbb{F}_q} \simeq B$ as abelian varieties over $\mathbb{F}_q$.

- (R) CM *lifting after finite residue field extension*: there is a local domain R with characteristic 0 and residue field κ of finite degree over $\mathbb{F}_q$, an abelian scheme A over R with relative dimension g equipped with an action (in the isogeny category over R) by a CM field L with $[L : \mathbb{Q}] = 2g$, and an isomorphism $\phi : A_\kappa \simeq B \times_{\mathrm{Spec}(\mathbb{F}_q)} \mathrm{Spec}(\kappa)$ as abelian varieties over κ.
- (I) CM *lifting up to isogeny*: there is a local domain R with characteristic 0 and residue field $\mathbb{F}_q$, an abelian scheme A over R with relative dimension g equipped with an action (in the isogeny category over R) by a CM field L with $[L : \mathbb{Q}] = 2g$, and an isogeny $A_{\mathbb{F}_q} \to B$ of abelian varieties over $\mathbb{F}_q$.
- (IN) CM *lifting to normal domains up to isogeny*: there is a normal local domain R with characteristic 0 and residue field $\mathbb{F}_q$ such that (I) is satisfied for B using R.
- (RIN) CM *lifting to normal domains up to isogeny after finite residue field extension*: there is a normal local domain R with characteristic 0 and residue field κ of finite degree over $\mathbb{F}_q$ such that (R) is satisfied for B using R except that ϕ is only required to be an isogeny rather than an isomorphism.
- (sCML) *strong* CM *lifting*: For every CM field $L \subset \mathrm{End}^0(B)$ with $[L : \mathbb{Q}] = 2g$ such that $O_L \subset \mathrm{End}(B)$, the abelian variety B satisfies (CML) using a lifting A such that the $\mathbb{Q}$-subalgebra $\mathrm{End}^0(A) \subset \mathrm{End}^0(B)$ contains L.

1.8.5.1. Remark. By expressing a local ring as a direct limit of local subrings essentially of finite type over $\mathbb{Z}$, in the formulation of (R) there is no loss of generality in replacing κ with an algebraic closure of $\mathbb{F}_q$ or allowing κ to vary over all extensions of $\mathbb{F}_q$. Likewise, the normality condition in (RIN) is irrelevant because it can be attained at the cost of a finite residue field extension (by a specialization argument that we will give in 2.1.1), and in (IN) we can assume R is complete since essentially finite type $\mathbb{Z}$-algebras are excellent (ensuring that normality is preserved under completion of such rings along an ideal). Even in (I) we can assume R is complete local noetherian since we may first descend to a local noetherian domain $R_0 \subset R$ of characteristic 0 with residue field $\mathbb{F}_q$, and then note that the completion $\widehat{R}_0$ of R_0 has a minimal prime of residue characteristic 0 (as $R_0 \to \widehat{R}_0$ is faithfully flat).

By Remark 1.6.5.2, (RIN) has an affirmative answer for any isotypic B over $\mathbb{F}_q$, and the CM lift can be chosen using *any* CM maximal commutative subfield $L \subset \mathrm{End}^0(B)$. There are several refinements we wish to answer:

(1) Is a residue field extension necessary? That is, does (IN) hold for every B?
(2) If (IN) does not hold for every B, can we characterize when it holds? And how about (I) in general (i.e., drop normality, but permit an isogeny without increasing the residue field)?
(3) Is an isogeny necessary? That is, does (R) hold for every B (requiring the local domain R to be normal is not a constraint, since we are allowing a finite extension on κ; cf. Remark 1.8.5.1), or does even (CML) hold for every B?

These questions can be made more specific in several respects. For example, since the $\mathbb{Q}$-simple $\mathrm{End}^0(B)$ is usually non-commutative, it generally contains more than one CM maximal commutative subfield L (up to conjugacy) and so we can pose the CM lifting questions requiring an order in a particular choice of L to lift to a CM structure over R. We will give examples to show that the choice of L can affect the the answer to some of the lifting questions. Even if we know for a given B and $L \subset \mathrm{End}^0(B)$ that there is a CM lift to characteristic 0 on which the action of an order in L also lifts, it could be that the CM order $L \cap \mathrm{End}(B)$ does not lift. We will give examples where this happens in 4.1.2 (see Theorem 4.1.1 and the non-algebraizable universal formal deformation in 4.1.2.3).

1.8.6. Answers to CM lifting problems. The proofs of the following answers form the backbone of subsequent chapters.

1. By [**93**, Thm. B], for any $g > 2$ there exist g-dimensional abelian varieties over an algebraic closure of $\mathbb{F}_p$ for which there is no CM lift to characteristic 0. These results are proved in a much stronger form in Chapter 3. Thus, (R) does not hold in general, so in particular (CML) sometimes fails to hold. Hence, *an isogeny is necessary*; that is, it is better to consider (I) than (CML).
2. Building on lifting results for p-divisible groups in Chapter 3, in Chapter 4 we prove that (I) holds for every B (so a strengthening of (RIN) holds, applying the isogeny before making a finite extension on the residue field). In fact, for any CM maximal commutative subfield $L \subset \mathrm{End}^0(B)$ we construct an isogeny $B \to B'$ to an abelian variety over $\mathbb{F}_q$ such that B' has a CM lift to characteristic 0 on which the action of the order $\mathbb{Z} + p\mathcal{O}_L$ in $\mathcal{O}_L$ also lifts.

   The abelian variety B' generally depends on L, but we can arrange that the isogeny to B' is a p-power at most p^{4g^2} (and examples show that we cannot arrange it to have degree not divisible by p in general). The CM lifting of B' can be found over an order in a p-adic field whose relative ramification degree over a specific p-adic reflex field is tightly controlled (usually 1).
3. In contrast with success for (I), if we require R to be normal and do not increase the residue field (but permit isogenies) then the answer is negative: in Chapter 2 we give examples for which (IN) fails. Hence, for the existence of a CM lifting to a normal domain of characteristic 0 we must allow a finite extension of the initial finite field (and an isogeny), as in Theorem 1.6.5.1.

   However, there is a salvage: for each B and choice of $L \subset \mathrm{End}^0(B)$ we will give (in Chapter 2) concrete *necessary and sufficient* conditions in terms of a $\overline{\mathbb{Q}}_p$-valued CM type Φ on L for (IN) to have an affirmative answer using a CM lifting to which the action of an order in L (in the isogeny category) also lifts and yields the specified CM type Φ. In Example 2.1.7 we will give B for which this necessary and sufficient condition is satisfied for one choice of $L \subset \mathrm{End}^0(B)$ (and a suitable Φ) but fails for another choice (and any Φ).

We expect that (sCML) in 1.8.5 does *not* hold in general, but this is a guess.

1.8.7. Remark. (CM types in (IN) and (I)). In the context of 1.8.5, for any CM structure $L \subset \mathrm{End}^0(B)$ on the abelian variety B the q-Frobenius endomorphism $\varphi \in \mathrm{End}(B)$ lies in $L^\times$ since L is its own centralizer in $\mathrm{End}^0(B)$. The combinatorial data of *L-slopes* of the L-linear isogeny class of $(B, L \subset \mathrm{End}^0(B))$ is the sequence of non-negative rational numbers $r_w = \mathrm{ord}_w(\varphi)/\mathrm{ord}_w(q)$ indexed by the set of places w of L above p. When L is a CM field, with complex conjugation ι, this satisfies the self-duality condition $r_w + r_{w\circ\iota} = 1$ for each w (since $\varphi \cdot \iota(\varphi) = q$ in L).

Fix an algebraic closure $\overline{\mathbb{Q}}_p$ of $\mathbb{Q}_p$ and an embedding of $\mathbb{F}_q$ into the residue field of the valuation ring of $\overline{\mathbb{Q}}_p$ (so $W(\mathbb{F}_q)[1/p]$ canonically embeds into $\overline{\mathbb{Q}}_p$). For any

CM lifting of $(B, L \subset \mathrm{End}^0(B))$ over the valuation ring of a subfield of $\overline{\mathbb{Q}}_p$ of finite degree over $W(\mathbb{F}_q)[1/p]$, the associated $\overline{\mathbb{Q}}_p$-valued CM type Φ on L must satisfy the following compatibility condition with the L-slopes:

$$\#\{\phi \in \Phi \mid \phi \text{ induces } w \text{ on } L\} = r_w$$

for all w; see 2.1.4.1–2.1.4.2. For any $\overline{\mathbb{Q}}_p$-valued CM type (L, Φ) satisfying this necessary condition, in Chapter 2 we prove a necessary and sufficient condition for the existence of a solution $\mathcal{B}$ to problem (IN) for (B, L) over a local normal domain $R \supseteq W(\mathbb{F}_q)$ with residue field $\mathbb{F}_q$ such that the L-action (in the isogeny category) lifts to $\mathcal{B}$ and the resulting CM structure over $K = \mathrm{Frac}(R)$ has $\overline{\mathbb{Q}}_p$-valued CM type Φ relative to some $W(\mathbb{F}_q)[1/p]$-algebra embedding of $\overline{\mathbb{Q}}_p$ into $\overline{K}$.

Without normality it is more difficult to determine which CM types on L arise from solutions to the lifting problem (I) for (B, L) when we require the L-action to lift. If L is a CM field, the answer is p-local for the maximal totally real subfield L^+: it can be analyzed separately for each p-adic place v of L^+. To be precise, a $\overline{\mathbb{Q}}_p$-valued CM type Φ for L corresponds to a sequence $\{(L_v^+ \otimes_{L^+} L, \Phi_v)\}_v$ where Φ_v is a set of $\mathbb{Q}_p$-algebra homomorphisms $L_v^+ \otimes_{L^+} L \to \overline{\mathbb{Q}}_p$, and via p-divisible groups the question reduces to determining (for each v) the family $\mathfrak{F}_v$ of all Φ_v that arise as the v-component of the $\overline{\mathbb{Q}}_p$-valued CM type of an affirmative solution to the CM lifting problem (I) for the CM structure $(B, L \subset \mathrm{End}^0(B))$.

When v is split in L, say with w and w' the two places over it on L, there turn out to be no restrictions on Φ_v beyond the above compatibility conditions

$$r_w = \#\{\phi \in \Phi_v \mid \phi \text{ induces } w \text{ on } L\}, \quad r_{w'} = \#\{\phi \in \Phi_v \mid \phi \text{ induces } w' \text{ on } L\}$$

(which can always be satisfied, since any $\phi \in \Phi_v$ induces w or w' on L). However when v is non-split in L, the proofs in Chapter 4, B.1, and B.2 provide only a non-empty subset $\mathfrak{F}'_v$ of $\mathfrak{F}_v$;[3] we do not know the discrepancy between $\mathfrak{F}_v$ and $\mathfrak{F}'_v$.[4]

[3] The set $\mathfrak{F}'_v$ is the family of all v-components of $\overline{\mathbb{Q}}_p$-valued CM types of affirmative solutions to (I) which can be constructed by the method in B.1 and B.2 together with the Serre tensor construction for p-divisible groups in 4.3. The proof in Chapter 4, especially 4.5.15 (iii)–4.5.17, which uses the Serre tensor construction and the existence of CM lifting of toy model p-divisible groups, gives a non-empty subset $\mathfrak{F}''_v$ of $\mathfrak{F}'_v$ which can be strictly smaller than $\mathfrak{F}'_v$.

[4] A complete solution of (sCML) should also provide an answer to this question.

CHAPTER 2

CM lifting over a discrete valuation ring

We trusted in the God who created integers and introduced what we called arithmetic invariants.

— Jun-Ichi Igusa

2.1. Introduction

In this chapter we consider some obstructions which can be used to study various aspects of the CM lifting problems from 1.8.5 over a normal base. The key observation is the *residual reflex condition* introduced in 2.1.5. In 2.2 we show that CM lifting up to isogeny is possible over $\mathbb{F}_q$ if the residual reflex obstruction is avoided in the strongest sense. In 2.3 we give counterexamples to the lifting condition (IN) from 1.8.5. The main result of this chapter (Theorem 2.1.6) is that the residual reflex condition is the only obstruction to the lifting problem (IN). Its proof is given by global methods in 2.5, building on preparations with algebraic Hecke characters in 2.4 (and the Main Theorem of Complex Multiplication, for which we provide a complete proof in A.2).

An alternative proof via local methods is given in 2.6 by replacing the global arithmetic theory of complex multiplication with local arguments resting on p-divisible groups and p-adic Hodge theory. We prefer the global argument because it is more explicit about the number field over which a given abelian variety over a finite field can be lifted. But the local methods will be used in Chapter 3, which in turn underlies the arguments in Chapter 4.

2.1.1. Some deformation ring arguments. In the formulation of the lifting problem (I) in 1.8.5, there is no loss of generality in requiring R to be a complete local noetherian domain with residue field $\mathbb{F}_q$ (see Remark 1.8.5.1). For such R, every maximal ideal $\mathfrak{m}$ of $R[1/p]$ has residue field $R[1/p]/\mathfrak{m}$ of finite degree over the fraction field of $W(\mathbb{F}_q)$ (by [**54**, 7.1.9]), and so by replacing R with its image in (the valuation ring of) $R[1/p]/\mathfrak{m}$ we see that R can even be taken to be an order with residue field $\mathbb{F}_q$ in a p-adic integer ring.

The finite residue field of R may increase after normalization. For this reason, on the one hand (IN) in 1.8.5 is a priori a stronger property than (I), but on the other hand (IN) is a natural condition to consider because abelian schemes over normal base schemes have the pleasant property that homomorphisms between generic fibers extend (uniquely) over the entire base (Lemma 1.8.4). In 2.2 we will give an elementary sufficient criterion for (I) to hold (and in Chapter 4 we will use deeper methods to prove that (I) always holds). In 2.3.5 this is combined with an arithmetic obstruction (explained in 2.1.5) to give absolutely simple examples with arbitrarily large dimension for which (I) holds but (IN) fails.

An interesting special case of (IN) is when the normal base ring R is a p-adic integer ring. We now explain why this is essentially the general case. Suppose (IN) holds for some B over $\mathbb{F}_q$ using a lifting $\mathcal{B}'$ of an isogenous B' over a characteristic-0 complete local normal noetherian domain R with residue field $\mathbb{F}_q$ (see Remark 1.8.5.1). There is a polarization of the generic fiber of $\mathcal{B}'$ over R, and by the normality of R this extends to an isogeny $\lambda : \mathcal{B}' \to \mathcal{B}'^t$ over R (Remark 1.8.4.2) that is necessarily symmetric. In fact, λ is a polarization due to:

2.1.1.1. Lemma. *Let $A \to S$ be an abelian scheme, and $\mathscr{L}$ a line bundle on A. The locus U of $s \in S$ such that $\mathscr{L}_s$ is ample on A_s is both open and closed in S, and $\mathscr{L}|_{A_U}$ is U-ample.*

PROOF. By [**34**, IV$_3$, 9.6.4], U is open and $\mathscr{L}|_{A_U}$ is U-ample. To prove that U is closed we can reduce to the case when S is the spectrum of a discrete valuation ring whose generic point η lies in U. We can also assume by passing to a positive power that $\mathscr{L}_\eta$ has a non-zero global section on A_η. The S-group map $\phi_{\mathscr{L}} : A \to A^t$ is an isogeny since it is so on the generic fiber (see Remark 1.8.4.2). Letting $s \in S$ be the closed point, the resulting finiteness of $\ker\phi_{\mathscr{L}_s}$ implies that $\mathscr{L}_s$ is ample provided that it has a non-zero global section [**82**, §6, App. I]. If $h^0(A_s, \mathscr{L}_s) = 0$ then the theorem on cohomology and base change implies $\mathrm{H}^0(A, \mathscr{L}) = 0$, a contradiction since $\mathrm{H}^0(A_\eta, \mathscr{L}_\eta) \neq 0$. ☐

The polarization property for λ yields a natural local $W(\mathbb{F}_q)$-algebra map $D \to R$ from the universal deformation ring D for B' equipped with its induced polarization and CM structure from $\mathcal{B}'$. (See Proposition 1.4.4.14 for the existence of D.) We claim that the image D_1 of D in R is finite flat over $W(\mathbb{F}_q)$, in which case its normalization $\widetilde{D}_1$ is also finite flat over $W(\mathbb{F}_q)$. Such a normalization must have residue field $\mathbb{F}_q$ because it lies inside R (due to the normality of R). Thus, granting the finiteness for D_1 in general, if (IN) has an affirmative answer for B then (IN) can be achieved using lifts over p-adic integer rings (such as $\widetilde{D}_1$).

Since R is a domain of characteristic 0, $\mathrm{Spec}(D_1)$ is a closed subscheme of the schematic closure of a generic point of $\mathrm{Spec}(D)$ with characteristic 0. Thus, the desired $W(\mathbb{F}_q)$-finiteness for D_1 is a consequence of:

2.1.1.2. Lemma. *Let $(A_0, \alpha_0 : \mathcal{O} \to \mathrm{End}(A_0))$ be a polarized* CM *abelian variety over a field K, and let ϕ_0 be a polarization of α_0. Let $\Lambda = K$ if $\mathrm{char}(K) = 0$ and let Λ be a Cohen ring for K if $\mathrm{char}(K) > 0$.*

Let D be the deformation ring for (A_0, ϕ_0, α_0) on the category of complete local noetherian Λ-algebras with residue field K. Every generic point of $\mathrm{Spec}(D)$ with characteristic 0 has Λ-finite closure.

PROOF. Let $g = \dim(A_0)$ and $d^2 = \deg(\phi_0)$. Choose an integer $n \geqslant 3$ not divisible by $\mathrm{char}(K)$, and let $\mathcal{G}$ be the unique finite étale Λ-group whose special fiber is $A_0[n]$. The proof of Proposition 1.4.4.14 identifies D with a completed local ring at a K-point on a locally finite type moduli scheme $\mathcal{M}$ over Λ that classifies 4-tuples (A, ϕ, α, τ) over Λ-schemes S, where $A \to S$ is an abelian scheme with relative dimension g, ϕ is a polarization of A with degree d^2, $\alpha : \mathcal{O} \to \mathrm{End}(A)$ is a ring homomorphism, and $\tau : A[n] \simeq \mathcal{G}_S$ is an S-group isomorphism.

It suffices to show the generic fiber of $\mathcal{M}$ over $F := \mathrm{Frac}(\Lambda)$ is 0-dimensional. Indeed, if $\mathrm{char}(K) = 0$ then $F = K$ and so it would follow that all local rings on $\mathcal{M}$ are artinian and K-finite. If instead $\mathrm{char}(K) > 0$ then the schematic closure of the generic fiber $\mathcal{M}_F$ would have relative dimension 0 due to its Λ-flatness and 0-dimensionality of the generic fiber, so the local rings at points of the special fiber would have K-finite quotients modulo the maximal ideal of Λ and hence would have Λ-finite completions. Consequently, our problem now concerns 0-dimensionality of a locally finite type moduli scheme of polarized CM abelian varieties with level structure over a field F of characteristic 0.

The idea is that by Theorem 1.7.2.1 there are no non-isotrivial families of CM abelian varieties in characteristic 0. To make this idea rigorous, we may assume $F = \overline{F}$ and just have to check that any point $m \in M$ (especially a generic point) has residue field $F(m)$ equal to F. Let $A \to M$ be the universal abelian scheme. By Theorem 1.7.2.1, the geometric fiber $A_{\overline{m}}$ descends as a CM abelian variety to a number field inside the algebraic closure $F(\overline{m})$ of $F(m)$, and hence it descends to a CM abelian variety over the algebraically closed subfield $F \subset F(m)$. The endomorphism and level structure on $A_{\overline{m}}$ both automatically descend to this F-descent of $A_{\overline{m}}$, due to Lemma 1.2.1.2 (and the constancy of n-torsion in an abelian variety over an algebraically closed field of characteristic 0). Thus, the F-morphism $\overline{m} \to M$ factors through an F-point. But its image is m, so $F(m) = F$. □

2.1.2. In 2.2 we use deformation theory to prove that (I) in 1.8.5 is satisfied using any CM subfield $L \subseteq \mathrm{End}^0(B)$ such that $[L : \mathbb{Q}] = 2g$ and all p-adic places of L have residue field $\mathbb{F}_p$ (provided that such a subfield L exists, which we exhibit in some interesting examples in 2.3.5). In Chapter 4 we prove (I) in general, but the special case in 2.2 is sufficient for some interesting examples in the present chapter.

The condition (IN) is subtle because there is an arithmetic obstruction to it coming from classical CM theory. To explain this obstruction, fix an isotypic abelian variety B over $\mathbb{F}_q$ with dimension $g > 0$. Let R be a local *normal* domain with characteristic 0 and residue field $\mathbb{F}_q$. Let A be an abelian scheme over R with relative dimension g, equipped with an action (in the isogeny category over R) by a CM field L with $[L : \mathbb{Q}] = 2g$. Assume that the special fiber $A_{\mathbb{F}_q}$ is isogenous to B. By direct limit arguments, we can arrange that R is essentially of finite type over $\mathbb{Z}$, so it is excellent. Thus, we can replace R with its completion without losing the normality hypothesis (and this does not change the residue field).

Since $K := \mathrm{Frac}(R)$ has characteristic 0, the action on A over R by an order in $\mathcal{O}_L$ uniquely extends to a K-linear action by L on the g-dimensional $\mathrm{Lie}(A_K)$. By Lemma 1.5.2, this action viewed over an algebraic closure $\overline{K}$ decomposes into eigenlines on which L acts through a set Φ of representatives for $\mathrm{Hom}_{\mathrm{ring}}(L, \overline{K})$ modulo the action on the CM field L by its intrinsic complex conjugation. We may of course view Φ as a subset of $\mathrm{Hom}_{\mathrm{ring}}(L, \overline{\mathbb{Q}}_p)$, where $\overline{\mathbb{Q}}_p$ is the algebraic closure of $\mathbb{Q}_p$ inside $\overline{K}$.

2.1.3. Definition. For a CM field L and a CM type Φ on L valued in an algebraically closed field K of characteristic 0, the *reflex field* $E \subset \overline{\mathbb{Q}} \subseteq K$ is the number field for which $\mathrm{Gal}(\overline{\mathbb{Q}}/E)$ is the subgroup of elements of $\mathrm{Gal}(\overline{\mathbb{Q}}/\mathbb{Q})$ stabilizing the subset $\Phi \subset \mathrm{Hom}_{\mathrm{ring}}(L, \overline{\mathbb{Q}})$.

Observe that whereas L is an abstract field, by definition the reflex field E associated to (L, Φ) is a subfield of $\overline{\mathbb{Q}}$.

2.1.3.1. Example. Reflex fields have an interpretation in terms of abelian varieties, as follows. Consider a CM type (L, Φ) arising from an abelian variety A over a field K of characteristic 0 equipped with a choice of algebraic closure $\overline{K}$ of K. By Galois descent, the reflex field E is the unique minimal subfield $F \subset \overline{K}$ such that the $L \otimes_{\mathbb{Q}} \overline{K}$-module $\mathrm{Lie}(A_{\overline{K}})$ descends to an $L \otimes_{\mathbb{Q}} F$-module. In particular, $E \subset K$. An equivalent description of E is that it is the subfield of K generated over $\mathbb{Q}$ by the K-linear trace of the action on $\mathrm{Lie}(A_K)$ by every $c \in L$.

Proposition 1.5.4.1 suggests that E should be viewed as a "field of definition" for the L-isogeny class of CM abelian varieties over $\overline{K}$ with $\overline{K}$-valued CM type (L, Φ). However, it can happen that no member of the isogeny class is actually defined over E. For example, the reflex field of a CM elliptic curve over $\overline{\mathbb{Q}}$ is equal to the associated imaginary quadratic field L (or rather, its canonical image in $\overline{\mathbb{Q}}$), but it follows from classical CM theory (and Proposition 1.7.4.5) that any field of definition for a member of the isogeny class must contain the Hilbert class field of L. Hence, if L does not have class number 1 then no such elliptic curve can be defined over the reflex field. Such class group obstructions are addressed in general for CM abelian varieties in A.4.6.1 and A.4.6.5 (using the Main Theorem of Complex Multiplication). We will soon see an alternative useful interpretation of E, as a genuine field of definition for a homomorphism between tori.

2.1.3.2. Remark. The reflex field E is always a CM field. Indeed, from the description via traces it is clear that E is a subfield of a CM field (such as the Galois closure of L over $\mathbb{Q}$), so it is either CM or totally real. But if it were totally real then a complex conjugation would lie in $\mathrm{Gal}(\overline{\mathbb{Q}}/E)$, contradicting the characteristization of this Galois group in terms of the CM type Φ.

Since we have introduced the reflex field E, we now briefly digress to introduce the concept of the *reflex norm* that will play an essential role in the Main Theorem of Complex Multiplication. This will be defined in two ways, first classically and then in a more modern form. Consider $\prod_{\varphi \in \Phi} \overline{\mathbb{Q}}$ as an $L \otimes_{\mathbb{Q}} \overline{\mathbb{Q}}$-module by making any $c \in L$ act via $c.(x_\varphi)_{\varphi \in \Phi} = (\varphi(c) x_\varphi)_{\varphi \in \Phi}$. This is $\mathrm{Gal}(\overline{\mathbb{Q}}/E)$-equivariant when an element $\sigma \in \mathrm{Gal}(\overline{\mathbb{Q}}/E)$ acts in the natural way on the right tensor factor of $L \otimes_{\mathbb{Q}} \overline{\mathbb{Q}}$ and acts on $\prod_{\varphi \in \Phi} \overline{\mathbb{Q}}$ via

$$\sigma(x_\varphi) = (\sigma(x_{\sigma^{-1} \circ \varphi})).$$

Applying Galois descent, this defines an $L \otimes_{\mathbb{Q}} E$-module t_Φ canonically attached to (L, Φ). An alternative characterization of t_Φ as an $L \otimes_{\mathbb{Q}} E$-module is that the E-linear action on it by each $c \in L$ has E-linear trace $\sum_{\varphi \in \Phi} \varphi(c) \in E$. This determines t_Φ up to isomorphism.

In the setup of Example 2.1.3.1, t_Φ is the $L \otimes_{\mathbb{Q}} E$-linear tangent space of a CM abelian variety over E with CM type (L, Φ) if such an abelian variety exists. However, we saw that usually no such abelian variety exists over E.

2.1.3.3. Definition. Let Φ be a $\overline{\mathbb{Q}}$-valued CM type on a CM field L, and let $E \subset \overline{\mathbb{Q}}$ be the reflex field. The *reflex norm*

$$\mathrm{N}_\Phi : \mathrm{Res}_{E/\mathbb{Q}}(\mathbb{G}_\mathrm{m}) \to \mathrm{Res}_{L/\mathbb{Q}}(\mathbb{G}_\mathrm{m})$$

is the L-linear determinant of the E-action on t_Φ. That is, for a $\mathbb{Q}$-algebra R and a unit $a \in E_R^\times$, $\mathrm{N}_\Phi(a) \in L_R^\times$ is the L_R-linear determinant of a-multiplication on $t_\Phi \otimes_\mathbb{Q} R$.

In Remark 2.1.3.5 we will describe the reflex norm in explicit Galois-theoretic terms, building on the following useful re-interpretation of the reflex field E and reflex norm N_Φ attached to a CM type (L, Φ). For any algebraically closed field $K \supset \overline{\mathbb{Q}}$ and any subset $\Sigma \subset \mathrm{Hom}(L, K) = \mathrm{Hom}(L, \overline{\mathbb{Q}})$, define the K-homomorphism

$$h_\Sigma : \mathbb{G}_\mathrm{m} \to \prod_{j \in \mathrm{Hom}(L,K)} \mathbb{G}_\mathrm{m} = \mathrm{Res}_{L/\mathbb{Q}}(\mathbb{G}_\mathrm{m})_K \tag{2.1.3.1}$$

to be the map $t \mapsto (t, \ldots, t, 1, \ldots, 1)$ whose jth component is trivial if $j \notin \Sigma$ and is the identity map if $j \in \Sigma$. This map clearly descends to the analogous such map defined over $\overline{\mathbb{Q}}$, and even over the finite Galois splitting field of L over $\mathbb{Q}$. Relative to the evident $\mathbb{Q}$-structures on its source and target, the minimal field of definition over $\mathbb{Q}$ for h_Σ is the number field in $\overline{\mathbb{Q}}$ corresponding to the open subgroup of $\mathrm{Gal}(\overline{\mathbb{Q}}/\mathbb{Q})$ whose elements are those σ such that $\sigma(\Sigma) = \Sigma$. In particular, h_Φ has its minimal field of definition over $\mathbb{Q}$ equal to the reflex field E of (L, Φ). The descent to E, also denoted h_Φ, encodes the reflex norm N_Φ:

2.1.3.4. Proposition. *For any finite separable extension of fields F'/F and any commutative affine F-group C of finite type, let $\mathrm{Nm}_{F'/F} : \mathrm{Res}_{F'/F}(C_{F'}) \to C$ denote the F-homomorphism that descends the F_s-homomorphism*

$$\mathrm{Res}_{F'/F}(C_{F'})_{F_s} = \mathrm{Res}_{(F_s \otimes_F F')/F_s}(C_{F_s \otimes_F F'}) = \prod_{j \in \mathrm{Hom}_F(F', F_s)} C_{F_s} \to C_{F_s}$$

defined by $(c_j) \mapsto \prod_j c_j$.

The reflex norm N_Φ is equal to the composite map

$$\mathrm{Res}_{E/\mathbb{Q}}(\mathbb{G}_\mathrm{m}) \xrightarrow{\mathrm{Res}_{E/\mathbb{Q}}(h_\Phi)} \mathrm{Res}_{E/\mathbb{Q}}(\mathrm{Res}_{L/\mathbb{Q}}(\mathbb{G}_\mathrm{m})_E) \xrightarrow{\mathrm{Nm}_{E/\mathbb{Q}}} \mathrm{Res}_{L/\mathbb{Q}}(\mathbb{G}_\mathrm{m}) .$$

PROOF. By unirationality of tori over fields, the set $E^\times$ of $\mathbb{Q}$-points in $\mathrm{Res}_{E/\mathbb{Q}}(\mathbb{G}_\mathrm{m})$ is Zariski-dense. Hence, it suffices to verify the identity on $\mathbb{Q}$-points. That is, for $x \in E^\times$, we claim that the element $\mathrm{N}_\Phi(x) \in L^\times$ equals the norm of $h_\Phi(x) \in \mathrm{Res}_{L/\mathbb{Q}}(\mathbb{G}_\mathrm{m})(E) = (E \otimes_\mathbb{Q} L)^\times$ relative to the finite flat ring homomorphism $L \to E \otimes_\mathbb{Q} L$. By definition, $\mathrm{N}_\Phi(x)$ is the E-linear determinant of multiplication by x on the $E \otimes_\mathbb{Q} L$-module t_Φ. But t_Φ is locally free over $E \otimes_\mathbb{Q} L$ with fibers of rank $\leqslant 1$, so it is invertible over a factor ring of $E \otimes_\mathbb{Q} L$. By construction, this factor ring as an E-algebra has its geometric support equal to the product of factors of $\overline{\mathbb{Q}} \otimes_E (E \otimes_\mathbb{Q} L) = \overline{\mathbb{Q}} \otimes_\mathbb{Q} L$ corresponding to the elements of $\Phi \subset \mathrm{Hom}(L, \overline{\mathbb{Q}})$.

Since $\mathrm{Gal}(\overline{\mathbb{Q}}/E)$ preserves the CM type Φ on L, $E \otimes_\mathbb{Q} L = R_\Phi \times R_{\Phi^*}$ where R_Φ is the E-algebra whose $\overline{\mathbb{Q}}$-points (over E) are precisely those of $E \otimes_\mathbb{Q} L$ that correspond to Φ and we define $\Phi^* := \mathrm{Hom}(L, \overline{\mathbb{Q}}) - \Phi$. Thus, $\mathrm{N}_\Phi(x) = \mathrm{Nm}_{R_\Phi/L}(x_\Phi)$

where $(x_\Phi, x_{\Phi^*}) = x \otimes 1$, so

$$\begin{aligned} \mathrm{N}_\Phi(x) = \mathrm{Nm}_{R_\Phi/L}(x_\Phi)\mathrm{Nm}_{R_{\Phi^*}/L}(1) &= \mathrm{Nm}_{(R_\Phi \times R_{\Phi^*})/L}(x_\Phi, 1) \\ &= \mathrm{Nm}_{(E\otimes_{\mathbb{Q}} L)/L}(x_\Phi, 1). \end{aligned}$$

It now suffices to prove that $h_\Phi(x) = (x_\Phi, 1)$ in $E \otimes_{\mathbb{Q}} L$ for $x \in E^\times$. Using extension of scalars along the canonical inclusion $E \hookrightarrow \overline{\mathbb{Q}}$, for any $x \in \overline{\mathbb{Q}}^\times$ the element $h_\Phi(x) \in (\overline{\mathbb{Q}} \otimes_{\mathbb{Q}} L)^\times = \prod_j \overline{\mathbb{Q}}^\times$ has j-component equal to 1 for $j \in \Phi^*$ and equal to x for $j \in \Phi$. Since the decomposition $\overline{\mathbb{Q}} \otimes_{\mathbb{Q}} L \simeq \prod_j \overline{\mathbb{Q}}$ carries the $\overline{\mathbb{Q}}$-algebra structure on the left over to the diagonal one on the right, the element $x \otimes 1$ on the left goes over to the diagonal embedding of x on the right. Hence, if $x \in E$ then the element $x_\Phi \in \overline{\mathbb{Q}} \otimes_E R_\Phi = \prod_\Phi \overline{\mathbb{Q}}$ is the diagonally embedded x, so we are done. □

2.1.3.5. Remark. Fix an embedding of a CM field L into $\overline{\mathbb{Q}}$, and a $\overline{\mathbb{Q}}$-valued CM type Φ on L. The reflex field $E = E(L, \Phi)$ and reflex norm N_Φ can be described in terms of the Galois group of the normal closure $L' \subset \overline{\mathbb{Q}}$ of $L/\mathbb{Q}$, recovering the definition of these concepts in classical CM theory as in [**113**, §5.5A].

To explain this, let $\Phi_{L'} \subset \mathrm{Gal}(L'/\mathbb{Q})$ be the inverse image of Φ arising via the isomorphism $\mathrm{Gal}(L'/\mathbb{Q})/\mathrm{Gal}(L'/L) \simeq \mathrm{Hom}_{\mathbb{Q}\text{-alg}}(L, L') = \mathrm{Hom}_{\mathbb{Q}\text{-alg}}(L, \overline{\mathbb{Q}})$. Let

$$\Phi^*_{L'} = \{s \in \mathrm{Gal}(L'/\mathbb{Q}) \mid s^{-1} \in \Phi_{L'}\} = \{s \in \mathrm{Gal}(L'/\mathbb{Q}) \mid s^{-1}|_L \in \Phi\},$$

and let H^* be the subgroup of $\mathrm{Gal}(L'/\mathbb{Q})$ consisting of all elements σ such that $\Phi^*_{L'} \cdot \sigma = \Phi^*_{L'}$. This condition on σ is equivalent to left multiplication by σ^{-1} on $\mathrm{Hom}_{\mathbb{Q}\text{-alg}}(L, L')$ carrying Φ into itself, so the reflex field $E = E(L, \Phi)$ equals the fixed field L'^{H^*}.

The Galois-theoretic description of the reflex norm requires a bit more work. Under the identification of $\mathrm{Gal}(L'/\mathbb{Q})/H^*$ with $\mathrm{Hom}_{\mathbb{Q}\text{-alg}}(E, L') = \mathrm{Hom}_{\mathbb{Q}\text{-alg}}(E, \overline{\mathbb{Q}})$, the H^*-stable subset $\Phi^*_{L'}$ defines a subset $\Phi^*_{L'}/H^* \hookrightarrow \mathrm{Hom}_{\mathbb{Q}\text{-alg}}(E, L')$ (the "reflex type"). Letting j denote the canonical $\mathbb{Q}$-embedding of L into L', we claim that N_Φ is given on $\mathbb{Q}$-points by

$$(j \circ \mathrm{N}_\Phi)(x) = \prod_{\tau \in \Phi^*_{L'}/H^*} \tau(x) \tag{2.1.3.2}$$

for all $x \in E(L, \Phi)^\times$. Define $\underline{L} = \mathrm{Res}_{L/\mathbb{Q}}(\mathbb{G}_{\mathrm{m}})$.

Letting $h : \mathbb{G}_{\mathrm{m}} \to \underline{L}^\times_E$ denote the E-descent of h_Φ (as in (2.1.3.1) with $\Sigma = \Phi$), by Proposition 2.1.3.4 our claim concerns the $\mathbb{Q}$-homomorphism $\underline{j} : \underline{L}^\times \to \underline{L'}^\times$, namely that the element $(\underline{j}_E \circ h)(x) \in (E \otimes_{\mathbb{Q}} L')^\times$ satisfies $\mathrm{Nm}_{E/\mathbb{Q}}((\underline{j}_E \circ h)(x)) = \prod_{\tau \in \Phi^*_{F'}/H^*} \tau(x)$ for all $x \in E^\times$. To prove this claim, via the identification

$$E \otimes_{\mathbb{Q}} L' = \prod_{\tau : E \hookrightarrow L'} L'$$

(product over all $\mathbb{Q}$-embeddings of E into L') it suffices to prove $(\underline{j}_E \circ h)(x)$ has τ-component $\tau(x)$ when $\tau \in \Phi^*_{L'}/H^*$ and has τ-component 1 otherwise.

Note that $\underline{j}_E \circ h$ is an E-descent of the composite geometric cocharacter

$$\underline{j}_{\overline{\mathbb{Q}}} \circ h_\Phi = h_{\Phi_{L'}} : \mathbb{G}_{\mathrm{m}} \to \underline{L'}^\times_{\overline{\mathbb{Q}}} = \prod_{s \in \mathrm{Gal}(L'/\mathbb{Q})} \mathbb{G}_{\mathrm{m}}$$

whose s-component is the identity map when $s \in \Phi_{L'}$ and is trivial otherwise. The natural identification of $\underline{L'}^*_E$ as an E-descent of $\underline{L'}^\times_{\overline{\mathbb{Q}}}$ carries the group $(E \otimes_{\mathbb{Q}} L')^\times = \prod_{\tau: E \to L'} L'^\times$ into $(\overline{\mathbb{Q}} \otimes_{\mathbb{Q}} L')^\times = \prod_{\sigma: L' \to \overline{\mathbb{Q}}} \overline{\mathbb{Q}}^\times$ via the map $(x_\tau) \mapsto (\sigma(x_{\sigma^{-1}|_E}))$ since for $a \in E$ and $b \in L'$ we have $a\sigma(b) = \sigma(\sigma^{-1}(a)b)$. In particular, for $a \in E^\times$ the trivial identity $a = s^{-1}(s(a))$ for each $s \in \mathrm{Gal}(L'/\mathbb{Q})$ implies that the point $h_{\Phi_{L'}}(a) = (\underline{j}_E \circ h)(a)$ comes from (x_τ) where $x_\tau = s(a)$ when τ lifts to $s \in \mathrm{Gal}(L'/\mathbb{Q})$ satisfying $s^{-1} \in \Phi_{L'}$ and $x_\tau = 1$ otherwise. This says exactly that $x_\tau = \tau(x)$ when $\tau \in \Phi^*_{L'}/H^*$ and $x_\tau = 1$ otherwise, as desired.

The proof of the Main Theorem of Complex Multiplication will rest on the fact that for any CM type (L, Φ), the image of the reflex norm N_Φ lies in a certain natural subtorus of $\mathrm{Res}_{L/\mathbb{Q}}(\mathbb{G}_\mathrm{m})$ that we now define. Let $\mathrm{Nm}_{L/L^+} : \mathrm{Res}_{L/\mathbb{Q}}(\mathbb{G}_\mathrm{m}) \to \mathrm{Res}_{L^+/\mathbb{Q}}(\mathbb{G}_\mathrm{m})$ denote the usual norm relative to the maximal totally real subfield $L^+ \subset L$; $\ker(\mathrm{Nm}_{L/L^+})$ is connected and hence is a $\mathbb{Q}$-torus (as one verifies via a calculation with character groups over $\overline{\mathbb{Q}}$). Thus, the preimage $T = \mathrm{Nm}_{L/L^+}^{-1}(\mathbb{G}_\mathrm{m})$ is a $\mathbb{Q}$-subtorus of $\mathrm{Res}_{L/\mathbb{Q}}(\mathbb{G}_\mathrm{m})$. This torus has the following important property.

2.1.3.6. Proposition. *The image of* N_Φ *is contained in* T.

PROOF. Since the locus of $\mathbb{Q}$-points is Zariski-dense in any $\mathbb{Q}$-torus (due to the unirationality of tori over any field), it suffices to prove that for any $a \in \mathrm{Res}_{E/\mathbb{Q}}(\mathbb{G}_\mathrm{m})(\mathbb{Q}) = E^\times$, the element $\mathrm{N}_\Phi(a) \in L^\times$ lies in $T(\mathbb{Q})$. In other words, we claim that $\mathrm{Nm}_{L/L^+}(\mathrm{N}_\Phi(a)) \in \mathbb{Q}^\times$. In fact, we will prove that this norm coincides with $\mathrm{Nm}_{E/\mathbb{Q}}(a)$.

By definition, $\mathrm{N}_\Phi(a) = \det_L(a : t_\Phi \to t_\Phi)$. But $\mathrm{Nm}_{L/L^+}(c) = cc^*$ where $c \mapsto c^*$ is the complex conjugation involution of L over L^+, and this involution carries the CM type Φ on L to its complement Φ^* in the set of embeddings of L into $\overline{\mathbb{Q}}$. Hence, inside L we have $\mathrm{Nm}_{L/L^+}(\mathrm{N}_\Phi(a)) = \mathrm{N}_\Phi(a)\mathrm{N}_{\Phi^*}(a)$. This product of L-linear determinants is the L-linear determinant of multiplication by a on the $E \otimes_{\mathbb{Q}} L$-module direct sum $t_\Phi \oplus t_{\Phi^*}$. But this direct sum is a free $E \otimes_{\mathbb{Q}} L$-module of rank 1, as we may check via scalar extension along the inclusion $E \hookrightarrow \overline{\mathbb{Q}}$, so we arrive at the L-linear determinant of the natural a-multiplication on $E \otimes_{\mathbb{Q}} L$. This is $\mathrm{Nm}_{E/\mathbb{Q}}(a) \in \mathbb{Q}^\times$. □

2.1.4. CM lifting obstruction. Returning to the topic of CM lifting problems in 2.1.2, observe that by the normality of R, the inclusion of fields $E \subset K = \mathrm{Frac}(R)$ yields the inclusion of rings $\mathcal{O}_E \subset R$. But R is local with residue characteristic p, so its maximal ideal contracts to a prime ideal $\mathfrak{p}$ of $\mathcal{O}_E$ over p. Note that this $\mathfrak{p}$-adic place on E is also induced by the inclusion of E into $\overline{\mathbb{Q}}_p$ arising from the definition of E in terms of (L, Φ).

By the construction of $\mathfrak{p}$ in terms of R, $\mathcal{O}_E/\mathfrak{p}$ occurs as a subfield of the residue field $\mathbb{F}_q$ of R. Thus, if (IN) in 1.8.5 holds for B over $\mathbb{F}_q$ using some CM field $L \subset \mathrm{End}^0(B)$ then there is a $\overline{\mathbb{Q}}_p$-valued CM type Φ on L whose associated reflex field $E \subset \overline{\mathbb{Q}}_p$ has residue field at the induced place over p that is not "too large" in the sense that it can be embedded into $\mathbb{F}_q$. This is an arithmetic restriction on Φ.

There is a further restriction on the set

$$\Phi \subset \mathrm{Hom}_{\mathrm{ring}}(L, \overline{\mathbb{Q}}_p) = \mathrm{Hom}_{\mathrm{ring}}(L, \overline{K})$$

arising from an L-linear CM abelian scheme over R that lifts the L-linear CM abelian variety B over $\mathbb{F}_q$. This restriction will not require normality of the complete local noetherian domain R, and so it will be a necessary condition on CM types arising in (I) in 1.8.5 when R there is noetherian and complete.

To explain this restriction, first note that since L is its own centralizer in $\mathrm{End}^0(B)$, and $\mathrm{End}(B) \cap L$ is an order in $\mathcal{O}_L$, there is a unique element $\mathrm{Fr}_{B,q} \in \mathcal{O}_L$ acting on B by the q-Frobenius endomorphism. Assume that there is an L-linear isogeny between B and $A_0 := A_{\mathbb{F}_q}$ for some A over R as in (I) with R noetherian and complete. This L-linear isogeny is compatible with q-Frobenius endomorphisms, so the element $\mathrm{Fr}_{A_0,q} \in \mathcal{O}_L$ is equal to $\mathrm{Fr}_{B,q}$.

2.1.4.1. The *Shimura–Taniyama formula* for the CM abelian variety A relates the factorization of $\mathrm{Fr}_{A_0,q}\mathcal{O}_L = \mathrm{Fr}_{B,q}\mathcal{O}_L$ to the CM type $\Phi \subset \mathrm{Hom}_{\mathrm{ring}}(L, \overline{\mathbb{Q}}_p) = \mathrm{Hom}_{\mathrm{ring}}(L, \overline{K})$ arising from the L-action on $\mathrm{Lie}(A_K)$: for each p-adic place v on L,

$$\frac{\mathrm{ord}_v(\mathrm{Fr}_{B,q})}{\mathrm{ord}_v(q)} = \frac{\#\{\phi \in \Phi : \phi \text{ induces } v \text{ on } L\}}{[L_v : \mathbb{Q}_p]}. \tag{2.1.4.1}$$

Note that the normalization of ord_v on the left side cancels out in the ratio, and the right side is independent of the choice of $\overline{\mathbb{Q}}_p$.

The ratios on the left side (as v varies) are called the *slopes* of B (over $\mathbb{F}_q$). More generally, if C is an abelian variety over $\mathbb{F}_q$ with $q = p^f$ and if $F_C \in \mathbb{Z}[T]$ is the associated characteristic polynomial of $\mathrm{Fr}_{C,q}$ then as λ_i varies through the zeros of F_C in an algebraic closure $\overline{\mathbb{Q}}_p$ of $\mathbb{Q}_p$, the numbers $\mathrm{ord}_p(\lambda_i)/f$ (usually counted with multiplicity) are called the *slopes* of C over $\mathbb{F}_q$; here, ord_p is normalized by the condition $\mathrm{ord}_p(p) = 1$. Obviously the formation of the slopes is invariant under isogeny and finite extension of the finite base field.

For R a p-adic integer ring, the formula (2.1.4.1) was first proved by Shimura and Taniyama using global arguments, and a local proof was given by Tate using p-divisible groups [**121**, §5]. See Corollary A.4.7.4 (and the subsequent Remark) for the proof of a more general statement for Frobenius elements coming from an algebraic Hecke character. The role of a Dedekind base in the proofs of Shimura–Taniyama and Tate is to make the CM order $\mathcal{O}$ become maximal upon passing to an isogenous abelian scheme A'. Indeed, the isogeny $f : A_K \to \mathcal{O}_L \otimes_{\mathcal{O}} A_K$ onto an abelian variety over K provided by Proposition 1.7.4.5 uniquely extends to an isogeny $A \to A'$ onto an abelian scheme over R by defining A' to be the quotient of A modulo the schematic closure in A of the K-finite $\ker(f) \subset A_K$, and visibly $\mathcal{O}_L \subset \mathrm{End}(A')$ inside $\mathrm{End}^0(A')$. (The closure of $\ker(f)$ in A is R-flat since R is Dedekind, so it is R-finite since A is proper.) Difficulties with R-flatness of scheme-theoretic closure over non-Dedekind R make it unclear if such an A' can be found for general R. Nonetheless, we have:

2.1.4.2. Proposition. *Formula* (2.1.4.1) *holds for any complete local noetherian domain R with fraction field of characteristic* 0.

PROOF. First we treat the case $\dim(R) = 1$. Let κ be the residue field of the R-finite normalization $\widetilde{R}$ of R and let $e = [\kappa : \mathbb{F}_q] \geqslant 1$. Clearly $\mathrm{Fr}_{B_\kappa,\#\kappa} = \mathrm{Fr}_{B,q}^e$ in $\mathcal{O}_L$ since κ has size q^e. Thus, the Shimura–Taniyama formula for $A_{\widetilde{R}}$ over the Dedekind $\widetilde{R}$ (equivalently, for B_κ) implies the same for A (equivalently, for B).

The general case can be reduced to the case of a 1-dimensional base by using an argument with universal deformation rings similar to what was done in 2.1.1, but here is a more concrete argument via direct limit and specialization arguments. Let K denote the fraction field of R, let $\mathcal{O} \subseteq \mathcal{O}_L$ be an order that acts on A over R, and let $\mathcal{O}' \subseteq \overline{\mathbb{Q}}_p$ be a p-adic integer ring that is large enough to contain $\varphi(\mathcal{O}_L)$ for all $\varphi \in \Phi$. Observe that not only is $\overline{K}$ the direct limit of finite separable (i.e., étale) extension fields of K, but it is also a direct limit of R-subalgebras that are finite étale over localizations $R[1/r_i]$. Thus, by applying direct limit arguments to the eigenline decomposition of $\mathrm{Lie}(A) \otimes_R \overline{K}$ over $\overline{K}$ via Φ, there is a non-zero $r \in R$ and an $\mathcal{O}'$-subalgebra $R' \subseteq \overline{K}$ that is finite étale over $R[1/r]$ and contains $\mathcal{O}'$ such that $\mathrm{Lie}(A) \otimes_R R'$ decomposes into a direct sum of free R'-modules of rank 1 on which $\mathcal{O}$ acts by the maps $\varphi : \mathcal{O} \to \mathcal{O}' \subseteq R'$. We can assume that $p|r$ in R, so $\mathrm{Spec}(R[1/r])$ is a dense open subscheme of $\mathrm{Spec}(R[1/p])$. Since $\mathrm{Spec}(R[1/p])$ is Jacobson (e.g., due to [**73**, Lemma 1], or [**54**, 7.1.9] and the Jacobson property of affinoid algebras [**9**, 5.2.6/3]), we can therefore choose a maximal ideal $\mathfrak{n} \in \mathrm{MaxSpec}(R[1/p])$ that does not contain r. Let $\mathfrak{p}$ be the corresponding prime ideal of R.

The local noetherian quotient domain $R_0 = R/\mathfrak{p}$ is clearly contained in the valuation ring of the finite extension $R_0[1/p] = R[1/r]/\mathfrak{n}$ of $\mathrm{Frac}(W(\mathbb{F}_q))$, so it is an order in a p-adic integer ring. The quotient $R'/\mathfrak{p}$ is finite étale over R_0, so upon choosing a local factor ring of $R'/\mathfrak{p}$ we see that $A \bmod \mathfrak{p}$ over R_0 equipped with its action by $\mathcal{O}$ has a CM type that is naturally identified with Φ and has special fiber A_0 over $\mathbb{F}_q$ (compatibly with the $\mathcal{O}$-actions). Hence, the Shimura–Taniyama formula for A over R follows from the settled case of the formula for $A \bmod \mathfrak{p}$ over the 1-dimensional R_0. □

2.1.4.3. An interesting feature of the Shimura–Taniyama formula (2.1.4.1) is that it determines the element $\mathrm{Fr}_{B,q} \in \mathcal{O}_L$ up to multiplication by a root of unity when B is L-linearly isogenous to the reduction of an L-linear A as above over R with $\overline{K}$-fiber of CM type (L, Φ). To see this, we first note that since the constant term of the characteristic polynomial in $\mathbb{Z}[T]$ associated to B is equal to q, $\mathrm{Fr}_{B,q}$ is a unit at all finite places of $\mathcal{O}_L$ away from p. The Shimura–Taniyama formula provides its valuation at the p-adic places of L in terms of the CM type Φ on L arising from $A_{\overline{K}}$. Thus, $\mathrm{Fr}_{B,q}$ is uniquely determined up to an $\mathcal{O}_L^\times$-multiple. The archimedean absolute values of $\mathrm{Fr}_{B,q}$ are known to be $\sqrt{q}$ by the Riemann Hypothesis for abelian varieties over finite fields, so the only possible unit multiplier ambiguity is by an element $u \in \mathcal{O}_L^\times$ whose archimedean absolute values are all equal to 1, which is to say that u lies in the finite group $\mu(L)$ of roots of unity in L.

The intervention of $\mu(L)$-multiplication cannot be eliminated in general, due to Galois twisting when $R = \mathcal{O}_K$ (with residue field $\mathbb{F}_q$), as we now explain. Adjust A_K (hence A) and B by L-linear isogenies as in Proposition 1.7.4.5 so that $\mathcal{O}_L \subset \mathrm{End}(B)$ and $\mathcal{O}_L \subset \mathrm{End}(A_K) = \mathrm{End}(A)$. For any $\zeta \in \mu(L)$ we can make the $\mathcal{O}_L$-linear twist B_ζ of B via the finite-order character $\chi : \mathrm{Gal}(\overline{\mathbb{F}}_p/\mathbb{F}_q) \to \mu(L) \subset \mathrm{Aut}(B)$ carrying the arithmetic Frobenius element to ζ (where $\overline{\mathbb{F}}_p$ is the residue field of $\overline{K}$). Clearly the q-Frobenius in $\mathcal{O}_L$ for B_ζ is $\zeta \cdot \mathrm{Fr}_{B,q}$. The Galois character χ lifts to a finite-order unramified character of $\mathrm{Gal}(\overline{K}/K)$ carrying the arithmetic Frobenius to ζ, and twisting by this defines an analogous $\mathcal{O}_L$-linear twist A_ζ of A over R whose $\overline{K}$-fiber has the same CM type Φ on L and whose reduction is L-linearly isogenous to B_ζ. Thus, just from the data of the CM type on the $\overline{K}$-fiber we can only determine $\mathrm{Fr}_{B,q}$ to within multiplication by a root of unity in L, and all elements of a $\mu(L)$-coset in $\mathcal{O}_L$ arise in this way using the same (L, Φ) when one such element does.

2.1.5. Residual reflex condition. The considerations in 2.1.4 show that if (IN) in 1.8.5 holds for an isotypic abelian variety B over $\mathbb{F}_q$ with dimension $g > 0$, then there is a CM subfield $L \subset \mathrm{End}^0(B)$ with $[L : \mathbb{Q}] = 2g$ and a CM type $\Phi \subset \mathrm{Hom}_{\mathrm{ring}}(L, \overline{\mathbb{Q}}_p)$ such that (L, Φ) satisfies the following *residual reflex condition*:

(i) The slopes of B are given in terms of (L, Φ) by the Shimura–Taniyama formula (2.1.4.1).
(ii) Let $E \subset \overline{\mathbb{Q}}_p$ be the reflex field attached to (L, Φ), and let v be the induced p-adic place of E. The residue field κ_v of $\mathcal{O}_{E,v}$ can be realized as a subfield of $\mathbb{F}_q$.

For example, if $g = 1$ then p splits in L if and only if the elliptic curve B is ordinary, so it is straightforward to check that the residual reflex condition is satisfied for one choice (resp. both choices) of Φ when B is ordinary (resp. supersingular). That is, for elliptic curves over finite fields the residual reflex condition can be satisfied for any imaginary quadratic subfield of the endomorphism algebra.

Condition (i) in the residual reflex condition is not a serious constraint because for *any* CM subfield $L \subset \mathrm{End}^0(B)$ with $[L : \mathbb{Q}] = 2g$ (which exists, by Corollary 1.6.2.5) there is a $\overline{\mathbb{Q}}_p$-valued CM type Φ on L for which (i) holds. Indeed, by Remark 1.6.5.2, there is a p-adic field $K \subset \overline{\mathbb{Q}}_p$ with residue field $\mathbb{F}_{q'} \supset \mathbb{F}_q$ and a member B' of the isogeny class of $B_{\mathbb{F}_{q'}}$ such that B' admits an L-linear lift to an abelian scheme $\mathcal{B}'$ over $\mathcal{O}_K$. Applying the Shimura–Taniyama formula to $\mathcal{B}'_K$, the $\overline{\mathbb{Q}}_p$-valued CM type Φ on L arising from $\mathcal{B}'_K$ satisfies (2.1.4.1) (as $q' \in q^{\mathbb{Z}}$).

Condition (ii) is more interesting because in 2.3.5 we will use Honda-Tate theory to give absolutely simple examples such that $L := \mathrm{End}^0(B)$ is a CM field of degree $2g$ and every $\overline{\mathbb{Q}}_p$-valued CM type Φ on L satisfying the Shimura–Taniyama formula violates (ii). Moreover, we will show by elementary methods (i.e., not requiring the general case in Chapter 4) that these examples satisfy (I) in 1.8.5.

Thus, the (second part of the) residual reflex condition is a non-trivial obstruction to the validity of (IN) in 1.8.5 for a given triple (B, L, Φ) over $\mathbb{F}_q$, and this obstruction is eliminated by passing to a finite extension $\mathbb{F}_{q'} \supset \mathbb{F}_q$ if and only if $\mathbb{F}_{q'}$ is large enough to contain the residue field κ_v as in (ii). In our absolutely simple counterexamples to (IN) that satisfy (I), B is an ℓ-dimensional abelian variety over $\mathbb{F}_p$ with two distinct slopes over $\mathbb{F}_p$, where $\ell \geqslant 5$ is an arbitrary prime and p lies in a certain non-empty set of congruence classes depending on ℓ.

It is natural to ask whether there are obstructions to (IN) other than the residual reflex condition. We show in 2.5 that the combinatorial residual reflex condition is the only obstruction for (IN). This amounts to a refinement of the general validity of (RIN) from 1.8.5 (see Remark 1.6.5.2), and it is the main result of this chapter:

2.1.6. Theorem. *Let B be an isotypic abelian variety of dimension $g > 0$ over $\mathbb{F}_q$ and let $L \subset \mathrm{End}^0(B)$ be a* CM *field with $[L : \mathbb{Q}] = 2g$. Let $\Phi \subset \mathrm{Hom}_{\mathrm{ring}}(L, \overline{\mathbb{Q}}_p)$ be a* CM *type, and let $E \subset \overline{\mathbb{Q}}_p$ be the associated reflex field. Assume that (L, Φ) satisfies the residual reflex condition in* 2.1.5.

There exists a finite extension E'/E inside $\overline{\mathbb{Q}}_p$, a g-dimensional abelian variety A over E' with good reduction at the p-adic place v' on E' induced by $\overline{\mathbb{Q}}_p$, and an inclusion $L \hookrightarrow \mathrm{End}^0(A)$ with associated $\overline{\mathbb{Q}}_p$-valued CM *type Φ such that the reduction of A at v' is L-linearly isogenous to B over an isomorphism of finite fields $\kappa_{v'} \simeq \mathbb{F}_q$. In particular, B satisfies* (IN) *from* 1.8.5 *using a lifting of the L-action over a p-adic integer ring with residue field $\mathbb{F}_q$.*

Note that the choice of isomorphism $\kappa_{v'} \simeq \mathbb{F}_q$ in Theorem 2.1.6 does not matter, since B is L-linearly isogenous to $B^{(p)}$ via the relative Frobenius isogeny (and the natural isomorphism $B^{(q)} \simeq B$ is L-linear). The main ingredient in the proof of Theorem 2.1.6 is a purely arithmetic result in Theorem 2.4.11 that gives a procedure to modify algebraic Hecke characters. (A review of algebraic Hecke characters is given in 2.4.) The arithmetic theory of complex multiplication will tell us which algebraic Hecke characters comes from abelian varieties with complex multiplication (see Theorem 2.5.2), so the algebraic Hecke character produced by Theorem 2.4.11 will give us a CM abelian variety over a number field E'/E. This will be a CM lifting of B (up to isogeny) as required in Theorem 2.1.6.

After assembling the necessary background in 2.2-2.4, the statement of Theorem 2.1.6 will be recalled again as Theorem 2.5.3, after which we give the proof. In 2.6 we give an alternative proof of Theorem 2.1.6 that uses a local analogue of Theorem 2.4.11 (see Lemma 2.6.1) and replaces the global arithmetic theory of complex multiplication with local arguments resting on p-divisible groups and p-adic Hodge theory.

The preceding considerations, including the formulation of the residual reflex condition, can be generalized to the case of abelian varieties with sufficiently many complex multiplications by a CM algebra (i.e., a product of finitely many CM fields). We leave it to the reader to give a precise statement of this generalization because the proofs are immediately reduced to the case when the CM algebra is a CM field.

2.1.7. Example. To illustrate the sensitivity of the residual reflex condition to the choice of $L \subset \mathrm{End}^0(B)$, let $B = E \times E$ over $\mathbb{F}_{p^2}$ where E is an elliptic curve with p^2-Frobenius $\pi_E = -p$. By Corollary 1.6.2.2(1), such an E exists and its geometric endomorphisms are defined over $\mathbb{F}_{p^2}$, with $D := \mathrm{End}^0(E)$ the unique quaternion division algebra over $\mathbb{Q}$ ramified at exactly p and ∞. In particular, B has endomorphism algebra $\mathrm{Mat}_2(D)$.

Let L_1 be a quartic CM field in which p is totally ramified with associated completion $(L_1)_p = \mathbb{Q}_p(p^{1/4})$. Such an L_1 is easy to construct via weak approximation. Let L_2 be a $\mathbb{Z}/(4)$-extension of $\mathbb{Q}$ that is CM and in which p is totally inert (i.e., Frob_p generates the Galois group). Such fields L_2 are easy to construct.

In both cases there is only one p-adic place on L_i, and D_{L_i} splits away from that place (the archimedean places being clear since L_i is totally complex), so $\mathrm{Mat}_2(D)$ is split by each L_i. Hence, by Proposition 1.2.3.1 each L_i embeds $\mathrm{Mat}_2(D)$ as a maximal commutative subfield. The only CM subfield of L_2 is itself, at which the p-adic place has residue field $\mathbb{F}_{p^4}$, so part (ii) of the residual reflex condition is violated and (IN) fails for (B, L_2). On the other hand, the Galois closure of L_1 over $\mathbb{Q}$ has p-adic completions $\mathbb{Q}_p(p^{1/4}, \sqrt{-1})$ that are at most quadratic over $(L_1)_p$, and this Galois closure contains the reflex field E of any CM type on L_1. Thus, the p-adic places of any such E embed into $\mathbb{F}_{p^2}$, so by Theorem 2.1.6 it follows that (B, L_1) satisfies (IN).

2.2. Existence of CM lifting up to isogeny

Let B be an isotypic abelian variety of dimension $g > 0$ over a finite field $\mathbb{F}_q$ with size q that is a power of a prime p. In this section we give a sufficient criterion for the condition (I) in 1.8.5 to be satisfied by B. (The general case is treated in Chapter 4.) By Corollary 1.6.2.5, there exists a CM subfield $L \subset \mathrm{End}^0(B)$ with $[L : \mathbb{Q}] = 2g$. Such an L is its own centralizer in $\mathrm{End}^0(B)$, so there is an element $\pi \in O_L$ whose action on B is the q-Frobenius endomorphism $\mathrm{Fr}_{B,q}$. As we noted in 2.1.5, there exists a $\overline{\mathbb{Q}}_p$-valued CM type Φ on L that satisfies the Shimura–Taniyama formula: for each p-adic place v of L,

$$\frac{\mathrm{ord}_v(\pi)}{\mathrm{ord}_v(q)} = \frac{\#\{\phi \in \Phi \,|\, \phi \text{ induces } v \text{ on } L\}}{[L_v : \mathbb{Q}_p]}.$$

Fix such a Φ.

We want to prove that (I) holds for B using the CM type (L, Φ), subject to a restriction on the behavior of p in L (given in Proposition 2.2.1 below). Let $O = L \cap \mathrm{End}(B)$ be the associated CM order in O_L. Since O_L is a finite O-algebra with the same fraction field as O, by Proposition 1.7.4.5 there is an L-linear isogeny $B \to B'$ where B' has CM order equal to O_L. Thus, upon replacing B with B' it suffices to consider the case when $O = O_L$.

2.2.1. Proposition. *Let B be an isotypic abelian variety of dimension $g > 0$ over a finite field $\mathbb{F}_q$ of characteristic p, and let L be a* CM *field of degree $2g$ equipped with an embedding of its maximal order O_L into* $\mathrm{End}(B)$. *Assume $O_L/\mathfrak{p} = \mathbb{F}_p$ for every prime $\mathfrak{p}$ of L over p.*

There exists an order R in a p-adic integer ring and an abelian scheme A over R equipped with an action by O_L having CM *type (L, Φ) as above such that R has residue field $\mathbb{F}_q$ and $A_{\mathbb{F}_q}$ is O_L-linearly isomorphic to B. In particular, B satisfies* (CML) *and hence satisfies* (I).

Some hypothesis on L at p is required in this proposition, since otherwise Remark 2.3.4 gives counterexamples via a tangential obstruction (the Kottwitz invariant). Beware that this does not imply that such examples cannot satisfy (I), since we may just need to pass to another member of the L-linear $\mathbb{F}_q$-isogeny class (perhaps with non-maximal CM order) to get the required lift over a possibly non-normal base; this is what will be done in Chapter 4.

In Corollary 4.6.5.4(ii) a mild generalization of Proposition 2.2.1 will be given.

PROOF. By Remark 1.6.5.2, we may find a number field $K \subset \overline{\mathbb{Q}}_p$ and a CM abelian variety A_1 over K with good reduction at the induced p-adic place w on K (and reduction at w denoted $\overline{A}_1$) such that: the CM type of (A_1, L) is Φ, the CM order is O_L, and there exists an embedding $\mathbb{F}_q \hookrightarrow \kappa_w$ and an O_L-linear isogeny $\phi_1 : \overline{A}_1 \to B_{\kappa_w}$. The kernel $\ker\phi_1$ is a finite O_L-submodule scheme of the g-dimensional $\overline{A}_1$.

We claim that there is a (unique) non-zero ideal $I \subseteq O_L$ such that $\ker\phi_1 = \overline{A}_1[I]$. This is obvious on ℓ-primary parts for $\ell \neq p$ since the ℓ-adic Tate module $T_\ell(\overline{A}_1)$ is an invertible module over $O_{L,\ell} = \prod_{\mathfrak{q}|\ell} O_{L,\mathfrak{q}}$. The case of p-primary parts requires more care because the (contravariant) Dieudonné module $\mathrm{M}^*(\overline{A}_1[p^\infty])$ is a module over $W(\kappa_w) \otimes_{\mathbb{Z}_p} \mathcal{O}_{L,p}$ rather than over $\mathcal{O}_{L,p}$. The natural map

$$L_p := \mathbb{Q}_p \otimes_{\mathbb{Q}} L \hookrightarrow \mathbb{Q}_p \otimes_{\mathbb{Q}} \mathrm{End}(\overline{A}_1) \to \mathrm{End}(\overline{A}_1[p^\infty])$$

is injective (Proposition 1.2.5.1), so by Proposition 1.4.3.9 (with $F = L_p$, $\kappa_1 = \kappa = \kappa_w$) we see that $\mathrm{M}^*(\overline{A}_1[p^\infty])$ is invertible over $W(\kappa) \otimes_{\mathbb{Z}_p} \mathcal{O}_{L,p}$.

Since $\mathrm{M}^*(\ker\phi_1)$ is a quotient of $\mathrm{M}^*(\overline{A}_1[p^\infty])$ as modules over the Dieudonné ring D_{κ_w} over $W(\kappa_w)$, the $W(\kappa_w) \otimes_{\mathbb{Z}_p} \mathcal{O}_{L,p}$-module invertibility of $\mathrm{M}^*(\overline{A}_1[p^\infty])$ implies that the kernel of the projection to this quotient is $J \cdot \mathrm{M}^*(\overline{A}_1[p^\infty])$ for a uniquely determined ideal $J \subseteq W(\kappa_w) \otimes_{\mathbb{Z}_p} \mathcal{O}_{L,p}$ that is invertible (since $\mathrm{M}^*(\ker\phi_1)$ has finite length over $W(\kappa_w)$). Each factor $\mathcal{O}_{L,\mathfrak{p}}$ of $\mathcal{O}_{L,p}$ has residue field $\mathbb{F}_p$, so each factor $W(\kappa_w) \otimes_{\mathbb{Z}_p} \mathcal{O}_{L,\mathfrak{p}}$ of $W(\kappa_w) \otimes_{\mathbb{Z}_p} \mathcal{O}_{L,p}$ is a local *unramified* extension of $\mathcal{O}_{L,\mathfrak{p}}$. Hence, J uniquely descends to an invertible ideal $I_p \subseteq \mathcal{O}_{L,p} = \prod_{\mathfrak{p}|p} \mathcal{O}_{L,\mathfrak{p}}$. This provides the p-part of the desired ideal $I \subseteq \mathcal{O}_L$, so we have constructed I.

Let $\mathcal{A}_1$ denote the Néron model of A_1 over the complete local ring $\mathcal{O}_{K,w}$ of $\mathcal{O}_K$ at w, so $\mathcal{A}_1$ is an abelian scheme over $\mathcal{O}_{K,w}$ and the $\mathcal{O}_L$-action on A_1 uniquely extends to an action on $\mathcal{A}_1$. Since the algebraic localization $\mathcal{O}_{L,(\ell)}$ is a principal ideal domain for every rational prime ℓ, functorial consideration of primary parts yields that the torsion subscheme $\mathcal{A}_1[I]$ is a finite flat $\mathcal{O}_{K,w}$-subgroup scheme of $\mathcal{A}_1$ (even if I is not globally principal). Thus, $\mathcal{A}_2 = \mathcal{A}_1/(\mathcal{A}_1[I])$ is an abelian scheme over $\mathcal{O}_{K,w}$ endowed with a natural action by $\mathcal{O}_L$. Moreover, the special fiber $\overline{A}_2$ of $\mathcal{A}_2$ is $\mathcal{O}_L$-linearly isomorphic to $\overline{A}_1/(\overline{A}_1[I]) = \overline{A}_1/(\ker\phi_1) \simeq B_{\kappa_w}$. Hence, by replacing A_1 with $A_1/(A_1[I])$ we can arrange that ϕ_1 is even an isomorphism. That is, $\mathcal{A}_1$ is an $\mathcal{O}_L$-linear deformation of B_{κ_w}. A polarization λ_1 of $\mathcal{A}_1$ over $\mathcal{O}_{K,w}$ induces a polarization on B_{κ_w}, but to get such a λ_1 whose reduction over κ_w descends to B over $\mathbb{F}_q$ it is convenient (and also useful for later purposes in the proof of Theorem 2.2.3) to construct polarizations that interact well with the $\mathcal{O}_L$-action in more general situations:

2.2.2. Lemma. *Let C be an abelian variety of dimension $g > 0$ over an arbitrary field K, and let L be a* CM *field of degree $2g$ over $\mathbb{Q}$ equipped with an action on C in the isogeny category over K. Let $L^+ \subset L$ be the maximal totally real subfield, and give C^t the dual* CM *structure as in* 1.5.3.1.

(1) *There exists an L-linear polarization $\lambda : C \to C^t$.*
(2) *Assume $\mathcal{O}_L \subset \mathrm{End}(C)$. The $\mathcal{O}_{L^+}$-module $\mathrm{Hom}^{\mathrm{sym}}_{\mathcal{O}_L}(C, C^t)$ of symmetric $\mathcal{O}_L$-linear homomorphisms over K is invertible, and the $\mathcal{O}_L$-module $\mathrm{Hom}_{\mathcal{O}_L}(C, C^t)$ of $\mathcal{O}_L$-linear homomorphisms over K is invertible. The formation of these modules is unchanged by scalar extension on K.*

As we explained after Definition 1.5.3.1, when $\mathrm{char}(K) = 0$ the dual CM structure on the dual abelian variety has the same CM type.

PROOF. By Proposition 1.7.4.5 and the functorial properties of polarizations, for the proof of (1) we may assume $\mathcal{O}_L \subset \mathrm{End}(C)$. Hence, we may and do make this assumption for the remainder of the argument. Also, once the invertibility properties in (2) are established, it follows that for any extension field K' of K the injective maps

$$\mathrm{Hom}^{\mathrm{sym}}_{\mathcal{O}_L}(C, C^t) \to \mathrm{Hom}^{\mathrm{sym}}_{\mathcal{O}_L}(C_{K'}, C^t_{K'}), \ \ \mathrm{Hom}_{\mathcal{O}_L}(C, C^t) \to \mathrm{Hom}_{\mathcal{O}_L}(C_{K'}, C^t_{K'})$$

are finite-index inclusions. Hence, if $f' : C_{K'} \to C^t_{K'}$ is an $\mathcal{O}_L$-linear map then nf' descends to a K-homomorphism for some integer $n > 0$. But this descent $C \to C^t$ kills $C[n]$ (as we may check over K'), so it has the form nf for a K-homomorphism $f : C \to C^t$. Obviously f is a descent of f', so it inherits the $\mathcal{O}_L$-linearity from f' (as well as symmetry if f' is symmetric).

Consider the $\mathcal{O}_L$-module $\mathrm{Hom}_{\mathcal{O}_L}(C, C^t)$ of $\mathcal{O}_L$-linear maps over K. The additive endomorphism $h \mapsto h' := h^t \circ \iota_C$ of $\mathrm{Hom}_{\mathcal{O}_L}(C, C^t)$ (where $\iota_C : C \simeq C^{tt}$ is the canonical isomorphism) is semi-linear over complex conjugation on $\mathcal{O}_L$ due to how the $\mathcal{O}_L$-action on C^t is defined, and its fixed set is the $\mathcal{O}_{L^+}$-module $\mathrm{Hom}^{\mathrm{sym}}_{\mathcal{O}_L}(C, C^t)$. Moreover, $h'' = h$ for all h because ι_{C^t} is inverse to ι^t_C, so to prove the two invertibility claims it suffices to prove that $\mathrm{Hom}_{\mathcal{O}_L}(C, C^t)$ is an invertible $\mathcal{O}_L$-module. It is clear that $\mathrm{Hom}_{\mathcal{O}_L}(C, C^t)$ is a finite flat $\mathcal{O}_L$-module, and by elementary rank considerations with ℓ-adic Tate modules for $\ell \neq \mathrm{char}(K)$ we see that its rank is at most 1. Thus, invertibility is reduced to being non-zero.

Step 1. Consider K with characteristic 0. For an algebraic closure $\overline{K}$ of K, we can descend the CM abelian variety $C_{\overline{K}}$ (with its CM structure) to a number field (Theorem 1.7.2.1), so there is a finite Galois extension K'/K such that $C_{K'}$ with its CM structure descends to a CM abelian variety (A, α) over a number field $F \subseteq K'$. The complex-analytic theory (Example 1.5.3) provides an L-linear polarization on a complex fiber of A (due to how we define the L-action on A^t in the dual CM structure; see 1.5.3.1). By Lemma 1.2.1.2 this descends to the corresponding $\overline{\mathbb{Q}}$-fiber of A, and so also to a finite Galois extension F'/F. Thus, by increasing K' to split F' over F, we can find an L-linear polarization $\lambda' : C_{K'} \to C^t_{K'}$ of $C_{K'}$. The sum of the $\mathrm{Gal}(K'/K)$-conjugates of the L-linear map λ' descends to the desired polarization over K, and in particular this is a non-zero symmetric $\mathcal{O}_L$-linear map.

Step 2. For K with characteristic $p > 0$ we will proceed in reverse, first proving that $\mathrm{Hom}_{\mathcal{O}_L}(C, C^t)$ is non-zero and then afterwards using this to infer the existence of an L-linear polarization. Since C must be isotypic (Theorem 1.3.1.1), $\mathrm{End}^0(C)$ is a simple $\mathbb{Q}$-algebra. Its center is identified with a subfield $Z \subseteq L$, and the image $Z^* \subseteq L$ of Z under complex conjugation on L is identified with the center of $\mathrm{End}^0(C^t)$ when using the action of L on C^t through duality and complex conjugation.

Let $h : C \to C^t$ be a polarization. The associated Rosati involution on $\mathrm{End}^0(C)$ restricts to an involution τ (perhaps trivial) on the center Z. By the proof of Lemma 1.3.5.4, either Z is totally real and τ is the identity or Z is a CM field and τ is its complex conjugation. Either way, $Z^* = Z$ inside L and the Rosati involution is complex conjugation on Z. This says that h is Z-linear via the definition of the L-action on C^t. It also follows that the center of $\mathrm{End}^0(C^t)$ is the subfield Z of $L \subseteq \mathrm{End}^0(C^t)$.

The finite-dimensional $\mathbb{Q}$-algebras $\mathrm{End}^0(C)$ and $\mathrm{End}^0(C^t)$ are simple $\mathbb{Q}$-algebras (by K-isotypicity of C) in which the center is $Z \subseteq L$, so the Z-linear h induces a Z-algebra isomorphism $\psi : \mathrm{End}^0(C) \simeq \mathrm{End}^0(C^t)$. This may not be L-linear, but we can use the Skolem-Noether theorem to find a unit $u \in \mathrm{End}^0(C)^\times$ such that composing ψ with conjugation by u is an L-algebra isomorphism. Thus, after replacing u with a non-zero integral multiple that lies in $\mathrm{End}(C)$, $h \circ u$ is an $\mathcal{O}_L$-linear isogeny $C \to C^t$ (perhaps not a polarization). So far, $\mathrm{char}(K)$ could have been arbitrary.

Step 3. Now we construct L-linear polarizations (over K) in characteristic $p > 0$. We can assume that K is finitely generated over $\mathbb{F}_p$, so K is the function field of a finite type integral affine scheme U over the finite field κ that is the algebraic closure of $\mathbb{F}_p$ in K. By shrinking U we can arrange that the CM abelian variety C over K spreads out to a CM abelian scheme $\mathcal{C}$ over U. Since $\mathrm{Hom}(C, C^t)$ is finitely generated, by further shrinking we may assume that every homomorphism $C \to C^t$ over $K = \kappa(U)$ extends (uniquely) to a U-homomorphism $\mathcal{C} \to \mathcal{C}^t$, and so for a choice of closed point $u \in U$ we get a specialization map

$$\mathrm{Hom}(C, C^t) \to \mathrm{Hom}(\mathcal{C}_u, \mathcal{C}_u^t)$$

that is injective (by the discussion at the start of 1.4.4). This induces an injection between the invertible $\mathcal{O}_L$-modules of $\mathcal{O}_L$-linear homomorphisms, so this latter injection is a finite-index inclusion. Hence, if the case of finite fields is settled then for a choice of L-linear polarization λ_u of $\mathcal{C}_u$ over the finite field $\kappa(u)$ then we can replace λ_u with a positive integral multiple so that it arises from an L-linear map $\lambda : C \to C^t$ over $K = \kappa(U)$. By Lemma 2.1.1.1 such a λ is a polarization because its extension over U specializes to a polarization on some fiber (namely, over u).

It remains to construct an L-linear polarization of C when K is a finite field κ, and it suffices to do this after replacing κ with a finite extension κ' (either due to the same finite-index inclusion trick as above, or by the more elementary Galois descent trick used in characteristic 0). By Remark 1.6.5.2, after making a finite extension on κ we can therefore assume that there is an L-linear κ-isogeny $f : C \to C'$ onto the reduction C' of an abelian scheme $\mathcal{C}'$ over the valuation ring of a p-adic field F with residue field κ such that $\mathcal{C}'_F$ has CM by L lifting that on the reduction (but its CM order may not be $\mathcal{O}_L$).

By Proposition 1.7.4.5 we can choose an L-linear isogeny $\mathcal{C}'_F \to \mathcal{C}''_F$ onto an abelian variety $\mathcal{C}''_F$ over F having CM order $\mathcal{O}_L$, and since $\mathcal{O}_F$ is Dedekind this extends to an L-linear isogeny $\mathcal{C}' \to \mathcal{C}''$ onto an abelian scheme $\mathcal{C}''$ over $\mathcal{O}_F$ with CM order $\mathcal{O}_L$. We may compose f with the reduction of this latter isogeny to get to the case when C' has CM order equal to $\mathcal{O}_L$. By functoriality properties of polarizations we can rename C' as C, which is to say that we can assume that C equipped with its $\mathcal{O}_L$-action is identified with the reduction of an abelian variety $\mathcal{C}_F$ equipped with an $\mathcal{O}_L$-action over a p-adic field F with residue field κ. Since $\mathrm{char}(F) = 0$ there exists an L-linear polarization of $\mathcal{C}_F$, and any such polarization has reduction on C that is the desired L-linear polarization over κ (again using Lemma 2.1.1.1). □

To apply Lemma 2.2.2 in the proof of Proposition 2.2.1, let $\lambda_1 : A_1 \to A_1^t$ be an $\mathcal{O}_L$-linear polarization as in the setup preceding the statement of Lemma 2.2.2. Its K_w-fiber extends to an $\mathcal{O}_L$-linear homomorphism $\lambda_{1,w} : \mathcal{A}_1 \to \mathcal{A}_1^t$ over $\mathcal{O}_{K,w}$ that is a polarization (Lemma 2.1.1.1). This reduces to an $\mathcal{O}_L$-linear polarization $\overline{\lambda}_1$ on B_{κ_w}. But the inclusion

$$\mathrm{Hom}_{\mathcal{O}_L}(B, B^t) \to \mathrm{Hom}_{\mathcal{O}_L}(B_{\kappa_w}, B^t_{\kappa_w})$$

between invertible $\mathcal{O}_L$-modules must be a finite-index inclusion, so by replacing λ_1 with a positive integral multiple we can assume that the polarization $\overline{\lambda}_1$ over κ_w descends to a polarization λ_B of B. (This λ_B is also $\mathcal{O}_L$-linear, but that will not be needed.)

Consider the deformation functor $\mathrm{Def}(B, i_B, \lambda_B)$ attached to the polarized abelian scheme (B, λ_B) equipped with the $\mathcal{O}_L$-action via the canonical inclusion $i_B : \mathcal{O}_L \to \mathrm{End}(B)$. This is the subfunctor of $\mathrm{Def}(B, \lambda_B)$ defined by requiring that the $\mathcal{O}_L$-action on B lifts (necessarily uniquely) to the deformation of B. By Proposition 1.4.4.14 (especially its final assertion), the deformation functor $\mathrm{Def}(B, i_B, \lambda_B)$ considered on the category of *all* complete local noetherian $W(\mathbb{F}_q)$-algebras (with residue field possibly larger than $\mathbb{F}_q$, such as κ_w) is represented by a complete local noetherian $W(\mathbb{F}_q)$-algebra $\mathcal{R}$ with residue field $\mathbb{F}_q$.

The polarized abelian scheme over $\mathcal{O}_{K,w}$ provided by $\mathcal{A}_1$ equipped with its $\mathcal{O}_L$-action and polarization $\lambda_{1,w}$ (and residual isomorphism ϕ_1 giving the deformation structure with respect to (B, i_B, λ_B) over $\mathbb{F}_q$) is classified by a local $W(\mathbb{F}_q)$-algebra homomorphism $h : \mathcal{R} \to \mathcal{O}_{K,w}$. Let $R = \mathcal{R}/(\ker h)$, so R is an order in a p-integer ring contained in $\mathcal{O}_{K,w}$ such that R has residue field $\mathbb{F}_q$ inside κ_w. The pullback to $\mathrm{Spec}(R)$ of the universal deformation over $\mathcal{R}$, equipped with its pullback $\mathcal{O}_L$-action and residual isomorphism to B over $\mathbb{F}_q$, is the desired CM lifting of (B, i_B) to characteristic 0 without increasing the residue field (and it has the same CM type Φ as does $\mathcal{A}_1$). This completes the proof of Proposition 2.2.1. □

We finish this section by using Lemma 2.2.2 to prove a tangential necessary and sufficient criterion for the algebraicity of a formal abelian scheme with complex multiplication. This criterion is used in our alternative local proof of Theorem 2.1.6 (in 2.6), as well as in the proof of the main result in Chapter 4 (to reduce the CM lifting problem for abelian varieties to the case of p-divisible groups without losing contact with algebraicity in the limit).

2.2.3. Theorem. *Let R be a $\mathbb{Z}_p$-flat 1-dimensional complete local noetherian ring with residue characteristic p, and let $\mathfrak{A}$ be a formal abelian scheme over R of relative dimension $g > 0$. Assume that the finite-dimensional $\mathbb{Q}$-algebra $\mathbb{Q} \otimes_{\mathbb{Z}} \mathrm{End}_R(\mathfrak{A})$ contains a* CM *subalgebra L with $[L : \mathbb{Q}] = 2g$. Then $\mathfrak{A}$ is algebraizable if and only if the $R[1/p]$-linear L-action on $\mathrm{Lie}(\mathfrak{A})[1/p]$ is given by a* CM *type over each geometric point of $\mathrm{Spec}(R[1/p])$. In such cases, the algebraization admits an ample line bundle giving rise to an L-linear polarization.*

Before we prove this theorem, we note that there are examples of p-adic integer rings R and pairs $(\mathfrak{A}, L)$ for which the tangential hypothesis is violated, so $\mathfrak{A}$ is not algebraizable. One way to make such examples is by using the deformation theory of formal modules, applied to the examples discussed in Remark 2.3.4. Also, in 4.1.2 we will build a non-algebraizable $\mathbb{Z}[\zeta_5]$-linear formal abelian surface $\mathfrak{A}$ over $W(\mathbb{F}_{p^4})$ for any $p \equiv \pm 2 \pmod 5$.

Any such $\mathfrak{A}$ that is non-algebraizable is interesting because the rigid-analytic generic fiber $G = \mathfrak{A}^{\mathrm{an}}$ is a smooth, proper, connected rigid-analytic group over a p-adic field such that G admits complex multiplication in the evident sense but is not algebraic (and does not become so after any finite extension on the base field). This is a striking contrast with the classical fact that every complex-analytic torus of dimension g equipped with an action by an order in a CM field of degree $2g$ is an abelian variety (i.e., is algebraic).

PROOF. The necessity of the tangential condition is a consequence of the complex-analytic theory of abelian varieties. Thus, we focus on sufficiency.

Step 1. Let us first reduce to the case when R is a domain. Observe that R_{red} is $\mathbb{Z}_p$-flat. If $\mathfrak{A}\widehat{\otimes}_R R_{\mathrm{red}}$ admits a formally ample line bundle $\mathfrak{L}$ then by positivity of the residue characteristic there is an $m \geqslant 1$ (depending on the order of nilpotence of the nilradical of R) such that $\mathfrak{L}^{\otimes p^m}$ lifts to a line bundle $\mathfrak{N}$ on $\mathfrak{A}$. Such a lift is necessarily formally ample, and the associated formal polarization $\phi_{\mathfrak{N}}$ is L-linear if $\phi_{\mathfrak{L}}$ is. Hence, we can assume R is reduced. Assuming the domain case is settled, we shall now induct on the number of irreducible components via a gluing argument along closed subschemes.

Let $\{\mathfrak{p}_1, \ldots, \mathfrak{p}_n\}$ be the minimal primes of the reduced R, so $\bigcap \mathfrak{p}_i = (0)$. We may assume $n > 1$, as otherwise we are in the domain case that we are assuming is settled. For $J = \mathfrak{p}_1$ and $J' = \bigcap_{j>1} \mathfrak{p}_j$, the natural map $R \to (R/J) \times_{R/(J+J')} (R/J')$ is an isomorphism. (Note that R/J and R/J' are $\mathbb{Z}_p$-flat since R is $\mathbb{Z}_p$-flat.) We now check rather generally that if R is an adic noetherian ring and $\mathfrak{V}$ is any proper flat formal R-scheme with algebraizable pullbacks to $\mathrm{Spf}(R/J)$ and $\mathrm{Spf}(R/J')$ for ideals $J, J' \subseteq R$ such that $R \simeq (R/J) \times_{R/(J+J')} (R/J')$ then $\mathfrak{V}$ is algebraizable.

Let V and V' be the respective proper flat algebraizations of $\mathfrak{V} \bmod J$ and $\mathfrak{V} \bmod J'$ over R/J and R/J', so by formal GAGA there is a unique isomorphism f between their pullbacks over $R/(J + J')$ respecting the identification of the formal completions of such pullbacks with $\mathfrak{V} \bmod (J + J')$. By Proposition 1.4.4.11(3), the R-flatness allows us to "glue" along f to get a proper flat R-scheme W such that the associated formal scheme $\widehat{W}$ over $\mathrm{Spf}(R)$ is the analogous "flat gluing" of $\widehat{V} = \mathfrak{V} \bmod J$ and $\widehat{V'} = \mathfrak{V} \bmod J'$. Since $\mathfrak{V}$ is R-flat, the "flat gluing" of $\widehat{V}$ and $\widehat{V'}$ is identified with $\mathfrak{V}$. Hence, W is an algebraization of $\mathfrak{V}$. This gluing argument also behaves well with respect to polarizations if $\mathfrak{V}$ is a formal abelian scheme.

Step 2. We next reduce to the case when the 1-dimensional $\mathbb{Z}_p$-flat complete local noetherian domain R is a discrete valuation ring with algebraically closed residue field. Let $\{\overline{A}_i\}$ be the set of isotypic factors of the special fiber $\overline{A}$ over the residue field k of R, so $L = \prod L_i$ for CM algebras L_i of degree $2\dim(\overline{A}_i)$ through which the L-action on $\overline{A}_i$ factors. The decomposition $L_i = \prod_{j \in J_i} L_{ij}$ into a direct product of CM fields gives rise to a collection of abelian subvarieties $\overline{A}_{ij} \subset \overline{A}_i$ such that $\prod_{j \in J_i} \overline{A}_{ij} \to \overline{A}_i$ is an isogeny and the L_i-action on $\overline{A}_i$ goes through an L_{ij}-action on $\overline{A}_{ij}$ for each j, thereby defining a CM structure on each $\overline{A}_{ij}$ via a CM field. By applying Lemma 2.2.2 to the CM abelian varieties $\overline{A}_{ij}$ equipped with the action by the CM field L_{ij}, we can construct an L-linear polarization $\phi_{\mathcal{L}} : \overline{A} \to \overline{A}^t$ arising from an ample line bundle $\mathcal{L}$ on $\overline{A}$.

Let I be an $\mathfrak{m}_R$-primary ideal admitting topologically nilpotent divided powers, such as pR if $p > 2$ or $4R$ if $p = 2$. Since R/I is 0-dimensional with residue characteristic p, if we replace $\mathcal{L}$ with a suitable p-power then we can assume that $\phi_{\mathcal{L}}$ lifts to a homomorphism ϕ_0 over R/I. Let $R_n = R/I^{n+1}$ and $A_n = \mathfrak{A} \otimes_R R_n$ for $n \geqslant 0$. Granting the case when R is a discrete valuation ring, we claim that ϕ_0 over R_0 lifts to an R-homomorphism $\phi_\infty : \mathfrak{A} \to \mathfrak{A}^t$, thereby providing a formally ample line bundle on $\mathfrak{A}$ (the $(1, \phi_\infty)$-pullback of the formal Poincaré bundle), so Grothendieck's algebraization theorems [**34**, III_1, 5.4.1, 5.4.5] apply to $\mathfrak{A}$. (See Example 1.4.5.4.)

Let $\Gamma = \mathfrak{A}[p^\infty]$ and $\Gamma' = \mathfrak{A}^t[p^\infty] \simeq \Gamma^t$, and for $n \geqslant 0$ let Γ_n and Γ'_n denote the reductions modulo I^{n+1}. Note that $R_n \twoheadrightarrow R_0$ has kernel I/I^{n+1} with canonical nilpotent divided powers. By the Serre-Tate deformation theorem (Theorem 1.4.5.3), to make the desired R-lift of ϕ_0 it is equivalent to construct an R-lifting of the L-linear k-isogeny $\Gamma_0 \to \Gamma'_0$ induced by ϕ_0 on p-divisible groups. By Grothendieck-Messing theory [**75**] and the divided power hypothesis on I, the data of the deformation Γ of Γ_0 is covariantly functorially encoded as a pair (D^0, D)

where D is a finite projective R-module of rank $2g$ and D^0 is a finite projective R-submodule of rank g that is a direct summand. Explicitly, if we let D_n be the Lie algebra of the universal vector extension $E(\Gamma_n)$ of Γ_n by a vector group over R_n then naturally $D_{n+1} \otimes_{R_{n+1}} R_n \simeq D_n$ for all $n \geqslant 0$ and D is the inverse limit of the D_n's. The functoriality of D in Γ_0 is determined by that of each D_n via the above specification of divided powers on I. The R-submodule $D^0 \subseteq D$ is the inverse limit of the subbundles $D_n^0 = \omega_{\Gamma_n^t} \subseteq D_n$ given by the cotangent spaces along the identity to the dual p-divisible groups Γ_n^t. There is a similar such pair (D'^0, D') for Γ' as a deformation of Γ'_0, and the k-map ϕ_0 induces an R-linear map $[\phi_0] : D \to D'$. The existence of ϕ_∞ is equivalent to the condition that $[\phi_0]$ carries D^0 into D'^0.

Since the R-submodules $D^0 \subseteq D$ and $D'^0 \subseteq D'$ are subbundles over $\mathrm{Spec}(R)$, and R is a $\mathbb{Z}_p$-flat domain, it is equivalent to check the subbundle compatibility of $[\phi_0]$ after extending scalars to an algebraic closure $\overline{K}$ of the field $K = \mathrm{Frac}(R)$ of characteristic 0. In particular, it is harmless to replace ϕ_0 with $p^r\phi_0$ for any $r \geqslant 1$. By functoriality, the K-linear map $[\phi_0]_K : D_K \to D'_K$ is also L-linear. We claim that D_K and D'_K are invertible $K \otimes_{\mathbb{Q}} L$-modules and that the submodules D^0_K and D'^0_K are spanned over $\overline{K}$ by the 1-dimensional eigenspaces for a common set of g embeddings $L \to \overline{K}$. This will force $[\phi_0]_K$ to carry D^0_K into D'^0_K, as required.

The definitions of (D^0, D) and (D'^0, D'), as well as the action on these pairs by CM orders in L, are intrinsic to the system of universal vector extensions of Γ_n and Γ'_n over $\mathrm{Spec}(R_n)$ for all $n \geqslant 0$. This eliminates the need to pay attention to functoriality of these vector extensions with respect to the k-morphism ϕ_0 that we do not yet know to have a lifting, since we have reduced ourselves to module-theoretic properties of (D^0, D) and (D'^0, D') that have nothing to do with ϕ_0.

By considering K-ranks, the desired invertibility of D_K over $K \otimes_{\mathbb{Q}} L$ is equivalent to its faithfulness as a $K \otimes_{\mathbb{Q}} L$-module. Since R is a noetherian local domain of dimension 1, we can choose a local injection $R \to \mathcal{O}$ into a complete discrete valuation ring $\mathcal{O}$. This injects K into $\mathrm{Frac}(\mathcal{O}) = \mathcal{O}[1/p]$. The formation of the universal vector extension of a p-divisible group (over a base on which p is locally nilpotent) commutes with any base change, so it suffices to consider our linear algebra problems after the base change $\mathrm{Spf}(\mathcal{O}) \to \mathrm{Spf}(R)$. That is, we may assume that R is a discrete valuation ring, and we can also assume that the residue field k of R is algebraically closed. (Observe that this reduction step preserves the geometric tangential hypothesis whose sufficiency for algebraization we are trying to prove.)

Step 3. We now focus on proving that D_K is an invertible $K \otimes_{\mathbb{Q}} L$-module (so we ignore D^0). This will rest on comparison isomorphisms between Dieudonné theory and crystalline cohomology after inverting p (to avoid restrictions on the absolute ramification degree). By [**75**, Ch. V, 2.1], there is a canonical $R/(p^{n+1})$-linear isomorphism between the Lie algebras of the universal vector extensions of $\Gamma \bmod p^{n+1}$ and $\mathfrak{A} \bmod p^{n+1}$ for all $n \geqslant 0$, and this is compatible with change in n. But [**74**, Ch. I, 2.6.7, 3.2.3, 4.1.7, 4.2.1] gives a canonical $R/(p^{n+1})$-linear isomorphism

$$\mathrm{Lie}(E(\mathfrak{A} \bmod p^{n+1})) \simeq \mathrm{H}^1_{\mathrm{dR}}((\mathfrak{A}^t \bmod p^{n+1})/(R/(p^{n+1})))$$

compatibly with change in n, so passage to the inverse limit gives $D \simeq \mathrm{H}^1_{\mathrm{dR}}(\mathfrak{A}^t/R)$ due to the theorem on formal functions for hypercohomology of bounded C-linear complexes of coherent sheaves on a proper formal schemes over a complete local noetherian ring C. (This version of the theorem on formal functions is easily deduced from the special case of cohomology of coherent sheaves on proper formal schemes given in [**34**, III_1, 3.4.4].) Hence, we have a canonical K-linear isomorphism

$$D_K \simeq \mathrm{H}^1_{\mathrm{dR}}(\mathfrak{A}^t/R)[1/p] \simeq K \otimes_W \mathrm{H}^1_{\mathrm{cris}}(\overline{A}^t/W), \tag{2.2.3.1}$$

where $W = W(k)$ and the final isomorphism comes from [**8**, (2.4.2)].

By canonicity with respect the action on $\mathfrak{A}$ by an order in L, the composite isomorphism (2.2.3.1) is $K \otimes_{\mathbb{Q}} L$-linear when using the L-action on $\mathrm{H}^1_{\mathrm{cris}}(\overline{A}^t/W)[1/p]$ defined via pullback of the dual of the L-action on $\overline{A}$ in the isogeny category over k. Thus, the invertibility of D_K over $K \otimes_{\mathbb{Q}} L$ is equivalent to that of $\mathrm{H}^1_{\mathrm{cris}}(\overline{A}^t/W)[1/p]$ over $K_0 \otimes_{\mathbb{Q}} L$, where $K_0 = W[1/p]$. For any abelian variety C over k, the comparison of classical contravariant Dieudonné theory and crystalline cohomology for abelian varieties [**6**, 2.5.5–2.5.7, 3.3.7, 4.2.14] naturally identifies the two W-module $\mathrm{H}^1_{\mathrm{cris}}(C/W)$ and $\mathrm{M}^*(C[p^\infty])^{(p)}$, so our problem reduces to proving that $\mathrm{M}^*(C[p^\infty])[1/p]$ is an invertible module over $K_0 \otimes_{\mathbb{Q}} L = K_0 \otimes_{\mathbb{Q}_p} L_p$ for any such C of dimension g endowed with a CM structure by L. Applying Proposition 1.4.3.9(1) thereby completes the proof of invertibility of D_K and D'_K over $K \otimes_{\mathbb{Q}} L$.

Step 4. It remains to determine the $K \otimes_{\mathbb{Q}} L$-module structures of D^0_K and D'^0_K. More precisely, when the invertible $K \otimes_{\mathbb{Q}} L$-modules D_K and D'_K are viewed over $\overline{K}$, each $\mathbb{Q}$-algebra map $L \to \overline{K}$ has a 1-dimensional eigenspace and the submodules $D^0_{\overline{K}}$ and $D'^0_{\overline{K}}$ are each spanned over $\overline{K}$ by the half of these eigenlines. We have to show that the eigencharacters arising in these submodules coincide.

By its construction, D^0 is the formal cotangent space of Γ^t along the identity, or equivalently of $\mathfrak{A}^t$ along the identity (see Example 1.4.3.6). This latter cotangent space is naturally identified with the R-linear dual of $\mathrm{H}^1(\mathfrak{A}, \mathscr{O}_{\mathfrak{A}})$ compatibly with the L-actions after inverting p (where L acts on $\mathrm{H}^1(\mathfrak{A}, \mathscr{O}_{\mathfrak{A}})[1/p]$ via pullback of its given action on $\mathfrak{A}$ in the isogeny category over R).

Likewise, via the double duality isomorphism $(\mathfrak{A}^t)^t \simeq \mathfrak{A}$ of formal abelian schemes over R, D'^0_K is identified with $\mathrm{H}^0(\mathfrak{A}, \Omega^1_{\mathfrak{A}/R})[1/p]$ as a $K \otimes_{\mathbb{Q}} L$-module where L acts through the complex conjugate of its given action on $\mathfrak{A}$ in the isogeny category over R (due to how L acts on $\mathfrak{A}^t$). Thus, it suffices to show that when we view $\mathrm{H}^1_{\mathrm{dR}}(\mathfrak{A}/R)[1/p]$ as an invertible $K \otimes_{\mathbb{Q}} L$-module using the given L-action on $\mathfrak{A}$ then complex conjugation on L swaps the sets of eigencharacters arising on the two graded pieces $\mathrm{H}^0(\mathfrak{A}, \Omega^1_{\mathfrak{A}/R})[1/p]$ and $\mathrm{H}^1(\mathfrak{A}, \mathscr{O}_{\mathfrak{A}})[1/p]$ of the Hodge filtration.

By invertibility over $K \otimes_{\mathbb{Q}} L$ and K-rank considerations, it is equivalent to check that there are no conjugate pairs arising in one of the graded parts (such as in $\mathrm{H}^0(\mathfrak{A}, \Omega^1_{\mathfrak{A}/R})[1/p]$). Passing to the K-linear dual, we are reduced to checking that the eigencharacters arising for the L-action on $\mathrm{Lie}(\mathfrak{A})[1/p]_{\overline{K}}$ constitute a ($\overline{K}$-valued) CM type of L. But this is exactly the initial tangential hypothesis. □

2.3. CM lifting to a normal domain up to isogeny: counterexamples

We now give two classes of counterexamples to (IN) in 1.8.5, the second of which is more satisfying but the first of which is easier to explain (and will arise again in 4.1.2 as the prototype for an important class of examples in Chapter 4).

2.3.1. Supersingular counterexamples. Choose a prime number p satisfying $p \equiv 2, 3 \pmod 5$, so p remains prime in $\mathbb{Q}(\mu_5)$. Consider an algebraic integer in $\mathbb{Q}(\mu_5)$ of the form $\pi = \pm p\zeta_5$, where ζ_5 is a primitive fifth root of unity in $\mathbb{Q}(\mu_5)$. This is a p^2-Weil integer. By Honda-Tate theory in 1.6.2, there is a simple abelian variety B over $\mathbb{F}_{p^2}$ such that the center of $\mathrm{End}^0(B)$ may be identified with $F := \mathbb{Q}[\pi] = \mathbb{Q}(\mu_5)$ via $\pi \mapsto \mathrm{Fr}_{B,p^2}$ and the central division algebra $D = \mathrm{End}^0(B)$ over F is split away from the unique p-adic place of F. It follows that D is globally split over F, which is to say $D = F$, so

$$2\dim(B) = [F : \mathbb{Q}]\sqrt{[D : F]} = 4$$

and hence $\dim(B) = 2$. Thus, up choosing an isomorphism $\mathbb{Q}(\mu_5) \simeq \mathrm{End}^0(B)$, the only CM structure on B is the canonical one provided by the action of $\mathbb{Q}(\mu_5)$.

2.3.2. Lemma. *The reflex field of every* CM *type on* $\mathbb{Q}(\mu_5)$ *is non-canonically isomorphic to* $\mathbb{Q}(\mu_5)$.

Proof. If L is a CM field that is Galois over $\mathbb{Q}$ then any CM type Φ on L has reflex field $E \subset \overline{\mathbb{Q}}$ that is contained in the canonical image of L in $\overline{\mathbb{Q}}$. But the reflex field of a CM type is a CM field as well (Remark 2.1.3.2), so if in addition L has no proper CM subfield then $E \simeq L$. An example of such an L is $\mathbb{Q}(\mu_\ell)$ for a Fermat prime ℓ, such as $\ell = 5$. □

2.3.3. Corollary. *For* $p \equiv \pm 2 \pmod 5$ *and* B *over* $\mathbb{F}_{p^2}$ *as above, no member of the isogeny class of* B *can be lifted to an abelian scheme* $\mathcal{B}$ *over a normal local domain* R *with characteristic* 0 *and residue field* $\mathbb{F}_{p^2}$ *such that the generic fiber of* $\mathcal{B}$ *has sufficiently many complex multiplications. In particular,* B *does not satisfy the condition* (IN) *in* 1.8.5.

Proof. Assume that such a $\mathcal{B}$ exists, and let K be the fraction field of R. By faithfulness of reduction to the special fiber, we get an injection

$$\mathrm{End}^0(\mathcal{B}_K) = \mathrm{End}^0(\mathcal{B}) \hookrightarrow \mathrm{End}^0(B) = \mathbb{Q}(\mu_5)$$

into a quartic number field. (The initial equality holds by Lemma 1.8.4, since R is normal.) Hence, since $\dim(\mathcal{B}_K) = \dim(B) = 2$, the only way that $\mathcal{B}_K$ can have sufficiently many complex multiplications is if it has a CM structure given by a lifting of the $\mathbb{Q}(\mu_5)$-action on its special fiber (upon fixing an isogeny between B and the special fiber of $\mathcal{B}$, the choice of which does not matter since $\mathrm{End}^0(B) = \mathbb{Q}(\mu_5)$ is commutative). Thus, we may assume there is such an action of $\mathbb{Q}(\mu_5)$ on $\mathcal{B}_K$ in the isogeny category over K (or equivalently on $\mathcal{B}$ in the isogeny category over R).

Choose an algebraic closure $\overline{K}$ of K and let Φ be the resulting $\overline{\mathbb{Q}}_p$-valued CM type on $\mathbb{Q}(\mu_5)$ (where $\overline{\mathbb{Q}}_p$ is the algebraic closure of $\mathbb{Q}_p$ in $\overline{K}$). The congruence condition on p implies that there is a unique p-adic place on $\mathbb{Q}(\mu_5)$ and it is unramified over p. It follows that Φ satisfies the Shimura–Taniyama formula (2.1.4.1) for B over $\mathbb{F}_{p^2}$. By Lemma 2.3.2, the reflex field of $(\mathbb{Q}(\mu_5), \Phi)$ is isomorphic to $\mathbb{Q}(\mu_5)$. Hence, this reflex field likewise admits a unique p-adic place which moreover has residue field of degree 4 over $\mathbb{F}_p$. By the necessity of the residual obstruction in 2.1.5 (since R is normal), the residue field of R contains $\mathbb{F}_{p^4}$. But the residue field of R is $\mathbb{F}_{p^2}$, so we have a contradiction and therefore no such $\mathcal{B}$ exists. □

2.3.4. Remark. If we replace $\mathbb{Q}(\mu_5)$ with $\mathbb{Q}(\mu_\ell)$ for any Fermat prime $\ell \geqslant 5$ (so there is no proper CM subfield of $\mathbb{Q}(\mu_\ell)$ and $[\mathbb{Q}(\mu_\ell) : \mathbb{Q}] > 2$), then the preceding construction works over $\mathbb{F}_{p^2}$ for any prime $p \neq \ell$ with $p \not\equiv \pm 1 \pmod \ell$ (so $p \pmod \ell$)

in $(\mathbb{Z}/\ell\mathbb{Z})^\times$ has order exceeding 2). In such examples, B is simple with dimension $(\ell-1)/2$ and has endomorphism algebra $\mathbb{Q}(\mu_\ell)$ in which $\mathrm{Fr}_{B,p^2} = \pi := \pm p\zeta_\ell$ for a primitive ℓth root of unity ζ_ℓ in $\mathbb{Q}(\mu_\ell)$. Thus, $\pi^\ell = \pm p^\ell$, so there is an isogeny between $B_{\mathbb{F}_{p^{2\ell}}}$ and a power of a supersingular elliptic curve over $\mathbb{F}_{p^{2\ell}}$. Hence, these B are supersingular abelian varieties.

Consider the special case when p is a generator of $(\mathbb{Z}/\ell\mathbb{Z})^\times$ and the CM order is $\mathbb{Z}[\mu_\ell]$. In this case the tangent space $\mathrm{Lie}(B)$ of dimension $(\ell-1)/2$ over $\mathbb{F}_{p^2}$ is 1-dimensional over $\mathbb{Z}[\mu_\ell]/(p) \simeq \mathbb{F}_{p^{\ell-1}}$. Since complex conjugation on $\mathbb{Z}[\mu_\ell]$ induces an $\mathbb{F}_{p^2}$-automorphism of $\mathbb{F}_{p^{\ell-1}}$, B over $\mathbb{F}_{p^2}$ cannot satisfy the condition (CML) in 1.8.5 (necessarily using $L = \mathbb{Q}(\mu_\ell)$) since the μ_ℓ-action on the tangent space in characteristic p forces the geometric tangential action in characteristic 0 to be given by a collection of eigencharacters that is stable under complex conjugation, contradicting the requirement to be a CM type.

2.3.5. Counterexamples with two slopes. The preceding supersingular counterexamples to condition (IN) in 1.8.5 are not absolutely simple. We now give absolutely simple counterexamples. To make it easier to violate the necessary residual reflex condition from 2.1.5 we will work over $\mathbb{F}_p$. Fix a prime $\ell \geqslant 5$ and integer r such that $1 < r < \ell/2$. We shall construct counterexamples to (IN) over $\mathbb{F}_p$ that satisfy the elementary criterion for (CML) in Proposition 2.2.1 (so they satisfy (I)) and are absolutely simple of dimension ℓ, with p satisfying some congruence conditions to be determined shortly. In these examples the slopes of B (in the sense of 2.1.4.1) will be r/ℓ and $1-r/\ell$, each with multiplicity ℓ.

Choose an imaginary quadratic field F whose discriminant is prime to ℓ (so in particular, F is linearly disjoint from $\mathbb{Q}(\mu_\ell)$ over $\mathbb{Q}$). Choose a prime $p \nmid \ell\cdot\mathrm{disc}(F)$ such that $p \bmod \ell$ is a generator of $(\mathbb{Z}/\ell\mathbb{Z})^\times$, which is to say that p is inert in $\mathbb{Q}(\mu_\ell)$. Assume that p splits in F and that the two prime ideals $\wp$ and $\overline{\wp}$ of $\mathcal{O}_F$ over p are principal ideals. For example, if we consider F with class number 1 (i.e., $F = \mathbb{Q}(\sqrt{-D})$ with $D = 1, 2, 3, 7, 11, 19, 43, 67, 163$, so at most one such F is ruled out by the condition $\ell \nmid \mathrm{disc}(F)$), then the conditions on p say that it lies in some non-empty set of congruence classes modulo $\ell\cdot\mathrm{disc}(F)$; in the special case $\ell = 5$ and $F = \mathbb{Q}(\sqrt{-1})$, the condition is $p \equiv 13, 17 \pmod{20}$.

Write $\wp = \alpha\mathcal{O}_F$. Let $L = F(u)$ be an extension field of F such that $u^\ell = \alpha^r\,\bar{\alpha}^{\ell-r} = p^r\overline{\alpha}^{\ell-2r}$. The extension L/F is unambiguous because the polynomial $X^\ell - \alpha^r\overline{\alpha}^{\ell-r} \in F[X]$ is irreducible (as it is even $\wp$-adically irreducible). Likewise, $F = \mathbb{Q}(u^\ell)$, so $L = \mathbb{Q}(u)$. Observe that the algebraic integer u is a Weil p-integer, so the field $L = \mathbb{Q}(u)$ containing the imaginary quadratic field F in which p is split is itself a CM field. Since $(p/u)^\ell = p^r\alpha^{\ell-2r}$, we see that L/F is totally ramified at both primes $\wp$ and $\overline{\wp}$ of $\mathcal{O}_F$ over p (so $\mathcal{O}_L$ has residue field $\mathbb{F}_p$ at all primes over p). By Honda-Tate theory in 1.6.2, there is a simple ℓ-dimensional abelian variety B over $\mathbb{F}_p$ endowed with an action by L in the isogeny category over $\mathbb{F}_p$ such that $\mathrm{Fr}_{B,p} = u$. This abelian variety has two distinct slopes, namely r/ℓ and $(\ell-r)/\ell$ (each with multiplicity ℓ), and its endomorphism algebra is $\mathbb{Q}(u) = L$.

We now check that B is absolutely simple. Consider $B_{\mathbb{F}_{p^e}}$ with $e \geqslant 1$. This is isotypic since B is $\mathbb{F}_p$-simple (Proposition 1.2.6.1), so the center of its endomorphism algebra is $\mathbb{Q}(u^e)$ because $u^e = \mathrm{Fr}_{B_{\mathbb{F}_{p^e}}/\mathbb{F}_{p^e},e}$. To prove absolute simplicity we may replace e with a multiple so that $e = \ell e'$ for some $e' \in \mathbb{Z}$. Since the element $u^e = \alpha^{re'}\overline{\alpha}^{(\ell-r)e'}$ in $\mathcal{O}_F$ generates the ideal $\wp^{re'}\overline{\wp}^{(\ell-r)e'}$ that does not come from $\mathbb{Z}$

(due to the unequal exponents), the inclusion $\mathbb{Q}(u^e) \subseteq \mathbb{Q}(u^\ell) = F$ is an equality for degree reasons. Hence, if A is the unique (up to isogeny) simple factor of $B_{\mathbb{F}_{p^e}}$ then the endomorphism algebra of A is a central division algebra D over F. The local invariants of D at the p-adic places of F are the slopes of $B_{\mathbb{F}_{p^e}}$, which are r/ℓ and $(\ell - r)/\ell$. These have denominator ℓ, and D is split at all places of F away from p, so $\sqrt{[D:F]} = \ell$. Thus, by Honda-Tate theory, $2\dim(A) = [F:\mathbb{Q}]\sqrt{[D:F]} = 2\ell$, so $\dim(A) = \ell = \dim(B_{\mathbb{F}_{p^e}})$. This establishes the desired absolute simplicity. (The reader can check that if $\ell \nmid e$ then $\mathbb{Q}(u^e) = \mathbb{Q}(u) = L$ and this is the endomorphism algebra of $B_{\mathbb{F}_{p^e}}$, as for $e = 1$.)

Let L' be a Galois closure of L over F. This is generated over $\mathbb{Q}$ by $L = \mathbb{Q}(u)$ and $\mathbb{Q}(\mu_\ell)$, and it is Galois over $\mathbb{Q}$ because u and p/u are respective ℓth roots of the $\mathbb{Q}$-conjugate elements $p^r\overline{\alpha}^{\ell-2r}$ and $p^r\alpha^{\ell-2r}$ in F that generate F over $\mathbb{Q}$. Since $[L:\mathbb{Q}] = 2\ell$, so F is the unique quadratic subfield of L, it follows from the linear disjointness of F and $\mathbb{Q}(\mu_\ell)$ over $\mathbb{Q}$ that L and $\mathbb{Q}(\mu_\ell)$ are linearly disjoint over $\mathbb{Q}$. Hence, $[L':L] = \ell - 1$ and $[L':\mathbb{Q}] = 2\ell(\ell-1)$. Since p splits in F and is inert in $\mathbb{Q}(\mu_\ell)$, $F(\mu_\ell)/F$ is inert at both primes over p. But L/F is totally ramified at both p-adic places of F, so by degree-counting we see that the composite field L' of degree $\ell(\ell-1)$ over F has exactly two p-adic places, each with residual degree $\ell - 1$ over $\mathbb{F}_p$. In particular, complex conjugation on the CM field L' switches the two p-adic places of L' and both such places are inert over L.

In addition to the preceding arithmetic properties of F and L at p, we need to record some group-theoretic properties before we can show that B does not satisfy (IN). Let $\Gamma := \mathrm{Gal}(L'/\mathbb{Q})$, $G := \mathrm{Gal}(L'/F)$, $H := \mathrm{Gal}(L'/L) = \mathrm{Gal}(L(\mu_\ell)/L) \simeq (\mathbb{Z}/\ell\mathbb{Z})^\times$, and $N := \mathrm{Gal}(L'/F(\mu_\ell)) \simeq \mathbb{Z}/\ell\mathbb{Z}$, so $\#N = \ell$, $\#H = \ell - 1$, and $G = N \rtimes H \simeq (\mathbb{Z}/\ell\mathbb{Z}) \rtimes (\mathbb{Z}/\ell\mathbb{Z})^\times$ with $(\mathbb{Z}/\ell\mathbb{Z})^\times$ acting on the additive group $\mathbb{Z}/\ell\mathbb{Z}$ via its canonical multiplicative scaling action. The group G has the following properties.

(i) Let ℓ' be a prime divisor of $\ell - 1$ and let P be the ℓ'-Sylow subgroup of H. The normalizer subgroup $\mathrm{N}_G(P)$ of P in G is equal to H. (In particular, $gHg^{-1} = H$ for $g \in G$ if and only if $g \in H$.)
(ii) The proper subgroups H' of G such that $N \cdot H' = G$ are the conjugates of H.
(iii) Any two distinct conjugates of H intersect trivially.
(iv) For any conjugate H' of H, there are exactly two orbits for the left H'-action on G/H. One orbit consists of a single coset and H' operates freely on the other H'-orbit.
(v) Let Σ be a subset of G/H with $1 < \#\Sigma < \ell - 1$ (i.e., Σ and its complement Σ^c are subsets of G/H with more than one element). The G-stabilizer $\mathrm{Stab}_G(\Sigma) \subseteq G$ of Σ has image in $G/N = \mathrm{Gal}(F(\mu_\ell)/F) = (\mathbb{Z}/\ell\mathbb{Z})^\times$ that is a proper subgroup of G/N.

The statement (i) follows from an easy calculation, and statements (ii) and (iii) follow from (i). (Statements (i)–(iii) are also easily seen in field-theoretic terms via the Galois correspondence.) The statement (iv) follows from (iii) (or is a simple calculation for $H' = H$, and for a general $H' = gHg^{-1}$ follows from the fact that left multiplication by g^{-1} on G/H carries H'-orbits to H-orbits). Finally, (v) follows from (ii) and (iv) because if $\mathrm{Stab}_G(\Sigma)$ maps onto G/N then by (ii) either $\mathrm{Stab}_G(\Sigma) = G$ or $\mathrm{Stab}_G(\Sigma)$ is a conjugate of H; the first option contradicts that Σ is a proper subset of G/H and (by (iv)) the second option contradicts the assumptions on $\#\Sigma$.

Let Φ be a CM type of L that is a collection of embeddings of L into its Galois closure L' over $\mathbb{Q}$. Such Φ are in bijective correspondence with H-invariant L'-valued CM types on L', which is to say subsets $\Phi_{L'} \subseteq \Gamma = \mathrm{Gal}(L'/\mathbb{Q})$ satisfying $\Phi_{L'} \cdot H = \Phi_{L'}$ and $\Phi_{L'} = \Psi \sqcup \iota \circ \Psi^c$ (where ι denotes the complex conjugation on the CM field L'), with Ψ a subset of G and Ψ^c the complement of Ψ in G. (Here we have used that $\Gamma = G \times C$, where C is the subgroup generated by the central involution given by complex conjugation on the CM field L'.)

The H-stability condition on $\Phi_{L'}$ says precisely that $\Psi \cdot H = \Psi$. If the above B over $\mathbb{F}_p$ is to satisfy (IN) using a particular CM lift (over a normal local domain R having characteristic 0 and residue field $\mathbb{F}_p$, with R taken to be noetherian and complete without loss of generality) then the Shimura–Taniyama formula in the residual reflex condition from 2.1.5 is exactly the requirement that the generic fiber of this lift has $\overline{\mathbb{Q}}_p$-valued CM type Φ on L corresponding to such a $\Phi_{L'}$ with $\#(\Psi/H) = r$ or $\#(\Psi/H) = \ell - r$ (depending on which of the two p-adic places of L' is induced by the choice of embedding of L' into $\overline{\mathbb{Q}}_p$ so as to view L'-valued CM types on L or L' as $\overline{\mathbb{Q}}_p$-valued CM types).

Hence, to prove that B does not satisfy (IN), it suffices to consider those L'-valued Φ such that $1 < \#(\Psi/H) < \ell - 1$ (such Φ do exist) and to show that the second part of the residual reflex condition is violated at all p-adic places of the reflex field E of (L, Φ). Since Ψ and Ψ^c have distinct cardinalities (as $r \neq \ell/2$) and complex conjugation is central in Γ, an element of Γ stabilizes $\Phi_{L'} \subseteq \Gamma = \mathrm{Gal}(L'/\mathbb{Q})$ under the left action if and only if it stabilizes the non-empty subset $\Psi \subseteq G = \mathrm{Gal}(L'/F)$. In particular, $F \subseteq E$ due to the definition of E. Hence, in terms of Galois theory, the subfield $E \subseteq L'$ corresponds to the subgroup $\mathrm{Gal}(L'/E) = \mathrm{Stab}_G(\Psi/H)$ using the left G-action on $G/H = \mathrm{Hom}_{\mathrm{ring}}(L, L')$.

Since the reflex field E of (L, Φ) is a CM field (as is the reflex field of any CM type, by Remark 2.1.3.2), complex conjugation acts non-trivially on E. But complex conjugation on L' switches the two p-adic places of L', and restriction to the imaginary quadratic field F in which p is split sets up a bijection between the p-adic places of L' and of F. Thus, since $F \subseteq E$, we see that E admits exactly two p-adic places, these places are switched by complex conjugation on E, and each has decomposition group in $\mathrm{Gal}(L'/E)$ equal to the full Galois group. The extension $L'/F(\mu_\ell)$ is totally ramified at both p-adic places whereas $F(\mu_\ell)/F$ is unramified with full residual degree $\ell - 1$ at both of its p-adic places (since p is inert in $\mathbb{Q}(\mu_\ell)$ and split in F), so the natural map from $\mathrm{Gal}(L'/E)$ to $G/N = \mathrm{Gal}(F(\mu_\ell)/F)$ has image equal to the Galois group of the residue field extension for L'/E at both p-adic places. Hence, the two p-adic places of E have residue field strictly larger than $\mathbb{F}_p$ if and only if $\mathrm{Gal}(L'/E)$ has proper image in G/N. But $\mathrm{Gal}(L'/E) = \mathrm{Stab}_G(\Psi/H)$, so property (v) above (applied to $\Sigma = \Psi/H \subseteq G/H$) implies that the p-adic places of E do indeed have non-trivial residual degree over $\mathbb{F}_p$. Thus, we have contradicted the second part of the residual reflex condition for the abelian variety B over $\mathbb{F}_p$, so it does not satisfy the property (IN).

To summarize, in these absolutely simple counterexamples to (IN) the endomorphism algebra over the finite field $\mathbb{F}_p$ is a CM field L, but its CM types that satisfy the requirement from the Shimura–Taniyama formula for a possible CM lifting (over a complete local noetherian domain as in the condition (I) from 1.8.5; see

Remark 1.8.5.1 and Proposition 2.1.4.2) do not satisfy the second part of the residual reflex condition. Moreover, since O_L has residue field $\mathbb{F}_p$ at all p-adic places, by Proposition 2.2.1 these examples all satisfy (CML) and hence satisfy (I) over $\mathbb{F}_p$.

2.3.6. The *Newton polygon* of an abelian variety B of dimension $g > 0$ over $\mathbb{F}_q$ is the concave-up polygon obtained by concatenating $2g$ segments with unit horizontal length and slopes (by definition) equal to the slopes of B in the sense defined in 2.1.4 (using suitably normalized p-adic ordinals of the roots of the characteristic polynomial F_B of the q-Frobenius of B); these slopes lie in $\mathbb{Q} \cap [0, 1]$. Such polygons $\mathfrak{N}$ are "symmetric" in the sense that for every slope λ occurring in $\mathfrak{N}$, the slopes λ and $1 - \lambda$ occur in $\mathfrak{N}$ with the same multiplicity; the symmetry holds because $\pi \mapsto q/\pi$ is a permutation of the set of zeros of F_B,

Denote by $\mathscr{N}\mathscr{P}_{\mathrm{NI}}$ (resp. $\mathscr{N}\mathscr{P}_{p,\mathrm{NI}}$) the set of all symmetric Newton polygons $\mathfrak{N}$ such that every *isotypic* abelian variety B over a finite field (resp. over any finite field of characteristic p) with Newton polygon $\mathfrak{N}$ satisfies (IN). For any symmetric Newton polygons $\mathfrak{N}_1, \ldots, \mathfrak{N}_r$ with no slopes in common and any non-negative integers $m_1, \ldots, m_r$, we write $\sum m_j \mathfrak{N}_j$ to denote the symmetric Newton polygon obtained by concatenation of the segments in m_j copies of each $\mathfrak{N}_j$ (with all segments arranged in order of increasing slope).

For each $n > 2$, let $(\frac{1}{n}, \frac{n-1}{n})$ denote the symmetric Newton polygon of length $2n$ with slopes $\frac{1}{n}$ and $\frac{n-1}{n}$ each occurring with multiplicity n. Analogously, we let $(\frac{1}{2}, \frac{1}{2})$ denote the symmetric Newton polygon with slope $\frac{1}{2}$ occurring with multiplicity 2, and we let $(0, 1)$ denote the symmetric Newton polygon with slopes 0 and 1 each occurring with multiplicity 1. In 2.3.1 and 2.3.5 we have seen examples of symmetric Newton polygons that are not in $\mathscr{N}\mathscr{P}_{\mathrm{NI}}$, namely $2(\frac{1}{2}, \frac{1}{2})$ and $(\frac{r}{\ell}, \frac{\ell-r}{\ell})$ for any odd prime $\ell \geqslant 5$ and $1 < r < \ell/2$. (In each case, the counterexamples were found in any characteristic $p > 0$ such that p satisfies a suitable non-vacuous congruence condition.) In the affirmative direction, the following are examples of symmetric Newton polygons $\mathfrak{N}$ in $\mathscr{N}\mathscr{P}_{\mathrm{NI}}$.

(i) An ordinary symmetric Newton polygon $\mathfrak{N} = g(0, 1)$, which is to say one whose only slopes (each with multiplicity $g > 0$) are 0 and 1, is in $\mathscr{N}\mathscr{P}_{\mathrm{NI}}$. This is exactly the case when the isotypic abelian variety B over a finite field κ of characteristic p has p-divisible group with étale part having the maximal possible height, namely $g = \dim(B)$, or in other words B is an ordinary abelian variety. To establish (IN) for any such B, choose a CM subfield $L \subseteq \mathrm{End}^0(B)$ with $[L : \mathbb{Q}] = 2g$ (Corollary 1.6.2.5). By Example 1.4.5.4, the Serre-Tate canonical lift of B over $W(\kappa)$ establishes property (IN) for B.

(ii) The Newton polygon $(\frac{1}{2}, \frac{1}{2})$ corresponds to a supersingular elliptic curve E over a finite field κ, and these satisfy (IN). (The counterexample to (IN) in 2.3.1 has Newton polygon $2(\frac{1}{2}, \frac{1}{2})$.) To prove this, choose an imaginary quadratic field $L \subseteq \mathrm{End}^0(E)$ and pass to a κ-isogenous elliptic curve if necessary so that $O_L \subseteq \mathrm{End}(E)$. Since $\mathbb{Z}_p \otimes_{\mathbb{Z}} O_L$ acts faithfully on the 1-dimensional $E[p^\infty]$ (Proposition 1.2.5.1) and this p-divisible group is local-local by supersingularity, there is a unique prime $\mathfrak{p}$ in O_L over p. Also, the action by O_L on the tangent space to E defines an embedding of $k = O_L/\mathfrak{p}$ into κ. Using this embedding to make κ into an $O_{L,\mathfrak{p}}$-algebra, $E[p^\infty]$ corresponds to a 1-dimensional formal $O_{L,\mathfrak{p}}$-module G over κ, and it has $O_{L,\mathfrak{p}}$-height 1 (since the $\mathbb{Z}_p$-height is 2 and $O_{L,\mathfrak{p}}$ is finite free of rank 2 over $\mathbb{Z}_p$).

Clearly $R := \mathcal{O}_{L,\mathfrak{p}} \otimes_{W(k)} W(\kappa)$ is a p-adic integer ring unramified over $\mathcal{O}_{L,\mathfrak{p}}$ with residue field κ. Since elliptic curves are canonically projective, to lift E with its $\mathcal{O}_L$-action to R (having tangential action via $\mathcal{O}_{L,\mathfrak{p}} \hookrightarrow R$) it is equivalent to lift G to a formal $\mathcal{O}_{L,\mathfrak{p}}$-module over R. By Lubin-Tate theory (or more generally the deformation theory of 1-dimensional formal modules with finite height over algebras over p-adic integer rings [**49**, 22.4.4]), for any p-adic integer ring Λ and complete local noetherian Λ-algebra $\mathcal{R}$, a height-1 formal Λ-module with dimension 1 over the residue field of $\mathcal{R}$ admits (up to unique isomorphism) a unique deformation to a formal Λ-module over $\mathcal{R}$. This provides the required lift of G over R as a formal $\mathcal{O}_{L,\mathfrak{p}}$-module.

(iii) Every Newton polygon of the form $\mathfrak{N} = (\frac{1}{n}, \frac{n-1}{n})$, $n > 2$, is in $\mathscr{N}\mathscr{P}_{\rm NI}$. To prove this, let B be an isotypic abelian variety over a finite field κ (with size q) such that it has this Newton polygon; in particular, $\dim(B) = n$. Let $L \subseteq \mathrm{End}^0(B)$ be a CM subfield with $[L : \mathbb{Q}] = 2n$, and use a κ-isogeny if necessary so that $\mathcal{O}_L \subseteq \mathrm{End}(B)$ (Proposition 1.7.4.5). The Frobenius element $\mathrm{Fr}_{B,q} \in \mathcal{O}_L$ has p-adic slopes $\mathrm{ord}_w(\mathrm{Fr}_{B,q})/\mathrm{ord}_w(q)$ given by only $1/n$ and $(n-1)/n = 1 - 1/n$. These are positive, so $G = B[p^\infty]$ is a connected p-divisible group with connected dual.

If w is a p-adic place of L with associated slope $\lambda_w = \mathrm{ord}_w(\mathrm{Fr}_{B,q})/\mathrm{ord}_w(q)$ then $\lambda_{\overline{w}} = 1 - \lambda_w \neq \lambda_w$, so $\overline{w} \neq w$. The faithful action of $\mathcal{O}_{L,p} \simeq \prod_{w|p} \mathcal{O}_{L,v}$ on G (Proposition 1.2.5.1) decomposes G into a product $\prod G_w$ where G_w is a non-zero connected p-divisible group equipped with an action by $\mathcal{O}_{L,w}$. Since

$$\sum_{w|p} \mathrm{height}(G_w) = \mathrm{height}(G) = 2n = [L : \mathbb{Q}] = \sum_{w|p} [L_w : \mathbb{Q}_p],$$

it follows by Proposition 1.4.3.9(1) that G_w has height $[L_w : \mathbb{Q}_p]$ for each w. Tate's method of proof of the Shimura–Taniyama formula in [**121**, §5] via p-divisible groups carries over to each G_w over the finite field κ to yield

$$\frac{\mathrm{ord}_w(\mathrm{Fr}_{B,q})}{\mathrm{ord}_w(q)} = \frac{\dim(G_w)}{\mathrm{height}(G_w)} = \frac{\dim(G_w)}{[L_w : \mathbb{Q}_p]}.$$

The left side is either $1/n$ or $1 - 1/n$, and hence has denominator n. This forces $n | [L_w : \mathbb{Q}_p]$ for all w, yet $[L : \mathbb{Q}] = 2n$. Hence, there are exactly two p-adic places on L, necessarily of the form v and $\overline{v}$, say with respective slopes $1/n$ and $1 - 1/n > 1/n$. We also get $\dim(G_v) = 1$ and $\dim(G_{\overline{v}}) = n - 1$.

The tangential action by $\mathcal{O}_{L,v}$ on the 1-dimensional G_v defines an embedding $\kappa_v \hookrightarrow \kappa$, so G_v may be identified with a 1-dimensional formal $\mathcal{O}_{L,v}$-module with $\mathcal{O}_{L,v}$-height 1 over the $\mathcal{O}_{L,v}$-algebra κ. Applying the same considerations to the dual abelian variety B^t over κ (equipped with the dual CM structure as in Definition 1.5.3.1), we see that via the canonical isomorphism $\mathcal{O}_{L,v} \simeq \mathcal{O}_{L,\overline{v}}$ induced by complex conjugation, $G_{\overline{v}}$ is identified with the Cartier dual of a 1-dimensional formal $\mathcal{O}_{L,v}$-module G'_v with $\mathcal{O}_{L,v}$-height 1 over κ.

Let $R = \mathcal{O}_{L,v} \otimes_{W(\kappa_v)} W(\kappa)$; this is a p-adic integer ring unramified over $\mathcal{O}_{L,v}$ with residue field κ. Using Lemma 2.2.2, choose an $\mathcal{O}_L$-linear polarization $\lambda : B \to B^t$. The associated $\mathcal{O}_{L,p}$-linear map $G_v \times (G'_v)^t \to G'_v \times (G_v)^t$ between p-divisible groups decomposes as the $\mathcal{O}_{L,p}$-linear product $\lambda_v \times \mu_v^t$ of an $\mathcal{O}_{L,v}$-linear isogeny $\lambda_v : G_v \to G'_v$ and the dual of another such isogeny $\mu_v : G_v \to G'_v$. To prove (IN) for B, by Grothendieck's algebraization theorem (as in Example 1.4.5.4) it suffices to deform λ_v and μ_v over R; we handle λ_v, and then μ_v will go in the same way.

Arguing via the deformation theory of 1-dimensional formal Λ-modules of Λ-height 1 over complete local noetherian algebras over p-adic integer rings Λ, exactly as in (ii) we may uniquely deform G_v and G'_v to respective formal groups $\mathcal{G}_v$ and $\mathcal{G}'_v$ over R. There is at most one lift $\mathcal{G}_v \to \mathcal{G}'_v$ of λ_v (by 1.4.4.3), and if $\overline{\kappa}/\kappa$ is an algebraic closure then we claim that it suffices to construct such a lift when κ is replaced with $\overline{\kappa}$. To prove this, let $R' = R \otimes_{W(\kappa)} W(\overline{\kappa})$, so the natural action by $\Gamma = \mathrm{Gal}(\overline{\kappa}/\kappa)$ on R' has R as its subring of invariant elements (as we see by computing relative to a finite $W(\kappa)$-basis of R). Hence, any finite flat R-module M likewise satisfies $M = M'^{\Gamma}$ for $M' := R' \otimes_R M$, so if M and N are two such R-modules then $\mathrm{Hom}_R(M, N) = \mathrm{Hom}_{R'}(M', N')^{\Gamma}$. It follows by consideration of the finite flat p^m-torsion subgroups of p-divisible groups ($m \geqslant 1$) that a Γ-equivariant R'-homomorphism $(\mathcal{G}_v)_{R'} \to (\mathcal{G}'_v)_{R'}$ descends uniquely to an R-homomorphism $\mathcal{G}_v \to \mathcal{G}'_v$. By uniqueness, if $(\lambda_v)_{\overline{\kappa}}$ admits a lift to an $\mathcal{O}_{L,v}$-linear R'-homomorphism $(\mathcal{G}_v)_{R'} \to (\mathcal{G}'_v)_{R'}$ then this lift is Γ-equivariant (due to 1.4.4.3 and the Γ-equivariance of the special map $(\lambda_v)_{\overline{\kappa}}$), so it descends to a lift of λ_v over R.

Now we may replace κ with $\overline{\kappa}$, so $\kappa = \overline{\kappa}$. Thus, G_v and G'_v are abstractly $\mathcal{O}_{L,v}$-linearly isomorphic. Using such an isomorphism, λ_v becomes multiplication by some non-zero element of $\mathcal{O}_{L,v}$. Uniqueness of deformations in the height-1 case thereby settles the lifting problem for λ_v.

It is easy to adapt the arguments underlying the preceding examples (i)–(iii) to show that any symmetric Newton polygon that can be written as a finite sum in the form

$$\mathfrak{N} = m_1\,(0,1) + \sum_{n \geqslant 2} m_n \left(\frac{1}{n}, \frac{n-1}{n}\right), \qquad m_1 \geqslant 0,\ m_n \in \{0,1\} \text{ for all } n \geqslant 2,$$

is in $\mathscr{N\!P}_{\mathrm{NI}}$. We do not know whether there is an element of $\mathscr{N\!P}_{\mathrm{NI}}$ that is not of the above form.

2.3.7. Remark. (i) The method in 2.3.5 using Galois extensions whose Galois group is the standard semi-direct product $G \simeq (\mathbb{Z}/n\mathbb{Z}) \rtimes (\mathbb{Z}/n\mathbb{Z})^{\times}$ for prime n fails dramatically when instead $n = 15$. In this case we have $N \simeq \mathbb{Z}/15\mathbb{Z}$ and $H = (\mathbb{Z}/15\mathbb{Z})^{\times} \simeq (\mathbb{Z}/3\mathbb{Z})^{\times} \times (\mathbb{Z}/5\mathbb{Z})^{\times}$. The left action of H on G/H is identified with the standard action of $(\mathbb{Z}/3\mathbb{Z})^{\times} \times (\mathbb{Z}/5\mathbb{Z})^{\times}$ on $(\mathbb{Z}/3\mathbb{Z}) \times (\mathbb{Z}/5\mathbb{Z})$, and this has 4 orbits (namely, $\{0\}$, $(\mathbb{Z}/3\mathbb{Z})^{\times}$, $(\mathbb{Z}/5\mathbb{Z})^{\times}$, and the rest) with respective cardinalities 1, 2, 4, and 8. Therefore, for any integer r with $1 \leqslant r \leqslant 14$ there exists a subset of G/H with r elements that is stable under the left action by H, and so likewise for the left action on G/H by any conjugate of H.

(ii) It would be interesting to determine which symmetric Newton polygons belong to $\mathscr{N\!P}_{\mathrm{NI}}$, and which ones belong to $\mathscr{N\!P}_{p,\mathrm{NI}}$ as the prime number p varies. For a given symmetric Newton polygon $\mathfrak{N}$, the question as to whether $\mathfrak{N}$ belongs to $\mathscr{N\!P}_{p,\mathrm{NI}}$ is easier when p does not divide the denominator of any slope of $\mathfrak{N}$. For instance it is not difficult to check that $(\frac{r}{15}, \frac{15-r}{15})$ is in $\mathscr{N\!P}_{p,\mathrm{NI}}$ for $r = 2, 4, 6, 7$ if $p \neq 3, 5$. We do not know any plausible statistics about $\mathscr{N\!P}_{\mathrm{NI}}$ or $\mathscr{N\!P}_{p,\mathrm{NI}}$ among all symmetric Newton polygons.

2.4. Algebraic Hecke characters

Our "global" proof of Theorem 2.1.6 will rest on encoding CM abelian varieties over number fields in terms of the notion of an algebraic Hecke character. In this section we will review the definition and basic properties of algebraic Hecke characters before establishing Theorem 2.4.11, which gives a procedure to modify an algebraic Hecke character by specifying its value on a uniformizer at a desired unramified place. The reader is referred to [**127**], [**109**, §7], [**105**, Chap. II], [**27**, §5], [**78**], and [**77**] for more information about algebraic Hecke characters. We sometimes write $\underline{L}^{\times}$ as shorthand for the $\mathbb{Q}$-torus $\mathrm{Res}_{L/\mathbb{Q}}(\mathbb{G}_{\mathrm{m}})$ over $\mathbb{Q}$, where $\mathrm{Res}_{L/\mathbb{Q}}$ denotes Weil restriction with respect to a finite extension $L/\mathbb{Q}$.

2.4.1. Definition. Let K be an infinite field. For a finite separable extension K'/K and an extension F/K, a homomorphism $\chi : K'^{\times} \to F^{\times}$ is *algebraic* (with respect to K) if one of the following equivalent conditions holds:

(i) The map χ is induced by a homomorphism of algebraic groups
$$\mathrm{Res}_{K'/K}(\mathbb{G}_{\mathrm{m}})_F \xrightarrow{\phi} \mathbb{G}_{\mathrm{m}}$$
over F. The relation between χ and ϕ is that the map $(K' \otimes_K F)^{\times} \to F^{\times}$ induced by ϕ on F-points restricts to χ on the subgroup $K'^{\times} \subseteq (K' \otimes_K F)^{\times}$ (inclusion defined by $t \mapsto t \otimes 1$).

(ii) (F/K finite separable) The map χ is induced by a K-homomorphism
$$\mathrm{Res}_{K'/K}(\mathbb{G}_{\mathrm{m}}) \to \mathrm{Res}_{F/K}(\mathbb{G}_{\mathrm{m}}).$$

(iii) Let $e_1, \ldots, e_n$ be a K-basis of K'. There exists a rational function
$$f(X_1, \ldots, X_n) \in F(X_1, \ldots, X_n)$$
such that every non-zero element of K^n is in the domain of definition of $f(X_1, \ldots, X_n)$ (i.e., is not a zero of its "reduced form" denominator) and
$$\chi\left(\sum_{i=1}^{n} x_i\, e_i\right) = f(x_1, \ldots, x_n)$$
for every non-zero element $(x_1, \ldots, x_n) \in K^n$.

(iv) Let $I = \mathrm{Hom}_K(K', F_s)$ be the finite set of K-embeddings of K' into a (fixed) separable closure F_s of F. Then there exists a $\mathrm{Gal}(F_s/F)$-invariant function $m : I \to \mathbb{Z}$, necessarily unique, such that
$$\chi(x) = \prod_{\tau \in I} \tau(x)^{m(\tau)}$$
for all $x \in K'^{\times}$.

The equivalence of (i) and (ii) is due to functorial adjointness, and the equivalence of (i) and (iv) is clear (due to Galois descent with respect to F_s/F; see Remark 2.4.2), as is the implication (iv) $\Rightarrow$ (iii). Finally, to deduce (i) from (iii) one uses that the set of K-points is Zariski-dense in any torus over K (such as $\mathrm{Res}_{K'/K}(\mathbb{G}_{\mathrm{m}})$), due to the unirationality of tori over fields, and that a rational map between smooth connected algebraic groups over a field is a morphism if it is generically a homomorphism (in an evident sense).

2.4.2. Remark. In practice the ground field K in Definition 2.4.1 is understood from context (such as $\mathbb{Q}$ or $\mathbb{Q}_p$). The Galois group $\mathrm{Gal}(F_s/F)$ operates naturally on I in condition (iv), and the displayed equation in (iv) implies that m is constant on each $\mathrm{Gal}(F_s/F)$-orbit of I. Conversely, for any function $m : I \to \mathbb{Z}$ that is constant on every $\mathrm{Gal}(F_s/F)$-orbit of I there is a unique algebraic homomorphism χ from $K'^{\times}$ to $F^{\times}$ giving rise to m as in (iv).

If $F \subseteq K'$ (over K) then the special case with m identically equal to 1 corresponds to $\chi = \mathrm{Nm}_{K'/F}$. If F/K splits K'/K then $\mathrm{Gal}(F_s/F)$ acts trivially on I, so in such cases the maps $K'^{\times} \to F^{\times}$ induced by field embeddings over K are a $\mathbb{Z}$-basis for the group of algebraic homomorphisms (with respect to K).

2.4.3. Definition. Let K be a number field, $\mathfrak{m}$ a non-zero ideal of the ring of integers $\mathcal{O}_K$ of K, $I_{\mathfrak{m}}$ the group of fractional ideals of $\mathcal{O}_K$ that are relatively prime to $\mathfrak{m}$, and $\mathbb{A}_K^{\times}$ the group of K-ideles. Let L be a field of characteristic 0.

(a) A homomorphism $\chi : I_{\mathfrak{m}} \to L^{\times}$ is an *algebraic Hecke character with conductor* $\mathfrak{m}$ if there exists a homomorphism $\chi_{\mathrm{alg}} : K^{\times} \to L^{\times}$ that is algebraic (with respect to $\mathbb{Q}$) and satisfies

$$\chi((x)) = \chi_{\mathrm{alg}}(x)$$

for every $x \in K^{\times}$ satisfying two conditions: $x \equiv 1 \pmod{\mathfrak{m}}$ and x has image in $K_{\infty}^{\times}$ lying in the identity component $(K_{\infty}^{\times})^0$ (i.e., x is positive at all real places). Here, $(x) := x\mathcal{O}_K$ is the fractional ideal of $\mathcal{O}_K$ generated by x. We say such a χ has *conductor* $\leqslant \mathfrak{m}$.

(b) A continuous homomorphism $\epsilon : \mathbb{A}_K^{\times} \to L^{\times}$ is an *algebraic Hecke character* if the restriction $\epsilon|_{K^{\times}}$ of ϵ to the diagonally embedded subgroup $K^{\times} \subset \mathbb{A}_K^{\times}$ is equal to an algebraic homomorphism $\epsilon_{\mathrm{alg}} : K^{\times} \to L^{\times}$ (with respect to $\mathbb{Q}$). The target $L^{\times}$ of ϵ is given the discrete topology, so the kernel of ϵ is an open subgroup of $\mathbb{A}_K^{\times}$.

An equivalence of these two definitions can be formulated and proved as follows. Suppose that $\epsilon : \mathbb{A}_K^{\times} \to L^{\times}$ is an algebraic Hecke character as in 2.4.3(b). Choose a non-zero ideal $\mathfrak{m}$ of $\mathcal{O}_K$ such that $\ker(\epsilon)$ contains the open subgroup

$$U_{\mathfrak{m}} := (K_{\infty}^{\times})^0 \times \prod_{v|\mathfrak{m}} (1 + \mathfrak{m}\mathcal{O}_{K,v}) \times \prod_{v \notin S} \mathcal{O}_{K,v}^{\times}$$

of $\mathbb{A}_K^{\times}$, where $S = S(\mathfrak{m})$ is the union of the set of archimedean places of K and the set of places of K dividing $\mathfrak{m}$. (We say ϵ has *conductor* $\leqslant \mathfrak{m}$.) Denote by $\mathbb{A}_K^S$ the factor ring of $\mathbb{A}_K$ consisting of factors away from S (i.e., the restricted product $\prod'_{v \notin S} K_v$), so its unit group is the restricted product $\mathbb{A}_K^{S,\times} = \prod'_{v \notin S} K_v^{\times}$. The restriction $\epsilon|_{\mathbb{A}_K^{S,\times}}$ of ϵ to $\mathbb{A}_K^{S,\times}$ factors through the canonical surjection $\pi_S : \mathbb{A}_K^{S,\times} \twoheadrightarrow I_{\mathfrak{m}}$ via a homomorphism $\chi : I_{\mathfrak{m}} \to L^{\times}$ satisfying the conditions in 2.4.3(a).

Conversely, given a homomorphism $\chi : I_{\mathfrak{m}} \to L^{\times}$ satisfying the conditions in 2.4.3(a), one defines a continuous homomorphism $\epsilon : \mathbb{A}_K^{\times} \to L^{\times}$ as follows. Every element $x \in \mathbb{A}_K^{\times}$ can be written as $x = u \cdot y \cdot z$ with $u \in U_{\mathfrak{m}}$, $y \in K^{\times}$, and $z \in \mathbb{A}_K^{S,\times}$, where S is as in the previous paragraph. Define $\epsilon : \mathbb{A}_K^{\times} \to L^{\times}$ by $\epsilon(x) = \chi_{\mathrm{alg}}(y)\chi(\pi_S(z))$ for $x = uyz$ as above. It is easy to see that ϵ is well-defined and satisfies the conditions in 2.4.3(b) (and has conductor $\leqslant \mathfrak{m}$).

If χ and ϵ correspond to each other as above, then the algebraic homomorphisms χ_{alg} and ϵ_{alg} are equal; this algebraic homomorphism is called the *algebraic part* of χ or ϵ. As an example, if K'/K is a finite extension and $\mathfrak{m}'$ is a sufficiently divisible modulus on K' lying over $\mathfrak{m}$ such that $\mathfrak{m}|\mathrm{Nm}_{K'/K}(\mathfrak{m}')$, then $\epsilon \circ \mathrm{Nm}_{K'/K}$ is an algebraic Hecke character with algebraic part $\epsilon_{\mathrm{alg}} \circ \mathrm{Nm}_{K'/K}$ and associated homomorphism $I_{\mathfrak{m}'} \to L^\times$ given by $\chi \circ \mathrm{Nm}_{K'/K}$.

Note that the concepts of algebraic character and algebraic Hecke character as discussed above make sense when $L = \prod L_i$ is a finite product of fields L_i of characteristic 0, an interesting example of which is $E_\ell := \mathbb{Q}_\ell \otimes_{\mathbb{Q}} E$ for a number field E. In such cases, the preceding definitions, assertions, and proofs concerning $\underline{L}^\times$-valued algebraic characters and $L^\times$-valued algebraic Hecke characters carry over essentially verbatim, and these "algebraicity" properties hold if and only if they do after composing with projection to each factor field L_i.

2.4.4. The concept of algebraic Hecke character can be expressed in terms of (possibly disconnected) linear algebraic $\mathbb{Q}$-groups $S_{\mathfrak{m}}$ defined in Chapter II of [**105**]. This will be very useful for our purposes, so we now review the definition of $S_{\mathfrak{m}}$ and, for any number field L, use it to formulate a bijective correspondence between $L^\times$-valued algebraic Hecke characters of a number field K and certain continuous representations $\mathrm{Gal}(K^{\mathrm{ab}}/K) \to L_\ell^\times$ (with a fixed rational prime ℓ).

The group $S_{\mathfrak{m}}$ is an extension of the finite group $C_{\mathfrak{m}} := \mathbb{A}_K^\times/(K^\times \cdot U_{\mathfrak{m}})$ by a certain $\mathbb{Q}$-torus $T_{\mathfrak{m}}$, defined as follows. By definition, $T_{\mathfrak{m}}$ is the quotient of $\underline{K}^\times := \mathrm{Res}_{K/\mathbb{Q}}(\mathbb{G}_m)$ by the Zariski closure of the image of the finitely generated $\mathbb{Z}$-module $K^\times \cap U_{\mathfrak{m}}$ in $\underline{K}^\times$. The $\mathbb{Q}$-torus $T_{\mathfrak{m}}$ stabilizes as a quotient of $\underline{K}^\times$ when $\mathfrak{m}$ is sufficiently divisible (since the closure of $K^\times \cap U_{\mathfrak{m}}$ in $\underline{K}^\times$ has identity component that is independent of $\mathfrak{m}$); we denote it by $\mathfrak{S}^K$, following the notation in [**77**].[1] By definition, $S_{\mathfrak{m}}$ is the middle term in the extension

$$1 \to T_{\mathfrak{m}} \to S_{\mathfrak{m}} \to C_{\mathfrak{m}} \to 1$$

that is the push-out of the short exact sequence

$$1 \to K^\times/(K^\times \cap U_{\mathfrak{m}}) \to \mathbb{A}_K^\times/U_{\mathfrak{m}} \to C_{\mathfrak{m}} \to 1$$

by the natural homomorphism $K^\times/(K^\times \cap U_{\mathfrak{m}}) \to T_{\mathfrak{m}}(\mathbb{Q})$. In other words, we have a commutative diagram of locally finite type $\mathbb{Q}$-groups

$$\begin{array}{ccccccccc} 1 & \longrightarrow & K^\times/K^\times \cap U_{\mathfrak{m}} & \longrightarrow & \mathbb{A}_K^\times/U_{\mathfrak{m}} & \longrightarrow & C_{\mathfrak{m}} & \longrightarrow & 1 \\ & & \downarrow & & \downarrow{\scriptstyle i_{K,\mathfrak{m}}} & & \downarrow{\scriptstyle =} & & \\ 1 & \longrightarrow & T_{\mathfrak{m}} & \longrightarrow & S_{\mathfrak{m}} & \longrightarrow & C_{\mathfrak{m}} & \longrightarrow & 1, \end{array}$$

with exact rows (and the top row consisting of constant groups). In particular, $i_{K,\mathfrak{m}}$ can equivalently be viewed as a map of abstract groups $\mathbb{A}_K^\times/U_{\mathfrak{m}} \to S_{\mathfrak{m}}(\mathbb{Q})$, and its image is Zariski-dense in $S_{\mathfrak{m}}$. These commutative diagrams form a projective system as $\mathfrak{m}$ runs through all non-zero ideals of $\mathcal{O}_K$.

By [**126**, 3.3], every Hopf algebra over a field is a direct limit of Hopf subalgebras that are flat and finitely generated, with faithfully flat transition maps. This property does not hold for flat Hopf algebras over general rings R, so for general R we call a flat affine R-group scheme $G = \mathrm{Spec}(A)$ *pro-algebraic* if A is such a direct

[1]The notation for the $\mathbb{Q}$-torus $\mathfrak{S}^K$ is S^K in [**78**] and is ${}^K S$ in [**28**].

limit over R. This property is preserved by base change; we will be interested in pro-algebraic groups arising by base change from affine group schemes over a field.

The notion of *short exact sequence* of affine group schemes relative to the fpqc topology is defined in 1.4.1.1. A direct limit of faithfully flat algebras is faithfully flat, so the inverse limit of the short exact sequences of finite type affine $\mathbb{Q}$-groups

$$1 \to T_{\mathfrak{m}} \to S_{\mathfrak{m}} \to C_{\mathfrak{m}} \to 1$$

indexed by non-zero ideals $\mathfrak{m}$ of $\mathcal{O}_K$ is a short exact sequence (for the fpqc topology)

$$1 \to \mathfrak{S}^K \to S_K \to \mathrm{Gal}(K^{\mathrm{ab}}/K) \to 1 \tag{2.4.4.1}$$

of affine $\mathbb{Q}$-groups.

The right term in (2.4.4.1) is a constant (profinite) $\mathbb{Q}$-group and the exact sequence is the pullback of the exact sequence $1 \to T_{\mathfrak{m}} \to S_{\mathfrak{m}} \to C_{\mathfrak{m}} \to 1$ by the canonical surjection $\mathrm{Gal}(K^{\mathrm{ab}}/K) \to C_{\mathfrak{m}}$ for $\mathfrak{m}$ sufficiently divisible so that $T_{\mathfrak{m}}$ does not change (as a quotient of $\underline{K}^\times$) when $\mathfrak{m}$ is replaced with any non-zero multiple. Here we have used class field theory to identify $\varprojlim C_{\mathfrak{m}}$ with $\mathrm{Gal}(K^{\mathrm{ab}}/K)$ as profinite groups. We have a natural commutative diagram

$$\begin{array}{ccccccccc}
1 & \longrightarrow & K^\times & \longrightarrow & \mathbb{A}_K^\times & \longrightarrow & \mathbb{A}_K^\times/K^\times & \longrightarrow & 1 \\
 & & \downarrow & & \downarrow{\scriptstyle i_K} & & \downarrow{\scriptstyle \mathrm{rec}_K} & & \\
1 & \longrightarrow & \mathfrak{S}^K(\mathbb{Q}) & \longrightarrow & S_K(\mathbb{Q}) & \longrightarrow & \mathrm{Gal}(K^{\mathrm{ab}}/K) & \longrightarrow & 1
\end{array}$$

where $\mathrm{rec}_K : \mathbb{A}_K^\times/K^\times \to \mathrm{Gal}(K^{\mathrm{ab}}/K)$ is the surjective Artin map with the classical (i.e., arithmetic) normalization.

By definition of S_K as a limit, any $\mathbb{Q}$-homomorphism from S_K to an affine finite type group over $\mathbb{Q}$ factors through the projection to some $S_{\mathfrak{m}}$. The following lemma gives another description of algebraic Hecke characters of K; see [**105**, §2.6, Chapter II] for details of its proof.

2.4.5. Lemma. *Let L be a number field.*

(i) *Let $\rho : S_K \to \underline{L}^\times$ (resp. $\rho_{\mathfrak{m}} : S_{\mathfrak{m}} \to \underline{L}^\times$) be a $\mathbb{Q}$-homomorphism. The composition*

$$\mathbb{A}_K^\times \xrightarrow{i_K} S_K(\mathbb{Q}) \xrightarrow{\rho} \underline{L}^\times(\mathbb{Q}) = L^\times$$

is an algebraic Hecke character (resp. the composition

$$I_{\mathfrak{m}} = \mathbb{A}_K^{S(\mathfrak{m}),\times} / \textstyle\prod_{v \notin S(\mathfrak{m})} \mathcal{O}_{K,v}^\times \longrightarrow \mathbb{A}_K^\times/U_{\mathfrak{m}} \xrightarrow{i_{K,\mathfrak{m}}} S_{\mathfrak{m}}(\mathbb{Q}) \xrightarrow{\rho_{\mathfrak{m}}} \underline{L}^\times(\mathbb{Q}) = L^\times$$

is an algebraic Hecke character with conductor $\leqslant \mathfrak{m}$).

(ii) *Conversely, every algebraic Hecke character $\epsilon : \mathbb{A}_K^\times \to L^\times$ (resp. algebraic Hecke character $\chi : I_{\mathfrak{m}} \to L^\times$ with conductor $\leqslant \mathfrak{m}$) comes from a unique $\mathbb{Q}$-homomorphism $\rho : S_K \to \underline{L}^\times$ (resp. $\rho_{\mathfrak{m}} : S_{\mathfrak{m}} \to \underline{L}^\times$).* □

2.4.6. Remark. (i) In the situation of Lemma 2.4.5, the composition of the natural (quotient) map $\underline{K}^\times \to \mathfrak{S}^K$ with the restriction $\rho|_{\mathfrak{S}^K}$ of ρ to $\mathfrak{S}^K \subseteq S^K$ is the algebraic part ϵ_{alg} of the algebraic Hecke character ϵ.

(ii) For later reference, we record the following fact: any two algebraic Hecke characters $\epsilon_1, \epsilon_2 : \mathbb{A}_K^\times \rightrightarrows L^\times$ with the same algebraic part coincide on an open subgroup of finite index containing $K^\times$. This is easily seen from either of the equivalent definitions of the concept of algebraic Hecke character in Definition 2.4.3, due to the finiteness of generalized ideal class groups (and their adelic descriptions).

2.4.7. Consider the exact sequence of $\mathbb{A}_f$-groups obtained from (2.4.4.1) by base change along $\operatorname{Spec}\mathbb{A}_f \to \operatorname{Spec}\mathbb{Q}$. There exists a natural continuous section ϕ : $\operatorname{Gal}(K^{\mathrm{ab}}/K) \to S_K(\mathbb{A}_f) = \varprojlim S_{\mathfrak{m}}(\mathbb{A}_f)$ over the subgroup of "constant points" in the "profinite constant" group scheme over $\operatorname{Spec}\mathbb{A}_f$ associated to $\operatorname{Gal}(K^{\mathrm{ab}}/K)$. We review its construction below; in Chapter II, §2.3 in [**105**] the associated continuous ℓ-adic splittings $\phi_\ell : \operatorname{Gal}(K^{\mathrm{ab}}/K) \to S_K(\mathbb{Q}_\ell) = \varprojlim S_{\mathfrak{m}}(\mathbb{Q}_\ell)$ over $\mathbb{Q}_\ell$ are constructed (or rather, the composite of this with projection to each $S_{\mathfrak{m}}(\mathbb{Q}_\ell)$ is constructed and is denoted ε_ℓ with $\mathfrak{m}$ fixed).

Beware that the splitting ϕ *cannot* extend to a splitting as group schemes over $\operatorname{Spec}(\mathbb{A}_f)$ since any map from an affine (flat) proalgebraic group scheme to a finitely presented affine group scheme factors through a finite type flat quotient of the source whereas the splitting maps $\operatorname{Gal}(K^{\mathrm{ab}}/K) \to S_{\mathfrak{m}}(\mathbb{Q}_\ell)$ that we shall construct will not factor through any finite quotient $C_{\mathfrak{m}'}$ of the source.

To construct ϕ it suffices to construct a compatible family of continuous homomorphisms $\phi_{\mathfrak{m}} : \operatorname{Gal}(K^{\mathrm{ab}}/K) \to S_{\mathfrak{m}}(\mathbb{A}_f)$, where $\phi_{\mathfrak{m}}$ lifts the canonical quotient map $\operatorname{Gal}(K^{\mathrm{ab}}/K) \twoheadrightarrow C_{\mathfrak{m}}$ (whose target is viewed as $\mathbb{Q}$-points within the $\mathbb{A}_f$-points of the finite constant $\mathbb{Q}$-group associated to $C_{\mathfrak{m}}$). Let $\pi_{\mathfrak{m}} : \underline{K}^\times \to T_{\mathfrak{m}}$ be the canonical quotient map over $\mathbb{Q}$. Define $\widetilde{\phi}_{\mathfrak{m}} : \mathbb{A}_K^\times \to S_{\mathfrak{m}}(\mathbb{A}_f)$ by

$$\widetilde{\phi}_{\mathfrak{m}}(x) = i_{K,\mathfrak{m}}(x \bmod U_{\mathfrak{m}}) \cdot \pi_{\mathfrak{m}}(x_f)^{-1},$$

where $x = (x_\infty, x_f)$ with $x_\infty \in K_\infty^\times$ and $x_f \in \mathbb{A}_{K,f}^\times$ denoting the archimedean and finite components of x respectively. The point $i_{K,\mathfrak{m}}(x) \in S_{\mathfrak{m}}(\mathbb{Q})$ is viewed as an element of $S_{\mathfrak{m}}(\mathbb{A}_f)$ in the natural manner, and $\pi_{\mathfrak{m}}(x_f) \in T_{\mathfrak{m}}(\mathbb{A}_f) \subseteq S_{\mathfrak{m}}(\mathbb{A}_f)$, so $i_{K,\mathfrak{m}}(x \bmod U_{\mathfrak{m}}) \cdot \pi_{\mathfrak{m}}(x_f)^{-1}$ makes sense as an element of $S_{\mathfrak{m}}(\mathbb{A}_f)$. By construction, the map $\widetilde{\phi}_{\mathfrak{m}}$ is continuous and its restriction to $K^\times$ is trivial. Thus, $\widetilde{\phi}_{\mathfrak{m}}$ factors continuously through the projection $\mathbb{A}_K^\times \twoheadrightarrow \mathbb{A}_K^\times/K^\times$. Since the target group of $\widetilde{\phi}_{\mathfrak{m}}$ is totally disconnected, $\widetilde{\phi}_{\mathfrak{m}}$ factors through the topological quotient $\mathbb{A}_K^\times \twoheadrightarrow \pi_0(\mathbb{A}_K^\times/K^\times) \simeq \operatorname{Gal}(K^{\mathrm{ab}}/K)$ to define a continuous homomorphism $\phi_{\mathfrak{m}} : \operatorname{Gal}(K^{\mathrm{ab}}/K) \to S_{\mathfrak{m}}(\mathbb{A}_f)$. The compatibility of $\phi_{\mathfrak{m}}$ with respect to change in $\mathfrak{m}$ is easy to check, so we have constructed ϕ.

For every prime number ℓ, define the continuous map

$$\phi_\ell := \pi_\ell \circ \phi : \operatorname{Gal}(K^{\mathrm{ab}}/K) \to S_K(\mathbb{Q}_\ell),$$

where $\pi_\ell : S_K(\mathbb{A}_f) \to S_K(\mathbb{Q}_\ell)$ is the homomorphism induced by the projection $\mathbb{A}_f \to \mathbb{Q}_\ell$. The associated ℓ-adic splitting $\operatorname{Gal}(K^{\mathrm{ab}}/K) \to S_{\mathfrak{m}}(\mathbb{Q}_\ell)$ is the ℓ-adic component of $\phi_{\mathfrak{m}}$.

2.4.8. Lemma. *Let $\epsilon : \mathbb{A}_K^\times \to L^\times$ be an algebraic Hecke character corresponding to a $\mathbb{Q}$-homomorphism $\rho : S_K \to \underline{L}^\times$, with L a number field. Let ℓ be a prime number. Then*

$$\psi_\ell := \rho \circ \phi_\ell : \operatorname{Gal}(K^{\mathrm{ab}}/K) \to \underline{L}^\times(\mathbb{Q}_\ell) = L_\ell^\times$$

is a continuous homomorphism with the following properties.

(i) *There exists a finite set Σ of places of K, including all archimedean places and all places above ℓ, such that ψ_ℓ is unramified outside Σ and such that*

$\psi_\ell(W_v) \subset L^\times$ for each place $v \notin \Sigma$ of K, where W_v is the Weil subgroup of the decomposition group of v in $\mathrm{Gal}(K^{\mathrm{ab}}/K)$. In fact, $\psi_\ell(\mathrm{Frob}_v) = \epsilon(\pi_v)$, where π_v is any local uniformizer at $v \notin \Sigma$ and $\mathrm{Frob}_v \in \mathrm{Gal}(K^{\mathrm{ab}}/K)$ is an arithmetic Frobenius element at v.

(ii) *The composition*

$$\psi_\ell \circ r_{K,\ell} : K_\ell^\times := \prod_{\lambda|\ell} K_\lambda^\times \to L_\ell^\times$$

of ψ_ℓ with the finite product of local Artin maps

$$r_{K,\ell} : K_\ell^\times := \prod_{\lambda|\ell} K_\lambda^\times \to \prod_{\lambda|\ell} \mathrm{Gal}(K_\lambda^{\mathrm{ab}}/K_\lambda) \to \mathrm{Gal}(K^{\mathrm{ab}}/K)$$

coincides with the algebraic homomorphism

$$\epsilon_{\mathrm{alg}}^{-1} : K_\ell^\times = \underline{K}^\times(\mathbb{Q}_\ell) \to \underline{L}^\times(\mathbb{Q}_\ell) = L_\ell^\times$$

on an open subgroup of $K_\ell^\times$. In particular, ψ_ℓ is locally algebraic at ℓ in the sense that $\psi_\ell \circ r_{K,\ell}$ coincides near 1 with a homomorphism that is algebraic with respect to $\mathbb{Q}_\ell$.

Moreover, if ρ factors canonically through a homomorphism $\rho_{\mathfrak{m}} : S_{\mathfrak{m}} \to \underline{L}^\times$ then $\psi_\ell = \rho_{\mathfrak{m}} \circ (\phi_{\mathfrak{m}})_\ell$, where $(\phi_{\mathfrak{m}})_\ell$ is the $\mathbb{Q}_\ell$-part of $\phi_{\mathfrak{m}}$.

PROOF. The assertion in (i) is immediate from the definitions. Note that if ϵ has conductor $\leqslant \mathfrak{m}$ then the set Σ in (i) can be taken to be the union of the set of all archimedean places of K and the set of all places of K dividing $\mathfrak{m}$. (See Chapter II, §2.5 in [**105**] for a related discussion.) Also, the validity of the proposed description of ψ_ℓ in terms of $\rho_{\mathfrak{m}}$ if ρ factors through $\rho_{\mathfrak{m}}$ is clear from the definitions.

As for (ii), let $r_K : \mathbb{A}_K^\times \to \mathrm{Gal}(K^{\mathrm{ab}}/K)$ be the composition of $\mathbb{A}_K^\times \to \mathbb{A}_K^\times/K^\times$ with the Artin map $\mathrm{rec}_K : \mathbb{A}_K^\times/K^\times \to \mathrm{Gal}(K^{\mathrm{ab}}/K)$. Observe that for $x \in K^\times$ we have $\psi_\ell(r_{K,\ell}(x)) = \psi_\ell(r_K((x)^{(\ell)}))^{-1}$, where $(x)^{(\ell)} \in \mathbb{A}_K^\times$ is the idele that is trivial at all $v|\ell$ and is given by x at all $v \nmid \ell$. But for any finite place $v \nmid \ell$ on K, the image of inertia at v in $\mathrm{Gal}(K^{\mathrm{ab}}/K)$ has finite pro-ℓ part. Thus, since ψ_ℓ is mapping continuously into $L_\ell^\times$, for x sufficiently near 1 at all $v \in \Sigma$ with $v \nmid \ell$ we have $\psi_\ell(r_K((x)^{(\ell)})) = \psi_\ell(r_K((x)^{(\Sigma)}))$ where the element $(x)^{(\Sigma)} \in \mathbb{A}_K^{\Sigma,\times}$ is the idele that is trivial at Σ and given by x at all places away from Σ.

By (i), $\psi_\ell(r_K((x)^{(\Sigma)})) = \epsilon((x)^{(\Sigma)})$. Taking x to also satisfy $x \equiv 1 \pmod{\mathfrak{m}}$ where $\mathfrak{m}$ is the modulus of ϵ, it follows from the algebraicity of ϵ that $\epsilon((x)^{(\Sigma)}) = \epsilon_{\mathrm{alg}}(x)$ for such x. Hence, $\psi_\ell(r_{K,\ell}(x)) = \epsilon_{\mathrm{alg}}(x)^{-1}$ for $x \in K^\times$ that is sufficiently close to 1 at all $v \in \Sigma$. Using these nearness conditions at $v|\ell$ gives an open neighborhood V_ℓ of the identity in $K_\ell^\times$ on which the desired identity holds because it does so on the subset $K^\times \cap V_\ell$ that is dense in V_ℓ. □

The following converse lemma includes a uniqueness result for ψ_ℓ as in Lemma 2.4.8, and it will be useful later when we need to check an equality of algebraic Hecke characters.

2.4.9. Lemma. *Let K and L be number fields. Let ℓ be a prime number, and let Σ be a finite set of places of K including all archimedean places and all places above ℓ. Let $\psi_\ell : \mathrm{Gal}(K^{\mathrm{ab}}/K) \to L_\ell^\times$ be a continuous homomorphism that is unramified outside Σ and for which $\psi_\ell(W_v) \subseteq L^\times$ for the Weil subgroup W_v in* *the decomposition group* $D_v \subset \mathrm{Gal}(K^{\mathrm{ab}}/K)$ *at each* $v \notin \Sigma$. *Assume moreover that there exists a* $\mathbb{Q}$*-homomorphism* $\chi_{\mathrm{alg}} : \underline{K}^\times \to \underline{L}^\times$ *such that* $\psi_\ell \circ r_{K,\ell}$ *coincides with* χ_{alg} *on an open subgroup* U_ℓ *of* $K_\ell^\times = \underline{K}^\times(\mathbb{Q}_\ell)$.

There exists a unique algebraic Hecke character $\epsilon : \mathbb{A}_K^\times \to L^\times$ *with algebraic part* χ_{alg}^{-1} *such that* ϵ *induces* ψ_ℓ *as in Lemma 2.4.8. In particular, if* K'/K *is a finite extension then* $\epsilon \circ \mathrm{Nm}_{K'/K}$ *corresponds to* $\psi_\ell|_{\mathrm{Gal}(K'^{\mathrm{ab}}/K')}$.

PROOF. Define $\epsilon : \mathbb{A}_K^\times \to L_\ell^\times$ by

$$\epsilon(x) = \psi_\ell(r_K(x)) \cdot \chi_{\mathrm{alg}}^{-1}(x_\ell), \tag{2.4.9.1}$$

where $x_\ell \in K_\ell^\times$ denotes the ℓ-adic component of $x \in \mathbb{A}_K^\times$. Choose an open subgroup W of $\prod_{p \in \Sigma,\ p \neq \ell, \infty} K_p^\times$ such that $(\psi_\ell \circ r_K)(W) = \{1\}$; such an open subgroup W exists because ψ_ℓ is continuous and takes values in a finite product of multiplicative groups of ℓ-adic fields. It is clear that ϵ is trivial on the open subgroup

$$U := (K_\infty^\times)^0 \times U_\ell \times W \times \prod_{v \notin \Sigma} \mathcal{O}_{K,v}^\times$$

of $\mathbb{A}_K^\times$. It is also clear that ϵ coincides with χ_{alg}^{-1} on $K^\times$. Since $\mathbb{A}_K^\times$ is generated by U, $K^\times$ and $(\mathbb{A}_K^\Sigma)^\times$, we have $\epsilon(\mathbb{A}_K^\times) \subseteq L^\times$. Thus, ϵ is an algebraic Hecke character with χ_{alg}^{-1} as its algebraic part. Moreover, ϵ is unramified away from Σ and for each $v \notin \Sigma$ we have $\psi_\ell(\mathrm{Frob}_v) = \epsilon(\pi_v)$, so by continuity this ψ_ℓ is the same as the map arising from ϵ in Lemma 2.4.8.

As for the uniqueness of ϵ, it has to be proved that if ϵ is an $L^\times$-valued finite-order Hecke character of K with $\rho_\epsilon \circ \phi_\ell = 1$ (where $\rho_\epsilon : S_K \to \underline{L}^\times$ corresponds to ϵ) then $\epsilon = 1$. But $\epsilon(\pi_v) = 1$ for all but finitely many v, so the result follows. The final part concerning $\epsilon \circ \mathrm{Nm}_{K'/K}$ follows from the uniqueness and the relationship in class field theory between Galois restriction and norm maps. □

The next lemma will play a technical role in the proof of the main arithmetic result in this section, Theorem 2.4.11.

2.4.10. Lemma. *Let* Γ_0 *be a normal subgroup of finite index in a group* Γ. *Let* γ *be an element of* Γ *of infinite order, and let* $\Gamma_1 := \Gamma_0 \cdot \gamma^{\mathbb{Z}}$ *be the subgroup of* Γ *generated by* Γ_0 *and* γ. *Let* n *be the unique positive integer such that* $\langle \gamma^n \rangle = \Gamma_0 \cap \gamma^{\mathbb{Z}}$.

Let $\rho_0 : \Gamma_0 \to H$ *be a homomorphism of groups and suppose* $h \in H$ *satisfies* $h^n = \rho_0(\gamma^n)$. *If* $\rho_0(\gamma x \gamma^{-1}) = h\, \rho_0(x)\, h^{-1}$ *for all* $x \in \Gamma_0$ *then there exists a unique extension of* ρ_0 *to a homomorphism* $\rho_1 : \Gamma_1 \to H$ *such that* $\rho_1(\gamma) = h$.

PROOF. Let Γ' be the semidirect product $\Gamma_0 \rtimes \mathbb{Z}$ defined by the group law

$$(x, a) \cdot (y, b) = (x\, \gamma^a\, y\, \gamma^{-a}, a + b)$$

for $x, y \in \Gamma_0$ and $a, b \in \mathbb{Z}$ (i.e., $1 \in \mathbb{Z}$ acts on Γ_0 by $y \mapsto \gamma y \gamma^{-1}$). Define a map $\rho' : \Gamma' \to H$ by $\rho'((x, a)) = \rho_0(x) h^a$ for all $(x, a) \in \Gamma'$. The assumption on ρ_0 guarantees that ρ' is a homomorphism of groups. The kernel of the natural surjection $\pi : \Gamma' \to \Gamma_1 = \Gamma_0 \cdot \gamma^{\mathbb{Z}}$ is the cyclic subgroup generated by the element $(\gamma^n, -n)$ since on the normal subgroup $\Gamma_0 \subseteq \Gamma'$ this surjection is the identity and modulo Γ_0 it is the natural projection $\mathbb{Z} \to \mathbb{Z}/n\mathbb{Z}$ (so its kernel cannot contain (γ_0, i) for $\gamma_0 \in \Gamma$ and $0 < i < n$, and for $i = 0$ necessarily $\gamma_0 = 1$). Since the restriction of ρ' to $\ker(\pi)$ is trivial, $\rho' = \rho_1 \circ \pi$ for a unique homomorphism $\rho_1 : \Gamma_1 \to H$. □

Let K and L be number fields. Let $\epsilon : \mathbb{A}_K^\times \to L^\times$ be an algebraic Hecke character, and let $\epsilon_{\mathrm{alg}} : \underline{K}^\times \to \underline{L}^\times$ be the algebraic part of ϵ. Let E be a subfield of K such that ϵ_{alg} factors through the norm map $\mathrm{Nm}_{K/E} : \underline{K}^\times \to \underline{E}^\times$; i.e., there exists a $\mathbb{Q}$-homomorphism $\chi_{\mathrm{alg}} : \underline{E}^\times \to \underline{L}^\times$ such that $\epsilon_{\mathrm{alg}} = \chi_{\mathrm{alg}} \circ \mathrm{Nm}_{K/E}$. (This is weaker than saying $\epsilon = \chi \circ \mathrm{Nm}_{K/E}$ for an algebraic Hecke character $\chi : \mathbb{A}_E^\times \to L^\times$ satisfying $\chi_{\mathrm{alg}} \circ \mathrm{Nm}_{K/E} = \epsilon_{\mathrm{alg}}$ since an algebraic Hecke character is only determined by its algebraic part up to a finite order character. Moreover, χ_{alg} through which ϵ_{alg} factors may not be the algebraic part of an algebraic Hecke character.)

We now construct a special kind of "local root extraction" of ϵ at the expense of replacing the extension K/E with another extension:

2.4.11. Theorem. *Using notation and hypotheses as above, let v be a finite place of K such that ϵ is unramified at v and let w be the place of E induced by v. Suppose that an element $\beta \in L^\times$ satisfies $\beta^{[\kappa_v:\kappa_w]} = \epsilon(\pi_v)$, where π_v is a local uniformizer of $\mathcal{O}_{K,v}$. There exists a finite extension $\widetilde{E}$ of E, a place $\widetilde{w}$ of $\widetilde{E}$ above w, and an algebraic Hecke character $\delta : \mathbb{A}_{\widetilde{E}}^\times \to L^\times$ with the following properties.*

(i) *The degree of the residue field extension for $\widetilde{w}$ over w is 1; i.e., $[\kappa_{\widetilde{w}} : \kappa_w] = 1$.*
(ii) *The algebraic part of δ is equal to $\chi_{\mathrm{alg}} \circ \mathrm{Nm}_{\widetilde{E}/E}$.*
(iii) *The algebraic Hecke character δ is unramified at $\widetilde{w}$, and $\delta(\pi_{\widetilde{w}}) = \beta$, where $\pi_{\widetilde{w}}$ is a uniformizer of $\mathcal{O}_{\widetilde{E},\widetilde{w}}$.*

PROOF. Let K'/K be a finite extension that is Galois over E, and choose a place v' on K' over v. The composition ϵ' of ϵ with $\mathrm{Nm}_{K'/K}$ on ideles is an algebraic Hecke character that is unramified at v' and has algebraic part $\epsilon_{\mathrm{alg}} \circ \mathrm{Nm}_{K'/K} = \chi_{\mathrm{alg}} \circ \mathrm{Nm}_{K'/E}$. If $\pi_{v'} \in \mathcal{O}_{K',v'}$ is a uniformizer then $\mathrm{Nm}_{K'_{v'}/K_v}(\pi_{v'})$ is an $\mathcal{O}_{K,v}^\times$-multiple of $\pi_v^{[\kappa_{v'}:\kappa_v]}$, so $\epsilon'(\pi_{v'}) = \beta^{[\kappa_{v'}:\kappa_w]}$. Thus, by replacing K, ϵ, and v with K', ϵ', and v' we may and do assume that K is Galois over E.

Choose an *arithmetic* Frobenius element σ in a decomposition group $D_{\widetilde{v}|w} \subset \mathrm{Gal}(K^{\mathrm{ab}}/E)$; i.e., σ induces $x \mapsto x^{q_w}$ on the residue field of an extension $\widetilde{v}$ of v to K^{ab}, with $q_w = \#\kappa_w$. Let ℓ be a prime number which is prime to v, and let $\psi_\ell : \mathrm{Gal}(K^{\mathrm{ab}}/K) \to L_\ell^\times$ be the ℓ-adic character attached to the algebraic Hecke character ϵ as in Lemma 2.4.8. As we saw in Lemma 2.4.8, ψ_ℓ is locally algebraic above ℓ with $\epsilon_{\mathrm{alg}}^{-1}$ as its algebraic part above ℓ. Choose $M \geqslant 1$ such that $\sigma^{f(v|w)M}$ lies in the decomposition group $D_{\widetilde{v}|v} \subset \mathrm{Gal}(K^{\mathrm{ab}}/K)$ at $\widetilde{v}$, with $f(v|w) := [\kappa_v : \kappa_w]$. This agrees modulo the inertia group $I_{\widetilde{v}|v}$ with the Mth power of an arithmetic Frobenius element in $D_{\widetilde{v}|v}$, so by unramifiedness of ψ_ℓ at v we compute $\psi_\ell(\sigma^{f(v|w)M}) = \psi_\ell(\mathrm{Frob}_v^M) = \epsilon(\pi_v)^M = \beta^{f(v|w)M}$.

Since K/E is Galois, the element $\sigma \in \mathrm{Gal}(K^{\mathrm{ab}}/E)$ acts on $\mathrm{Gal}(K^{\mathrm{ab}}/K)$ by conjugation. Consider the conjugate of ψ_ℓ by σ; i.e., the continuous homomorphism ${}^\sigma\psi_\ell : \mathrm{Gal}(K^{\mathrm{ab}}/K) \to L_\ell^\times$ defined by $\tau \mapsto \psi_\ell(\sigma^{-1}\tau\sigma)$. Clearly ${}^\sigma\psi_\ell$ is unramified at all but finitely many places and is locally algebraic above ℓ, with algebraic part above ℓ given by composing the algebraic part of ψ_ℓ above ℓ with the action by $\sigma^{-1}|_K$ on $K_\ell^\times$. Because ϵ_{alg} factors through $\mathrm{Nm}_{K/E}$, it is invariant under the action of $\sigma|_K \in \mathrm{Gal}(K/E)$. Hence, the algebraic part of ${}^\sigma\psi_\ell$ above ℓ is also equal to $\epsilon_{\mathrm{alg}}^{-1}$. Therefore, by Remark 2.4.6(ii), there exists a finite abelian extension field K_1 of K which is Galois over E such that ${}^\sigma\psi_\ell$ coincides with ψ_ℓ on $\mathrm{Gal}(K^{\mathrm{ab}}/K_1)$. Replacing

K_1 by a finite extension that is abelian over K and Galois over E with sufficiently divisible degree over K_1 (e.g., a sufficiently large ray class field of K), we may assume (by the relationship between finite-index subgroups of $\mathbb{Z}$ and open subgroups of $\widehat{\mathbb{Z}}$) that there exists a positive integer N_1 such that $\sigma^{\mathbb{Z}} \cap \mathrm{Gal}(K^{\mathrm{ab}}/K_1) = \sigma^{f(v|w)MN_1\mathbb{Z}}$.

Let $\widetilde{E} \subseteq K_1$ be the subfield over E that is the fixed field of the cyclic group $\langle \sigma|_{K_1} \rangle \subseteq \mathrm{Gal}(K_1/E)$ with order $f(v|w)MN_1$. Since $\psi_\ell(\sigma^{f(v|w)MN_1}) = \beta^{f(v|w)MN_1}$, by applying Lemma 2.4.10 with $\Gamma_0 = \mathrm{Gal}(K^{\mathrm{ab}}/K_1)$, $\Gamma = \mathrm{Gal}(K^{\mathrm{ab}}/E)$, $\gamma = \sigma$, and $n = f(v|w)MN_1$, we see that there is a unique continuous homomorphism $\widetilde{\psi}_\ell : \mathrm{Gal}(K^{\mathrm{ab}}/\widetilde{E}) \to L_\ell^\times$ that coincides with ψ_ℓ on $\mathrm{Gal}(K^{\mathrm{ab}}/K_1)$ and satisfies $\widetilde{\psi}_\ell(\sigma) = \beta$. In particular, $\widetilde{\psi}_\ell$ is unramified at all but finitely many places of $\widetilde{E}$ since ψ_ℓ is unramified at all but finitely many places of K and the finite extension $K_1/\widetilde{E}$ is unramified at all but finitely many places.

Let $\widetilde{w}$ be the restriction to $\widetilde{E}$ of the place $\widetilde{v}$ on K^{ab}, so $\sigma \in D_{\widetilde{v}|\widetilde{w}}$. Since the subgroup $\mathrm{Gal}(K_1/\widetilde{E}) \subseteq \mathrm{Gal}(K_1/E)$ is the subgroup generated by $\sigma|_{K_1}$, we have that $\kappa_w \subseteq \kappa_{\widetilde{w}}$. But $\kappa_{\widetilde{w}}$ has trivial action by the automorphism $x \mapsto x^{q_w}$ induced by $\sigma|_{K_1}$. Hence, $\kappa_w = \kappa_{\widetilde{w}}$, so (i) holds. Thus, $q_{\widetilde{w}} = q_w$, so the element $\sigma \in D_{\widetilde{v}|\widetilde{w}} \subset \mathrm{Gal}(K^{\mathrm{ab}}/\widetilde{E})$ is an arithmetic Frobenius element in this decomposition group. Beware that even though $\sigma|_{K_1}$ generates $\mathrm{Gal}(K_1/\widetilde{E})$, we do not see how to avoid the possibility that $K_1/\widetilde{E}$ is ramified at the restriction v_1 of $\widetilde{v}$ to K_1 (especially if K/E is ramified at v). Thus, we will have to do some work to check that the algebraic Hecke character δ that we shall construct is unramified at $\widetilde{w}$.

By construction, the continuous $L_\ell^\times$-valued character $\widetilde{\psi}'_\ell$ of $\mathrm{Gal}(\widetilde{E}^{\mathrm{ab}}/\widetilde{E})$ arising from the continuous homomorphism $\widetilde{\psi}_\ell : \mathrm{Gal}(K^{\mathrm{ab}}/\widetilde{E}) \to L_\ell^\times$ is locally algebraic above ℓ, with algebraic part $\chi_{\mathrm{alg}}^{-1} \circ \mathrm{Nm}_{\widetilde{E}/E}$ because the relative norm map goes over to "restriction" on Galois groups via class field theory and we can compute the algebraic part by working near the identity (such as on the open image of $\mathrm{Gal}(K^{\mathrm{ab}}/K_1)$). Since the continuous $\widetilde{\psi}'_\ell$ is unramified at all but finitely many places, the proof of Lemma 2.4.9 (especially the analysis of open kernels) therefore shows that the continuous homomorphism $\delta : \mathbb{A}_{\widetilde{E}}^\times \to L_\ell^\times$ defined by the formula

$$\delta(x) = \widetilde{\psi}'_\ell(r_{\widetilde{E}}(x)) \cdot \chi_{\mathrm{alg}}(\mathrm{Nm}_{\widetilde{E}/E}(x_\ell))$$

(where the ℓ-component of $x \in \mathbb{A}_{\widetilde{E}}^\times$ is denoted $x_\ell \in \widetilde{E}_\ell^\times$) is an algebraic Hecke character of $\widetilde{E}$ with values in the multiplicative group of L_ℓ, a finite product of fields. We likewise see that $\delta_{\mathrm{alg}} = \chi_{\mathrm{alg}} \circ \mathrm{Nm}_{\widetilde{E}/E}$. By algebraicity δ has open kernel, so it is continuous with respect to the discrete topology on $L_\ell^\times$.

Let U be the open subgroup $\widetilde{E}^\times \cdot \mathrm{Nm}_{K_1/\widetilde{E}}(\mathbb{A}_{K_1}^\times)$ of finite index in $\mathbb{A}_{\widetilde{E}}^\times$, and let $\pi_{\widetilde{w}}$ be a uniformizer of $\mathcal{O}_{\widetilde{E},\widetilde{w}}$ chosen such that the arithmetic Frobenius element $r_{\widetilde{E}}(\pi_{\widetilde{w}}) \in \mathrm{Gal}(\widetilde{E}^{\mathrm{ab}}/\widetilde{E})$ at $\widetilde{w}$ is equal to the arithmetic Frobenius element $\sigma|_{\widetilde{E}^{\mathrm{ab}}}$. Such a $\pi_{\widetilde{w}}$ can be found because the profinite unit group $\mathcal{O}_{\widetilde{w}}^\times$ surjects onto the inertia subgroup of $\mathrm{Gal}(\widetilde{E}_{\widetilde{w}}^{\mathrm{ab}}/\widetilde{E}_{\widetilde{w}})$. The idele group $\mathbb{A}_{\widetilde{E}}^\times$ is generated by U and $\pi_{\widetilde{w}}$ because $\mathrm{Gal}(K_1/\widetilde{E})$ is generated by $\sigma|_{K_1}$. Thus, to show that δ takes its values in $L^\times \subset L_\ell^\times$ it suffices to prove that $\delta(\pi_{\widetilde{w}}) \in L^\times$ and $\delta|_U$ is valued in $L^\times$.

Since $\ell \nmid \widetilde{w}$ (as $\ell \nmid v$), the definitions of δ, $\widetilde{\psi}_\ell$, and $\pi_{\widetilde{w}}$ imply that

$$\delta(\pi_{\widetilde{w}}) = \widetilde{\psi}'_\ell(r_{\widetilde{E}}(\pi_{\widetilde{w}})) = \widetilde{\psi}'_\ell(\sigma|_{\widetilde{E}^{\mathrm{ab}}}) = \widetilde{\psi}_\ell(\sigma) = \beta \in L^\times.$$

For any element $x \in U$, written in the form $x = \mathrm{Nm}_{K_1/\widetilde{E}}(y) \cdot z$ with $y \in \mathbb{A}_{K_1}^\times$ and $z \in \widetilde{E}^\times$, the triviality of $r_{\widetilde{E}}(\widetilde{E}^\times)$ implies

$$\begin{aligned} \delta(x) &= \widetilde{\psi}'_\ell(r_{\widetilde{E}}(x)) \cdot \chi_{\mathrm{alg}}(\mathrm{Nm}_{\widetilde{E}/E}(x_\ell)) \\ &= \psi_\ell(r_K(\mathrm{Nm}_{K_1/K}(y))) \cdot \chi_{\mathrm{alg}}(\mathrm{Nm}_{K/E}(\mathrm{Nm}_{K_1/K}(y_\ell))) \cdot \chi_{\mathrm{alg}}(\mathrm{Nm}_{\widetilde{E}/E}(z_\ell)) \\ &= \epsilon(\mathrm{Nm}_{K_1/K}(y)) \cdot \chi_{\mathrm{alg}}(\mathrm{Nm}_{\widetilde{E}/E}(z)) \in L^\times, \end{aligned}$$

where the final equality uses the construction of ϵ from ψ_ℓ as in the proof of Lemma 2.4.9 (since $\epsilon_{\mathrm{alg}} = \chi_{\mathrm{alg}} \circ \mathrm{Nm}_{K/E}$). Thus, indeed $\delta(\mathbb{A}_{\widetilde{E}}^\times) \subseteq L^\times$.

We have shown that δ is an algebraic Hecke character satisfying the required properties (i)–(iii), except for verifying that δ is unramified at $\widetilde{w}$. The key to settling this remaining issue even if $K_1/\widetilde{E}$ is ramified at $\widetilde{w}$ is to exploit the cyclicity of $K_1/\widetilde{E}$, and more specifically the Hasse Norm Theorem for cyclic extensions of number fields. For a local unit $u \in \mathcal{O}_{\widetilde{E},\widetilde{w}}^\times$, we have to prove $\delta(u) = 1$ when u is viewed in $\mathbb{A}_{\widetilde{E}}^\times$ in the natural manner.

We can write $u = z \cdot \mathrm{Nm}_{K_1/\widetilde{E}}(y) \cdot \pi_{\widetilde{w}}^i$ for some $i \in \mathbb{Z}$, $z \in \widetilde{E}^\times$, and $y \in \mathbb{A}_{K_1}^\times$. In particular, since the $\widetilde{E}$-idele $u\pi_{\widetilde{w}}^{-i}$ has trivial local component at every place away from $\widetilde{w}$, it follows that the element $z \in \widetilde{E}^\times$ is a local norm from K_1 at every place away from $\widetilde{w}$. By the triviality of $r_{\widetilde{E}}$ on $\widetilde{E}^\times$, z is a local norm at $\widetilde{w}$. Hence, by the Hasse Norm Theorem for cyclic extensions, z is a global norm: $z = \mathrm{Nm}_{K_1/E}(z_1)$ for some $z_1 \in K_1^\times$. Thus, by replacing y with $z_1 y$ we can assume $z = 1$. There is a unique place v_1 of K_1 over $\widetilde{w}$ because $\sigma \in D_{\widetilde{v}|\widetilde{w}} \subset \mathrm{Gal}(K^{\mathrm{ab}}/\widetilde{E})$ and $\mathrm{Gal}(K_1/\widetilde{E})$ is generated by $\sigma|_{K_1}$. We therefore have $u = \mathrm{Nm}_{K_{1,v_1}/\widetilde{E}_{\widetilde{w}}}(y_1)\pi_{\widetilde{w}}^i$, where y_1 is the v_1-component of the K_1-idele y.

Let $\pi_{v_1} \in \mathcal{O}_{K_1,v_1}^\times$ be a uniformizer, so there are unique $j \in \mathbb{Z}$ and $u_1 \in \mathcal{O}_{K_1,v_1}^\times$ such that $y_1 = \pi_{v_1}^j u_1$. Since $\mathrm{Nm}_{K_{1,v_1}/\widetilde{E}_{\widetilde{w}}}(\pi_{v_1})$ is an $\mathcal{O}_{\widetilde{E},\widetilde{w}}^\times$-multiple of $\pi_{\widetilde{w}}^{f(v_1|\widetilde{w})}$, the condition $\mathrm{ord}_{\widetilde{w}}(u) = 0$ forces $i = -f(v_1|\widetilde{w})j = -f(v_1|w)j = -f(v_1|v)f(v|w)j$. Hence,

$$\begin{aligned} \delta(u) = \delta(\mathrm{Nm}_{K_1/\widetilde{E}}(y_1))\delta(\pi_{\widetilde{w}})^i &= \epsilon(\mathrm{Nm}_{K_1/K}(y_1))\beta^{-f(v_1|v)f(v|w)j} \\ &= \epsilon(\mathrm{Nm}_{K_1/K}(y_1))\epsilon(\pi_v)^{-f(v_1|v)j} \end{aligned}$$

where y_1 is viewed as an K_1-idele supported at v_1 (so its K_1/K-norm is

$$\mathrm{Nm}_{K_1/K}(y_1) = \mathrm{Nm}_{K_{1,v_1}/K_v}(y_1)$$

as an K-idele supported at v). Since ϵ is unramified at v, $\epsilon(\mathrm{Nm}_{K_{1,v_1}/K_v}(y_1)) = \epsilon(\mathrm{Nm}_{K_{1,v_1}/K_v}(\pi_{v_1}))^j$. Thus, we just have to show that

$$\epsilon(\pi_v)^{f(v_1|v)} = \epsilon(\mathrm{Nm}_{K_{1,v_1}/K_v}(\pi_{v_1})).$$

This equality follows from the unramifiedness of ϵ at v and the fact that the (normalized) v-adic ordinal of $\mathrm{Nm}_{K_{1,v_1}/K_v}(\pi_{v_1})$ is $f(v_1|v)$. □

2.4.12. Remark. In the final part of the proof of Theorem 2.4.11, to prove the unramifiedness of δ at $\widetilde{w}$, we had to do some extra work with the Hasse Norm Theorem to overcome the possibility that $K_1/\widetilde{E}$ could be ramified at its unique place above $\widetilde{w}$. Let us indicate an alternative procedure to bypass this difficulty by using the Grunwald-Wang theorem [**4**, Ch. X, Thm. 6].

Let v_1 be the unique place of K_1 above $\widetilde{w}$, and let $e(v_1|\widetilde{w})$ be the ramification index. By the Grunwald-Wang theorem, there exists an cyclic extension K_2/K_1 which is unramified at v_1 such that the residual degree $f(v_1, K_2/K_1)$ is a multiple of $e(v_1|\widetilde{w})$. Let K_3/K_2 be a Galois closure of K_2 over $\widetilde{E}$, so K_3/K_1 is abelian and unramified over v_1. Let v_2 be a place of K_2 above v_1, and let v_3 be a place of K_3 above v_2.

Consider the short exact sequence

$$1 \to I_{v_1|\widetilde{w}} \to D_{v_1|\widetilde{w}} \to \mathrm{Gal}(\kappa_{v_1}/\kappa_{\widetilde{w}}) \to 1. \tag{2.4.12.1}$$

The short exact sequence $1 \to I_{v_3|\widetilde{w}} \to D_{v_3|\widetilde{w}} \to \mathrm{Gal}(\kappa_{v_3}/\kappa_{\widetilde{w}}) \to 1$ is the pullback of (2.4.12.1) by the natural surjection $\mathrm{Gal}(\kappa_{v_3}/\kappa_{\widetilde{w}}) \twoheadrightarrow \mathrm{Gal}(\kappa_{v_1}/\kappa_{\widetilde{w}})$ because $e(v_3|\widetilde{w}) = e(v_1|\widetilde{w})$. This pullback sequence splits (as a semi-direct product) because $\mathrm{Gal}(\kappa_{v_3}/\kappa_{\widetilde{w}})$ is a cyclic group whose cardinality is a multiple of the cardinality of $D_{v_1|\widetilde{w}}$; here we use that K_2/K_1 was chosen so that $e(v_1|\widetilde{w})|[\kappa_{v_2} : \kappa_{v_1}]$. Thus, there exists $\sigma_3 \in D_{v_3|\widetilde{w}}$ that maps to the arithmetic Frobenius element in $\mathrm{Gal}(\kappa_{v_3}/\kappa_{\widetilde{w}})$ and for which the subgroup generated by σ_3 is a lifting of $\mathrm{Gal}(\kappa_{v_3}/\kappa_{\widetilde{w}})$ in $D_{v_3|\widetilde{w}}$.

Let E_3 be the subfield of K_3 fixed by σ_3, so v_3 is the only place on K_3 over its restriction w_3 in E_3. The extension K_3/E_3 is unramified over w_3, by consideration of field degrees (residually and generically). The statement of Theorem 2.4.11 holds with $(\widetilde{E}, \widetilde{w})$ replaced by (E_3, w_3); the unramifiedness of δ at w_3 is clear because K_3/E_3 is unramified there.

2.5. Theory of complex multiplication

We now review part of the theory of complex multiplication due to Shimura and Taniyama, and use it to deduce from Theorem 2.4.11 a result concerning CM liftings of abelian varieties up to isogeny. We will make extensive use of the reflex field $E \subset \overline{\mathbb{Q}}$ associated to a $\overline{\mathbb{Q}}$-valued CM type (L, Φ), as well as the reflex norm

$$\mathrm{N}_\Phi : \mathrm{Res}_{E/\mathbb{Q}}(\mathbb{G}_\mathrm{m}) \to \mathrm{Res}_{L/\mathbb{Q}}(\mathbb{G}_\mathrm{m})$$

(see Definitions 2.1.3 and 2.1.3.3). For our present purposes, the following version of the Main Theorem of Complex Multiplication will suffice. A refined version is stated and proved in A.2.

2.5.1. Theorem (Main Theorem of CM). *Let $(A, \alpha : L \to \mathrm{End}^0(A))$ be a* CM *abelian variety over a number field $K \subset \overline{\mathbb{Q}}$ with L a* CM *field, and let $\Phi \subset$* $\mathrm{Hom}(L, \overline{\mathbb{Q}})$ *be the associated* CM *type.*

(i) *The reflex field $E \subset \overline{\mathbb{Q}}$ of (L, Φ) is contained in K.*
(ii) *There exists a unique algebraic Hecke character $\epsilon : \mathbb{A}_K^\times \to L^\times$ such that for every prime number ℓ, the continuous homomorphism $\psi_\ell : \mathrm{Gal}(K^{\mathrm{ab}}/K) \to L_\ell^\times$ attached to ϵ in Lemma 2.4.8 is equal to the ℓ-adic representation of $\mathrm{Gal}(K^{\mathrm{ab}}/K)$ attached to the ℓ-power torsion points of A.*
(iii) *The algebraic part of the algebraic Hecke character ϵ is the map $\mathrm{N}_\Phi \circ \mathrm{Nm}_{K/E}$, where $\mathrm{N}_\Phi : \mathrm{Res}_{E/\mathbb{Q}}(\mathbb{G}_\mathrm{m}) \to \mathrm{Res}_{L/\mathbb{Q}}(\mathbb{G}_\mathrm{m})$ is the reflex norm.*
(iv) *Let v be a finite place of K. The abelian variety A has good reduction at v if and only if the algebraic Hecke character ϵ is unramified at v.*
(v) *If v is a finite place of K where A has good reduction, say with reduction $\overline{A}$ over κ_v, then $\epsilon(\pi_v) = \mathrm{Fr}_{\overline{A}, q_v}$, where π_v is a uniformizer of $\mathcal{O}_{K_v}$ and $q_v = \#\kappa_v$.*

Part (i) is immediate from the definition of the reflex field (see Example 2.1.3.1), and in A.2 we use the torus from Proposition 2.1.3.6 to state and prove an alternative version of the Main Theorem (see Theorem A.2.4.4), from which we deduce Theorem 2.5.1. More precisely, parts (ii) and (iii) correspond to Theorem A.2.5.4 and Remark A.2.5.5 (the uniqueness in (ii) being obvious), and parts (iv) and (v) correspond to the two respective parts of Theorem A.2.5.9.

2.5.1.1. Remark. Since Theorem 2.5.1 is insensitive to L-linear isogenies in A over K, by Proposition 1.7.4.5 there is no loss of generality in assuming that the CM order is $\mathcal{O}_L$, as is sometimes done in the literature on the Main Theorem. Parts (ii) and (iii) of Theorem 2.5.1 amount to a reformulation of the Main Theorem as stated in [**111**]. Part (iv) is proved in §7 of [**109**] as a consequence of this Main Theorem (and (v) follows from (ii) and (iv)). See [**105**, (II-25)–(II-28)] and [**65**, §1, Ch. 4] for further discussion of the Main Theorem and related constructions.

The following converse to Theorem 2.5.1 is a consequence of the motivic theory of complex multiplication (see [**28**, §E, Prop. E.1]). It was first proved in [**114**, §2, Thm. 6] and attributed to Casselman, using an older style of algebraic geometry and the traditional analytic formulation of the Main Theorem of Complex Multiplication. (This analytic version of the Main Theorem is stated in [**65**, Thm. 6.1, Ch. 3], and is stated and proved in a coordinate-free form in A.2.8.) In A.3 we provide a proof of this converse result using the constructions in the proof of the Main Theorem in A.2. See also [**65**, Theorem 5.1] and [**115**, Theorem 21.4] for treatments of Casselman's theorem along the same lines as in [**114**].

2.5.2. Theorem. *Let L be a* CM *field with $[L:\mathbb{Q}] = 2g$. Let Φ be a* CM *type for L valued in a fixed algebraic closure $\overline{\mathbb{Q}}$ and let $E \subset \overline{\mathbb{Q}}$ be the reflex field of (L,Φ). Let K/E be a finite extension. Let $\epsilon : \mathbb{A}_K^\times \to L^\times$ be an algebraic Hecke character whose algebraic part is the composite map $\mathrm{N}_\Phi \circ \mathrm{Nm}_{K/E}$, where N_Φ is the reflex norm associated to (L,Φ).*

There exists a g-dimensional abelian variety A over K and a ring homomorphism $\alpha : L \to \mathrm{End}^0(A)$ such that Φ is the CM *type of (A,α) and ϵ is the algebraic Hecke character attached to (A,α) as in Theorem* 2.5.1(ii)*. Moreover (A,α) is unique up to L-linear K-isogeny.*

We are now in position to prove Theorem 2.1.6. Let us first recall the statement.

2.5.3. Theorem. *Let L be a* CM *field with degree $2g$ over $\mathbb{Q}$. Let $q = p^r$, where p is a prime number. Let B be an isotypic abelian variety of dimension g over a finite field $\mathbb{F}_q$ with size q, and let $\alpha_B : L \to \mathrm{End}^0(B)$ be a ring homomorphism. Suppose that there exists a* CM *type $\Phi \subset \mathrm{Hom}_{\mathrm{ring}}(L,\overline{\mathbb{Q}}_p)$ of L such that the residual reflex condition in* 2.1.5 *is satisfied for (L,Φ). Let $E \subset \overline{\mathbb{Q}}_p$ be the reflex field of (L,Φ), and let w be the induced p-adic place on E.*

There exists a finite extension $\widetilde{E}/E$ inside $\overline{\mathbb{Q}}_p$, an abelian variety A over $\widetilde{E}$, and a ring homomorphism $\alpha : L \to \mathrm{End}^0(A)$ such that the following hold.

(i) *The residue field $\kappa_{\widetilde{w}}$ of $\widetilde{E}$ at the place $\widetilde{w}$ induced from $\overline{\mathbb{Q}}_p$ is isomorphic to $\mathbb{F}_q$.*
(ii) *The $\overline{\mathbb{Q}}_p$-valued* CM *type of (A,α) is Φ.*
(iii) *The abelian variety A over $\widetilde{E}$ has good reduction at $\widetilde{w}$.*

(iv) *Via a suitable isomorphism* $\mathbb{F}_q \simeq \kappa_{\widetilde{w}}$, B *is* L*-linearly isogenous to the reduction of* A *at* $\widetilde{w}$.

Note that in (iv) the choice of isomorphism $\mathbb{F}_q \simeq \kappa_{\widetilde{w}}$ does not matter, since B and $B^{(p)}$ are L-linearly isogenous over $\mathbb{F}_q$ via the relative Frobenius map for B.

PROOF. This proposition is a consequence of Theorems 2.4.11, 2.5.1, and 2.5.2, as we now explain.

By Proposition 1.5.4.1 (and Proposition 1.7.4.5), there exists an abelian variety C of dimension g over a finite extension F of $\mathbb{Q}_p$ inside $\overline{\mathbb{Q}}_p$ and a homomorphism $\xi : \mathcal{O}_L \to \mathrm{End}(C)$ such that (C, ξ) has CM type Φ. In particular, the reflex field $E \subset \overline{\mathbb{Q}}_p$ is a subfield of F such that the canonical p-adic absolute value on F induces the p-adic place w on E. By replacing F with a finite extension if necessary, we can assume that C has good reduction over $\mathcal{O}_F$ and that the size q' of the residue field κ of F is a power of q. Since the reduction $\overline{C}$ has CM by a field, it is isotypic due to Theorem 1.3.1.1.

Let $\pi' \in \mathcal{O}_L$ be the element whose action on the reduction $\overline{C}$ over κ is the q'-Frobenius endomorphism (so it is a Weil q'-integer). The Shimura–Taniyama formula (2.1.4.1) gives

$$\frac{\mathrm{ord}_v(\pi')}{\mathrm{ord}_v(q')} = \frac{\#\{\phi \in \Phi \mid \phi \text{ induces } v \text{ on } L\}}{[L_v : \mathbb{Q}_p]}$$

for all p-adic places v of L. The first part of the residual reflex condition in 2.1.5 is the hypothesis on the $\overline{\mathbb{Q}}_p$-valued CM type Φ that the right side of this identity equals $\mathrm{ord}_v(\mathrm{Fr}_{B,q})/\mathrm{ord}_v(q)$ for each v (where $\mathrm{Fr}_{B,q}$ is identified with a non-zero element of $\mathcal{O}_L$ via the CM structure on B over $\mathbb{F}_q$). Since q' is a power of q, we can choose an embedding $\mathbb{F}_q \hookrightarrow \kappa$ to make sense of $\mathrm{Fr}_{B_\kappa,q'}$, and as an element of L this q'-Frobenius endomorphism is $\mathrm{Fr}_{B,q}^{[\mathbb{F}_{q'}:\mathbb{F}_q]}$. Hence,

$$\frac{\mathrm{ord}_v(\pi')}{\mathrm{ord}_v(q')} = \frac{\mathrm{ord}_v(\mathrm{Fr}_{B,q})}{\mathrm{ord}_v(q)} = \frac{\mathrm{ord}_v(\mathrm{Fr}_{B_\kappa,q'})}{\mathrm{ord}_v(q')},$$

so the Weil q'-integers π' and $\mathrm{Fr}_{B_\kappa,q'}$ in $\mathcal{O}_L$ have the same order at each p-adic place of $\mathcal{O}_L$. Their ratio is therefore a root of unity, so by replacing F with a suitable unramified extension (so as to replace q' with a suitable power) we can arrange that $\pi' = \mathrm{Fr}_{B_\kappa,q'}$ inside $\mathcal{O}_L$.

By Theorem 1.6.2.1, the g-dimensional isotypic abelian varieties $\overline{C}$ and B_κ over κ are isogenous. Although a choice of such an isogeny $\overline{C} \to B_\kappa$ may not be L-linear, it at least carries $\pi' = \mathrm{Fr}_{\overline{C},q'}$ to $\mathrm{Fr}_{B_\kappa,q'}$. In other words, the induced isomorphism

$$\theta : \mathrm{End}^0(\overline{C}) \simeq \mathrm{End}^0(B_\kappa)$$

of simple $\mathbb{Q}$-algebras may not be L-linear but it is linear over the subfield $Z \subseteq L$ generated by the q'-Frobenius endomorphisms in each algebra. But Z is the center of these simple $\mathbb{Q}$-algebras (Corollary 1.6.2.2) in which L is a maximal commutative subfield, so the Z-linearity allows us to use the Skolem-Noether theorem to find a unit $u \in \mathrm{End}^0(B_\kappa)^\times$ such that composing θ with conjugation by u is an L-algebra map. In other words, if we compose the initial choice of isogeny with a suitable isogeny $B_\kappa \to B_\kappa$ (corresponding to a non-zero $\mathbb{Z}$-multiple of u) then we can arrange that the isogeny $\overline{C} \to B_\kappa$ is L-linear. Finally, by replacing F with a further finite extension if necessary we can ensure (Theorem 1.7.2.1) that the pair (C, ξ) descends to a number field $K \subset F$, and by increasing K we can ensure that F is identified with the completion K_v at the induced p-adic place v on K. In particular, $E \subseteq K$ and v on K restricts to w on E.

To summarize, by using the first part of the residual reflex condition we have constructed a finite extension K/E inside $\overline{\mathbb{Q}}_p$ and a CM abelian variety (A_1, α_1) over K with good reduction at the induced p-adic place v of K such that the associated $\overline{\mathbb{Q}}_p$-valued CM type is (L, Φ) and the reduction $\overline{A}_1$ at v is L-linearly isogenous to B_{κ_v} via a choice of embedding $\mathbb{F}_q \hookrightarrow \kappa_v$. Such an isogeny carries $\mathrm{Fr}_{\overline{A}_1, q_v}$ to $\mathrm{Fr}_{B_{\kappa_v}, q_v}$ as endomorphisms, and so the corresponding elements of $\mathcal{O}_L$ (via the CM structures) are equal, due to L-linearity of the isogeny.

Pick a finite extension E_1 of E that is linearly disjoint from K over E and which has a place w_1 above w such that κ_{w_1} is isomorphic to $\mathbb{F}_q$. Choose an E-embedding $E_1 \to \overline{\mathbb{Q}}_p$ inducing w_1. By replacing K with the composite field $E_1 \otimes_E K \subset \overline{\mathbb{Q}}_p$, we may assume that K contains a subfield E'/E on which the induced place w' from v satisfies $\kappa_{w'} \simeq \mathbb{F}_q$. By Theorem 2.5.1, we get an algebraic Hecke character $\epsilon : \mathbb{A}_K^\times \to L^\times$ attached to (A_1, α_1) which has $\mathrm{N}_\Phi \circ \mathrm{Nm}_{K/E}$ as its algebraic part and is unramified at v with $\epsilon(\pi_v) = \mathrm{Fr}_{(B, \alpha_B)_{\kappa_v}}$.

Apply Theorem 2.4.11 to (ϵ, E', β) with $\chi_{\mathrm{alg}} := \mathrm{N}_\Phi \circ \mathrm{Nm}_{E'/E}$ and β equal to the element of $\mathcal{O}_L$ corresponding to $\mathrm{Fr}_{B,q}$ via α_B (so $\beta^{[\kappa_v : \kappa_{w'}]} = \beta^{[\kappa_v : \mathbb{F}_q]} = \mathrm{Fr}_{(B, \alpha_B)_{\kappa_v}} = \epsilon(\pi_v)$). This provides a finite extension $\widetilde{E}/E'$, a place $\widetilde{w}$ above w', and an algebraic Hecke character $\delta : \mathbb{A}_{\widetilde{E}}^\times \to L^\times$ with the properties stated in Theorem 2.4.11. Hence, $\delta_{\mathrm{alg}} = \chi_{\mathrm{alg}} \circ \mathrm{Nm}_{\widetilde{E}/E_1} = \mathrm{N}_\Phi \circ \mathrm{Nm}_{\widetilde{E}/E}$, δ is unramified at $\widetilde{w}$, $\delta(\pi_{\widetilde{w}}) = \beta = \mathrm{Fr}_{B,q}$, and $(\widetilde{E}, \widetilde{w})$ satisfies property (i).

Fix an E'-embedding of $\widetilde{E}$ into $\overline{\mathbb{Q}}_p$ inducing $\widetilde{w}$ on $\widetilde{E}$. By Theorem 2.5.2 (and Theorem 2.5.1(iv),(v)), the algebraic Hecke character δ comes from a pair (A, α) over $\widetilde{E}$ with the required properties (ii)–(iv). Let us briefly explain how to establish (iv). Let $\overline{A}$ denote the reduction of A at $\widetilde{w}$ and let $\pi \in \mathcal{O}_L$ denote $\delta(\pi_{\widetilde{w}})$, so the respective CM structures $\overline{\alpha}$ and α_B on $\overline{A}$ and B satisfy $\overline{\alpha}(\pi) = \mathrm{Fr}_{\overline{A}, q}$ and $\alpha_B(\pi) = \mathrm{Fr}_{B,q}$. Since $\overline{A}$ and B are isotypic (due to the existence of CM by fields), it follows from Honda-Tate theory (Theorem 1.6.2.1) that there is an isogeny between $\overline{A}$ and B over any isomorphism $\kappa_{\widetilde{w}} \simeq \mathbb{F}_q$. We want to find such an isogeny that is L-linear. At least any isogeny respects the q-Frobenius endomorphisms, and so (via the CM structures $\overline{\alpha}$ and α_B) is linear over the subfield $\mathbb{Q}(\pi) \subseteq L$. But via $\overline{\alpha}$ and α_B the subfield $\mathbb{Q}(\pi)$ is the center of each of the endomorphism algebras of $\overline{A}$ and B (Corollary 1.6.2.2), so we can use the Skolem-Noether theorem as we did earlier in this proof to compose the isogeny with a suitable isogeny $B \to B$ to get an L-linear isogeny. □

2.6. Local methods

It is possible to prove Theorem 2.5.3 by replacing the global Theorem 2.5.2 with purely local methods (but still using the classical theory of complex multiplication, to construct a suitable abelian variety with good reduction over a local field). This rests on some basic concepts in p-adic Hodge theory, such as Tate's work on p-divisible groups over discrete valuation rings (Theorem 1.4.5.1), crystalline representations, and the relationship between these two topics. (We will provide references for what we need.)

After using local methods to prove a local version of Theorem 2.5.3 (in which the number field $\widetilde{E}$ is replaced with a p-adic field) we will formally deduce the global version (over number fields) from the local version. The local version of Theorem 2.5.3 is given at the end of 2.6.4 below; the formal deduction of the global result from the local version is carried out in 2.6.5.

First we establish a local analogue of Theorem 2.4.11, using the argument in Remark 2.4.12.

2.6.1. Lemma. *Let L be a number field, and ϕ an element of $L^\times$. Let E_w be a finite extension field of $\mathbb{Q}_p$, K_v a finite extension of E_w, and $\epsilon_v : K_v^\times \to L^\times$ an unramified character such that $\epsilon_v(\pi_v^n) = \phi^{n[\kappa_v:\kappa_w]}$ for some fixed $n \geqslant 1$, with π_v a chosen uniformizer of K_v.*

There exists a finite extension $\widetilde{K}_{\widetilde{v}}/K_v$ and a subfield $\widetilde{E}_{\widetilde{w}} \subseteq \widetilde{K}_{\widetilde{v}}$ over E_w such that the following conditions hold.

(i) *The extension $\widetilde{K}_{\widetilde{v}}/\widetilde{E}_{\widetilde{w}}$ is unramified.*
(ii) *The residue field extension $\kappa_{\widetilde{w}}/\kappa_w$ is trivial; i.e., $\widetilde{E}_{\widetilde{w}}/E_w$ is totally ramified.*
(iii) *The value $\epsilon_v(\mathrm{Nm}_{\widetilde{K}_{\widetilde{v}}/K_v}(\pi_{\widetilde{v}}))$ is equal to $\phi^{[\kappa_{\widetilde{v}}:\kappa_w]}$.*
(iv) *For any uniformizer $\pi_{\widetilde{w}}$ of $\widetilde{E}_{\widetilde{w}}$, there is a unique character $\epsilon_{\widetilde{w}} : \widetilde{E}_{\widetilde{w}}^\times \to L^\times$ such that $\epsilon_v \circ \mathrm{Nm}_{\widetilde{K}_{\widetilde{v}}/K_v} = \epsilon_{\widetilde{w}} \circ \mathrm{Nm}_{\widetilde{K}_{\widetilde{v}}/\widetilde{E}_{\widetilde{w}}}$ and $\epsilon_{\widetilde{w}}(\pi_{\widetilde{w}}) = \phi$. Moreover, $\epsilon_{\widetilde{w}}$ is unramified.*

PROOF. Let K_1 be the unramified extension field of K_v of degree n, and let K_2 be the Galois closure of K_1 over E_w. Consider the short exact sequence of finite groups

$$1 \to I_{K_2/E_w} \to \mathrm{Gal}(K_2/E_w) \to \mathrm{Gal}(\kappa_{K_2}/\kappa_w) \to 1$$

where I_{K_2/E_w} is the inertia subgroup in $\mathrm{Gal}(K_2/E_w)$. This is not necessarily a split extension, but it would be split if we replace K_2 by the unramified extension K_3 of K_2 of degree $e(K_2/E_w)$, or more generally by any finite unramified extension K_3 such that $[K_3 : K_2]$ is a multiple of the ramification index $e(K_2/E_w)$, because the short exact sequence $1 \to I_{K_3/E_w} \to \mathrm{Gal}(K_3/E_w) \to \mathrm{Gal}(\kappa_{K_3}/\kappa_w) \to 1$ is the pullback of the exact sequence $1 \to I_{K_2/E_w} \to \mathrm{Gal}(K_2/E_w) \to \mathrm{Gal}(\kappa_{K_2}/\kappa_w) \to 1$ by the natural surjection $\mathrm{Gal}(\kappa_{K_3}/\kappa_w) \twoheadrightarrow \mathrm{Gal}(\kappa_{K_2}/\kappa_w)$.

Define $\widetilde{K}_{\widetilde{v}} = K_3$. Choose a splitting $j : \mathrm{Gal}(\kappa_{\widetilde{v}}/\kappa_w) \to \mathrm{Gal}(\widetilde{K}_{\widetilde{v}}/E_w)$ of the short exact sequence

$$1 \to I_{\widetilde{K}_{\widetilde{v}}/E_w} \to \mathrm{Gal}(\widetilde{K}_{\widetilde{v}}/E_w) \to \mathrm{Gal}(\kappa_{\widetilde{v}}/\kappa_w) \to 1\,,$$

and let $\widetilde{E}_{\widetilde{w}}$ be the subfield of $\widetilde{K}_{\widetilde{v}}$ fixed by $j(\mathrm{Gal}(\kappa_{\widetilde{v}}/\kappa_w))$. The statements (i)–(iv) are all easy to check (for (iv) we use that the local norm map is surjective on local unit groups in the unramified case, and for (iii) we use that $n|[\kappa_{\widetilde{v}} : \kappa_v]$ by construction of K_1). □

2.6.1.1. Our aim is to remove the global Theorem 2.5.2 from the proof of Theorem 2.5.3. The construction of a CM abelian variety (A_1, α_1) with $\overline{\mathbb{Q}}_p$-valued CM type (L, Φ) over a finite extension K/E inside $\overline{\mathbb{Q}}_p$ goes as near the start of the proof of Theorem 2.5.3 (where it is called (C, ξ)). In particular, A_1 has good reduction at the place v on K over w induced by the inclusion $K \subset \overline{\mathbb{Q}}_p$ and moreover the residual extension κ_v/κ_w has $\mathbb{F}_q$ as an intermediate extension such that the reduction $\overline{A}_1$ over κ_v is L-linearly isogenous to B_{κ_v}.

Let $E'_{w'}/E_w$ be the intermediate unramified extension of E_w corresponding to $\mathbb{F}_q/\kappa_w$, and let ϵ_v be the restriction to $K_v^\times$ of the algebraic Hecke character ϵ attached to the CM abelian variety (A_1, α_1) over K with CM order $\mathcal{O}_L$. The character

$$\psi_{p,v} : \mathrm{Gal}(K_v^{\mathrm{ab}}/K_v) \to \mathcal{O}_{L,p}^\times$$

attached to the p-divisible group $(A_1)_{K_v}[p^\infty]$ equipped with its L_p-action via α_1 is related to the character ϵ_v via the formula

$$\epsilon_v(x) = \psi_{p,v}(r_K(x)) \cdot \mathrm{N}_\Phi(\mathrm{Nm}_{K_v/E_w}(x))$$

for all $x \in K_v^\times$ by Lemma 2.4.8, Lemma 2.4.9 (see (2.4.9.1)), and Theorem 2.5.1(iii). Apply Theorem 2.5.1(iv) and Lemma 2.6.1 to $(E'_{w'}, K_v, \epsilon_v, \phi, n)$, with $n = 1$ and $\phi = \alpha_B(\mathrm{Fr}_{B,q}) \in L^\times$. We obtain a finite extension $\widetilde{K}_{\widetilde{v}}/K_v$ inside $\overline{\mathbb{Q}}_p$, an intermediate extension $\widetilde{E}_{\widetilde{w}}$ in $\widetilde{K}_{\widetilde{v}}$ over $E'_{w'}$, and an $L^\times$-valued unramified character $\epsilon_{\widetilde{w}}$ of $\widetilde{E}_{\widetilde{w}}^\times$ satisfying properties (i)–(iv) in Lemma 2.6.1. In particular, $\widetilde{K}_{\widetilde{v}}/\widetilde{E}_{\widetilde{w}}$ is unramified, $\kappa_{\widetilde{w}} = \kappa_{w'} = \mathbb{F}_q$, and $\epsilon_{\widetilde{w}}(\pi_{\widetilde{w}}) = \alpha_B(\mathrm{Fr}_{B,q}) \in L^\times$.

Denote by $\mathcal{A}_1$ the abelian scheme over $\mathcal{O}_{\widetilde{K}_{\widetilde{v}}}$ extending the base change $(A_1)_{\widetilde{K}_{\widetilde{v}}}$, and endow it with the unique L-action (in the isogeny category) extending that defined on its generic fiber via α_1. The p-divisible group $\mathcal{G}_1 = \mathcal{A}_1[p^\infty]$ over $\mathcal{O}_{\widetilde{K}_{\widetilde{v}}}$ is thereby endowed with a natural action by L_p in the isogeny category, and its generic fiber is given by the restriction to $\mathrm{Gal}(\widetilde{K}_{\widetilde{v}}^{\mathrm{ab}}/\widetilde{K}_{\widetilde{v}})$ of the Galois character $\psi_{p,v}$ of K_v as above.

There is a unique continuous character $\psi_{p,\widetilde{w}} : \mathrm{Gal}(\widetilde{E}_{\widetilde{w}}^{\mathrm{ab}}/\widetilde{E}_{\widetilde{w}}) \to \mathcal{O}_{L,p}^\times$ such that

$$\epsilon_{\widetilde{w}}(x) = \psi_{p,\widetilde{w}}(r_{\widetilde{E}_{\widetilde{w}}}(x)) \cdot \mathrm{N}_\Phi(\mathrm{Nm}_{\widetilde{E}_{\widetilde{w}}/E_w}(x)) \tag{2.6.1.1}$$

for all $x \in \widetilde{E}_{\widetilde{w}}^\times$. Indeed, uniqueness is clear, and for existence it suffices to check that the continuous homomorphism

$$\epsilon_{\widetilde{w}} \cdot (\mathrm{N}_\Phi \circ \mathrm{Nm}_{\widetilde{E}_{\widetilde{w}}/E_w})^{-1} : \widetilde{E}_{\widetilde{w}}^\times \to L_p^\times$$

has image contained in $\mathcal{O}_{L,p}^\times$ (so it is of "Galois type": factors through the profinite completion of its source). This is a problem of analyzing the image of a single uniformizer $\pi_{\widetilde{w}}$ up to $\mathcal{O}_{L,p}^\times$-multiple. Thus, it suffices to check that the product $\mathrm{Fr}_{B,q} \cdot \mathrm{N}_\Phi(\pi_w)^{-[\kappa_{\widetilde{w}}:\kappa_w]}$ in $L_p^\times$ lies in $\mathcal{O}_{L,p}^\times$ for some uniformizer π_w in E_w. It suffices to check this after raising to the $[\kappa_v : \kappa_{\widetilde{w}}]$th-power. Since $\kappa_{\widetilde{w}} = \kappa_{w'} = \mathbb{F}_q$, the desired result follows from the fact that $\epsilon_v(\pi_v)$ is equal to $\mathrm{Fr}_{B,q}^{[\kappa_v:\kappa_{w'}]}$ and is an $\mathcal{O}_{L,p}^\times$-multiple of $\mathrm{N}_\Phi(\pi_w^{[\kappa_v:\kappa_w]})$ (due to the Galois character $\psi_{p,v}$ taking values in $\mathcal{O}_{L,p}^\times$).

By using Lemma 2.6.1(iv), we get the identity

$$(\psi_{p,v} \circ r_{K_v}) \circ \mathrm{Nm}_{\widetilde{K}_{\widetilde{v}}/K_v} = (\psi_{p,\widetilde{w}} \circ r_{\widetilde{E}_{\widetilde{w}}}) \circ \mathrm{Nm}_{\widetilde{K}_{\widetilde{v}}/\widetilde{E}_{\widetilde{w}}}.$$

Thus, the character $\psi_{p,\widetilde{w}}$ defines a descent G' over $\widetilde{E}_{\widetilde{w}}$ of the generic fiber G_1 of $\mathcal{G}_1$ over $\widetilde{K}_{\widetilde{v}}$ equipped with its natural $\mathcal{O}_{L,p}$-action. The p-divisible group G' over $\widetilde{E}_{\widetilde{w}}$ acquires "good reduction" over the finite unramified extension $\widetilde{K}_{\widetilde{v}}$ in the sense that $G'_{\widetilde{K}_{\widetilde{v}}} \simeq G_1$ over $\widetilde{K}_{\widetilde{v}}$ and this extends to the p-divisible group $\mathcal{G}_1$ over $\mathcal{O}_{\widetilde{K},\widetilde{v}}$. The resulting Galois descent data relative to $\widetilde{K}_{\widetilde{v}}/\widetilde{E}_{\widetilde{w}}$ on the generic fiber of $\mathcal{G}_1$ over $\widetilde{K}_{\widetilde{v}}$

can be viewed as a collection of isomorphisms between G_1 and its $\mathrm{Gal}(\widetilde{K}_{\widetilde{v}}/\widetilde{E}_{\widetilde{w}})$-twists (satisfying a cocycle condition).

These isomorphisms to the Galois twists uniquely extend to isomorphisms from $\mathcal{G}_1$ to its Galois twists, by Tate's full faithfulness result in Theorem 1.4.5.1. Since $\widetilde{K}_{\widetilde{v}}$ is unramified over $\widetilde{E}_{\widetilde{w}}$, we can use the equivalence of finite étale descent and Galois descent for finite unramified extensions of p-adic integer rings [**10**, 6.2B] (applied at each torsion level over $\mathcal{O}_{\widetilde{E}_{\widetilde{w}}}$) to construct a unique p-divisible group $\mathcal{G}'$ over $\mathcal{O}_{\widetilde{E}_{\widetilde{w}}}$ having generic fiber G' compatibly with the descent data. Again using Tate's full faithfulness theorem, the $\mathcal{O}_{L,p}$-action on G' extends uniquely to one on $\mathcal{G}'$.

2.6.2. Lemma. *The q-Frobenius endomorphism of the reduction $G'_0 := \mathcal{G}'_{\kappa_{\widetilde{w}}}$ is given by the action of $\epsilon_{\widetilde{w}}(\pi_{\widetilde{w}}) \in \mathcal{O}_L$.*

This lemma (along with some of the ingredients in its proof) is a local version of Theorem 2.5.1(v) for p-divisible groups with complex multiplication.

PROOF. By decomposing G'_0 up to isogeny according to the factor fields of L_p, our task is a special case of the following problem. Let Γ be a p-divisible group of height d over a p-adic integer ring $\mathcal{O}_K$ with finite residue field κ of size q and fraction field K. Let $L/\mathbb{Q}_p$ be an extension of degree d and assume that Γ is equipped with an action by L in the isogeny category over $\mathcal{O}_K$. By Proposition 1.4.3.9, L is its own centralizer in the endomorphism algebra of the reduction Γ_0 over κ. Hence, some non-zero element $\lambda \in L$ induces the q-Frobenius endomorphism of Γ_0. We need to relate λ to the continuous Galois character $\psi : \mathrm{Gal}(K^{\mathrm{ab}}/K) \to \mathcal{O}_L^\times$ associated to the generic fiber Γ_K.

It suffices to prove that there is a unique homomorphism $\chi : K^\times \to L^\times$ algebraic with respect to $\mathbb{Q}_p$ such that $\chi, \psi \circ r_K : K^\times \rightrightarrows L^\times$ coincide on $\mathcal{O}_K^\times$ and the character $(\psi \circ r_K) \cdot \chi^{-1} : K^\times \to L^\times$ that is trivial on $\mathcal{O}_K^\times$ carries uniformizers to λ. Indeed, this would give the desired result concerning G'_0 due to (2.6.1.1), since $\epsilon_{\widetilde{w}}$ is unramified. The general assertion concerning χ refines the local part of Serre's theory of locally algebraic abelian representations in [**105**, III.1] since we are asserting the algebraicity of $\psi \circ r_K$ on the entire local unit group $\mathcal{O}_K^\times$ of $K^\times$. The existence and uniqueness of χ is a special case of a general property of abelian semisimple crystalline p-adic representations of Galois groups of p-adic fields, and it is proved in [**26**, App. B]. □

Fix an isomorphism $\mathbb{F}_q \simeq \kappa_{\widetilde{w}}$ so as to view B as an abelian variety over $\kappa_{\widetilde{w}}$. The p-divisible groups $B[p^\infty]$ and G'_0 over $\kappa_{\widetilde{w}}$ are endowed with actions by L_p in the isogeny category. Since $\alpha_B(\pi) = \mathrm{Fr}_{B,q}$ and $\pi = \epsilon_{\widetilde{w}}(\pi_{\widetilde{w}})$, by Lemma 2.6.2 the element $\pi \in \mathcal{O}_L$ acts as the q-Frobenius endomorphism on each of these p-divisible groups. This is crucial in the proof of the following lemma.

2.6.3. Lemma. *There is an L_p-linear isogeny $B[p^\infty] \to G'_0$ over $\kappa_{\widetilde{w}}$.*

PROOF. Let $k = \kappa_{\widetilde{w}}$. We will use contravariant Dieudonné theory. Let $D_k = W(k)[\mathcal{F}, \mathcal{V}]$ be the Dieudonné ring over the finite field k (so D_k is non-commutative if $k \neq \mathbb{F}_p$). The category of left $D_k[1/p]$-modules that have finite dimension over

$W(k)[1/p]$ and admit a D_k-stable $W(k)$-lattice is anti-equivalent to the isogeny category of p-divisible groups over k. Note that the center of $D_k[1/p] = W(k)[1/p][\mathcal{F}]$ is $\mathbb{Q}_p[\mathcal{F}_q]$ where $\mathcal{F}_q = \mathcal{F}^{[k:\mathbb{F}_p]}$.

The Dieudonné modules $\mathrm{M}^*(B[p^\infty])$ and $\mathrm{M}^*(G'_0)$ are left D_k-modules that are finite free over $W(k)$ with rank $2g$ that is equal to the $\mathbb{Z}_p$-rank of $\mathcal{O}_{L,p}$, and each is equipped with a D_k-linear action by $\mathcal{O}_{L,p}$. The actions by $\mathcal{O}_{L,p}$ on both p-divisible groups are faithful. Indeed, for $B[p^\infty]$ the faithfulness is due to Proposition 1.2.5.1 with $\ell = p$, and for G'_0 the faithfulness can be checked after the ground field extension $\kappa_{\widetilde{v}}/k$ where it becomes the p-divisible group of an abelian variety (the reduction of the CM abelian variety (A_1, α_1) over $\widetilde{K}_{\widetilde{v}}$). In view of this faithfulness, by Proposition 1.4.3.9(2) both Dieudonné modules are invertible as $W(k) \otimes_{\mathbb{Z}_p} \mathcal{O}_{L,p}$-modules.

Let $Z = \mathbb{Q}(\pi)$, so if $f \in \mathbb{Q}[T]$ is the minimal polynomial of π then $\mathbb{Q}[T]/(f) \simeq Z$ via $T \mapsto \pi$. Let $\prod_{v'|p} f_{v'}$ be the monic irreducible factorization of f in $\mathbb{Q}_p[T]$, corresponding to the decomposition $Z_p \simeq \prod_{v'|p} Z_{v'}$. Here, $Z_{v'} \simeq \mathbb{Q}_p[T]/(f_{v'})$ in which $\pi \in Z \subset Z_{v'}$ is the image of T. Since $\mathcal{F}_q$ is central in $D_k[1/p]$, it makes sense to form the $Z_{v'}$-algebra $C_{v'} = D_k[1/p]/D_k[1/p]f_{v'}(\mathcal{F}_q)$ in which the element $\pi \in Z_{v'}$ acts as $\mathcal{F}_q$. By Theorem A.1.2.1, $C_{v'}$ is a central simple $Z_{v'}$-algebra. Thus, the isomorphism class of a finitely generated left $C_{v'}$-module is determined by its $Z_{v'}$-dimension.

Both $\mathrm{M}^*(B[p^\infty])[1/p]$ and $\mathrm{M}^*(G'_0)[1/p]$ are invertible $W(k) \otimes_{\mathbb{Z}_p} L_p$-modules, and their $Z_{v'}$-factors are left $C_{v'}$-modules because the q-Frobenius endomorphisms of $B[p^\infty]$ and G'_0 are induced by the element $\pi \in \mathcal{O}_Z \subseteq \mathcal{O}_L$ whose image in $Z_{v'}$ is a root of $f_{v'} \in \mathbb{Q}_p[T]$. When viewing these Dieudonné modules as left $C_{v'}$-modules, they have the same $Z_{v'}$-dimension, namely that of the $Z_{v'}$-part of the Z_p-algebra $W(k) \otimes_{\mathbb{Z}_p} L_p$. Hence, the $Z_{v'}$-factors are isomorphic as $C_{v'}$-modules, and forming the direct product of such isomorphisms over all v' gives a Z_p-linear isomorphism between the underlying $D_k[1/p]$-modules. This provides a Z_p-linear isogeny $B[p^\infty] \to G'_0$ over k, but it may not be L_p-linear. Hence, the resulting isomorphism of endomorphism algebras

$$\mathrm{End}^0(B[p^\infty]) \simeq \mathrm{End}^0(G'_0)$$

is Z_p-linear but perhaps not L_p-linear.

These Z_p-isomorphic endomorphism algebras are central simple Z_p-algebras with L_p a maximal commutative subalgebra in each because Theorem 1.6.1.1 gives $\mathbb{Q}_p \otimes_{\mathbb{Q}} \mathrm{End}^0(B) \simeq \mathrm{End}^0(B[p^\infty])$ with $\mathrm{End}^0(B)$ a central simple Z-algebra having L as a maximal commutative Z-subalgebra. Hence, the Skolem-Noether theorem ensures that if we compose a choice of Z_p-linear isogeny $B[p^\infty] \to G'_0$ with a suitable Z_p-linear isogeny $G'_0 \to G'_0$ then we get an L_p-linear isogeny. □

2.6.4. Using Lemma 2.6.3, we may choose an L_p-linear $\kappa_{\widetilde{w}}$-isogeny $B[p^\infty] \to G'_0$. The kernel is identified with a finite subgroup scheme of B over $\kappa_{\widetilde{w}} = \mathbb{F}_q$, so if we replace B with its quotient modulo this kernel then we gain the property that there is an L_p-linear isomorphism $B[p^\infty] \simeq G'_0$ of p-divisible groups (not just an isogeny). To summarize, the p-divisible group $\mathcal{G}'$ over $\mathcal{O}_{\widetilde{E}_{\widetilde{w}}}$ equipped with its L-action (in the isogeny category) is identified with a deformation of the p-divisible group of the abelian variety B over $\kappa_{\widetilde{w}}$ equipped with its L-action (in the isogeny category). Hence, by working with CM orders we deduce via Theorem 1.4.5.3 that there is a unique formal abelian scheme $\mathfrak{A}'$ over $\mathrm{Spf}(\mathcal{O}_{\widetilde{E}_{\widetilde{w}}})$ equipped with an L-action (in the isogeny category) that compatibly deforms B and has p-divisible group $\mathcal{G}'$ (respecting the L-actions).

The tangent space of $\mathfrak{A}'$ along its identity section coincides with that of $\mathcal{G}'$ over $\mathcal{O}_{\widetilde{E}_{\widetilde{w}}}$ (Example 1.4.3.6), which in turn descends the tangent space $\mathrm{Lie}(\mathcal{G}_1) \simeq \mathrm{Lie}(\mathcal{A}_1)$ over $\mathcal{O}_{\widetilde{K}_{\widetilde{v}}}$, all respecting the actions by an order in $\mathcal{O}_L$. After inverting p this recovers $\mathrm{Lie}(A_1)_{\widetilde{K}_{\widetilde{v}}}$, on which the L-action is given by the $\overline{\mathbb{Q}}_p$-valued CM type Φ of $(A_1, \alpha_1)_{K_v}$. Hence, by Theorem 2.2.3, $\mathfrak{A}'$ algebraizes to an abelian scheme $\mathcal{A}'$ over $\mathcal{O}_{\widetilde{E}_{\widetilde{w}}}$ endowed with an L-action (in the isogeny category) giving rise to the CM type Φ. This algebraization has reduction B and its p-divisible group is the descent $\mathcal{G}'$ of $\mathcal{G}_1 = \mathcal{A}_1[p^\infty]$ compatibly with L-actions and the residual isomorphism $B[p^\infty] \simeq G'_0$.

Let (A', α') be the CM generic fiber of $\mathcal{A}'$ over $\widetilde{E}_{\widetilde{w}}$ with its L-action, so (A', α') has good reduction that is L-linearly isomorphic to (B, α_B) over the initial choice of isomorphism $\kappa_{\widetilde{w}} \simeq \mathbb{F}_q$ and moreover the $\overline{\mathbb{Q}}_p$-valued CM type of (A', α') is (L, Φ). We conclude that (B, α_B) satisfies (IN) in 1.8.5, with a lift of type (L, Φ) over the valuation ring of the subfield $\widetilde{E}_{\widetilde{w}} \subset \overline{\mathbb{Q}}_p$. This is a local version of Theorem 2.5.3, using $\widetilde{E}_{\widetilde{w}}$ rather than a number field.

2.6.5. To complete the local proof of Theorem 2.5.3, it remains to carry out one global step: descend (A', α') to a number field inside $\widetilde{E}_{\widetilde{w}}$ (necessarily containing the reflex field $E \subset \overline{\mathbb{Q}}_p$ of the CM type (L, Φ)); we can then increase such a number field inside $\widetilde{E}_{\widetilde{w}}$ so that its completion at the place induced by $\widetilde{w}$ is $\widetilde{E}_{\widetilde{w}}$; in particular, this place has residue field $\kappa_{\widetilde{w}}$ and the descent of A' has good reduction there (since the formation of Néron models over a discrete valuation ring R commutes with scalar extension to $\widehat{R}$). The desired descent to a number field is:

2.6.5.1. Lemma. *Any* CM *abelian variety A with good reduction over a p-adic field F descends with its* CM *structure to a* CM *abelian variety over a number field inside F.*

PROOF. Fix a positive integer $n \geqslant 3$ not divisible by p, and let $\mathcal{A}$ over $R = \mathcal{O}_F$ be the Néron model of A. The finite étale group scheme $\mathcal{A}[n]$ over R is uniquely determined by its special fiber. Let $K \subset F$ be a dense subfield that is a number field and let v be the induced p-adic place of K, so $K_v = F$. The special fiber of $\mathcal{A}[n]$ can be uniquely lifted to a finite étale group scheme $\mathcal{G}$ over the henselization $\mathcal{O}^{\mathrm{h}}_{K,(v)}$ of the algebraic local ring of $\mathcal{O}_K$ at v, and by uniqueness of finite étale liftings over henselian local rings this descends $\mathcal{A}[n]$ via the canonical local inclusion $\mathcal{O}^{\mathrm{h}}_{K,(v)} \to R$. The generic fiber of $\mathcal{G}$ lives over an algebraic extension of $\mathbb{Q}$, so it descends to a finite étale group G over a number field $K' \subset F$ containing K.

Choose a polarization of A, say with degree d^2. Let $g = \dim(A)$, and let $\mathcal{O}$ denote the CM order on A arising from the CM structure. Consider the moduli scheme $\mathcal{M}$ of finite type over K' classifying isomorphism classes of polarized abelian schemes of relative dimension g (over K'-schemes) endowed with the following extra structure: a degree-d^2 polarization, an isomorphism of the n-torsion with the pullback of the K'-group G, and an action by $\mathcal{O}$. Every geometric point of this moduli scheme $\mathcal{M}$ descends to $\overline{\mathbb{Q}}$ (Theorem 1.7.2.1), so $\mathcal{M}$ is 0-dimensional. Hence, the K'-map $\mathrm{Spec}(F) \to \mathcal{M}$ corresponding to A with its additional structure factors through some point of $\mathcal{M}$ that is necessarily defined over a finite extension of K'. This is a number field inside F to which A with its CM structure descends. □

2.6.5.2. Remark. The local version of 2.5.3 given at the end of 2.6.4 will be strengthened in Proposition 3.7.11: the lifted CM p-divisible group can be defined over the p-adic reflex field of the chosen $\overline{\mathbb{Q}}_p$-valued CM type Φ on L, which is the minimal possible field of definition over $\mathbb{Q}_p$ (given Φ for the lift). In particular, the lift can be built without any ramification beyond that arising from the p-adic reflex field. However, the method of proof in this local approach is not entirely "p-adic", since it rests on Casselman's Theorem 2.5.2 and the Main Theorem of Complex Multiplication in 2.5.1, both of whose proofs use complex-analytic uniformization.

An advantage of the local method is that it shows the degree of the isogeny between B and the special fiber of A in 2.6.4(iv) can be arranged to be a power of p when the CM order of B has index in $\mathcal{O}_L$ not divisible by p. Examples in Chapter 3 (see 3.8.3–3.8.4) show that it is sometimes impossible to find a CM lifting of $B[p^\infty]$, and so the isogeny between B and the special fiber of A cannot induce an isomorphism between p-divisible groups; i.e., the degree of the isogeny must be divisible by p. Thus, we get "best possible" control on the degree of the isogeny in cases with p not dividing the index of the CM order in $\mathcal{O}_L$.

In the proof of the main result of Chapter 4 (corresponding to problem (I) in 1.8.5) we will use local methods to show that if B has CM order whose index in $\mathcal{O}_L$ is not divisible by p then the field of definition of the desired CM lift can be arranged to be the p-adic reflex field when $1/2$ is not a slope of B and can be arranged to be *tamely ramified* over the p-adic reflex field when $1/2$ is a slope of B; see Theorem 4.1.1.

CHAPTER 3

CM lifting of p-divisible groups

Although p-divisible groups are interesting enough in their own right, our main motivation for studying them has been their applications to abelian varieties.

— John Tate

3.1. Motivation and background

In this chapter we study the questions (CML) and (R) in 1.8.5, and want to determine if there exist abelian varieties over finite fields that do not admit a CM lift, even after an extension of the base field (but not applying any non-trivial isogeny). This problem was mainly settled in [**93**], where it was shown that there are "many" abelian varieties over an algebraic closure $\overline{\mathbb{F}_p}$ of $\mathbb{F}_p$ that do not admit a CM lifting. However, that analysis was restricted to abelian varieties over $\overline{\mathbb{F}_p}$ whose endomorphism algebra is a commutative field. We seek to make non-liftable examples without this restriction on the endomorphism algebra, and also to get effective[1] control on the finite fields over which such examples can be constructed.

3.1.1. Why p-divisible groups? In 1.4 we reviewed some basic notions and theorems related to p-divisible groups and Dieudonné modules. To improve upon CM lifting results in [**93**] (which we review below), we will use the deformation theory of p-divisible groups via Theorem 1.4.5.3 (illustrated in Example 1.4.5.4). This encodes deformations of homomorphisms as well, so p-divisible groups are especially well-suited to the study of CM lifting problems for abelian varieties even though most p-divisible groups do not arise from abelian varieties.

Among the advantages of working with p-divisible groups (apart from their "computability" via semi-linear objects such as Dieudonné modules) is that the isogeny classes are easily understood over an algebraically closed field (Theorem 3.1.3) and their endomorphism algebras are more accessible than for abelian varieties. Since a simple abelian variety in characteristic $p > 0$ can have a p-divisible group that is isogenous to a non-trivial product, leaving the framework of abelian varieties opens the door to more operations. In this way, we can remove the restrictive hypothesis in [**93**] that the endomorphism algebra of the residual abelian variety in characteristic p is a commutative field, leading to many more non-liftable examples and the solution of some CM lifting problems for supersingular abelian surfaces (thereby answering [**93**, Question C]). Moreover, we will provide an effective procedure to construct isogenies to non-liftable abelian varieties over specific

[1]The effectiveness assertion here and in 3.1.1 below, and later in 3.5.4, is illustrated in 3.6.3.

finite fields (see 3.8.3). In 3.8.4 and 3.8.5 we illustrate how our method constructs explicit examples for which no CM lifting exists.

To prove that certain abelian varieties do not admit a CM lifting to characteristic 0 in various senses (as in 1.8.5), it will suffice to show this for the associated p-divisible group (in a sense defined in 3.7.1). It will not be necessary to keep track of polarizations when proving such negative results. In contrast, when we allow isogenies and prove affirmative liftability results via the Serre–Tate deformation theorem (Theorem 1.4.5.3) in Chapter 4 then polarizations must be taken into account to ensure the algebraizability of the formal abelian schemes that we construct. (In 4.1.2.3 we will exhibit *non-algebraizable* CM formal abelian schemes over p-adic integer rings.) Below we shall recall some definitions related to p-divisible groups and their Serre duals (see 1.4.3.3), with an emphasis on the relative setting (which we need in deformation theory).

3.1.2. Slopes. In 2.1.4.1 we defined the concept of *slopes* for abelian varieties over finite fields. More generally, for a p-divisible group G over a finite field K of size $q = p^f$, the numbers $\mathrm{ord}_p(\lambda)/f$ arising from the eigenvalues λ (counted with multiplicity) of the $W(K)$-linear q-Frobenius on the (contravariant) Dieudonné module $\mathrm{M}^*(G)$ define the *slopes* of G in a manner that is consistent with the case of abelian varieties over K and is invariant under finite extension of the finite ground field. However, for infinite ground fields this viewpoint does not make sense. We need a way to define the slopes of a general p-divisible groups over any field of characteristic $p > 0$, especially algebraically closed fields, and this will also define the slopes of an abelian variety A over any field of characteristic p (as the slopes of $A[p^\infty]$). The definition in such generality rests on some basic concepts in Dieudonné theory, so we first review some aspects of that theory.

Recall from 1.4.3.1 that for a perfect field κ of characteristic $p > 0$, the *Dieudonné ring* D_κ is the associative ring generated by the p-adic Witt ring $W(\kappa)$ and two generators $\mathcal{F}$ and $\mathcal{V}$ subject to the relations

$$\mathcal{F}\cdot\mathcal{V} = p = \mathcal{V}\cdot\mathcal{F}, \qquad \mathcal{F}\cdot a = \sigma(a)\cdot\mathcal{F},\ \ a\cdot\mathcal{V} = \mathcal{V}\cdot\sigma(a) \quad \text{for all } a \in W(k),$$

where σ is the ring automorphism of $W(\kappa)$ given by

$$a = (a_0, a_1, a_2, \ldots) \mapsto (a_0^p, a_1^p, a_2^p, \ldots).$$

A basic result in classical contravariant Dieudonné theory is that there are two compatible contravariant equivalences of categories:

(i) an anti-equivalence of categories between p-divisible groups over κ and left D_κ-modules which are free of finite rank over $W(\kappa)$,
(ii) an anti-equivalence of categories between finite commutative κ-group schemes of p-power order and left D_κ-modules finitely generated over $W(\kappa)$ and killed by a power of p.

These results were reviewed in 1.4; see [**29**, III, §8–9] or B.3.5 for a summary of classical Dieudonné theory.

To define the slopes of a p-divisible group G over any field of characteristic p, we shall use the classification of isogeny classes of p-divisible groups over an algebraic closure k of the ground field. This is really a problem in semilinear algebra, using left $D_k[1/p]$-modules of finite dimension over $W(k)[1/p]$. Not every such $D_k[1/p]$-module arises from a D_k-module that is finitely generated over $W(k)$ (i.e., arises from the isogeny class of a p-divisible group over k). The existence of a $W(k)$-lattice stable under $\mathcal{F}$ and $\mathcal{V}$ imposes non-trivial constraints on the actions of $\mathcal{F}$ and $\mathcal{V}$.

3.1.3. Theorem (Dieudonné–Manin). *Let k be an algebraically closed field of characteristic $p > 0$.*

(1) *The isogeny category of p-divisible groups over k is semisimple.*
(2) *For any $\lambda = r/s \in \mathbb{Q} \cap [0, 1]$ in reduced form (with $s \geqslant 1$), the left D_k-module $D_k/D_k(\mathcal{F}^{s-r} - \mathcal{V}^r)$ is finite free over $W(k)$ of rank s and the corresponding p-divisible group $\mathcal{G}_\lambda$ of height s has dimension r and is simple in the isogeny category.*
(3) *The $\mathcal{G}_\lambda$ are pairwise non-isogenous, and every simple object in the isogeny category of p-divisible groups over k is isogenous to some $\mathcal{G}_\lambda$.*
(4) *The Serre dual $(\mathcal{G}_\lambda)^t$ of $\mathcal{G}_\lambda$ is $\mathcal{G}_{1-\lambda}$.*
(5) *For any p-divisible group X there is a unique sequence of "slopes" $\lambda_i \in \mathbb{Q} \cap [0, 1]$ and "multiplicities" $e_i \geqslant 1$ for which there is an isogeny*

$$X \sim \prod_i \mathcal{G}_{\lambda_i}^{e_i}.$$

PROOF. See [**71**, Chapter 2, §4] and [**29**, IV, §4, Theorem]. □

For a p-divisible group G over a perfect field κ, the Frobenius isogeny $G \to G^{(p)}$ corresponds to the semilinear endomorphism $\mathcal{F}$ of the contravariant Dieudonné module $\mathrm{M}^*(G)$. Since this is not $W(\kappa)$-linear (unless $G = 0$ or $\kappa = \mathbb{F}_p$), it does not have a well-defined characteristic polynomial and so has no notion of "eigenvalue" in general (whereas over a finite field of size $q = p^f$ we can use the $W(K)$-linear fth power of $\mathcal{F}$, as noted at the start of 3.1.2).

If K is a general field of characteristic p then by definition the *slopes* of G are the numbers λ_i in Theorem 3.1.3(5) applied to $X = G_{\overline{K}}$, with respective *multiplicities* $m_i := e_i \cdot \mathrm{ht}(\mathcal{G}_{\lambda_i}) = e_i s_i$ where r_i/s_i is the reduced form of λ_i (so $m_i \lambda_i \in \mathbb{Z}$ for all i). In particular, a non-zero G has only one slope λ precisely when $G_{\overline{K}}$ is isotypic in the sense that it is isogenous to a power of a simple object $\mathcal{G}_\lambda$; such G are called *isoclinic* (of slope λ). We leave it to the reader to check that for finite K of size q, this notion of slope agrees with the definition using eigenvalues of the $W(K)$-linear q-Frobenius as in the case of abelian varieties over finite fields in 2.1.4.1.

3.1.4. Newton polygons. For a p-divisible group X over a field K of characteristic p, we apply 3.1.3 to $X_{\overline{K}}$ to obtain slopes λ_i with multiplicities e_i. The lower convex polygon $\mathcal{N}(X)$ obtained by taking the slopes with these multiplicities and arranging them in non-decreasing order is the *Newton polygon* of X. By construction, $\mathcal{N}(X)$ only depends on the isogeny class of X, and $\mathcal{N}(X^t)$ is obtained from $\mathcal{N}(X)$ by replacing the set of slopes $\{\lambda_i\}$ (with multiplicity) with the set of slopes $\{1 - \lambda_i\}$ (with multiplicity), due to 3.1.3(4).

For an abelian variety A over K we write $\mathcal{N}(A)$ instead of $\mathcal{N}(A[p^\infty])$. For any such A, $\mathcal{N}(A)$ is symmetric in the sense that any λ and $1 - \lambda$ appear with the same multiplicity since A is isogenous to A^t; see [**71**, Theorem 4.1], [**87**, III.19]. When K is finite this recovers 2.3.6.

3.1.5. Definition. For any $\lambda = r/s \in \mathbb{Q} \cap [0,1]$ in reduced form (with $s \geqslant 1$), the p-divisible group $\mathcal{G}_\lambda$ over $\mathbb{F}_p$ is defined by the condition

$$\mathrm{M}^*(\mathcal{G}_\lambda) = D_{\mathbb{F}_p}/D_{\mathbb{F}_p}(\mathcal{F}^{s-r} - \mathcal{V}^r) =: M_\lambda$$

Over any field K of characteristic p we write $\mathcal{G}_\lambda$ to denote $\mathcal{G}_{\lambda,K}$ when the ground field K is understood from the context.

It is elementary to check (as in 3.1.3(2)) that M_λ is finite free of rank s over $\mathbb{Z}_p$ and $M_\lambda/\mathcal{F}M_\lambda$ is r-dimensional over $\mathbb{F}_p$, so $\mathcal{G}_\lambda$ has dimension r and height s. For $\lambda \in \mathbb{Q} \cap [0,1]$, the reason we say (over any field K of characteristic p) that $\mathcal{G}_\lambda$ is "isoclinic of slope λ" is that for all positive integral multiples n of the denominator s of λ, the n-fold relative Frobenius isogeny $\mathcal{G}_\lambda \to \mathcal{G}_\lambda^{(p^n)}$ is approximately a unit multiple of $[p^{\lambda n}]_{\mathcal{G}_\lambda}$ in the sense that their kernels are "uniformly comparable" as n varies. To make this precise, we switch to the language of contravariant Dieudonné theory over $\overline{K}$ (since the operator $\mathcal{F}$ on the contravariant Dieudonné module corresponds to the relative Frobenius morphism on the p-divisible group). The Dieudonné module $M = \mathrm{M}^*(\mathcal{G}_{\lambda,\overline{K}})$ has a presentation with a generator y and the relation $(\mathcal{F}^{s-r} - \mathcal{V}^r)\cdot y = 0$. Multiplying on the left by $\mathcal{F}^r$ gives $\mathcal{F}^s y = p^{\lambda s} y$, so the cokernel of $\mathcal{F}^n$ is comparable to the cokernel of $p^{\lambda n}$ uniformly in $n \in \mathbb{Z}_{>0} \cdot s$ in the sense that for some $n_0 \geqslant 0$ and all $n > n_0$, $p^{\lfloor (n+n_0)\lambda \rfloor} M \subset \mathcal{F}^n M \subset p^{\lfloor (n-n_0)\lambda \rfloor} M$.

3.1.6. Example. A p-divisible group Y over K is isoclinic of slope 0 if and only if Y coincides with its étale part $Y^{\text{ét}}$ in the sense of Example 1.4.3.6; such Y are called *étale*. This condition is equivalent to $Y_{\overline{K}}$ being a power of the unique constant p-divisible group $\mathbb{Q}_p/\mathbb{Z}_p$ over $\overline{K}$ of height 1 and dimension 0. Likewise, Y is isoclinic of slope 1 precisely when it has étale Serre dual; such Y are said to be *of multiplicative type*.

For any p-divisible group X over K, we defined its connected component X^0 in Example 1.4.3.6 (and its étale part $X^{\text{ét}}$ is X/X^0). Clearly X^0 is the smallest p-divisible subgroup of X such that X/X^0 is isoclinic of slope 0, and the p-divisible subgroup $X^{\text{mult}} \subset X^0$ whose Serre dual corresponds to the quotient $(X^0)^{\text{ét}}$ of X^0 is the largest p-divisible subgroup that is of multiplicative type. We write

$$X_{(0,1)} := X^0/X^{\text{mult}};$$

this is called the *local-local part* of X.

The p-divisible group X is *ordinary* if $X_{(0,1)} = 0$, and is *local-local* if $X = X_{(0,1)}$ (i.e., X is its own local-local part). We claim that X is ordinary precisely when X^0 has an étale Serre dual, or equivalently $X_{\overline{K}}$ has no slopes different from 0 and 1, and that X is local-local precisely when X and X^t are connected, or equivalently all slopes of $X_{\overline{K}}$ lie strictly between 0 and 1 (hence the notation "$X_{(0,1)}$"). These assertions are invariant under isogeny, and it suffices to check them over an algebraically closed extension of K, so they are immediate from Theorem 3.1.3.

3.1.7. The slope filtration. Let X be a p-divisible group over a field K of characteristic p. (We allow that K may be imperfect.) By a theorem of Grothendieck and Zink (see [**138**, Thm. 7]), there is a unique filtration by p-divisible subgroups

$$Y_0 = 0 \subset Y_1 \subset \cdots \subset Y_t = X$$

over K such that for any $1 \leqslant i \leqslant t$ the quotient Y_{i-1}/Y_i is isoclinic of slope λ_i and $\lambda_1 > \cdots > \lambda_t$. This is called the *slope filtration*, and it is a refinement of the Dieudonné–Manin theorem (which applies over $\overline{K}$). We will only use it to clarify some definitions and observations; it will not be used in our proofs.

In general the slope filtration is not a split filtration; e.g., the connected-étale sequence

$$0 \to X^0 \to X \to X^{\text{ét}} \to 0$$

usually does not split as a direct sum decomposition of X over K. However, if the ground field K is *perfect* then the connected-étale sequence is uniquely split (see [**87**, I.2]), so over such K we obtain a canonical decomposition

$$X = X^{\text{mult}} \times X_{(0,1)} \times X^{\text{ét}}.$$

This direct product decomposition cannot be made finer, in the sense that the extensions defined by successive terms of the slope filtration of $X_{(0,1)}$ are generally non-trivial even when K is perfect.

3.1.8. Lemma. *Let X be a non-zero p-divisible group over a field of characteristic p. The $\mathbb{Z}_p$-algebra* End(X) *is finite free as a $\mathbb{Z}_p$-module.*

We will write $\text{End}^0(X)$ to denote $\mathbb{Q}_p \otimes_{\mathbb{Z}_p} \text{End}(X)$. We do not claim that this is a semisimple $\mathbb{Q}_p$-algebra; such semisimplicity holds when the ground field K is algebraically closed (by Theorem 3.1.3(1)), but in general the isogeny category of p-divisible groups over K is not semisimple (e.g., if K is finite then semisimplicity fails for the isogeny category of étale p-divisible groups).

PROOF. It suffices to prove the lemma after replacing the ground field by an algebraic closure. Also it suffices to prove the lemma after we replace the p-divisible group by an isogenous one. Thus, by the Dieudonné–Manin classification (Theorem 3.1.3) it suffices to treat the simple objects $X = \mathcal{G}_\lambda$ with slope $\lambda \in \mathbb{Q} \cap [0,1]$ over an algebraically closed field k. Equivalently, it suffices to prove $\mathbb{Z}_p$-finiteness for the D_k-linear endomorphism ring of each Dieudonné module $M_\lambda := D_k/D_k(\mathcal{F}^{s-r} - \mathcal{V}^r)$ where $\lambda = r/s$ with relatively prime integers $0 \leqslant r \leqslant s$ (and $s \geqslant 1$).

The $\mathbb{Q}_p$-finiteness of the endomorphism algebra of the left $D_k[1/p]$-module $M_\lambda[1/p]$ is a classical fact; see the self-contained computation with cyclic algebras in Example 3.6.3 for an explicit description, and see [**29**, IV, §3] and [**18**, 4.52] for explicit descriptions of both $\text{End}(M_\lambda[1/p])$ and $\text{End}(M_\lambda)$. (In A.1.2 we prove an analogous fact over finite ground fields.) Since $\text{End}(M_\lambda)$ is p-adically separated and complete, and $\text{End}(M_\lambda)[1/p] = \text{End}(M_\lambda[1/p])$, it remains to prove that if N is a p-adically separated and complete flat $\mathbb{Z}_p$-module then it is finitely generated over $\mathbb{Z}_p$ if $\dim_{\mathbb{Q}_p}(N[1/p])$ is finite. This is proved by contradiction, as follows.

By p-power scaling, we can choose a $\mathbb{Q}_p$-basis $\{n_1, \ldots, n_d\}$ of $N[1/p]$ consisting of elements $n_j \in N$. Let N' be the $\mathbb{Z}_p$-span of the n_j's in N. We seek an integer $\nu > 0$ so that $N \subseteq p^{-\nu}N'$. If no such ν exists then there exists a strictly increasing sequence of positive integers $j_1 < j_2 < \ldots$ such that $p^{-j_i}N'$ contains an element v_i not in $p^{-(j_i-1)}N'$. Thus, $n'_i := p^{j_i}v_i \in N'$ and this element has a $\mathbb{Z}_p^\times$-coefficient relative to the $\mathbb{Z}_p$-basis $\{n_1, \ldots, n_d\}$ of N'. By passing to a subsequence and rearranging this $\mathbb{Z}_p$-basis, we can assume that each n'_i has its n_1-coefficient in $\mathbb{Z}_p^\times$ and that the sequence $\{n'_i\}$ in N' has a limit n' in N', so $n' \neq 0$ (as its n_1-coefficient is a unit). But visibly $n'_i \to 0$ in N relative to the p-adic topology, so it suffices to check that the p-adic topology on N induces the p-adic topology on N'.

By a classic result of Chevalley (see [**73**, Exer. 8.7]), any decreasing sequence of submodules of a finitely generated module over a complete local noetherian ring $(R, \mathfrak{m})$ tends to 0 for the $\mathfrak{m}$-adic topology if its intersection is 0. In our case, the submodules $N'_i = N' \bigcap p^i N$ in N' satisfy $\bigcap N'_i \subseteq \bigcap p^i N = 0$, so each $p^i N'$ contains $N'_{e(i)}$ some $e(i) > 0$. Visibly N'_i contains $p^i N'$ for all i, so $\{N'_i\}$ defines the p-adic topology on N'. □

3.1.9. Definition. A finite commutative group scheme N over a field K of characteristic $p > 0$ is an *α-group scheme* if it is of p-power order and

$$\ker(\mathrm{Fr}_{N/K} : N \to N^{(p)}) = 0, \quad \ker(\mathrm{Ver}_{N/K} : N^{(p)} \to N) = 0.$$

Clearly α_p is an α-type group scheme, so any power of it is also such a group scheme. These are the only examples:

3.1.10. Proposition. *A finite commutative group scheme N over a field K of characteristic $p > 0$ is an α-group scheme if and only if there exists an integer r and a K-isomorphism*

$$N \simeq (\alpha_p)^r.$$

PROOF. The statement is immediate from Dieudonné theory when the base field K is perfect; the corresponding (contravariant) Dieudonné module is an r-dimensional vector space over K on which both $\mathcal{F}$ and $\mathcal{V}$ are zero. We give two proofs for a general field K of characteristic p.

The first proof is based on the following general fact. The category of commutative K-group schemes of finite type on which the relative Frobenius homomorphism is zero is equivalent to the category of commutative p-Lie algebras over K via the functor $G \rightsquigarrow \mathrm{Lie}_p(G)$ where $\mathrm{Lie}_p(G)$ is $\mathrm{Lie}(G)$ equipped with the p-operation arising from the p-power map $D \mapsto D^p$ on invariant derivations on $\mathcal{O}_G$. A proof can be found in [**82**, Ch. II, §14, Thm.].[2] For example, the p-operation on α_p vanishes since the same clearly holds for $\mathbb{G}_{\mathrm{a}}$ (and the p-operation is functorial with respect to K-homomorphisms, such as $\alpha_p \hookrightarrow \mathbb{G}_{\mathrm{a}}$). Under this equivalence, an α-group scheme N over K has commutative p-Lie algebra on which the p-operator is zero (since we can check this over the perfect field $\overline{K}$). Clearly such a p-Lie algebra is isomorphic to the p-Lie algebra of $(\alpha_p)^r$, where $r = \dim_K(\mathrm{Lie}(N))$, so $N \simeq (\alpha_p)^r$.

The strategy of the second proof is to reduce to the case when K is perfect by fpqc descent.[3] A direct calculation proves that the endomorphism functor of α_p on the category of K-schemes is represented by the ring scheme $\mathbb{G}_{\mathrm{a}}$, so the automorphism functor of $(\alpha_p)^r$ on the category of K-schemes is represented by

---

[2] The proof in [**82**] works verbatim over any field of characteristic $p > 0$, despite the blanket assumption in [**82**] that the base field is algebraically closed. The generalization to non-commutative G and non-commutative p-Lie algebras is given in [**32**, II, 7.4] and [**30**, $\mathrm{VII_A}$, 7.4].

[3] Here is a variant that avoids the formalism of fpqc descent. Let $V = \mathrm{Hom}_{K\text{-gp}}(N, \mathbb{G}_{\mathrm{a}})$, a K-vector space via the K-linear structure on $\mathbb{G}_{\mathrm{a}}$. This is the K-vector space of $f \in \Gamma(N, \mathcal{O}_N) =: R$ such that $\mu^*(f) = f \otimes 1 + 1 \otimes f \in R \otimes_K R$, where $\mu^* : R \to R \otimes_K R$ corresponds to the group law of N. We claim that V has K-dimension r where $p^r = \mathrm{rk}(N)$, and that the natural Hopf K-algebra map $h : \mathrm{Sym}^\bullet(V) \to R$ is a surjection whose kernel is generated by all p-th power of elements of V in the symmetric algebra $\mathrm{Sym}^\bullet(V)$. The formation of V and h commute with arbitrary extensions of the base field (e.g., $\overline{K}/K$), so we are reduced to the obvious case when K is perfect.

GL_r. Since $N_{\overline{K}} \simeq (\alpha_p)^r_{\overline{K}}$ for some $r \geqslant 0$, we obtain a descent datum describing N as a twisted form of $(\alpha_p)^r$ relative to $\overline{K}/K$. This descent datum defines a class in the cohomology set $\mathrm{H}^1(\overline{K}/K, \mathrm{GL}_r)$. This cohomology set vanishes (as it also classifies descent through $\mathrm{Spec}\,\overline{K} \to \mathrm{Spec}\,K$ for rank-r vector bundles, and K-vector spaces have bases), so the descent datum can be trivialized; i.e., $N \simeq (\alpha_p)^r$. □

3.1.11. Definition. Let K be a field of characteristic $p > 0$, and let G be a finite commutative K-group scheme of p-power order. The unique maximal α-group scheme in G is denoted $G[\mathcal{F}, \mathcal{V}]$. Similarly, for a p-divisible group X over K we denote by $X[\mathcal{F}, \mathcal{V}]$ the unique maximal α-subgroup scheme of X.

To see that the above definitions make sense, first note that $X[\mathcal{F}, \mathcal{V}] \subset X[p]$, so $X[\mathcal{F}, \mathcal{V}] = (X[p])[\mathcal{F}, \mathcal{V}]$. For finite commutative G of p-power order, if H and H' are K-subgroups of G that are α-group schemes then so is the K-subgroup that they generate (since $H \times H'$ is faithfully flat over that K-subgroup via multiplication in G). Hence, order considerations ensure the existence and uniqueness of of $G[\mathcal{F}, \mathcal{V}]$.

3.2. Properties of a-numbers

The construction of counterexamples to some of the CM lifting questions in 1.8.5 will rest on a close study of subgroups of the form $(\alpha_p)^r$, so we now investigate such subgroups.

3.2.1. Let K be a field of characteristic $p > 0$, and let G be an fppf commutative group sheaf on the category of K-schemes (e.g., a finite commutative K-group, or a p-divisible group over K). We make the subset $\mathrm{Hom}(\alpha_p, G) \subset G(\alpha_p)$ into a K-vector space by using the K-action on α_p. By a computation with Dieudonné modules of infinitesimal Frobenius kernels it is easy to check that if K is perfect and K'/K is a perfect extension then when $G[p]$ is represented by a finite type K-group (e.g., G a finite type K-group, or a p-divisible group), the natural map

$$K' \otimes_K \mathrm{Hom}((\alpha_p)_K, G) \to \mathrm{Hom}((\alpha_p)_{K'}, G_{K'})$$

is an isomorphism. Thus, if $G[p]$ is a finite type K-group, it is unambiguous for arbitrary fields K of characteristic p to define the *a-number* $a(G)$ to be

$$a(G) := \dim_{K'}(\mathrm{Hom}(\alpha_p, G_{K'}))$$

for any perfect extension K'/K (e.g., $K' = \overline{K}$). The use of a perfect extension is unavoidable, since over any imperfect field K of characteristic p there is an infinitesimal commutative finite K-group G of order p^3 such that

$$\dim_K(\mathrm{Hom}(\alpha_p, G)) < \dim_{\overline{K}}(\mathrm{Hom}(\alpha_p, G_{\overline{K}}));$$

see [**18**, 5.8].

Having defined the a-number for rather general commutative group functors, we now introduce an important concept in the relative theory of p-divisible groups. The group schemes N that arise as the p^n-power torsion of a p-divisible group over an arbitrary scheme S satisfy some non-trivial conditions that can be expressed entirely in terms of N. (We do not need this explicit list of conditions, so we refer the reader to [**51**, 1.1] and [**75**, 1.2–1.3] for details.) A finite locally free commutative group scheme $N \to S$ satisfying these conditions is called a *truncated Barsotti–Tate group scheme of level* n, or simply a BT_n group scheme. It is a remarkable fact (which we do not need) that over a perfect field of characteristic

p, any BT_n group scheme is the p^n-torsion of a p-divisible group over the ground field; see [**51**, 1.7]. Over general fields of characteristic p, the notion of a BT_1 group scheme can be expressed in explicit terms, which we take as our definition:

3.2.2. Definition. Let K be a field of characteristic $p > 0$. A finite commutative K-group N is called a BT_1 *group scheme* when it is killed by p and the complex

$$N \xrightarrow{\mathrm{Fr}} N^{(p)} \xrightarrow{\mathrm{Ver}} N$$

is exact (as fppf group sheaves, or equivalently Fr induces $N/\ker(\mathrm{Fr}) \simeq \ker(\mathrm{Ver})$).

As an example, if X is a p-divisible group over K then $N = X[p]$ is a BT_1. Indeed, to check this we may apply scalar extension to an algebraically closed extension of K to reduce to the case that K is perfect. In such cases the (contravariant) Dieudonné module $M = \mathrm{M}^*(X)$ is finite free over $W(K)$ and the commuting semi-linear endomorphisms $\mathcal{F}$ and $\mathcal{V}$ of M have composition equal to p, so $\mathcal{F}$ is injective on M. Hence, the sequence of abelian groups

$$M/pM \xrightarrow{\mathcal{V}} M/pM \xrightarrow{\mathcal{F}} M/pM$$

is exact, and this is precisely the Dieudonné module version of the desired exact sequence in Definition 3.2.2 for $N = X[p]$.

Letting $\#(G)$ denote the order of a finite K-group scheme G, for any finite commutative K-group N with p-power order we have the following obvious relations among orders of finite K-groups:

$$\#(N[\mathrm{Fr}]) \cdot \#(\mathrm{Fr}(N)) = \#(N) = \#(N^{(p)}[\mathrm{Ver}]) \cdot \#(\mathrm{Ver}(N^{(p)})),$$

where $N[f]$ denotes the kernel of a homomorphism $f : N \to N'$ and $f(N)$ denotes the closed K-subgroup $N/\ker(f) \hookrightarrow N'$. Thus, in the definition of a BT_1 group scheme it is equivalent to say that the complex

$$N^{(p)} \xrightarrow{\mathrm{Ver}} N \xrightarrow{\mathrm{Fr}} N^{(p)}$$

is exact.

It follows that for a BT_1 group scheme N over a *perfect* field, the Dieudonné module $M = \mathrm{M}^*(N)$ satisfies

$$\mathcal{F}M = M[\mathcal{V}], \quad \mathcal{V}M = M[\mathcal{F}],$$

where $M[\mathcal{F}] := \ker(\mathcal{F} : M \to M)$ and similarly for $\mathcal{V}$.

3.2.3. Proposition. *For a* BT_1 *group scheme* N *over a field* K *of characteristic* $p > 0$, $a(N) = a(N^D)$.

PROOF. It suffices to prove the theorem after extension of the ground field, so we may assume that K is perfect. Let $M = \mathrm{M}^*(N)$ be the Dieudonné module. Clearly

$$a(N) = \dim_K(M/(\mathcal{F}M + \mathcal{V}M)), \quad a(N^D) = \dim_K(\mathcal{F}M \cap \mathcal{V}M).$$

Since $\mathcal{F}M = M[\mathcal{V}]$ and $\mathcal{V}M = M[\mathcal{F}]$ (due to the BT_1 property), and

$$\dim_K(\mathcal{F}M) + \dim_K(M[\mathcal{F}]) = \dim_K(M),$$

we see that

$$a(N) = \dim_K M - \dim_K(\mathcal{F}M + \mathcal{V}M) = \dim_K(\mathcal{F}M \cap \mathcal{V}M) = a(N^D).$$

□

3.2.4. Corollary. *For a p-divisible group X over a field K of characteristic p, the a-numbers $a(X)$ and $a(X^t)$ coincide.*

PROOF. Clearly $a(X) = a(X[p])$, and $X[p]$ is a BT_1 group scheme. By definition, we have $X^t[p] = (X[p])^D$. Hence, by 3.2.3,

$$a(X) = a(X[p]) = a((X[p])^D) = a(X^t[p]) = a(X^t).$$

□

3.2.5. Remark. The condition "N is a BT_1 group scheme" in 3.2.3 is essential. Here is an example, over any field of characteristic p, consisting of a p-torsion commutative finite group scheme N such that $a(N) \neq a(N^D)$.

Let N be the finite commutative $\mathbb{F}_p$-group scheme of rank p^3 whose Dieudonné module is

$$M = \mathbb{F}_p e \oplus \mathbb{F}_p e' \oplus \mathbb{F}_p e''$$

with $\mathcal{F}(e) = e'$, $\mathcal{F}(e') = \mathcal{F}(e'') = 0$, $\mathcal{V}(e) = e''$, $\mathcal{V}(e') = \mathcal{V}(e'') = 0$. Thus,

$$a(N) = \dim_{\mathbb{F}_p}(M/(\mathcal{F}M + \mathcal{V}M)) = 1 \quad \text{and} \quad a(N^D) = \dim_{\mathbb{F}_p}(M[\mathcal{F}] \cap M[\mathcal{V}]) = 2.$$

For any field K of characteristic p we likewise have $a(N_K) = a(N) = 1$ and $a((N_K)^D) = a((N^D)_K) = a(N^D) = 2$.

Later we will study quotients by parametric families of α_p-subgroups, so it is useful to have some control on the behavior of a-numbers in a family:

3.2.5.1. Proposition. *Let $G \to S$ be a p-torsion finite locally free commutative group scheme over an $\mathbb{F}_p$-scheme S. The function $s \mapsto a(G_s)$ is upper semi-continuous; i.e., for any integer n, the set of $s \in S$ such that $a(G_s) \geqslant n$ is closed.*

PROOF. By standard limit arguments, we may assume S is noetherian. Since closedness for a subset of S is equivalent to constructibility and stability under specialization, it suffices to show that the condition $a(G_s) \geqslant n$ defines a constructible set in S that is moreover stable under specialization.

For constructibility, by noetherian induction it suffices to prove that if S is integral with generic point η then $a(G_s) = a(G_\eta)$ for all s in a dense open subset of S. It is harmless to replace S with a dense open subset, so by generic flatness over S for the kernel of $\mathrm{Fr}_{G/S} : G \to G^{(p)}$ we may and do assume $\mathrm{Fr}_{G/S} = 0$. Since $a(G_s) = a(G_s^{(p)})$, it is equivalent to show that $a(G_s^{(p)}) = a(G_\eta^{(p)})$ for all s in a dense open subset of S. But $p^{a(G_s)}$ is the order of the kernel of the s-fiber of $\mathrm{Ver}_{G/S} : G^{(p)} \to G$, so by applying generic flatness to this kernel over S we get the desired constancy near η. The constructibility of the a-number is now proved.

For the behavior under specialization, we may assume $S = \mathrm{Spec}(R)$ for a discrete valuation ring R of equicharacteristic p, say with closed point s and generic point η. We just have to show that $a(G_\eta) \leqslant a(G_s)$. We may replace R with its completion, and then with its integral closure in a sufficiently large finite extension of the fraction field so that G_η contains an η-subgroup scheme of the form $(\alpha_p)^m$ with $m = a(G_\eta)$. The Zariski closure of this subgroup in G is a finite *flat* R-subgroup H of G with order p^m, and $\mathrm{Fr}_{H/S}$ and $\mathrm{Ver}_{H/S}$ vanish since we may check the vanishing on the η-fibers. Thus, the closed subgroup scheme $H_s \subset G_s$ is an α-group scheme and has order p^m, so $a(G_s) \geqslant m = a(G_\eta)$. □

3.3. Isogenies and duality

To go further, we need a systematic theory of isogenies for p-divisible groups over any (noetherian) base scheme. This is motivated in part by the analogous theory for abelian schemes, which we briefly introduced in 1.4.2.3–1.4.2.5 and now revisit before we turn our attention back to p-divisible groups.

3.3.1. Proposition. *Let $\varphi : A \to B$ be a homomorphism between abelian schemes over a scheme S. The set of $s \in S$ such that φ_s is an isogeny is open and closed.*

PROOF. Since the functions $s \mapsto \dim A_s, \dim B_s$ are Zariski-locally constant on S, we may assume that they are equal to a constant g. We may also assume that these fiber dimensions coincide, as otherwise there is nothing to do, so for any $s \in S$ the map φ_s is an isogeny if and only if $\ker(\varphi_s)$ is finite. Since $\ker(\varphi)$ is a finitely presented S-scheme, by semi-continuity of fiber dimension [**34**, IV_3, 13.1.3] the locus of points in $\ker(\varphi)$ at which the fiber of $\ker(\varphi) \to S$ has positive dimension is closed. By S-properness of $\ker(\varphi)$, that closed set has closed image in S. The open complement of that closed image is precisely the locus of $s \in S$ such that φ_s is an isogeny.

Let Z be the closed set of $s \in S$ such that φ_s is not an isogeny. We want to show that Z is open. By working Zariski-locally on S, we may arrange that there is a prime ℓ that is a unit on S. Choose $s_0 \in S$, so the kernel $\ker(\varphi_{\overline{s}_0})$ between geometric fibers over s_0 has positive dimension. Hence, $(\ker(\varphi_{\overline{s}_0}))^0_{\mathrm{red}}$ is a non-zero abelian variety, so for any $n > 0$ its ℓ^n-torsion is not killed by ℓ^{n-1}. The map $\varphi_n : A[\ell^n] \to B[\ell^n]$ between finite étale S-groups must have kernel $\ker(\varphi_n)$ that is finite étale. The s_0-fiber of this kernel is not killed by ℓ^{n-1}.

If S is connected then the endomorphism of ℓ^{n-1}-multiplication on the finite étale S-group $\ker(\varphi_n)$ is non-zero on all fibers since it is non-zero on one fiber. Hence, if S is connected then for every $s \in S$ there is no single power of ℓ killing the ℓ^n-torsion in $\ker(\varphi_s)$ for all n, whence $\ker(\varphi_s)$ cannot be finite for any $s \in S$. For general S we may apply this to the pullback of φ over the connected scheme $\mathrm{Spec}(\mathscr{O}_{S,s_0})$, so Z is stable under generization. In particular, Z is open if S is locally noetherian. (It is not true that closed sets stable under generization are open without a noetherian hypothesis. A counterexample is $S = \mathrm{Spec}(R)$ with $R = \prod_{n=1}^{\infty} k$ for a field k and Z the closed complement of the union of the set of isolated points corresponding to the evident factor fields of R.)

To prove that Z is open in general, we need to relate the general case to the noetherian case. We may assume that $S = \mathrm{Spec}(R)$ is affine, and then by expressing R as the direct limit of its noetherian subrings R_i we may find a sufficiently large i_0 so that the finitely presented R-schemes A, B descend to abelian schemes A_{i_0}, B_{i_0} of relative dimension g over R_{i_0}. By increasing i_0 if necessary, we may also arrange that φ descends to a homomorphism $\varphi_{i_0} : A_{i_0} \to B_{i_0}$ over R_{i_0}. The set Z is the preimage under $\mathrm{Spec}(R) \to \mathrm{Spec}(R_{i_0})$ of the analogous set $Z_{i_0} \subset \mathrm{Spec}(R_{i_0})$. But Z_{i_0} is open since R_{i_0} is noetherian, so we are done. □

The preceding result suggests that we try to define the concept of isogeny for p-divisible groups over S via a condition on (geometric) fibers over S, but this is inconvenient for developing the relative theory when S is arbitrary (since non-zero p-divisible groups are not finitely presented, unlike abelian schemes). Thus, we shall use another definition (see 3.3.5) which looks stronger and suffices for our purposes, and in 3.3.8 and 3.10 we will revisit its relation with a fibral definition. (Our avoidance of noetherian hypotheses on S is for aesthetic reasons. In the later applications, S is noetherian.)

The proof of the analogue of Theorem 1.4.2.5 for p-divisible groups is different from the proof for abelian schemes. Before stating the result, we first develop some basic notions related to p-divisible groups over a general scheme.

3.3.2. Inductive and projective limits. Recall that a p-divisible group $X \to S$ over a scheme S can be defined as an inductive system of finite locally free commutative S-groups $\{X_i \mid \iota_j^i\}$ fitting into exact sequences

$$0 \to X_j \xrightarrow{\iota^j_{j+j'}} X_{j+j'} \xrightarrow{\pi^{j+j'}_{j'}} X_{j'} \to 0$$

with properties as stated in 1.4.3.3 (or see [**51**]). The direct limit sheaf $X = \varinjlim X_j$ (relative to the ι_j^i's) for the fppf topology on the category of S-schemes recovers the initial inductive system via the identifications $X_j = X[p^j]$, under which the map $\iota^j_{j'}$ corresponds to the inclusion $X[p^j] \hookrightarrow X[p^{j'}]$. We also call X the *p-divisible group* associated to this set of exact sequences and maps.

The inductive limit X is a convenient way of encoding the data of a p-divisible group; these are precisely the fppf abelian sheaves whose points are locally killed by a power of p and whose p-torsion subgroup sheaf is represented by a finite locally free S-scheme (forcing the same for each $X[p^j]$ by induction on j due to Proposition 1.4.1.3 since $0 \to X[p] \to X[p^{j+1}] \to X[p^j] \to 0$ is exact for the fppf topology for every j).

Inspired by the study of étale sheaves via Tate modules, it is natural to wonder if we can efficiently study p-divisible groups using the viewpoint of projective limits. This turns out to be much less convenient than inductive limits when using sheaf-theoretic methods (as we must do when p is not a unit on the base).

The "asymmetry" between the usefulness of inductive and projective limits in the relative setting is due to the nature of the sheaf axioms. Consider the projective system of fppf abelian sheaves $\{\pi_j^{j'} : X_{j'} \longrightarrow X_j\}$. In contrast with the case of inductive limits, the projective limit sheaf $\varprojlim X_j$ is not useful because we cannot reconstruct X from it. For example, consider the p-divisible group $\{\mu_{p^n}\}_{n\geqslant 1}$ over $\mathbb{Q}$. The associated inductive limit sheaf assigns to any number field K the (finite) group of all p-power roots of unity in K, whereas the associated projective limit sheaf assigns the trivial group to every number field (since $\varprojlim \mu_{p^n}(K) = 1$ for such fields K). Thus, the right object to consider (over quasi-compact S) is the "essential" projective limit $\mathrm{T}_p(X_\bullet)$: this is the projective system $\{X_j\}$ viewed in the category of projective systems $\{Y_j\}$ for which Y_j is killed by p^j for all j and morphisms $f : \{Y'_j\} \to \{Y_j\}$ are maps of projective systems taken up to the equivalence relation $f \sim h$ when $f_j \circ \pi_j^{j+n} = h_j \circ \pi_j^{j+n}$ for all j and some $n \geqslant 0$. (The intervention of the maps π_j^{j+n} is not necessary if they are all epimorphisms, such as when Y' and Y are p-divisible groups.)

For a p-divisible group X over S, we call $\mathrm{T}_p(X_\bullet)$ the *Tate-p-group* attached to X (or to $\{X_j\}$). When S is a quasi-compact scheme on which p is a unit, this coincides with the notion of a lisse étale $\mathbb{Z}_p$-sheaf, so when such an S is non-empty and connected it is the same as a representation of the étale fundamental group on a finite free $\mathbb{Z}_p$-module. [4] Any of the following three structures determine the other two:

- the inductive system $\{\iota^i_{i'} : X_i \longrightarrow X_{i'}\}_{i' \geqslant i}$,
- the p-divisible group X,
- the projective system $\{\pi^{i'}_i : X_{i'} \longrightarrow X_i\}_{i' \geqslant i}$,

Explicitly,

$$X = \varinjlim X_n, \quad X_n = X[p^n], \quad \pi^{n'+n}_n : X_{n'+n} \to X_{n'+n}/X_{n'}.$$

We will not use the notion of Tate-p-groups, and will not use the notation $\mathrm{T}_p(X_\bullet)$. However, in order to understand duality for p-divisible groups it is useful to keep in mind both the inductive and the projective aspects.

As a preliminary step towards the definition of "isogeny" for p-divisible groups, we now construct quotients by suitable finite subgroup schemes. The starting point is the observation that if X is a p-divisible group over S and if $N \subset X$ is a subgroup sheaf that is represented by a finite locally free S-scheme then the fibral order of N is constant Zariski-locally on S, so Zariski-locally N is killed by a power of p (by Theorem 1.4.2.4). That is, S has a covering by Zariski-open sets $\{U_i\}$ such that $N|_{U_i} \subset X[p^{r_i}]$ for integers $r_i \geqslant 1$.

3.3.3. Lemma. *Let X be a p-divisible group over a scheme S and let $N \subset X$ be a finite locally free subgroup scheme.*

(1) *Assume N is killed by p^r with $r > 0$; equivalently, $N \subset X_r = X[p^r]$. For any $n \in \mathbb{Z}_{\geqslant 0}$ and any $m \geqslant n$ the natural map*

$$(X_{n+r}/N)\,[p^n] \to (X_{m+r}/N)\,[p^n]$$

is an isomorphism.

(2) *For any $n \geqslant 1$, define Y_n by the evident gluing of the group schemes defined Zariski-locally over S by $(X_{m+r}/N)[p^n]$ for any $m \geqslant n$ and $r \geqslant 1$ such that p^r kills N. Each Y_n is finite locally free as an S-scheme, and the resulting inductive system $Y = \{Y_n\}$ is a p-divisible group.*

The evident homomorphism $\varphi : X \to Y$ has functorial kernel N and it identifies the fppf sheaf Y with the fppf quotient sheaf X/N. For any morphism of schemes $S' \to S$ the natural map $X_{S'}/N_{S'} \to (X/N)_{S'}$ is an isomorphism.

Proof. It clearly suffices to work Zariski-locally on S, so we may and do assume that $N \subset X[p^r]$ for some $r \geqslant 1$. We define $Y_n := (X[p^{n+r}]/N)[p^n]$, which obviously is independent of the choice of r and is naturally compatible with base change on S. To see that Y_n is finite locally free as an S-scheme, consider the faithfully flat map $X[p^{n+r}] \to X[p^r]$ induced by multiplication by p^n on $X[p^{n+r}]$. This carries N into itself, and (upon passing to the quotient modulo N) the p^n-multiplication endomorphism of $X[p^{n+r}]/N$ has image $X[p^r]/N$ in terms of fppf abelian sheaves. But an fppf-sheaf surjection between finite locally free commutative group schemes is faithfully flat (as we may check on geometric fibers, due to the fibral flatness criterion), so the induced map $X[p^{n+r}]/N \to X[p^r]/N$ is faithfully flat. Thus, its kernel Y_n is a finite locally free commutative S-group.

[4] If the base scheme is not connected then one needs to use the fundamental groupoid instead.

To prove (1) and (2), we have to show that the natural maps $Y_n \to Y_m[p^n]$ for $m \geqslant n$ are isomorphisms and the inductive system $\{Y_n\}$ forms a p-divisible group. Since each Y_n is finite locally free over S, for this purpose we can pass to geometric fibers. That is, we can assume $S = \mathrm{Spec}(K)$ for an algebraically closed field K. The case $\mathrm{char}(K) \neq p$ is an elementary computation with finitely generated $\mathbb{Z}_p$-modules, and the case $\mathrm{char}(K) = p$ goes similarly via Dieudonné modules (using $W(K)$ in place of $\mathbb{Z}_p$).

Returning to a general base, by construction the natural map

$$\varphi_n : X_n = X[p^n] \to (X[p^{n+r}]/N)[p^n] = Y_n$$

has kernel $N[p^n]$. Thus, taking $n \geqslant r$ gives $\ker(\varphi_n) = N$, so $\varphi : X \to Y$ has kernel (represented by) N. It is now clear that the fppf sheaf defined by Y is X/N. □

3.3.4. Corollary. *Let X and Y be p-divisible groups of the same height over a scheme S. Let $\varphi : X \to Y$ be a homomorphism such that its kernel $N = \ker(\varphi)$ is finite locally free over S. The induced homomorphism $X/N \to Y$ is an isomorphism. Moreover if S is quasi-compact or N has fibral order that is bounded over S then there exists a homomorphism $\psi : Y \to X$ and an integer $m \geqslant 0$ such that*

$$\varphi \circ \psi = [p^m]_Y, \quad \psi \circ \varphi = [p^m]_X,$$

and $N' := \ker(\psi)$ is finite locally free over S with $Y/N' \to X$ an isomorphism.

PROOF. For the isomorphism property we may work locally on the base. Thus, to prove the corollary we may and do assume either that S is quasi-compact or that N has bounded order. Either way, N is killed by some p^r (use Theorem 1.4.2.4 in the latter case), and we choose $m = r$. Once we show that φ identifies Y with X/N, we define ψ to be

$$Y = X/N \to X/X[p^r] = X$$

to arrange that $\psi \circ \varphi = [p^r]_X$. Then $N' := \ker(\psi) = X[p^r]/N$ is finite locally free with $Y/N' \to X$ an isomorphism and

$$(\varphi \circ \psi) \circ \varphi = \varphi \circ (\psi \circ \varphi) = \varphi \circ [p^r]_X = [p^r]_Y \circ \varphi,$$

so the universal mapping property of $\varphi : X \to X/N$ forces $\varphi \circ \psi = [p^r]_Y$ as well.

Now our problem is to show that the induced map $X/N \to Y$ between p-divisible groups over S is an isomorphism. It is equivalent to check the isomorphism condition on p^i-torsion levels for all $i \geqslant 1$. Since the formation of X/N commutes with base change on S (by Lemma 3.3.3), by the fibral isomorphism criterion we may assume $S = \mathrm{Spec}(k)$ for an algebraically closed field k. The case $\mathrm{char}(k) \neq p$ is an elementary computation with finite free $\mathbb{Z}_p$-modules of the same rank, and a similar argument works when $\mathrm{char}(k) = p$ by using Dieudonné modules that are finite free over $W(k)$ of the same rank. □

3.3.5. Definition. Let X and Y be p-divisible groups over a scheme S. A homomorphism $\varphi : X \to Y$ is an *isogeny* when the heights coincide and $\ker(\varphi)$ is represented by a finite locally free S-scheme.

Equivalently, by the two preceding results, an isogeny between p-divisible groups is the projection to the quotient modulo a finite locally free S-subgroup scheme.

By 3.3.4, if S is quasi-compact or the S-group $\ker(\varphi)$ has bounded fibral order then there also exists an isogeny $Y \to X$.

3.3.5.1. Example. Let $\varphi : A \to B$ be a homomorphism between abelian schemes over a scheme S, with all fibers A_s and B_s of dimension $g > 0$, and let $f : A[p^\infty] \to B[p^\infty]$ be the associated homomorphism between p-divisible groups of height $2g$. Then we claim that φ is an isogeny (see 1.4.2.3) if and only if f is an isogeny. It suffices to show that $\ker(\varphi)$ is finite locally free if and only if $\ker(f)$ is.

Certainly if f is an isogeny then for every geometric fiber $\overline{s}$ of S the abelian variety $\ker(\varphi_{\overline{s}})^0_{\text{red}}$ has finite p-power torsion and hence must vanish. That is, $\ker(\varphi)$ has finite fibers, so φ is a fiberwise isogeny and hence (by definition) is an isogeny. Conversely, suppose φ is an isogeny, so $\ker(\varphi)$ is a finite locally free S-group. Zariski-locally on the base, this kernel is killed by an integer $n > 0$ (see 1.4.2.4), which is to say it is a group functor valued in $\mathbb{Z}/(n)$-modules. Hence, it has a functorial decomposition into primary parts indexed by the prime factors of n (with vanishing primary parts for primes not dividing n), and its p-part is $\ker(f)$.

It remains to show that if $G \to S$ is a finite locally free commutative group scheme killed by n then for any factorization $n = dd'$ with $d, d' > 0$ and $\gcd(d, d') = 1$, the finite and finitely presented S-group $G[d]$ is S-flat. We have functorially $G = G[d] \times G[d']$, so if $S = \operatorname{Spec}(R)$ for a ring R (as we may assume) then the coordinate rings satisfy $\mathcal{O}(G) \simeq \mathcal{O}(G[d]) \otimes_R \mathcal{O}(G[d'])$ as R-algebras. The co-unit of $\mathcal{O}(G[d'])$ identifies R as a direct summand of $\mathcal{O}(G[d'])$, so $\mathcal{O}(G[d])$ is a direct summand of the R-flat $\mathcal{O}(G)$ and thus $\mathcal{O}(G[d])$ is also R-flat.

It is not obvious if the dual of an isogeny is an isogeny. As motivation for how to analyze this problem, consider a finite-index inclusion $f : M' \to M$ between finite free $\mathbb{Z}_p$-modules (the "projective limit" viewpoint on an isogeny between p-divisible groups), say with M/M' killed by p^r. The induced map

$$f \bmod p^n : M'/p^n M' \to M/p^n M$$

has cokernel $(M/M')/(p^n)$ identified with M'/M for $n \geqslant r$, and the kernel is

$$\operatorname{Tor}_1^{\mathbb{Z}_p}(\mathbb{Z}_p/(p^n), M/M') = (M/M')[p^n] = M/M'$$

when $n \geqslant r$. We now establish an analogous result for isogenies between general p-divisible groups.

3.3.6. Proposition. *Let $\varphi : X \longrightarrow Y$ be an isogeny between p-divisible groups over a scheme S, and let $N = \ker(\varphi)$. Assume N is killed by p^r with $r \geqslant 0$. For any $n \geqslant r$, consider the exact sequence of fppf abelian sheaves*

$$0 \to N_n = N \longrightarrow X_n \xrightarrow{\varphi_n} Y_n \longrightarrow Q_n \to 0.$$

The Q_n are represented by finite locally free commutative S-groups of the same order as N. Moreover:

(1) *for any $n \geqslant r$ and $m \geqslant 0$, the inclusion $X_n \hookrightarrow X_{n+m}$ restricts to the identity on N and the quotient map $X_{n+m} \twoheadrightarrow X_m$ induced by p^n kills N;*

(2) *for any $m \geqslant r$ and $n \geqslant 0$ there are unique isomorphisms $Q_{n+m} \simeq Q_m$ fitting into commutative diagrams*

$$\begin{array}{ccc} Y_n & \twoheadrightarrow & Q_n \\ \big\downarrow & & \big\downarrow{\scriptstyle 0} \\ Y_{n+m} & \twoheadrightarrow & Q_{n+m} \\ \big\downarrow & & \big\downarrow{\scriptstyle \simeq} \\ Y_m & \twoheadrightarrow & Q_m \end{array}$$

(3) *for any $m, n \geqslant r$, the maps φ_n, φ_{n+m}, and φ_m naturally define isomorphisms $\delta_{n,m} : N \simeq Q_n$ independent of m and respecting the isomorphism $Q_n \simeq Q_r$ from* (2).

PROOF. The assertions in (1) are obvious. To prove (2), we first analyze the structure of the group sheaves Q_n. For each $n \geqslant r$, Q_n is the cokernel of a monic homomorphism $X_n/N \to Y_n$. The source X_n/N is finite locally free over S because N is, so it is a closed subgroup of Y_n. Thus, the quotient Q_n is indeed represented by a finite locally free S-group. Since X_n and Y_n have the same order (as X and Y have the same height), we see that Q_n has the same order as N for all $n \geqslant r$. In particular, the map $Q_n \to Q_r$ between cokernels induced by the commutative diagram

$$\begin{array}{ccc} X_n & \xrightarrow{\varphi_n} & Y_n \\ \big\downarrow & & \big\downarrow \\ X_r & \xrightarrow[\varphi_r]{} & Y_r \end{array}$$

is a quotient map and thus an isomorphism (for order reasons).

We conclude that the natural map $Q_{n+m} \to Q_m$ is an isomorphism for all $m \geqslant r$ and $n \geqslant 0$. Thus, to show that the map $Q_n \to Q_{n+m}$ induced by $Y_n \hookrightarrow Y_{n+m}$ vanishes for $m \geqslant r$ and $n \geqslant 0$ it is harmless to compose it with the isomorphism $Q_{n+m} \to Q_m$ (induced by p^n on Y) and precompose it with the quotient map $Y_n \to Q_n$. This resulting composite map

$$Y_n \to Q_n \to Q_{n+m} \to Q_m$$

vanishes since the composition

$$Y_n \to Y_{n+m} \to Y_m$$

is zero (as Y_n is killed by $[p^n]_Y$ that induces $Y_{n+m} \to Y_m$). This proves (2).

To prove (3), we use the snake lemma for fppf abelian sheaves. For $n, m \geqslant r$, consider the commutative diagram

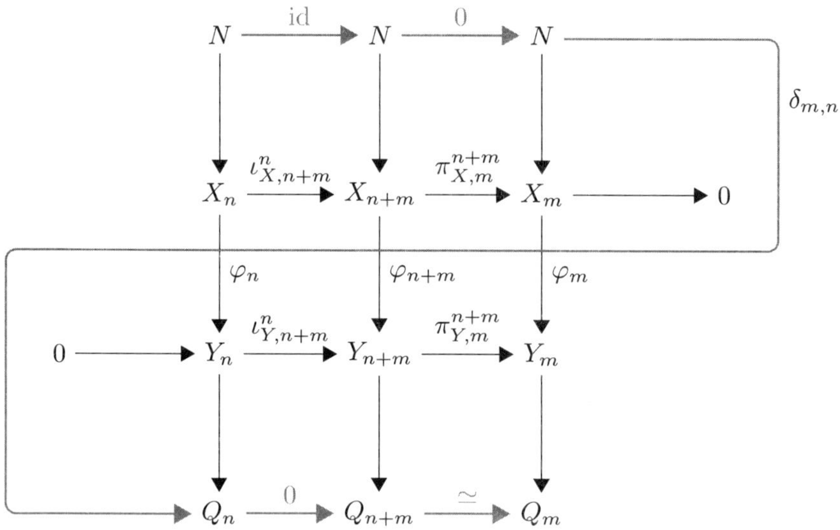

with exact rows and columns. The vertical exact sequences are given by the isogeny φ, so the first arrow (respective the last arrow) of each column is the kernel (respectively the cokernel) of the middle arrow in that column. By the snake lemma we get an exact sequence

$$N \xrightarrow{\text{id}} N \xrightarrow{0} N \xrightarrow{\delta_{m,n}} Q_n \xrightarrow{0} Q_{n+m} \xrightarrow{\simeq} Q_m \,.$$

Functoriality of the snake lemma implies that the isomorphisms $\delta_{m,n} : N \xrightarrow{\sim} Q_n$ for $m, n \geqslant r$ are independent of m (by comparing m and $m+1$ for a common n, using $X_{m'+m} \hookrightarrow X_{m'+(m+1)}$ for any $m' \geqslant 0$, and likewise for Y) and compatible with the natural identifications $Q_n \simeq Q_r$ (by comparing n and $n+1$ for a common m, using $X_{n'+(n+1)} \twoheadrightarrow X_{n'+n}$ for any $n' \geqslant 0$ and likewise for Y). □

3.3.7. Theorem. *Let $\varphi : X \to Y$ be an isogeny of p-divisible groups over a scheme S, and let $N = \ker(\varphi)$. The dual map $\varphi^t : Y^t \to X^t$ is an isogeny with $\ker(\varphi^t)$ naturally isomorphic to N^D.*

Note that in 1.4.2.5 we write A^t for the dual abelian scheme, and in 3.3.7 the notation X^t stands for the Serre dual of X; see 1.4.3.3.

PROOF. The projective system $\{Y_n \mid \pi_i^j\}$ gives the inductive system

$$\{(Y_n)^D \mid (\pi_i^j)^D\}$$

which defines the Serre dual p-divisible group Y^t. We have a map of projective systems

$$\{Y_n \twoheadrightarrow Q_n \simeq N \mid n \geqslant r\}$$

using the identity map on N at every stage, so applying Cartier duality yields a map of inductive systems $\{N^D \hookrightarrow (Y_n)^D = Y^t[p^n]\}$ using the identity map on N^D.

We conclude that the projective system of right-exact sequences $X_n \to Y_n \to Q_n \to 0$ of finite locally free commutative S-groups (with flat kernel on the left) for $n \geqslant r$ dualizes to an inductive system of left-exact sequences $0 \to N^D \to (Y_n)^D \to$

$(X_n)^D$. Hence, by working with p-divisible groups as fppf abelian sheaves we obtain a left-exact sequence

$$0 \to N^D \longrightarrow Y^t \xrightarrow{\varphi^t} X^t.$$

In other words, φ^t is a homomorphism between p-divisible groups of the same height and its kernel is represented by the finite locally free S-group N^D, so φ^t is an isogeny. (In particular, φ^t identifies X^t with the quotient sheaf Y^t/N^D.) □

We have established that the notion of isogeny for p-divisible groups satisfies several good properties, and now we turn attention to fibral aspects. In view of Proposition 3.3.1, it is natural to ask if the isogeny property for p-divisible groups in the relative setting can be detected on fibers. Under a noetherian hypothesis, we can access infinitesimal fibers over artin local rings via formal group techniques to get a satisfactory answer:

3.3.8. Proposition. *If S is locally noetherian and $\varphi : X \to Y$ is a homomorphism of p-divisible groups over S then the set U of $s \in S$ such that φ_s is an isogeny is open and $\varphi_U : X_U \to Y_U$ is an isogeny.*

The locus U is also closed, but this is deep; we address it near the end of §3.3.

PROOF. We may certainly assume that X and Y have the same height. Let's first treat the case when $S = \mathrm{Spec}(R)$ for an artin local ring R. In this case the map φ_0 between special fibers is an isogeny and we want to show that $\ker(\varphi)$ is represented by a finite flat R-group.

We may assume the residue field κ has characteristic p, as otherwise every $X[p^n]$ and $Y[p^n]$ is finite étale over R (so everything is easy). Let X^0 and $X^{\text{ét}}$ respectively denote the connected and étale parts of X, and similarly for Y, so we have a commutative diagram

(3.3.8.1)
$$\begin{array}{ccccccccc} 0 & \longrightarrow & X^0 & \longrightarrow & X & \longrightarrow & X^{\text{ét}} & \longrightarrow & 0 \\ & & \downarrow{\scriptstyle \varphi^0} & & \downarrow{\scriptstyle \varphi} & & \downarrow{\scriptstyle \varphi^{\text{ét}}} & & \\ 0 & \longrightarrow & Y^0 & \longrightarrow & Y & \longrightarrow & Y^{\text{ét}} & \longrightarrow & 0 \end{array}$$

whose formation commutes with passage to the special fiber.

Since φ_κ is an isogeny, the induced map between connected parts has finite kernel and the height of X^0 is at most that of Y^0. But there is an isogeny $Y_\kappa \to X_\kappa$ by the theory over fields, so likewise the height of Y^0 is at most that of X^0. Hence, the heights of X^0 and Y^0 coincide, so the heights of $X^{\text{ét}}$ and $Y^{\text{ét}}$ also coincide, and φ^0_κ is an isogeny. Thus, snake lemma considerations with the special fiber of (3.3.8.1) imply that $\varphi^{\text{ét}}_\kappa$ is also an isogeny. If φ^0 and $\varphi^{\text{ét}}$ are isogenies (over R) then snake lemma considerations over R with (3.3.8.1) imply that $\ker(\varphi)$ is an fppf abelian sheaf extension of the finite flat group $\ker(\varphi^{\text{ét}})$ by the finite flat group $\ker(\varphi^0)$. In such cases, $\ker(\varphi)$ is finite flat over R by Proposition 1.4.1.3. Hence, over the artin local R it suffices to separately treat the connected and étale cases.

The étale case is easy, so assume X and Y are connected. As we noted in 1.4.3.6, $\mathcal{O}(X) := \varprojlim \mathcal{O}(X[p^n])$ is a formal power series ring over R equipped with a formal group law whose p^n-torsion is $X[p^n]$. Since R is artin local with residue characteristic p, all points of $\mathcal{O}(X)$ valued in a finite local R-algebras are p-power torsion (as shown in the proof of Proposition 1.4.4.3), so the formal Hopf algebra $\mathcal{O}(X)$ pro-represents X on that category. The same holds with $\mathcal{O}(Y)$ for Y.

It suffices to show that the map $\varphi^* : \mathcal{O}(Y) \to \mathcal{O}(X)$ is finite flat. Indeed, in that case the finite flat R-group G whose coordinate ring is the quotient of $\mathcal{O}(X)$ by the ideal generated by the φ^*-image of the augmentation ideal of $\mathcal{O}(Y)$ would be killed by some power of p, hence contained in $X[p^n]$ for sufficiently large n, so for such n the maps $X[p^n] \to Y[p^n]$ have scheme-theoretic kernel G that is finite flat and independent of n. This implies that the finite flat R-group G represents $\ker(\varphi)$ on the category of all S-schemes.

Since $\mathcal{O}(X)$ and $\mathcal{O}(Y)$ are formal power series rings over R, φ^* is finite flat if and only if the map $(\varphi^*)_\kappa$ between special fibers is finite flat. The special fibers of $\mathcal{O}(X)$ and $\mathcal{O}(Y)$ coincide with the analogous formal Hopf algebras associated to X_κ and Y_κ, so $(\varphi^*)_\kappa = (\varphi_\kappa)^*$. Hence, we are reduced to showing that if $S = \mathrm{Spec}(\kappa)$ for a field κ of characteristic p and if φ is an isogeny then φ^* is finite flat.

We have a factorization $[p^n]_X = \psi \circ \varphi$ for some $\psi : Y \to X$ and some $n \geqslant 1$, so $\varphi^* \circ \psi^* = [p^n]_X^*$ with $[p]_X^*$ finite flat, and hence φ^* is finite. The equality

$$\varphi \circ \psi \circ \varphi = \varphi \circ [p^n]_X = [p^n]_Y \circ \varphi$$

with φ an fppf sheaf surjection implies $\varphi \circ \psi = [p^n]_Y$. Hence, the map $\psi^* \circ \varphi^* = [p^n]_Y^*$, and this is finite flat as an endomorphism of $\mathcal{O}(Y)$, so it is injective and hence φ^* is injective. Any finite injection between formal power series rings over a field satisfies the "dimension formula" and so regularity of such power series rings it must be flat [**73**, 23.1]. This settles the case when S is artin local.

Next we consider the case when $S = \mathrm{Spec}(R)$ for a local noetherian ring $(R, \mathfrak{m})$ with residue field κ. Assuming only that the map φ_κ between special fibers is an isogeny, we will show that φ is an isogeny. For $m \geqslant 1$ let $R_m = R/\mathfrak{m}^{m+1}$ and let $\varphi_m : X_{R_m} \to Y_{R_m}$ denote the induced map over R_m. By the settled artin local case, $\ker(\varphi_m)$ is represented by a finite flat closed subgroup $N_m \subset X_{R_m}$ whose order coincides with that of N_0, say p^e. By Theorem 1.4.2.4, $N_m \subset X_{R_m}[p^e] = X[p^e]_{R_m}$. The formation of these subgroup schemes respects change in m, so since $X[p^e]$ is finite flat over R, the N_m's arise from a finite flat subgroup scheme $\widehat{N} \subset X[p^e]_{\widehat{R}}$ that is killed by $\varphi_{\widehat{R}}$.

The map $\varphi_{\widehat{R}}$ therefore factors through a homomorphism $\varphi' : X_{\widehat{R}}/\widehat{N} \to Y_{\widehat{R}}$ between p-divisible groups over $\widehat{R}$ that is an *isomorphism* over every R_m, whence the map induced by φ' between the $\widehat{R}$-finite p-power torsion levels are isomorphisms modulo every power of the maximal ideal. Thus, φ' is an isomorphism, so $(\ker(\varphi))_{\widehat{R}}$ is represented by $\widehat{N} \subset X[p^e]_{\widehat{R}}$. The fppf sheaves defined by p-divisible groups are easily checked to be fpqc sheaves, so $\ker(\varphi)$ is an fpqc sheaf on the category of R-schemes. We have just shown that its restriction over the category of $\widehat{R}$-schemes represented by a finite flat closed subscheme of $X[p^e]_{\widehat{R}}$, so by fpqc descent through $\mathrm{Spec}(\widehat{R}) \to \mathrm{Spec}(R)$ (which generally leaves the noetherian setting, via the intervention of $\widehat{R} \otimes_R \widehat{R}$) we see that $\widehat{N}$ descends to a finite flat closed R-subgroup of $X[p^e]$ that represents $\ker(\varphi)$.

Finally, consider a general locally noetherian base. Fix a choice of $s \in U$ and let $R := \mathcal{O}_{S,s}$, so there is a finite flat subgroup $N(s) \subset X[p^{e(s)}]_R$ that represents $\ker(\varphi_R)$. Since $X[p^{e(s)}]$ is finite flat over S, by replacing S with a Zariski-open neighborhood of s (as we may do) it can be arranged that $N(s)$ "spreads out" to a finite flat S-subgroup $N \subset X[p^{e(s)}]$. By further shrinking we can arrange that the composite map $N \hookrightarrow X[p^{e(s)}] \to Y[p^{e(s)}]$ vanishes (as it vanishes over the local ring $R = \mathcal{O}_{S,s}$). It suffices to show that N represents $\ker(\varphi)$ over an open neighborhood of s in S. Replacing X with X/N, we may arrange that φ_R is an isomorphism, so by shrinking S around s we may assume that $\varphi : X[p] \to Y[p]$ is an isomorphism. Snake lemma arguments for the fppf topology with the diagrams

$$\begin{array}{ccccccccc} 0 & \longrightarrow & X[p] & \longrightarrow & X[p^{n+1}] & \longrightarrow & X[p^n] & \longrightarrow & 0 \\ & & \downarrow & & \downarrow & & \downarrow & & \\ 0 & \longrightarrow & Y[p] & \longrightarrow & Y[p^{n+1}] & \longrightarrow & Y[p^n] & \longrightarrow & 0 \end{array}$$

then imply that φ is an isomorphism. □

3.3.9. Question. Suppose $\varphi : X \to Y$ is a homomorphism of p-divisible groups over an arbitrary base scheme S. If $\varphi_s : X_s \to Y_s$ is an isogeny for all $s \in S$ (or equivalently for all geometric points of S), is φ is an isogeny?

The preceding methods that provide an affirmative answer for locally noetherian S do not easily carry over to the case of general S; in contrast with abelian schemes (as in the proof of Proposition 3.3.1), one cannot reduce problems over a general base to the case of a (locally) noetherian base since (non-zero) p-divisible groups are not finitely presented. Nonetheless, Gabber has proved an affirmative answer for an arbitrary base scheme S.

Over a normal base scheme it is natural to seek a stronger result:

3.3.10. Example. Consider a homomorphism $\varphi : X \to Y$ between p-divisible groups over a normal integral base scheme S such that the map $\varphi_\eta : X_\eta \to Y_\eta$ between generic fibers is an isogeny. We may choose a map $\psi_\eta : Y_\eta \to X_\eta$ whose composition with φ_η in either order is multiplication by p^n for some $n \geqslant 0$. When S is locally noetherian, it follows from theorems of Tate [**119**, §4, 4.2] and de Jong [**55**, 1.2] that ψ_η extends to a homomorphism $\psi : Y \to X$ whose composition with φ in either order is multiplication by p^n, so φ is an isogeny (since we can check the isogeny property on fibers over a locally noetherian base); see Proposition 3.10.4.1 for further discussion. This can fail if the noetherian hypothesis is removed; a counterexample is given in Example 3.10.2 over the integral closure of $\mathbb{Z}_p$ in an algebraic closure of $\mathbb{Q}_p$.

An interesting consequence of the theorems of Tate and de Jong is that in Proposition 3.3.8 the open locus U is closed. It is equivalent to show that the open immersion $U \hookrightarrow S$ is proper, and by the valuative criterion we reduce to the case $S = \mathrm{Spec}(R)$ for a discrete valuation ring R. This case is settled because such an S is normal.

3.3.11. Question. Let S be a normal integral $\mathbb{F}_p$-scheme (not assumed to be locally noetherian). Suppose $\varphi : X \to Y$ is a homomorphism between p-divisible groups over S such that the map $\varphi_\eta : X_\eta \to Y_\eta$ between generic fibers is an isogeny. Is $\varphi : X \to Y$ an isogeny?

Gabber has proved that the answer is affirmative. There are further questions related to 3.3.9 and 3.3.11. We will return to this topic in 3.10.

3.4. Some p-divisible groups with small a-number

In this section we study variation of the a-number within the isogeny class of a p-divisible group, with an emphasis on members of the isogeny class (if any) with a-number $\leqslant 2$. We work over a field K of characteristic $p > 0$.

3.4.1. Definition. For an abelian variety A over K, the *p-rank* of A is the integer $f(A)$ such that

$$\#(A(\overline{K})[p]) = p^{f(A)}.$$

For a p-divisible group X over K, the *p-rank* of X is the height of $X^{\text{ét}}$.

Note that the p-rank of A equals the p-rank of $A[p^\infty]$ since the connected-étale sequence of $A_{\overline{K}}[p^\infty]$ splits, and that in general $f(A) \leqslant \dim(A)$. The case $f(A) = \dim(A)$ corresponds exactly to the case that $A[p^\infty]$ has vanishing local-local part, which is to say that A is ordinary (see Example 1.4.5.4).

3.4.2. Definition. A p-divisible group X over K is *of extended Lubin-Tate type* if its local-local part $X_{(0,1)}$ is either trivial, has dimension 1, or has dual of dimension 1. (Equivalently, X is not of extended Lubin–Tate type if and only if $X_{(0,1)}$ and $X^t_{(0,1)}$ each has dimension $\geqslant 2$.)

By inspection of the Dieudonné–Manin classification (Theorem 3.1.3) over $\overline{K}$, being of extended Lubin–Tate type amounts to all slopes in $\mathbb{Q} \bigcap (0, 1)$ having the reduced form $1/n$ or $1 - 1/n$ with multiplicity n for exactly one $n \geqslant 2$. Also, since the height of a p-divisible group X over K is equal to $\dim(X) + \dim(X^t)$, if X is non-zero and isogenous to its own dual (such as the p-divisible group of a non-zero abelian variety) then X is of extended Lubin–Tate type if and only if $X^{\text{ét}}$ has height $\dim(X)$ or $\dim(X) - 1$. In the first of these two cases all slopes are 0 or 1, and in the second case $1/2$ is the only additional slope and it occurs with multiplicity 2. As a special case, the p-divisible group of a non-zero abelian variety A over K is of extended Lubin-Tate type if and only if the p-rank of A is either $\dim(A)$ (i.e. A is ordinary) or $\dim(A) - 1$ (called *almost ordinary*).

3.4.3. Let k be a perfect field of characteristic $p > 0$. We shall study p-divisible groups X over k such that:

(i) $a(X) = 2$,

(ii) there exists a k-subgroup scheme $\alpha_p \subset X$ with $a(X/\alpha_p) = 1$.

The ubiquity of such X within isogeny classes will be explained in 3.4.7. Here is an example showing that (i) does not imply (ii):

3.4.3.1. Example. Let $\mathcal{G}_{1/2}$ be the p-divisible group as in 3.1.5 of height 2 and dimension 1 that is isoclinic of slope $1/2$. We consider it over a perfect field k of characteristic $p > 0$ and assume that $k \neq \mathbb{F}_p, \mathbb{F}_{p^2}$. (Over an algebraically closed base field, or even over $\mathbb{F}_{p^2}$, $\mathcal{G}_{1/2}$ is isomorphic to the p-divisible group $E[p^\infty]$ attached to a supersingular elliptic curve E.)

Define $M = \mathrm{M}^*((\mathcal{G}_{1/2})^3)$; this has an evident $\mathbb{Z}_p$-descent arising from the $\mathbb{F}_p$-descent of $\mathcal{G}_{1/2}$. For $m \in M$, let $\overline{m}$ denote $m \bmod \mathcal{F}M \in M/\mathcal{F}M$. Clearly

$M = D_k e_1 + D_k e_2 + D_k e_3$ with $\mathcal{F}e_i = \mathcal{V}e_i$, so $\mathcal{F}M = \mathcal{V}M$ and the quotient $M/\mathcal{F}M = \oplus k\overline{e}_i$ is 3-dimensional over k with vanishing $\mathcal{V}$-operator. Choose $x, y \in M$ such that the k-subspace

$$k\overline{x} + k\overline{y} \subset M/\mathcal{F}M$$

is 2-dimensional and is not defined over $k \cap \mathbb{F}_{p^2}$ relative to the canonical $\mathbb{F}_p$-descent of the k-vector space $M/\mathcal{F}M$. (In terms of the $\mathbb{P}^2$ classifying planes in 3-space, this amounts to choosing a point in $\mathbb{P}^2(k)$ that is not rational over $k \cap \mathbb{F}_{p^2}$.)

Let X be the p-divisible group over k with $\mathrm{M}^*(X) = M' := D_k x + D_k y + \mathcal{F}M \subset M$. By construction, M' has the D_k-module quotient $k\overline{x} + k\overline{y} \subset M/\mathcal{F}M$ on which $\mathcal{F}$ and $\mathcal{V}$ vanish, so this quotient corresponds to some $(\alpha_p)^2 \hookrightarrow X$. Thus, $a(X) \geqslant 2$.

We claim that $a(X) = 2$, so $M'/\mathcal{F}M' \twoheadrightarrow k\overline{x} + k\overline{y}$ is an isomorphism. In particular, by the topological nilpotence of $\mathcal{F}$ on the finite free $W(k)$-module M' (inherited by that of $\mathcal{F}$ on M), it would follow that $M' = D_k x + D_k y$. Of course, conversely if $M' = D_k x + D_k y$ then $M'/(\mathcal{F}M' + \mathcal{V}M')$ has k-dimension at most 2 and hence equal to 2 (since $a(X) \geqslant 2$), so the condition $a(X) = 2$ is equivalent to the condition that M' is spanned by x and y over D_k, or equivalently $\mathcal{F}M \subset D_k x + D_k y$.

To prove $a(X) = 2$, we carry out a computation by replacing (x, y) with a lift in M of a convenient k-basis of the 2-dimensional span of $\overline{x}$ and $\overline{y}$ inside $M/\mathcal{F}M$; any such lift can be used to define the same M', so such a change in (x, y) is harmless for our purposes. Certainly the k-plane $k\overline{x} + k\overline{y}$ inside $M/\mathcal{F}M = \oplus k\overline{e}_i$ is distinct from the span of $\{\overline{e}_2, \overline{e}_3\}$, so at least one of $\overline{x}$ or $\overline{y}$ has a non-zero $\overline{e}_1$-coefficient. Thus, by swapping x and y if necessary, and also swapping e_2 and e_3 if necessary, we may arrange that $\overline{y}$ has $\overline{e}_1$-coefficient equal to 1 and $\overline{x} = \overline{e}_2 + b\overline{e}_3$ for some $b \in k$. We can therefore also assume $\overline{y} = \overline{e}_1 + c\overline{e}_3$ for some $c \in k$, so we may and do assume $x = e_2 + [b]e_3$ and $y = e_1 + [c]e_3$, where $[a] \in W(k)$ is the Teichmüller lift of $a \in k$.

First suppose $c = 0$, so $b \notin k \bigcap \mathbb{F}_{p^2}$ and $y = e_1$. Hence, $D_k x + D_k y$ contains the elements $\mathcal{F}e_1$, $\mathcal{F}(x) = \mathcal{F}e_2 + [b^p]\mathcal{F}e_3$, and $\mathcal{V}(x) = \mathcal{F}e_2 + [b^{1/p}]\mathcal{F}e_3$. Since $b^p - b^{1/p} \in k^\times$, it follows that $D_k x + D_k y$ contains all $\mathcal{F}e_j$'s, so this D_k-submodule of M' contains $\mathcal{F}M$ and hence exhausts M'. That is, if $c = 0$ then $a(X) = 2$.

Suppose instead that $c \neq 0$. In this case the D_k-span of x and y inside M' contains $\mathcal{F}(x)$, $\mathcal{V}(x)$, $\mathcal{F}(y)$, and $\mathcal{V}(y)$. The differences $\mathcal{F}(x) - \mathcal{V}(x)$ and $\mathcal{F}(y) - \mathcal{V}(y)$ respectively equal $([b^p] - [b^{1/p}])\mathcal{F}(e_3)$ and $([c^p] - [c^{1/p}])\mathcal{F}(e_3)$. If $b \notin k \bigcap \mathbb{F}_{p^2}$ then $[b^p] - [b^{1/p}] \in W(k)^\times$, so in such cases $D_k x + D_k y$ contains $\mathcal{F}(e_3)$ and hence the containment of $\mathcal{F}(x) = \mathcal{F}(e_2) + [b^p]\mathcal{F}(e_3)$ and $\mathcal{F}(y) = \mathcal{F}(e_1) + [c^p]\mathcal{F}(e_3)$ in $D_k x + D_k y$ implies that this D_k-span contains $\mathcal{F}M$. Thus, once again $a(X) = 2$. If instead $b \in k \bigcap \mathbb{F}_{p^2}$ then necessarily $c \notin k \bigcap \mathbb{F}_{p^2}$ so we may again conclude that $a(X) = 2$.

We have shown that X satisfies condition (i) in 3.4.3 and now we show that condition (ii) is not satisfied for X: for any $\alpha_p \subset X$, we claim that $a(X/\alpha_p) \geqslant 2$. To prove this, first note that the quotient $Y = X/\alpha_p$ has Dieudonné module $\mathrm{M}^*(Y)$ of the form $M'' = D_k z + \mathcal{F}M = Wz + \mathcal{F}M \subset M'$ for some $z \in Wx + Wy$ (recall that $\mathcal{V}M = \mathcal{F}M$) since the quotient M'/M'' of M' corresponds to the chosen embedding $\alpha_p \subset X$. Note also that $\mathcal{F}M'' + \mathcal{V}M'' = W\mathcal{F}(z) + W\mathcal{V}(z) + pM$. Since the quotient $\mathcal{F}M/pM = \mathcal{F}M/\mathcal{F}^2M$ has dimension 3 over k and $z \notin \mathcal{F}M$, clearly the quotient $M''/pM = (Wz + \mathcal{F}M)/pM$ has dimension 4 over k. But $M''/(\mathcal{F}M'' + \mathcal{V}M'')$ is the quotient of this modulo the k-span of the classes of $\mathcal{F}(z)$ and $\mathcal{V}(z)$, so it has k-dimension at least 2. We conclude that $a(Y) \geqslant 2$.

For any p-divisible group X over a perfect field k of characteristic p, consider the largest α-type subgroup $X[\mathcal{F},\mathcal{V}]$ in X (or equivalently, in $X[p]$) as in 3.1.11. Explicitly, this is the intersection of the kernel of $\mathrm{Fr}_{X/k}$ and the kernel of the twist $\mathrm{Ver}_{X/k}^{(1/p)} : X \to X^{(1/p)}$ of $\mathrm{Ver}_{X/k}$ by the inverse of the Frobenius automorphism of k. In terms of Dieudonné modules we have $\mathrm{M}^*(X[\mathcal{F},\mathcal{V}]) = \mathrm{M}^*(X)/(\mathcal{F},\mathcal{V})$ and it has k-dimension $a(X)$. This link with the a-number is useful in the following generalization of Example 3.4.3.1.

3.4.3.2. Example. Let k be algebraically closed of characteristic $p > 0$, and let X' be a p-divisible group with $a(X') \geqslant 3$. In particular, $\dim(X') \geqslant 3$. We claim that there exists Y isogenous to X' such that $a(Y) = 2$ and Y does not satisfy condition (ii) of 3.4.3.

First note that by the direct sum decomposition of p-divisible groups over k into étale, multiplicative, and local-local parts (as k is perfect), the p-divisible groups over k with vanishing a-number are precisely the ones with vanishing local-local part. Hence, for every X'' in the isogeny class of X' we have $1 \leqslant a(X'') \leqslant \dim(X'') = \dim(X')$. By similar reasoning, we may assume that X' is local-local, so all members of its isogeny class are local-local.

By Corollary 3.4.7 below, there is an isogeny $X' \to X''$ with $a(X'') = 2$. We wish to refine this choice of X'' so that it is dominated by degree-p isogeny from a quotient of X with a-number equal to 3. The kernel $N = \ker(X' \to X'')$ is a non-trivial local-local group, so the Frobenius and Verschiebung operators on its Dieudonné module are nilpotent. Hence, N admits a composition series $N = N_0 \supset N_1 \supset \cdots \supset N_r = 1$ over k such that each N_i/N_{i+1} is isomorphic to α_p.

Consider the quotients $X'_i = X'/N_i$. We have $a(X'_r) \geqslant 3$, $a(X'_0) = 2$, and $a(X'_i) \geqslant 1$ for all i. Since $X'_i \to X'_{i-1}$ is an isogeny with kernel α_p, clearly $a(X'_{i-1}) \geqslant a(X'_i) - 1$ for $1 \leqslant i \leqslant r$, so if we choose i_0 maximal such that $a(X'_{i_0}) = 2$ then $i_0 < r$ and $a(X'_{i_0+1}) \in \{1, 3\}$. The case $a(X'_{i_0+1}) = 1$ cannot occur. Indeed, if this happens then $i_0 + 1 < r$ and $1 \leqslant a(X'_{i_0+2}) \leqslant 2$ yet $a(X'_{i_0+2}) \neq 2$ due to the maximality of i_0. In other words, necessarily $a(X'_{i_0+2}) = 1$ and we can continue by induction to eventually show that $a(X'_r) = 1$, a contradiction. Thus, for $X := X'_{i_0+1}$ and $Y = X'_{i_0} = X/\alpha_p$ in the isogeny class of X', we have $a(X) = 3$ and $a(Y) = 2$.

We claim that Y violates condition (ii) in 3.4.3. That is, for any $b : \alpha_p \hookrightarrow Y$, necessarily $a(Y/b(\alpha_p)) \geqslant 2$. Let $q : X \to Y$ be the natural isogeny, so $\ker(q) = \alpha_p$. The inclusion of group schemes $X[\mathcal{F},\mathcal{V}]/\ker(q) \hookrightarrow Y[\mathcal{F},\mathcal{V}]$ is an isomorphism for order reasons (as $X[\mathcal{F},\mathcal{V}]$ has order $p^{a(X)} = p^3$ and $Y[\mathcal{F},\mathcal{V}]$ has order $p^{a(Y)} = p^2$), so the natural map $X[\mathcal{F},\mathcal{V}] \twoheadrightarrow Y[\mathcal{F},\mathcal{V}]$ is faithfully flat. It follows that the preimage of $b(\alpha_p)$ in the α-group scheme $X[\mathcal{F},\mathcal{V}]$ has order p^2 and hence is isomorphic to $(\alpha_p)^2$. In other words, the isogeny $X \to Y_b := Y/b(\alpha_p)$ has kernel $(\alpha_p)^2$, so the dual isogeny has kernel $((\alpha_p)^2)^D = (\alpha_p)^2$ (by Theorem 3.3.7). This defines an inclusion $(\alpha_p)^2 \hookrightarrow (Y_b)^t$, so $a((Y_b)^t) \geqslant 2$. But $a(Y_b) = a((Y_b)^t)$ by 3.2.4, so $a(Y_b) \geqslant 2$ as claimed.

3.4.4. Fix an X that satisfies (i) and (ii) in 3.4.3, and choose a k-isomorphism

$$(\alpha_p)^2 \simeq X[\mathcal{F},\mathcal{V}].$$

For any extension K of k, the set of K-subgroups

$$N \subset X_K[\mathcal{F},\mathcal{V}] = (\alpha_p)^2$$

that are isomorphic to α_p is in bijective correspondence with the set of lines in $\mathrm{Lie}((\alpha_p)^2)$ via the map $N \mapsto \mathrm{Lie}(N) \subset \mathrm{Lie}((\alpha_p)^2) = K^2$, so the set of such N is naturally parameterized by $\mathbb{P}^1(K)$. For any $b \in \mathbb{P}^1(K)$ we write $b(\alpha_p)$ to denote the corresponding K-subgroup $N \subset X[\mathcal{F}, \mathcal{V}]_K \subset X_K$, and we define the p-divisible group quotient $X_b := X_K/b(\alpha_p)$ over K using the quotient formalism in 3.3.3. Since the a-number is defined in terms of a geometric fiber, if $b_0 \in \mathbb{P}^1(k(b))$ denotes the physical image point of b in $\mathbb{P}^1_k$ then $a(X_{b_0}) = a(X_b)$. For this reason, the notation $a(X_b)$ is insensitive to extension of K.

Under conditions (i) and (ii) in 3.4.3, the set of $b \in \mathbb{P}^1_k$ with $a(X_b) = 1$ is a dense open subset of $\mathbb{P}^1_k$. Indeed, it is non-empty by condition (ii), and $a(X_b) \geqslant 1$ for all b (since the a-number is defined in terms of the geometric fiber, and over a *perfect* field the a-number vanishes if and only if the local-local part vanishes). By upper semi-continuity of the a-number (see Proposition 3.2.5.1), it follows that "$a(X_b) = 1$" is an open condition on b.

3.4.5. Proposition. *Let X be a p-divisible group over k satisfying conditions* (i) *and* (ii) *in* 3.4.3, *and let K be an extension of k. Let $b \in \mathbb{P}^1(K)$ be a point such that $a(X_b) = 1$, where $X_b := X_K/b(\alpha_p)$. Up to unique K-isomorphism over X_b, the only K-isogeny $X' \to X_b$ of p-divisible groups with kernel isomorphic to α_p is the canonical quotient map $q : X_K \to X_b$.*

PROOF. By duality, it is equivalent to show that the quotient $(X_K)^t$ of $(X_b)^t$ via q^t is uniquely K-isomorphic to the quotient X'^t of $(X_b)^t$ via f^t. The uniqueness is clear, and each is the quotient by a K-subgroup of $(X_b)^t$ isomorphic to $(\alpha_p)^D \simeq \alpha_p$ (by Proposition 3.3.7), so it suffices to show that $(X_b)^t$ has a unique K-subgroup isomorphic to α_p. There is certainly at least one such K-subgroup, and to prove the uniqueness it suffices to check after replacing K with $\overline{K}$. Since $a((X_b)^t) = a(X_b) = 1$ by Corollary 3.2.4, we are done. □

Now we show that over an algebraically closed field of characteristic $p > 0$, the isogeny class of a non-zero local-local p-divisible group contains a member with a-number equal to 1. This was used in Example 3.4.3.2 to show that any local-local p-divisible group X over such a field with $a(X) \geqslant 3$ has within its isogeny class a p-divisible group with a-number 2.

3.4.6. Proposition. *Let X be a p-divisible group over a field κ of characteristic p with $X_{(0,1)} \neq 0$. There exists a finite extension κ' of κ and a p-divisible group Y over κ' in the κ'-isogeny class of $X_{\kappa'}$ such that $a(Y) = 1$.*

This result and its corollary were proved in [**93**, 3.2] for abelian varieties over finite fields, and the method of proof works for p-divisible groups over an arbitrary field of characteristic p. Below we give a different proof.

PROOF. First we show the sufficiency of carrying out the construction over a single algebraically closed extension of κ (with X fixed). Let K be an algebraically closed extension of κ such that there exists an isogeny $f : X_K \to X_K/N =: Y$ with $a(Y) = 1$. Let $p^r = \deg(f)$ with $r \geqslant 1$, so $N \subset X_K[p^r]$.

By working over the Grassmannian of rank-p^r quotient bundles of the coordinate ring of $X[p^r]$, we obtain a moduli scheme $\mathcal{M}$ of finite type over κ classifying finite flat closed subgroup schemes of $X[p^r]$ of order p^r. (We do not know any results concerning the irreducibility of $\mathcal{M}$.) Letting $H \subset X[p^r]_{\mathcal{M}}$ be the universal such subgroup scheme over $\mathcal{M}$, we get a point $\xi \in \mathcal{M}(K)$ corresponding to N. Consider the p-torsion $G = (X_{\mathcal{M}}/H)[p]$ in the p-divisible group $X_{\mathcal{M}}/H$ over $\mathcal{M}$. Since X has non-trivial local-local part, $a(G_m) \geqslant 1$ for all $m \in \mathcal{M}$. Thus, by the upper semi-continuity of the a-number (Proposition 3.2.5.1), the condition "$a(G_m) = 1$" defines an open subscheme $U \subset \mathcal{M}$ through which ξ factors. Hence, $U \neq \emptyset$, so any closed point of U provides a finite extension κ' of κ and a point $u \in U(\kappa')$ such that $Y = X_{\kappa'}/G_u$ has a-number equal to 1.

Now we may and do assume κ is algebraically closed. The isogeny category of p-divisible groups is therefore semisimple, and since it is harmless to replace X with another member X' of its isogeny class, and we can therefore assume $X = \prod X_i'$ for X_i' that are simple in the isogeny category. The decomposition over κ into a direct product of étale, multiplicative, and local-local parts allows us to assume that all X_i' are local-local. For each i there is an isogeny $X_i' \to X_i$ where $a(X_i) = 1$. Indeed, by the Dieudonné–Manin classification (Theorem 3.1.3) each X_i' is isogenous to one of the standard isosimple p-divisible groups $\mathcal{G}_{\lambda_i}$ with slope $\lambda_i \in \mathbb{Q} \bigcap (0,1)$, and in general if $\lambda = r/s$ with $1 \leqslant r < s$ and $\gcd(r,s) = 1$ (so $r, s - r \geqslant 1$) then $a(\mathcal{G}_\lambda) = 1$ because the quotient of $\mathrm{M}^*(\mathcal{G}_\lambda) = D_\kappa / D_\kappa(\mathcal{F}^{s-r} - \mathcal{V}^r)$ modulo $\mathcal{F}$ and $\mathcal{V}$ is 1-dimensional over κ by inspection.

We choose an enumeration of the set of factors X_i and thereby define an increasing filtration on X by the p-divisible subgroups $\mathrm{Fil}^j(X) = \prod_{i \leqslant j} X_i$. Each successive quotient $\mathrm{Fil}^i(X)/\mathrm{Fil}^{i-1}(X) = X_i$ has a-number equal to 1 and $\kappa = \overline{\kappa}$, so by [**94**, Lemma 2.7] (proved via displays) there exists a deformation $(\mathcal{X}, \mathrm{Fil}^\bullet(\mathcal{X}))$ of $(X, \mathrm{Fil}^\bullet(X))$ (as a filtered p-divisible group) over a formal power series ring $\kappa[\![t_1, \ldots, t_r]\!]$ such that (i) the geometric generic fiber of each successive quotient $\mathrm{Fil}^i(\mathcal{X})/\mathrm{Fil}^{i-1}(\mathcal{X})$ is isoclinic with the same slope as the geometric special fiber X_i and (ii) the geometric generic fiber $\mathcal{X}_{\overline{\eta}}$ of $\mathcal{X}$ has a-number equal to 1. To summarize, the p-divisible groups $\mathcal{X}_{\overline{\eta}}$ and $X_{\overline{\eta}}$ have the same slopes (with multiplicity) and $a(\mathcal{X}_{\overline{\eta}}) = 1$. This equality of slopes implies that $X_{\overline{\eta}}$ is isogenous to $\mathcal{X}_{\overline{\eta}}$ (by the Dieudonné–Manin classification, applied over $\overline{\eta}$). Thus, we found an algebraically closed extension of the ground field (namely, corresponding to $\overline{\eta}$) over which the base change of X has an isogenous quotient with a-number equal to 1. □

In the preceding proof we produced an isogeny from each X_i' to an X_i with $a(X_i) = 1$ by using the Dieudonné–Manin classification. The construction of such an X_i can also be given by a deformation-theoretic method, as in the proof of [**56**, Thm. 4.1] (a non-trivial result for which there is still no easy conceptual proof).

3.4.7. Corollary. *Let X be a p-divisible group over a field κ of characteristic p, and assume that X is not of extended Lubin-Tate type. There exists a finite extension κ' of κ and an isogeny $X_{\kappa'} \to Y$ where Y satisfies conditions* (i) *and* (ii) *in* 3.4.3.

PROOF. Since $X_{\overline{\kappa}}$ is not of extended Lubin–Tate type, the local-local part of X is non-trivial. Thus, by 3.4.6, there exists a finite extension κ' of κ and an isogeny $X_{\kappa'} \to X'$ with $a(X') = 1$. We apply the following lemma to the kernel N of this isogeny over the field κ':

3.4.7.1. Lemma. *Let N be a finite local-local group scheme over a field κ of characteristic $p > 0$. There exists a filtration $0 \subset N_1 \subset \cdots \subset N_i \subset \cdots \subset N_r = N$ by κ-subgroups such that N_{i+1}/N_i is isomorphic to α_p for all $1 \leqslant i < r$.*

PROOF. It suffices to show that for a local-local finite commutative κ-group $N \neq 0$, the maximal α-subgroup $N[\mathcal{F}, \mathcal{V}]$ is nontrivial (as then this κ-group is a power of α_p and $N/N[\mathcal{F}, \mathcal{V}]$ is local-local with smaller order). Equivalently, if N is a local-local finite commutative κ-group and $N[\mathcal{F}, \mathcal{V}] = 0$ then we claim $N = 0$. We may certainly assume $\kappa = \overline{\kappa}$, so κ is perfect. The Dieudonné module $\mathrm{M}^*(N)$ has nilpotent $\mathcal{F}$ and $\mathcal{V}$ operators yet $\mathrm{M}^*(N)/(\mathcal{F}, \mathcal{V}) = \mathrm{M}^*(N[\mathcal{F}, \mathcal{V}]) = 0$, so $\mathrm{M}^*(N) = 0$ and hence $N = 0$. □

Returning to the proof of Corollary 3.4.7, by taking successive quotients via the composition series in the lemma we obtain a sequence of isogenies

$$Y_0 := X_{\kappa'} \to Y_1 := X_{\kappa'}/N_1 \to \cdots \to Y_i := X_{\kappa'}/N_i \to \cdots \to X_{\kappa'}/N_r = X' =: Y_r$$

with all $a(Y_i) \geqslant 1$ (since all Y_i have non-trivial local-local part) and $a(Y_{i+1}) - 1 \leqslant a(Y_i) \leqslant a(Y_{i+1}) + 1$ for all $i = 0, 1, \ldots, r-1$.[5] The final term in this sequence of isogenies has a-number equal to 1 and the initial term has a-number $\geqslant 2$ (since X is not of extended Lubin–Tate type), so there is at least one index s with $0 \leqslant s \leqslant r-1$ such that

$$a(Y_s) = 2, \quad a(Y_{s+1}) = 1.$$

The choice $Y = Y_s$ satisfies the conditions in 3.4.3. □

3.4.8. Remark. Let k be an algebraically closed field of characteristic $p > 0$. For $\mathcal{G}_\lambda$ with $\lambda \in (0,1)$ we have seen that $a(\mathcal{G}_\lambda) = 1$. It is a fact that the set of a-numbers of members in the k-isogeny class of $\mathcal{G}_\lambda$ is the set of all integers between 1 and $\min(r, s-r)$ (the a-priori maximum possible value, in view of the invariance of the a-number under duality and the upper bound on it provided by the dimension). Thus, a p-divisible group X over k is of extended Lubin-Tate type if and only if all members of its isogeny class have a-number at most 1. It follows via Corollary 3.4.7 that the k-isogeny class of X contains a member satisfying (i) and (ii) in 3.4.3 if and only if X is not of extended Lubin-Tate type.

3.5. Earlier non-liftability results and a new proof

In [**93**, Thm. B] we find the following theorem:

3.5.1. Theorem. *Let B be an abelian variety of dimension $g \geqslant 2$ over an algebraic closure k of $\mathbb{F}_p$. Suppose that the p-rank of B is at most $g-2$ (i.e. $B[p^\infty]$ is not ordinary and not almost ordinary). If the endomorphism algebra $\mathrm{End}^0(B)$ is a (commutative) field then there exists an isogeny $B \to B'$ such that B' does not admit a* CM *lift to characteristic* 0.

Below, for the convenience of the reader, we will sketch the idea behind the original proof of Theorem 3.5.1 (which uses CM theory over $\mathbb{C}$). Later in this chapter we will generalize the theorem (e.g., remove the condition that the endomorphism algebra is a field) and give a proof which does not use CM theory over $\mathbb{C}$. Before we take

[5]The first inequality follows from the invariance of a-numbers under duality and the fact that the dual isogeny $(Y_{i+1})^t \to (Y_i)^t$ has kernel $(\alpha_p)^D = \alpha_p$.

up the arguments, we give an easy example that may seem trivial and unimportant but in fact contains the basic idea behind the construction of examples showing that an isogeny may be necessary in order to obtain a CM lift to characteristic 0:

3.5.2. Example. Let A be an abelian variety over a finite field κ of size q such that $A[p^\infty]$ satisfies conditions (i) and (ii) in 3.4.3. One such class of abelian varieties is $E \times E$ for a supersingular elliptic curve E over κ. To make other such A (with q large), first note that via Honda–Tate theory there exist simple abelian varieties B over κ of dimension $g \geqslant 2$ such that B has p-rank at most $g-2$ (equivalently, B is neither ordinary nor almost ordinary); this is seen by constructing Weil q-integers with enough p-adic ordinals strictly between 0 and 1. For such a B the p-divisible group $B[p^\infty]$ is not of extended Lubin–Tate type, so by Corollary 3.4.7 we can replace κ with a finite extension so that there is a κ-isogeny $B[p^\infty] \to X$ with X satisfying (i) and (ii) in 3.4.3. Then we may take A to be the p-isogenous quotient of B such that $A[p^\infty] = X$ as quotients of $B[p^\infty]$.

Fix an isomorphism $(\alpha_p)^2 \simeq A[\mathcal{F}, \mathcal{V}]$ as κ-groups. Over a field $K \supset \kappa(t)$, consider the inclusion

$$(t, 1) : \alpha_p \hookrightarrow (\alpha_p)^2 = A_K[\mathcal{F}, \mathcal{V}] \subset A_K.$$

Let A_t denote the quotient A_K modulo this α_p subgroup. We claim that A_t does not admit a CM lift to characteristic 0 (over a local domain with residue field K, which we may and do take to be noetherian). Since any extension field of the residue field of a local noetherian domain R is realized by a faithfully flat local extension of local noetherian domains $f : R \to R'$ [**34**, 0_{III}, 10.3.1], and we can sacrifice faithful flatness while retaining injectivity for f and arranging for R' to be a complete local domain by passing to the quotient of $\widehat{R'}$ by a minimal prime, we may reduce to the case that K is algebraically closed and R is complete. In particular, the normalization of R is module-finite over R (as all complete local noetherian rings are excellent), so we can also arrange that R is normal.

By 1.7.2.1 and 1.7.2.3 we can replace $\mathrm{Frac}(R)$ with a finite extension (and R with its module-finite integral closure in such an extension) so that a hypothetical CM lift (if one exists) descends to the ring of integers of a number field $F \subset \mathrm{Frac}(R)$. The reduction of such a descent at the maximal ideal of $\mathcal{O}_F$ contained in the maximal ideal of R is a descent of A_t to a finite subfield of K. Hence, to rule out the CM lift in characteristic 0 it suffices to show that A_t as an abstract abelian variety over K does not descend to the subfield $k = \overline{\kappa} \subset K$.

Assuming such a descent B exists, consider the quotient map

$$(A_{\overline{\kappa}})_K \twoheadrightarrow A_t \simeq B_K = (B_{\overline{\kappa}})_K.$$

By Lemma 1.2.1.2, this map descends to a $\overline{\kappa}$-homomorphism $A_{\overline{\kappa}} \to B_{\overline{\kappa}}$, so the kernel over K is defined over $\overline{\kappa}$ inside A_K. But by construction of A_t, the kernel inside $A_K[\mathcal{F}, \mathcal{V}] = (\alpha_p)^2$ (equality respecting the κ-descent on both sides!) has minimal field of definition over κ equal to $\kappa(t)$. This is a contradiction.

The point of Theorem 3.5.1 is that instead of making examples of the failure of CM lifting for an isogenous quotient over an extension of $\kappa(t)$ (such as A_t over $\kappa(t)$ above), examples can be found using an isogenous quotient over a finite extension of κ that is "large enough".

3.5.3. Now we sketch the original proof of Theorem 3.5.1. The condition $f(B) \leqslant \dim(B) - 2$ implies that $B[p^\infty]$ is not of extended Lubin-Tate type. By 3.4.7, we may find a finite subfield $\kappa \subset k$ and an abelian variety A over κ such that A_k is k-isogenous to B and $X := A[p^\infty]$ satisfies conditions (i) and (ii) in 3.4.3. (The existence of such an A is also explained in [**93**, 3.2].) Since we require $\mathrm{End}^0(B)$ to be a commutative field (so B is simple), by descending B to a sufficiently large finite field and using Tate's results on the structure of endomorphism algebras of simple abelian varieties over finite fields we have $\mathrm{End}^0(A) = \mathrm{End}^0(A_k) \simeq \mathrm{End}^0(B)$. (In particular, $\mathrm{End}(A) = \mathrm{End}(A_k)$.)

Fix a κ-isomorphism $(\alpha_p)^2 \simeq A[\mathcal{F}, \mathcal{V}]$, so for every $b \in \mathbb{P}^1(k)$ we get a quotient

$$A_{\kappa(b)} \to A_b := A_{\kappa(b)}/b(\alpha_p).$$

By hypothesis $a(A_b) = 1$ for some b, so upper semi-continuity of the a-number implies that $a(A_b) = 1$ for all but finitely many b. Choose such a b. In [**93**, §4] it is shown that if $(A_b)_k$ can be CM lifted to characteristic zero then $\kappa(b)$ is contained in a pro-p extension of the extension of κ of degree Q' for an integer Q' determined by A. Once Q' is known, one easily produces explicit closed points $b \in \mathbb{P}^1_\kappa$ such that $\kappa(b)$ is not contained in any such extension, so $(A_b)_k$ cannot be CM lifted to characteristic zero. This ends our sketch of the original proof of 3.5.1.

3.5.4. Instead of $\kappa(b)$ avoiding containment in the maximal pro-p extension of the degree-Q' extension of κ, we will prove a much finer result in 3.8.3: it suffices for $\kappa(b)$ to not be contained in a specific *finite* extension of κ, the degree of which can be effectively computed. (This involves the residue fields of a certain finite collection of p-adic fields, as in 3.8.2 applied to $A[p^\infty]_k$, and also an integer $Q = Q(A)$ defined in 3.6.1–3.6.2 and computed in some examples in 3.6.3.) In terms of the notion of CM *lifting to characteristic* 0 for p-divisible groups, which is defined in 3.7.1.8, we will prove:

3.5.5. Theorem. *Let X be a p-divisible group over $k = \overline{\mathbb{F}}_p$. Assume X is not of extended Lubin–Tate type. There exists a p-divisible group X' over k isogenous to X that cannot be* CM *lifted to characteristic* 0.

A more precise result will be given in Theorem 3.8.3: we can arrange that X' is defined over a finite field, though possibly of large size. (See 3.8.4 and 3.8.5 for examples.) A consequence of Theorem 3.5.5 is that the restrictions on $\mathrm{End}^0(B)$ can be eliminated from Theorem 3.5.1, thereby answering [**93**, Question C]:

3.5.6. Corollary. *Let B be an abelian variety over $k = \overline{\mathbb{F}}_p$ of dimension $g \geqslant 2$ such that the p-rank of B is at most $g - 2$. There exists an isogeny $B \to B'$ over k such that B' does not admit a* CM *lifting to characteristic* 0.

PROOF. Apply 3.5.5 to $B[p^\infty]$. □

3.5.7. Remark. The result in Corollary 3.5.6 is optimal in the following sense: if B is an abelian variety over a finite field κ and $f(B) \geqslant \dim(B) - 1$ (i.e., B is ordinary or almost ordinary, so $B[p^\infty]$ is of extended Lubin-Tate type) then B admits a CM lift over a p-adic discrete valuation ring with residue field κ.

By the Serre–Tate theorem on canonical lifts, ordinary CM p-divisible groups admit a canonical CM lift over $W(\kappa)$. For ordinary B we apply this to $B[p^\infty]$ and argue as in 1.4.5.4 to prove algebraicity for the associated formal abelian scheme deformation of B.

In the almost ordinary case, $B[p^\infty]$ is canonically a direct product of an ordinary CM p-divisible group over κ and a local-local CM p-divisible group that is isoclinic of slope 1/2 and height 2. A CM p-divisible group of this latter local-local type can be viewed as a 1-dimensional formal module over a quadratic ramified extension R of $W(\kappa)$, so it admits a canonical R-linear CM lift over R (as we explained in 2.3.6(ii), or see [**91**, Lemmas 14.7, 14.8] for an another proof without using the viewpoint of formal modules). These two cases can be combined to handle the almost ordinary case; an additional argument is required to prove algebraicity of the associated formal abelian scheme deformation. (See [**93**, §2] for further details.)

3.6. A lower bound on the field of definition

In this section we fix a finite field κ, and we study p-divisible groups over κ satisfying conditions (i) and (ii) in 3.4.3. Such p-divisible groups exist in abundance at the cost of replacing κ with a finite extension: by 3.4.7, for every p-divisible group over κ not of extended Lubin–Tate type there is a finite extension κ' of κ for which there is a κ'-isogeny onto a p-divisible group satisfying (i) and (ii) in 3.4.3.

3.6.1. A construction. Let k be an algebraic closure of κ, and let X be a p-divisible group over κ that satisfies (i) and (ii) in 3.4.3. It will also be convenient to consider an abelian variety A over κ such that $A[p^\infty]$ satisfies (i) and (ii) in 3.4.3. (In 3.5.3 we saw how such A arise in the course of proving Theorem 3.5.1.) A technical difference between the cases of abelian varieties and p-divisible groups is that every endomorphism of A_k descends to an endomorphism of $A_{\kappa'}$ for a finite extension κ' of κ inside k, whereas in general endomorphisms of X_k do not admit a field of definition of finite degree over κ.

Since $\mathrm{End}(X_k)$ is finitely generated over $\mathbb{Z}_p$ (Lemma 3.1.8) and $X[\mathcal{F},\mathcal{V}] = (\alpha_p)^2$, the image of $\mathrm{End}(X_k)$ in $\mathrm{End}(X_k[\mathcal{F},\mathcal{V}]) = \mathrm{Mat}_2(k)$ lies in $\mathrm{Mat}_2(\kappa')$ for some finite extension κ' of κ inside k. Thus, by replacing κ with a suitable finite extension we may and do arrange that $\mathrm{End}(X_k)$ and $\mathrm{End}(X)$ have the same image in $\mathrm{End}(X_k[\mathcal{F},\mathcal{V}])$. In the abelian variety case we may and do replace κ with a finite extension so that $\mathrm{End}(A_k) = \mathrm{End}(A)$. Thus,

$$\Gamma_X = \mathrm{image}(\mathrm{Aut}(X_k) \to \mathrm{Aut}(X_k[\mathcal{F},\mathcal{V}])) \subset \mathrm{GL}_2(k)$$

and

$$\Gamma_A = \mathrm{image}(\mathrm{Aut}(A_k) \to \mathrm{Aut}(A_k[\mathcal{F},\mathcal{V}])) \subset \mathrm{GL}_2(k)$$

lie inside $\mathrm{GL}_2(\kappa)$ and so are finite (with $\Gamma_A \subset \Gamma_{A[p^\infty]}$). We define $Q(X)$ to be the least common multiple of the orders of the elements in Γ_X, and we similarly define $Q(A)$ using Γ_A. (In what follows we can use $\#\Gamma_X$ instead of $Q(X)$, but for constructing examples it is better to use an optimized definition such as $Q(X)$.)

The set of all $b \in \mathbb{P}^1(k)$ with $a(X_b) > 1$ is finite (due to the hypotheses on X and the upper semi-continuity of the a-number), so there exists a finite extension κ_0 of κ inside k such that $a(X_b) = 1$ for all $b \in \mathbb{P}^1(k)$ with $b \notin \mathbb{P}^1(\kappa_0)$. The same holds for A in place of X since $a(A) = a(A[p^\infty])$.

3.6.2. Proposition. *For X over κ as above (so $\mathrm{End}(X_k) \to \mathrm{End}(X[\mathcal{F},\mathcal{V}]_k) = \mathrm{Mat}_2(k)$ has the same image as $\mathrm{End}(X)$) and any $b \in \mathbb{P}^1(k)$ such that $a(X_b) = 1$, if X_b over k descends to a p-divisible group X' over a finite extension κ' of κ inside k then $[\kappa'(b) : \kappa']$ divides $Q(X)$. In particular,*

$$[\kappa' : \kappa] = \frac{[\kappa'(b) : \kappa]}{[\kappa'(b) : \kappa']} \geqslant \frac{[\kappa(b) : \kappa]}{Q(X)}.$$

Likewise, for A as above (so $\mathrm{End}(A_k) = \mathrm{End}(A)$) and any $b \in \mathbb{P}^1(k)$ such that $a(A_b) = 1$, the same conclusion holds in terms of $Q(A)$ for finite extensions κ' of κ to which A_b descends.

The main point of this result is that by choosing b such that $[\kappa(b) : \kappa]$ is sufficiently large (relative to $Q(X)$ and $[\kappa_0 : \kappa]$, the latter to ensure $b \notin \mathbb{P}^1(\kappa_0)$, so $a(X_b) = 1$), the abstract p-divisible group X_b cannot be defined over a finite extension of κ whose κ-degree is small.

PROOF. We treat the case of p-divisible groups, and the case of abelian varieties goes in the same way. Since $a(X_b) = 1$ and the formation of the a-number is invariant under extension of a perfect ground field, we have $a(X') = 1$. Thus, $a({X'}^t) = 1$ by Proposition 3.2.4.

The quotient map $\xi' : {X'}^t \to {X'}^t/({X'}^t[\mathcal{F},\mathcal{V}])$ is a κ'-descent of the quotient of ${X'}^t_k = (X_b)^t$ by its unique k-subgroup isomorphic to α_p. But as we saw in the proof of 3.4.5, this latter quotient map over k is the dual of the natural quotient map $X_k \twoheadrightarrow X_b$. Hence, the dual of ξ' is an isogeny of p-divisible groups $\xi : Y' \to X'$ over κ' that is a κ'-descent of the quotient map $f_b : X_k \to X_b$.

The $\mathrm{Gal}(k/\kappa')$-twisting that constructs Y' from X_k is through k-automorphisms of X_k that preserve $\ker f_b$. Thus, $Y'[\mathcal{F},\mathcal{V}]$ is a $\mathrm{Gal}(\kappa'/\kappa)$-twist of $X[\mathcal{F},\mathcal{V}]_{\kappa'}$ through the image $\Gamma_X \subset \mathrm{Aut}(X[\mathcal{F},\mathcal{V}]) = \mathrm{GL}_2(\kappa) \subset \mathrm{GL}_2(k)$ of $\mathrm{Aut}(X_k)$ in $\mathrm{Aut}(X[\mathcal{F},\mathcal{V}]_k) = \mathrm{GL}_2(k)$ (and even just the subgroup of elements in Γ_X whose action on $X[\mathcal{F},\mathcal{V}]_k$ preserves $\ker f_b$). This latter Galois twist is classified by an element of

$$\mathrm{H}^1(\mathrm{Gal}(k/\kappa'), \Gamma_X) = \mathrm{Hom}_{\mathrm{cont}}(\mathrm{Gal}(k/\kappa'), \Gamma_X).$$

Since $\mathrm{Gal}(k/\kappa')$ is pro-cyclic, this cohomology class is split by a finite extension κ'' of κ' of degree dividing $Q(X)$ (due to how $Q(X)$ was defined). In particular, the κ''-structure imposed on $X[\mathcal{F},\mathcal{V}]_k$ from the κ'-structure $Y'[\mathcal{F},\mathcal{V}]$ coincides with the one arising from the κ-structure $X[\mathcal{F},\mathcal{V}]$. But as we have already noted, the twisting is even through k-automorphisms that preserve the k-subgroup $\ker f_b \subset X[\mathcal{F},\mathcal{V}]_k$, and the k-subgroup $\ker f_b$ has field of definition over κ (relative to the initial κ-structure $X[\mathcal{F},\mathcal{V}]$) equal to $\kappa(b)$. Thus, the field of definition over κ' for the k-subgroup $\ker f_b$ relative to the κ'-structure $X[\mathcal{F},\mathcal{V}]_{\kappa'}$ is $\kappa'(b)$, so κ'' contains $\kappa'(b)$ over κ' inside k. This proves that $[\kappa'(b) : \kappa']$ divides $[\kappa'' : \kappa']$, which divides $Q(X)$. □

3.6.3. Example. In any given situation it is usually not difficult to find an explicit finite extension κ_0 of κ such that $a(X_b) = 1$ for all $b \in \mathbb{P}^1_\kappa(k) - \mathbb{P}^1_\kappa(\kappa_0)$, and $Q(X)$ is effectively computable. Here is an example for which we compute κ_0 and Q.

Fix relatively prime integers $m \geqslant n \geqslant 1$, and let $h = m + n \geqslant 2$. (We will soon take $n = 1$ and $m \geqslant 2$, but for now we work more generally.) Let κ be a finite field of size p^h, $k = \overline{\kappa}$, and $\lambda = m/h > 1 - \lambda$. For the p-divisible group $\mathcal{G}_\lambda$ over $\mathbb{F}_p$ of height h and dimension n (see 3.1.5), the description of $\mathrm{M}^*(\mathcal{G}_\lambda)$ and commutativity of the Dieudonné ring $D_{\mathbb{F}_p}$ imply that

$$\mathrm{End}(\mathcal{G}_\lambda) \simeq \mathbb{Z}_p[\mathcal{F},\mathcal{V}]/(\mathcal{F}\mathcal{V}-p,\mathcal{F}^m-\mathcal{V}^n) = \mathbb{Z}_p[\![\mathcal{F},\mathcal{V}]\!]/(\mathcal{F}\mathcal{V}-p,\mathcal{F}^m-\mathcal{V}^n)$$

(with $\mathcal{F}^h = p^n$ and $\mathcal{V}^h = p^m$). This ring is an order in $\mathbb{Z}_p[p^{1/h}]$ via $\mathcal{F} \mapsto (p^{1/h})^n$ and $\mathcal{V} \mapsto (p^{1/h})^m$ (so it is the valuation ring when $n=1$). More generally, for $K_0 = W(k)[1/p]$ and the localized Dieudonné ring $D_k[1/p] = K_0\{\mathcal{F}\}$ (with commutation relations $\mathcal{F}c = \sigma(c)\mathcal{F}$ for all $c \in K_0$ and the absolute Frobenius automorphism σ of K_0), $\mathrm{End}^0((\mathcal{G}_\lambda)_k)$ is the opposite algebra of the endomorphism algebra of the left $K_0\{\mathcal{F}\}$-module $\mathrm{M}^*((\mathcal{G}_\lambda)_k)[1/p] = K_0\{\mathcal{F}\}/K_0\{\mathcal{F}\}(\mathcal{F}^h - p^n)$.

Such endomorphisms are induced by $x \mapsto xf$ for precisely those

$$f = c_0 + c_1\mathcal{F} + \cdots + c_{h-1}\mathcal{F}^{h-1} \in K_0\{\mathcal{F}\}$$

such that $x \mapsto xf$ preserves the left ideal $K_0\{\mathcal{F}\}(\mathcal{F}^h - p^n)$. This is exactly the condition $\sigma^h(c_i) = c_i$ for all i, which is to say $c_i \in W(\kappa)[1/p] =: F$. Thus, $\mathrm{End}((\mathcal{G}_\lambda)_k) = \mathrm{End}((\mathcal{G}_\lambda)_\kappa)$ and it is an order in the cyclic algebra $(F/\mathbb{Q}_p, \sigma, p^n)$ over $\mathbb{Q}_p$ attached to the element $p^n \in \mathbb{Q}_p^\times$ and the absolute Frobenius generator σ of $\mathrm{Gal}(F/\mathbb{Q}_p)$. This cyclic algebra is a division algebra since the normalized valuation $\mathrm{ord}_p(p^n) = n$ is relatively prime to h (and $\mathrm{End}((\mathcal{G}_\lambda)_k)$ is the maximal order in this division algebra [**49**, 20.2.14] when $n=1$, but we do not use this fact). Similarly, for the Serre dual $\mathcal{G}_{1-\lambda} = \mathcal{G}_{n/h}$ of $\mathcal{G}_\lambda$ we have $\mathrm{End}((\mathcal{G}_{1-\lambda})_k) = \mathrm{End}((\mathcal{G}_{1-\lambda})_\kappa) = \mathrm{End}((\mathcal{G}_\lambda)_\kappa)^{\mathrm{opp}}$.

Assume $m > n$ (so $m \geqslant 2$ and $h \geqslant 3$), and let

$$Y = (\mathcal{G}_\lambda)_\kappa, \quad Y' = (\mathcal{G}_{1-\lambda})_\kappa, \quad X = Y \times Y'.$$

A computation with Dieudonné modules shows that $a(Y) = 1$ and $a(Y') = 1$, so we may and do identify the κ-group scheme $X[\mathcal{F},\mathcal{V}] = Y[\mathcal{F},\mathcal{V}] \times Y'[\mathcal{F},\mathcal{V}]$ with $(\alpha_p)^2$. Now we assume $n=1$, so $h = m+1$. For $b \in \mathbb{P}^1_\kappa(k)$, we find via a straightforward computation with Dieudonné modules that $a(X_k/b(\alpha_p)) = 1$ whenever $b \notin \{(1,0),(0,1)\}$. We may therefore choose κ_0 to be κ. Conditions (i) and (ii) in 3.4.3 are satisfied by X, and $\mathrm{End}(X) = \mathrm{End}(X_k)$, so the group

$$\Gamma_X = \mathrm{image}(\mathrm{Aut}(X_k) \to \mathrm{Aut}(X_k[\mathcal{F},\mathcal{V}])) \subset \mathrm{GL}_2(k)$$

is $\kappa^\times \times \kappa^\times$ in the diagonal subgroup $k^\times \times k^\times \subset \mathrm{GL}_2(k)$ and hence $Q(X) = p^h - 1$.

More generally, for $m \geqslant 2$ and *any* p-divisible group X over κ with slopes $1/(m+1)$ and $m/(m+1)$ each having (minimal possible) multiplicity $m+1$ (not just direct products X as above), we may find an isogenous quotient over a finite extension of κ such that (i) and (ii) in 3.4.3 are satisfied and κ_0 is the ground field and $Q = p^{m+1} - 1$. If instead X has additional slopes then it is generally more involved to find an explicit p-divisible group X' over (an explicit finite extension of) κ such that X_k and X'_k are isogenous and X' satisfies (i) and (ii) in 3.4.3.

3.7. Complex multiplication for p-divisible groups

3.7.1. Let R be a complete local noetherian domain with fraction field K of characteristic 0 and residue field of characteristic $p > 0$. For any non-zero p-divisible $\mathcal{X}$ over R, the endomorphism ring is a $\mathbb{Z}_p$-subalgebra of $\mathrm{End}(T_p(\mathcal{X}_{\overline{K}}))$. Thus, the

$\mathbb{Z}_p$-algebra $\mathrm{End}(\mathcal{X})$ is finite free as a $\mathbb{Z}_p$-module and $\mathrm{End}^0(\mathcal{X}) := \mathbb{Q}_p \otimes_{\mathbb{Z}_p} \mathrm{End}(\mathcal{X})$ is a finite-dimensional $\mathbb{Q}_p$-algebra. We are interested in situations when this contains large commutative semisimple $\mathbb{Q}_p$-subalgebras. The following lemma is straightforward to prove, by adapting the proof of Theorem 1.3.1.1.

3.7.1.1. Lemma. *Let $h > 0$ be the height of $\mathcal{X}$. For any commutative semisimple $\mathbb{Q}_p$-subalgebra $F \subseteq \mathrm{End}^0(\mathcal{X})$, $[F : \mathbb{Q}_p] \leqslant h$ with equality if and only if $V_p(\mathcal{X}_{\overline{K}})$ is free of rank 1 as an F-module.* □

3.7.1.2. Definition. We say that $\mathcal{X}$ admits *sufficiently many complex multiplications*, or $\mathcal{X}$ is a CM *p-divisible group*, if there exists a commutative semisimple algebra F over $\mathbb{Q}_p$ and an injective $\mathbb{Q}_p$-algebra homomorphism $i : F \to \mathrm{End}^0(\mathcal{X})$ such that $[F : \mathbb{Q}_p] = \mathrm{height}(\mathcal{X})$.

In such situations, we also say that $\mathcal{X}$ *has* CM *by F*. A natural source of examples is abelian schemes: if the complete local noetherian domain R is normal and A is an abelian scheme of relative dimension g over R such that $\mathrm{End}^0(A_K)$ contains a commutative semisimple $\mathbb{Q}$-algebra L of rank $2g$ then the equality $\mathrm{End}(A_K) = \mathrm{End}(A)$ (Lemma 1.8.4) identifies the commutative semisimple $\mathbb{Q}_p$-algebra $L_p := \mathbb{Q}_p \otimes_{\mathbb{Q}} L$ with a $\mathbb{Q}_p$-subalgebra of $\mathrm{End}^0(A[p^\infty])$ since the map $\mathbb{Z}_p \otimes_{\mathbb{Z}} \mathrm{End}(A_K) \to \mathrm{End}(A_K[p^\infty])$ is injective. Hence, the p-divisible group $A[p^\infty]$ over R has CM by L_p.

For an algebraic closure $\overline{\mathbb{Q}}_p$ of $\mathbb{Q}_p$ and a finite-dimensional commutative semisimple $\mathbb{Q}_p$-algebra F, denote by $\mathrm{Hom}_{\mathbb{Q}_p\text{-alg}}(F, \overline{\mathbb{Q}}_p)$ the set of all $\mathbb{Q}_p$-algebra homomorphisms from F to $\overline{\mathbb{Q}}_p$. We have a natural decomposition $F = \prod F_i$ for finite extension fields F_i of $\mathbb{Q}_p$, so $\mathrm{Hom}_{\mathbb{Q}_p\text{-alg}}(F, \overline{\mathbb{Q}}_p)$ is the disjoint union of the sets $\mathrm{Hom}_{\mathbb{Q}_p\text{-alg}}(F_i, \overline{\mathbb{Q}}_p)$. Although elements of $\mathrm{Hom}_{\mathbb{Q}_p\text{-alg}}(F, \overline{\mathbb{Q}}_p)$ are injections of F into $\overline{\mathbb{Q}}_p$ only when F is a field, we sometimes abuse language and call an element of $\mathrm{Hom}_{\mathbb{Q}_p\text{-alg}}(F, \overline{\mathbb{Q}}_p)$ a "$\mathbb{Q}_p$-embedding of F into $\overline{\mathbb{Q}}_p$".

In general, since F is a $\mathbb{Q}_p$-algebra and $F \cap \mathrm{End}(\mathcal{X})$ is a $\mathbb{Z}_p$-order in F, there is a natural $F \otimes_{\mathbb{Q}_p} K$-module structure on the generic fiber $\mathrm{Lie}(\mathcal{X})_K$ of the Lie algebra of $\mathcal{X}$ over R. Much like in the case of abelian varieties, we have:

3.7.1.3. Lemma. *Let d denote the relative dimension of $\mathcal{X}$ over R. The $\overline{K}$-linear action of F on $\mathrm{Lie}(\mathcal{X})_{\overline{K}}$ is given by a collection $\Phi = \Phi(\mathcal{X}, F)$ of d distinct $\mathbb{Q}_p$-embeddings of F into $\overline{K}$ (or equivalently, into $\overline{\mathbb{Q}}_p$). The dual action of F on the Serre dual $\mathcal{X}^t$ defines an embedding $F \hookrightarrow \mathrm{End}(\mathcal{X}^t)$, and $\Phi(\mathcal{X}^t, F)$ is the complement of $\Phi(\mathcal{X}, F)$ in $\mathrm{Hom}_{\mathbb{Q}_p\text{-alg}}(F, \overline{K})$.*

PROOF. Since $\mathrm{Lie}(\mathcal{X})_{\overline{K}}$ is a module over the semisimple $\overline{K}$-algebra $F \otimes_{\mathbb{Q}} \overline{K}$, the first problem is to prove that the $\mathbb{Q}_p$-embeddings $F \to \overline{K}$ whose associated eigenspace in the Lie algebra is non-zero actually have a 1-dimensional eigenspace. That is, we have to prove that there are no non-trivial multiplicities for the eigencharacters of the F-action. For the analogous fact concerning CM types arising from abelian varieties over fields of characteristic 0, the only known proofs use Hodge theory (or at least the Hodge filtration of de Rham cohomology). Likewise, for p-divisible groups we will use the Hodge-Tate decomposition.

First, we pass to the case of a simpler base ring, as follows. It is well-known that there is a local injective map of R into a complete discrete valuation ring.[6] The formation of the identity component of a p-divisible group over a complete local noetherian ring (of residue characteristic p) commutes with *local* extension of scalars, so we may apply such a scalar extension to reduce to the case when R is a discrete valuation ring. We may likewise arrange that its residue field is algebraically closed.

Let $\mathbb{C}_K$ denote the completion of $\overline{K}$.[7] The Hodge-Tate decomposition (1.4.5.1) provides a functorial $\mathbb{C}_K$-linear isomorphism

$$\mathbb{C}_K \otimes_{\mathbb{Q}_p} V_p(\mathcal{X}_{\overline{K}}) \simeq (\mathbb{C}_K(1) \otimes_K \mathrm{Lie}(\mathcal{X})_K) \oplus (\mathbb{C}_K \otimes_K \mathrm{Lie}(\mathcal{X}^t)_K^\vee),$$

where $V_p(\mathcal{X}_{\overline{K}}) := \mathbb{Q}_p \otimes_{\mathbb{Z}_p} T_p(\mathcal{X}_{\overline{K}})$. Consider its functoriality with respect to the F-action (in the isogeny category over R). Since $V_p(\mathcal{X}_{\overline{K}})$ is free of rank 1 over F (Lemma 3.7.1.1), the left side is free of rank 1 over $\mathbb{C}_K \otimes_{\mathbb{Q}_p} F$. Hence, there is a 1-dimensional eigenspace over $\mathbb{C}_K$ for each $\mathbb{Q}_p$-embedding of F into $\mathbb{C}_K$ (or equivalently, into $\overline{\mathbb{Q}}_p$). Passing to the right side, both Lie algebras over K have no repeated eigencharacters for the F-actions on their geometric fibers, and the sets of maps $F \to \overline{K}$ arising as eigencharacters on these respective Lie algebras are complements of each other in $\mathrm{Hom}_{\mathbb{Q}_p\text{-alg}}(F, \overline{K})$. □

The (*p-adic*) CM *type* of $(\mathcal{X}, F)$ is the collection of embeddings in Lemma 3.7.1.3; note that these embeddings are valued in the algebraic closure $\overline{\mathbb{Q}}_p$ of $\mathbb{Q}_p$ in $\overline{K}$. It is equivalent to work with the $\mathbb{Q}_p$-linear trace $F \to K$ that classifies the K-linear F-action on $\mathrm{Lie}(\mathcal{X})_K$ up to isomorphism. More generally, we have:

3.7.1.4. Definition. Fix an algebraic closure $\overline{\mathbb{Q}}_p$ of $\mathbb{Q}_p$. For a finite-dimensional commutative semisimple $\mathbb{Q}_p$-algebra F, a *p-adic* CM *type* for F valued in $\overline{\mathbb{Q}}_p$ is a subset $\Phi \subseteq \mathrm{Hom}_{\mathbb{Q}_p\text{-alg}}(F, \overline{\mathbb{Q}}_p)$.

This is linked to the notion of a $\overline{\mathbb{Q}}_p$-valued CM type on a CM field L as at the end of 2.1.2 because $\mathrm{Hom}(L, \overline{\mathbb{Q}}_p) = \mathrm{Hom}_{\mathbb{Q}_p\text{-alg}}(L_p, \overline{\mathbb{Q}}_p)$ for $L_p := \mathbb{Q}_p \otimes_{\mathbb{Q}} L$.

3.7.1.5. Remark. In Definition 3.7.2.3 we will define the notion of *reflex field* for a p-adic CM type. An immediate consequence of Lemma 3.7.1.3 will be that the fraction field of the base ring R of the p-divisible group $\mathcal{X}$ in 3.7.1.3 contains the reflex field of the p-adic CM type $(F, \Phi(\mathcal{X}, F))$.

Now we consider p-divisible groups in characteristic p. By using Dieudonné modules, we obtain an analogue of Lemma 3.7.1.1:

3.7.1.6. Proposition. *Let X be a p-divisible group of height $h > 0$ over a field of characteristic p, and let $F \subseteq \mathrm{End}^0(X)$ be a commutative semisimple $\mathbb{Q}_p$-subalgebra.*

---

[6]Here is a geometric proof. Let $f : Z \to \mathrm{Spec}(R)$ be the (surjective) blow-up at the closed point, and let z be a generic point of the special fiber of f. The local ring $\mathcal{O}_{Z,z}$ is domain (as Z is an integral scheme), and $R \to \mathcal{O}_{Z,z}$ is a local injection. A local generator at z for the ideal of the exceptional divisor on Z generates an $\mathfrak{m}_z$-primary ideal in $\mathcal{O}_{Z,z}$, so $\mathcal{O}_{Z,z}$ is 1-dimensional. Its normalization is a semi-local Dedekind domain (Krull-Akizuki [**73**, Thm. 11.7]), so the completion at a maximal ideal of the normalization does the job.

[7]Strictly speaking $\mathbb{C}_K$ depends on the choice of an algebraic closure $\overline{K}$ of K; it is determined by K up to non-unique topological isomorphism.

Relative to the decomposition $F = \prod F_i$ as a finite product of fields, let X_i denote the non-zero isogeny factor of X corresponding to F_i.

Then $[F : \mathbb{Q}_p] \leqslant h$, and when equality holds each X_i is isoclinic.

PROOF. It suffices to prove this when the base field is algebraically closed, which we assume to be the case. Since $h = \sum h_i$ where h_i is the height of X_i, it suffices to treat the X_i separately. Thus, we may and do assume that F is a field. By the Dieudonné-Manin classification (Theorem 3.1.3), we have a *functorial* decomposition in the isogeny category $X \sim \prod Y_j$ for non-zero isoclinic Y_j with pairwise distinct slopes. By functoriality, F preserves each Y_j. By Proposition 1.4.3.9(1) each Y_j has height at least $[F : \mathbb{Q}_p] = h$, so there is only one such factor. That is, X is isoclinic. □

Over fields of characteristic p it is now natural to adapt the definition of a CM p-divisible group in 3.7.1.2 to the case of non-zero p-divisible groups Z: we say that Z admits *sufficiently many complex multiplications* if $\mathrm{End}^0(Z)$ contains a commutative semisimple $\mathbb{Q}_p$-subalgebra F with $[F : \mathbb{Q}_p] = \mathrm{height}(Z)$. We then say that Z is a CM *p-divisible group*, and that it has CM by F.

3.7.1.7. Example. A CM p-divisible group $\mathcal{X}$ (in the sense of Definition 3.7.1.2) over a complete noetherian local domain $(R, \mathfrak{m})$ with generic characteristic 0 and residue characteristic p has special fiber X that is a CM p-divisible group over the residue field. Also, if B is a CM abelian variety over a field of characteristic $p > 0$ then its p-divisible group is a CM p-divisible group, since $\mathbb{Z}_p \otimes_{\mathbb{Z}} \mathrm{End}(B) \to \mathrm{End}(B[p^\infty])$ is injective (Proposition 1.2.5.1 with $\ell = p$).

3.7.1.8. Definition. Let X be a p-divisible group over a field κ of characteristic p. A CM *lifting* of X over a complete local noetherian domain $(R, \mathfrak{m})$ with generic characteristic 0 is a triple $(\mathcal{X}, \phi, f)$ where $\mathcal{X}$ is a CM p-divisible group over R, $\phi : R/\mathfrak{m} \simeq \kappa$ is an isomorphism, and $f : \mathcal{X}_\kappa \simeq X$ is an isomorphism.

In the study of lifting problems for CM p-divisible groups, we will find it useful to apply the deformation theory of p-divisible groups (much as the deformation theory of abelian varieties is useful in the study of CM lifting problems for abelian varieties, illustrated in 2.1.1). This deformation theory was reviewed in 1.4.4.

Consider a p-divisible group X of height h and dimension d over a *perfect* field k of characteristic $p > 0$. Let D be the deformation ring for X (as in 1.4.4.7), so D is a formal power series ring over $W(k)$ in $h(d-h)$ variables. If X admits a CM lifting over a complete local noetherian domain R of characteristic 0 with residue field k then by working with a suitable quotient of D (as in Theorem 1.4.5.5(1)) the same deformation ring argument as with CM lifting for abelian varieties in 2.1.1 provides a finite extension k' of k such that $X_{k'}$ admits a CM lifting over a complete discrete valuation ring of characteristic 0 with residue field k'. Moreover, as in the case of abelian varieties, we can require $k' = k$ if we allow R to be an order in finite totally ramified extension of $W(k)$ rather than to be the valuation ring in such an extension. In particular, if k is finite then the problem of existence of a CM lifting over orders in p-adic integer rings is essentially the general case.

3.7.2. Slopes, CM types, and reflex fields. Let (F, Φ) be the p-adic CM type of a CM p-divisible group $\mathcal{X}$ over R as in 3.7.1. Let X denote the special fiber. The product decomposition $F = \prod F_j$ with fields F_j induces product decompositions of p-divisible groups $X \sim \prod X_j$ and $\mathcal{X} \sim \prod \mathcal{X}_j$ in the respective isogeny categories. Each $\mathcal{X}_j$ has height $[F_j : \mathbb{Q}_p]$ and CM by F_j, so the same holds for each X_j. In particular, the X_j are isoclinic (by Proposition 3.7.1.6).

Renaming $\mathcal{X}_j$ as $\mathcal{X}$, let us focus on the case when F is a field, so X is isoclinic. Is there any relationship between the unique slope and the CM type? Indeed there is a very simple relationship:

3.7.2.1. Lemma. *When the F-action on X is obtained by reduction from characteristic 0 then the slope is $\#\Phi/[F : \mathbb{Q}_p]$.*

PROOF. Since $[F : \mathbb{Q}_p]$ is equal to the height and

$$\#\Phi = \dim_K \mathrm{Lie}(\mathcal{X})_K = \dim \mathcal{X} = \dim X,$$

the assertion reduces to the claim that the slope of an isoclinic p-divisible group over a field of characteristic p is its dimension divided by its height. Extending the ground field does not change the slopes of a p-divisible group, so Theorem 3.1.3(2) gives the result. □

This motivates:

3.7.2.2. Definition. Let X be a p-divisible group of height $h > 0$ over a field characteristic $p > 0$, and suppose X is equipped with an action by a $\mathbb{Z}_p$-order in a commutative semisimple $\mathbb{Q}_p$-algebra $F = \prod F_j$ of rank h, with each F_j a field. Choose a p-adic CM type $\Phi = \coprod \Phi_j$ of $F = \prod F_j$; we sometimes write $\coprod(F_j, \Phi_j)$ to denote (F, Φ).

The p-adic CM type (F, Φ) is *compatible* with the F-action on X if $\#(\Phi_j) = \dim(X_j)$ for each j, or equivalently if the slope of the isoclinic X_j is $\#\Phi_j/[F_j : \mathbb{Q}_p]$ for each j.

Our interest in this definition is due to the fact (proved above) that if there is a CM lift $\mathcal{X}$ of X over a mixed characteristic local noetherian domain $(R, \mathfrak{m})$ and it has CM type (F, Φ), then (F, Φ) is compatible with the F-action on X.

We need one more definition. As motivation, fix an algebraic closure $\overline{K}$ of $K = \mathrm{Frac}(R)$, and view the CM types as being $\overline{K}$-valued. The CM type Φ of F acting on $\mathcal{X}$ above decomposes into the disjoint union of the CM types Φ_j for the $\mathcal{X}_j$, with $\Phi_j \subseteq \mathrm{Hom}_{\mathbb{Q}_p\text{-alg}}(F_j, \overline{K})$ for each j. Thus, the following definition applies.

3.7.2.3. Definition. Let F be a finite-dimensional non-zero commutative semisimple $\mathbb{Q}_p$-algebra, and $\{F_j\}$ its set of factor fields. Fix an algebraic closure $\overline{\mathbb{Q}}_p$ of $\mathbb{Q}_p$, and let $\Phi = \coprod \Phi_j$ be a $\overline{\mathbb{Q}}_p$-valued p-adic CM type on $F = \prod F_j$. The *reflex field* $E_j = E(F_j, \Phi_j) \subset \overline{\mathbb{Q}}_p$ over $\mathbb{Q}_p$ is the subfield corresponding to the open subgroup of elements $\sigma \in \mathrm{Gal}(\overline{\mathbb{Q}}_p/\mathbb{Q}_p)$ such that $\sigma(\Phi_j) = \Phi_j$ inside $\mathrm{Hom}_{\mathbb{Q}_p\text{-alg}}(F_j, \overline{\mathbb{Q}}_p)$. The *reflex field* $E = E(F, \Phi) \subset \overline{\mathbb{Q}}_p$ is the compositum over $\mathbb{Q}_p$ of the reflex fields $E_j = E(F_j, \Phi_j)$ inside $\overline{\mathbb{Q}}_p$.

In the special case that (F, Φ) arises from a CM p-divisible group over R as above and we take $\overline{\mathbb{Q}}_p$ to be the algebraic closure of $\mathbb{Q}_p$ inside a choice of algebraic closure $\overline{K}$ of K, it is clear that $E(F, \Phi) \subseteq K$. There is also a compatibility with the notion of reflex field for abelian varieties. To formulate this, let A be an abelian scheme over a complete discrete valuation ring R with mixed characteristic $(0, p)$ and fraction field K, and fix an algebraic closure $\overline{K}$ of K. Let $\overline{\mathbb{Q}}_p$ be the algebraic closure of $\mathbb{Q}_p$ inside $\overline{K}$, and let $\overline{\mathbb{Q}} \subset \overline{\mathbb{Q}}_p$ be the algebraic closure of $\mathbb{Q}$. Assume that A_K is equipped with a CM structure using a CM field L. Via the identification of $\mathrm{Lie}(A_K)$ with $\mathrm{Lie}(A[p^\infty])_K$ (see 1.4.3.6), the associated CM type

$$\Phi \subset \mathrm{Hom}_{\mathrm{ring}}(L, \overline{K}) = \mathrm{Hom}_{\mathbb{Q}_p\text{-alg}}(\mathbb{Q}_p \otimes_{\mathbb{Q}} L, \overline{\mathbb{Q}}_p) = \coprod_{v|p} \mathrm{Hom}_{\mathbb{Q}_p\text{-alg}}(L_v, \overline{\mathbb{Q}}_p)$$

coincides with the p-adic CM type for $A[p^\infty]$ and the p-adic reflex field $E(\mathbb{Q}_p \otimes_{\mathbb{Q}} L, \Phi)$ is easily checked to be the completion of $E(L, \Phi) \subset \overline{\mathbb{Q}}$ relative to the p-adic place induced by the inclusion $\overline{\mathbb{Q}} \hookrightarrow \overline{\mathbb{Q}}_p$.

3.7.2.4. Remark. Let F be a non-zero finite-dimensional commutative semisimple $\mathbb{Q}_p$-algebra and let Φ be a p-adic CM type of F. Fix an algebraic closure $\overline{\mathbb{Q}}_p$ of $\mathbb{Q}_p$. Denote by $\underline{F}^\times$ the Weil restriction torus $\mathrm{Res}_{F/\mathbb{Q}_p} \mathbb{G}_{\mathrm{m}}$; its group of $\mathbb{Q}_p$-points is canonically identified with $F^\times$ (and this group of points is Zariski-dense in the $\mathbb{Q}_p$-torus). The (geometric) character group of $\underline{F}^\times$ is naturally identified with the set $\mathrm{Hom}_{\mathbb{Q}_p\text{-alg}}(F, \overline{\mathbb{Q}}_p)$ of $\mathbb{Q}_p$-algebra homomorphisms from F to $\overline{\mathbb{Q}}_p$ via the conditions

$$\xi^\sigma(a) = \sigma(a) \qquad \text{for } a \in \underline{F}^\times(\mathbb{Q}_p) = F^\times, \sigma \in \mathrm{Hom}_{\mathbb{Q}_p\text{-alg}}(F, \overline{\mathbb{Q}}_p),$$

where ξ^σ is the geometric character of the $\mathbb{Q}_p$-torus $\underline{F}^\times$ corresponding to σ.

The $\mathbb{Z}$-basis of the (geometric) cocharacter group of $\underline{F}^\times$ dual to the $\mathbb{Z}$-basis $\{\xi^\sigma\}$ of the character group of $\underline{F}^\times$ is denoted by $\{\nu_\sigma\}$. The subset $\Phi \subset \mathrm{Hom}_{\mathbb{Q}_p\text{-alg}}(F, \overline{\mathbb{Q}}_p)$ defines a cocharacter μ_Φ of $\underline{F}^\times$ over $\overline{\mathbb{Q}}_p$ via

$$\langle \xi^\sigma, \mu_\Phi \rangle = \begin{cases} 1 & \text{if } \sigma \in \Phi \\ 0 & \text{if } \sigma \notin \Phi \end{cases}$$

under the pairing of the character group and the cocharacter group (i.e., $\mu_\Phi = \sum_{\sigma \in \Phi} \nu_\sigma$). An alternative definition of the reflex field $E(F, \Phi)$, in the spirit of (2.1.3.1), is that it is the minimal field of definition over $\mathbb{Q}_p$ for the geometric cocharacter μ_Φ of $\underline{F}^\times$. Equivalently, $E(F, \Phi)$ is the largest subfield of $\overline{\mathbb{Q}}_p$ over $\mathbb{Q}_p$ fixed by the stabilizer subgroup of μ_Φ in $\mathrm{Gal}(\overline{\mathbb{Q}}_p/\mathbb{Q}_p)$.

Motivated by Proposition 2.1.3.4, we make the following definition.

3.7.2.5. Definition. Using notation as above, with $E = E(F, \Phi)$ the reflex field of the p-adic CM type (F, Φ) and $\mu : \mathbb{G}_{\mathrm{m}} \to (\underline{F}^\times)_E$ the E-descent of μ_Φ, the *reflex norm* $\mathrm{N}\mu_\Phi : \underline{E}^\times \to \underline{F}^\times$ of (F, Φ) is the composite $\mathbb{Q}_p$-homomorphism

$$\mathrm{Res}_{E/\mathbb{Q}_p}(\mathbb{G}_{\mathrm{m}}) \xrightarrow{\mathrm{Res}_{E/\mathbb{Q}_p}(\mu)} \mathrm{Res}_{E/\mathbb{Q}_p}((\underline{F}^\times)_E) \xrightarrow{\mathrm{Nm}_{E/\mathbb{Q}_p}} \underline{F}^\times .$$

(The "norm" homomorphism $\mathrm{Nm}_{K'/K} : \mathrm{Res}_{K'/K}(C_{K'}) \to C$ used above is as in Proposition 2.1.3.4 for any finite separable extension of fields K'/K and commutative affine K-group C of finite type.)

The reflex norm is characterized by the condition that the associated Galois-equivariant map between geometric cocharacter groups sends the geometric cocharacter ν_{τ_0} of $\underline{E}^\times$ indexed by the inclusion map $\tau_0 : E \hookrightarrow \overline{\mathbb{Q}}_p$ to the geometric cocharacter μ_Φ of $\underline{F}^\times$. Indeed, this is an analogue of Proposition A.4.8.1 and is proved in exactly the same way up to some changes in notation (taking the ground field to be $\mathbb{Q}_p$ rather than $\mathbb{Q}$, and allowing F to be a product of several fields rather than a field).

3.7.2.6. Remark. When F is a subfield of $\overline{\mathbb{Q}}_p$ of finite degree over $\mathbb{Q}_p$, the reflex field $E(F,\Phi)$ and reflex norm $\mathrm{N}\mu_\Phi$ admit Galois-theoretic descriptions similar to the classical CM case in 2.1.3.5 by using the Galois group of the normal closure $F' \subset \overline{\mathbb{Q}}_p$ of $F/\mathbb{Q}_p$, as follows.

Replacing $(L/\mathbb{Q}, \overline{\mathbb{Q}}, \Phi, L')$ in 2.1.3.5 with $(F/\mathbb{Q}_p, \overline{\mathbb{Q}}_p, \Phi, F')$, we can define the subsets $\Phi_{F'}, \Phi^*_{F'} \subset \mathrm{Gal}(F'/\mathbb{Q}_p)$ and the subgroup $H^* \subset \mathrm{Gal}(F'/\mathbb{Q}_p)$ similarly to the classical case, yielding the description $E(F,\Phi) = F'^{H^*}$ by exactly the same argument. Use the identification of $\mathrm{Gal}(F'/\mathbb{Q}_p)/H^*$ with $\mathrm{Hom}_{\mathbb{Q}_p\text{-alg}}(E,F') = \mathrm{Hom}_{\mathbb{Q}_p\text{-alg}}(E,\overline{\mathbb{Q}}_p)$ to define the injection $\Phi^*_{F'}/H^* \hookrightarrow \mathrm{Hom}_{\mathbb{Q}_p\text{-alg}}(E,F')$. When the reflex norm (now *defined* in terms of the E-descent of μ_Φ in the role of the descent of h_Φ in the classical CM case) is composed with the inclusion of $\mathbb{Q}_p$-tori $\underline{F}^\times \hookrightarrow \underline{F'}^\times$, the resulting map of $\mathbb{Q}_p$-tori $\underline{E}^\times \to \underline{F'}^\times$ is given on $\mathbb{Q}_p$-points by the formula

$$x \mapsto \prod_{\tau \in \Phi^*_{F'}/H^*} \tau(x) \in F'^\times$$

for all $x \in E(F,\Phi)^\times$; this is proved exactly as in the classical CM case (except now working throughout with $\mathbb{Q}_p$-algebra maps).

The following two results (Propositions 3.7.3 and 3.7.4)[8] record some facts for CM p-divisible groups over discrete valuation rings, analogous to results discussed earlier for CM abelian varieties (as we review in the subsequent Remark below). These will be used in the proof of a necessary condition for CM lifting of p-divisible groups in Proposition 3.8.2; their proofs are respectively given in 3.7.10 and 3.7.8.

3.7.3. Proposition. *Let F be a non-zero finite-dimensional commutative semisimple $\mathbb{Q}_p$-algebra. Let Φ be a $\overline{\mathbb{Q}}_p$-valued p-adic* CM *type for F. Let $E = E(F,\Phi) \subset \overline{\mathbb{Q}}_p$ be the reflex field for (F,Φ).*

(1) *There is an O_F-linear* CM *p-divisible group $\mathcal{Y}$ over O_E with* CM *type (F,Φ).*

(2) *Given any O_F-linear* CM *p-divisible group $\mathcal{Y}$ over O_E with* CM *type (F,Φ), there is an unramified twist $\mathcal{Y}'$ of $\mathcal{Y}$ such that the local Galois representation*
$$\rho_{\mathcal{Y}'} : \mathrm{Gal}(\overline{E}/E) \to O_F^\times$$
attached to $\mathcal{Y}'$ has the property that $\rho_{\mathcal{Y}'}(I_E) = \rho_{\mathcal{Y}'}(\mathrm{Gal}(\overline{E}/E))$.

In Proposition 3.7.3(2), "unramified twist" means a p-divisible group related through a twist by an unramified $O_F^\times$-valued character of the Galois group of E (see 1.4.5.2).

[8]There is little doubt that 3.7.3 and 3.7.4 were known to the experts already in the 1970's, but we could not find these statements in the literature. We give "modern" proofs using some basic facts in p-adic Hodge theory.

3.7.4. Proposition. *Let F be a non-zero finite-dimensional commutative semisimple $\mathbb{Q}_p$-algebra, and let k be an algebraically closed field of characteristic p. Choose an algebraic closure $\overline{K}_0$ of $K_0 = W(k)[1/p]$, and let $\overline{\mathbb{Q}}_p$ denote the algebraic closure of $\mathbb{Q}_p$ in $\overline{K}_0$. Let K/K_0 be a finite extension inside $\overline{K}_0$, and let $R = \mathcal{O}_K$.*

(1) *Let $\mathcal{Y}_1$ and $\mathcal{Y}_2$ be p-divisible groups over R with height $h := [F : \mathbb{Q}_p]$, each equipped with an F-action defining the same* CM *type (F, Φ). Then $\mathcal{Y}_1$ and $\mathcal{Y}_2$ are F-linearly isogenous over R. If in addition the subring $\mathcal{O}_F$ in each endomorphism algebra* $\mathrm{End}^0(\mathcal{Y}_j)$ *lies in the endomorphism ring* $\mathrm{End}(\mathcal{Y}_j)$ *then $\mathcal{Y}_1$ and $\mathcal{Y}_2$ are $\mathcal{O}_F$-linearly isomorphic.*
(2) *Let $\mathcal{K}$ be a subfield of $\overline{\mathbb{Q}}_p$ of finite degree over $\mathbb{Q}_p$. Let $\mathcal{Y}_1$ and $\mathcal{Y}_2$ be p-divisible groups over $\mathcal{O}_{\mathcal{K}}$ and R respectively, each equipped with a* CM *structure by F having the same $\overline{K}$-valued* CM *type Φ.*

   There exists a finite extension $\mathcal{K}'/\mathcal{K}$ inside $\overline{\mathbb{Q}}_p$, a finite flat $\mathcal{O}_{\mathcal{K}'}$-subgroup C in $(\mathcal{Y}_1)_{\mathcal{O}_{K'}}$, and an F-linear isomorphism

$$((\mathcal{Y}_1)_{\mathcal{O}_{K'}}/C)_{\mathcal{O}_{K'}} \simeq (\mathcal{Y}_2)_{\mathcal{O}_{K'}}$$

   where $K' := \mathcal{K}'K$ is the compositum of $\mathcal{K}'$ and K over $\mathbb{Q}_p$ inside $\overline{K}_0$.

Remark. Before we prove Propositions 3.7.3 and 3.7.4, we record the analogies with the CM theory of abelian varieties. In view of Tate's full faithfulness theorem for p-divisible groups over mixed-characteristic $(0, p)$ discrete valuation rings (see Theorem 1.4.5.1), note that in Proposition 3.7.4 (1) it is equivalent to make the isogeny between the generic fibers and in Proposition 3.7.4 (2) it is equivalent (using schematic closure from the generic fiber) to say that there is an F-linear isogeny $(\mathcal{Y}_1)_{K'} \to (\mathcal{Y}_2)_{K'}$ whose kernel is defined over $\mathcal{K}'$.

Hence, Proposition 3.7.4 (1) is analogous to the isogeny aspect of Proposition 1.5.4.1, and Proposition 3.7.4 (2) is analogous to a hybrid of Theorem 1.7.2.1 and Theorem 1.7.2.5. Likewise, Proposition 3.7.3(1) is analogous to Theorem 2.5.2 and Corollary A.4.6.5. In fact the proof of Proposition 3.7.3 uses Corollary A.4.6.5, which depends on Theorem 2.5.2 and Theorem A.4.5.1; the latter is an existence statement for algebraic Hecke characters having a specified algebraic part over the field of moduli and having good reduction at all places above p.

We will prove Proposition 3.7.4 before Proposition 3.7.3 because a corollary of the former is used in the proof of the latter. Readers are encouraged to skip both proofs and the rest of §3.7 and go directly to the main results of this chapter in §3.8. To begin the proofs of the two preceding results, we record several lemmas.

3.7.5. Lemma. *Let F be a finite extension field of $\mathbb{Q}_p$. There exists a totally real number field L^+ with a unique place $v|p$ and a totally imaginary quadratic extension L of L^+ in which v splits such that the v-adic completion L_v^+ of L^+ is $\mathbb{Q}_p$-isomorphic to F.*

PROOF. By weak approximation in $\mathbb{Q}$, we may pick a monic polynomial $f \in \mathbb{Q}[t]$ that p-adically approximates the minimal polynomial in $\mathbb{Q}_p[t]$ for a primitive element of $F/\mathbb{Q}_p$ and in $\mathbb{R}[t]$ approximates a totally split monic polynomial of the same degree. By Krasner's Lemma, choosing the p-adic approximation closely enough ensures that f is irreducible over $\mathbb{Q}_p$, hence over $\mathbb{Q}$, so $L^+ := \mathbb{Q}[t]/(f)$ is a totally real number field equipped with a p-adic place v at which the completion is $\mathbb{Q}_p$-isomorphic to F and has $\mathbb{Q}_p$-degree equal to $[L^+ : \mathbb{Q}]$ (so v is the unique p-adic place of L^+). Weak approximation in L^+ provides a further quadratic extension L of L^+ that is totally complex and is split at v. □

3.7.6. Lemma. *Let R be a mixed characteristic $(0, p)$ complete discrete valuation ring with fraction field K and a perfect residue field. Let $\mathcal{X}$ and $\mathcal{Y}$ be p-divisible groups over R.*

For any totally ramified finite extension K' of K, any homomorphism

$$\beta' : \mathcal{X}_{R'} \to \mathcal{Y}_{R'}$$

over the valuation ring R' of K' descends uniquely to a homomorphism $\beta : \mathcal{X} \to \mathcal{Y}$ over R, and β is an isogeny if and only if β' is an isogeny.

PROOF. It suffices to verify the descent condition for morphisms, relative to the finite flat ring extension R'/R. Let $R'' = R' \otimes_R R'$. Let $\mathcal{X}''$ (respectively $\mathcal{Y}''$) be the pullback of $\mathcal{X}$ (respectively $\mathcal{Y}$) to $\mathrm{Spec}(R'')$. We have two pullbacks $\mathrm{pr}_1^*(\beta')$ and $\mathrm{pr}_2^*(\beta')$ over R'' of the isogeny β' over R', via the two projections $\mathrm{pr}_i : \mathrm{Spec}(R'') \to \mathrm{Spec}(R')$. We need to show that $\mathrm{pr}_1^*(\beta') = \mathrm{pr}_2^*(\beta')$; then β' descends to a homomorphism β over R by faithfully flat descent, and it is obvious that β is an isogeny if and only if β' is an isogeny (consider heights and kernels).

The ring R'' is a complete noetherian *local* ring because R' is totally ramified over R. The closed fiber of the two homomorphisms $\mathrm{pr}_1^*(\beta')$ and $\mathrm{pr}_2^*(\beta')$ over R'' are equal because both are naturally identified with the closed fiber of β'. We know from Proposition 1.4.4.3 that passage to the special fiber is a faithful functor on p-divisible groups over complete local noetherian rings (such as R'') with residue characteristic p, so $\mathrm{pr}_1^*(\beta') = \mathrm{pr}_2^*(\beta')$ and hence β' descends to R. □

3.7.6.1. Remark. Using Tate's Theorem 1.4.5.1 on extending homomorphisms between p-divisible groups, one can deduce the following from Lemma 3.7.6, with R and K as in that lemma, and we also give a more "modern" proof:

> *Let $\rho_1, \rho_2 : \mathrm{Gal}(\overline{K}/K) \rightrightarrows \mathrm{GL}_h(\mathbb{Q}_p)$ be continuous homomorphisms isomorphic to respective Galois representations $V_p(\mathcal{X}_1)$ and $V_p(\mathcal{X}_2)$ for p-divisible groups $\mathcal{X}_1$ and $\mathcal{X}_2$ of height h over R. Suppose that the restrictions of $\rho_1|_U$ and $\rho_2|_U$ to an open subgroup U of the inertia subgroup $I_K \subset \mathrm{Gal}(\overline{K}/K)$ are isomorphic. Then the restrictions $\rho_1|_{I_K}$ and $\rho_2|_{I_K}$ to I_K are isomorphic.*

To reprove this assertion, first note that (by a theorem of Fontaine [**43**, 6.2]) both ρ_i are crystalline. (We refer to [**42**, 3.2.5], [**43**, §5], [**44**, 5.1.4] for the notion of a crystalline p-adic representation.) Thus, the assertion is a special case of the general result in p-adic Hodge theory that if an element in the representation space of a crystalline p-adic representation (such as an element in the tensor product $\rho_1^\vee \otimes \rho_2 = \mathrm{Hom}_{\mathbb{Q}_p}(\rho_1, \rho_2)$) is fixed by an open subgroup of the inertia subgroup then it is fixed by the whole inertia subgroup. In other words, a finitely ramified crystalline representation is unramified; cf. [**42**, 3.3.1].

3.7.7. Lemma. *Let K and F be finite extensions of $\mathbb{Q}_p$, and let $\rho : \mathrm{Gal}(K^{\mathrm{ab}}/K) \to F^\times$ be a crystalline linear representation on a 1-dimensional F-vector space. Let $\chi : \underline{K}^\times \to \underline{F}^\times$ be a $\mathbb{Q}_p$-homomorphism such that on $\mathbb{Q}_p$-points its restriction to a* *homomorphism* $\mathcal{O}_K^\times \to \mathcal{O}_F^\times$ *between maximal compact subgroups agrees near* 1 *with* $\rho \circ r_K$. *Then* $\chi|_{\mathcal{O}_K^\times}$ *agrees with* $\rho \circ r_K$ *on the entirety of* $\mathcal{O}_K^\times$.

PROOF. It is harmless to increase F so that it splits K over $\mathbb{Q}_p$. That is, we may assume that the natural map $F \otimes_{\mathbb{Q}_p} K \to \prod_{\tau: K \to F} F$ defined by $a \otimes b \mapsto (\tau(b)a)$ is an isomorphism, where τ varies through the $\mathbb{Q}_p$-embeddings of K into F. It is also enough to show that $\chi|_{\mathcal{O}_K^\times}$ extends to *some* crystalline representation $\rho' : \mathrm{Gal}(K^{\mathrm{ab}}/K) \to F^\times$, as then ρ and ρ' are crystalline representations whose restrictions to I_K agree on an open subgroup and thus are equal (see Remark 3.7.6.1).

To extend $\chi|_{\mathcal{O}_K^\times}$ to an $F^\times$-valued crystalline representation on $\mathrm{Gal}(K^{\mathrm{ab}}/K)$, we first note that since F splits K over $\mathbb{Q}_p$, the universal property of the functor $\mathrm{Res}_{F/\mathbb{Q}_p}$ implies that the group of $\mathbb{Q}_p$-homomorphisms $\underline{K}^\times \to \underline{F}^\times$ has as a basis the maps $[\tau] : \underline{K}^\times \to \underline{F}^\times$ induced by the $\mathbb{Q}_p$-algebra homomorphisms $\tau : K \to F$. Hence, it is enough to extend each $[\tau] : \mathcal{O}_K^\times \to F^\times$ to a crystalline representation $\mathrm{Gal}(K^{\mathrm{ab}}/K) \to F^\times$. In fact, it is enough to treat the case $F = K$ and τ the identity map, as then applying $F \otimes_{\tau, K} (\cdot)$ gives the desired result for $[\tau]$ (since the crystalline condition only depends on the underlying $\mathbb{Q}_p$-linear representation space).

Consider a Lubin–Tate formal group G_π over $\mathcal{O}_K$ attached to a uniformizer π of $\mathcal{O}_K$. The representation $\rho_\pi = V_p(G_\pi)$ defines a homomorphism $\mathrm{Gal}(K^{\mathrm{ab}}/K) \to \mathcal{O}_K^\times$ whose composition with r_K restricts to inversion on $\mathcal{O}_K^\times$ [**103**, §3.4, Thm. 3(c)], so its K-linear dual extends the identity map on $\mathcal{O}_K^\times$. In view of the link between connected p-divisible groups and commutative formal groups over $\mathcal{O}_K$ on which $[p]$ is an isogeny (see Example 1.4.3.6), ρ_π arises from the generic fiber of a p-divisible group over $\mathcal{O}_K$ and hence is crystalline [**43**, 6.2]. □

3.7.8. PROOF OF PROPOSITION 3.7.4. Part (2) is clearly a consequence of part (1) (by using a schematic closure argument with finite group schemes to construct C), so we may and do focus on part (1). As a preliminary step, we check it suffices to show that the continuous homomorphisms $\rho_1, \rho_2 : \mathrm{Gal}(K^{\mathrm{ab}}/K) \rightrightarrows F^\times$ coincide on an open subgroup U of the abelianized Galois group.

Grant equality on such a U, and let K'/K be the finite totally ramified abelian extension corresponding to U, so the p-divisible groups $(\mathcal{Y}_1)_{\mathcal{O}_{K'}}$ and $(\mathcal{Y}_2)_{\mathcal{O}_{K'}}$ have F-linearly isogenous generic fibers. Hence, Tate's theorem on extending homomorphisms of p-divisible groups (Theorem 1.4.5.1) provides an F-linear isogeny

$$\beta' : (\mathcal{Y}_1)_{\mathcal{O}_{K'}} \to (\mathcal{Y}_2)_{\mathcal{O}_{K'}}$$

between p-divisible groups over $\mathcal{O}_{K'}$. If $\mathcal{O}_F$ lies inside $\mathrm{End}(\mathcal{Y}_1)$ and $\mathrm{End}(\mathcal{Y}_2)$ then the generic fiber Tate modules over K' are $\mathcal{O}_F$-linearly isomorphic (not merely isogenous), so in such cases β' may be chosen to be an isomorphism. By Lemma 3.7.6 the isogeny β' descends to an F-linear isogeny β from $\mathcal{Y}_1$ to $\mathcal{Y}_2$ over R, with β an isomorphism when β' is an isomorphism.

It remains to check the desired equality of $F^\times$-valued characters on an open subgroup of $\mathrm{Gal}(K^{\mathrm{ab}}/K)$. For this purpose it is harmless to replace K with a finite extension. It is also harmless to change each $\mathcal{Y}_i$ by an F-linear isogeny, so we may assume that each decomposes as a direct product according to the primitive idempotents of F. Hence, by treating the factor fields of F separately, we may assume F is a field. By replacing K with a finite extension, we may assume that K contains the Galois closure of F over $\mathbb{Q}_p$ inside $\overline{\mathbb{Q}}_p \subset \overline{K}$.

Each $\mathbb{C}_K$-semilinear representation $\mathbb{C}_K \otimes_{\mathbb{Q}_p} \rho_i$ is invertible as a module over $\mathbb{C}_K \otimes_{\mathbb{Q}_p} F = \prod_{\tau:F\to\overline{\mathbb{Q}}_p} \mathbb{C}_K(\tau)$ with $\mathbb{C}_K(\tau)$ a copy of $\mathbb{C}_K$ on which F acts through the $\mathbb{Q}_p$-embedding τ. By naturality of the Hodge–Tate decomposition of each $\mathbb{C}_K \otimes_{\mathbb{Q}_p} \rho_i$ into parts with Hodge–Tate weights 0 and 1 (via Theorem 1.4.5.1), the weight-1 part corresponds to the embeddings $\tau : F \to \overline{\mathbb{Q}}_p$ in the CM type for F on $\mathrm{Lie}(\mathcal{Y}_i)[1/p]_{\overline{\mathbb{Q}}_p}$. Since $\mathcal{Y}_1$ and $\mathcal{Y}_2$ each have the same CM type (F, Φ), it follows from [**105**, Ch. III, A.5, Thm. 2(i)⇔(iii)] that the p-adic Galois representations ρ_i are given by the same formula in terms of (F, Φ) on an open subgroup of $\mathrm{Gal}(K^{\mathrm{ab}}/K)$. □

The preceding proof used the fact that the Galois representation arising from a CM p-divisible group over R (in case of algebraically closed residue field) is given on an open subgroup of $\mathrm{Gal}(K^{\mathrm{ab}}/K)$ by an explicit formula in terms of the p-adic CM type. This can be strengthened to a formula in terms of the p-adic CM type on the entirety of $\mathrm{Gal}(\overline{K}/K)$:

3.7.9. Corollary. *Let R and F be as in Proposition 3.7.4, and let $\mathcal{Y}$ be a CM p-divisible group over R of CM type (F, Φ). The representation $\rho : \mathrm{Gal}(\overline{K}/K) \to F^\times$ attached to $\mathcal{Y}$ is equal to the composition*

$$\mathrm{Gal}(\overline{K}/K)^{\mathrm{ab}} \longrightarrow I'_E \xleftarrow[\simeq]{r_E} \mathcal{O}_E^\times \xrightarrow{1/\mathrm{N}\mu_\Phi} F^\times ,$$

where

- *$E = E(F, \Phi) \subset K$ is the reflex field attached to the p-adic CM type (F, Φ),*
- *I'_E is the inertia subgroup of $\mathrm{Gal}(E^{\mathrm{ab}}/E)$,*
- *r_E is the arithmetically normalized local Artin map,*
- *$\mathrm{N}\mu_\Phi : \mathcal{O}_E^\times \to \mathcal{O}_F^\times$ is the restriction to $\mathcal{O}_E^\times \subset E^\times$ of the reflex norm attached to the CM type (F, Φ).*

The containment $E \subset K$ was noted immediately after Definition 3.7.2.3, and the reflex norm was defined in 3.7.2.5 (using a formula motivated by the alternative description of the reflex norm in the classical global CM setting in 2.1.3.4).

PROOF. We first wish to reduce to treating the factor fields F_i of F separately, but this requires paying attention to the distinction between $E = E(F, \Phi)$ and its subfield $E_i = E(F_i, \Phi_i)$. In general, if $E' \subset E$ is a subfield over $\mathbb{Q}_p$ then the functoriality of local class field theory gives the commutativity of the diagram

$$\begin{array}{ccc} \mathcal{O}_E^\times & \xrightarrow[\simeq]{r_E} & I'_E \\ {\scriptstyle \mathrm{Nm}_{E/E'}}\downarrow & & \downarrow{\scriptstyle \mathrm{res}} \\ \mathcal{O}_{E'}^\times & \xrightarrow[r_{E'}]{\simeq} & I'_{E'} \end{array}$$

Also, the i_0th component of reflex norm $\mathrm{N}\mu_\Phi : \mathcal{O}_E^\times \to F^\times = \prod F_i^\times$ is the composition of $\mathrm{Nm}_{E/E_{i_0}}$ with the reflex norm $\mathrm{N}\mu_{\Phi_{i_0}}$ (as is easily verified from the definitions), so it suffices to work separately with each (F_i, Φ_i). That is, we may and do assume F is a field.

Choose a CM field L and the unique p-adic place v of L^+ as in Lemma 3.7.5 for F. Let $\widetilde{v}$ be one of the two places of L above v, let $\wp_{\widetilde{v}}$ be the maximal ideal of $\mathcal{O}_L$ corresponding to $\widetilde{v}$, and fix a $\mathbb{Q}_p$-isomorphism γ from the $\widetilde{v}$-adic completion of L to the given p-adic field F. We choose and fix a $\mathbb{Q}_p$-embedding $F \hookrightarrow \overline{\mathbb{Q}}_p$, so $L \subset \overline{\mathbb{Q}}_p$ in this way.

Since v is split in L, so complex conjugation on L swaps the two places of L over v, we can find a $\overline{\mathbb{Q}}_p$-valued CM type Ψ of L which induces the given p-adic CM type Φ of F through the chosen $\mathbb{Q}_p$-isomorphism $\gamma\colon L_{\widetilde{v}} \simeq F$. For the CM reflex field $E(L,\Psi) \subset \overline{\mathbb{Q}} \subset \overline{\mathbb{Q}}_p$, the embedding $E(L,\Psi) \hookrightarrow \overline{\mathbb{Q}}_p$ defines a p-adic place w of $E(L,\Psi)$. The w-adic completion of $E(L,\Psi)$ inside $\overline{\mathbb{Q}}_p$ is the reflex field E of the p-adic CM type Φ since Ψ induces Φ and is *uniquely determined* by Φ (because Ψ is a CM type and L^+ has v as its unique p-adic place, with complex conjugation swapping the two p-adic places of L).

Let the subfield $\mathcal{K} \subset \overline{\mathbb{Q}} \subset \overline{\mathbb{Q}}_p$ be the field of moduli of the $L^\times$-valued reflex norm N_Ψ as defined in A.4.6. This is an *everywhere unramified* abelian extension of $E(L,\Psi)$. Corollary A.4.6.5 (with N there equal to p here) provides an $\mathcal{O}_L$-linear abelian variety $(A, \mathcal{O}_L \to \mathrm{End}(A))$ over $\mathcal{K}$ with CM type Ψ and good reduction at all p-adic places (in particular, all places above w). Pick a place $\widetilde{w}$ of $\mathcal{K}$ above w, so the $\widetilde{w}$-adic completion $\mathcal{K}_{\widetilde{w}}$ of $\mathcal{K}$ is an unramified extension of $E(L,\Psi)_w = E$.

Let $\mathscr{A}$ be the abelian scheme over the ring of integers $\mathcal{O}_{\mathcal{K}_{\widetilde{w}}}$ of $\mathcal{K}_{\widetilde{w}}$ that extends $A_{\mathcal{K}_{\widetilde{w}}}$. The CM p-divisible group $\mathscr{A}[\wp_{\widetilde{v}}^\infty]$ over $\mathcal{O}_{\mathcal{K}_{\widetilde{w}}}$ is defined to be the direct factor of $\mathscr{A}[p^\infty]$ with respect to the primitive idempotent of $\mathbb{Z}_p \otimes_{\mathbb{Z}} \mathcal{O}_L$ associated to $\widetilde{v}$, and it is an $\mathcal{O}_F$-linear p-divisible group via the isomorphism $F \simeq L_{\widetilde{v}}$ chosen above; as such, it has CM type Φ due to how Ψ was constructed.

The classical theory of complex multiplication (see Theorem 2.5.1 and Lemma 2.4.8(ii)) tells us that the restriction to the inertia subgroup $I_{\mathcal{K}_{\widetilde{w}}} \subset \mathrm{Gal}(\mathcal{K}_{\widetilde{w}}^{\mathrm{ab}}/\mathcal{K}_{\widetilde{w}})$ of the Galois representation attached to the $\widetilde{v}$-adic part of $V_p(A_{\mathcal{K}_{\widetilde{w}}})$ agrees on an open subgroup with the *reciprocal* of the composition

$$I'_{\mathcal{K}_{\widetilde{w}}} \longrightarrow I'_E \xleftarrow[r_E]{\simeq} \mathcal{O}_E^\times \xrightarrow{\mathrm{N}\mu_\Phi} F^\times .$$

By Proposition 3.7.4(2), the restriction of this reciprocal p-adic Galois representation to the abelianized Galois group $\mathrm{Gal}(K^{\mathrm{ab}}/K)$ agrees on an open subgroup with the $F^\times$-valued Galois representation attached to the given p-divisible group $\mathcal{Y}$ over R since $\mathcal{Y}$ also has CM type (F,Φ). By Lemma 3.7.6, these representations must therefore coincide on the entirety of $\mathrm{Gal}(K^{\mathrm{ab}}/K)$.

It now suffices to show that the reciprocal of the algebraic homomorphism $\mathcal{K}_{\widetilde{w}}^\times \to F^\times$ induced by $\mathrm{N}\mu_\Phi \circ \mathrm{Nm}_{\mathcal{K}_{\widetilde{w}}/E}$ agrees on the entirety of $\mathcal{O}_{\mathcal{K}_{\widetilde{w}}}^\times$ (and not just on an open subgroup) with the inertial restriction of the Galois representation $\mathrm{Gal}(\mathcal{K}_{\widetilde{w}}^{\mathrm{ab}}/\mathcal{K}_{\widetilde{w}}) \to F^\times$ defined by $V_p(\mathscr{A}[\wp_{\widetilde{v}}^\infty])$. Since $\mathscr{A}[\wp_{\widetilde{v}}^\infty]$ is a p-divisible group over $\mathcal{O}_{\mathcal{K}_{\widetilde{w}}}$, its generic fiber Galois representation is crystalline [**43**, 6.2]. Hence, we can conclude by applying Lemma 3.7.7 to the p-adic field $K = \mathcal{K}_{\widetilde{w}}$ and the homomorphism of $\mathbb{Q}_p$-tori $1/(\mathrm{N}\mu_\Phi \circ \mathrm{Nm}_{\mathcal{K}_{\widetilde{w}}/E})$. □

3.7.10. Proof of Proposition 3.7.3. We first establish part (2), granting part (1). For every unramified continuous homomorphism $\chi : \mathrm{Gal}(\overline{E}/E) \to \mathcal{O}_F^\times$, by Corollary 1.4.5.2 there is a (unique) $\mathcal{O}_F$-linear p-divisible group $\mathcal{Y}(\chi)$ over $\mathcal{O}_E$ whose associated Galois representation $\rho_{\mathcal{Y}(\chi)}$ is the χ-twist of $\rho_{\mathcal{Y}}$. The abelianization $\mathrm{Gal}(E^{\mathrm{ab}}/E)$ of $\mathrm{Gal}(\overline{E}/E)$ is the direct product of its inertia subgroup and its unramified quotient, so there exists an unramified continuous homomorphism

$\chi : \mathrm{Gal}(\overline{E}/E) \to \mathcal{O}_F^\times$ such that

$$(\rho_{\mathcal{Y}} \cdot \chi)(I_E) = (\rho_{\mathcal{Y}} \cdot \chi)(\mathrm{Gal}(\overline{E}/E))$$

Thus, (2) is proved by taking $\mathcal{Y}' = \mathcal{Y}(\chi)$.

To prove part (1), we may assume that F is a field. As in the proof of Corollary 3.7.9, we can find a CM field L such that there is a unique place $v|p$ on L^+ and it splits in L, and a $\overline{\mathbb{Q}}_p$-valued CM type Ψ on L such that $F \simeq L_v^+ = L_{\widetilde{v}}$ (for a place $\widetilde{v}$ of L over v) and the p-adic CM type of F thereby induced by Ψ is Φ. Let w be the p-adic place on the global reflex field $E(L,\Psi) \subset \overline{\mathbb{Q}}_p$ induced by the valuation on $\overline{\mathbb{Q}}_p$, so its completion $E(L,\Psi)_w \subset \overline{\mathbb{Q}}_p$ is the p-adic reflex field $E = E(F,\Phi)$.

We also saw in the proof of Corollary 3.7.9 that there is a subextension $\mathcal{K}$ of the Hilbert class field of $E(L,\Psi)$ and an $\mathcal{O}_L$-linear abelian variety A over $\mathcal{K}$ with CM type (L,Ψ) and good reduction at all places over w. Pick a place $\widetilde{w}$ of $\mathcal{K}$ above w. The $\widetilde{w}$-adic completion $\mathcal{K}_{\widetilde{w}}$ of $\mathcal{K}$ is a unramified extension of $E(L,\Psi)_w = E$. Let $\mathscr{A}$ be the abelian scheme over the ring of integers $\mathcal{O}_{\mathcal{K}_{\widetilde{w}}}$ of $\mathcal{K}_{\widetilde{w}}$ that extends $A_{\mathcal{K}_{\widetilde{w}}}$, so the CM p-divisible group $\mathscr{A}[\wp_{\widetilde{v}}^\infty]$ over $\mathcal{O}_{\mathcal{K}_{\widetilde{w}}}$ is an $\mathcal{O}_F$-linear p-divisible group with p-adic CM type (F,Φ).

By Corollary 3.7.9, the abelian representation $\mathrm{Gal}(\mathcal{K}_{\widetilde{w}}^{\mathrm{ab}}/\mathcal{K}_{\widetilde{w}}) \to F^\times$ arising from $\mathscr{A}[\wp_{\widetilde{v}}^\infty]$ has inertial restriction which is given on $\mathcal{O}_{\mathcal{K}_{\widetilde{w}}}^\times$ (via the local reciprocity map for $\mathcal{K}_{\widetilde{w}}$) by the composition

$$\mathcal{O}_{\mathcal{K}_{\widetilde{w}}}^\times \xrightarrow{\mathrm{Nm}_{\mathcal{K}_{\widetilde{w}}/E}} \mathcal{O}_E^\times \xrightarrow{1/\mathrm{N}\mu_\Phi} F^\times.$$

A section of $\mathrm{Gal}(E^{\mathrm{ab}}/E) \twoheadrightarrow \mathrm{Gal}(\overline{\mathbb{F}}_p/\kappa_E)$ defines an identification

$$\mathrm{Gal}(E^{\mathrm{ab}}/E) \simeq \mathcal{O}_E^\times \times \mathrm{Gal}(\overline{\mathbb{F}}_p/\kappa_E),$$

and we use this to define a continuous homomorphism $\rho : \mathrm{Gal}(E^{\mathrm{ab}}/E) \to F^\times$ to be trivial on $\mathrm{Gal}(\overline{\mathbb{F}}_p/\kappa_E)$ and to be the reciprocal $1/\mathrm{N}\mu_\Phi$ of the reflex norm on $\mathcal{O}_E^\times$.

Viewing ρ as a representation of $\mathrm{Gal}(\overline{\mathbb{Q}}_p/E)$ into $\mathcal{O}_F^\times$, its inertial restriction is the same as that of $V_p(A_{\mathcal{K}_{\widetilde{w}}}[\wp_{\widetilde{v}}^\infty])$ since $\mathcal{K}_{\widetilde{w}}/E$ is unramified. Hence, since $A_{\mathcal{K}_{\widetilde{w}}}[\wp_{\widetilde{v}}^\infty]$ extends to a p-divisible group over $\mathcal{O}_{\mathcal{K}_{\widetilde{w}}}$, by Corollary 1.4.5.2 it follows that ρ arises from a $\mathcal{O}_F$-linear CM p-divisible group $\mathcal{Y}$ over $\mathcal{O}_E$. The CM type of $\mathcal{Y}$ is Φ, and $\mathcal{Y}$ and $\mathscr{A}[\wp_{\widetilde{v}}^\infty]$ have the same pullback over the completion of the valuation ring of the common maximal unramified extension of $\mathcal{K}_{\widetilde{w}}$ and E. □

We now record a lifting result that will be deduced from the proof of Proposition 3.7.3 and strengthens the main work in 2.6. For instance, when κ is the residue field of the p-adic reflex field $E(F,\Phi)$, the CM p-divisible group produced in the following lifting result is defined over $E(F,\Phi)$ rather than over some totally ramified extension of this reflex field. However the construction of this lift uses CM abelian varieties over $\mathbb{C}$, because the proof of Proposition 3.7.4 uses such abelian varieties.

3.7.11. Proposition. *Let Y be a p-divisible group of height $h > 0$ over a finite field κ of characteristic p, and let $F \subset \mathrm{End}^0(Y)$ be a commutative semisimple $\mathbb{Q}_p$-subalgebra with $\mathbb{Q}_p$-degree h. Let Φ be a p-adic CM type of F that is compatible with the F-action on Y in the sense of Definition 3.7.2.2, and let $E' \subset \overline{\mathbb{Q}}_p$ be a finite extension of $\mathbb{Q}_p$ containing the reflex field $E = E(F,\Phi)$ such that the residue field κ' of E' admits an embedding $\kappa \hookrightarrow \kappa'$; fix such an embedding.*

There exists an $\mathscr{O}_F$-linear p-divisible group $\mathcal{Y}$ over $\mathscr{O}_{E'}$ of CM *type* (F, Φ) *such that its special fiber is F-linearly isogenous to $Y_{\kappa'}$.*

PROOF. We may and do assume that F is a field. In the proof of Proposition 3.7.3, for a given p-adic CM type Φ of F we used classical CM theory for abelian varieties and an existence result for algebraic Hecke characters to construct an $\mathscr{O}_F$-linear p-divisible group $\mathcal{X} := \mathscr{A}[\wp_{\tilde{v}}^{\infty}]$ of CM type (F, Φ) over the ring of integers of a finite unramified extension field $\mathcal{K}/E$ inside $\overline{\mathbb{Q}}_p$ (where $\mathscr{A}$ is an abelian scheme over $\mathscr{O}_{\mathcal{K}}$ built using classical CM theory over number fields).

Let $K' := \mathcal{K} \cdot E' \subset \overline{\mathbb{Q}}_p$ be the compositum of $\mathcal{K}$ with E' over E, so K' is a finite unramified extension of E'. Let $\overline{\mathbb{F}}_p$ denote the residue field of $\overline{\mathbb{Q}}_p$. For every continuous unramified character

$$\chi\colon \mathrm{Gal}(K'^{\mathrm{ab}}/K') \twoheadrightarrow \mathrm{Gal}(\overline{\mathbb{F}}_p/\kappa_{K'}) \longrightarrow \mathscr{O}_F^{\times},$$

the twist $\mathcal{X}_{K'}(\chi)$ of $\mathcal{X}_{K'}$ by χ is a p-divisible group with the following properties:

- the p-adic Galois representations $\rho_{\mathcal{X}_{K'}}$ and $\rho_{\mathcal{X}_{K'}}(\chi)$ attached to the generic fibers are related by
$$\rho_{\mathcal{X}_{K'}(\chi)} = \rho_{\mathcal{X}_{K'}} \cdot \chi,$$
- the $q_{K'}$-Frobenius endomorphisms $\mathrm{Fr}_{\mathcal{X}_{K'}}$ and $\mathrm{Fr}_{\mathcal{X}_{K'}(\chi)}$ of the closed fibers of these two $\mathscr{O}_F$-linear p-divisible groups over $\mathscr{O}_{K'}$ are related by the equality
$$\mathrm{Fr}_{\mathcal{X}_{K'}(\chi)} = \mathrm{Fr}_{\mathcal{X}_{K'}} \cdot \chi(\mathrm{Fr}_{\kappa_{K'}})$$
inside $\mathscr{O}_F$, where $\mathrm{Fr}_{\kappa_{K'}}$ is the arithmetic $q_{K'}$-Frobenius automorphism of the residue field of $\overline{\mathbb{Q}}_p$.

By Corollary 1.4.5.2, an extension ρ of

$$\rho_{\mathcal{X}_{K'}(\chi)}\colon \mathrm{Gal}(K'^{\mathrm{ab}}/K') \to \mathscr{O}_F^{\times}$$

to $\mathrm{Gal}(E'^{\mathrm{ab}}/E')$, if one exists, defines an $\mathscr{O}_F$-linear descent $\mathcal{Y}_\rho$ over $\mathscr{O}_{E'}$ of the p-divisible group $\mathcal{X}_{K'}(\chi)$ over $\mathscr{O}_{K'}$. Granting the existence of ρ (which we will construct later), we now wish to compute the element of $F^\times$ that gives the action of the $q_{E'}$-Frobenius endomorphism $\mathrm{Fr}_{Y_\rho/\kappa_{E'}}$ of the closed fiber Y_ρ over $\kappa_{E'}$. We claim that

$$\mathrm{Fr}_{Y_\rho/\kappa_{E'}} = ((\rho \circ r_{E'})\cdot(\mathrm{N}\mu_\Phi \circ \mathrm{Nm}_{E'/E}))(\pi_{E'})$$

for any generator $\pi_{E'}$ of the maximal ideal of $\mathscr{O}_{E'}$, where $r_{E'}\colon (E')^\times \to \mathrm{Gal}(E'^{\mathrm{ab}}/E')$ is the arithmetically normalized local Artin map. This expression for $\mathrm{Fr}_{Y_\rho/\kappa_{E'}}$ is unambiguous because the abelian character

$$(\rho \circ r_{E'})\cdot(\mathrm{N}\mu_\Phi \circ \mathrm{Nm}_{E'/E})\colon (E')^\times \to F^\times$$

is unramified (by Corollary 3.7.9). Such a formula for the $q_{E'}$-Frobenius operation in terms of the generic fiber crystalline Galois representation and its "algebraic part" is proved in general in [**26**, B.4, B.5]. (Since the algebraic part is determined by the inertial restriction, $\rho \circ r_{E'}$ has algebraic part that is the reciprocal of $\mathrm{N}\mu_\Phi \circ \mathrm{Nm}_{E'/E}$ due to how $\mathscr{A}$ is constructed from a CM abelian variety over a number field.)

We use local class field theory to identify the character $\rho_{\mathcal{X}_{K'}}$ of $\mathrm{Gal}(K'^{\mathrm{ab}}/K')$ with a character $\rho_{K'}\colon (K')^\times \to F^\times$ of $(K')^\times$. We know from the construction of $\mathcal{X}$

as the p-divisible group of an abelian scheme $\mathscr{A}$ over $\mathcal{O}_K$ that (as in the proof of Corollary 3.7.9) $\rho_{_{K'}}$ coincides with $1/(\mathrm{N}\mu_\Phi \circ \mathrm{Nm}_{K'/E})$ on $\mathcal{O}_{K'}^\times$:

$$\rho_{_{K'}}|_{\mathcal{O}_{K'}^\times} = 1/\left(\mathrm{N}\mu_\Phi \circ \mathrm{Nm}_{K'/E}\right)|_{\mathcal{O}_{K'}^\times}.$$

Since $\ker(\mathrm{Nm}_{K'/E'}\colon (K')^\times \to (E')^\times)$ is a closed subgroup of $\mathcal{O}_{K'}^\times$, the homomorphism $\rho_{_{K'}}$ factors through $\mathrm{Nm}_{K'/E'}$.

Let U be the image of $\mathrm{Nm}_{K'/E'} : (K')^\times \to (E')^\times$, and let $\rho\colon U \to F^\times$ be the continuous character such that $\rho_{_{K'}} = \rho \circ \mathrm{Nm}_{K'/E'}$. Note that $U \supset \mathcal{O}_{E'}^\times$ since K'/E' is unramified, so U has finite index in $(E')^\times$. The $(\#\kappa)$-Frobenius endomorphism $\mathrm{Fr}_{Y/\kappa}$ for the F-linear p-divisible group Y is an element of $F^\times$. Suppose we can find an unramified character $\chi\colon U \to \mathcal{O}_F^\times$ such that $\rho \cdot \chi : U \to \mathcal{O}_F^\times$ extends to a character $\widetilde{\rho} : (E')^\times \to F^\times$ with the property that

$$(3.7.11.1) \qquad (\widetilde{\rho} \cdot (\mathrm{N}\mu_\Phi \circ \mathrm{Nm}_{E'/E}))(\pi_{_{E'}}) = (\mathrm{Fr}_{Y/\kappa})^{[\kappa':\kappa]}.$$

This character $\widetilde{\rho}$ agrees on the inertia subgroup of $\mathrm{Gal}(\overline{\mathbb{Q}}_p/E')$ with the restriction of $\rho_{K'}$ (recall that K'/E' is unramified). The representation $\rho_{K'}$ is the Galois representation attached to the generic fiber of a p-divisible group $\mathcal{X}$ over $\mathcal{O}_{K'}$, so by Corollary 1.4.5.2 such a $\widetilde{\rho}$ would have to arise from an $\mathcal{O}_F$-linear p-divisible group over $\mathcal{O}_{E'}$ and for $q' = \#\kappa'$ the q'-Frobenius endomorphism of its special fiber would have to equal the element of $F^\times$ given by the left side of (3.7.11.1) (again, by [**26**, B.4, B.5]). The right side of (3.7.11.1) shows that this coincides (as an element of $F^\times$) with the q'-Frobenius endomorphism of the $\mathcal{O}_{F'}$-linear p-divisible group $Y_{\kappa'}$. We would then be done, due to:

3.7.12. Lemma. *Let F be a field of finite degree over $\mathbb{Q}_p$. An F-linear* CM *p-divisible group $X \neq 0$ over a finite field κ of size q is determined up to F-linear isogeny by the element of $F^\times$ that gives the effect of the q-Frobenius endomorphism.*

PROOF. The isogeny category of p-divisible groups over κ is not semisimple, but we claim X is nonetheless isogenous to a power of a simple object. To prove this, let X_0 be a simple quotient of X in the isogeny category, so $\mathrm{M}^*(X_0)[1/p]$ is a simple subobject of $\mathrm{M}^*(X)[1/p]$ in the isogeny category of Dieudonné modules over κ. The resulting F-linear map $F \otimes_{\mathbb{Q}_p} \mathrm{M}^*(X_0)[1/p] \to \mathrm{M}^*(X)[1/p]$ has image that corresponds to a non-zero F-linear quotient Y of X. But F is a field, so $F \to \mathrm{End}^0(Y)$ is injective and hence $[F : \mathbb{Q}_p] \leqslant \mathrm{height}(Y)$ by 3.7.1.1. Since $[F : \mathbb{Q}_p] = \mathrm{height}(X)$, we conclude that $Y = X$. Thus, in the isogeny category of p-divisible groups over κ, X is a quotient of $\mathcal{O}_F \otimes_{\mathbb{Z}_p} X_0$. But $\mathcal{O}_F \otimes_{\mathbb{Z}_p} X_0$ is a power of X_0, so X is isogenous to a power X_0^n by general nonsense with semisimplicity.

The action of the q-Frobenius on X_0 must satisfy an *irreducible* polynomial g in $\mathbb{Q}_p[T]$ (since the q-Frobenius is central in the endomorphism algebra), and this minimal polynomial is certainly not T. Hence, by Theorem A.1.2.1, $C_g := D_\kappa[1/p]/D_\kappa[1/p]g(\mathcal{F}^{[\kappa:\mathbb{F}_p]})$ is a central simple algebra over the p-adic field generated by the q-Frobenius. Any central simple algebra over a field has a unique simple left module up to isomorphism, and in the case of C_g this simple left module is $\mathrm{M}^*(X_0)[1/p]$. Thus, X_0 is determined up to κ-isogeny in terms of g,

The multiplicity $n = \mathrm{height}(X)/\mathrm{height}(X_0) = [F : \mathbb{Q}_p]/\mathrm{height}(X_0)$ of X_0 in X (up to κ-isogeny) is determined by g (and F). Since the minimal polynomial over $\mathbb{Q}_p$ for the q-Frobenius endomorphism of X is the same as the minimal polynomial over

$\mathbb{Q}_p$ for the corresponding element of F, from the existence of the CM structure by F and the knowledge of the q-Frobenius element in $F^\times$ we can determine $\mathrm{M}^*(X)[1/p]$ up to isomorphism as a left $D_\kappa[1/p]$-module.

Consider another F-linear CM p-divisible group X' over κ with the same q-Frobenius in F, so $\mathrm{M}^*(X)[1/p]$ and $\mathrm{M}^*(X')[1/p]$ are isomorphic as $D_\kappa[1/p]$-modules compatibly with the q-Frobenius endomorphisms. This provides an isogeny $X \to X'$ as p-divisible groups over κ, but it may not be F-linear. However, the resulting isomorphism between the endomorphism algebras does respect the q-Frobenius endomorphisms, and these algebras are central simple over the p-adic field generated by the common q-Frobenius element in F. Hence, we can apply the Skolem-Noether theorem to change the choice of isogeny (thereby changing the isomorphism of endomorphism algebras by a unit conjugation) in order that it becomes F-linear. □

At this point we are reduced to the task of extending a given $\mathcal{O}_F^\times$-valued character ρ on the finite-index open subgroup $U \subset (E')^\times$ to an $\mathcal{O}_F^\times$-valued character on $(E')^\times$ after applying a preliminary unramified $\mathcal{O}_F^\times$-valued twist. (The main point is to keep the values of our characters inside $\mathcal{O}_F^\times$.) Let $m = [(E')^\times : U] = [K' : E']$. Choose and fix a generator $\pi_{E'}$ of the maximal ideal of $\mathcal{O}_{E'}$. Clearly $\pi_{E'}^m \in U$, and $U/\mathcal{O}_{E'}^\times$ is generated by $\pi_{E'}^m$, so we can define $\chi : U \to \mathcal{O}_F^\times$ to be the unramified character such that

$$\chi(\pi_{E'}^m) = \left(\rho \cdot (\mathrm{N}\mu_\Phi \circ \mathrm{Nm}_{E'/E})\right)(\pi_{E'}^m)^{-1} \cdot \mathrm{Fr}_{Y/\kappa}^{[\kappa':\kappa]m}.$$

We claim that $(\rho\cdot(\mathrm{N}\mu_\Phi \circ \mathrm{Nm}_{E'/E}))(\pi_{E'}^m)$ and $\mathrm{Fr}_{Y/\kappa}^{[\kappa':\kappa]m}$ have the same p-adic valuation (so their ratio lies in $\mathcal{O}_F^\times$).

To see this, first recall that the slope of the CM p-divisible group Y over κ is $\#\Phi/[F : \mathbb{Q}_p]$ (by definition of what it means for the F-action to be compatible with Φ in 3.7.2.2), so for the residue field κ'' of K' of degree $[\kappa' : \kappa]m$ over κ the same holds for $Y_{\kappa''}$. Since $\pi_{E'}^m$ is an $\mathcal{O}_{E'}^\times$-multiple of $\mathrm{Nm}_{K'/E'}(\pi_{K'})$ (for any uniformizer $\pi_{K'}$ of K'), it follows by the definition of ρ and consideration of maximal compact subgroups that the element $(\rho \cdot (\mathrm{N}\mu_\Phi \circ \mathrm{Nm}_{E'/E}))(\pi_{E'}^m)$ of F is an $\mathcal{O}_F^\times$-multiple of $(\rho_{K'} \cdot (\mathrm{N}\mu_\Phi \circ \mathrm{Nm}_{K'/E}))(\pi_{K'})$. But this element of $F^\times$ computes the effect of the $(\#\kappa'')$-Frobenius endomorphism of the special fiber of $\mathcal{X}_{\mathcal{O}_{K'}}$, or equivalently of $\mathscr{A}_{\mathcal{O}_{K'}}$. This latter Frobenius endomorphism has slope $\#\Phi/[F : \mathbb{Q}_p]$ by Lemma 3.7.2.1, so the desired equality of p-adic valuations in F is proved.

Observe that $(E')^\times$ is an amalgamated product of U with the cyclic group $\langle \pi_{E'} \rangle := \pi_{E'}^{\mathbb{Z}}$ over their common cyclic subgroup $\langle \pi_{E'}^m \rangle := \pi_{E'}^{m\mathbb{Z}}$; that is,

$$(E')^\times = U \times^{\langle \pi_{E'}^m \rangle} \langle \pi_{E'} \rangle.$$

In other words, $(E')^\times$ is the quotient of $U \times \langle \pi_{E'} \rangle$ by the cyclic subgroup generated by $(\pi_{E'}^m, \pi_{E'}^{-m})$. The definition of the unramified character χ on U guarantees that the restriction of $\rho \cdot \chi$ to the cyclic subgroup $\langle \pi_{E'}^m \rangle$ coincides with the restriction to $\langle \pi_{E'}^m \rangle$ of the character on $\langle \pi_{E'} \rangle$ such that $\pi_{E'} \mapsto (\mathrm{N}\mu_\Phi \circ \mathrm{Nm}_{E'/E})(\pi_{E'})^{-1} \cdot \mathrm{Fr}_{Y/\kappa}^{[\kappa':\kappa]}$. Thus, these glue to define a character $\widetilde{\rho}$: $(E')^\times \to \mathcal{O}_F^\times$ of $(E')^\times$ such that

$$\left(\widetilde{\rho} \cdot (\mathrm{N}\mu_\Phi \circ \mathrm{Nm}_{E'/E})\right)(\pi_{E'}) = \mathrm{Fr}_{Y/\kappa}^{[\kappa':\kappa]}.$$

□

3.8. An upper bound for a field of definition

3.8.1. Let Z be a p-divisible group over $\overline{\mathbb{F}}_p$ with $h := \text{height}(Z) > 0$. As we noted in the proof of Lemma 3.1.8, $\text{End}^0(Z)$ is a finite-dimensional semisimple $\mathbb{Q}_p$-algebra. In fact, $\dim(\text{End}^0(Z)) \leqslant h^2$. To prove this, by multiplicativity and the Dieudonné–Manin classification (Theorem 3.1.3) it suffices to treat the case $Z = \mathcal{G}_\lambda$ for $\lambda \in \mathbb{Q} \bigcap [0,1]$. The cases λ and $1-\lambda$ are interchangeable via duality, and the case $\lambda = 0$ is obvious, so we can assume $0 < \lambda < 1$. Writing $\lambda = r/s$ with $1 \leqslant r < s$ and $\gcd(r,s) = 1$, the height h is equal to s and the computations in Example 3.6.3 show that $\text{End}^0(\mathcal{G}_\lambda)$ is a central division algebra over $\mathbb{Q}_p$ of rank s^2.

It is a classical fact that there are only finitely many extensions of $\mathbb{Q}_p$ of a given degree up to isomorphism, due to finiteness of the residue field, Krasner's Lemma, and the compactness of the space of Eisenstein polynomials over a p-adic field (all of which are separable). Thus, up to isomorphism there are only a finite number of commutative semisimple $\mathbb{Q}_p$-algebras that can be embedded in $\text{End}^0(Z)$; this is bounded in terms of h (and p), since $\dim_{\mathbb{Q}_p}(\text{End}^0(Z)) \leqslant h^2$.

We just saw that each simple factor of $\text{End}^0(Z)$ is a central simple $\mathbb{Q}_p$-algebra (by computation of the endomorphism algebras $\text{End}^0(\mathcal{G}_\lambda)$ for the simple objects in the isogeny category over $\overline{\mathbb{F}}_p$). Thus, every maximal commutative semisimple $\mathbb{Q}_p$-subalgebra of $\text{End}^0(Z)$ must be its own centralizer (and hence is a direct product of maximal commutative subfields of the simple factors of $\text{End}^0(Z)$). It follows from Proposition 1.4.3.9(2) and the Skolem-Noether theorem there are only a finite number of $\text{End}^0(Z)^\times$-conjugacy classes of pairs

$$(F \subset \text{End}^0(Z), \Phi)$$

where F is a (necessarily maximal) commutative semisimple $\mathbb{Q}_p$-subalgebra of $\text{End}^0(Z)$ of $\mathbb{Q}_p$-degree h and Φ is a CM type of F compatible with Z (in the sense of Definition 3.7.2.2) and valued in a fixed algebraic closure $\overline{\mathbb{Q}}_p$ (whose residue field we identify with $\overline{\mathbb{F}}_p$).

Let $(F_j \hookrightarrow \text{End}^0(Z), \Phi_j)$ $(1 \leqslant j \leqslant N)$ be a complete set of representatives of these conjugacy classes. Let $E_j = E(F_j, \Phi_j) \subset \overline{\mathbb{Q}}_p$ be the reflex field of (F_j, Φ_j). For each j, let $\mathcal{Y}_j$ be an $\mathcal{O}_{F_j}$-linear p-divisible group over $\mathcal{O}_{E_j}$ such that

(i) the CM type of $\mathcal{Y}_j$ is (F_j, Φ_j),
(ii) the Galois representation $\rho_{\mathcal{Y}_j} : \text{Gal}(\overline{\mathbb{Q}}_p/E_j) \to \mathcal{O}_{F_j}^\times$ attached to $\mathcal{Y}_j$ carries the inertia subgroup I_{E_j} onto the entire image.

The existence of $\mathcal{Y}_j$ is guaranteed by Proposition 3.7.3. Denote by κ_j the residue field of E_j, and let Y_j be the closed fiber of $\mathcal{Y}_j$ over the residue field κ_j of $\mathcal{O}_{E_j}$.

Now we will deduce some bounds on finite fields of definition for CM p-divisible groups over $\overline{\mathbb{F}}_p$ that admit a CM lift to characteristic 0. Consider a p-divisible group Y over $\overline{\mathbb{F}}_p$ in the isogeny class of a fixed Z, and assume Y admits a CM lift $\mathcal{Y}$ over the ring of integers of a finite extension field K of $K_0 = W(\overline{\mathbb{F}}_p)[1/p]$. There exists an integer j with $1 \leqslant j \leqslant N$ such that the CM type of $\mathcal{Y}$ is (F_j, Φ_j). By Proposition 3.7.4, there exists a finite extension field E_j' of E_j, a finite flat subgroup scheme $C \subset (\mathcal{Y}_j)_{\mathcal{O}_{E_j'}}$ over $\mathcal{O}_{E_j'}$, and an F_j-linear isomorphism

$$\xi : ((\mathcal{Y}_j)_{\mathcal{O}_{E_j'}}/C)_{\mathcal{O}_{K'}} \simeq \mathcal{Y}_{\mathcal{O}_{K'}}$$

where $K' := E_j'K$ is a compositum of E_j' and K over $\mathbb{Q}_p$.

Denote by $\mathcal{K}_j$ the unique minimal subfield of E'_j over E_j over which the étale generic fiber C_η of C is defined relative to the E_j-structure $(\mathcal{Y}_j)_{E_j}$ on $(\mathcal{Y}_j)_{E'_j}$, so $\mathcal{K}_j$ is a subfield of the splitting field over E_j inside $\overline{\mathbb{Q}}_p$ of the group of geometric points of C_η (viewed as geometric points of a torsion-level subgroup of $(\mathcal{Y}_j)_{E_j}$). This splitting field is totally ramified over E_j because $\rho_{\mathcal{Y}_j}$ carries I_{E_j} onto the image of $\mathrm{Gal}(\overline{\mathbb{Q}}_p/E_j)$, so $\mathcal{K}_j$ is totally ramified over E_j.

Via schematic closure considerations over a Dedekind base we see that C descends to a finite flat subgroup scheme $C' \subset (\mathcal{Y}_j)_{\mathcal{O}_{\mathcal{K}_j}}$ over $\mathcal{O}_{\mathcal{K}_j}$. Let

$$\mathcal{Y}' := (\mathcal{Y}_j)_{\mathcal{O}_{\mathcal{K}_j}}/C';$$

this is a p-divisible group over $\mathcal{O}_{\mathcal{K}_j}$ of CM type (F_j, Φ_j). By construction, there is an F_j-linear $\mathcal{O}_{K'}$-isomorphism

$$\mathcal{Y}' \times_{\mathrm{Spec}(\mathcal{O}_{\mathcal{K}_j})} \mathrm{Spec}(\mathcal{O}_{K'}) \simeq \mathcal{Y} \times_{\mathrm{Spec}(\mathcal{O}_K)} \mathrm{Spec}(\mathcal{O}_{K'}).$$

In particular the closed fiber Y' of $\mathcal{Y}'$ is a p-divisible group over the residue field $\kappa_{\mathcal{K}_j}$ of $\mathcal{K}_j$ and its geometric fiber $Y'_{\overline{\mathbb{F}}_p}$ is isomorphic to Y.

As Y varies over all p-divisible groups over $\overline{\mathbb{F}}_p$ that are in the isogeny class of Z and admit a CM lift corresponding to a fixed j as above, the subfields $\mathcal{K}_j \subset E'_j$ vary. However, their residue fields $\kappa_{\mathcal{K}_j}$ remain constant. In fact, since we have shown that $\mathcal{K}_j$ is totally ramified over E_j, clearly $\kappa_{\mathcal{K}_j}$ is equal to the residue field κ_j of E_j. We summarize our conclusions in the following proposition.

3.8.2. Proposition. *Using notation as above, let κ be the compositum over $\mathbb{F}_p$ of the residue fields κ_j of the reflex fields $E_j = E(F_j, \Phi_j)$, $j = 1, \ldots, N$. Let Y be a p-divisible group over $\overline{\mathbb{F}}_p$ isogenous to Z.*

If Y admits a CM lifting to characteristic 0 then for some j there exists a totally ramified finite extension $\mathcal{K}_j$ of E_j and a CM lifting of Y over $\mathcal{O}_{\mathcal{K}_j}$ of CM type (F_j, Φ_j). In particular there exists a p-divisible group Y' over κ such that $Y'_{\overline{\mathbb{F}}_p}$ is isomorphic to Y.

We now give many counterexamples to (CML) and (R) from 1.8.5. To describe these, we require some preliminary hypotheses. Let X be a p-divisible group over a finite field $\kappa \subset \overline{\mathbb{F}}_p$ such that $X_{\overline{\mathbb{F}}_p}$ satisfies conditions (i) and (ii) in 3.4.3. (In every isogeny class of p-divisible groups over κ not of extended Lubin-Tate type, after making a preliminary finite extension on κ we see via Corollary 3.4.7 that there are p-divisible groups X which satisfy these conditions. In particular, we can make such examples in the isogeny class of any abelian variety that is not ordinary or almost ordinary, at the cost of a preliminary finite extension of the ground field. This answers [**93**, Question C] affirmatively.) Increase κ inside $\overline{\mathbb{F}}_p$ if necessary so that it contains the residue fields κ_j of the p-adic reflex fields of all pairs (F_j, Φ_j) arising from $Z = X_{\overline{\mathbb{F}}_p}$ in Proposition 3.8.2. Let $Q = Q(X_{\overline{\mathbb{F}}_p})$ be the integer as defined in 3.6.1. The following result is immediate from Propositions 3.6.2 and 3.8.2.

3.8.3. Theorem. *For X and κ as above, increase κ by a finite amount if necessary so $\mathrm{End}(X_{\overline{\mathbb{F}}_p}) \to \mathrm{End}(X_{\overline{\mathbb{F}}_p}[\mathcal{F}, \mathcal{V}])$ has the same image as $\mathrm{End}(X)$ (with $X[\mathcal{F}, \mathcal{V}]$ defined as in 3.1.11). Let $\mathbf{a} \subset X_{\overline{\mathbb{F}}_p}$ be a subgroup isomorphic to α_p such that*

$a(X_{\overline{\mathbb{F}}_p}/\mathbf{a}) = 1$ and the minimal field of definition κ' over κ for $\mathbf{a} \subset X_{\overline{\mathbb{F}}_p}$ has κ-degree not dividing Q. Let $\mathbf{a}_0 \subset X_{\kappa'}$ be the associated descent over κ'.

The quotient $Y = X_{\kappa'}/\mathbf{a}_0$ has no CM lifting to characteristic 0, and if $X = A[p^\infty]$ is the p-divisible group attached to an abelian variety A over κ then it suffices to impose the initial condition on endomorphism rings for $\mathrm{End}(A_{\overline{\mathbb{F}}_p})$ and $\mathrm{End}(A)$. In these latter cases, the abelian variety $A_{\kappa'}/\mathbf{a}_0$ does not admit a CM lift to characteristic 0, even after extension of scalars to an arbitrary field containing κ'.

Away from finitely many points in $\mathbb{P}^1_\kappa(\overline{\mathbb{F}}_p)$, there are many such choices of $\mathbf{a}$. We now illustrate Theorem 3.8.3 with two examples.

3.8.4. Consider the imaginary quadratic field $L := \mathbb{Q}(\sqrt{-7})$ that has class number 1. Let p be a prime number which is split in L, or equivalently $p \equiv 1, 2, 4 \pmod 7$. Let $\wp_1$ and $\wp_2$ be the two prime ideals in $\mathcal{O}_L$ above p. Because L has class number 1, there exists an element $\pi \in L^\times$ such that $\pi\mathcal{O}_L = \wp_1 \cdot \wp_2^2 = (p)\wp_2$.

By the product formula, π is a Weil p^3-integer, so by Honda-Tate theory in 1.6.2 there is a simple abelian variety B over $\mathbb{F}_{p^3}$ with π as its Weil number, and it has dimension 3. By Proposition 1.7.4.5, we may choose B such that $\mathcal{O}_L \subseteq \mathrm{End}(B)$, so the p-divisible group of B decomposes as the product of $B[\wp_1^\infty]$ and $B[\wp_2^\infty]$ with respective slopes 1/3 (of dimension 1 and height 3) and 2/3 (with height 3). The dual of $B[\wp_2^\infty]$ is isogenous to $B[\wp_1^\infty]$ and so has dimension 1. The 1-dimensionality properties ensure that the local-local $B[\wp_i^\infty]$ are each unique in their isogeny class over any extension of $\mathbb{F}_{p^3}$. We conclude that $a(B) = 2$ and the p-divisible groups $B[\wp_i^\infty]_{\overline{\mathbb{F}}_p}$ are simple and non-isogenous in the isogeny category over $\overline{\mathbb{F}}_p$. (This B is a *hypersymmetric* abelian variety in the sense of [**17**].) The endomorphism algebra $\mathrm{End}^0(B)$ is a 9-dimensional central division algebra over $L = \mathbb{Q}(\sqrt{-7})$.

For any CM structure F on $B[p^\infty]$ and compatible CM type Φ in the sense of Definition 3.7.2.2, (F, Φ) has the form $(F_1, \Phi_1) \coprod (F_2, \Phi_2)$, where F_1 and F_2 are cubic extension fields of $\mathbb{Q}_p$, $\#\Phi_1 = 1$, and $\#\Phi_2 = 2$. Thus, $E(F_j, \Phi_j) = F_j$ and by Example 3.6.3 (and the uniqueness of each $B[\wp_i^\infty]$ in its isogeny class) we can take $Q = p^3 - 1$ and $\kappa = \mathbb{F}_{p^3}$ in the statement of Theorem 3.8.3. Upon choosing an isomorphism $B[\mathcal{F}, \mathcal{V}] \simeq \alpha_p \times \alpha_p$ over κ, for any extension κ' of $\mathbb{F}_{p^3}$ not contained in $\mathbb{F}_{p^{3(p^3-1)}}$ and any $b_0 \in \mathbb{P}^1(\kappa')$ not in $\mathbb{P}^1(\mathbb{F}_{p^3})$ we conclude via Theorem 3.8.3 that the quotient $B_{\kappa'}/b_0(\alpha_p)$ does not admit a CM lifting to characteristic 0 (even after scalar extension of the ground field).

3.8.5. Here is an example that also gives an explicit answer to a 2-dimensional instance of [**93**, Question C]. The Weil p-number $\sqrt{-p}$ corresponds to a supersingular elliptic curve E' over $\mathbb{F}_p$, and $E'' := E'_{\mathbb{F}_{p^2}}$ has all geometric endomorphisms defined over $\mathbb{F}_{p^2}$. One shows by a calculation with Dieudonné modules (in the same style as Example 3.4.3.1) that for $X'' := (E'' \times E'')[p^\infty]$ and any $b \notin \mathbb{F}_{p^2}$, the corresponding α_p-quotient Y_b of $X''_{\kappa(b)}$ satisfies $a(Y_b) = 1$. Let $E = E'_\kappa$ where $\kappa = \mathbb{F}_{p^8}$. For $A = E \times E$, define $X = A[p^\infty]$ (so $X_{\overline{\mathbb{F}}_p}$ satisfies conditions (i) and (ii) in 3.4.3).

The reason for making a base change to κ is to ensure that the hypotheses in Theorem 3.8.3 are satisfied (especially that κ is "large enough"). To explain this, first observe that $\mathrm{End}^0(X) \simeq \mathrm{Mat}_2(\Delta)$ where $\Delta = \mathrm{End}^0(E)$ is the unique (up to isomorphism) quaternion division algebra over $\mathbb{Q}_p$. The maximal commutative semisimple $\mathbb{Q}_p$-subalgebras $F \subset \mathrm{Mat}_2(\Delta)$ have $\mathbb{Q}_p$-degree 4 since such F are in bijective correspondence with the maximal $\mathbb{Q}_p$-tori T in the associated unit group $G = \mathrm{GL}_2(\Delta)$ over $\mathbb{Q}_p$: F is the $\mathbb{Q}_p$-subalgebra generated by $T(\mathbb{Q}_p)$, and $T = \underline{F}^\times \subset G$. Clearly the number of factor fields of F is the rank of the split part of T, so since the maximal split $\mathbb{Q}_p$-tori in G have rank 2 we see that when F is not a field it is a product of exactly two fields. Since every maximal commutative subfield of a central simple algebra of rank d^2 has rank d, we conclude that either F is a quartic field extension or a product of two quadratic field extensions of $\mathbb{Q}_p$.

If $F = F_1 \times F_2$ for quadratic fields F_j over $\mathbb{Q}_p$, any p-adic CM type $\Phi = \Phi_1 \coprod \Phi_2$ on F has reflex fields $E(F_j, \Phi_j) \subset F_j$, so the residue fields κ_j are at most quadratic over $\mathbb{F}_p$. If instead F is a quartic field then we claim that its Galois closure F' over $\mathbb{Q}_p$ (which contains the p-adic reflex field of any p-adic CM type Φ on F) has $\mathbb{Q}_p$-degree dividing 8. It would then follow that the residue fields of the p-adic reflex fields $E(F, \Phi)$ are subfields of κ, as required for Theorem 3.8.3. (Note that $\mathrm{End}^0(A_{\overline{\mathbb{F}}_p}) = \mathrm{End}^0(A)$, since the same holds for for $E = E'_{\kappa}$.) To control the degree of the Galois closure F' over $\mathbb{Q}_p$ in a way that treats all p on an equal footing, we note that if $F' \neq F$ then $\mathrm{Gal}(F'/\mathbb{Q}_p)$ is a transitive subgroup of S_4 with size > 4, so it is either D_4, A_4, or S_4. For any finite Galois extension of non-archimedean local fields, the Galois group has a solvability series whose terms are normal in the entire group; i.e., such groups are "supersolvable", not just solvable. But A_4 is not supersolvable, so S_4 is not either. Hence, $[F' : \mathbb{Q}_p]$ divides 8 as claimed.

The next step is to compute the integer $Q = Q(A)$, or at least a non-zero multiple of it. The $\mathbb{Z}_p$-finite $\mathrm{End}(E)$ lies in the maximal order $\mathcal{O}_\Delta$ of Δ (in fact these are equal, but we do not need this). The unique maximal 2-sided ideal $\mathfrak{p}$ in $\mathcal{O}_\Delta$ must kill the unique order-p subgroup α_p inside $E'' = E'_{\mathbb{F}_{p^2}}$, so $\mathrm{End}(A) \subset \mathrm{Mat}_2(\mathcal{O}_\Delta)$ and its image in $\mathrm{End}(A[\mathcal{F}, \mathcal{V}])$ factors through the quotient $\mathrm{Mat}_2(\mathcal{O}_\Delta/\mathfrak{p}) = \mathrm{Mat}_2(\mathbb{F}_{p^2})$. Hence, the image of $\mathrm{Aut}(A)$ in $\mathrm{Aut}(A[\mathcal{F}, \mathcal{V}])$ is a quotient of a subgroup of $\mathrm{GL}_2(\mathbb{F}_{p^2})$ (in fact it is isomorphic to $\mathrm{GL}_2(\mathbb{F}_{p^2})$, but we do not need this), so Q divides the least common multiple of the orders of the elements of $\mathrm{GL}_2(\mathbb{F}_{p^2})$.

By the Jordan decomposition, the order of any element of $\mathrm{GL}_2(\mathbb{F}_{p^2})$ either divides $p^4 - 1$ or divides $p(p^2 - 1)$, so Q divides $p(p^4 - 1)$. It follows that any $b \notin \mathbb{F}_{p^{8 \cdot p(p^4-1)}}$ provides a p-divisible group isogenous to $X_{\overline{\mathbb{F}}_p}$ that does not admit a CM lift to characteristic 0 (even after increasing the ground field). Thus, for any finite extension κ' of $\mathbb{F}_{p^8}$ not contained in $\mathbb{F}_{p^{8 \cdot p(p^4-1)}}$ and any generator b of κ' over $\mathbb{F}_{p^8}$, we have constructed an α_p-quotient B of $A_{\kappa'}$ such that $B_{\overline{\mathbb{F}}_p}$ does not admit a CM lift to characteristic 0.

3.9. Appendix: algebraic abelian p-adic representations of local fields

3.9.1. Locally algebraic representations. Let R be a complete discrete valuation ring of mixed characteristic $(0, p)$ with a perfect residue field and fraction field K. A theorem of Tate, proved in [**105**, III, A.7] for the case when K is locally compact and in [**108**, §2.2, Thm. 4] for the general case, asserts that a continuous abelian semisimple representation ρ: $\mathrm{Gal}(K^{\mathrm{ab}}/K) \to \mathrm{GL}(V)$ on a finite-dimensional vector space V over $\mathbb{Q}_p$ is Hodge-Tate if and only if it is *locally algebraic*; i.e., there exists a finite extension K'/K inside $\overline{K}$, a subfield $E \subset K'$ of finite degree over $\mathbb{Q}_p$, an algebraic $\mathbb{Q}_p$-homomorphism $h : \underline{E}^\times = \mathrm{Res}_{E/\mathbb{Q}}(\mathbb{G}_\mathrm{m}) \longrightarrow \mathrm{GL}(V)$, and an open subgroup $U \subset \mathcal{O}_E^\times$ such that $r_E(u)$ lies in the image the restriction homomorphism

$\delta\colon \mathrm{Gal}(K'^{\mathrm{ab}}/K') \to \mathrm{Gal}(E^{\mathrm{ab}}/E)$ for all $x \in U$ and

$$\rho\left(\delta^{-1}(r_E(x))\right) = 1/h(x) \qquad \text{for all } x \in U,$$

where $r_E\colon E^\times \to \mathrm{Gal}(E^{\mathrm{ab}}/E)$ is the arithmetically normalized local Artin map and $\delta^{-1}(r_E(x))$ denotes the entire (non-empty) preimage of $r_E(x)$ in $\mathrm{Gal}(K'^{\mathrm{ab}}/K')$. It is a standard fact (proved by ramification considerations) that the closed image of the inertia subgroup of $\mathrm{Gal}(\overline{K}/K')$ in $\mathrm{Gal}(\overline{E}/E)$ is an open subgroup of the inertia subgroup of $\mathrm{Gal}(\overline{E}/E)$, so all sufficiently small U lie in the image of δ. Note also that if ρ is locally algebraic then the restriction of ρ to the inertia subgroup of $\mathrm{Gal}(K^{\mathrm{ab}}/K)$ is semisimple.

If a locally algebraic commutative representation ρ is simple then $\mathrm{End}_{(G,\rho)}(V)$ is a finite extension field F of $\mathbb{Q}_p$, V is a 1-dimensional vector space over F, and there exists a continuous homomorphism $\rho_1\colon \mathrm{Gal}(\overline{K}/K)^{\mathrm{ab}} \to F^\times$ such that ρ is the composition of ρ_1 with the injection $F^\times \subset \mathrm{GL}(V)$ coming from the action of F on V. Moreover there is a unique $\mathbb{Q}_p$-homomorphism $h_1\colon \underline{E}^\times \to \underline{F}^\times$ such that

$$\rho_1\left(\delta^{-1}(r_E(x))\right) = 1/h_1(x) \qquad \text{for all } x \in U,$$

with some U as above. We call h_1 *an algebraic part* of the locally algebraic ρ_1.

Note that E in general is not uniquely determined by ρ_1. However, as is explained at the end of A.4.2, there is a unique smallest subfield $E(\rho_1)$ of E over $\mathbb{Q}_p$ such that h_1 factors through $\mathrm{Nm}_{E/E(\rho_1)}$, so $h_1 = h(\rho) \circ \mathrm{Nm}_{E/E(\rho_1)}$ for a unique $\mathbb{Q}_p$-homomorphism $h(\rho)\colon \underline{E(\rho_1)}^\times \to \underline{F}^\times$. In view of the uniqueness of $E(\rho_1)$, it is unaffected by replacing E with a finite extension and hence is independent of the initial choice of $E/\mathbb{Q}_p$ inside $\overline{K}$. We may therefore denote it $E(\rho)$ and call $h(\rho)$ the *algebraic part* of ρ (or of ρ_1).

3.9.2. Remark. In the definition of the algebraic part of a locally algebraic homomorphism, we used the reciprocal of h_1 instead of h_1 itself. This was done for local-global compatibility in the following sense. Suppose that $\epsilon : \mathbb{A}_K^\times \to L^\times$ is an algebraic Hecke character for number fields K and L, $\chi : \underline{K}^\times \to \underline{L}^\times$ is the algebraic part of ϵ, and $\psi\colon \mathrm{Gal}(K^{\mathrm{ab}}/K) \to \mathbb{A}_L^\times$ is the adelic Galois representation attached to ϵ as in 2.4.7. If E is identified with the completion $K_\wp$ of K at a p-adic place $\wp$ of K, F is identified with the completion L_v of L at a p-adic place v of L, and ρ_1 is identified with the v-adic component $\mathrm{pr}_v \circ (\psi\,|_{\mathrm{Gal}(K_\wp^{\mathrm{ab}}/K_\wp)})$ of the restriction of ψ to the decomposition group at $\wp$, then the algebraic part h_1 of ρ_1 is identified (on $\mathbb{Q}_p$-points) with the composition

$$K_\wp^\times \longrightarrow (\mathbb{A}_{K,f})^\times \xrightarrow{\ \chi\ } (\mathbb{A}_{L,f})^\times \xrightarrow{\ \mathrm{pr}_v\ } L_v^\times$$

where the first map is the inclusion as the $\wp$-factor.

3.9.3. Let F be a non-zero finite-dimensional commutative semisimple $\mathbb{Q}_p$-algebra. The structure of the character group of the torus $\underline{F}^\times = \mathrm{Res}_{F/\mathbb{Q}_p}\mathbb{G}_{\mathrm{m}}$ over $\mathbb{Q}_p$ is given in A.4.2, and we will follow the notation in A.4. The geometric character group $\mathrm{X}^*(\underline{F}^\times)$ is naturally identified with the group $\mathbb{Z}^{\mathrm{Hom}_{\mathbb{Q}_p\text{-alg}}(F,\overline{\mathbb{Q}}_p)}$ of $\mathbb{Z}$-valued functions on the set $\mathrm{Hom}_{\mathbb{Q}_p\text{-alg}}(F,\overline{\mathbb{Q}}_p)$ of all $\mathbb{Q}_p$-algebra homomorphisms $F \to \overline{\mathbb{Q}}_p$, where the character ξ^f corresponding to a function $f \in \mathbb{Z}^{\mathrm{Hom}_{\mathbb{Q}_p\text{-alg}}(F,\overline{\mathbb{Q}}_p)}$ is determined by

$$\xi^f(x) = \prod_{\sigma \in \mathrm{Hom}_{\mathbb{Q}_p\text{-alg}}(F,\overline{\mathbb{Q}}_p)} \sigma(x)^{f(\sigma)}$$

for all $x \in \underline{F}^\times(\mathbb{Q}_p) = F^\times$.

For any $\sigma \in \mathrm{Hom}_{\mathbb{Q}_p\text{-alg}}(F, \overline{\mathbb{Q}}_p)$, denote by ξ^σ the geometric character of $\underline{F}^\times$ corresponding to the δ-function at σ. Explicitly (as we see by Zariski-density considerations over $\mathbb{Q}_p$), the isomorphism $\underline{F}^\times(\overline{\mathbb{Q}}_p) \simeq (\overline{\mathbb{Q}}_p \otimes_{\mathbb{Q}_p} F)^\times \simeq \prod_\tau \overline{\mathbb{Q}}_p^\times$ carries ξ^σ to the map $(c_\tau) \mapsto c_\sigma$. Clearly $\{\xi^\sigma\}_\sigma$ is a $\mathbb{Z}$-basis of $\mathrm{X}^*(\underline{F}^\times)$. The natural $\mathrm{Gal}(\overline{\mathbb{Q}}_p/\mathbb{Q}_p)$-action on $\mathrm{X}^*(\underline{F}^\times)$ is given by $\gamma^*(\xi^\sigma) = \xi^{\gamma\circ\sigma}$ for all $\gamma \in \mathrm{Gal}(\overline{\mathbb{Q}}_p/\mathbb{Q}_p)$; see A.4.2 for a detailed discussion of this computation (carried out over a general field k in place of $\mathbb{Q}_p$).

We let $\{\nu_\sigma\}$ be the $\mathbb{Z}$-basis of the cocharacter group $\mathrm{X}_*(\underline{F}^\times)$ of $\underline{F}^\times$ dual to $\{\xi^\sigma\}$. This cocharacter group is identified with the group of $\mathbb{Z}$-valued measures on $\mathrm{Hom}_{\mathbb{Q}_p\text{-alg}}(F, \overline{\mathbb{Q}}_p)$. The natural $\mathrm{Gal}(\overline{\mathbb{Q}}_p/\mathbb{Q}_p)$-actions on $\mathrm{X}^*(\underline{F}^\times)$ and $\mathrm{X}_*(\underline{F}^\times)$ are compatible via $\mathbb{Z}$-duality, so $\gamma^*(\nu_\sigma) = \nu_{\gamma\circ\sigma}$ for all $\gamma \in \mathrm{Gal}(\overline{\mathbb{Q}}_p/\mathbb{Q}_p)$. Hence, the Galois action transitively permutes the elements of both bases $\{\xi^\sigma\}$ and $\{\nu_\sigma\}$. For a subfield $F' \subset F$ over $\mathbb{Q}_p$, composition with $\mathrm{Nm}_{F/F'} : \underline{F}^\times \to \underline{F'}^\times$ pulls $\xi^{\tau'}$ back to $\sum_{\tau|_{F'}=\tau'} \xi^\tau$, so it is easy to check that $\mathrm{Nm}_{F/F'} \circ \nu_\tau = \nu_{\tau|_{F'}}$ as geometric cocharacters of $\underline{F'}^\times$.

For any geometric cocharacter $\mu \in \mathrm{X}_*(\underline{F}^\times)$ of $\underline{F}^\times$, denote by $E(\mu)$ the (minimal) field of definition for μ over $\mathbb{Q}_p$. Thus, $E(\mu)$ is the largest subfield of $\overline{\mathbb{Q}}_p$ over $\mathbb{Q}_p$ fixed by the stabilizer in $\mathrm{Gal}(\overline{\mathbb{Q}}_p/\mathbb{Q}_p)$ of the cocharacter μ. We call the subfield $E(\mu)$ of $\overline{\mathbb{Q}}_p$ the *reflex field* of the geometric cocharacter μ. For example, since $\gamma^*(\xi^\sigma) = \xi^{\gamma\sigma}$ for all $\gamma \in \mathrm{Gal}(\overline{\mathbb{Q}}_p/\mathbb{Q}_p)$ we see that $E(\xi^\sigma) = \sigma(F)$ inside $\overline{\mathbb{Q}}_p$, and likewise ν_σ has minimal field of definition over $\mathbb{Q}_p$ equal to $\sigma(F)$ inside $\overline{\mathbb{Q}}_p$.

There exists a unique $\mathbb{Q}_p$-homomorphism

$$\mathrm{N}\mu \colon \underline{E(\mu)}^\times \to \underline{F}^\times$$

such that $\mathrm{N}\mu_*(\nu_{\tau_0}) = \mu$, where ν_{τ_0} is the cocharacter of $\underline{E(\mu)}^\times$ corresponding to the embedding $\tau_0 : E(\mu) \to \overline{\mathbb{Q}}_p$ given by the definition inclusion. The construction of $\mathrm{N}\mu$ is explained in detail in A.4.8, where the base field is $\mathbb{Q}$ instead of $\mathbb{Q}_p$, but the method applies over any ground field, so A.4.8.1 gives the existence and uniqueness of $\mathrm{N}\mu$.[9] The $\mathbb{Q}_p$-homomorphism $\mathrm{N}\mu$ is called the *reflex norm* of the geometric cocharacter μ. For instance, a study of definition of the reflex norm shows that $\mathrm{N}\nu_\sigma : \underline{\sigma(F)}^\times \to \underline{F}^\times$ is induced by the inverse isomorphism of $\mathbb{Q}_p$-algebras $\sigma^{-1} : \sigma(F) \simeq F$. The following example confirms the consistency with an earlier notion of "reflex norm" (also cf. 3.7.2.4):

3.9.4. Example. Consider a p-adic CM type Φ for F, and let μ_Φ be the geometric cocharacter of $\underline{F}^\times$ such that

$$\langle \xi^\sigma, \mu_\Phi \rangle = \begin{cases} 1 & \text{if } \sigma \in \Phi, \\ 0 & \text{if } \sigma \notin \Phi. \end{cases}$$

[9]The notation in A.4.8 is $\chi(\mu)$, which is different from the notation $\mathrm{N}\mu$ used here.

In other words, under the isomorphism

$$\underline{F}^\times(\overline{\mathbb{Q}}_p) = (\overline{\mathbb{Q}}_p \otimes_{\mathbb{Q}_p} F)^\times \simeq \prod_\sigma \overline{\mathbb{Q}}_p^\times$$

$\mu_\Phi(c)$ goes over to the tuple (a_σ) with $a_\sigma = c$ if $\sigma \in \Phi$ and $a_\sigma = 1$ otherwise. Thus, the field of definition $E(\mu_\Phi)$ of μ_Φ over $\mathbb{Q}_p$ is the reflex field $E(\Phi)$ of Φ defined in 3.7.2.3, and by A.4.8.1 (adapted to the ground field $\mathbb{Q}_p$ rather than $\mathbb{Q}$) the reflex norm $\mathrm{N}\mu_\Phi$ of μ_Φ coincides with the reflex norm of Φ defined in 3.7.2.5.

In the situation of 3.9.1 with h_1 equal to an algebraic part of some ρ_1, if we take $\overline{\mathbb{Q}}_p$ to be the algebraic closure of $\mathbb{Q}_p$ in $\overline{K}$ and μ to be the composition of h_1 with the geometric cocharacter ν_{τ_0} of $\underline{E}^\times$ corresponding to the canonical $\mathbb{Q}_p$-embedding $\tau_0 : E \hookrightarrow \overline{\mathbb{Q}}_p$ then we claim that $E(\mu)$ is the field $E(\rho_1)$ defined in 3.9.1 and $\mathrm{N}\mu$ is the algebraic part of ρ_1 defined in 3.9.1. To prove this, consider more generally a $\mathbb{Q}_p$-homomorphism $h : \underline{E}^\times \to \underline{F}^\times$ and a subextension $E' \subseteq E$ over $\mathbb{Q}_p$ such that h factors as $h' \circ \mathrm{Nm}_{E/E'}$ for a $\mathbb{Q}_p$-homomorphism $h' : \underline{E'}^\times \to \underline{F}^\times$. Letting $\tau_0' = \tau_0|_{E'}$, the equality $\mathrm{Nm}_{E/E'} \circ \nu_{\tau_0} = \nu_{\tau_0'}$ of geometric cocharacters of $\underline{E'}^\times$ yields an equality

$$\mu := h \circ \nu_{\tau_0} = h' \circ \mathrm{Nm}_{E/E'} \circ \nu_{\tau_0} = h' \circ \nu_{\tau_0'}$$

of geometric cocharacters of $\underline{F}^\times$. It follows that $E(\mu) \subseteq E'$ since $\nu_{\tau_0'}$ is defined over E', so the minimal E' coincides with $E(\mu)$ (and hence the corresponding h' equals $\mathrm{N}\mu$) because the factorization $\mu = h \circ \nu_{\tau_0}$ is unique and μ factors as $\mathrm{N}\mu \circ \nu_{\tau_\mu} = (\mathrm{N}\mu \circ \mathrm{Nm}_{E/E(\mu)}) \circ \nu_{\tau_0}$ for the canonical $\mathbb{Q}_p$-embedding $\tau_\mu : E(\mu) \to \overline{\mathbb{Q}}_p$.

The proof of the following existence result for locally algebraic representations with a given algebraic part uses Proposition 3.7.3, so it depends on CM theory for abelian varieties over $\mathbb{C}$.

3.9.5. Proposition. *Let F be a non-zero finite-dimensional commutative semisimple $\mathbb{Q}_p$-algebra. Let $\mu \in \mathrm{X}_*(\underline{F}^\times)$ be a geometric cocharacter of the $\mathbb{Q}_p$-torus $\underline{F}^\times$. Let $E(\mu)$ be the reflex field of μ over $\mathbb{Q}_p$, a subfield of a fixed algebraic closure $\overline{\mathbb{Q}}_p$.*

(1) *There exist p-adic* CM *types $\Phi_1, \ldots, \Phi_a$ of F and pairwise distinct non-zero integers $n_1, \ldots, n_a$ such that the following properties hold:*
   (a) *the subsets $\Phi_i \subset \mathrm{Hom}_{\mathbb{Q}_p\text{-alg}}(F, \overline{\mathbb{Q}}_p)$ are mutually disjoint,*
   (b) *for all i, $E_i := E(F_i, \Phi_i) \subset E(\mu)$ and the reflex field $E(\mu)$ of μ is the smallest subfield of $\overline{\mathbb{Q}}_p$ which contains E_i for all i.*
   (c) $\mu = \sum_{i=1}^a n_i \cdot \mu_{\Phi_i}$.

(2) *For all i, there is an $\mathcal{O}_F$-linear* CM *p-divisible group $\mathcal{Y}_i$ over $\mathcal{O}_{E_i}$ with* CM *type (F, Φ_i).*

(3) *Let $\rho_{\mathcal{Y}_i} : \mathrm{Gal}(E_i^{\mathrm{ab}}/E_i) \to \mathcal{O}_F^\times$ be the p-adic Galois representation attached to the $\mathcal{O}_F$-linear* CM *p-divisible group $\mathcal{Y}_i$ for each i. For $E = E(\mu)$ we have*

$$\mathrm{N}\mu(x) = \prod_{i=1}^a (\mathrm{N}\mu_{\Phi_i} \circ \mathrm{Nm}_{E/E_i})(x)^{n_i} = \prod_{i=1}^a (\rho_{\mathcal{Y}_i} \circ r_{E_i} \circ \mathrm{Nm}_{E/E_i})(x)^{-n_i}$$

for all $x \in \mathcal{O}_E^\times$.

PROOF. Express the given cocharacter μ as a unique $\mathbb{Z}$-linear combination

$$\mu = \sum_{\sigma \in \mathrm{Hom}_{\mathbb{Q}_p\text{-alg}}(F, \overline{\mathbb{Q}}_p)} m_\sigma \cdot \nu_\sigma$$

of the $\mathbb{Z}$-basis $\{\nu_\sigma\}$. Let $f : \mathrm{Hom}_{\mathbb{Q}_p\text{-alg}}(F, \overline{\mathbb{Q}}_p) \to \mathbb{Z}$ be the function $f(\sigma) = n_\sigma$. Denote the distinct non-zero elements in the image of f as $\{n_1, \ldots, n_a\}$. Decompose $\mathrm{Hom}_{\mathbb{Q}_p\text{-alg}}(F, \overline{\mathbb{Q}}_p)$ into a disjoint union of subsets $\Phi_0, \Phi_1, \ldots, \Phi_a$ such that $\Phi_0 = f^{-1}(0)$ and $\Phi_i = f^{-1}(n_i)$.

It is easy to verify that the conditions (a)–(c) in (1) are satisfied, so (1) follows. The assertion in (2) is immediate from Proposition 3.7.3 (1). The first equality in (3) is an easy consequence of the properties in (1) and the definition of the reflex norm (and the functoriality of the local Artin map with respect to Galois restriction in terms of field norms, which is relevant because the definitions of $\mathrm{N}\mu$ and $\mathrm{N}\mu_{\Phi_i}$ use the minimal fields of definition $E(\mu)$ and E_i that need not coincide). The second equality in (3) follows from Corollary 3.7.9. □

3.9.6. From the point of view of p-adic Hodge theory, the significance of Proposition 3.9.5 is its consequence that for a field K as in 3.9.1, the restriction to the inertia subgroup of $\mathrm{Gal}(K^{\mathrm{ab}}/K)$ of any *crystalline* abelian p-adic representation $\rho : \mathrm{Gal}(K^{\mathrm{ab}}/K) \to F^\times$ is *algebraic*. We now precisely formulate this assertion and explain its proof.

Since ρ is crystalline, it is Hodge–Tate. Thus, by Tate's theorem recalled in 3.9.1, we know that ρ is locally algebraic. Let $h : \underline{E}^\times \to \underline{F}^\times$ be an algebraic part of ρ as in 3.9.1, and let $\overline{\mathbb{Q}}_p$ be the algebraic closure of $\mathbb{Q}_p$ in an algebraic closure $\overline{K}$ of K. Let ν_{τ_0} be the geometric cocharacter of $\underline{E}^\times$ corresponding to the inclusion τ_0 of E into $\overline{\mathbb{Q}}_p$, and let $\mu \in \mathrm{X}_*(\underline{F}^\times)$ be $h \circ \nu_{\tau_0}$. Clearly the minimal field of definition $E(\mu)$ of μ is contained in E, and by Example 3.9.4 we see that $E(\mu) = E(\rho)$ and $\mathrm{N}\mu$ is the algebraic part of ρ.

Write μ as a sum $\mu = \sum_{i=1}^a n_i \cdot \mu_{\Phi_i}$ for pairwise disjoint p-adic CM types $\Phi_1, \ldots, \Phi_a$ of F as in 3.9.5. Let Φ_0 denote the elements of $\mathrm{Hom}_{\mathbb{Q}_p\text{-alg}}(F, \overline{\mathbb{Q}}_p)$ not in any of $\Phi_1, \ldots, \Phi_a$, and define $n_0 = 0$. Let $\mathbb{C}_K$ denote the completion of $\overline{K}$, and let V be the representation space for ρ_1 over $\mathbb{Q}_p$. Because ρ is Hodge–Tate, we have a Hodge–Tate decomposition

$$\mathbb{C}_K \otimes_{\mathbb{Q}_p} V_\rho \simeq \bigoplus_{j \in \mathbb{Z}} (\mathbb{C}_K \otimes_K V[j]),$$

where

$$V[j] := \left\{ v \in \mathbb{C}_K \otimes_{\mathbb{Q}_p} V_\rho \mid \gamma \cdot v = \chi_{\mathrm{cycl}}^j(\gamma) v \text{ for all } \gamma \in \mathrm{Gal}(\overline{K}/K) \right\}$$

for every $j \in \mathbb{Z}$, with χ_{cycl} denoting the p-adic cyclotomic character on $\mathrm{Gal}(K^{\mathrm{ab}}/K)$. Clearly the action of F on V_ρ induces a natural action of F on the finite dimensional K-vector spaces $V[j]$.

We claim that $V[j] \neq 0$ precisely when either $j = n_i$ with $1 \leqslant i \leqslant a$ or when $j = 0$ and $\Phi_0 \neq \emptyset$, and that in general $\dim_K V[n_i] = \#\Phi_i$ for $0 \leqslant i \leqslant a$. To prove this we may reduce to the case when F is a field that is moreover Galois over $\mathbb{Q}_p$, and then we may replace K with a finite extension so that it contains a Galois closure of F over $\mathbb{Q}_p$ inside $\overline{\mathbb{Q}}_p \subset \overline{K}$. By tensorial considerations this reduces the problem to the special case $\mu = \nu_\sigma$, in which case $E(\mu) = \sigma(F)$. The results in [**105**, III, A.5] then settle this special case.

Moreover, for every $i = 0, \ldots, a$ the $\overline{K} \otimes_{\mathbb{Q}_p} F$-module $\overline{K} \otimes_K V[n_i]$ is isomorphic to the submodule of $\overline{K} \otimes_{\mathbb{Q}_p} F$ corresponding to Φ_i (as we may check this after applying $\mathbb{C}_K \otimes_{\overline{K}} (\cdot)$). Thus, the reflex field $E(F, \Phi_i) \subset \overline{\mathbb{Q}}_p \subset \overline{K}$ is contained in K for each $i = 1, \ldots, a$, so $E(\mu)$ is contained in K as well.

Let

$$\delta\colon \operatorname{Gal}(K^{\mathrm{ab}}/K) \to \operatorname{Gal}(E(\mu)^{\mathrm{ab}}/E(\mu))$$

be the natural restriction map. This carries the inertia subgroup onto an open subgroup of the inertia subgroup of $\operatorname{Gal}(E(\mu)^{\mathrm{ab}}/E(\mu))$.

3.9.7. Corollary. *Consider an abelian crystalline representation*

$$\rho\colon \operatorname{Gal}(K^{\mathrm{ab}}/K) \to F^\times$$

and the associated geometric cocharacter μ of $\underline{F}^\times$ as above with reflex norm $\mathrm{N}\mu$ equal to the algebraic part of ρ.

The reflex field $E(\mu)$ of μ is contained in K, and the restriction of ρ to the inertia subgroup of $\operatorname{Gal}(K^{\mathrm{ab}}/K)$ is algebraic; i.e.

$$\rho(\delta^{-1}(r_{E(\mu)}(x))) = 1/\mathrm{N}\mu(x) \qquad \text{for all } x \in U\,,$$

where $U \subset \mathcal{O}_{E(\mu)}^\times$ is the open subgroup corresponding (via the local Artin map) to the image of the inertia subgroup of $\operatorname{Gal}(K^{\mathrm{ab}}/K)$.

PROOF. We have seen that $E(\mu) \subset K$. By parts (1)(c) and (3) in Proposition 3.9.5 and [**105**, III, A.4, Prop. 4], it follows from [**105**, III, A.5, Theorem 2] that there exists an open subgroup U' of U such that $\rho(\delta^{-1}(r_{E(\mu)}(x))) = 1/\mathrm{N}\mu(x)$ for all $x \in U'$. On the other hand, Proposition 3.9.5 tells us that there exists a crystalline representation

$$\rho' = \prod \rho_{\mathcal{Y}_i}^{n_i} : \operatorname{Gal}(E(\mu)^{\mathrm{ab}}/E(\mu)) \to F^\times$$

such that $\rho'(r_{E(\mu)}(x)) = 1/\mathrm{N}\mu(x)$ for all $x \in \mathcal{O}_{E(\mu)}^\times$. In particular ρ and ρ' coincide on an open subgroup of the inertia subgroup of $\operatorname{Gal}(K^{\mathrm{ab}}/K)$. Therefore, ρ and ρ' coincide on the whole inertia subgroup of $\operatorname{Gal}(K^{\mathrm{ab}}/K)$ since a finitely ramified crystalline representation is unramified [**42**, 3.3.1]. □

The following summarizes the main points of 3.9.5 and 3.9.7.

3.9.8. Corollary. *For a finite-dimensional non-zero commutative semisimple $\mathbb{Q}_p$-algebra F, a continuous representation $\rho\colon \operatorname{Gal}(K^{\mathrm{ab}}/K) \to F^\times$ is crystalline if and only if the restriction of ρ to the inertia subgroup I'_K is algebraic in the following sense: there exists a geometric cocharacter μ of $\underline{F}^\times$ such that the reflex field $E(\mu) \subset \overline{K}$ of μ is contained in K over $\mathbb{Q}_p$ and*

$$\rho(\delta^{-1}(r_{E(\mu)}(x))) = 1/\mathrm{N}\mu(x) \qquad \text{for all } x \in U\,,$$

where $\delta : \operatorname{Gal}(K^{\mathrm{ab}}/K) \to \operatorname{Gal}(E(\mu)^{\mathrm{ab}}/E)$ is the natural homomorphism and $U \subseteq \mathcal{O}_{E(\mu)}^\times$ is the open subgroup corresponding to $\delta(I'_K)$ via $r_{E(\mu)}$. Moreover, for every $\mu \in \mathrm{X}^*(\underline{F}^\times)$ there exists a crystalline $\rho : \operatorname{Gal}(E(\mu)^{\mathrm{ab}}/E(\mu)) \to F^\times$ with algebraic part $\mathrm{N}\mu$.

PROOF. The "if" part and the existence of ρ given μ are both immediate from 3.9.5 since all $\rho_{\mathcal{Y}_i}$ are crystalline (as each $\mathcal{Y}_i$ is a p-divisible group over $\mathcal{O}_{E_i}^{\times}$). The "only if" part is 3.9.7. □

3.10. Appendix: questions and examples on extending isogenies

In this appendix we discuss several questions which naturally arise when contemplating p-divisible groups over a general base scheme, including some easy examples and some problems which appear to be open. We are especially grateful to Ofer Gabber for providing answers to some questions (see 3.3.9 and 3.3.11). Since the material in this appendix will not be used elsewhere in this book, proofs will be largely omitted; we hope to come back to this circle of ideas in the near future.

3.10.1. We use the following notation. Let R be a local domain R with residue field κ of characteristic $p > 0$ and fraction field K, and write

$$\kappa \xleftarrow{\ s\ } R \xrightarrow{\ \eta\ } K, \quad S = \mathrm{Spec}(R), \quad s \in S, \ \ \eta \in S,$$

for the residue class map s and the inclusion map η into the field of fractions. We take R to be a valuation ring (possibly not noetherian) unless we say otherwise.

3.10.2. Example. There exist examples of a homomorphism $\varphi : X \to Y$ of p-divisible groups over such a scheme S with $\varphi_\eta : X_\eta \to Y_\eta$ an isogeny but φ_s not an isogeny. For example, let R be the integral closure of $\mathbb{Z}_p$ in an algebraic closure K of $\mathbb{Q}_p$, and consider the isomorphism

$$\underline{\mathbb{Q}_p/\mathbb{Z}_p}(R) \simeq \mathbb{Q}_p/\mathbb{Z}_p \simeq \mu_{p^\infty}(R)$$

defined by a compatible system of p-power roots of unity in K. This defines a homomorphism of p-divisible groups

$$\varphi : \underline{\mathbb{Q}_p/\mathbb{Z}_p} \to \mu_{p^\infty}$$

over R that is an isomorphism between K-fibers but vanishes between κ-fibers.

3.10.3. Question. Is there a homomorphism $\varphi : X \to Y$ over R with $\mathrm{char}(K) = 0$ such that φ_η is an isomorphism and $\varphi_s = 0$ (as in the example above) with X_s and Y_s both local-local?

3.10.4. Over an irreducible normal base scheme S, any homomorphism between generic fibers of abelian schemes extends to a homomorphism over the entire base. In view of the analogy between p-divisible groups and abelian schemes, it is natural to wonder if the same is true for p-divisible groups. In the noetherian case, the answer is affirmative:

3.10.4.1. Proposition. *Let R be normal local domain and X and Y two p-divisible groups over $S = \mathrm{Spec}(R)$. If R is noetherian then any isogeny $\varphi_\eta : X_\eta \to Y_\eta$ over the generic point η of S extends uniquely to a homomorphism $\varphi : X \to Y$ over S that is moreover an isogeny.*

PROOF. Once the extension result is proved in general, by applying it to $\psi_\eta : Y_\eta \to X_\eta$ such that $\psi_\eta \circ \varphi_\eta = [p^r]_{X_\eta}$ and $\varphi_\eta \circ \psi_\eta = [p^r]_{Y_\eta}$ with $r \geqslant 0$ we see that the extended homomorphism over S is an isogeny on fibers and hence is an isogeny (since S is noetherian). Thus, we focus on extending φ_η over S.

Since R is the intersection of its localizations at height-1 prime ideals, we may assume R is a discrete valuation ring and that the residue characteristic is p (as otherwise the p-divisible groups are étale and the assertion is trivial). Let $K = \mathrm{Frac}(R)$. In this case the extension over R is a theorem of Tate (see 1.4.5.1) when $\mathrm{char}(K) = 0$, and is a theorem of de Jong [**55**, 1.2] when $\mathrm{char}(K) = p$. □

3.10.4.2. Example. If we drop the noetherian condition then the conclusion of the preceding proposition can fail even when S is local and the p-divisible groups have local-local special fiber (to avoid trivial examples such as φ_η^{-1} in Example 3.10.2). In fact, there are p-divisible groups X and Y over a base scheme $S = \mathrm{Spec}(R)$ as in 3.10.1 and a homomorphism (even an isomorphism) $f : X_\eta \to Y_\eta$ between the generic fibers such that f does not extend to a homomorphism from X to Y and the special fibers X_s and Y_s are local-local. The ring R can be taken to be the valuation ring of $\overline{\mathbb{Q}}_p$ or an algebraic closure of $\overline{\mathbb{F}}_p((t))$.

Let R' be either $\overline{\mathbb{F}}_p[\![t]\!]$ or the valuation ring of a finite extension of $W(\overline{\mathbb{F}}_p)[1/p]$, and let R be the valuation ring of an algebraic closure of $\mathrm{Frac}(R')$. We first note that there are examples of p-divisible groups X' and Y' of the same height over R' whose geometric generic fibers are isomorphic and whose closed fibers X'_s and Y'_s are not isogenous. For instance, when $\mathrm{char}(R') = 0$ we can pick any local-local p-divisible groups over $\overline{\mathbb{F}}_p$ having the same height $\geqslant 2$ but no slopes in common: these lift over R', due to the unobstructedness of the deformation theory of p-divisible groups (Proposition 1.4.4.7), and the geometric generic fibers are constant of the same height. For the case $\mathrm{char}(R') = p$, consider $X'_s = \mathcal{G}_{1/2} \times \mathcal{G}_1$ and $Y'_s = \mathcal{G}_{2/3}$ of height 2. These admit deformations over R' whose geometric generic fibers $X'_{\eta'}$ and $Y'_{\eta'}$ are both ordinary of dimension 1.

Given such X' and Y', let $X = X'_R$ and $Y = Y'_R$. Since $\mathrm{Frac}(R)$ is algebraically closed, we may pick an isomorphism $f : X_\eta \xrightarrow{\simeq} Y_\eta$. If f and f^{-1} extend to R-homomorphisms between X and Y then these extended maps over R must be inverse to each other (as we may check over η), contradicting that the special fibers are not isomorphic. Hence, either f or f^{-1} does not extend to an R-homomorphism.

3.10.5. Instead of extending homomorphisms between generic fibers of two p-divisible groups over R, it is also natural to seek to extend a p-divisible subgroup of the generic fiber. As motivation, we note that by the Néron–Ogg-Shafarevich criterion, if $\mathscr{A}$ is an abelian scheme over a discrete valuation ring R and B is an abelian subvariety of the generic fiber $A = \mathscr{A}_\eta$ then B extends to an abelian scheme $\mathscr{B}$ over R that is naturally closed in $\mathscr{A}$.

In Tate's original work on p-divisible groups there was given an example due to Serre (using a product of two elliptic curves with supersingular reduction) for which there is a map between p-divisible groups over a discrete valuation ring such that the induced map between generic fibers is a closed immersion but the induced map between special fibers has non-zero finite kernel. Such examples are recalled below. Here is a loosely formulated general question about this situation.

3.10.5.1. Question. Let Y be a p-divisible group over a valuation ring R of residue characteristic p, and let $G \subset Y_\eta$ be a closed p-divisible subgroup of the generic fiber. Does the inclusion $G \hookrightarrow Y_\eta$ between p-divisible groups over η extend to a homomorphism of p-divisible groups $\varphi : X \to Y$ over $S = \mathrm{Spec}(R)$ (with φ not necessarily a closed immersion)?

3.10.5.2. Under some finiteness conditions, the answer to 3.10.5.1 is affirmative. If we assume that R is a discrete valuation ring then $\varphi : X \to Y$ asked for in 3.10.5.1 does exist (and is unique up to unique isomorphism by Proposition 3.10.4.1). For example, in [**37**] it is shown that when R a discrete valuation ring of characteristic 0 then $X \to Y$ can be constructed as follows. Let X'_m be the Zariski closure of $G[p^m]$ in $Y[p^m]$ for each $m \geqslant 1$. Using discriminant arguments, there is an m_0 such that the directed system of finite flat group schemes $X_n := X'_{n+m_0}/X'_{m_0}$ over R forms a p-divisible group with an evident homomorphism to Y that does the job.

As we have already noted, in such a situation *it can happen that the map* $\varphi_s : X_s \to Y_s$ *has a non-zero finite kernel* even though φ_η is a closed immersion. An equi-characteristic example is given in 3.10.5.3, and a mixed characteristic example is given in 3.10.5.4. In both constructions we choose a closed immersion $\rho : X \hookrightarrow Y'$ and a finite flat R-subgroup scheme $N \subset Y'$ fitting into a diagram

$$\begin{array}{ccccc} X & \overset{\rho}{\hookrightarrow} & Y' & \qquad & \rho_\eta(X_\eta) \cap N_\eta = 0, \\ & \searrow^{\varphi} & \downarrow & & \\ & & Y := Y'/N & & \rho_s(X_s) \cap N_s \neq 0. \end{array}$$

3.10.5.3. Example. For $\kappa = \mathbb{F}_p$, let $R = \kappa[t]_{(t)}$ (local ring at the origin of the affine line). Let $X = (\mathcal{G}_{1/2})_R$ and $Y' = X \times X$ with $\rho : X \to Y'$ the map $x \mapsto (x, 0)$. Let N be the closed subgroup $(1, t) : \alpha_p \to \alpha_p \times \alpha_p \subset Y'$ over R. Thus, $\varphi : X \to Y'/N =: Y$ is an example asked for in the equi-characteristic case.

3.10.5.4. Example. Let R be a complete discrete valuation ring of mixed characteristic $(0, p)$, and let X be a p-divisible group over R with special fiber $\mathcal{G}_{1/3}$ (height 3, dimension 1). Likewise, let X' be a p-divisible group over R with special fiber $\mathcal{G}_{2/3}$ (e.g., $X' = X^t$). It is easy to make such (R, X, X'). We let $Y'' = X \times X'$. By increasing K by finite amount if necessary, we can arrange that $X_\eta[p^2]$ and $X'_\eta[p^2]$ are constant (i.e., $X[p^2](K) = (\mathbb{Z}/p^2\mathbb{Z})^3 = X'[p^2](K)$). Thus, $Y''[p^2](K) = (\mathbb{Z}/p^2\mathbb{Z})^6$.

Choose non-zero $P \in X[p](K)$ and non-zero $P' \in X'[p](K)$, and let $P'' = (P, P') \in Y''[p](K)$. Let $N'' \subset Y''$ be the closure over R of the subgroup generated by P'' over K. Since $P' \neq 0$, it follows that $N''_s \cap (X_s \times \{0\}) = 0$. Hence, for $Y' := Y''/N''$ the natural map $\rho : X \to Y''/N'' = Y'$ has kernel whose special fiber is trivial, so it is a closed immersion (as we may check on each p-power torsion level). Note that $Y'[p](K) \simeq (\mathbb{Z}/p\mathbb{Z})^6$ since $Y''[p^2](K) = (\mathbb{Z}/p^2\mathbb{Z})^6$. Moreover, as $P \neq 0$ and $P' \neq 0$ it follows that $a(Y'_s) = 1$.

We choose $Q \in Y'[p](K)$ such that $Q \notin X[p](K)$. Let $N \subset Y$ be the finite flat closed R-subgroup whose generic fiber is the constant cyclic subgroup generated by Q. We see that $N_\eta \cap \rho_\eta(X_\eta) = 0$. Since N_s has rank p, it is the unique α-group contained in Y'_s. Hence, $N_s \cap \rho_s(X_s) \neq 0$. This is a mixed characteristic example of the desired type over a discrete valuation ring.

3.10.5.5. Suppose R is a discrete valuation ring and $\varphi : X \to Y$ is a homomorphism such that φ_η an inclusion (and then φ_s has a finite kernel). Does every such example arise, possibly after a finite extension of R, from the general construction $X \to Y'/N$ as considered above? In other words, does φ factor as a closed immersion of p-divisible groups over R followed by an isogeny over R?

3.10.6. Question. Consider p-divisible groups X and Y over R as in 3.10.1 with $\mathrm{char}(K) = 0$, and suppose there is given a homomorphism $f : X_\eta \hookrightarrow Y_\eta$ that is either a closed immersion or isomorphism. Are there reasonable conditions (weaker than requiring R to be noetherian) which ensure f can be extended? For example, if every slope in the Newton polygon of X_s is strictly smaller than any slope in the Newton polygon of Y_s, then does f extend to a homomorphism $\varphi : X \to Y$?

CHAPTER 4

CM lifting of abelian varieties up to isogeny

I also see along which lines to look for a generalization of Tate's theorem to crystals in the unequal characteristic case... the description of this functor remains however the most mysterious point... Maybe these statements will even turn out to be provable!

— Alexandre Grothendieck

4.1. Introduction

In this chapter we prove the following strong form of an affirmative answer to Problem (I) in 1.8.5, CM lifting up to isogeny without increasing the finite base field. (The details are given in 4.6.4, building on preparations developed below.) This is the main result of the book.

4.1.1. Theorem (CM lifting up to isogeny). *Let B be an abelian variety of dimension $g > 0$ over a finite field κ of characteristic p, and let $\alpha : L \to \mathrm{End}^0(B)$ be an embedding of a* CM *field L of degree $2g$ over $\mathbb{Q}$. Let $\mathcal{O}' = \mathbb{Z} + p\mathcal{O}_L$.*

There exists an abelian variety B' over κ and a κ-isogeny $f : B \to B'$ such that $\mathcal{O}_L \subset \mathrm{End}(B')$ via the identification $\mathrm{End}^0(B') = \mathrm{End}^0(B)$ and such that B' equipped with its $\mathcal{O}'$-action lifts to an abelian scheme $\mathcal{B}'$ with $\mathcal{O}'$-action over a 1-dimensional complete local noetherian domain D with generic characteristic 0 and residue field κ. If $\mathcal{O}_L \subset \mathrm{End}(B)$ then we can arrange that $\deg(f)|p^{4g^2}$.

For each $\mathbb{Q}_p$-embedding $j : D \hookrightarrow \overline{\mathbb{Q}}_p$, let Φ_j be the $\overline{\mathbb{Q}}_p$-valued CM *type on L associated to the base change $\mathcal{B}'_{\overline{\mathbb{Q}}_p}$ along j and let w_j be the induced p-adic place on the reflex field $E(L, \Phi_j) \subset D[1/p]$. The triple $(D, \mathcal{B}', j)$ can be chosen so that the extension of p-adic fields $E(L, \Phi_j)_{w_j} \to D[1/p]$ is tame with relative ramification degree dividing $p^4 - 1$ (and equal to 1 when B has no slope equal to $1/2$).*

It is explained in [**93**] and in Chapter 3 (see 3.1.10(iii) and 3.8.3) how to make examples of B having a negative solution to the CM lifting problem if we do not permit κ-isogenies; the obstruction is related to sizes of fields of definition associated to the p-divisible group $B[p^\infty]_{\overline{\kappa}}$. Thus, it is necessary to replace the given abelian variety B with a κ-isogenous B' as in the statement of Theorem 4.1.1 to ensure the existence of a CM lifting. In the examples of failure of the CM lifting without allowing a κ-isogeny, even the p-divisible group fails to have a CM lifting (in the sense of 3.7.1.8). Hence, it is unavoidable that p may divide the degree of the isogeny as in Theorem 4.1.1. We do not know which CM types Φ on L arise from L-linear lifts $\mathcal{B}'$ of members of the L-linear κ-isogeny class of B (see 1.8.7).

It may be impossible to find the κ-isogenous B' so that the action of the entire ring of integers $\mathcal{O}_L$ on B' lifts to characteristic 0. As an example, consider the 2-dimensional C_0 over $\mathbb{F}_{p^{2n}}$ in 4.1.2 below (so $p \equiv \pm 2 \pmod 5$ with $L = \mathbb{Q}(\zeta_5)$ and $\mathcal{O}_L$ operating on C_0). We shall see in 4.1.2.3 that for *odd* n no member C of the $\mathcal{O}_L$-linear $\mathbb{F}_{p^{2n}}$-isogeny class of C_0 admits a lift to an abelian scheme over a local domain with characteristic 0 and residue field $\mathbb{F}_{p^{2n}}$ such that the action of the entire ring of integers $\mathcal{O}_L$ on the special fiber lifts to the abelian scheme. (There is a formal $\mathcal{O}_L$-linear lift of each C over $W(\mathbb{F}_{p^{2n}})$, but it is not algebraizable.)

This 2-dimensional example is the key. Through a series of intermediate steps involving localization to p-divisible groups, the classification of isomorphism classes and isogeny classes of $(\mathcal{O}_L \otimes_{\mathbb{Z}} \mathbb{Z}_p)$-linear p-divisible groups over finite fields and algebraically closed fields of characteristic p, and Galois descent, the proof of Theorem 4.1.1 is reduced to the study of a mild generalization of $C_0[p^\infty]$ (see 4.1.3)!

4.1.2. An instructive example. Choose a prime $p \equiv \pm 2 \pmod 5$. For $q = p^{2n}$, let C_0 be a simple abelian surface over $\mathbb{F}_q$ such that its q-Frobenius is $\pm p^n \cdot \zeta_5$. Intrinsically, the q-Frobenius endomorphism π is p^n times either a primitive 5th root of unity (denotes as ζ_5) or a primitive 10th root of unity (denoted as $-\zeta_5$). This defines an embedding of $\mathbb{Q}(\zeta_5)$ into $\mathrm{End}^0(C_0)$. Since $\mathrm{M}^*(C_0[p^\infty])[1/p]$ is 1-dimensional over the quartic extension $\mathbb{Q}_p(\zeta_5)$ of $\mathbb{Q}_p$, necessarily $\mathrm{M}^*(C_0[p^\infty])$ is free of rank 1 over $\mathbb{Z}_p[\zeta_5]$. Thus, we see that π kills $C_0[p^n]$, so p^n divides π in $\mathrm{End}(C_0)$ and hence there is a preferred embedding $\mathbb{Z}[\zeta_5] \subset \mathrm{End}(C_0)$ (via $\pm\zeta_5 \mapsto \pi/p^n$).

These CM abelian surfaces are a generalization of the supersingular abelian surfaces over $\mathbb{F}_{p^2}$ in 2.3.1. A calculation with Honda-Tate theory as in 2.3.1 shows that $L = \mathrm{End}^0(C_0)$, so there is a preferred isomorphism $\mathbb{Z}[\zeta_5] \simeq \mathrm{End}(C_0)$. In general C_0 is not unique in its $\mathbb{Z}[\zeta_5]$-linear $\mathbb{F}_q$-isogeny class, but our results concerning C_0 apply (with the same proofs) to all members of its $\mathbb{Z}[\zeta_5]$-linear $\mathbb{F}_q$-isogeny class

By the following general result, the infinitesimal deformation theory of C_0 equipped with its $\mathbb{Z}[\zeta_5]$-action is formally étale. In particular, C_0 admits a unique formal deformation $\mathfrak{C}_0$ as a $\mathbb{Z}[\zeta_5]$-linear formal abelian scheme over $W(\mathbb{F}_q)$.

4.1.2.1. Proposition. *Let Λ be a complete local noetherian ring with residue field k of characteristic $p > 0$.*

(1) *Let A_0 be an abelian variety of dimension $g > 0$ over k, and F a commutative semisimple $\mathbb{Q}$-algebra such that $[F : \mathbb{Q}] = 2g$ and p is unramified in every factor field of F. Let $\mathcal{O}$ be an order in the maximal order $\mathcal{O}_F$ such that $p \nmid [\mathcal{O}_F : \mathcal{O}]$, and let $\alpha_0 : \mathcal{O} \to \mathrm{End}(A_0)$ be an injective ring map. The deformation functor $\mathrm{Def}_\Lambda(A_0, \alpha_0)$ is formally étale; equivalently, its deformation ring is Λ.*
(2) *Let X_0 be a p-divisible group over k of height $h > 0$. Let F be a commutative semisimple $\mathbb{Q}_p$-algebra such that $[F : \mathbb{Q}_p] = h$ and every factor field of F is unramified over $\mathbb{Q}_p$. Let $\mathcal{O}_F \subset F$ be the maximal $\mathbb{Z}_p$-order. For any injective $\mathbb{Z}_p$-algebra map $\alpha_0 : \mathcal{O}_F \to \mathrm{End}(X_0)$, the deformation functor $\mathrm{Def}_\Lambda(X_0, \alpha_0)$ is formally étale; equivalently, its deformation ring is Λ.*

PROOF. Since $\mathbb{Z}_p \otimes_{\mathbb{Z}} \mathcal{O} = \mathbb{Z}_p \otimes_{\mathbb{Z}} \mathcal{O}_F$ in (1), by the Serre-Tate deformation theorem (Theorem 1.4.5.3) the assertion in (1) is reduced to the assertion in (2). Thus, we now focus on (2). By Theorem 1.4.5.5(2), if W is a Cohen ring for k and $W \to \Lambda$ is a local homomorphism lifting the identity map on residue fields then the deformation ring for $\mathrm{Def}_\Lambda(X_0, \alpha_0)$ is obtained from the deformation ring $\mathcal{R}$ for $\mathrm{Def}_W(X_0, \alpha_0)$ by applying $\Lambda \widehat{\otimes}_W(\cdot)$. Hence, we may assume $\Lambda = W$. We seek to prove that the natural map $f : W \to \mathcal{R}$ is an isomorphism.

Let us reduce to the case of a perfect ground field so that Dieudonné theory may be used. Consider an extension of fields $j : \mathsf{k} \to \mathsf{k}'$ and a local map of Cohen rings $W \to W'$ lifting j. Let $(X'_0, \alpha'_0) = (X_0, \alpha_0)_{\mathsf{k}'}$. The deformation ring for $\mathrm{Def}_{W'}(X'_0, \alpha'_0)$ is naturally identified with $W' \widehat{\otimes}_W \mathcal{R}$ (by Theorem 1.4.5.5(2)). If the structure map $f' : W' \to W' \widehat{\otimes}_W \mathcal{R}$ is an isomorphism then reducing modulo p shows that $\mathcal{R}/(p)$ is 1-dimensional as a vector space over k and hence the structure map $f : W \to \mathcal{R}$ is surjective. The isomorphism property for f' then forces f to be an isomorphism. Hence, it suffices to work over k' rather than k, so we may and do assume k is perfect. We aim to show that the infinitesimal deformation theory of (X_0, α_0) (on artinian local k-algebras with residue field k) is formally étale.

Let $(R, \mathfrak{m})$ be an artinian local ring with residue field k. Choose a square-zero ideal $J \subset R$, and let X be a p-divisible group over R/J equipped with an action by $\mathcal{O}_F$ and an $\mathcal{O}_F$-linear isomorphism of its special fiber with X_0. It is necessary and sufficient to prove that X admits a unique $\mathcal{O}_F$-linear deformation over R. To analyze this deformation problem via crystalline methods, we equip $J = \ker(R \twoheadrightarrow R/J)$ with the trivial divided power structure.

The Lie algebra of the universal vector extension $\mathrm{E}(X)$ of X is naturally an $\mathcal{O}_F \otimes_{\mathbb{Z}} (R/J)$-linear extension of $\mathrm{Lie}(X)$ by $\mathrm{Lie}(X^t)^\vee$ [**75**, IV, 1.14], and its crystalline nature [**75**, IV, 2.5] yields a canonical $\mathcal{O}_F \otimes_{\mathbb{Z}_p} R$-module N that lifts $\mathrm{Lie}(\mathrm{E}(X))$ and is finite free over R. By [**75**, V, 1.6], to give an isomorphism class of $\mathcal{O}_F$-linear deformations of X over R is the same as to give an $\mathcal{O}_F \otimes_{\mathbb{Z}_p} R$-submodule of E' that is a direct summand over R and lifts $\mathrm{Lie}(X^t)^\vee \subseteq \mathrm{Lie}(\mathrm{E}(X)) = N/JN$. Hence, it is necessary and sufficient to show that there is a unique such submodule.

By Proposition 1.4.3.9(2) applied to the action of $\mathcal{O}_F$ on X_0, the Dieudonné module $\mathrm{M}^*(X_0)$ is free of rank 1 over $\mathcal{O}_F \otimes_{\mathbb{Z}_p} W(\mathsf{k})$. Thus, passing to the quotient modulo p, $\mathrm{M}^*(X_0[p])$ is free of rank 1 over $P_0 := (\mathcal{O}_F/p\mathcal{O}_F) \otimes_{\mathbb{F}_p} \mathsf{k}$. The unramifiedness of the factor fields of F implies that $P_0 = \prod \mathsf{k}_i$ for finite extension fields k_i of k, and in the exact sequence of P_0-modules

$$\begin{aligned} 0 \to \ker(\mathcal{V} : \mathrm{M}^*(X_0[p]) \to \mathrm{M}^*(X_0[p])^{(p)}) \to \mathrm{M}^*(X_0[p]) \\ \to \mathrm{M}^*(X_0[p])/\mathcal{F}(\mathrm{M}^*(X_0[p])) \to 0 \end{aligned}$$

the first term is identified with $\mathrm{Lie}(X_0[p]^D) = \mathrm{Lie}(X_0^t)$ (as its k-linear dual is the cokernel of the linearized Frobenius of $\mathrm{M}^*(X_0[p]^D)$), whereas the final term is identified with the k-linear dual $\mathrm{Lie}(X_0[p])^\vee = \mathrm{Lie}(X_0)^\vee$ (see Theorem 1.4.3.2(4)). The k-linear dual of k_i is 1-dimensional over k_i, so under the decomposition of P_0-modules according to the factor fields k_i of P_0, for each i exactly one of $\mathrm{Lie}(X_0^t)$ or $\mathrm{Lie}(X_0)$ has a non-zero k_k-factor and that factor module has k_i-dimension 1.

Consider the isomorphism

$$\mathcal{O}_F \otimes_{\mathbb{Z}_p} (R/J) \simeq \prod W(k_i) \otimes_{W(\mathsf{k})} (R/J).$$

For any $\mathcal{O}_F \otimes_{\mathbb{Z}_p} (R/J)$-module M, we thereby get a decomposition $M = \prod M_i$ into a direct product of modules over the local factor rings. In particular, if M is free over the local ring R/J then each M_i is free as a module over R/J, so such an M

is free of rank 1 over $\mathcal{O}_F \otimes_{\mathbb{Z}_p} (R/J)$ if and only if its fiber M_0 over the residue field k of R/J is free of rank 1 over $(\mathcal{O}_F/p\mathcal{O}_F) \otimes_{\mathbb{F}_p} \mathsf{k} = P_0$.

An analogous conclusion holds with R/J replaced by R. Applying this to the R-free N, it follows that N is free of rank 1 over $\mathcal{O}_F \otimes_{\mathbb{Z}_p} R$ and that $M \mapsto M/\mathfrak{m}M$ is a bijection between the set of submodules of N that are R-linear direct summands and the set of P_0-submodules of $N/\mathfrak{m}N = \mathrm{Lie}(\mathrm{E}(X_0))$. The same applies to $\mathrm{Lie}(\mathrm{E}(X))$ in place of N when using R/J in place of R, so $M \mapsto M/JM$ is a bijection between the set of $\mathcal{O}_F \otimes_{\mathbb{Z}_p} R$-submodules of N that are R-linear direct summands and the set of $\mathcal{O}_F \otimes_{\mathbb{Z}_p} (R/J)$-submodules of $\mathrm{Lie}(\mathrm{E}(X))$ that are R/J-linear direct summands. □

4.1.2.2. Remark. In the preceding proof, the role of unramifiedness at p in the coefficient algebra F is to ensure (for perfect k) that $M := \mathrm{M}^*(X_0[p])$ is free of rank 1 over $(\mathcal{O}_F/p\mathcal{O}_F) \otimes_{\mathbb{F}_p} \mathsf{k}$ and that the submodule $\ker \mathcal{V} = \mathcal{F}M$ and the quotient $M/\mathcal{F}M$ are each direct summands as $(\mathcal{O}_F/p\mathcal{O}_F) \otimes_{\mathbb{F}_p} \mathsf{k}$-modules.

For general $F = \prod_{i \in I} F_i$ with fields F_i, these module-theoretic properties hold whenever X_0 arises as a Serre tensor construction $\mathcal{O}_F \otimes_{\mathcal{O}_{F_0}} X'_0$ where $F_0 := \prod F'_i$ for the maximal unramified subfields $F'_i \subset F_i$ and X'_0 is a p-divisible group over k equipped with an action by the maximal $\mathbb{Z}_p$-order $\mathcal{O}_{F_0}$ in F_0. (This instance of the Serre tensor construction is easy to build, since $\mathcal{O}_F$ is a finite free $\mathcal{O}_{F_0}$-module; see 4.3.1 for a more general context. The key point is that on finite flat k-groups this construction becomes ordinary extension of scalars after passing to Dieudonné modules, so we get the module-theoretic properties over $\mathcal{O}_F$ from the known case of $\mathcal{O}_{F_0}$.) When $\mathsf{k} = \overline{\mathsf{k}}$, necessary and sufficient conditions for X_0 to arise as $\mathcal{O}_F \otimes_{\mathcal{O}_{F_0}} X'_0$ are given in Lemma 4.3.4(iii) and Proposition 4.3.5.

4.1.2.3. We now show that if n is odd (i.e., $q = p^{4m+2}$) then the unique $\mathbb{Z}[\zeta_5]$-linear formal abelian scheme deformation $\mathfrak{C}_0$ of C_0 over $W(\mathbb{F}_q)$ is not algebraizable, due to the tangential criterion in Theorem 2.2.3. Fix an algebraic closure $\overline{\mathbb{Q}}_p$ of $W(\mathbb{F}_q)[1/p]$, and let $\kappa = \mathbb{Z}[\zeta_5]/(p)$ (a finite field of size p^4). The action of $\mathbb{Z}[\zeta_5]$ on $\mathrm{Lie}(\mathfrak{C}_0)_{\overline{\mathbb{Q}}_p}$ corresponds to a subset

$$\Phi \subseteq \mathrm{Hom}_{\mathbb{Q}\text{-alg}}(\mathbb{Q}(\zeta_5), \overline{\mathbb{Q}}_p) = \mathrm{Hom}_{\mathbb{Z}\text{-alg}}(\mathbb{Z}[\zeta_5], \overline{\mathbb{Z}}_p).$$

Since p is inert in $\mathbb{Q}(\zeta_5)$, so $\mathrm{Hom}(\mathbb{Z}[\zeta_5], \overline{\mathbb{Z}}_p) = \mathrm{Hom}(\kappa, \overline{\mathbb{F}}_p)$, the subset Φ is determined by the $\mathbb{F}_q$-linear action of κ on $\mathrm{Lie}(\mathfrak{C}_0)/(p) = \mathrm{Lie}(C_0)$. More specifically, the $\mathbb{F}_q$-vector space $\mathrm{Lie}(C_0)$ is a module over

$$\kappa \otimes_{\mathbb{F}_p} \mathbb{F}_q = \prod_j \kappa \otimes_{j, \mathbb{F}_{p^2}} \mathbb{F}_q,$$

where the product is taken over the two $\mathbb{F}_p$-embeddings $\mathbb{F}_{p^2} \rightrightarrows \kappa$. But $q = p^{4m+2}$, so both local factor rings in this direct product ring are *fields* (of degree 2 over $\mathbb{F}_q$).

Since $\dim_{\mathbb{F}_q} \mathrm{Lie}(C_0) = 2$, there is a unique embedding $j_0 : \mathbb{F}_{p^2} \to \kappa$ such that $\mathrm{Lie}(C_0)$ is free of rank 1 over the field $\kappa \otimes_{j_0, \mathbb{F}_{p^2}} \mathbb{F}_q$. Passing to $C_0^{(p)}$ has the effect of replacing j_0 with the other embedding (so $C_0^{(p)}$ is *not* $\mathbb{Z}[\zeta_5]$-linearly isomorphic to C_0, as even their Lie algebras are not linearly isomorphic over $\kappa \otimes_{\mathbb{F}_p} \mathbb{F}_q$). We use j_0 to identify $\mathbb{F}_{p^2}$ as a subfield of κ, and conclude that Φ is stable under the action of the element τ in $\mathrm{Gal}(\mathbb{Q}(\zeta_5)/\mathbb{Q})$ that induces the unique involution of κ (i.e., the unique non-trivial $\mathbb{F}_{p^2}$-automorphism $x \mapsto x^{p^2}$ of κ). But τ is complex conjugation on $\mathbb{Q}(\zeta_5)$, so Φ violates the necessary and sufficient algebraizability criterion in 2.2.3 that it be disjoint from its complex conjugate.

For any perfect extension field k of $\mathbb{F}_q$, $(\mathfrak{C}_0)_{W(k)}$ is the universal formal deformation of $(C_0)_k$ to a $\mathbb{Z}[\zeta_5]$-linear formal abelian scheme over a complete local noetherian ring with residue field k. The tangential criterion 2.2.3 implies that $(\mathfrak{C}_0)_{W(k)}$ is not algebraizable, and likewise for $(\mathfrak{C}_0)_D$ for any domain D that is finite flat over $W(k)$. In particular, $(C_0)_k$ has no $\mathbb{Z}[\zeta_5]$-linear lifting to characteristic 0 (since any such lift gives one over some D, by a slicing argument with formal deformation rings as explained early in 2.1.1).

Remark. The above argument also shows that if n is odd then the CM structure

$$(C_0, \mathbb{Z}[\zeta_5] \to \mathrm{End}(C_0)) \times_{\mathrm{Spec}(\mathbb{F}_{p^{2n}})} \mathrm{Spec}\,\overline{\mathbb{F}}_p$$

cannot be lifted to a pair $(C, \mathbb{Z}[\zeta_5] \to \mathrm{End}(C))$ over a local domain R of characteristic 0 with residue field $\overline{\mathbb{F}}_p$, where C is an abelian scheme over R. One can also prove this statement directly: the key point is that for an algebraic closure F of $\mathrm{Frac}(R)$, the $(\mathbb{Z}[\zeta_5] \otimes_{\mathbb{F}_{p^{2n}}} \overline{\mathbb{F}}_p)$-module structure for the Lie algebra $\mathrm{Lie}(C_0) \otimes_{\mathbb{F}_{p^2}} \overline{\mathbb{F}}_p$ determines the $(\mathbb{Z}[\zeta_5] \otimes_{\mathbb{Z}} \overline{F})$-module structure for the Lie algebra $\mathrm{Lie}(C) \otimes_R \overline{F}$ of the geometric generic fiber of C (if such a lifting exists).[1] See 4.6.3.2 for a generalization to higher-dimensional abelian varieties.

4.1.2.4. We continue to assume that n is odd (i.e., $q = p^{4m+2}$). The above non-algebraizability result in such cases shows that there does not exist a $\mathbb{Z}[\zeta_5]$-linear lifting of C_0 to characteristic 0, but a CM lifting that lifts the action of the order $\mathcal{O}' = \mathbb{Z} + p\mathbb{Z}[\zeta_5]$ without increasing the residue field can be constructed. The rest of 4.1.2 is devoted to proving this, via methods which will be generalized in 4.1.3.

The first key point is to show that the kernel H_0 of the Frobenius isogeny $C_0 \to C_0^{(p)}$ is isomorphic to $\alpha_p \times \alpha_p$. The q-Frobenius isogeny of C_0 is the endomorphism $\pm p^n \cdot \zeta_5 : C_0 \to C_0$, so all slopes of $C_0[p^\infty]$ are positive and hence the étale part vanishes. But $C_0[p^\infty]$ is isogenous to its dual, so it also has vanishing multiplicative part and thus is local-local (with dimension $\dim(C_0) = 2$). We conclude that $C_0[p]$ is local-local, so H_0 of order $p^{\dim C_0} = p^2$ is local-local. By using the $\mathbb{F}_q$-linear action of the finite field $\kappa = \mathbb{Z}[\zeta_5]/(p)$ on $\mathrm{Lie}(C_0)$ we shall prove that the Verschiebung morphism of H_0 vanishes, so $H_0 \simeq \alpha_p \times \alpha_p$.

In 4.1.2.3 we saw that the κ-action on H_0 makes $\mathrm{Lie}(H_0) = \mathrm{Lie}(C_0)$ a 1-dimensional vector space over a factor field $\kappa \otimes_{j_0, \mathbb{F}_{p^2}} \mathbb{F}_q$, thereby defining a preferred embedding j_0 of $\mathbb{F}_{p^2}$ into κ (recovering the $\mathbb{F}_{p^2}$-linear structure induced by the $\mathbb{F}_q$-linear structure on this Lie algebra). The $\mathbb{F}_{p^2}$-action on H_0 arising from the κ-action induces an $\mathbb{F}_{p^2}$-action on the Dieudonné module $\mathrm{M}^*(H_0)$ coinciding with the action arising from the $\mathbb{F}_q$-linear structure, due to the natural $\mathbb{F}_q$-linear isomorphism $\mathrm{M}^*(H_0) \simeq \mathrm{Lie}(H_0)^\vee$ (as for any commutative infinitesimal group scheme with vanishing Frobenius morphism; see 1.4.3.2(4)). But the κ-action on this Dieudonné module commutes the Verschiebung operator $\mathcal{V}$, so it follows that $\mathcal{V}$ is *linear* for the natural $\mathbb{F}_{p^2}$-linear structure on $\mathrm{M}^*(H_0)$ arising from the fact that H_0 is a group over $\mathbb{F}_q \supseteq \mathbb{F}_{p^2}$. Since $\mathcal{V}$ is always *semilinear* over the absolute Frobenius automorphism on $\mathbb{F}_q$, it must therefore vanish. That is, $H_0 \simeq \alpha_p \times \alpha_p$ as desired.

---

[1] More generally, in the notation of 4.2.2, the map $\mathbf{d} : \mathrm{R}_K(\mathcal{O}_F) \to \mathrm{R}_k(\mathcal{O}_F)$ in 4.2.2.1 is an isomorphism for any unramified finite extension F of $\mathbb{Q}_p$, such as $F = \mathbb{Q}_p(\zeta_5)$ for $p \equiv \pm 2 \pmod 5$.

4.1.2.5. Now we consider the base change $C_0' = (C_0)_k$ and its Frobenius kernel $H' \simeq \alpha_p \times \alpha_p$, where k is a perfect extension of $\mathbb{F}_q = \mathbb{F}_{p^{4m+2}}$ containing a subfield of order p^4 (so $\kappa \otimes_{\mathbb{F}_p} k$ splits as a finite étale k-algebra of rank 4). We will find an abelian variety A_0 over k with a-number 1 (i.e., a unique α_p-subgroup) and a $\mathbb{Z}[\zeta_5]$-action on A_0 such that C_0' equipped with its $\mathbb{Z}[\zeta_5]$-action is the quotient of A_0 by its unique (hence $\mathbb{Z}[\zeta_5]$-stable) α_p-subgroup. The construction of A_0 will be designed so that the unique $\mathbb{Z}[\zeta_5]$-linear formal abelian scheme deformation $\mathfrak{A}$ of A_0 over $W(k)$ (via Proposition 4.1.2.1) satisfies the tangential algebraizability criterion in Theorem 2.2.3.

A suitable p-isogenous quotient of this algebraization (after a preliminary finite extension on $W(k)[1/p]$) will be used to construct a CM lifting of C_0 with its action by the order $\mathbb{Z} + p\mathbb{Z}[\zeta_5]$. The method involves constructing a lift after replacing C_0' with $(C_0')_{k'}$ for a finite extension k' of k, and then using an argument with deformation rings to carry out the construction for C_0 over $\mathbb{F}_q$ (i.e., avoiding the intervention of non-trivial extensions of $\mathbb{F}_q$).

The first key step is to understand the κ-action on the k-group H'. The Lie functor defines an additive isomorphism $\mathrm{End}(H') \simeq \mathrm{End}(\mathrm{Lie}(H'))$ (since H' is a power of α_p and $\mathrm{End}(\alpha_p) \to \mathrm{End}(\mathrm{Lie}(\alpha_p))$ is bijective), so there is a unique k-linear structure on H' lifting that on its Lie algebra. The k-linear action of κ on $\mathrm{Lie}(C_0') = \mathrm{Lie}(H')$ makes this Lie algebra into a module over $\kappa \otimes_{\mathbb{F}_p} k$. But $\mathrm{M}^*((C_0')[p])$ is a free $\kappa \otimes_{\mathbb{F}_p} k$-module of rank 1 (due to Proposition 1.4.3.9(2)), and the 2-dimensional k-linear dual $\mathrm{Lie}(C_0')^\vee$ is a quotient module, so the κ-action on $\mathrm{Lie}(C_0')$ decomposes into eigenlines over k for a subset $\overline{\Phi} \subset \mathrm{Hom}_{\mathbb{F}_p\text{-alg}}(\kappa, k)$ of size 2 (i.e., there is not a repeated eigencharacter).

The set $\overline{\Phi}$ is stable under precomposition with the action of the unique involution τ of κ because the canonical k-linear isomorphism $\mathrm{Lie}(C_0) \otimes_{\mathbb{F}_q} k \simeq \mathrm{Lie}(C_0')$ is also κ-linear (and the $\mathbb{F}_q$-linear κ-action on $\mathrm{Lie}(C_0)$ makes it a 1-dimensional vector space over $\kappa \otimes_{j_0, \mathbb{F}_{p^2}} \mathbb{F}_q$ for an $\mathbb{F}_p$-embedding $j_0 : \mathbb{F}_{p^2} \to \kappa$). Thus, $\overline{\Phi} = \{\varphi, \varphi \circ \tau\}$ for some $\mathbb{F}_p$-embedding $\varphi : \kappa \to k$.

For any $\mathbb{F}_p$-embedding $\chi : \kappa \to k$, let $\alpha_p(\chi)$ denote the k-group α_p equipped with the κ-action composing χ with the natural k-action on α_p. The identification $\mathrm{Lie}(H') = \mathrm{Lie}(C_0')$ defines a κ-linear product decomposition of k-groups

$$H' = H'_\varphi \times H'_{\varphi \circ \tau}$$

with $H'_\varphi \simeq \alpha_p(\varphi)$ and $H'_{\varphi\circ\tau} \simeq \alpha_p(\varphi \circ \tau)$.

Let $B_0 = C_0'/H'_{\varphi\circ\tau}$. The Frobenius kernel of this $\mathbb{Z}[\zeta_5]$-linear abelian surface contains H'_φ and its Lie algebra is equal to the 2-dimensional $\mathrm{Lie}(B_0)$. When this Lie algebra is considered as a $\kappa \otimes_{\mathbb{F}_p} k$-module it has two eigenlines over k for the κ-action, corresponding to a pair of distinct embeddings $\kappa \rightrightarrows k$. One of these embeddings is φ (with associated eigenline $\mathrm{Lie}(H'_\varphi)$), and we now determine the other embedding by a computation with Dieudonné modules (and this computation will be applied later to a certain class of p-divisible groups).

4.1.2.6. Lemma. *The embeddings $\kappa \rightrightarrows k$ arising from the k-linear κ-action on the Lie algebra of the Frobenius kernel of B_0 are $\{\varphi, \varphi^{1/p}\}$. Moreover, $a(B_0) = 1$ (so H'_φ is the unique subgroup of B_0 isomorphic to α_p).*

PROOF. The $W(\kappa)$-linear Dieudonné module M of $C_0'[p^\infty]$ is free of rank 1 over $W(\kappa) \otimes_{\mathbb{Z}_p} W(k)$ (by 1.4.3.9(2)). Letting $\sigma : k \simeq k$ be the absolute Frobenius automorphism $c \mapsto c^p$, we pick a $W(\kappa) \otimes_{\mathbb{Z}_p} W(k)$-basis of M to identify $\mathcal{F} : M \to M$ with a $1 \otimes \sigma$-semilinear endomorphism $\mathcal{F}_M$ of $W(\kappa) \otimes_{\mathbb{Z}_p} W(k)$ given by $x \mapsto t \cdot (1 \otimes \sigma)(x)$ for some $t \in W(\kappa) \otimes_{\mathbb{Z}_p} W(k)$ that divides p (since $\mathcal{F} \circ \mathcal{V} = p$). Thus, the Verschiebung morphism for $C_0'[p^\infty]$ corresponds to the $1 \otimes \sigma^{-1}$-semilinear map $\mathcal{V}_M : x \mapsto (p/(1 \otimes \sigma^{-1})(t)) \cdot (1 \otimes \sigma^{-1})(x)$.

Under the canonical decomposition

$$W(\kappa) \otimes_{\mathbb{Z}_p} W(k) \simeq \prod_{j:\kappa \to k} W(k)_j$$

into a product of copies of $W(k)$ indexed by the embeddings $j : \kappa \to k$, t corresponds to a 4-tuple (t_j) and $\mathcal{F}_M$ carries the direct factor $W(k)_j$ into the direct factor $W(k)_{\sigma \circ j}$. Thus, as a $W(\kappa) \otimes_{\mathbb{Z}_p} W(k)$-module,

$$\mathrm{M}^*(\ker \mathrm{Fr}_{C_0'[p^\infty]/k}) = \prod_j (W(k)/\sigma(t_{\sigma^{-1} \circ j})).$$

This is p-torsion and 2-dimensional over k since $C_0'[p^\infty]$ has Frobenius kernel $\alpha_p \times \alpha_p$, so exactly two of the t_j's are in $W(k)^\times$ and the other two are in $p \cdot W(k)^\times$. Since the Frobenius kernel of $C_0'[p^\infty]$ has the same Lie algebra as C_0', and we know that the κ-action on the Lie algebra is given by a τ-orbit of embeddings $\kappa \to k$, the j's such that t_j is a non-unit are a τ-orbit $\{\varphi, \varphi \circ \tau\}$ and so likewise for the set $\{\varphi^p, \varphi^p \circ \tau = \varphi^{1/p}\}$ of j's such that t_j is a unit.

We conclude that for some embedding $\varphi : \kappa \to k$,

$$t = (pu_\varphi, u_{\varphi^p}, pu_{\varphi \circ \tau}, u_{\varphi^p \circ \tau})$$

for units $u_j \in W(k)_j^\times$. In particular, $\mathcal{F}_M^2$ is p times a $1 \otimes \tau$-semilinear automorphism, so $\mathrm{Fr}_{C_0'[p^\infty]/k, p^2}$ is p times an isomorphism. Hence, the Frobenius kernel of $C_0'[p^\infty]/H'_{\varphi \circ \tau}$ is contained in $G := C_0'[p]/H'_{\varphi \circ \tau}$, so it suffices to prove two properties of the κ-vector group G:

(i) the eigenlines over k for the κ-action on $\mathrm{Lie}(\ker \mathrm{Fr}_{G/k})$ are given by the embeddings $\varphi, \varphi^{1/p} : \kappa \rightrightarrows k$,
(ii) $a(G) = 1$.

Both (i) and (ii) will follow from a direct computation with $\mathrm{M}^*(C_0'[p]) = M/pM$.

The Frobenius operator on $M/pM = k_\varphi \times k_{\varphi^p} \times k_{\varphi \circ \tau} \times k_{\varphi^p \circ \tau}$ is the σ-semilinear endomorphism

$$(c_\varphi, c_{\varphi^p}, c_{\varphi \circ \tau}, c_{\varphi^p \circ \tau}) \mapsto (0, \overline{u}_{\varphi^p} c_\varphi^p, 0, \overline{u}_{\varphi^p \circ \tau} c_{\varphi \circ \tau}^p)$$

where $\overline{u}_j = u_j \pmod p$. Likewise, the Verschiebung operator on M/pM is the σ^{-1}-semilinear endomorphism

$$(c_\varphi, c_{\varphi^p}, c_{\varphi \circ \tau}, c_{\varphi^p \circ \tau}) \mapsto (0, (c_{\varphi \circ \tau}/\overline{u}_{\varphi \circ \tau})^{1/p}, 0, (c_\varphi/\overline{u}_\varphi)^{1/p}).$$

The inclusion of $\ker \mathrm{Fr}_{C_0'[p]/k} = H'_\varphi \circ H'_{\varphi \circ \tau}$ into $C_0'[p]$ corresponds to the canonical projection from M/pM onto the cokernel of $\mathcal{F}$; i.e., it is the canonical projection onto $k_\varphi \times k_{\varphi \circ \tau}$ (on which $\mathcal{F}$ vanishes by design, and $\mathcal{V}$ must vanish either because $a(C_0'[p^\infty]) = 2$ or by direct computation). Thus, the inclusion of $H'_{\varphi \circ \tau}$ into $C_0'[p]$ corresponds to the canonical projection of M/pM onto $k_{\varphi \circ \tau}$, so the Dieudonné submodule $\mathrm{M}^*(G) \subset M/pM$ is $k_\varphi \times k_{\varphi^p} \times k_{\varphi^p \circ \tau}$ on which $\mathcal{F}$ is the σ-semilinear endomorphism

$$(c_\varphi, c_{\varphi^p}, c_{\varphi^p \circ \tau}) \mapsto (0, \overline{u}_{\varphi^p} c_\varphi^p, 0)$$

and $\mathcal{V}$ is the σ^{-1}-semilinear endomorphism

$$(c_\varphi, c_{\varphi^p}, c_{\varphi^p\circ\tau}) \mapsto (0, 0, (c_\varphi/\overline{u}_\varphi)^{1/p}).$$

The $\mathcal{F}$-cokernel $k_\varphi \times k_{\varphi^p\circ\tau}$ of $\mathrm{M}^*(G)$ is the Dieudonné module of $\ker \mathrm{Fr}_{G/k}$. Thus, the κ-action on $\mathrm{Lie}(G)$ is through the embeddings φ and $\varphi^p\circ\tau = \varphi^{1/p}$ of κ into k, as desired. The $\mathcal{V}$-operator on $\mathrm{M}^*(\ker \mathrm{Fr}_{G/k})$ is $(c_\varphi, c_{\varphi^p\circ\tau}) \mapsto (0, (c_\varphi/\overline{u}_\varphi)^{1/p})$, so $a(G) = 1$. □

Apply the preceding considerations with C_0' replaced by the $\mathbb{Z}[\zeta_5]$-linear $C_0'^{(p)}$ (whose Lie algebra supports the embeddings $\varphi^p, \varphi^p \circ \tau : \kappa \rightrightarrows k$ via its k-linear κ-action) and with $H'_{\varphi\circ\tau}$ replaced by $H''_{\varphi^p\circ\tau} := (H'_{\varphi\circ\tau})^{(p)} \simeq \alpha_p(\varphi^p \circ \tau)$. This proves that the $\mathbb{Z}[\zeta_5]$-linear abelian variety $B_0^{(p)}$ satisfies $a(B_0^{(p)}) = 1$ and its unique α_p has the form $\alpha_p(\varphi^p)$. The quotient of $B_0^{(p)}$ modulo this subgroup is the quotient of $C_0'^{(p)}$ modulo its entire Frobenius kernel, so the projection to this quotient is a $\mathbb{Z}[\zeta_5]$-linear degree-p isogeny $B_0^{(p)} \twoheadrightarrow C_0'^{(p^2)}$. Undoing the 2-fold Frobenius twist on C_0' over the perfect field k, $A_0 := B_0^{(1/p)}$ is a $\mathbb{Z}[\zeta_5]$-linear abelian variety with a-number 1 that admits C_0' as a $\mathbb{Z}[\zeta_5]$-linear quotient via a degree-p isogeny.

4.1.2.7. Lemma. *There is a unique $\mathbb{Z}[\zeta_5]$-linear abelian scheme A over $W(k)$ lifting A_0.*

PROOF. The infinitesimal deformation theory of A_0 with its $\mathbb{Z}[\zeta_5]$-action is formally étale (Proposition 4.1.2.1), so there is a unique $\mathbb{Z}[\zeta_5]$-linear formal abelian scheme $\mathfrak{A}$ over $W(k)$ lifting A_0. Our problem is precisely to prove that $\mathfrak{A}$ is algebraic.

By the tangential criterion in 2.2.3, this amounts to the action of $\mathbb{Q}(\zeta_5)$ on a geometric fiber of the $W(k)[1/p]$-vector space $\mathrm{Lie}(\mathfrak{A})[1/p]$ being a CM type. Since $\mathbb{Z}[\zeta_5] \subset \mathrm{End}(A_0)$ and complex conjugation on $\mathbb{Q}(\zeta_5)$ induces the involution τ on $\kappa = \mathbb{Z}[\zeta_5]/(p)$, it is equivalent to show that the κ-action on $\mathrm{Lie}(A_0) = \mathrm{Lie}(B_0)^{(1/p)}$ is disjoint from its τ-conjugate. By Lemma 4.1.2.6, the set of embeddings $\kappa \rightrightarrows k$ occurring on $\mathrm{Lie}(A_0)$ is $\{\varphi^{1/p}, \varphi^{1/p^2} = \varphi\circ\tau\}$; this is disjoint from its τ-conjugate. □

Pick the minimal extension K of $W(k)[1/p]$ such that the 1-dimensional κ-vector group $A[p]_K$ over K is constant (a cyclic extension of degree dividing p^4-1), and let k' be the residue field of K. Pick an $\mathbb{F}_p$-line in $A[p](K)$, and let $\mathcal{G}$ be its finite flat Zariski closure inside $A[p]$ over $\mathcal{O}_K$. This is killed by $p\mathbb{Z}[\zeta_5]$, so it is stable under the action of the order $\mathcal{O}' = \mathbb{Z} + p\mathbb{Z}[\zeta_5]$. Thus, the abelian scheme $A/\mathcal{G}$ is equipped with an evident action by $\mathcal{O}'$. The special fiber $\mathcal{G}_0$ of $\mathcal{G}$ and the kernel of $(A_0)_{k'} \to (C_0')_{k'}$ are both order-p subgroups of $(A_0)_{k'}$, so each is isomorphic to α_p. These subgroups coincide since $a(A_0) = 1$. We conclude that $C := A/\mathcal{G}$ is an $\mathcal{O}'$-linear lift of $(C_0)_{k'}$ over $\mathcal{O}_K$.

By Proposition 1.4.4.5, there is a formal deformation ring $\mathcal{R}$ of the abelian variety C_0 *equipped with* its action by the order $\mathcal{O}'$; it is complete local noetherian with residue field $\mathbb{F}_q$ (and a quotient of the formal deformation ring of C_0). The compatibility of this deformation ring with respect to "change of coefficients" (as recorded in Proposition 1.4.4.14) implies that the deformation ring is universal relative to formal deformations over complete local noetherian rings with any perfect residue field k containing $\mathbb{F}_q$ (e.g., $W(k)\widehat{\otimes}_{W(\mathbb{F}_q)}\mathcal{R}$ is the analogous formal deformation ring with residue field k). Thus, C as an $\mathcal{O}'$-linear lift of C_0 is classified by a local map

$\mathcal{R} \to \mathcal{O}_K$ lifting the inclusion of residue fields $\mathbb{F}_q \hookrightarrow k'$, so $\mathcal{R}[1/p] \neq 0$. The quotient of $\mathcal{R}[1/p]$ modulo any maximal ideal is of finite degree over $W(\mathbb{F}_q)[1/p]$ [**54**, 7.1.9], so the image D of $\mathcal{R}$ in such a quotient is a characteristic 0 (*non*-normal) $\mathbb{Z}_p$-finite local domain with residue field $\mathbb{F}_q$ over which the universal formal deformation defines an $\mathcal{O}'$-linear lift of C_0 to a *formal* abelian scheme over D.

4.1.2.8. The preceding argument with formal deformations will adapt nicely to the setting of p-divisible groups (since their formal deformation theory involves no "algebraization" problem, in contrast with the case of abelian varieties). However, we seek an $\mathcal{O}'$-linear lift of C_0 (not merely $(C_0)_{k'}$) that is an abelian scheme rather than just a formal abelian scheme. This requires incorporating polarizations into the deformation problem that we consider.

A more general version of this algebraization problem will arise in the final step of the proof of Theorem 4.1.1 (see 4.6.4.3). The present discussion serves solely as motivation for the general arguments in 4.6.4.3, so here we just sketch the ideas behind the deformation theory technique by which the algebraization problem (without increasing the residue field) is solved in general in 4.6.4.3; a simpler and complete argument specific to the above class of abelian surfaces will be given at the end of 4.1.3 via the algebraization criterion in 2.2.3.

Pick an L-linear polarization ϕ of $(A/\mathcal{G})_K$. Use 2.2.2 and 2.1.1.1 to show that after multiplying it by a sufficiently divisible positive integer, the extended homomorphism $A/\mathcal{G} \to (A/\mathcal{G})^t$ over $\mathcal{O}_K$ has special fiber that is a polarization of $(C_0)_{k'}$ which is *defined over* $\mathbb{F}_q$. The degree of this polarization and the $\mathcal{O}'$-action define a formal deformation ring $\mathcal{R}$ for C_0 with residue field $\mathbb{F}_q$. The $\mathcal{O}'$-linear lift over $\mathcal{O}_K$ of the polarized C_0 is classified by a $W(\mathbb{F}_q)$-algebra map $\mathcal{R} \to \mathcal{O}_K$. By Lemma 2.1.1.2, this factors through a quotient D' of $\mathcal{R}$ that is a $W(\mathbb{F}_q)$-order with residue field $\mathbb{F}_q$. The pullback of the universal (algebraized) structure over $\mathcal{R}$ is the desired $\mathcal{O}'$-linear lift of C_0 over a p-adic order without increasing the residue field.

4.1.3. Toy models. The basic features of the pair $(C_0, \mathbb{Z}[\zeta_5] \to \mathrm{End}(C_0))$ over $\mathbb{F}_{p^{4m+2}}$ that was studied in 4.1.2 are encoded in the following definition.

4.1.3.1. Definition. A *toy model* abelian variety over a field k of characteristic $p > 0$ is a CM abelian surface $(C, \mathcal{O} \hookrightarrow \mathrm{End}(C))$ over k such that:

- $\mathcal{O}$ is an order in a quartic CM field L, $p \nmid [\mathcal{O}_L : \mathcal{O}]$, and p is inert (i.e., p is unramified and remains prime in $\mathcal{O}_L$, or equivalently $\mathcal{O} \otimes_{\mathbb{Z}} \mathbb{Z}_p \simeq W(\mathbb{F}_{p^4})$),
- for the finite field $\kappa = \mathcal{O}/p\mathcal{O} = \mathcal{O}_L/p\mathcal{O}_L$ with p^4 elements, the subset $\overline{\Phi} \subset \mathrm{Hom}_{\mathbb{F}_p\text{-alg}}(\kappa, \overline{k})$ corresponding to the κ-action on $\mathrm{Lie}(C)_{\overline{k}}$ is stable under the action of the unique involution τ of κ.

In 4.2.10 the toy model abelian varieties over finite fields and algebraically closed fields of characteristic $p > 0$ will be classified up to prime-to-p isogeny. For every p there will be examples arising from the isogeny class of $E \times E$ for supersingular elliptic curves E with all geometric endomorphisms defined over the ground field, and the only other examples over finite fields turn out to be the ones studied in 4.1.2 over the fields $\mathbb{F}_{p^{2n}}$ for $p \equiv \pm 2 \pmod 5$ (subject to an additional condition when n is even). This classification is not needed for our purposes; it will be more convenient to work with the following analogue of the notion of toy model that is adapted to the setting of p-divisible groups.

4.1.3.2. Definition. Let κ be a finite field of characteristic p with size p^4, and let k be a field of characteristic p. A *toy model* p-divisible group over k is a p-divisible group X_0 over k of dimension 2 and height 4 equipped with a $W(\kappa)$-linear action such that the subset of $\mathrm{Hom}_{\mathbb{F}_p\text{-alg}}(\kappa, \overline{\mathsf{k}})$ encoding the $\overline{\mathsf{k}}$-linear action of $W(\kappa)/pW(\kappa) \simeq \kappa$ on the 2-dimensional $\mathrm{Lie}(X_0)_{\overline{\mathsf{k}}}$ is stable under the action of the unique involution τ of κ.

If $(C, \mathcal{O} \to \mathrm{End}(C))$ is a toy model abelian variety over k then $C[p^\infty]$ with its action by $\mathbb{Z}_p \otimes_{\mathbb{Z}} \mathcal{O} = \mathbb{Z}_p \otimes_{\mathbb{Z}} \mathcal{O}_L \simeq W(\mathbb{F}_{p^4})$ is a toy model p-divisible group over k because naturally $\mathrm{Lie}(C[p^\infty]) \simeq \mathrm{Lie}(C)$. In 4.2.10 we will show that every toy model p-divisible group over an algebraically closed field of characteristic p arises from a toy model abelian variety.

If there is a toy model p-divisible group X_0 (or toy model abelian variety C) over k then k *must* contain a subfield $\mathbb{F}_{p^2}$ of order p^2 and the κ-action on $\mathrm{Lie}(X_0)_{\overline{\mathsf{k}}} = \mathrm{Lie}(X_0[p])_{\overline{\mathsf{k}}}$ must be given by a pair of *distinct* embeddings of κ into $\overline{\mathsf{k}}$ (and likewise for $\mathrm{Lie}(C) = \mathrm{Lie}(C[p^\infty])$ for toy model abelian varieties C over k).

To prove that k contains a subfield $\mathbb{F}_{p^2}$ of size p^2, we may replace k with its perfect closure k. The Dieudonné module $\mathrm{M}^*(X_0[p])$ is free of rank 1 over $\kappa \otimes_{\mathbb{F}_p} k$ (since $\mathrm{M}^*(X_0)$ is free of rank 1 over $W(\kappa) \otimes_{\mathbb{Z}_p} W(k)$ by 1.4.3.9(2)), so the κ-action on the 2-dimensional quotient $\mathrm{Lie}(X_0[p])^\vee_{\overline{k}}$ of $\mathrm{M}^*(X_0[p])_{\overline{k}} = \mathrm{M}^*(X_0[p]_{\overline{k}})$ is given by two distinct embeddings of κ into $\overline{k}$. The Lie algebra $\mathrm{Lie}(X_0)$ is a 2-dimensional k-vector space with a compatible structure of $\kappa \otimes_{\mathbb{F}_p} k$-module and $[\kappa : \mathbb{F}_p] > 2$, so $\kappa \otimes_{\mathbb{F}_p} k$ cannot be a field and thus k contains a subfield $\mathbb{F}_{p^2}$ of order p^2.

Now we return to working over a general field k. The τ-stability condition in the second part of 4.1.3.2 implies that the action of κ on $\mathrm{Lie}(X_0)$ makes the quadratic subfield of κ act through a single isomorphism onto $\mathbb{F}_{p^2} \subset \mathsf{k}$, so we may and do unambiguously view $\mathbb{F}_{p^2}$ as a subfield of κ. Equivalently, the requirement for the κ-action on $\mathrm{Lie}(X_0)_{\overline{\mathsf{k}}}$ in the definition of a toy model p-divisible group says exactly that k contains a subfield $\mathbb{F}_{p^2}$ of order p^2 and that the $\kappa \otimes_{\mathbb{F}_p} \mathsf{k}$-module structure on $\mathrm{Lie}(X_0)$ factors through an invertible module structure over $\kappa \otimes_{j_0, \mathbb{F}_{p^2}} \mathsf{k}$ for a unique embedding $j_0 : \mathbb{F}_{p^2} \to \kappa$.

The reason we restricted to the case $q = p^{2n}$ with odd n in 4.1.2 is because in such cases the Lie algebra is automatically a module over $\kappa \otimes_{j_0, \mathbb{F}_{p^2}} \mathbb{F}_q$ (a field when $2 \nmid n$) for some j_0. Since such a condition has just been seen to follow formally from the definition of a toy model, many of the arguments for C_0 with its action by $\mathbb{Z}[\zeta_5]$ over $\mathbb{F}_{p^{2n}}$ in 4.1.2 with odd n carry over to any toy model abelian variety $(C, \mathcal{O} \to \mathrm{End}(C))$ or toy model p-divisible group X_0 over any perfect field k of characteristic $p > 0$. For example:

4.1.3.3. Proposition. *The 2-dimensional X_0 is local-local (so $C[p^\infty]$ is local-local for any toy model abelian variety C over k) and $\ker \mathrm{Fr}_{X_0/\mathsf{k}} \simeq \alpha_p \times \alpha_p$ as k-groups (equivalently, $a(X_0) = 2$). In particular, any toy model abelian variety over k has a-number equal to 2.*

PROOF. We may replace k with its perfect closure k. To establish the local-local property, consider the Dieudonné model $\mathrm{M}^*(X_0)$. This is free of rank 1 over $W(\kappa) \otimes_{\mathbb{Z}_p} W(k)$ (by Proposition 1.4.3.9(2)), so there is no non-zero proper $W(\kappa) \otimes_{\mathbb{Z}_p} W(k)$-submodule that is stable under the absolute Frobenius operator (since the Frobenius automorphism of $W(k)$ transitively permutes the local factors of $W(\kappa) \otimes_{\mathbb{Z}_p} W(k)$). This forces the vanishing of the étale and multiplicative parts of the 2-dimensional p-divisible group X_0 of height 4, so X_0 is local-local as claimed.

To show that $a(X_0) = 2$, first note that as in the study of C_0 over $\mathbb{F}_{p^{4m+2}}$, the Frobenius kernel H of X_0 has order p^2 and we just have to show that the Verschiebung operator $\mathcal{V}$ on $\mathrm{M}^*(H)$ vanishes. The κ-action on $\mathrm{M}^*(H)$ commutes with $\mathcal{V}$, so the same holds for the action on $\mathrm{M}^*(H)$ by the subfield $\mathbb{F}_{p^2} \subset \kappa$. The natural isomorphism $\mathrm{M}^*(H) \simeq \mathrm{Lie}(H)^\vee$ is k-linear, so the κ-action on $\mathrm{M}^*(H)$ induces the $\mathbb{F}_{p^2}$-linear structure *arising from* the k-linear structure (due to how the embedding of $\mathbb{F}_{p^2}$ into κ has been defined using the Lie algebra $\mathrm{Lie}(H) = \mathrm{Lie}(X_0)$). It follows that $\mathcal{V}$ acts linearly for the $\mathbb{F}_{p^2}$-linear structure on $\mathrm{M}^*(H)$ arising from the k-linear structure. But relative to the k-linear structure $\mathcal{V}$ also acts semilinearly over the absolute Frobenius on k, so $\mathcal{V}$ acts semilinearly over (the inverse of) the absolute Frobenius on $\mathbb{F}_{p^2}$. The linearity of $\mathcal{V}$ over the *same* $\mathbb{F}_{p^2}$-action therefore forces $\mathcal{V} = 0$. □

4.1.3.4. Let Λ be a Cohen ring for k, and $K_0 = \Lambda[1/p]$. The $W(\kappa)$-linear deformation theory of X_0 over Λ works out similarly to our earlier study of C_0: by 4.1.2.1 there is a unique $W(\kappa)$-linear deformation X of X_0 over Λ, and inspection of the κ-action on the geometric fiber $\mathrm{Lie}(X_0)_{\overline{k}}$ shows that the $W(\kappa)[1/p]$-action on the geometric fiber $\mathrm{Lie}(X)[1/p]_{\overline{K}_0}$ is given by a pair of $\mathbb{Q}_p$-embeddings $W(\kappa)[1/p] \rightrightarrows \overline{K}_0$ that are permuted by τ. Similarly, if $(C, \mathcal{O})$ is a toy model abelian variety over k then C has a unique $\mathcal{O}$-linear formal abelian surface deformation $\mathfrak{C}$ over Λ and the action of $L = \mathrm{Frac}(\mathcal{O})$ on $\mathrm{Lie}(\mathfrak{C})[1/p]_{\overline{K}_0}$ is stable under the complex conjugation of L, so it is *not* a CM type. In particular, $\mathfrak{C}$ is non-algebraizable.

Of the utmost importance is that the procedure for passing to a $\mathbb{Z}[\zeta_5]$-linearly p-isogenous abelian variety over $\mathbb{F}_{p^{4m+2}}$ with a-number equal to 1 via 4.1.2.6ff carries over without change to the $W(\kappa)$-linear X_0 over a perfect field k except that the appeal to 1.4.4.14 has to be replaced with an appeal to 1.4.5.5(1). (In particular, the Dieudonné module computations in the proof of Lemma 4.1.2.6 apply verbatim with X_0 in the role of $C_0[p^\infty]$.)

This yields a $(\mathbb{Z}_p + pW(\kappa))$-linear deformation X of X_0 over $W(k)$ such that the pair of $\mathbb{Q}_p$-embeddings $W(\kappa)[1/p] \rightrightarrows \overline{K}_0$ encoding the action of $W(\kappa)[1/p]$ on $\mathrm{Lie}(X)[1/p]_{\overline{K}_0}$ is *disjoint* from its τ-conjugate. (There is no need to include a quasi-polarization in the deformation problem, since there is no "algebraization" problem for formal deformations of p-divisible groups, in contrast with the case of abelian schemes.) In the same way, for any toy model abelian variety $(C, \mathcal{O})$ over a perfect field k we can build a $(\mathbb{Z} + p\mathcal{O})$-linear formal abelian surface deformation $\mathcal{C}$ over $W(k)$ such that $\mathcal{C}$ satisfies the tangential algebraization criterion in 2.2.3. In particular, $\mathcal{C}$ algebraizes to an abelian surface over $W(k)$ that is a $(\mathbb{Z} + p\mathcal{O})$-linear lift of C. This completes our preliminary study of toy models. They play a key role in the proof of Theorem 4.1.1; see 4.1.7(b).

4.1.4. Preliminaries. The concepts and tools needed in the proof of Theorem 4.1.1 are defined and developed in §4.2–§4.5, and these are brought together to give the proof in §4.6. Since it is a long story, as a convenience to the reader we provide an overview the ingredients in 4.1.4–4.1.6, and in 4.1.7 we sketch how they are combined to give the proof (followed by some remarks and open questions in

4.1.8–4.1.9). The reader may skip all of this and go straight to §4.2 without missing anything from a purely logical point of view.

Changing B by an $\mathbb{F}_q$-isogeny in accordance with 1.7.4.5 enables us to *arrange that $\mathcal{O}_L$ operates on B*, so for any order $\mathcal{O}$ in $\mathcal{O}_L$ we may view B as an $\mathcal{O}$-linear abelian variety. For any such $\mathcal{O}$, in 4.6.4.3 we will show via a formal deformation theory argument with $\mathcal{O}$-linear abelian schemes equipped with an $\mathcal{O}$-linear polarization that the existence of a CM lifting is a *geometric* property: a pair $(A_0, L \to \mathrm{End}^0(A_0))$ consisting of an abelian variety over a finite field $\mathbb{F}_q$ equipped with complex multiplication by a CM field L can be lifted to characteristic 0 if (and only if) its base change to an algebraic closure k of $\mathbb{F}_q$ can be CM lifted. Thus, in 1.8.5 the statement (CML) is equivalent to the apparently weaker statement (R).

We conclude that to prove Theorem 4.1.1 it suffices to show that if $\mathcal{O}_L \subset \mathrm{End}(B)$ then there is an $\mathbb{F}_q$-isogeny $B \to B'$ such that for the induced CM structure $L \hookrightarrow \mathrm{End}^0(B) = \mathrm{End}^0(B')$ the base change B'_k with its L-action is the specialization of an abelian variety with CM by L over a finite extension K of the fraction field K_0 of $W(k)$. We will arrange the construction of this B' so that its CM order contains $\mathcal{O}' := \mathbb{Z} + p\mathcal{O}_L$; the order $\mathcal{O}'$ is well-suited to lifting via deformation theory. We also need to control the ramification over a reflex field for a CM lifting of B' as well as the degree of the isogeny from B onto B', since Theorem 4.1.1 requires it.

Consider all $\mathcal{O}_L$-linear abelian varieties B' over $\mathbb{F}_q$ that are $\mathcal{O}_L$-linearly isogenous to B by an isogeny of p-power degree. We have to show that at least one among them admits an L-linear lift to characteristic 0, at least after extending the ground field to k. For a finite extension K of $K_0 = W(k)[1/p]$, by the Serre-Tate theorem (Theorem 1.4.5.3) the construction of a formal abelian scheme $\mathfrak{B}$ over $\mathcal{O}_K$ lifting some B'_k such that the action of an order $\mathcal{O} \subseteq \mathcal{O}_L$ lifts to $\mathfrak{B}$ is equivalent to the construction of an $\mathcal{O}$-linear p-divisible group $\mathcal{Y}$ over $\mathcal{O}_K$ that $\mathcal{O}$-linearly lifts the p-divisible group $B'[p^\infty]_k$.

For the resulting formal abelian scheme to be algebraizable, by 2.2.3 it is sufficient (and also necessary) that the K-linear action of L on $\mathrm{Lie}(\mathcal{Y})[1/p]_{\overline{K}}$ is given by a $\overline{K}$-valued CM type of L. This is a "*self-duality*" condition on the $(L\otimes_{\mathbb{Q}}K)$-module $\mathrm{Lie}(\mathcal{Y})[1/p]$ with respect to the involution for the quadratic extension L/L^+, where L^+ is the maximal totally real subfield of L. The "self-duality" terminology is motivated by the intervention of complex conjugation in the definition of the induced CM structure on the dual of a CM abelian variety in characteristic 0 (in order to retain the same CM type and hence preserve the geometric linear isogeny class with respect to the CM field; see 1.5.3.1 and 1.5.4.1).

For any p-adic place v of L^+, the preceding self-duality condition can be localized at v as follows. Write $L_v := L\otimes_{L^+} L^+_v$, and let $\mathcal{O}_{L,v} := \mathcal{O}_L \otimes_{\mathcal{O}_{L^+}} \mathcal{O}_{L^+,v}$ be the maximal order in L_v; the complex conjugation for the CM field L induces the unique involution of the L^+_v-algebra L_v. For every L_v-linear CM p-divisible group $\mathcal{X}_v$ over $\mathcal{O}_K$, the *p-adic* CM *type* of $\mathcal{X}_v$ is the subset Φ_v of $\mathrm{Hom}_{\mathbb{Q}_p\text{-alg}}(L_v, \overline{K})$ which corresponds to the linear L_v-action on $\mathrm{Lie}(\mathcal{X}_v)[1/p]_{\overline{K}}$; we say Φ_v is *self-dual* if it is disjoint from its "complex conjugate".

4.1.5. Lie types of $\mathcal{O}_{L,v}$-linear CM p-divisible groups in characteristic p. Let v be a place of L^+ above p. An $\mathcal{O}_{L,v}$-linear p-divisible group Z_v over a field $\mathcal{K}$ of characteristic p is called a CM *p-divisible group* if its $\mathcal{O}_{L,w}$-part has height $[L_w : \mathbb{Q}_p]$ for each place w of L over v. The *Lie type* of such a Z_v is the class $[\mathrm{Lie}(Z_v)]$

of the $(\mathcal{O}_{L,v} \otimes_{\mathbb{Z}_p} \mathcal{K})$-module $\mathrm{Lie}(Z_v)$ in the Grothendieck group $\mathrm{R}_{\mathcal{K}}(\mathcal{O}_{L,v})$ of the category of finitely generated $(\mathcal{O}_{L,v} \otimes_{\mathbb{Z}_p} \mathcal{K})$-modules. This amounts to recording the length of the factor module of $\mathrm{Lie}(Z_v)$ over each local factor ring of $\mathcal{O}_{L,v} \otimes_{\mathbb{Z}_p} \mathcal{K} = (\mathcal{O}_{L,v}/(p)) \otimes_{\mathbb{F}_p} \mathcal{K}$.

The slopes of Z_v can be read off from its Lie type when $\mathcal{K}$ is algebraically closed (see 4.2.5–4.2.6, applied to each factor field of L_v), so the Lie type is a useful tool for classifying $\mathcal{O}_{L,v}$-linear CM p-divisible groups over fields of characteristic p. It is also useful for constructing isogenies and $\mathcal{O}_{L,v}$-linear liftings to characteristic 0; see (L1)–(L3) below. The idea that the Lie type can be fruitfully exploited for CM lifting questions is due to C.-F. Yu. In particular, the methods (L1) and (L3) below appeared in [**133**, §4-§5].

(L1) Classification of CM p-divisible groups in characteristic p via Lie types.
Let B be as in Theorem 4.1.1, and assume $\mathcal{O}_L \subset \mathrm{End}(B)$. The p-divisible group $B[p^\infty]$ decomposes into the product of $\mathcal{O}_{L,v}$-linear CM p-divisible groups $B[v^\infty]$, where v runs through all places of L^+ above p, and we will localize our problem accordingly. (We work locally on L^+ rather than locally on L since we need to retain contact with the action of the unique involution of L over L^+ in order to use the tangential algebraicity criterion 2.2.3 for applications to abelian varieties.) A classification result in [**133**] says that an $\mathcal{O}_{L,v}$-linear CM p-divisible group over an algebraically closed field of characteristic p is determined up to isomorphism by its Lie type; see Proposition 4.2.6 for a proof.

(L2) Galois descent—constructing isogenies from Lie types.
There is an evident natural (left) action of the Galois group $\mathrm{Gal}(\overline{\mathbb{F}}_p/\mathbb{F}_p)$ on the Grothendieck group $\mathrm{R}_{\overline{\mathbb{F}}_p}(\mathcal{O}_{L,v})$. The Lie type of any $\mathcal{O}_{L,v}$-linear CM p-divisible group Z_v over a finite subfield $\mathbb{F}_q \subset \overline{\mathbb{F}}_p$ is a $\mathbb{F}_q$-rational element of $\mathrm{R}_{\overline{\mathbb{F}}_p}(\mathcal{O}_{L,v})$ in the sense that it is fixed by $\mathrm{Gal}(\overline{\mathbb{F}}_p/\mathbb{F}_q)$. Moreover, the following weak Galois descent statement holds (see Proposition 4.2.8 for a proof):

> if X_v is an $\mathcal{O}_{L,v}$-linear CM p-divisible group over a finite subfield $\mathbb{F}_q \subset \overline{\mathbb{F}}_p$ and δ_v is an $\mathbb{F}_q$-rational element of $\mathrm{R}_{\overline{\mathbb{F}}_p}(\mathcal{O}_{L,v})$ whose set of slopes (in the sense defined in 4.2.5) counted with multiplicity coincides with those of the Lie type of X_v then there exists an $\mathcal{O}_{L,v}$-linear p-divisible group Y_v over $\mathbb{F}_q$ with Lie type δ_v such that Y_v is $\mathcal{O}_{L,v}$-linearly isogenous to X_v.

Building on this, in 4.2.9.2 we prove a localized version of the classification of $\mathcal{O}_{L,v}$-linear isogeny and isomorphism classes of CM p-divisible groups over finite fields and algebraically closed fields of characteristic p. (The "localized" aspect is related to the fact that $\mathcal{O}_{L,v}$ may only be semi-local rather than local.)

(L3) Lifting by the $\mathcal{O}_{L,v}$-action and self-duality.
For every place v of L^+ above p, an $\mathcal{O}_{L,v}$-linear CM p-divisible group $\mathcal{X}_v$ over the valuation ring of an unspecified finite extension of $K_0 = W(\overline{\mathbb{F}}_p)[1/p]$ inside $\overline{K}_0$ is determined up to isomorphism by its *p-adic CM type* Φ_v (i.e. the subset of $\mathrm{Hom}_{\mathbb{Q}_p-\mathrm{alg}}(L_v, \overline{K}_0)$ that corresponds to the class in the Grothendieck group $\mathrm{R}_{\overline{K}_0}(L_v)$ of the $(L_v \otimes_{\mathbb{Q}_p} \overline{K}_0)$-module $\mathrm{Lie}(\mathcal{X}_v)[1/p]_{\overline{K}_0}$). This is an application of Proposition 3.7.4(1).

With this in mind, the classification result (L1) becomes a method to construct $\mathcal{O}_{L,v}$-linear CM liftings, as follows. To lift an $\mathcal{O}_{L,v}$-linear p-divisible group X_v

over $\overline{\mathbb{F}}_p$ to an $\mathcal{O}_{L,v}$-linear p-divisible group $\mathcal{X}_v$ over the ring of integers of some unspecified finite extension of K_0 with a self-dual p-adic CM type, it suffices to find a *self-dual* p-adic CM type Φ_v for L_v that is carried to the Lie type of X_v under the natural reduction map $\mathrm{R}_{\overline{K}_0}(L_v) \to \mathrm{R}_{\overline{\mathbb{F}}_p}(\mathcal{O}_{L,v})$ between Grothendieck groups. (See Proposition 4.4.6 for the precise statement.) An element of $\mathrm{R}_{\overline{\mathbb{F}}_p}(\mathcal{O}_{L,v})$ obtained by reduction of a self-dual p-adic CM type of L_v is necessarily a self-dual Lie type in the sense of 4.4.3. There is a simple combinatorial description for self-dual Lie types in characteristics 0 and p; see 4.4.3 and 4.4.5.

4.1.6. Given a CM abelian variety $(B, \mathcal{O}_L \hookrightarrow \mathrm{End}(B))$ over a finite field $\mathbb{F}_q$ of characteristic p, the task of constructing a CM lifting after performing an $\mathbb{F}_q$-isogeny is easier for some p-adic places of L and not as easy for others. The following definition gives the exact conditions.

Definition. Let L be a CM field, let L^+ be the maximal totally real subfield of L, and let $\mathbb{F}_q$ be a finite field of characteristic p. Let $\Delta_{L^+} = \Delta(L/L^+, \mathbb{F}_q)$ be the set of all p-adic places v of L^+ such that the following hold:

(i) L/L^+ is unramified and inert above v (denote by w the place of L above v);
(ii) the common ramification indices $e(L/\mathbb{Q}, w)$ and $e(L^+/\mathbb{Q}, v)$ are odd;
(iii) the residual degree $f_w = [\kappa_w : \mathbb{F}_p]$ is divisible by 4;
(iv) $[\kappa_w : (\kappa_w \cap \mathbb{F}_q)]$ is even.

A p-adic place v of L^+ in Δ_{L^+} is *bad* (with respect to L/L^+ and $\mathbb{F}_q$); p-adic places of L^+ not in Δ_{L^+} are *good* (with respect to L/L^+ and $\mathbb{F}_q$).[2] Denote by Δ_L the set of all p-adic places of L above Δ_{L^+}. A p-adic place w of L is *bad* (with respect to $\mathbb{F}_q$) if $w \in \Delta_L$; otherwise w is *good* (with respect to $\mathbb{F}_q$).

Remark. Slightly more general definitions of badness and goodness are given in 4.5.7 in a "semi-local" context, where we consider a non-zero finite-dimensional commutative semisimple $\mathbb{Q}_p$-algebra E with an involution ι such that the subring E^+ of E fixed by ι is a field.

4.1.7. Sketch of the proof of Theorem 4.1.1

(a) Construction of a suitable $\mathcal{O}_L$-linear isogeny $B' \to B$ over $\mathbb{F}_q$.
We have arranged that the ring of integers $\mathcal{O}_L$ of the CM field L operates on the given abelian variety B over $\mathbb{F}_q$. We use (L1)–(L3) to find an abelian variety B' over $\mathbb{F}_q$ that is $\mathcal{O}_L$-linearly $\mathbb{F}_q$-isogenous to B such that $B'[p^\infty]$ is "as close to being self-dual as possible". More precisely, consider all places v of L^+ lying above p. Whenever it is possible to change the $\mathcal{O}_{L,v}$-linear p-divisible group $B'[v^\infty]$ by an $\mathcal{O}_{L,v}$-linear isogeny over $\mathbb{F}_q$ to make it self-dual, we do so; we show that this can be done precisely when v is good with respect to L/L^+ and $\mathbb{F}_q$ (i.e., $v \notin \Delta_{L^+}$).

For every place $w \in \Delta_L$ above a bad $v \in \Delta_{L^+}$, choose an $\mathcal{O}_{L,w}$-linear CM p-divisible group in the $\mathcal{O}_{L,w}$-linear isogeny class of $B'[w^\infty] = B'[v^\infty]$ such that its Lie type is "maximally symmetric" for the action of $\mathrm{Gal}(\kappa_w/\mathbb{F}_p) = \mathrm{Hom}_{\mathbb{F}_p\text{-alg}}(\kappa_w, \overline{\mathbb{F}}_p)$. The precise meaning of this condition is called *striped* in 4.5.3 (also see 4.5.4(i)).

[2] The problem of constructing a CM lifting of $B[v^\infty]$ (up to an $\mathbb{F}_q$-isogeny) with a self-dual p-adic CM type is easier if $v \notin \Delta_{L^+}$ and more difficult if $v \in \Delta_{L^+}$. We need self-duality to ensure algebraicity of formal deformations, by 2.2.3.

Combing the good and bad cases, we obtain an $\mathcal{O}_L$-linear $\mathbb{F}_q$-isogeny $B' \to B$ such that $B'[v^\infty]$ is self-dual for every good place $v|p$ of L^+ and is striped for every bad place $v|p$ of L^+, where badness and goodness are taken relative to $(L/L^+, \mathbb{F}_q)$.

(b) Reduction of the CM lifting problem to toy model p-divisible groups. Let v be a bad p-adic place of the totally real field L^+ and let w be the place of L above v. In particular, κ_w contains a subfield $\mathbb{F}_{p^4}$ of size p^4, so $W(\mathbb{F}_{p^4}) \subseteq W(\kappa_w) \subseteq \mathcal{O}_{L,w}$. We will show that the $\mathcal{O}_{L,w}$-linear p-divisible group $B'[w^\infty]_{\overline{\mathbb{F}}_p}$ has the form $\mathcal{O}_{L,w} \otimes_{W(\mathbb{F}_{p^4})} Y_{\text{toy}}$ (a Serre tensor construction), where Y_{toy} is a $W(\mathbb{F}_{p^4})$-linear toy model p-divisible group over $\overline{\mathbb{F}}_p$. The tensor product for this Serre tensor construction is taken in the category of sheaves for the fppf topology and is explained in 4.3.1–4.3.2; it is an elementary notion, since $\mathcal{O}_{L,w}$ is finite free over $W(\mathbb{F}_{p^4})$. See 4.5.11 and parts (ii) and (iii) in 4.5.15 for further details.

For each good p-adic place v of L^+, we know by (L3) how to lift the $\mathcal{O}_{L,v}$-linear p-divisible group $B'[v^\infty]_{\overline{\mathbb{F}}_p}$ to an $\mathcal{O}_{L,v}$-linear CM p-divisible group (over a mixed characteristic complete discrete valuation ring) with a self-dual p-adic CM type. By the Serre-Tate theorem and the tangential algebraicity criterion in 2.2.3, to prove that B' admits a CM lifting to characteristic 0 it suffices to produce, for each bad p-adic place w of L, a lifting $\mathcal{X}_w$ of $B'[w^\infty]_{\overline{\mathbb{F}}_p}$ to characteristic 0 such that the action of a $\mathbb{Z}_p$-order of $\mathcal{O}_{L,w}$ on $B'[w^\infty]_{\overline{\mathbb{F}}_p}$ extends to $\mathcal{X}_w$ and the resulting p-adic CM type of $\mathcal{X}_w$ is self-dual.

We have seen in 4.1.3 how to construct a CM lifting with self-dual p-adic CM type for the p-divisible group of certain toy model abelian surfaces. That method will be generalized to any toy model p-divisible group over $\overline{\mathbb{F}}_p$ in 4.5.17. For any bad place w of L, the $\mathcal{O}_{L,w}$-linear p-divisible group $B'[w^\infty]_{\overline{\mathbb{F}}_p}$ is isomorphic to the Serre tensor product of $\mathcal{O}_{L,w}$ over $W(\mathbb{F}_{p^4})$ against a $W(\mathbb{F}_{p^4})$-linear toy model p-divisible group. Thus, the $W(\mathbb{F}_{p^4})$-linear lifting of such p-divisible groups with a self-dual p-adic CM type provides a CM lifting of $B'[w^\infty]_{\overline{\mathbb{F}}_p}$ with a self-dual p-adic CM type.

(c) From p-divisible groups back to abelian varieties
It suffices to construct, for each place v of L^+ above p, a p-divisible group $\mathcal{X}_v$ over the ring of integers $\mathcal{O}_K$ of a finite extension field K of $K_0 = W(\overline{\mathbb{F}}_p)[1/p]$ such that $\mathcal{X}_v$ has closed fiber isomorphic to $B'[v^\infty]_{\overline{\mathbb{F}}_p}$ and the action of the $\mathbb{Z}_p$-order $\mathcal{O}'_v = \mathbb{Z}_p + p\mathcal{O}_{L,v}$ of $\mathcal{O}_{L,v}$ on $B'[v^\infty]_{\overline{\mathbb{F}}_p}$ lifts to an action on $\mathcal{X}_v$ having a self-dual p-adic CM type. Indeed, by arranging (as we may) for $\mathcal{O}_K$ to be large enough that it works for all such v, since the $\mathcal{O}'$-linear p-divisible group over $\mathcal{O}_K$ has special fiber $B'[p^\infty]_{\overline{\mathbb{F}}_p}$ it follows from the Serre–Tate deformation theorem that $\prod_v \mathcal{X}_v$ arises from an $\mathcal{O}'$-linear formal abelian scheme deformation of $B'_{\overline{\mathbb{F}}_p}$ over $\mathcal{O}_K$. This algebraizes to an $\mathcal{O}'$-linear abelian scheme deformation over $\mathcal{O}_K$ by 4.6.3.1.

To find $\mathcal{X}_v$, there is no problem if v is a good place of L^+ above p: we simply take $\mathcal{X}_v$ to be an $\mathcal{O}_{L,v}$-linear lifting of $B'[v^\infty]_{\overline{\mathbb{F}}_p}$ with self-dual p-adic CM type Φ_v (whose existence is deduced from the self-duality of $B'[v^\infty]_{\overline{\mathbb{F}}_p}$ at v). For a bad place v of L^+ and the unique place w of L above v, something wonderful happens: we will show that $B'[w^\infty]_{\overline{\mathbb{F}}_p}$ is isomorphic to $\mathcal{O}_{L,w} \otimes_{W(\mathbb{F}_{p^4})} Y_{\text{toy}}$, so we can construct $\mathcal{X}_v$

by applying the Serre tensor construction for $\mathcal{O}_{L,w}$ over $W(\mathbb{F}_{p^4})$ against a solution in the case of toy model p-divisible groups over $\overline{\mathbb{F}}_p$ (for which the replacement of the algebraization problem in the abelian variety case is the self-duality of the p-adic CM type of the lift). Details are given in 4.6.4.

4.1.8. The original proof of Theorem 4.1.1 was more complicated than the proof sketched above, in that the construction of a CM lifting of $B'[w^\infty]_{\overline{\mathbb{F}}_p}$ with a self-dual p-adic CM type for a bad prime was first reduced, using a Serre-tensor product construction, to the case when L_w is unramified over $\mathbb{Q}_p$ (so $\mathcal{O}_{L,w} = W(\kappa_w)$). Under this unramifiedness property, the required CM lifting of $B'[w^\infty]_{\overline{\mathbb{F}}_p}$ was constructed directly via p-adic Hodge theory; later we replaced p-adic Hodge theory with results of Raynaud [**99**] on finite flat group schemes. These two versions of the direct construction of CM lifting can be found in Appendices B.2 and B.1 respectively; cf. 4.5.18. (In B.3 we summarize the relationship between different flavors of Dieudonné theory over a perfect field of characteristic p.)

This chapter is notation-heavy, partly because of the need to decode the CM data explicitly. Summaries of notation are given in 4.2.1 (local case), 4.4.1 (semi-local case), and 4.6.1 (global case).

4.1.9. There remains a significant open question on CM lifting: can one kill all obstructions to the existence of a CM lifting by requiring that the whole ring of integers of the CM field operates on the abelian variety? This question, stated precisely in 1.8.5 as (sCML) and recalled below, is essentially geometric in the sense that it is equivalent to the apparently weaker version when $\mathbb{F}_q$ is replaced by $\overline{\mathbb{F}}_p$. (This equivalence follows from deformation theory for p-divisible groups, via the "change of coefficients" isomorphism in Theorem 1.4.5.5(2), similarly to the argument used for abelian varieties at the end of 4.1.2.)

Question (sCML) *strong* CM *lifting.*
Choose a CM field L and a CM abelian variety B of dimension $[L:\mathbb{Q}]/2$ over a finite field $\mathbb{F}_q$, equipped with $\mathcal{O}_L \hookrightarrow \mathrm{End}(B)$. Does there exist an abelian scheme $\mathcal{B}$ over a complete local noetherian domain R of generic characteristic 0 with residue field $\mathbb{F}_q$ and an embedding $L \to \mathrm{End}^0(\mathcal{B})$ such that the closed fiber of $\mathcal{B}$ is L-linearly isomorphic to B?

This problem amounts to avoiding an ($\mathcal{O}_L$-linear) $\mathbb{F}_q$-isogeny, and in the $\mathcal{O}_L$-linear case our proof of Theorem 4.1.1 only requires such isogenies during the study of $B[v^\infty]$ for a p-adic place v on L^+ that is *bad* (with respect to L/L^+ and $\mathbb{F}_q$). Here is a closely related question.

Question (LTI) *Lie types of closed fibers of* CM *abelian schemes isogenous to a* CM *lift.*
Let L be a CM field, and $\mathcal{B}$ be an abelian scheme of relative dimension $[L:\mathbb{Q}]/2$ over a mixed characteristic complete discrete valuation ring R with algebraically closed residue field of characteristic $p > 0$, equipped with an embedding $L \hookrightarrow \mathrm{End}^0(\mathcal{B})$.

Consider the family $\mathscr{I}(\mathcal{B})$ of all abelian schemes $\mathcal{B}'$ over R isogenous to $\mathcal{B}$ such that the whole ring of integers $\mathcal{O}_L$ of L operates on the closed fiber B' of $\mathcal{B}'$ via the induced CM structure $L \to \mathrm{End}^0(B')$. Let $\mathrm{LTI}(\mathcal{B})$ be the set of all Lie types of the closed fibers of CM abelian schemes in $\mathscr{I}(\mathcal{B})$. What can one say about the subset $\mathrm{LTI}(\mathcal{B})$ of the set $\mathrm{LT}(\mathcal{O}_L, p)$ of all Lie types for the CM field L?

We expect (sCML) does not hold in general. There are examples where $\mathrm{LTI}(\mathcal{B}) \neq \mathrm{LT}(\mathcal{O}_L, p)$, but we do not know any structure or statistics for $\mathrm{LTI}(\mathcal{B})$.

4.2. Classification and Galois descent by Lie types

We review in this section the classification up to both isomorphism (see 4.2.6, 4.2.7) and isogeny (see 4.2.8) for $\mathcal{O}_F$-linear CM p-divisible groups over an algebraically closed field k of characteristic p, where $\mathcal{O}_F$ is the ring of integers of a finite extension F of $\mathbb{Q}_p$.[3] The field F will play a role akin to that of a CM field of degree $2g$ in the study of CM abelian varieties of dimension g. For completeness we also address the classification up to $\mathcal{O}_F$-linear isomorphism and isogeny over finite fields of characteristic p; see Remark 4.2.9.2.

The isomorphism classification is a special case of results in [**133**]. We begin by setting up some notation and terminology, as well as some elementary lemmas.

4.2.1. Notation in the local case. We shall work throughout with a finite extension F of $\mathbb{Q}_p$ and an algebraically closed field k of characteristic $p > 0$. The maximal ideal and residue field of F are respectively denoted $\mathfrak{m}_F$ and κ_F. The absolute ramification index $e(F/\mathbb{Q}_p)$ is denoted e, and the residual degree $[\kappa_F : \mathbb{F}_p]$ is denoted f (or f_F). Likewise, $q = q_F$ denotes $p^f = \#\kappa_F$. The maximal unramified subfield $W(\kappa_F)[1/p] \subseteq F$ is denoted F_0, so we often write $\mathcal{O}_{F_0}$ rather than $W(\kappa_F)$.

Dieudonné modules of p-divisible groups over k equipped with an action by $\mathcal{O}_F$ are modules over $\mathcal{O}_F \otimes_{\mathbb{Z}_p} W(k)$, so we will frequently use the canonical decomposition of this ring into a direct product of local rings:

$$\mathcal{O}_F \otimes_{\mathbb{Z}_p} W(k) \simeq \mathcal{O}_F \otimes_{\mathcal{O}_{F_0}} (\mathcal{O}_{F_0} \otimes_{\mathbb{Z}_p} W(k)) \simeq \prod_{i \in I} (\mathcal{O}_F \otimes_{\mathcal{O}_{F_0}, i} W(k)) \tag{4.2.1.1}$$

where

$$I := \mathrm{Hom}_{\mathbb{Z}_p\text{-alg}}(\mathcal{O}_{F_0}, W(k)) \simeq \mathrm{Hom}_{\mathbb{Z}_p\text{-alg}}(\mathcal{O}_{F_0}, k) \simeq \mathrm{Hom}_{\mathbb{F}_p\text{-alg}}(\kappa_F, k).$$

(Note that each local factor in (4.2.1.1) is a complete discrete valuation ring with ramification index e over $W(k)$.) We have an analogous decomposition for $\mathcal{O}_F \otimes_{\mathbb{Z}_p} k$ using the local factor rings $\mathcal{O}_F \otimes_{\mathcal{O}_{F_0}, i} k$ that are each k-isomorphic to $k[t]/(t^e)$. (In contrast, if κ is a finite subfield of k then the set of local factor rings of $\mathcal{O}_F \otimes_{\mathbb{Z}_p} W(\kappa)$ depends on the composite field $\kappa_F \kappa$.)

The set I has a natural structure of torsor for the finite cyclic group $\mathrm{Gal}(\kappa_F/\mathbb{F}_p)$ acting through pre-composition. The Galois group $\mathrm{Gal}(\kappa_F/\mathbb{F}_p)$ is identified with $\mathbb{Z}/f\mathbb{Z}$ using the arithmetic Frobenius generator $\sigma : x \mapsto x^p$, so for $i \in I$ and $n \in \mathbb{Z}$ we shall write $i + n$ to denote $i \circ \sigma^n$.

When working with finitely generated modules over $\mathcal{O}_F \otimes_{\mathbb{Z}_p} k$ or $\mathcal{O}_F \otimes_{\mathbb{Z}_p} \kappa$ for a finite subfield $\kappa \subset k$, it will be very convenient to use the language of Grothendieck groups. To that end, for any field $\mathcal{K}$ of characteristic $p > 0$ (such as k or one of its finite subfields) we denote by $\mathrm{R}_{\mathcal{K}}(\mathcal{O}_F)$ the Grothendieck group of the category of finitely generated $(\mathcal{O}_F \otimes_{\mathbb{Z}_p} \mathcal{K})$-modules. Likewise,

$$\mathrm{R}^+_{\mathcal{K}}(\mathcal{O}_F) \subset \mathrm{R}_{\mathcal{K}}(\mathcal{O}_F)$$

denotes the subset of *effective* elements in $\mathrm{R}_{\mathcal{K}}(\mathcal{O}_F)$; i.e., the classes $[M]$ of finitely generated $(\mathcal{O}_F \otimes_{\mathbb{Z}_p} \mathcal{K})$-modules M. The analogous Grothendieck group for the

[3] Only the case $k = \overline{\mathbb{F}}_p$ is needed for the proof of Theorem 4.1.1; see 4.6.4.

coefficient ring $\mathcal{O}_F \otimes_{\mathcal{O}_{F_0},i} k$ is denoted $\mathrm{R}_k(\mathcal{O}_F, i)$, and its subset of effective elements is denoted $\mathrm{R}_k^+(\mathcal{O}_F, i)$.

Since $\mathcal{O}_F \otimes_{\mathcal{O}_{F_0},i} k$ is artin local with residue field k, the forgetful functor from $\mathcal{O}_F \otimes_{\mathcal{O}_{F_0},i} k$-modules to finite-dimensional k-vector spaces defines canonical isomorphisms

$$\epsilon_i \colon \mathrm{R}_k(\mathcal{O}_F, i) \xrightarrow{\simeq} \mathrm{R}(k) \simeq \mathbb{Z},$$

where $\mathrm{R}(k)$ denotes the Grothendieck group of the category of finite-dimensional k-vector spaces, identified with $\mathbb{Z}$ via dimension. In particular, $\epsilon_i(\mathrm{R}_k^+(\mathcal{O}_F, i)) = \mathbb{Z}_{\geqslant 0}$ and for a finitely generated $\mathcal{O}_F \otimes_{\mathcal{O}_{F_0},i} k$-module M the integer $\epsilon_i([M])$ is the length of M over the discrete valuation ring $\mathcal{O}_F \otimes_{\mathcal{O}_{F_0},i} k$. The canonical decomposition of the k-algebra $\mathcal{O}_F \otimes_{\mathcal{O}_{F_0}} k$ as a direct product of its local factors induces a decomposition

$$\mathrm{R}_k(\mathcal{O}_F) \xrightarrow{\simeq} \bigoplus_{i \in I} \mathrm{R}(\mathcal{O}_F, i) \xrightarrow[\simeq]{\oplus_i \epsilon_i} \bigoplus_{i \in I} \mathbb{Z}$$

under which the subset $\mathrm{R}_k^+(\mathcal{O}_F)$ of effective classes corresponds to $\bigoplus_{i \in I} \mathbb{Z}_{\geqslant 0}$.

Finally, we denote by $\epsilon \colon \mathrm{R}_k(\mathcal{O}_F) \to \mathbb{Z}$ the homomorphism

$$\epsilon : \delta \mapsto \sum_{i \in I} \epsilon_i(\delta)$$

and call $\epsilon(\delta)$ the *dimension* of δ (since on an effective class $[M] \in \mathrm{R}_k^+(\mathcal{O}_F)$ it is $\dim_k M$). For a finite subfield $\kappa \subset k$ we sometimes abuse notation and also denote by $\epsilon \colon \mathrm{R}_\kappa(\mathcal{O}_F) \to \mathbb{Z}$ the composition of ϵ with the natural map $j_\kappa : \mathrm{R}_\kappa(\mathcal{O}_F) \to \mathrm{R}_k(\mathcal{O}_F)$ induced by the functor $M \rightsquigarrow k \otimes_\kappa M$.

4.2.2. Terminology in the local case. An $\mathcal{O}_F$-*linear* (respectively F-*linear*) p-divisible group over a ring R is a p-divisible group X over R equipped with a ring homomorphism $\alpha : \mathcal{O}_F \to \mathrm{End}(X)$ (respectively $\alpha : F \to \mathrm{End}^0(X) := \mathbb{Q}_p \otimes_{\mathbb{Z}_p} \mathrm{End}(X)$). We say that (X, α) is an $\mathcal{O}_F$-*linear* CM p-*divisible group* if $[F : \mathbb{Q}_p] = \mathrm{height}(X)$; the notion of an F-linear CM p-divisible group is defined similarly. For an $\mathcal{O}_F$-linear CM p-divisible group (X_0, α_0) over k, the *Lie type* of (X_0, α_0) is the class

$$[\mathrm{Lie}(X_0)] := [\mathrm{Lie}(X_0, \alpha_0)] \in \mathrm{R}_k(\mathcal{O}_F)$$

of the $(\mathcal{O}_F \otimes_{\mathbb{Z}_p} k)$-module $\mathrm{Lie}(X_0)$.

For an algebraic closure $\overline{K}_0$ of $K_0 = W(k)[1/p]$ and a finite extension K of K_0 contained in $\overline{K}_0$, the $\overline{K}_0$-valued p-*adic* CM *type* of an F-linear p-divisible group $(\mathcal{Y}, \alpha)$ over $\mathcal{O}_K$ is the subset

$$\Psi \subset \mathrm{Hom}_{\mathbb{Q}_p\text{-alg}}(F, \overline{K}_0)$$

of size $\dim(\mathcal{Y})$ consisting of the $\mathbb{Q}_p$-embeddings $F \to \overline{K}_0$ arising from the $\overline{K}_0$-linear F-action on $\mathrm{Lie}(\mathcal{Y})[1/p]_{\overline{K}_0}$. (This set has size $\dim(\mathcal{Y}) = \dim_K \mathrm{Lie}(\mathcal{Y})[1/p]$, which is to say that there are no repeated embeddings in the list, because the Hodge-Tate decomposition 1.4.5.1 naturally identifies $\mathbb{C}_K(1) \otimes_K \mathrm{Lie}(\mathcal{Y})[1/p]$ as a direct summand of the $\mathbb{C}_K$-vector space $\mathbb{C}_K \otimes_{\mathbb{Q}_p} V_p(\mathcal{Y}_K)$ that is free of rank 1 over $\mathbb{C}_K \otimes_{\mathbb{Q}_p} F$.)

For every subset $\Phi \subseteq \mathrm{Hom}_{\mathbb{Q}_p\text{-alg}}(F, \overline{K}_0)$, the *reduction*

$$\xi(\Phi) \in \mathrm{R}_k(\mathcal{O}_F)$$

is the element of $\mathrm{R}_k(\mathcal{O}_F)$ characterized by the condition that

$$\epsilon_i(\xi(\Phi)) = \#\{\sigma \in \Phi \mid \overline{\sigma} = i\}$$

for all $i \in I$, where $\overline{\sigma}$ denotes the embedding $\kappa_F \to k$ induced by $\sigma\colon F \to \overline{K}_0$. An element $\delta \in \mathrm{R}_k(\mathcal{O}_F)$ is a *Lie type* if $0 \leqslant \epsilon_i(\delta) \leqslant e$ for all $i \in I$.

It is easy to see that $\epsilon(\xi(\Phi)) = \#\Phi$ for any $\Phi \subseteq \mathrm{Hom}_{\mathbb{Q}_p\text{-alg}}(F, \overline{K}_0)$, and that the reductions $\xi(\Phi)$ for varying Φ are precisely the Lie types in $\mathrm{R}_k(\mathcal{O}_F)$. For an $\mathcal{O}_F$-linear p-divisible group X_0 over k, we will see in Proposition 4.2.6 that the "dimension" $\epsilon([\mathrm{Lie}(X_0)])$ is equal to $\dim(X_0)$ and that $[\mathrm{Lie}(X_0)]$ is a Lie type (thereby justifying our terminology).

4.2.2.1. Remark. Here is another way to think about the reduction map $\Phi \mapsto \xi(\Phi)$. This map defines a homomorphism

$$\mathbf{d} : \mathrm{R}_K(\mathcal{O}_F) \to \mathrm{R}_k(\mathcal{O}_F),$$

where K is any subextension of $\overline{K}_0/K_0$ that contains a Galois closure of $F/\mathbb{Q}_p$ and $\mathrm{R}_K(\mathcal{O}_F)$ is the Grothendieck group of the category of finitely generated modules over $\mathcal{O}_F \otimes_{\mathbb{Z}_p} K$. Each $\mathbb{Q}_p$-embedding $\sigma : F \to K$ defines an $(\mathcal{O}_F \otimes_{\mathbb{Z}_p} K)$-module structure $K(\sigma)$ on K and the classes of the $K(\sigma)$'s form a $\mathbb{Z}$-basis of $\mathrm{R}_K(\mathcal{O}_F)$. The homomorphism $\mathbf{d}$ sends the class of $K(\sigma)$ to the generator of the semi-group $\mathrm{R}_k^+(\mathcal{O}_F, \overline{\sigma}) \simeq \mathbb{Z}_{\geqslant 0}$.

Now consider the Grothendieck groups $\mathrm{P}_k(\mathcal{O}_F)$ and $\mathrm{P}_{\mathcal{O}_K}(\mathcal{O}_F)$ of the respective categories of finitely generated projective modules over $\mathcal{O}_F \otimes_{\mathbb{Z}_p} \mathcal{O}_K$ and $\mathcal{O}_F \otimes_{\mathbb{Z}_p} k$. There is a natural isomorphism $\mathrm{P}_{\mathcal{O}_K}(\mathcal{O}_F) \simeq \mathrm{P}_k(\mathcal{O}_F)$ defined by "tensoring with k". Notice that $\mathrm{P}_k(\mathcal{O}_F)$ is a free abelian group with a free basis given by the classes of the $(\mathcal{O}_F \otimes_{\mathbb{Z}_p} k)$-modules $\mathcal{O}_F \otimes_{\mathcal{O}_{F_0}, i} k$ for $i \in I$.

The map "tensoring with K" defines a homomorphism $\mathrm{P}_{\mathcal{O}_K}(\mathcal{O}_F) \to \mathrm{R}_K(\mathcal{O}_F)$; pre-composing this homomorphism with the natural isomorphism from $\mathrm{P}_k(\mathcal{O}_F)$ to $\mathrm{P}_{\mathcal{O}_K}(\mathcal{O}_F)$ gives a homomorphism

$$\mathbf{e} : \mathrm{P}_k(\mathcal{O}_F) \to \mathrm{R}_K(\mathcal{O}_F).$$

It is not difficult to check that the composition $\mathbf{d} \circ \mathbf{e} : \mathrm{P}_k(\mathcal{O}_F) \to \mathrm{R}_k(\mathcal{O}_F)$ is equal to the homomorphism

$$\mathbf{c} : \mathrm{P}_k(\mathcal{O}_F) \to \mathrm{R}_k(\mathcal{O}_F)$$

that sends the class of a finitely generated projective $\mathcal{O}_F \otimes_{\mathbb{Z}_p} k$-module M to the class of M in $\mathrm{R}_k(\mathcal{O}_F)$.[4]

We also have a duality relation between $\mathbf{d}$ and $\mathbf{e}$, similar to [**106**, 15.4 (b)], as follows. The groups $\mathrm{P}_k(\mathcal{O}_F)$ and $\mathrm{R}_k(\mathcal{O}_F)$ are dual to each other via a perfect pairing $\langle \cdot, \cdot \rangle_k \colon \mathrm{P}_k(\mathcal{O}_F) \times \mathrm{R}_k(\mathcal{O}_F) \to \mathbb{Z}$ defined by

$$\langle\, [P],\ [N]\, \rangle_k := \dim_k \mathrm{Hom}_{\mathcal{O}_F \otimes_{\mathbb{Z}_p} k}(P, N)$$

[4]We have chosen the symbols $\mathbf{c}, \mathbf{d}, \mathbf{e}$ to underline the similarity with [**106**, Ch. 15], titled *The cde triangle.* Of course, our situation is substantially simpler than the set-up in [**106**] for modular representations, and the homomorphisms $\mathbf{c}, \mathbf{d}, \mathbf{e}$ can be described explicitly. The homomorphism $\mathbf{c}$ is represented by the diagonal matrix with all diagonal entries equal to e if we use the respective natural basis indexed by I for the source and the target of $\mathbf{c}$. The homomorphism $\mathbf{e}$ sends the class $[\mathcal{O}_F \otimes_{\mathcal{O}_{F_0}, i} k] \in \mathrm{P}_k(\mathcal{O}_F)$ to the sum of classes $[K(\sigma)] \in \mathrm{R}_K(\mathcal{O}_F)$, where σ runs through all $\mathbb{Q}_p$-embeddings $\sigma : F \to K$ such that $\overline{\sigma} = i$, or equivalently $\epsilon_i(\xi(\{\sigma\})) = 1$.

for every projective $(\mathcal{O}_F \otimes_{\mathbb{Z}_p} k)$-module P of finite type and every $(\mathcal{O}_F \otimes_{\mathbb{Z}_p} k)$-module N of finite type. Similarly we have a perfect pairing of $\mathrm{R}_K(\mathcal{O}_F)$ with itself, given by

$$\langle\, [M_1],\ [M_2]\,\rangle_K := \dim_K \mathrm{Hom}_{\mathcal{O}_F \otimes_{\mathbb{Z}_p} K}(M_1, M_2)$$

for all finitely generated $(\mathcal{O}_F \otimes_{\mathbb{Z}_p} K)$-modules M_1 and M_2. The homomorphisms $\mathbf{d}$ and $\mathbf{e}$ are adjoint to each other with respect to the above pairings in the sense that

$$\langle x, \mathbf{d}(y)\rangle_k = \langle \mathbf{e}(x), y\rangle_K$$

for all $x \in \mathrm{P}_k(\mathcal{O}_F)$ and $y \in \mathrm{R}_K(\mathcal{O}_F)$. In other words, $\mathbf{d}$ can be identified with the transpose of $\mathbf{e}$ (and vice-versa).

The link between p-adic CM types and Lie types via the "reduction" operation ξ is given by:

4.2.3. Lemma. *Let $\mathcal{X}$ be an $\mathcal{O}_F$-linear CM p-divisible group over the valuation ring of a finite extension K of $K_0 = W(k)[1/p]$ contained in $\overline{K}_0$. Let $\Phi \subseteq \mathrm{Hom}_{\mathbb{Q}_p\text{-alg}}(F, \overline{K}_0)$ be the p-adic CM type associated to $\mathrm{Lie}(\mathcal{X})[1/p]$. The effective class $[\mathrm{Lie}(X_0)]$ associated to the CM special fiber satisfies*

$$[\mathrm{Lie}(X_0)] = \xi(\Phi)$$

inside $\mathrm{R}_k(\mathcal{O}_F)$.

The proof of this lemma is a straightforward verification left to the reader (using that $\mathrm{Lie}(X_0) = \mathrm{Lie}(\mathcal{X}) \otimes_{\mathcal{O}_K} k$ as $\mathcal{O}_F \otimes_{\mathbb{Z}_p} k$-modules), as is the proof of the following lemma.

4.2.4. Lemma. *Let $\kappa \subset k$ be a finite subfield.*

(i) *The canonical homomorphism*

$$j_\kappa\colon \mathrm{R}_\kappa(\mathcal{O}_F) \longrightarrow \mathrm{R}_k(\mathcal{O}_F)$$

is injective.

(ii) *The inverse image of $\mathrm{R}_k^+(\mathcal{O}_F)$ under j_κ is equal to $\mathrm{R}_\kappa^+(\mathcal{O}_F)$.*

(iii) *For $\gamma \in \mathrm{Aut}(k)$ and a finitely generated $\mathcal{O}_F \otimes_{\mathbb{Z}_p} k$-module M, let ${}^\gamma M$ denote $M \otimes_{k,\gamma} k$ viewed as an $\mathcal{O}_F \otimes_{\mathbb{Z}_p} k$-module with $\mathcal{O}_F$ operating on the first tensor factor. Then $\epsilon_i([{}^\gamma M]) = \epsilon_{\gamma^{-1} \circ i}([M])$ for all $i \in I = \mathrm{Hom}_{\mathbb{F}_p\text{-alg}}(\kappa_F, k)$.*

(iv) *The canonical morphism j_κ identifies $\mathrm{R}_\kappa(\mathcal{O}_F)$ with the set of elements of $\mathrm{R}_k(\mathcal{O}_F) = \oplus_{i \in I}\mathbb{Z}$ fixed by the action of the subgroup*

$$\mathrm{Gal}(\kappa_F/(\kappa_F \cap \kappa)) \subset \mathrm{Gal}(\kappa_F/\mathbb{F}_p)$$

via the natural action of $\mathrm{Gal}(\kappa_F/\mathbb{F}_p)$ on I.

When $k = \overline{\mathbb{F}}_p$, Lemma 4.2.4(iii) says that the identification of $\mathrm{R}_{\overline{\mathbb{F}}_p}(\mathcal{O}_F)$ with $\oplus_{i \in I}\mathbb{Z}$ carries the natural action of $\mathrm{Gal}(\overline{\mathbb{F}}_p/\mathbb{F}_p)$ on $\mathrm{R}_{\overline{\mathbb{F}}_p}(\mathcal{O}_F)$ over to its translation action on I composed with inversion on the abelian group $\mathrm{Gal}(\overline{\mathbb{F}}_p/\mathbb{F}_p)$.

4.2.5. Definition. The *slope* of a Lie type $\delta \in \mathrm{R}_k(\mathcal{O}_F)$ is

$$\epsilon(\delta)/(ef) = \epsilon(\delta)/[F : \mathbb{Q}_p].$$

Two Lie types in $\mathrm{R}_k(\mathcal{O}_F)$ are *isogenous* if they have the same slope.

An element $\delta \in \mathrm{R}_k(\mathcal{O}_F)$ is *rational* over a finite subfield $\kappa \subset k$ if δ belongs to the image of the natural map $j_\kappa\colon \mathrm{R}_\kappa(\mathcal{O}_F) \to \mathrm{R}_k(\mathcal{O}_F)$.

By Lemma 4.2.4(iv), the rationality condition on δ coincides with the apparently weaker condition that δ is fixed under the natural action of $\mathrm{Aut}(k/(\kappa \cap \kappa_F))$ on $\mathrm{R}_k(\mathcal{O}_F)$. More explicitly, let $p^r = \#\kappa$ and define $s = \gcd(r, f_F)$ (so $\#(\kappa_F \cap \kappa) = p^s$). We identify the Galois group $\mathrm{Gal}(\kappa_F/\mathbb{F}_p)$ with the cyclic group $\mathbb{Z}/f_F\mathbb{Z}$ via the arithmetic Frobenius generator $(x \mapsto x^p)$ and regard I as a $(\mathbb{Z}/f_F\mathbb{Z})$-torsor as usual.

The rationality condition on δ over κ is precisely that

$$\epsilon_i(\delta) = \epsilon_{i+s}(\delta)$$

for all $i \in I$. In particular, the possibilities for the slope of a class in $\mathrm{R}_\kappa(\mathcal{O}_F)$ are precisely the rational numbers

$$\frac{\sum_{0 \leqslant j < s}(f_F/s)a_j}{[F:\mathbb{Q}_p]} = \frac{\sum_{0 \leqslant j < s} a_j}{s e_F}$$

where $0 \leqslant a_j \leqslant e_F$. These are the rational numbers in $[0,1]$ with denominator dividing $se_F = \mathrm{ord}_F(\#(\kappa_F \cap \kappa))$. It is also clear via Lemma 4.2.4(ii) that the κ-rational Lie types in $\mathrm{R}_k(\mathcal{O}_F)$ correspond to the classes arising from cyclic $(\mathcal{O}_F/(p)) \otimes_{\mathbb{F}_p} \kappa$-modules.

The next two propositions are proved in [**133**, §4]; we give alternative proofs.

4.2.6. Proposition. *Let X_0 be an $\mathcal{O}_F$-linear* CM *p-divisible group over an algebraically closed field k of characteristic p and let $[\mathrm{Lie}(X_0)]$ be the associated class in $\mathrm{R}_k^+(\mathcal{O}_F)$.*

(i) *For each $i \in I$, the torsion module $\mathrm{Lie}(X_0) \otimes_{\mathcal{O}_{F_0},i} k$ over the discrete valuation ring $\mathcal{O}_F \otimes_{\mathcal{O}_{F_0},i} W(k)$ is cyclic.*
(ii) *We have $0 \leqslant \epsilon_i([\mathrm{Lie}(X_0)]) \leqslant e$ for all $i \in I$, and $\epsilon([\mathrm{Lie}(X_0)]) = \dim(X_0)$.*
(iii) *The p-divisible group X_0 is isoclinic. Its unique slope is the slope of $[\mathrm{Lie}(X_0)]$.*
(iv) *If Y_0 is an $\mathcal{O}_F$-linear* CM *p-divisible group over k such that $[\mathrm{Lie}(Y_0)] = [\mathrm{Lie}(X_0)]$ in $\mathrm{R}_k(\mathcal{O}_F)$ then Y_0 is $\mathcal{O}_F$-linearly isomorphic to X_0.*

PROOF. Let $M = \mathrm{M}^*(X_0)$ be the contravariant Dieudonné module of X_0. The $\mathbb{Z}_p$-algebra $\mathcal{O}_F$ operates $W(k)$-linearly on M, commuting with the actions of $\mathcal{F}$ and $\mathcal{V}$, and by Proposition 1.4.3.9(2) the $\mathcal{O}_F \otimes_{\mathbb{Z}_p} W(k)$-module structure on M is free of rank 1. Thus, the decomposition

$$\mathcal{O}_F \otimes_{\mathbb{Z}_p} W(k) = \prod_{i \in I} \mathcal{O}_F \otimes_{\mathcal{O}_{F_0},i} W(k)$$

induces a decomposition

$$M = \bigoplus_{i \in I} M_i,$$

where M_i is a free module of rank 1 over the discrete valuation ring $\mathcal{O}_F \otimes_{\mathcal{O}_{F_0},i} W(k)$. Clearly

$$\mathcal{V}\cdot M_i \subset M_{i-1}, \quad \mathcal{F}\cdot M_i \subset M_{i+1}$$

for all $i \in I$, using the natural structure on I of a torsor for the group $\mathrm{Gal}(\kappa_F/\mathbb{F}_p) = \mathbb{Z}/f_F\mathbb{Z}$ (via the arithmetic Frobenius $x \mapsto x^p$ as the generator).

For each $i \in I$, $\mathcal{F} \cdot M_{i-1}$ is a non-zero $(\mathcal{O}_F \otimes_{\mathcal{O}_{F_0},i} W(k))$-submodule of M_i, so it equals $\mathfrak{m}_F^{e_i} M_i$ for some integer $e_i \geqslant 0$. We have $0 \leqslant e_i \leqslant e$ for all i because $\mathcal{F}\mathcal{V} = \mathcal{V}\mathcal{F} = p \cdot \mathrm{id}_M$, and $\sum e_i = \dim_k \mathrm{Lie}(X_0) = \dim(X_0)$. The k-linear dual $\mathrm{Lie}(X_0)^\vee$ is naturally isomorphic to $M/\mathcal{F}M$ (see 1.4.3.2(4)) and the k-linear dual of $(\mathcal{O}_F/\mathfrak{m}_F^n) \otimes_{\mathcal{O}_{F_0},i} k = k[t]/(t^n)$ is free of rank 1 over $k[t]/(t^n)$ for any $n \geqslant 1$, so the cyclicity in (i) holds and e_i is equal to $\epsilon_i([\mathrm{Lie}(X_0)])$ for each $i \in I$.

Since $\sum e_i = \dim(X_0)$, (ii) is also proved. The assertions in (iii) are immediate from (ii) and the results on slopes of p-divisible groups in characteristic p given in 3.7.1.6 and 3.1.3(2).

To prove (iv), fix a uniformizer π of $\mathcal{O}_F$. We will show that the $\mathcal{O}_F$-linear isomorphism class of the Dieudonné module of X_0 is determined by the integers e_i, which is to say by the Lie type $[\mathrm{Lie}(X_0)]$ (as this corresponds to the I-tuple (e_i) under the isomorphism $\mathrm{R}_k(\mathcal{O}_F) \simeq \oplus_{i\in I}\mathbb{Z}$). This will establish (iv). For each $i \in I$, choose a generator m_i of M_i over $\mathcal{O}_F \otimes_{\mathcal{O}_{F_0},i} W(k)$. By the definition of e_i, there exists a unit $u_i \in (\mathcal{O}_F \otimes_{\mathcal{O}_{F_0},i} W(k))^\times$ such that

$$\mathcal{F} \cdot m_{i-1} = u_i \, \pi^{e_i} \, m_i. \tag{4.2.6.1}$$

The action of $\mathcal{V}$ is uniquely determined by these structural equations since $\mathcal{F}\mathcal{V} = \mathcal{V}\mathcal{F} = p \cdot \mathrm{id}_M$. Thus, it suffices to show that we can choose the m_i so that $u_i = 1$ for all i. This will use that $k = \overline{k}$.

The choices of bases $\{m_i'\}$ are $m_i' = a_i m_i$ for $a_i \in (\mathcal{O}_F \otimes_{\mathcal{O}_{F_0},i} W(k))^\times$, with system of unit multipliers $\{u_i'\}$ in the structural equations given by the formula

$$u_i' = (\sigma_i(a_{i-1})/a_i) u_i$$

where $\sigma_i : \mathcal{O}_F \otimes_{\mathcal{O}_{F_0},i-1} W(k) \simeq \mathcal{O}_F \otimes_{\mathcal{O}_{F_0},i} W(k)$ is the isomorphism $z \otimes w \mapsto z \otimes \sigma(w)$ induced by the absolute Frobenius automorphism σ of $W(k)$. The condition $u_i' = 1$ for all i is the system of equations $a_i = \sigma_i(a_{i-1}) u_i$ for all i, so by choosing an embedding i_0 and an undetermined unit $a_{i_0} \in R := \mathcal{O}_F \otimes_{\mathcal{O}_{F_0},i_0} W(k)$ we successively solve for the remaining a_i and arrive at the necessary and sufficient self-consistency condition

$$a_{i_0} = (\sigma_{i_0} \circ \sigma_{i_0-1} \circ \cdots \circ \sigma_{i_0+1})(a_{i_0}) u$$

for a certain unit $u \in R^\times$ defined in terms of products involving the u_i.

This condition on a_{i_0} is exactly the equation $x = (1 \otimes \sigma_q)(x) u$ using the $\mathcal{O}_{F_0}$-algebra q-Frobenius automorphism $\sigma_q = \sigma^f$ of $W(k)$. Hence, it suffices to show that the endomorphism $x \mapsto (1 \otimes \sigma_q)(x)/x$ of $R^\times$ is surjective. We proceed by successive approximation relative to the powers of the maximal ideal $\mathfrak{m}_F R$ of R, the base case being the surjectivity of the $(q-1)$th-power endomorphism of $k^\times$. To carry out the approximation calculation, we use a uniformizer π of $\mathcal{O}_F$ to identify R with $W(k)[\![\pi]\!]$. For a unit of the form $1 + [c]\pi^j$ with $j > 0$ and $[c]$ denoting the Teichmüller lift of $c \in k$, the equation

$$\frac{(1 \otimes \sigma_q)(1 + [b]\pi^j)}{1 + [b]\pi^j} \equiv 1 + [c]\pi^j \pmod{\pi^{j+1}}$$

is the Artin-Schreier equation $b^q - b = c$. Thus, the successive approximation succeeds since k is separably closed. □

Proposition 4.2.6 says that $\mathcal{O}_F$-linear CM p-divisible groups over an algebraically closed field k of characteristic p are determined up to isomorphism by their Lie type. The next result says that all Lie types arise in this way.

4.2.7. Proposition. *For an algebraically closed field k of characteristic $p > 0$ and any Lie type $\delta_0 \in \mathrm{R}_k(\mathcal{O}_F)$, there exists an $\mathcal{O}_F$-linear p-divisible group X_0 over k such that* $[\mathrm{Lie}(X_0)] = \delta_0$.

PROOF. We use the computations that arose in the proof of 4.2.6. Let $e_i := \epsilon_i(\delta_0)$ and consider an $\mathcal{O}_F$-linear Dieudonné module M with generators m_i for M_i satisfying (4.2.6.1) with $a_i = 1$ for all $i \in I$. This makes sense because the condition $e_i \leqslant e$ ensures that we can define $\mathcal{V}$ to be "$p\mathcal{F}^{-1}$" without requiring denominators. The corresponding $\mathcal{O}_F$-linear p-divisible group X_0 does the job. □

The following result is a weak form of Galois descent for $\mathcal{O}_F$-linear CM p-divisible groups over k in terms of their Lie type. In the proof of Theorem 4.1.1 (CM lifting up to isogeny), we will take k to be an algebraic closure of a finite field $\mathbb{F}_q$ and use this weak Galois descent to construct an abelian variety B' over $\mathbb{F}_q$ isogenous to a given CM abelian variety $(B, L \hookrightarrow \mathrm{End}^0(B))$ over $\mathbb{F}_q$ such that $(B', L \hookrightarrow \mathrm{End}^0(B) = \mathrm{End}^0(B'))$ admits a CM-lifting to characteristic 0.

4.2.8. Proposition. *Let k be algebraically closed of characteristic $p > 0$, and let κ be a finite subfield of k. Every κ-rational Lie type in $\mathrm{R}_k(\mathcal{O}_F)$ has the form $[\mathrm{Lie}((Y_0)_k)]$ for an $\mathcal{O}_F$-linear* CM *p-divisible group Y_0 over κ.*

If X_0 is an $\mathcal{O}_F$-linear CM *p-divisible group over κ and δ_0 is a κ-rational Lie type in $\mathrm{R}_k(\mathcal{O}_F)$ with the same slope as X_0 (equivalently, δ_0 and $[\mathrm{Lie}((X_0)_k)]$ are isogenous) then the $\mathcal{O}_F$-linear κ-isogeny class of X_0 contains a member Y_0 with $[\mathrm{Lie}((Y_0)_k)] = \delta_0$. The degree of the κ-isogeny between X_0 and Y_0 may be chosen to be at most $p^{f_F \dim(X_0) + [F:\mathbb{Q}_p](f_F - 1)/2}$.*

We are freely using the terminology introduced in 4.2.5.

PROOF. Without loss of generality, k is an algebraic closure of $\mathbb{F}_p$. We use the notation as in the proofs of 4.2.6 and 4.2.7 (so $q = \#\kappa = p^f$), and first show that every κ-rational Lie type in $\mathrm{R}_k(\mathcal{O}_F)$ arises as the Lie type of an $\mathcal{O}_F$-linear CM p-divisible group over κ. In other words, for any cyclic $(\mathcal{O}_F/(p)) \otimes_{\mathbb{F}_p} \kappa$-module M_0 we seek a Dieudonné module M over κ equipped with an $\mathcal{O}_F$-action making M free of rank 1 over $\mathcal{O}_F \otimes_{\mathbb{Z}_p} W(\kappa)$ such that $M/\mathcal{F}M \simeq M_0$ as $(\mathcal{O}_F/(p)) \otimes_{\mathbb{F}_p} \kappa$-modules.

Since the local factor rings of $\mathcal{O}_F \otimes_{\mathbb{Z}_p} W(\kappa)$ are discrete valuation rings, obviously $M_0 \simeq (\mathcal{O}_F \otimes_{\mathbb{Z}_p} W(\kappa))/(a)$ for some $a \in \mathcal{O}_F \otimes_{\mathbb{Z}_p} W(\kappa)$ that divides p. We may use $M = \mathcal{O}_F \otimes_{\mathbb{Z}_p} W(\kappa)$ with Frobenius and Verschiebung operators

$$\mathcal{F}(x) := a(1 \otimes \sigma)(x), \quad \mathcal{V}(x) := \frac{p}{(1 \otimes \sigma^{-1})(a)} \cdot (1 \otimes \sigma^{-1})(x),$$

where σ denotes the absolute Frobenius automorphism of $W(\kappa)$.

Next, we show that for a given X_0 over κ, each κ-rational δ_0 with the same slope as X_0 arises as the Lie type of a member Y_0 of the $\mathcal{O}_F$-linear κ-isogeny class of X_0. Let M be the $\mathcal{O}_F$-linear Dieudonné module of $(X_0)_k$ and let $\oplus_{i \in I} M_i$ be the decomposition of M. The κ-descent X_0 of $(X_0)_k$ is encoded in terms of an $\mathcal{O}_F$-linear Galois descent datum on M relative to k/κ: a $W(k)$-semilinear action of $\mathrm{Gal}(k/\kappa)$ that is discrete on each $M/(p^n)$ and equivariant with respect to the operator $\mathcal{F}$ (and hence with respect to the operator $\mathcal{V} = p\mathcal{F}^{-1}$).

Choose an element $i_0 \in I$ and an $\mathcal{O}_F \otimes_{\mathcal{O}_{F_0}, i_0} W(k)$-submodule $N_{i_0} \subset M_{i_0}$ that is free of rank 1 (e.g., $N_{i_0} = M_{i_0}$). We claim that the κ-rational element $\delta_0 \in \mathrm{R}_\kappa(\mathcal{O}_F)$, together with N_{i_0}, determines for every $i \in I$ a free $\mathcal{O}_F \otimes_{(\mathcal{O}_{F_0}, i_0)} W(k)$-submodule $N_i \subset M_i[1/p]$ of rank 1 such that

$$\mathcal{F} N_{i-1} = \mathfrak{m}_F^{\epsilon_i(\delta_0)} N_i . \tag{4.2.8.1}$$

To see this, by starting with the chosen N_{i_0} the condition (4.2.8.1) determines

$$N_{i_0-1}, N_{i_0-2}, \ldots, N_{i_0-(f-1)}$$

successively, so we just need to verify a self-consistency condition "$N_{(i_0-(f-1))-1} = N_{i_0}$" that amounts to the equality

$$\mathcal{F}^f N_{i_0} \stackrel{?}{=} \mathfrak{m}_F^{\sum_{i \in I} \epsilon_i(\delta_0)} N_{i_0}.$$

Since $\sum_{i \in I} \epsilon_i(\delta_0) = \epsilon(\delta_0) = \epsilon([\mathrm{Lie}(X_0)]) = \dim(X_0)$ (by our hypothesis that δ_0 has the same slope as X_0), we want that the endomorphism $\mathcal{F}^f$ of M_{i_0} carries N_{i_0} onto $\mathfrak{m}_F^{\dim(X_0)} N_{i_0}$. The endomorphism $\mathcal{F}^f$ of M is the semilinear scalar extension of the q-Frobenius endomorphism of $\mathrm{M}^*(X_0)$, and so must arise from the action of a non-zero element $c \in \mathcal{O}_F$ (as F is its own centralizer in $\mathrm{End}^0(X_0)$, by Proposition 1.4.3.9(2)). It therefore suffices to show that $c \cdot \mathcal{O}_F = \mathfrak{m}_F^{\dim(X_0)}$.

Clearly $c \cdot \mathcal{O}_F = \mathfrak{m}_F^{e \cdot \mathrm{ord}_p(c)}$, so it is equivalent to prove that $e \cdot \mathrm{ord}_p(c) = \dim(X_0)$; i.e., $\mathrm{ord}_p(c)/f = \dim(X_0)/[F : \mathbb{Q}_p]$. By 4.2.6(iii), X_0 is isoclinic with unique slope equal to the slope of $[\mathrm{Lie}(X_0)]$, which is the slope of δ_0; this slope is $\epsilon(\delta_0)/[F : \mathbb{Q}_p] = \dim(X_0)/[F : \mathbb{Q}_p]$. Hence, the self-consistency condition is reduced to verifying that the CM p-divisible group X_0 over the finite field κ of size q has its unique slope equal to $\mathrm{ord}_p(c)/f = \mathrm{ord}_p(c)/\mathrm{ord}_p(q)$ (where the action of $c \in \mathcal{O}_F - \{0\}$ is the q-Frobenius endomorphism of X_0). This is exactly the consistency between the general definition of slopes for p-divisible groups (using the Dieudonné-Manin classification 3.1.3) and the definition in terms of q-Frobenius eigenvalues on the Dieudonné module for p-divisible groups over finite fields of size q, a consistency that is left to the reader as an exercise.

By changing N_{i_0} to a smaller free submodule of rank 1 if necessary (e.g., $\mathfrak{m}_F^{\dim(X_0)} M_{i_0}$ is sufficient) we may arrange that $N_i \subset M_i$ for all $i \in I$. The fact that δ_0 is stable under $\mathrm{Gal}(\kappa_F/(\kappa_F \cap \kappa))$ implies that the submodule $N := \oplus_{i \in I} N_i$ of M is stable under the natural action of $\mathrm{Gal}(k/\kappa)$ on M. The Dieudonné module N corresponds to an $\mathcal{O}_F$-linear CM p-divisible group over k that is an isogenous quotient of $(X_0)_k$ (corresponding to the inclusion of N into M).

The Galois-equivariance condition on N says exactly that this quotient of $(X_0)_k$ is the quotient by a finite k-subgroup of $(X_0)_k$ that is defined over κ (inside X_0). In other words, we obtain an $\mathcal{O}_F$-linear CM p-divisible group quotient Y_0 of X_0 over κ such that the induced map $\mathrm{M}^*(Y_0) \to \mathrm{M}^*(X_0)$ recovers $N \hookrightarrow M$ after scalar extension to $W(k)$. By the construction of $N = \mathrm{M}^*((Y_0)_k)$ inside M, we see that the image $[\mathrm{Lie}((Y_0)_k)]$ of $[\mathrm{Lie}(Y_0)]$ under the injection $\mathrm{R}_\kappa(\mathcal{O}_F) \hookrightarrow \mathrm{R}_k(\mathcal{O}_F)$ is δ_0.

Finally, we give an upper bound on the smallest possible degree for such an isogeny. Using the choice $N_{i_0} = \mathfrak{m}_F^{\dim(X_0)} M_{i_0}$,

$$N_{i_0+1} = \mathfrak{m}_F^{\dim(X_0) - \epsilon_{i_0+1}(\delta_0)} \mathcal{F} M_{i_0} = \mathfrak{m}_F^{\dim(X_0) + \epsilon_{i_0+1}([\mathrm{Lie}(X_0)]) - \epsilon_{i_0+1}(\delta_0)} M_{i_0+1}.$$

Each time we pass to the next index, we add another difference $\epsilon_i([\mathrm{Lie}(X_0)]) - \epsilon_i(\delta_0)$ to the exponent of $\mathfrak{m}_F$. These differences lie between e_F and $-e_F$, so for $0 \leqslant i \leqslant f_F - 1$ we have $N_{i_0+i} \supset \mathfrak{m}_F^{\dim(X_0) + i e_F} M_{i_0+i}$. Hence, $\ell_{W(k)}(M/N) \leqslant f_F \dim(X_0) + e_F f_F (f_F - 1)/2$. □

The arguments for 4.2.6–4.2.8 also prove another property of Lie types:

4.2.9. Proposition. *Let k_0 be a subfield of the algebraically closed field k with characteristic $p > 0$. Let $\xi : X_0 \to Y_0$ be an $\mathcal{O}_F$-linear isogeny between $\mathcal{O}_F$-linear* CM *p-divisible groups over k_0. Suppose that the Lie types of Y_0 and X_0 are equal. Then there exists a non-zero element $u \in \mathcal{O}_F$ such that* $\ker(\xi) = X_0[u]$. *Consequently Y_0 and X_0 are $\mathcal{O}_F$-linearly isomorphic over k_0.*

PROOF. We may and do extend the ground field from k_0 to k so that $k_0 = k$. Let $N \hookrightarrow M$ be the homomorphism between the contravariant Dieudonné modules associated to Y_0 and X_0 induced by ξ, so M/N is killed by a power of p. Let $N = \oplus_{i\in I} N_i$ and $M = \oplus_{i\in I} M_i$ be the decompositions of N and M respectively, so $N_i \subset M_i$ for each $i \in I$. Define non-negative integers c_i and n_i for $i \in I$ by

$$c_i := \epsilon_i([\mathrm{Lie}(M)]) = \epsilon_i([\mathrm{Lie}(N)])\,, \qquad N_i = \mathfrak{m}_F^{n_i} M_i,$$

so

$$\mathcal{F} N_{i-1} = \mathfrak{m}_F^{c_i} N_i \quad \text{and} \quad \mathcal{F} M_{i-1} = \mathfrak{m}_F^{c_i} M_i$$

for all $i \in I$. Thus, all the integers n_i are equal to a fixed integer n, so $N = \mathfrak{m}_F^n M$ and hence $\ker(\xi) = X_0[\pi_F^n]$ for any uniformizer π_F of $\mathcal{O}_F$. □

4.2.9.1. Remark. In contrast with algebraically closed fields, over a finite field κ of characteristic $p > 0$ the Lie type does not determine the $\mathcal{O}_F$-linear κ-isogeny class. To see this, let Y_0 be an $\mathcal{O}_F$-linear CM p-divisible group over κ, and let $q = \#\kappa$, so the central q-Frobenius endomorphism of Y_0 must arise from the action of a non-zero $\lambda \in \mathcal{O}_F$ (by 1.4.3.9(2)), with $\mathrm{ord}_F(\lambda)/\mathrm{ord}_F(q)$ equal to the unique slope of Y_0. Clearly λ is an invariant of the $\mathcal{O}_F$-linear κ-isogeny class of Y_0. We can change λ without changing the Lie type, as follows.

For a unit $u \in (\mathcal{O}_F \otimes_{\mathbb{Z}_p} W(\kappa))^\times$, replace the Frobenius and Verschiebung operators on $\mathrm{M}^*(Y_0)$ with $\mathcal{F}' = u\mathcal{F}$ and $\mathcal{V}' = (1 \otimes \sigma^{-1})(1/u)\mathcal{V}$ respectively, where σ is the absolute Frobenius automorphism of $W(\kappa)$. This defines an $\mathcal{O}_F$-linear CM p-divisible group Y_0' over κ whose central q-Frobenius endomorphism is given by the action of

$$\lambda' = \mathrm{Nm}_{(\mathcal{O}_F\otimes_{\mathbb{Z}_p} W(\kappa))/\mathcal{O}_F}(u) \cdot \lambda.$$

The norm map on units for the finite étale cover $\mathrm{Spec}(\mathcal{O}_F \otimes_{\mathbb{Z}_p} W(\kappa)) \to \mathrm{Spec}(\mathcal{O}_F)$ is surjective (since $\mathcal{O}_F$ has finite residue field), so by varying u we can vary λ' through all elements of $\mathcal{O}_F$ with the same valuation as λ. Since u is a unit, clearly

$$\mathrm{Lie}(Y_0')^\vee = \mathrm{M}^*(Y_0')/\mathcal{F}'\mathrm{M}^*(Y_0') \simeq \mathrm{M}^*(Y_0)/\mathcal{F}\mathrm{M}^*(Y_0) \simeq \mathrm{Lie}(Y_0)^\vee$$

as $(\mathcal{O}_F \otimes_{\mathbb{Z}_p} \kappa)$-modules, so $[\mathrm{Lie}(Y_0')] = [\mathrm{Lie}(Y_0)]$ in $\mathrm{R}_\kappa(\mathcal{O}_F)$.

4.2.9.2. Remark. As we noted above Proposition 4.2.7, over an algebraically closed field k of characteristic p the Lie type in $\mathrm{R}_k(\mathcal{O}_F)$ is a complete $\mathcal{O}_F$-linear isomorphism invariant of an $\mathcal{O}_F$-linear CM p-divisible group over k (i.e., it determines the isomorphism class, and all Lie types occur). Thus, the $\mathcal{O}_F$-linear isogeny classes over such fields are the same as the $\mathcal{O}_F$-linear isomorphism classes, by 4.2.9. Over a finite field κ of characteristic p, 4.2.8 implies that all κ-rational Lie types occur and by 4.2.9 the Lie type determines the $\mathcal{O}_F$-linear κ-isomorphism class within a given $\mathcal{O}_F$-linear κ-isogeny class. Moreover, the κ-rational Lie types that occur within an isogeny class are those having a fixed slope (by 4.2.8). However, the classification of $\mathcal{O}_F$-linear isogeny classes over κ is more subtle.

Let $q = \#\kappa$, so the q-Frobenius endomorphism of an $\mathcal{O}_F$-linear CM p-divisible group over κ is the action of some $\lambda \in \mathcal{O}_F - \{0\}$ and the slope is $\mathrm{ord}_F(\lambda)/\mathrm{ord}_F(q)$. The slope is determined by the Lie type, and all κ-rational Lie types occur, so the twisting argument in 4.2.9.1 shows that the possibilities for λ are precisely those $\lambda | q$ such that $\mathrm{ord}_F(\lambda)$ is divisible by $[\kappa : \kappa_F \cap \kappa] = [\kappa_F\kappa : \kappa_F]$. We claim that λ is a complete $\mathcal{O}_F$-linear κ-isogeny class invariant.

Let X_0 and X_0' be $\mathcal{O}_F$-linear CM p-divisible groups over κ. Their Dieudonné modules are free of rank 1 over $\mathcal{O}_F \otimes_{\mathbb{Z}_p} W(\kappa)$. Choose respective bases v and v', so $\mathcal{F}(v) = av$ and $\mathcal{F}(v') = a'v'$ for $a, a' \in \mathcal{O}_F \otimes_{\mathbb{Z}_p} W(\kappa)$ that divide p (so likewise $\mathcal{V}(v) = (p/(1 \otimes \sigma^{-1})(a))v$ and similarly for v' using a'). The existence of an $\mathcal{O}_F$-linear κ-isogeny between X_0 and X_0' is equivalent to that of an $\mathcal{F}$-equivariant and $F \otimes_{\mathbb{Q}_p} K_0$-linear isomorphism $\mathrm{M}^*(X_0)[1/p] \simeq \mathrm{M}^*(X_0')[1/p]$ for $K_0 := W(\kappa)[1/p]$. Such an isomorphism is $v \mapsto cv'$ for $c \in (F \otimes_{\mathbb{Q}_p} K_0)^\times$ such that $a = (((1 \otimes \sigma)(c))/c)a'$. Since $\mathrm{Spec}(F \otimes_{\mathbb{Q}_p} K_0) \to \mathrm{Spec}(F)$ is a finite étale $\mathrm{Gal}(K_0/\mathbb{Q}_p)$-torsor and σ generates this Galois group, by Hilbert's Theorem 90 the existence of c is equivalent to the condition that a and a' have the same norm in F. These norms are the respective elements $\lambda, \lambda' \in F^\times$ that compute the q-Frobenius endomorphisms of X_0 and X_0', so we are done.

4.2.10. Classification of toy model abelian varieties. The proof of Theorem 4.1.1 only uses toy model p-divisible groups rather than toy model abelian varieties. Nonetheless, it is of interest to know that every such p-divisible group over an algebraically closed field of characteristic p does arise from a toy model abelian variety over that field. (Such a result is not true over a finite field $\mathbb{F}_q$. For example, twisting by $W(\mathbb{F}_{p^4})^\times$-valued characters of $\mathrm{Gal}(\overline{\mathbb{F}}_q/\mathbb{F}_q)$ yields toy model p-divisible groups over $\mathbb{F}_q$ whose $W(\mathbb{F}_q)$-linear q-Frobenius does not have its eigenvalues on the Dieudonné module given by Weil q-integers.)

It has been shown in 4.1.3 that when there exists a toy model abelian variety or p-divisible group over the ground field then this field necessarily contains a subfield $\mathbb{F}_{p^2}$ of order p^2, so we now assume such a containment holds. Fix a finite field κ of order p^4, and assume for a moment that there exists a toy model abelian variety $(C, \mathcal{O})$ over $\mathbb{F}_{p^2}$ for a prime p. (When $p \equiv \pm 2 \pmod 5$, we have exhibited examples as C_0 in 4.1.2.) This yields a $W(\kappa)$-linear toy model p-divisible group X_0 over every field containing $\mathbb{F}_{p^2}$ upon choosing an isomorphism of $\mathcal{O}/p\mathcal{O}$ with κ in order to identify $\mathcal{O} \otimes_{\mathbb{Z}} \mathbb{Z}_p$ with $W(\kappa)$.

By changing this isomorphism of finite fields via composition with a generator of $\mathrm{Gal}(\kappa/\mathbb{F}_p)$ if necessary, we can arrange that over an algebraic closure k of the ground field the resulting action of κ on $\mathrm{Lie}(X_0)_k$ is changed to the other $\mathrm{Gal}(\kappa/\mathbb{F}_{p^2})$-stable pair of embeddings $\kappa \rightrightarrows k$. Such a pair of $\mathbb{F}_p$-embeddings of κ amounts to specifying the Lie type of C (or $C[p^\infty]$). There are exactly two such Lie types that can occur, and the classification of $W(\kappa)$-linear CM p-divisible groups over algebraically closed fields in terms of Lie types (see 4.2.9.2) shows that each does arise from a $W(\kappa)$-linear CM p-divisible group over $\mathbb{F}_{p^2}$.

To summarize, the task of proving that every toy model p-divisible group over an algebraically closed field of characteristic p arises from a toy model abelian variety over the same field is reduced to the problem of constructing a toy model abelian variety over $\mathbb{F}_{p^2}$ for every prime p. We now give such a construction over any finite field of even degree over $\mathbb{F}_p$, using the unique quaternion division algebra $D_{p,\infty}$ over $\mathbb{Q}$ ramified at exactly p and ∞.

4.2.10.1. Example. TOY MODELS ABELIAN SURFACES C WITH $\mathrm{End}^0(C)$ ISOMORPHIC TO $\mathrm{Mat}_2(D_{p,\infty})$.

Let E_0 be a supersingular elliptic curve over $\mathbb{F}_{p^{2n}}$ whose associated Weil p^{2n}-integer is $\pm p^n$ (see 1.6.2.2(1)). Equivalently, this is a member of one of the two isogeny classes of supersingular elliptic curves over $\mathbb{F}_{p^{2n}}$ whose endomorphism algebra is isomorphic to $D_{p,\infty}$; see 1.6.2.4.

The abelian surface $E_0 \times E_0$ has endomorphism algebra $\mathrm{Mat}_2(D_{p,\infty})$. It is elementary to construct a quartic CM field L that is inert at p (e.g., a suitable totally imaginary quadratic extension of a real quadratic field), and any such L splits $\mathrm{Mat}_2(D_{p,\infty})$ (as we see by local calculations at the p-adic and archimedean places of $\mathbb{Q}$). Thus, upon choosing such an L we may also choose an embedding $L \hookrightarrow \mathrm{Mat}_2(D_{p,\infty})$, and then via 1.7.4.5 we obtain an L-linearly isogenous quotient C of $E_0 \times E_0$ over $\mathbb{F}_{p^{2n}}$ such that $\mathcal{O}_L \subset \mathrm{End}(C)$.

Let $\kappa = \mathcal{O}_L/(p) \simeq \mathbb{F}_{p^4}$. For the $\mathcal{O}_L$-linear C to be a toy model, we need to check that its Lie type is $\mathrm{Gal}(\kappa/\mathbb{F}_{p^2})$-stable. By 4.2.8, if this condition fails then we can pass to an $\mathcal{O}_L$-linearly isogenous abelian surface over $\mathbb{F}_{p^2}$ to adjust the Lie type to satisfy the desired condition. This completes the construction of a toy model abelian variety over $\mathbb{F}_{p^{2n}}$ for any $n \geqslant 1$ (and in particular over $\mathbb{F}_{p^2}$).

4.2.10.2. To classify toy model abelian surfaces $(C, \mathcal{O} \to \mathrm{End}(C))$ over finite fields of characteristic p up to prime-to-p $\mathcal{O}$-linear isogeny, by (the proof of) 1.7.4.5 we may and do apply a preliminary such isogeny so that $\mathcal{O}_L \subseteq \mathrm{End}(C)$ (where L is the quartic CM fraction field of $\mathcal{O}$). Thus, we now fix L and focus on the cases with $\mathcal{O}_L \subseteq \mathrm{End}(C)$ over a finite field $\mathbb{F}_q$ with $q = p^{2n}$ (as there are no examples if $q = p^{2m+1}$). Let $\mathbb{F}_{p^2}$ be the subfield of $\mathbb{F}_q$ of size p^2 and let κ denote the finite field $\mathcal{O}_L/(p)$ of size p^4. By the definition of a toy model, there are exactly two possibilities for the Lie type (one for each embedding of $\mathbb{F}_{p^2}$ into κ), and by passing to $C^{(p)}$ we pass to the other possible Lie type. In particular, if one Lie type occurs then they both occur.

We claim that the examples arising from Example 4.2.10.1 are determined up to prime-to-p $\mathcal{O}_L$-linear isogeny by their abstract $\mathbb{F}_q$-isogeny class and Lie type. (We do not go beyond an $\mathcal{O}_L$-linear isogeny classification to an $\mathcal{O}_L$-linear isomorphism classification in general, due to class group obstructions in L.) To prove this, suppose C and C' are two such CM abelian surfaces in the same abstract $\mathbb{F}_q$-isogeny class. The simple $\mathbb{Q}$-algebra $\mathrm{End}^0(C) \simeq \mathrm{End}^0(C') \simeq \mathrm{Mat}_2(D_{p,\infty})$ has center $\mathbb{Q}$, so by the Skolem-Noether theorem we can compose an initial choice of $\mathbb{F}_q$-isogeny $\phi : C' \to C$ with a suitable isogenous endomorphism of C so that ϕ is L-linear. Then by 4.2.9 the $\mathcal{O}_L$-linear finite $\mathbb{F}_q$-subgroup $\ker(\phi) \subset C'$ has p-primary part $C'[p^\infty][u]$ for a non-zero $u \in \mathcal{O}_{L_p}$. Any element of $\mathcal{O}_{L_p}$ sufficiently close to u is a unit multiple of u, so by weak approximation we can arrange that $u \in \mathcal{O}_L$. Hence, $C'/C'[u]$ has the same $\mathcal{O}_{L_p}$-linear p-divisible group as C, yet $C' \simeq C'/C'[u]$ as $\mathcal{O}_L$-linear abelian varieties, so we can use this latter isomorphism to pass to the case when $\ker(\phi)$ has trivial p-part. That is, C and C' are related by an $\mathcal{O}_L$-linear isogeny of degree not divisible by p, as desired.

4.2.10.3. Apart from the toy models constructed in Example 4.2.10.1, which we have just classified up to prime-to-p isogeny, we shall now show that other toy model abelian varieties C over a finite field $\mathbb{F}_q$ of characteristic $p > 0$ exist if and only if $q = p^{2n}$ with $p \equiv \pm 2 \pmod 5$, and $L = \mathbb{Q}(\zeta_5)$. We will also describe the possibilities for these additional C, and classify them up to $\mathcal{O}_L$-linear *isomorphism* when C is simple. (It is possible to classify these additional examples up to isomorphism and not just isogeny because $\mathbb{Q}(\zeta_5)$ has class number 1, in contrast with a general quartic CM field L as in 4.2.10.2.)

Consider an arbitrary toy model abelian variety C over $\mathbb{F}_q$, so $q = p^{2n}$. Since C is a CM abelian surface, it is isotypic (by 1.3.1.1). Also, by definition its endomorphism algebra contains the quartic CM field L in which p is inert. This field L is its own centralizer in $\mathrm{End}^0(C)$, so L contains $\mathbb{Q}(\pi)$ where π is the Weil q-integer that corresponds to the unique simple isogeny factor of C. All slopes of C must be positive (as $C[p^\infty]$ is local-local), so either C is simple or it is $\mathbb{F}_q$-isogenous to the square of a supersingular elliptic curve. Since p must be inert in $\mathbb{Q}(\pi)$ yet $q = p^{2n}$, the list of supersingular possibilities in Example 1.6.2.4 allows only the Weil q-integers $\pm p^n$. (This entails ruling out the possibilities $\pi = \pm p^n \zeta$ for an imaginary quadratic root of unity $\zeta \neq \pm 1$ for which p is inert in $\mathbb{Q}(\zeta)$. That is, we have to rule out the case $\pi = \pm p^n i$ when $p \equiv -1 \pmod 4$ and $\pi = \pm p^n \zeta_3$ when $p \equiv -1 \pmod 3$. In any such case $\mathrm{End}^0(E_0) = \mathbb{Q}(\zeta)$, so $\mathrm{End}^0(E_0^2) = \mathrm{Mat}_2(\mathbb{Q}(\zeta))$. But this matrix algebra contains *no* quartic CM field L in which p is inert, since necessarily L is the compositum of its distinct quadratic subfields L^+ and $\mathbb{Q}(\zeta)$, forcing L to be biquadratic.) We conclude that C must be one of the toy model abelian varieties that we have built in Example 4.2.10.1 and classified up to $\mathcal{O}_L$-linear prime-to-p isogeny in 4.2.10.2. This takes care of non-simple C, so now we may and do assume that C is simple over $\mathbb{F}_q$.

An inspection of the Honda-Tate classification in 1.6.2.2 shows that $\mathbb{Q}(\pi)$ must be a CM field and the central simple algebra $\mathrm{End}^0(C)$ over $\mathbb{Q}(\pi)$ is split at all places away from the unique p-adic place of $\mathbb{Q}(\pi)$, so in fact $\mathrm{End}^0(C) = \mathbb{Q}(\pi)$. This forces $L = \mathbb{Q}(\pi)$, and the Weil q-integer π must then generate a power of the unique prime $p\mathcal{O}_L$ over p. Thus, $\pi = p^n u$ for some $u \in \mathcal{O}_L^\times$, and all archimedean absolute values of u are 1, so $u = \zeta$ is a root of unity. In other words, $L = \mathbb{Q}(\zeta)$ is a quartic cyclotomic field. This forces $\zeta = \pm\zeta_5$, ζ_8, or ζ_{12}. The existence of an inert rational prime forces $\mathrm{Gal}(L/\mathbb{Q})$ to be cyclic, so the only possibility is $L = \mathbb{Q}(\zeta_5)$ with $\pi = \pm p^n \cdot \zeta_5$. Since p is inert in L, necessarily $p \equiv \pm 2 \pmod 5$.

We conclude that a simple C must be a $\mathbb{Z}[\zeta_5]$-linear abelian surface over $\mathbb{F}_{p^{2n}}$ as introduced at the start of 4.1.2, so there is a preferred isomorphism $\mathbb{Z}[\zeta_5] \simeq \mathrm{End}(C)$ via $\zeta_5 \mapsto \pm\pi/p^n$ and C is uniquely determined up to $\mathbb{Z}[\zeta_5]$-linear isogeny by its Weil p^{2n}-integer. The remaining problem is to determine, up to $\mathbb{Z}[\zeta_5]$-linear isomorphism, which members of this isogeny class are toy models. We will show that there are exactly two members, related by absolute Frobenius twisting, and so each is determined up to $\mathbb{Z}[\zeta_5]$-linear isomorphism by its Lie type, for which there are exactly two possibilities. (For the quartic CM field $L = \mathbb{Q}(\zeta_5)$ we can make additional $\mathcal{O}_L$-linear toy model abelian surfaces by composing the canonical isomorphism $\mathbb{Z}[\zeta_5] \simeq \mathrm{End}(C)$ with the non-trivial automorphisms of $\mathbb{Z}[\zeta_5]$. These are then clearly the only additional examples.)

Since $\mathbb{Z}[\zeta_5]$ is a principal ideal domain and we have specified a canonical isomorphism of $\mathbb{Z}[\zeta_5]$ with $\mathrm{End}(C)$, the same argument with 4.2.9 as in our prime-to-p

isogeny classification of the toy models arising from Example 4.2.10.1 shows that any such C that is a toy model is uniquely determined up to $\mathbb{Z}[\zeta_5]$-linear isomorphism by its Lie type. There are only two possible Lie types, and passing to $C^{(p)}$ changes the Lie type, so it remains to show that there is *some* toy model within the $\mathbb{Z}[\zeta_5]$-linear isogeny class of simple abelian surfaces over $\mathbb{F}_{p^{2n}}$ with each Weil number $\pm p^n\zeta_5$.

Twisting by the unique quadratic character of the Galois group of $\mathbb{F}_q$ has the effect of negating the Weil q-number, so it suffices to consider the Weil q-number $p^n\zeta_5$. The case of odd n was handled in 4.1.2. For even n we write $n = 5^e m$ with $5 \nmid m$, so there is an example A over the subfield $\mathbb{F}_{p^{2\cdot 5^e}} \subseteq \mathbb{F}_q$ with Weil number $p^{5^e}\zeta_5$. Applying scalar extension to $\mathbb{F}_q$ raises the Weil number to the mth power, so since $5 \nmid m$ we thereby get an example over $\mathbb{F}_q$ with a Weil number that is p^n times a primitive 5th root of unity (though we may need to change the induced isomorphism $\mathbb{Z}[\zeta_5] \simeq \mathrm{End}(C)$ in order to make it the "canonical" one that carries $p^n\zeta_5$ to the p^{2n}-Frobenius endomorphism).

Conclusion. Suppose that $(C, \mathcal{O} \to \mathrm{End}(C))$ is a toy model abelian surface over a finite field $\mathbb{F}_{p^{2n}}$, where $\mathcal{O}$ is an order in an imaginary quadratic field L with $p \nmid [\mathcal{O}_L : \mathcal{O}]$.

(a) If $\mathrm{End}^0(C) \simeq \mathrm{Mat}_2(D_{p,\infty})$ then there exists a supersingular elliptic curve E_0 over $\mathbb{F}_{p^{2n}}$ with Weil number $\pm p^n$ and a prime-to-p isogeny $C \to E_0 \times E_0$ over $\mathbb{F}_{p^{2n}}$.

(b) If $\mathrm{End}^0(C)$ is not isomorphic to $\mathrm{Mat}_2(D_{p,\infty})$ then $p \equiv \pm 2 \pmod 5$, $\mathrm{End}^0(C) \simeq \mathbb{Q}(\zeta_5)$, and the Weil integer associated to C is $\pm p^n$ times a primitive 5th root of 1.

4.2.10.4. Remark. For an arbitrary field $\mathcal{K}$ of characteristic $p > 0$ and a toy model abelian surface $(A, \mathcal{O} \to \mathrm{End}(A))$ over $\mathcal{K}$ we claim $\mathrm{End}^0(A)$ is isomorphic to either $\mathbb{Q}(\zeta_5)$ or $\mathrm{Mat}_2(D_{p,\infty})$, with the former only possible if $p \equiv \pm 2 \pmod 5$. This fact will not be used in what follows, and our proof of it rests on the refined structure of endomorphism algebras of simple abelian varieties in positive characteristic as in 1.3.8.2.

Since A is a CM abelian surface, if it is not simple then by 1.3.2.1 it is isogenous to the square of a CM elliptic curve over $\mathcal{K}$. That is, either $\mathrm{End}^0(A)$ is a division algebra or it is a 2×2 matrix algebra over the endomorphism algebra of a CM elliptic curve E over $\mathcal{K}$. We first treat the latter case, as it is easier. In the non-simple case we have $\mathrm{End}^0(A) = \mathrm{Mat}_2(D)$ for $D := \mathrm{End}^0(E) \neq \mathbb{Q}$. If D is commutative then it is a quadratic subfield of $L = \mathrm{Frac}(\mathcal{O})$. But the real quadratic subfield L^+ of the quartic CM field L is the only quadratic subfield, as otherwise L would be biquadratic, which is inconsistent with the existence of the rational prime p that remains prime in L. We conclude that if D is commutative then $L^+ = \mathrm{End}^0(E)$, and this is impossible since a CM elliptic curve over a field never has real quadratic endomorphism algebra. Hence, the endomorphism algebra $D = \mathrm{End}^0(E)$ must be non-commutative, so by $\mathbb{Q}$-dimension reasons and an inspection of the list of endomorphism algebras of elliptic curves over algebraically closed fields of positive characteristic we see that $D \simeq D_{p,\infty}$. This settles the case of non-simple A.

Now suppose that A is simple, so $D := \mathrm{End}^0(A)$ is a division algebra. By 1.3.8.2, either D is a central quaternion division algebra over a totally real field (type III) or D is a central division algebra over a CM field (type IV). The center of D is a CM subfield of L, and we have already noted that the real quadratic subfield L^+ is the unique proper subfield of L distinct from $\mathbb{Q}$. Hence, in the type IV case we have $D = L$ and in the type III case the center of D is either $\mathbb{Q}$ or L^+. In particular, for the type III case the center Z is inert at p. But in the type III case the center is always either $\mathbb{Q}$ or $\mathbb{Q}(\sqrt{p})$ by 1.3.6.4 (which rests on Zarhin's proof of Tate's isogeny theorem over function fields), so in such cases the center of D must be $\mathbb{Q}$. But D cannot be a quaternion division algebra over $\mathbb{Q}$ since it contains the quartic number field L. We conclude that $\mathrm{End}^0(A) = L$. It remains to show that in such cases $L = \mathbb{Q}(\zeta_5)$ and $p \equiv \pm 2 \pmod 5$.

We may assume $\mathcal{O}_L \subset \mathrm{End}(A)$ and that $\mathcal{K}$ is finitely generated over $\mathbb{F}_{p^2}$. The toy model property specifies a unique embedding $j_0 : \mathbb{F}_{p^2} \to \kappa$ such that $\mathrm{Lie}(A)$ is free of rank 1 as a $\kappa \otimes_{j_0, \mathbb{F}_{p^2}} \mathcal{K}$-module. We may spread out A to an $\mathcal{O}_L$-linear abelian scheme $\mathcal{A} \to T$ over a smooth connected affine $\mathbb{F}_{p^2}$-scheme T with function field $\mathcal{K}$ such that $\mathrm{End}^0(\mathcal{A}) = \mathrm{End}(A)$ and $\mathrm{Lie}(\mathcal{A})$ is free of rank 1 as a $\kappa \otimes_{j_0, \mathbb{F}_{p^2}} \mathcal{O}_T$-module. The fiber $\mathcal{A}_t$ at any closed point $t \in T$ is an $\mathcal{O}_L$-linear toy model abelian variety over a finite field $\mathbb{F}_{q_t}$.

Specialization defines an injection of rings

$$L = \mathrm{End}^0(A) = \mathrm{End}^0(\mathcal{A}) \subset \mathrm{End}^0(\mathcal{A}_t).$$

By our isogeny classification of toy model abelian surfaces over finite fields, the endomorphism algebra $\mathrm{End}^0(\mathcal{A}_t)$ is either $\mathrm{Mat}_2(D_{p,\infty})$ or $\mathbb{Q}(\zeta_5)$ with the former precisely the case when the q_t-Frobenius lies in $\mathbb{Q}$ and the latter only occurring when $p \equiv \pm 2 \pmod 5$. Thus, we just have to show the existence of t for which the q_t-Frobenius is not in $\mathbb{Q}$. For every t, the q_t-Frobenius endomorphism of $\mathcal{A}_t$ arises from an element $\varphi_t \in \mathcal{O}_L$ since L is its own centralizer in $\mathrm{End}^0(\mathcal{A}_t)$. Let $Z' \subseteq L$ denote the subfield generated by the φ_t as we vary t. For a prime $\ell \neq p$, it follows from the Chebotarev Density Theorem and Tate's isogeny theorem over function fields (exactly as in the proof of 1.3.8.2) that the natural map

$$L_\ell = \mathbb{Q}_\ell \otimes_{\mathbb{Q}} \mathrm{End}^0(A) \to \mathrm{End}_{Z'_\ell}(V_\ell(A))$$

is an isomorphism. This rules out the case $Z' = \mathbb{Q}$, so we are done.

4.3. Tensor construction for p-divisible groups

We have used the Serre tensor product construction for abelian varieties via applications of 1.7.4.5, and we now require an analogue for p-divisible groups over fields and for Lie types. These matters are addressed in 4.3.1–4.3.6.

4.3.1. Lemma. *Let $\mathcal{K}$ be an arbitrary field, and let X be either an abelian variety over $\mathcal{K}$ or a p-divisible group over $\mathcal{K}$, with $\mathrm{char}(\mathcal{K}) = p$ in the latter case. Let R be a subring of $\mathrm{End}(X)$ that is a $\mathbb{Z}_p$-subalgebra in the case of p-divisible groups, and let M be a finitely generated right R-module.*

(a) *The fppf sheafification of $U \rightsquigarrow M \otimes_R X(U)$ on the category of $\mathcal{K}$-schemes is represented by an R-linear abelian variety when X is an abelian variety, and is represented by an R-linear p-divisible group when X is a p-divisible group. This representing object is denoted $M \otimes_R X$.*

(b) *If M has a compatible structure of associative R-algebra with identity then $M \otimes_R X$ has a natural left M-linear action and the natural map $\gamma \colon X \to M \otimes_R X$ is initial among R-linear homomorphisms from X to M-linear abelian* *varieties over $\mathcal{K}$ when X is an abelian variety, and likewise with M-linear p-divisible groups over $\mathcal{K}$ when X is a p-divisible group.*

(c) *In the setting of* (b), *if the structural morphism $R \to M$ is injective and the quotient M/R is a finite abelian group then $\gamma\colon X \to M \otimes_R X$ is an isogeny.*

PROOF. The case of abelian varieties was handled in 1.7.4.5, so we focus on the case that X is a p-divisible group (so $\mathrm{char}(\mathcal{K}) = p$).

The statement (b) follows immediately from (a). To prove (c) conditional on (a), suppose that M viewed as a $\mathbb{Z}_p$-module has finite length. There exists a positive integer N which is a power of p such that the map "multiplication by N" on M factors as the composition of an R-linear map $r\colon M \to R$ and the inclusion of (the image of) R in M. The map r induces a homomorphism $\delta\colon M \otimes_R X \longrightarrow X$ such that $\gamma \circ \delta = N \cdot \mathrm{id}_{M \otimes_R X}$ and $\delta \circ \gamma = N \cdot \mathrm{id}_X$. Thus, (c) also follows from (a). It remains to prove the existence statement (a) for X.

Pick an exact sequence $R^n \xrightarrow{h} R^m \longrightarrow M$ of R-modules, so we have an exact sequence

$$X^n = R^n \otimes_R X \xrightarrow{h\otimes 1} X^m = R^m \otimes_R X \longrightarrow M \otimes_R X := \mathscr{M} \longrightarrow 0$$

of fppf sheaves on $\mathrm{Spec}(\mathcal{K})$. Clearly the fppf sheaf $\mathscr{M}$ is p-power torsion as a sheaf and it is p-divisible. Thus, by the definition of a p-divisible group, it suffices to show that the kernel $\mathscr{M}[p]$ of $[p]\colon \mathscr{M} \to \mathscr{M}$ is represented by a finite group scheme over $\mathcal{K}$.

Let $\mathscr{K}$ and $\mathscr{F}$ be the respective kernel and the image of $h \otimes 1\colon X^n \to X^m$ (as sheaves). We have short exact sequences

$$0 \longrightarrow \mathscr{K} \longrightarrow X^n \longrightarrow \mathscr{F} \longrightarrow 0 \quad \text{and} \quad 0 \longrightarrow \mathscr{F} \longrightarrow X^m \longrightarrow \mathscr{M} \longrightarrow 0\,.$$

The sheaf $\mathscr{F}$ is p-divisible because it is a quotient of the p-divisible sheaf X^m, so we have a short exact sequence

$$0 \longrightarrow \mathscr{F}[p] \longrightarrow X^m[p] \longrightarrow \mathscr{M}[p] \longrightarrow 0\,.$$

Thus, to show that $\mathscr{M}[p]$ is represented by a finite group scheme over $\mathcal{K}$, it suffices to show that $\mathscr{F}[p]$ is represented by a finite group scheme over $\mathcal{K}$.

From the short exact sequence $0 \longrightarrow \mathscr{K} \longrightarrow X^n \longrightarrow \mathscr{F} \longrightarrow 0$ we get a 4-term exact sequence

$$0 \longrightarrow \mathscr{K}[p] \longrightarrow X^n[p] \longrightarrow \mathscr{F}[p] \longrightarrow \mathrm{coker}([p]\colon \mathscr{K} \to \mathscr{K}) \longrightarrow 0\,.$$

Clearly $\mathscr{K}[p^j]$ is represented by a finite group scheme over $\mathcal{K}$ for each j, because it is the kernel of the homomorphism $X^n[p^j] \to X^m[p^j]$ induced by $h \otimes 1\colon X^n \to X^m$. Thus, it suffices to show that the fppf sheaf $\mathrm{coker}([p]\colon \mathscr{K} \to \mathscr{K})$ on $\mathrm{Spec}(\mathcal{K})$ is represented by a finite group scheme over $\mathcal{K}$.

Because the fppf sheaf $\mathrm{coker}([p]\colon \mathscr{K} \to \mathscr{K})$ is the direct limit of the cokernels

$$\mathscr{C}_j := \mathrm{coker}(u_j\colon \mathscr{K}[p^{j+1}] \longrightarrow \mathscr{K}[p^j])$$

induced by "multiplication by p", where the transition maps $\iota_j\colon \mathscr{C}_j \to \mathscr{C}_{j+1}$ are induced by the inclusions $\mathscr{K}[p^j] \hookrightarrow \mathscr{K}[p^{j+1}]$, it suffices to show that the directed system $(\mathscr{C}_j)_{j \geqslant 1}$ of commutative finite group schemes over $\mathcal{K}$ terminates for large j (i.e., the transition maps are isomorphisms for large j).

Consider the commutative diagram

$$\begin{array}{ccccccccc}
0 & \longrightarrow & \mathscr{K}[p^{j+1}] & \longrightarrow & \mathscr{K}[p^{j+2}] & \longrightarrow & \mathscr{K}[p^{j+2}]/\mathscr{K}[p^{j+1}] & \hookrightarrow & X[p^{j+2}]/X[p^{j+1}] \\
 & & \downarrow{\scriptstyle u_j} & & \downarrow{\scriptstyle u_{j+1}} & & \downarrow{\scriptstyle v_j} & & {\scriptstyle \simeq}\downarrow{\scriptstyle w_j} \\
0 & \longrightarrow & \mathscr{K}[p^{j}] & \longrightarrow & \mathscr{K}[p^{j+1}] & \longrightarrow & \mathscr{K}[p^{j+1}]/\mathscr{K}[p^{j}] & \hookrightarrow & X[p^{j+1}]/X[p^{j}]
\end{array}$$

where the horizontal arrows are induced by inclusions and the vertical arrows are induced by "multiplication by p". Clearly $\ker(v_j) = 0$ for all $j \geqslant 1$ (because w_j is an isomorphism), so

$$\ker\left(\iota_j : \mathscr{C}_j \longrightarrow \mathscr{C}_{j+1}\right) = 0$$

for all $j \geqslant 1$. The trivial bound on $\mathcal{K}$-ranks

$$\mathrm{rk}(\mathscr{C}_j) \leqslant \mathrm{rk}(\mathscr{K}[p^j]) \leqslant \mathrm{rk}(X^n[p]) = p^{n\cdot \mathrm{height}(X)}$$

of the finite $\mathcal{K}$-group schemes $\mathscr{C}_j$ implies that there exists a $j_0 > 0$ such that

$$\mathrm{coker}([p] \colon \mathscr{K} \to \mathscr{K}) \simeq \mathscr{C}_j$$

for all $j \geqslant j_0$, so ι_j is an isomorphism for all $j \geqslant j_0$. □

Remark. Assume $\mathcal{K}$ is a perfect field of characteristic $p > 0$. In such cases, the existence of the cokernel of $h \otimes 1 \colon X^n \longrightarrow X^m$ for p-divisible groups X over $\mathcal{K}$ is an easy consequence of classical Dieudonné theory. Moreover, the isogeny assertion in (c) for p-divisible groups can be proved in such cases by a "saturation" procedure with Dieudonné modules, exactly as is done in the case of abelian varieties using ℓ-adic Tate modules for primes ℓ dividing $\#(M/R)$ with $\ell \in \mathcal{K}^\times$.

4.3.2. Remark. In 4.3.1(a), if we assume that X is an abelian scheme over a base scheme T and M is finitely generated projective right R-module, then we claim that the functor $T' \rightsquigarrow M \otimes_R A(T')$ on T-schemes is an fppf sheaf that is represented by an M-linear abelian scheme $M \otimes_R A$ over T. Similarly, if X is a p-divisible group over T and M is a finitely generated projective R-module then we claim that the functor $T' \rightsquigarrow M \otimes_R X(T')$ on T-schemes is an fppf sheaf that is represented by an M-linear p-divisible group $M \otimes_R X$ over T.

The proofs of these refinements over a general base are obvious when M is a free R-module, and the general case is deduced from that case by expressing a finitely generated projective module as a direct summand of a finite free module. Details on this existence result and various refinements (for Lie algebras, Cartier duality, etc.) are given in [**22**, 7.2–7.9].

When the base T is not the spectrum of a field, we will use the Serre tensor construction only in situations where M is a finitely generated projective R-module.

4.3.3. Let $F/F'/\mathbb{Q}_p$ be a tower of finite extensions of fields. In 4.3.5 we give a necessary and sufficient condition for an $\mathcal{O}_F$-linear CM p-divisible group over an algebraically closed field k of characteristic p to come from an $\mathcal{O}_{F'}$-linear CM p-divisible group over k via the Serre tensor construction (relative to $\mathcal{O}_{F'} \to \mathcal{O}_F$).

We carry over to $F'/\mathbb{Q}_p$ the same notation that we have used for $F/\mathbb{Q}_p$ in 4.2, but now we write e', f', and I' rather than e, f, and I. We also write $e_{F/F'}$ and $f_{F/F'}$ for the relative ramification index and residual degree (so $e = e_{F/F'}e'$ and

$f = f_{F/F'}f')$. By restriction of embeddings from κ_F to its subfield $\kappa_{F'}$ we get a natural surjection

$$\rho_{F/F'}\colon I \twoheadrightarrow I'$$

that is equivariant with respect to the natural surjective homomorphism

$$\mathrm{Gal}(\kappa_F/\mathbb{F}_p) \twoheadrightarrow \mathrm{Gal}(\kappa_{F'}/\mathbb{F}_p)$$

and the actions of $\mathrm{Gal}(\kappa_F/\mathbb{F}_p)$ on I and of $\mathrm{Gal}(\kappa_{F'}/\mathbb{F}_p)$ on I'.

The exact functor $M \rightsquigarrow \mathcal{O}_F \otimes_{\mathcal{O}_{F'}} M$ from the category of finitely generated $(\mathcal{O}_{F'} \otimes_{\mathbb{Z}_p} k)$-modules to the category of finitely generated $(\mathcal{O}_F \otimes_{\mathbb{Z}_p} k)$-modules induces a homomorphism

$$\tau_{F/F'}\colon \mathrm{R}_k(\mathcal{O}_{F'}) \longrightarrow \mathrm{R}_k(\mathcal{O}_F)$$

between the Grothendieck groups.

4.3.4. Lemma. *The map $\tau_{F/F'}$ satisfies the following properties.*

(i) *For every $\delta' \in \mathrm{R}_k(\mathcal{O}_{F'})$ and $i \in I$,*

$$\epsilon_i(\tau_{F/F'}(\delta)) = e_{F/F'} \cdot \epsilon_{\rho_{F/F'}(i)}(\delta).$$

An element $\delta' \in \mathrm{R}_k(\mathcal{O}_{F'})$ is in $\mathrm{R}_k^+(\mathcal{O}_{F'})$ if and only if $\tau_{F/F'}(\delta) \in \mathrm{R}_k^+(\mathcal{O}_F)$, and δ' is a Lie type in $\mathrm{R}_k(\mathcal{O}_{F'})$ if and only if $\tau_{F/F'}(\delta)$ is a Lie type in $\mathrm{R}_k(\mathcal{O}_F)$.

(ii) *The homomorphism $\tau_{F/F'}\colon \mathrm{R}_k(\mathcal{O}_{F'}) \longrightarrow \mathrm{R}_k(\mathcal{O}_F)$ is injective.*

(iii) *The image $\tau_{F/F'}(\mathrm{R}_k(\mathcal{O}_{F'}))$ consists of the elements $\delta \in \mathrm{R}_k(\mathcal{O}_F)$ such that $e_{F/F'} | \epsilon_i(\delta)$ for all $i \in I$ and the function $I \to \mathbb{Z}$ defined by $i \mapsto \epsilon_i(\delta)$ factors through $\rho_{F/F'}$. In particular, if F' is the maximal unramified subextension of F then the divisibility condition alone characterizes the image of $\tau_{F/F'}$.*

PROOF. The formula in (i) is elementary, and the rest of (i) follows from the formula. Parts (ii) and (iii) are immediate consequences of (i). □

4.3.5. Proposition. *An $\mathcal{O}_F$-linear* CM *p-divisible group Y over a field $k = \overline{k}$ of characteristic p is $\mathcal{O}_F$-linearly isomorphic to $\mathcal{O}_F \otimes_{\mathcal{O}_{F'}} X$ for an $\mathcal{O}_{F'}$-linear* CM *p-divisible group X over k if and only if its Lie type $[\mathrm{Lie}(Y)] \in \mathrm{R}_k(\mathcal{O}_F)$ belongs to $\tau_{F/F'}(\mathrm{R}_k(\mathcal{O}_F))$. In such cases, the $\mathcal{O}_{F'}$-linear p-divisible group X is uniquely determined by Y up to $\mathcal{O}_{F'}$-linear isomorphism.*

PROOF. The Lie algebra of $\mathcal{O}_F \otimes_{\mathcal{O}_{F'}} X$ is naturally isomorphic to $\mathcal{O}_F \otimes_{\mathcal{O}_{F'}} \mathrm{Lie}(X)$ because we can replace X with $X[p]$ and then compute the Lie algebras functorially in terms of dual numbers. Hence, the implication "$\Rightarrow$" is proved. Conversely, suppose that the Lie type $[\mathrm{Lie}(Y)] \in \mathrm{R}_k(\mathcal{O}_F)$ belongs to $\tau_{F/F'}(\mathrm{R}_k(\mathcal{O}_F))$, so $[\mathrm{Lie}(Y)] = \tau_{F/F'}(\delta')$ for a unique Lie type $\delta' \in \mathrm{R}_k(\mathcal{O}_{F'})$ by 4.3.4. By 4.2.7, there exists an $\mathcal{O}_{F'}$-linear CM p-divisible group X over k with Lie type δ'. It then follows from 4.3.4 and 4.2.6 that Y is $\mathcal{O}_F$-linearly isomorphic to $\mathcal{O}_F \otimes_{\mathcal{O}_{F'}} X$ since $k = \overline{k}$. The final assertion in the proposition is an immediate consequence of 4.3.4 and 4.2.6. □

4.3.6. Remark. If in 4.3.5 we replace field k with a finite field κ of characteristic p then the assertions are generally false. For example, consider an $\mathcal{O}_{F'}$-linear CM p-divisible group X over a finite field κ. Twisting by a continuous homomorphism $\rho\colon \mathrm{Gal}(\overline{\kappa}/\kappa) \longrightarrow \mathcal{O}_F^\times$ defines an $\mathcal{O}_F$-linear κ-form of $\mathcal{O}_F \otimes_{\mathcal{O}_{F'}} X$. We claim that only those twists attached to a homomorphism ρ with image in $\mathcal{O}_{F'}^{\times}$ are of the form $\mathcal{O}_F \otimes_{\mathcal{O}_{F'}} X'$ for an $\mathcal{O}_{F'}$-linear p-divisible group X' over κ.

To prove this claim, we first note that for $q = \#\kappa$, the q-Frobenius $\lambda \in \mathcal{O}_{F'}$ for X is recovered from $X_{\overline{\kappa}}$ as the composite isogeny

$$X_{\overline{\kappa}} \overset{\mathrm{Fr}_q}{\to} X_{\overline{\kappa}}^{(q)} \simeq X_{\overline{\kappa}}$$

(where the final isomorphism encodes the Galois descent to κ). Thus, the ρ-twist $\widetilde{X}$ over κ has q-Frobenius $\widetilde{\lambda} = \rho(\sigma_q)\lambda \in \mathcal{O}_F$ (where σ_q is the arithmetic q-Frobenius in $\mathrm{Gal}(\overline{\kappa}/\kappa)$). This is because (by definition) the Galois descent datum for $\widetilde{X}$ as a descent of $\mathcal{O}_F \otimes_{\mathcal{O}_{F'}} X_{\overline{\kappa}}$ is scaling by the $\mathcal{O}_F^{\times}$-valued ρ against the Galois descent datum for X as a descent of $X_{\overline{\kappa}}$. Consequently, if $\widetilde{X} \simeq \mathcal{O}_F \otimes_{\mathcal{O}_{F'}} X'$ for an $\mathcal{O}_{F'}$-linear X' over κ then necessarily multiplication by $\widetilde{\lambda}$ on $\mathcal{O}_F \otimes_{\mathcal{O}_{F'}} X'$ must preserve X' (in the isogeny category over κ). This is precisely the condition $\widetilde{\lambda} \in F'$ inside F, which in turn is equivalent to $\rho(\sigma_q)$ lying in $\mathcal{O}_{F'}^{\times}$.

4.4. Self-duality and CM lifting

The set-up of this section is semi-local, intermediate between the local situation in 4.2 and the global situation in 4.6. As before, k is an algebraically closed field of characteristic p and κ denotes a finite subfield of k unless otherwise specified.

4.4.1. Notation in the semi-local case. We are interested in situations that arise from a p-adic completion of a CM field viewed as an extension of its maximal totally real subfield. To that end, we now let F^+ be a finite extension of $\mathbb{Q}_p$ and F be a commutative semisimple F^+-algebra of F^+-dimension 2. Thus, either F is a quadratic extension field of F^+ or $F = F^+ \times F^+$. We refer to the two possibilities as the *non-split* and *split* cases respectively. (In applications in 4.6, F^+ will be the completion of a totally real field L^+ at a place v above p, and F will be the v-adic completion $L_v := L_v^+ \otimes_{L^+} L$ of a quadratic totally imaginary extension L of L^+.) There is a unique non-trivial F^+-algebra automorphism of F (swapping the factor fields in the split case), and it is denoted ι.

We now carry over a lot of the notation from the local case into the semi-local situation. The main point is to make appropriate definitions in the split case so that it can be treated simultaneously with the non-split case. The absolute residual degree and ramification index for F^+ are respectively denoted $f^+ = [\kappa_{F^+} : \mathbb{F}_p]$ and $e^+ = e(F^+/\mathbb{Q}_p)$, and in the split case we define $\mathcal{O}_F = \mathcal{O}_{F^+} \times \mathcal{O}_{F^+}$. Likewise, in the split case we define the "residue field" $\kappa_F := \kappa_{F^+} \times \kappa_{F^+}$ and the residual degree $f = f_F := [\kappa_F : \mathbb{F}_p]$ and ramification degree $e = e^+$, so $ef = [F : \mathbb{Q}_p]$ as in the non-split case. In the split case we also define $F_0 := F_0^+ \times F_0^+$ and $\mathcal{O}_{F^+} := \mathcal{O}_{F_0^+} \times \mathcal{O}_{F_0^+}$, where $F_0^+ := W(\kappa_{F^+})[1/p]$ is the maximal unramified subfield of F^+.

In terms of this notion of F_0 in the split case, we continue to use the notation

$$I = I_F = \mathrm{Hom}_{\mathbb{Z}_p\text{-alg}}(\mathcal{O}_{F_0}, W(k)) \xrightarrow{\simeq} \mathrm{Hom}_{\mathbb{Z}_p\text{-alg}}(\mathcal{O}_{F_0}, k) \xrightarrow{\simeq} \mathrm{Hom}_{\mathbb{F}_p\text{-alg}}(\kappa_F, k)$$

in all cases. In the split case there is a natural bijection $I_F \xrightarrow{\simeq} I_{F^+} \times I_{F^+}$ which is equivariant with respect to the natural action of $\mathrm{Gal}(\kappa_{F^+}/\mathbb{F}_p) \times \mathrm{Gal}(\kappa_{F^+}/\mathbb{F}_p)$, and it makes I_F a torsor for this product of Galois groups. The involution ι on F induces a natural involution of I_F given by interchanging the two factors in the split case.

For any field $\mathcal{K}$ of characteristic p, the Grothendieck group $\mathrm{R}_{\mathcal{K}}(\mathcal{O}_F)$ is defined in the split case exactly as in the non-split case (using finitely generated $\mathcal{O}_F \otimes_{\mathbb{Z}_p} \mathcal{K}$-modules). We likewise define the subset $\mathrm{R}^+_{\mathcal{K}}(\mathcal{O}_F)$ of *effective* classes. Thus, in the split case we have

$$\mathrm{R}_{\mathcal{K}}(\mathcal{O}_F) := \mathrm{R}_{\mathcal{K}}(\mathcal{O}_{F^+}) \times \mathrm{R}_{\mathcal{K}}(\mathcal{O}_{F^+}).$$

In all cases (split and non-split), the F^+-algebra involution ι of F induces an involution on the Grothendieck group $\mathrm{R}_{\mathcal{K}}(\mathcal{O}_F)$.

The "justification" for our definition of I in the split case is that we always have a product decomposition

$$\mathcal{O}_F \otimes_{\mathbb{Z}_p} W(k) \simeq \prod_{i\in I} \left(\mathcal{O}_F \otimes_{\mathcal{O}_{F_0},i} W(k)\right).$$

Moreover, for each $i \in I$ the Grothendieck group $\mathrm{R}(\mathcal{O}_F, i)$ of the category of finitely generated $\mathcal{O}_F \otimes_{\mathcal{O}_{F_0},i} k$-modules admits a natural isomorphism

$$\epsilon_i \colon \mathrm{R}(\mathcal{O}_F, i) \xrightarrow{\simeq} \mathbb{Z}$$

under which the subset of effective classes corresponds to $\mathbb{Z}_{\geqslant 0}$. We also carry over the definition of the *dimension* function $\epsilon = \epsilon_F \colon \mathrm{R}_k(\mathcal{O}_F) \to \mathbb{Z}$ via the formula $\delta \mapsto \sum_{i\in I} \epsilon_i(\delta)$, so in the split case the identification $\mathrm{R}_k(\mathcal{O}_F) = \mathrm{R}_k(\mathcal{O}_{F^+}) \times \mathrm{R}_k(\mathcal{O}_{F^+})$ yields the formula

$$\epsilon_F(\delta_1, \delta_2) = \epsilon_{F^+}(\delta_1) + \epsilon_{F^+}(\delta_2).$$

The product decomposition for $\mathcal{O}_F \otimes_{\mathbb{Z}_p} W(k)$ as a product of local factor rings indexed by I yields an isomorphism

$$\mathrm{R}_k(\mathcal{O}_F) \xrightarrow{\simeq} \bigoplus\nolimits_{i\in I} \mathrm{R}(\mathcal{O}_F, i) \xrightarrow{\oplus_i \epsilon_i} \bigoplus\nolimits_{i\in I} \mathbb{Z}$$

in general (even in the split case), under which the subset of effective classes corresponds to $\oplus_{i\in I}\mathbb{Z}_{\geqslant 0}$.

As in 4.2.4, for a finite subfield $\kappa \subset k$ the natural map

$$j_\kappa \colon \mathrm{R}_\kappa(\mathcal{O}_F) \longrightarrow \mathrm{R}_k(\mathcal{O}_F)$$

is an injection and identifies $\mathrm{R}_\kappa(\mathcal{O}_F)$ with the set of elements of $\mathrm{R}_k(\mathcal{O}_F)$ fixed by the natural action of $\mathrm{Aut}(k/\kappa)$. These elements are called *κ-rational.* In the split case, an element

$$(\delta_1, \delta_2) \in \mathrm{R}_k(\mathcal{O}_{F^+}) \times \mathrm{R}_k(\mathcal{O}_{F^+}) = \mathrm{R}_k(\mathcal{O}_F)$$

is κ-rational if and only if δ_1 and δ_2 are κ-rational in $\mathrm{R}_k(\mathcal{O}_{F^+})$.

Finally, an element δ of $\mathrm{R}_k(\mathcal{O}_F)$ is called a *Lie type* if $0 \leqslant \epsilon_i(\delta) \leqslant e$ for all $i \in I$ (where we recall that $e = e^+$ in the split case). Clearly every Lie type is effective; i.e., it belongs to the subset $\mathrm{R}^+_k(\mathcal{O}_F)$.

4.4.2. The notions of *$\mathcal{O}_F$-linear* (respectively *F-linear*) p-divisible group (X, α) over a ring R are defined exactly as in the local case in 4.2.2 except that we impose the additional requirement (automatic in the local case) that $\alpha : \mathcal{O}_F \to \mathrm{End}(X)$ is injective. Such a pair (X, α) is called an $\mathcal{O}_F$-linear (respectively F-linear) CM *p-divisible group* over R if the height of X is equal to $[F : \mathbb{Q}_p]$.

If $\mathcal{X}$ is an F-linear CM p-divisible group over the valuation ring $\mathcal{O}_K$ of a finite extension K of $K_0 = W(k)[1/p]$ contained in a fixed algebraic closure $\overline{K}_0$ of K_0 then the ($\overline{K}_0$-valued) *p-adic* CM *type* of $\mathcal{X}$ is the subset $\Psi \subset \mathrm{Hom}_{\mathbb{Q}_p\text{-alg}}(F, \overline{K}_0)$ of size $\dim(\mathcal{X})$ giving the $\overline{K}_0$-linear action of F on $\mathrm{Lie}(\mathcal{X})[1/p]_{\overline{K}_0}$. (The set Ψ has size

$\dim(\mathcal{X})$ due to the Hodge–Tate decomposition of $V_p(\mathcal{X}_K)$, exactly as in the local case in 4.2.2 since $\alpha : F \to \mathrm{End}^0(\mathcal{X})$ is *injective.*) The *Lie type* of an $\mathcal{O}_F$-linear CM p-divisible group (X, α) over a field $\mathcal{K}$ of characteristic p is the class in $\mathrm{R}_{\mathcal{K}}(\mathcal{O}_F)$ of the $\mathcal{K}$-linear action of $\mathcal{O}_F$ on $\mathrm{Lie}(X)$. This is a Lie type in the sense defined at the end of 4.4.1, due to the analogue in the local case (see 4.2.6(ii)).

Exactly as in 4.2.2, we define a "reduction" operation ξ assigning to any subset $\Phi \subseteq \mathrm{Hom}_{\mathbb{Q}_p\text{-alg}}(F, \overline{K}_0)$ a Lie type in $\mathrm{R}_k(\mathcal{O}_F)$ by the condition that

$$\epsilon_i(\xi(\Phi)) = \#\{\sigma \in \Phi \mid \overline{\sigma} = i\}$$

for all $i \in I$, where $\overline{\sigma}$ is the map $\kappa_F \to k$ induced by σ.

The *reflex field* $E(\Phi) \subset \overline{K}_0$ attached to any subset Φ of $\mathrm{Hom}_{\mathbb{Q}_p\text{-alg}}(F, \overline{K}_0)$ is defined as follows. In the non-split case it is defined in the usual way (using Galois theory, as for abelian varieties). In the split case with $F = F^+ \times F^+$ we have

$$\mathrm{Hom}_{\mathbb{Q}_p\text{-alg}}(F, \overline{K}_0) = \mathrm{Hom}_{\mathbb{Q}_p\text{-alg}}(F^+, \overline{K}_0) \coprod \mathrm{Hom}_{\mathbb{Q}_p\text{-alg}}(F^+, \overline{K}_0),$$

so Φ is the disjoint union of its components $\Phi_1, \Phi_2 \subseteq \mathrm{Hom}_{\mathbb{Q}_p\text{-alg}}(F^+, \overline{K}_0)$. In such cases we define the reflex field of Φ in $\overline{K}_0$ to be the subfield generated over $\mathbb{Q}_p$ by the reflex fields of Φ_1 and Φ_2.

Notions of self-duality can be defined for p-adic CM types and for Lie types:

4.4.3. Definition. A Lie type $\delta \in \mathrm{R}_k(\mathcal{O}_F)$ is *self-dual* if

$$\delta + \iota(\delta) = [\mathcal{O}_F \otimes_{\mathbb{Z}_p} k]$$

in $\mathrm{R}_k(\mathcal{O}_F)$. A subset $\Phi \subseteq \mathrm{Hom}_{\mathbb{Q}_p\text{-alg}}(F, \overline{K}_0)$ is *self-dual* if $\mathrm{Hom}_{\mathbb{Q}_p\text{-alg}}(F, \overline{K}_0)$ is the disjoint union of Φ and $\Phi \circ \iota$.

For a finite extension K of K_0 inside $\overline{K}_0$, an F-linear CM p-divisible group $\mathcal{X}$ over $\mathcal{O}_K$ is *self-dual up to isogeny* if its p-adic CM type $\Phi(\mathcal{X}) \subseteq \mathrm{Hom}_{\mathbb{Q}_p\text{-alg}}(F, \overline{K}_0)$ is self-dual. (By 3.7.1.3 and 3.7.4(1), it is equivalent to say that $\mathcal{X}$ is F-linearly isogenous to its Serre dual $\mathcal{X}^t$ equipped with the dual CM structure.)

It is clear that if Φ is a self-dual p-adic CM type on F then its reduction $\xi(\Phi) \in \mathrm{R}_k(\mathcal{O}_F)$ is self-dual. For Lie types, self-duality can be described explicitly, depending on whether we are in the split or non-split cases. In the non-split case it means that

$$\epsilon_i(\delta) + \epsilon_{i \circ \overline{\iota}}(\delta) = e$$

for all $i \in I$, where $\overline{\iota}$ is the involution on κ_F induced by ι. In particular, if F is a field and is ramified over F^+ then $e = e(F/\mathbb{Q}_p)$ is even and the self-duality condition means that $\epsilon_i(\delta) = e/2$ for all $i \in I_F$. In the split case with $\delta = (\delta_1, \delta_2)$ for $\delta_1, \delta_2 \in \mathrm{R}_k(\mathcal{O}_{F^+})$, self-duality for δ means that

$$\epsilon_i(\delta_1) + \epsilon_i(\delta_2) = e^+ = e(F^+/\mathbb{Q}_p)$$

for all $i \in I_{F^+}$; in particular, the sum of the slopes δ_1 and δ_2 is 1.

The following three results (4.4.4–4.4.6) relating p-adic CM types and Lie types to lifting and isomorphism problems for CM p-divisible groups, especially incorporating a self-duality condition, are due to C.-F. Yu in [**133**, §5].

4.4.4. Lemma. *Let K be a finite extension of K_0 contained in $\overline{K}_0$.*

(i) *Let $\mathcal{Y}$ be an $\mathcal{O}_F$-linear* CM *p-divisible group over $\mathcal{O}_K$, with p-adic* CM *type Φ. The Lie type of the closed fiber Y_0 is the reduction $\xi(\Phi)$ of Φ. Moreover, K contains the reflex field of (F, Φ), and $\mathcal{Y}$ is determined by Φ (and $\mathcal{O}_K$) up to $\mathcal{O}_F$-linear isomorphism.*

(ii) *Let Φ be a subset of* $\mathrm{Hom}_{\mathbb{Q}_p\text{-alg}}(F, \overline{K})$. *If K contains the reflex field of Φ then there exists an $\mathcal{O}_F$-linear* CM *p-divisible group $\mathcal{Y}$ over $\mathcal{O}_K$ whose p-adic* CM *type is Φ.*

PROOF. The first part of (i) is a reformulation of 4.2.3 in the present set-up, and the containment of the reflex field inside K is clear. The final part of (i) is an application of Proposition 3.7.4(1), and (ii) is an application of Proposition 3.7.3(1) (where E there is our $E(\Phi)$). $\square$

As we noted after 4.4.3, if $\Psi \subset \mathrm{Hom}_{\mathbb{Q}_p\text{-alg}}(F, \overline{K}_0)$ is a self-dual subset then its reduction $\xi(\Psi)$ is a self-dual element in $\mathrm{R}_k(\mathcal{O}_F)$. Here is a converse:

4.4.5. Lemma. *For every self-dual element $\delta \in \mathrm{R}_k(\mathcal{O}_F)$ there exists a self-dual subset $\Psi \subset \mathrm{Hom}_{\mathbb{Q}_p\text{-alg}}(F, \overline{K}_0)$ such that $\xi(\Psi) = \delta$.*

PROOF. We will handle the case when F is a field; the split case is similar but easier. Let $\delta \in \mathrm{R}_k(\mathcal{O}_F)$ be a self-dual element. For each $i \in I_F = \mathrm{Hom}_{\mathbb{F}_p\text{-alg}}(\kappa_F, k)$, define the subset $S_i \subseteq \mathrm{Hom}_{\mathbb{Q}_p\text{-alg}}(F, \overline{K}_0) = \mathrm{Hom}_{\mathbb{Z}_p\text{-alg}}(\mathcal{O}_F, \overline{K}_0)$ by

$$S_i := \left\{\tau \in \mathrm{Hom}_{\mathbb{Z}_p\text{-alg}}(\mathcal{O}_F, \overline{K}) \mid \tau \text{ induces } i \text{ on } \kappa_F\right\}.$$

We separately treat the cases when the quadratic extension of fields F/F^+ is ramified or unramified.

In the ramified case (i.e., $\kappa_F = \kappa_{F^+}$), $\epsilon_i(\delta) = \epsilon_i(\iota \cdot \delta) = 1/2$ for each $i \in I$. Pick a subset Ψ_i of representatives of $S_i/\{1, \iota\}$ in S_i for each $i \in I_{F^+}$. The union Ψ of the Ψ_i has the required properties.

Suppose now that F/F^+ is unramified. Pick a subset J of representatives for $I/\{1, \iota\}$ in I. For each $j \in J$, pick a subset $\Psi_j \subset S_j$ with $\#\Psi_j = \epsilon_j(\delta)$. The union Ψ of the Ψ_j has the required properties. $\square$

4.4.6. Proposition. *Let Y_0 be an $\mathcal{O}_F$-linear* CM *p-divisible group over k whose Lie type is self-dual.*

(i) *There exists a self-dual subset $\Phi \subset \mathrm{Hom}_{\mathbb{Q}_p\text{-alg}}(F, \overline{K}_0)$ whose reduction $\xi(\Phi)$ is equal to the Lie type of Y_0.*

(ii) *For any self-dual subset $\Phi \subset \mathrm{Hom}_{\mathbb{Q}_p\text{-alg}}(F, \overline{K}_0)$ such that $\xi(\Phi) = [\mathrm{Lie}(Y_0)]$ and any finite subextension K in $\overline{K}_0/K_0$ which contains the reflex field of (F, Φ) in $\overline{K}_0$, there exists an $\mathcal{O}_F$-linear* CM *p-divisible group $\mathcal{Y}$ over $\mathcal{O}_K$ with Φ as its p-adic* CM *type such that the closed fiber of $\mathcal{Y}$ is $\mathcal{O}_F$-linearly isomorphic to Y_0.*

In particular, there exists a finite extension K of K_0 and an $\mathcal{O}_F$-linear CM *p-divisible group $\mathcal{Y}$ over $\mathcal{O}_K$ whose closed fiber is $\mathcal{O}_F$-linearly isomorphic to Y_0.*

PROOF. Assertion (i) follows from 4.4.5. Let K be a finite extension of K_0 containing the reflex field of (F, Φ) as in (ii). By 4.4.4 (ii), there exists an $\mathcal{O}_F$-linear CM p-divisible group $\mathcal{Y}$ over $\mathcal{O}_K$ whose p-adic CM type is equal to Φ. In the split case, every $\mathcal{O}_F$-linear p-divisible group has a canonical decomposition as a direct product of two p-divisible groups according to the decomposition $\mathscr{O}_{F^+} \times \mathscr{O}_{F^+}$ for $\mathscr{O}_F$. Thus, by 4.2.6(iv), the closed fiber of $\mathcal{Y}$ is $\mathscr{O}_F$-linearly isomorphic to Y_0. $\square$

4.4.7. Lemma. *In the split case, for any finite field $\kappa \subset k$ and any κ-rational Lie type $\delta_1 \in \mathrm{R}_k(\mathscr{O}_{F^+})$ there exists a unique κ-rational Lie type $\delta_2 \in \mathrm{R}_k(\mathscr{O}_{F^+})$ such that the element*

$$\delta := (\delta_1, \delta_2) \in \mathrm{R}_k(\mathscr{O}_{F^+}) \times \mathrm{R}_k(\mathscr{O}_{F^+}) = \mathrm{R}_k(\mathscr{O}_F)$$

is a self-dual κ-rational Lie type in $\mathrm{R}_k(\mathscr{O}_F)$. Moreover, $\epsilon(\delta_1) + \epsilon(\delta_2) = [F^+ : \mathbb{Q}_p]$.

The proof of this lemma is straightforward and left to the reader. This result is useful when we are given an $\mathscr{O}_F$-linear CM p-divisible group over a finite field κ in the split case, such as in:

4.4.8. Corollary. *Suppose that $F = F^+ \times F^+$, and let X_0 be an $\mathscr{O}_F$-linear* CM *p-divisible group over a finite subfield $\kappa \subset k$. For the class $[\mathrm{Lie}(X_0)] = (\delta_1, \delta_2) \in \mathrm{R}_k(\mathscr{O}_{F^+}) \times \mathrm{R}_k(\mathscr{O}_{F^+})$, assume that the sum of the slopes of δ_1 and δ_2 is* 1.

There exists an $\mathscr{O}_F$-linear CM *p-divisible group over κ that is $\mathscr{O}_F$-linearly isogenous to X_0 and admits an $\mathscr{O}_F$-linear* CM *lift X over a* 1*-dimensional complete local noetherian domain with residue field κ and generic characteristic* 0 *such that the geometric generic fiber of X has a self-dual p-adic* CM *type. In particular, X_0 is $\mathscr{O}_F$-linearly isogenous to its Serre dual equipped with the dual* CM *structure relative to the F^+-involution of F.*

The isogenies in this corollary are constructed using 4.2.8, so in terms of the absolute ramification degree and residual degree invariants e^+ and f^+ for F^+ over $\mathbb{Q}_p$ we can arrange that the degree of the $\mathscr{O}_F$-linear κ-isogeny is at most $p^{2(f^+ \dim(X_0) + [F^+:\mathbb{Q}_p](f^+-1)/2)}$.

PROOF. The final assertion concerning the Serre dual of X_0 follows from the rest due to 3.7.1.3. Also, the initial hypothesis on the slopes ensures that the "self-dual complement" to δ_1 in the sense of 4.4.7 has its other component with the same slope as δ_2. Thus, by 4.2.8, X_0 is $\mathscr{O}_F$-linearly isogenous to an $\mathscr{O}_F$-linear p-divisible group Y_0 over κ whose Lie type is self-dual. We apply 4.4.6 to see that $(Y_0)_k$ can be $\mathscr{O}_F$-linearly lifted to an $\mathscr{O}_F$-linear CM p-divisible group $\mathcal{Y}$ in characteristic 0 with a self-dual p-adic CM type. The remaining task is to make such a lifting construction without increasing the residue field.

By Theorem 1.4.5.5, the infinitesimal deformation functor of Y_0 equipped with its $\mathscr{O}_F$-action admits a formal deformation ring $\mathcal{R}$ with residue field κ and it is universal with respect to formal deformations after any extension of the ground field. Thus, the existence of $\mathcal{Y}$ provides a local map from $\mathcal{R}$ to a complete local noetherian domain with generic characteristic 0, so $\mathcal{R}[1/p] \neq 0$ and the pullback of the universal deformation over $\mathcal{R}[1/p]$ admits geometric fibers at which the p-adic CM type is self-dual.

The p-adic CM type is a discrete invariant. To be precise, if K/K_0 is a finite extension that splits F over $\mathbb{Q}_p$ then by viewing the Lie algebra of the universal deformation as a vector bundle on $\mathrm{Spec}(\mathcal{R}_{\mathscr{O}_K})$ we see that the isomorphism class of the p-adic CM type at geometric points of $\mathrm{Spec}(\mathcal{R}[1/p]_K)$ is Zariski-locally constant on the base and there is a connected component on which it is self-dual. The same therefore holds on $\mathrm{Spec}(\mathcal{R})$. Now we can perform a slicing argument on such a connected component as at the start of 2.1.1 (using [**54**, 7.1.9]) to construct the desired 1-dimensional domain as a quotient of $\mathcal{R}$. □

The hypothesis on the sum of the slopes in Corollary 4.4.8 is satisfied in applications to p-divisible groups arising from CM abelian varieties over finite fields due to the Shimura–Taniyama formula in 2.1.4.1 (which computes the p-adic slopes of the q-Frobenius π viewed as an element of the CM field) and the Riemann Hypothesis (which ensures that $v(\pi)/\mathrm{ord}_v(q) + \overline{v}(\pi)/\mathrm{ord}_{\overline{v}}(q) = 1$ for every p-adic place v of the CM field).

A more general class of examples satisfying this slope hypothesis arises from any CM p-divisible group Y_0 over κ with CM by $\mathbb{Q}_p \otimes_{\mathbb{Q}} L$ for a CM field L, provided that the p-adic CM type Φ on $\mathbb{Q}_p \otimes_{\mathbb{Q}} L$ arises from a genuine CM type on L and the slope of $Y_0[w^\infty]$ is equal to $\#\Phi_w/[L_w : \mathbb{Q}_p]$ for every p-adic place w of L (exactly as in the Shimura–Taniyama formula). Given any such (Y_0, L, Φ), we can take $F = L_v$ and $X_0 = Y_0[v^\infty]$ for any p-adic place v of L^+ that splits in L. Indeed, in such cases the places of L over v are a distinct pair of the form $\{w, \overline{w}\}$ and $\Phi_{\overline{w}} = \overline{\Phi}_w$, so $\#\Phi_w + \#\Phi_{\overline{w}} = [L_w : \mathbb{Q}_p] = [L_{\overline{w}} : \mathbb{Q}_p]$ as required.

4.5. Striped and supersingular Lie types

In addition to self-duality for Lie types, another aspect we need to understand (allowing the split case) is a notion of "slope" for Lie types that encodes the notion of slope for p-divisible groups in characteristic p. This will lead to special classes of Lie types called "supersingular" and "striped", which respectively generalize the Lie types of supersingular elliptic curves and the toy models (abelian surfaces and p-divisible groups), as in 4.1.3.1 and 4.1.3.2. Such abelian varieties and p-divisible groups exhibit exceptional behavior for CM lifting problems (cf. 4.1.2.3).

Let k be an algebraically closed field of characteristic $p > 0$, and let F be as in 4.4.1. Since F may be a product of two fields, certain definitions initially made only for F a field (such as Definition 4.2.5) need to be revisited:

4.5.1. Definition. For a Lie type $\delta \in \mathrm{R}_k(\mathcal{O}_F)$, the *slopes* of δ and their *multiplicities* are defined as follows.

(a) In the non-split case, $\epsilon(\delta)/[F : \mathbb{Q}_p]$ is the only slope of δ (agreeing with 4.2.5), and it is assigned multiplicity $[F : \mathbb{Q}_p]$.
(b) In the split case, write $\delta = (\delta_1, \delta_2)$ with $\delta_j \in \mathrm{R}_k(\mathcal{O}_{F^+})$. Then $\epsilon(\delta_1)/[F^+ : \mathbb{Q}_p]$ and $\epsilon(\delta_2)/[F^+ : \mathbb{Q}_p]$ are the slopes of δ. If these are distinct then the multiplicity of each slope is $[F^+ : \mathbb{Q}_p]$. If they coincide then the multiplicity of this unique slope is $[F : \mathbb{Q}_p] = 2[F^+ : \mathbb{Q}_p]$.

An element $\delta \in \mathrm{R}_k(\mathcal{O}_F)$ is *supersingular* if it is a Lie type with $1/2$ as its only slope. Two Lie types in $\mathrm{R}_k(\mathcal{O}_F)$ are *isogenous* if they have the same slopes and the same multiplicities. (The requirement to have the same multiplicities is redundant.)

4.5.2. Remark. If F is a field and δ is a self-dual Lie type then δ is supersingular (since $\epsilon(\iota\delta) = \epsilon(\delta)$). If F is a ramified quadratic extension field of F^+ and δ is a self-dual Lie type then $\epsilon_i(\delta) = e^+ = e/2$ for all $i \in I$.

4.5.3. Definition. Suppose that F is a field; i.e., we are in the non-split case. A Lie type $\delta \in \mathrm{R}_k(\mathcal{O}_F)$ is *uniform* if the integers $\epsilon_i(\delta)$ for $i \in I$ are all the same. If

$f = [\kappa_F : \mathbb{F}_p]$ is even then a Lie type $\delta \in \mathrm{R}_k(\mathcal{O}_F)$ is *striped* if δ is $\mathbb{F}_{p^2}$-rational and $\epsilon_i(\delta) \in \{0, e\}$ for all $i \in I$.

Supersingular uniform Lie types δ exist if only if $e = e(F/\mathbb{Q}_p)$ is even. In such cases, $\epsilon_i(\delta) = e/2$ for all $i \in I$ and the unique supersingular uniform Lie type δ_{unif} is self-dual and rational over $\mathbb{F}_p$. By 4.1.3.2 and the subsequent discussion there, we see that any toy model p-divisible group has a striped Lie type.

4.5.4. Remark. Assume that F is a field and $f = f_F$ is even.

(i) Every striped Lie type is supersingular. There are exactly two striped Lie types in $\mathrm{R}_k(\mathcal{O}_F)$; each is of the form $\delta_{\mathrm{stp},J}$ for a unique $\mathrm{Gal}(\kappa_F/\mathbb{F}_{p^2})$-orbit J in I, characterized by the following property:

$$\epsilon_i(\delta_{\mathrm{stp},J}) = \begin{cases} e & \text{if } i \in J, \\ 0 & \text{if } i \notin J. \end{cases}$$

(This description explains the terminology "striped", as follows. Using the identification $\mathrm{Gal}(\kappa_F/\mathbb{F}_p) = \mathbb{Z}/f\mathbb{Z}$ via the arithmetic Frobenius generator with $f = f_F$ even, the $\mathbb{Z}/f\mathbb{Z}$-torsor I can be visualized as a roulette wheel with f sectors and a thin black annulus all the way around so that each sector meets the annulus in e equally-sized slots into which white marbles may land. The Galois orbit J is one of the two ways to make an alternating choice of $f/2$ sectors, and a Lie type δ is an assignment of $\epsilon_i(\delta) \leqslant e$ black slots to be filled with white marbles in the ith sector (beginning from the inner side in each sector). The striped Lie type associated to J is the one in which the ith sector is completely filled with white marbles for $i \in J$ and empty for $i \notin J$, forming a stripe-like pattern of black and white along the annulus.)

(ii) If $f \equiv 2 \pmod 4$ and F/F^+ is unramified then ι lifts the non-trivial element of $\mathrm{Gal}(\mathbb{F}_{p^2}/\mathbb{F}_p)$ and so both striped Lie types in $\mathrm{R}_k(\mathcal{O}_F)$ are self-dual. Otherwise neither striped Lie type in $\mathrm{R}_k(\mathcal{O}_F)$ is self-dual.

(iii) The Lie type $[\mathrm{Lie}(Y_0)]$ of the p-divisible group Y_0 in B.1.4, represented by the class of the $\mathcal{O}_F$-linear k-module $N/\mathcal{V}N$ there, is striped. We will need the notion of striped Lie types only when e is odd and $f \equiv 0 \pmod 4$.

(iv) We will only use uniform Lie types that are supersingular.

4.5.5. Lemma. *Suppose that F is a field and f_F is even. Let F_2 be the subfield of F that is is unramified of degree 2 over $\mathbb{Q}_p$. A Lie type $\delta \in \mathrm{R}_k(\mathcal{O}_F)$ is striped if and only if there exist a Lie type $\delta' \in \mathrm{R}_k(\mathcal{O}_{F_2})$ such that δ' is striped and $\delta = \tau_{F/F_2}(\delta')$, with τ_{F/F_2} as in* 4.3.3.

PROOF. The "if" part is easy to see from 4.5.3 and 4.3.4. Conversely, suppose that $\delta \in \mathrm{R}_k(\mathcal{O}_F)$ is a striped Lie type. By 4.5.4(i), there exists a $\mathrm{Gal}(\kappa/\mathbb{F}_{p^2})$-orbit $J \subset I$ such that δ is equal to the Lie type δ_J defined in 4.5.4(i). For exactly one of the two $\mathbb{Z}_p$-embeddings i of $\mathcal{O}_{F_2}$ into $W(k)$, the elements of I have restriction to $\mathcal{O}_{F_2}$ equal to i. Let $\delta_{\{i\}}$ be the striped Lie type in $\mathrm{R}_k(\mathcal{O}_{F_2})$ attached to $\{i\} \subset \mathrm{Hom}_{\mathbb{Z}_p\text{-alg}}(\mathcal{O}_{F_2}, k)$. We see from 4.3.4(i) that $\tau_{F/F_2}(\delta_{\{i\}}) = \delta$. □

4.5.6. Corollary. *For every striped Lie type $\delta \in \mathrm{R}_k(\mathcal{O}_F)$ and every subextension F' of $F/\mathbb{Q}_p$ with $f_{F'}$ even, there exists a unique striped Lie type $\delta' \in \mathrm{R}_k(\mathcal{O}_{F'})$ such that $\delta = \tau_{F/F'}(\delta')$.*

4.5.7. Given a finite field κ of characteristic p and a κ-rational supersingular Lie type $\delta \in \mathrm{R}_k^+(\mathcal{O}_F)$, we want to know if there is a κ-rational self-dual (supersingular) Lie type that is isogenous to δ. By 4.4.7, this holds when F is split. Likewise, it holds when F is a field and e is even, by using the unique supersingular uniform Lie type δ_{unif}. Thus, by 4.5.9 below, such a Lie type can be found unless the following purely field-theoretic properties (having nothing to do with Lie types) all hold:

(1) F is a field,
(2) e is odd (so F/F^+ is unramified and $2|f$),
(3) $4|f$,
(4) $[\kappa_F : \kappa_F \cap \kappa]$ is even.

We say $(F/F^+, \kappa)$ is *bad* if the conditions (1)–(4) above all hold; otherwise we say that the pair is *good.*

Remark. The definition in 4.5.7 generalizes 4.1.6, where the extension $L_v^+ \to L_v$ there (for a p-adic place v of L^+) plays the role F/F^+ here.

A p-divisible group over a field of characteristic p is *supersingular* when it is isoclinic of slope 1/2. The Lie type of a supersingular $\mathcal{O}_F$-linear CM p-divisible group X over κ is clearly supersingular (as a Lie type). For any δ as above, if $(F/F^+, \kappa)$ is good then there exists a self-dual κ-rational Lie type δ' isogenous to δ (see 4.5.10). Proposition 4.2.8 (applied to the local factors of $\mathcal{O}_F$ in the split case) gives us an $\mathcal{O}_F$-linear CM p-divisible group X' over κ that is $\mathcal{O}_F$-linearly isogenous to X and has Lie type δ'. By 4.4.6, X' can be $\mathcal{O}_F$-linearly lifted to a p-divisible group in characteristic 0 with self-dual p-adic CM type, so we are done with the CM lifting problem (I) for the CM p-divisible group X in such "good" cases.

On the other hand, suppose that $(F/F^+, \kappa)$ is bad. In such cases, to find a CM lift we modify X by an isogeny: Lemma 4.5.14 shows that $\kappa \supset \mathbb{F}_{p^2}$ and there exists a κ-rational striped Lie type in $\mathrm{R}_\kappa(\mathcal{O}_F)$, so it follows from 4.2.8 and 4.5.4(1) that X is $\mathcal{O}_F$-linearly κ-isogenous to an $\mathcal{O}_F$-linear p-divisible group over κ whose Lie type is striped.

4.5.8. Lemma. *Suppose that F is a field and the absolute ramification index e of F is even. Let δ_{unif} be the unique $\mathbb{F}_p$-rational supersingular uniform element in $\mathrm{R}_k^+(\mathcal{O}_F)$ (so δ_{unif} is self-dual and $\epsilon_i(\delta_{\mathrm{unif}}) = e/2$ for all $i \in I$). There exists a self-dual subset $\Phi_0 \subset \mathrm{Hom}(F, \overline{K}_0)$ such that $\xi(\Phi_0) = \delta_{\mathrm{unif}}$.*

PROOF. The result follows from 4.4.5, but here is a more explicit proof.

Suppose first that F is ramified over F^+. For each $\mathbb{Q}_p$-embedding $\sigma\colon F^+ \hookrightarrow \overline{K}_0$, choose an extension $\widetilde{\sigma}\colon F \hookrightarrow \overline{K}_0$ of σ. If Φ_0 is the set of all such $\widetilde{\sigma}$ then $\xi(\Phi_0) = \delta_{\mathrm{unif}}$ and Φ_0 is self-dual.

Assume F is unramified over F^+, so $F = F^+ \otimes_{F_0^+} F_0$ and ι arises from the non-trivial automorphism of F_0 over F_0^+. For each $\mathbb{Q}_p$-embedding $\sigma\colon F_0^+ \hookrightarrow \overline{K}_0$, let $\tau_1, \ldots, \tau_e$ be the embeddings of F^+ into $\overline{K}_0$ extending σ, and let $\widetilde{\sigma}, \widetilde{\sigma}'$ be the embeddings of F_0 into $\overline{K}_0$ extending σ. For $j = 1, \ldots, e/2$, let $\tau(\sigma)_j\colon F \hookrightarrow \overline{K}_0$ be the $\mathbb{Q}_p$-embedding given by $c^+ \otimes c_0 \mapsto \tau_j(c^+)\widetilde{\sigma}(c_0)$. For $j = (e/2)+1, , \ldots, e$, let $\tau(\sigma)_j\colon F \hookrightarrow \overline{K}_0$ be the $\mathbb{Q}_p$-embedding given by $\widetilde{\sigma}'$ and τ_j. Let Φ_0 be the subset of $\mathrm{Hom}(F, \overline{K}_0)$ consisting of all $\tau(\sigma)_j$, where σ runs through $\mathrm{Hom}(F_0^+, \overline{K}_0)$ and $j = 1, \ldots, e$. Then Φ_0 is self-dual and $\xi(\Phi_0) = \delta_{\mathrm{unif}}$. □

4.5.9. Lemma. *Assume that F is a field and e is odd (so f is even and F is unramified over F^+). Let κ be a finite field. If $f \equiv 2 \pmod 4$ and there exists a κ-rational supersingular Lie type in $\mathrm{R}_k(\mathcal{O}_F)$ then $[\kappa_F : \kappa_F \cap \kappa]$ is odd. Whenever this field degree is odd, the following properties hold:*

(i) *there exists a subset $\Psi \subset \mathrm{Hom}_{\mathbb{Q}_p\text{-alg}}(F_0, \overline{K}_0)$ such that*
   - *$\mathrm{Hom}_{\mathbb{Q}_p\text{-alg}}(F_0, \overline{K}_0)$ is the disjoint union of Ψ and $\Psi \circ \iota$,*
   - *Ψ is stable under the action of the subgroup of $\mathrm{Gal}(F_0/\mathbb{Q}_p)$ which corresponds to $\mathrm{Gal}(\kappa_F/(\kappa_F \cap \kappa))$ under the natural isomorphism $\mathrm{Gal}(F_0/\mathbb{Q}_p) \simeq \mathrm{Gal}(\kappa_F/\mathbb{F}_p)$;*

(ii) *the preimage $\Phi \subset \mathrm{Hom}(F, \overline{K}_0)$ of Ψ is self-dual and $\xi(\Phi)$ is a κ-rational self-dual supersingular element in $\mathrm{R}_k(\mathcal{O}_F)$.*

PROOF. Let $d = [\kappa_F : \kappa_F \cap \kappa]$. If d is odd then $\#(\mathrm{Gal}(\kappa_F/(\kappa_F \cap \kappa))\backslash I)$ must be even, so the combinatorial assertions (i) and (ii) hold because $\mathrm{R}_k(\mathcal{O}_F) = \mathrm{R}_{\kappa_F}(\mathcal{O}_F)$ (ensuring that κ-rational Lie types are the same as $(\kappa \cap \kappa_F)$-rational Lie types).

Now assume that $f = 2f^+$ for an odd integer f^+ and that there exists a supersingular Lie type $\delta \in \mathrm{R}_\kappa(\mathcal{O}_F)$. To prove the oddness of d in such cases, we argue by contradiction. Suppose d is even, so each orbit of $\mathrm{Gal}(\kappa_F/(\kappa_F \cap \kappa))$ on I has even cardinality. Let δ be a supersingular element of $\mathrm{R}_\kappa(\mathcal{O}_F)$. In the sum $\sum_{i\in I} \epsilon_i(\delta)$ the terms with i in a common orbit are equal, so the sum over each orbit is even and hence $\sum_{i\in I} \epsilon_i(\delta)$ is even. But $\sum_{i\in I} \epsilon_i(\delta) = ef^+$ because δ is supersingular, and ef^+ is odd by assumption. This is a contradiction. □

4.5.10. Lemma. *Suppose that F is a field. Let κ be a finite field and δ be a κ-rational Lie type in $\mathrm{R}_k(\mathcal{O}_F)$. There exists a self-dual κ-rational Lie type in $\mathrm{R}_k(\mathcal{O}_F)$ isogenous to δ provided that δ is supersingular and $(F/F^+, \kappa)$ is good.*

PROOF. This is a consequence of 4.5.8 and 4.5.9. □

4.5.11. Proposition. *Suppose that F is a field and f_F is even. Let F_2 be the subfield of F that is quadratic unramified over $\mathbb{Q}_p$. Let Y_0 be an $\mathcal{O}_F$-linear CM p-divisible group over an algebraically closed field $k \supset \mathbb{F}_p$. Assume that the Lie type of Y_0 is striped.*

There exists an $\mathcal{O}_F$-linear k-isomorphism

$$Y_0 \simeq \mathcal{O}_F \otimes_{\mathcal{O}_{F_2}} Y_0'$$

between $\mathcal{O}_F$-linear CM p-divisible groups, where Y_0' is an $\mathcal{O}_{F_2}$-linear CM p-divisible group over k whose Lie type is striped. In particular, if $4|f_F$ and F' is the subfield of F that is unramified of degree 4 over $\mathbb{Q}_p$ then $Y_0 \simeq \mathcal{O}_F \otimes_{\mathcal{O}_{F'}} Y_{\mathrm{toy}}$ for an $\mathcal{O}_{F'}$-linear CM p-divisible group Y_{toy} over k whose Lie type is striped.

PROOF. This is an immediate consequence of 4.3.5 and 4.5.6, but we now give a direct proof of the weaker assertion that $Y_0 \simeq \mathcal{O}_F \otimes_{\mathcal{O}_{F_0}} Y_0'$ for an $\mathcal{O}_{F_0}$-linear CM p-divisible group Y_0' over k whose Lie type is striped. (This "descent to unramified coefficients" in such cases is of interest for reasons explained in Remark 4.1.2.2.)

By the assumptions, $[\kappa_F : \mathbb{F}_p] = [\kappa_{F_0} : \mathbb{F}_p]$ is even. Let J be the $\mathrm{Gal}(\kappa_F/\mathbb{F}_{p^2})$-orbit in $I = I_F$ such that $[\mathrm{Lie}(Y_0)]$ is the striped Lie type $\delta_{\mathrm{stp},J}$ attached to J. Note that I_{F_0} is canonically identified with I_F, so J can be also regarded as a $\mathrm{Gal}(\kappa_{F_0}/\mathbb{F}_{p^2})$-orbit in I_{F_0}. Let Y_0' be an $\mathcal{O}_{F_0}$-linear CM p-divisible group over k

whose Lie type is the striped Lie type in $\mathrm{R}_k(\mathcal{O}_{F_0})$ attached to the subset $J \subset I_{F_0}$; such an $\mathcal{O}_{F_0}$-linear CM p-divisible group Y_0' exists, by 4.2.7. The Lie type of $\mathcal{O}_F \otimes_{\mathcal{O}_{F_0}} Y_0'$ is the class of $\mathcal{O}_F \otimes_{\mathcal{O}_{F_0}} \mathrm{Lie}(Y_0')$ in $\mathrm{R}_k(\mathcal{O}_F)$, which is easily seen to be $\delta_{\mathrm{stp},J}$. By 4.2.6(iv), $\mathcal{O}_F \otimes_{\mathcal{O}_{F_0}} Y_0'$ is $\mathcal{O}_F$-linearly isomorphic to Y_0. □

4.5.12. Remark. The reason for the notation "Y_{toy}" in 4.5.11 when $4|f_F$ is that from the explicit description of striped Lie types in 4.5.4(i) and the dimension formula in terms of Lie types in 4.2.6(ii), such Y_{toy} are precisely the toy model p-divisible groups over k in the sense of 4.1.3.2.

4.5.13. Remark. Suppose that X_0' is an $\mathcal{O}_{F_0}$-linear p-divisible group over k, and define the $\mathcal{O}_F$-linear p-divisible group

$$X_0 = \mathcal{O}_F \otimes_{\mathcal{O}_{F_0}} X_0'.$$

Clearly X_0 has height $[F : \mathbb{Q}_p]$ if and only if X_0' has height $[F_0 : \mathbb{Q}_p]$, and in such cases the Lie type of X_0' is striped if and only if the Lie type of X_0 is striped. In the proof of 4.5.11 above, the $\mathcal{O}_{F_2}$-linear p-divisible group Y_0' is known to CM with a striped Lie type by construction, before the relationship between Y_0' and Y_0 is proved. However, these properties of Y_0' are also a consequence of the $\mathcal{O}_F$-linear isomorphism $Y_0 \simeq \mathcal{O}_F \otimes_{\mathcal{O}_{F_0}} Y_0'$.

4.5.14. Lemma. *Assume that F is a field and e is odd (so f is even). Let J be an orbit for the action $\mathrm{Gal}(\kappa_F/\mathbb{F}_{p^2})$ of on I, and let $\delta_{\mathrm{stp},J}$ be the element of $\mathrm{R}_k(\mathcal{O}_F) = \mathrm{R}_{\kappa_F}(\mathcal{O}_F)$ such that*

$$\epsilon_i(\delta_{\mathrm{stp},J}) = \begin{cases} e & \text{if } i \in J, \\ 0 & \text{if } i \notin J \end{cases}$$

(so $\delta_{\mathrm{stp},J}$ is supersingular). Suppose that κ is a finite field and there exists a supersingular Lie type δ in $\mathrm{R}_\kappa(\mathcal{O}_F)$. Then $\kappa \supset \mathbb{F}_{p^2}$, so $\delta_{\mathrm{stp},J}$ is κ-rational in $\mathrm{R}_k(\mathcal{O}_F)$

PROOF. Let δ be a κ-rational element of $\mathrm{R}_k(\mathcal{O}_F)$. Write $f = 2^a b$ where b is an odd integer and $a \geqslant 1$. Fix an isomorphism $\mathrm{Gal}(\kappa_F/\mathbb{F}_p) \simeq \mathbb{Z}/2^a b\mathbb{Z}$. The condition $\kappa \supset \mathbb{F}_{p^2}$ (or equivalently, $\kappa \cap \kappa_F \supset \mathbb{F}_{p^2}$) means that the Galois group $\mathrm{Gal}(\kappa_F/(\kappa_F \cap \kappa))$ is contained in the subgroup $2\mathbb{Z}/2^a b\mathbb{Z}$ of $\mathbb{Z}/2^a b\mathbb{Z}$. If this does not hold then $\mathrm{Gal}(\kappa_F/(\kappa_F \cap \kappa))$ contains the 2-Sylow subgroup $b\mathbb{Z}/2^a b\mathbb{Z}$ of $\mathbb{Z}/2^a b\mathbb{Z}$. But the κ-rational δ is κ_F-rational and hence $(\kappa \cap \kappa_F)$-rational, so it is fixed by the action of $\mathrm{Gal}(\kappa_F/(\kappa_F \cap \kappa))$. Thus,

$$\begin{aligned} \textstyle\sum_{i \in I} \epsilon_i(\delta) &\equiv 0 \pmod{\# \mathrm{Gal}(\kappa_F/(\kappa_F \cap \kappa))}, \\ &\equiv 0 \pmod{2^a}. \end{aligned}$$

This is absurd because $\sum_{i \in I} \epsilon_i(\delta) = ef/2 = 2^{a-1}be$ is not divisible by 2^a. □

The following theorem summarizes many results in 4.2 and 4.4, and it reduces the proof of CM lifting up to isogeny for abelian varieties over finite fields to the following problem: for $\mathcal{O}_F$-linear p-divisible groups Y of the form $\mathcal{O}_F \otimes_{\mathcal{O}_{F_0}} Y_0$, where Y_0 is an $\mathcal{O}_{F_0}$-linear p-divisible group over $\overline{\mathbb{F}}_p$ whose Lie type is striped, construct an $\mathcal{O}_F$-linear CM lifting of Y whose p-adic CM type is self-dual.

4.5.15. Theorem. *Let (F, F^+) be as in 4.4.1, and Y be an $\mathcal{O}_F$-linear CM p-divisible group over a finite field κ. Assume that either Y has two distinct slopes*

μ_1, μ_2 with $\mu_1 + \mu_2 = 1$ (which happens only when $F = F^+ \times F^+$ with both slopes of multiplicity $[F^+ : \mathbb{Q}_p]$) or that $1/2$ is the only slope of Y.

(i) *Suppose either that $F = F^+ \times F^+$, or that F is a field and $(F/F^+, \kappa)$ is good. Then there exists*
   - *an $\mathcal{O}_F$-linear p-divisible group Y' over κ whose Lie type is self-dual,*
   - *an $\mathcal{O}_F$-linear κ-isogeny $Y' \to Y$,*
   - *a finite extension K of $K_0 = W(\overline{\kappa})[1/p]$ contained in $\overline{K}_0$,*
   - *an $\mathcal{O}_F$-linear p-divisible group $\mathcal{Y}$ over $\mathcal{O}_K$*

   such that the $\overline{K}_0$-valued p-adic CM *type Φ of $\mathcal{Y}$ is self-dual and the closed fiber $\mathcal{Y}_0$ is $\mathcal{O}_F$-linearly $\overline{\kappa}$-isomorphic to $Y'_{\overline{\kappa}}$. The p-adic* CM *type $\Phi \subset \mathrm{Hom}(F, \overline{K}_0)$ can be taken to be any self-dual one on F, and K can be chosen arbitrarily containing the reflex field of (F, Φ).*

(ii) *Suppose that $(F/F^+, \kappa)$ is* bad. *Then there exists*
   - *an $\mathcal{O}_F$-linear* CM *p-divisible group Y' over κ whose Lie type is striped,*
   - *an $\mathcal{O}_F$-linear κ-isogeny $Y' \to Y$,*
   - *an $\mathcal{O}_{F_0}$-linear* CM *p-divisible group Y'' over $\overline{\kappa}$ whose Lie type is striped,*
   - *an $\mathcal{O}_F$-linear $\overline{\kappa}$-isomorphism*

$$Y'_{\overline{\kappa}} \simeq \mathcal{O}_F \otimes_{\mathcal{O}_{F_0}} Y''.$$

(iii) *Continue to assume that $(F/F^+, \kappa)$ is bad, and let F' be the unramified subextension of $F/\mathbb{Q}_p$ such that $[F' : \mathbb{Q}_p] = 4$. Then there exists*
   - *an $\mathcal{O}_F$-linear* CM *p-divisible group Y' over κ whose Lie type is striped,*
   - *an $\mathcal{O}_F$-linear κ-isogeny $Y' \to Y$,*
   - *an $\mathcal{O}_{F'}$-linear* CM *p-divisible group Y_{toy} of height 4 and dimension 2 over $\overline{\kappa}$ whose Lie type is striped,*
   - *an $\mathcal{O}_F$-linear $\overline{\kappa}$-isomorphism*

$$Y'_{\overline{\kappa}} \simeq \mathcal{O}_F \otimes_{\mathcal{O}_{F'}} Y_{\mathrm{toy}}.$$

In all cases, we can arrange that the $\mathcal{O}_F$-linear $Y' \to Y$ has degree at most $p^{f\dim(X_0)+[F:\mathbb{Q}_p](f-1)/2}$ when F is a field and at most $p^{2(f^+\dim(X_0)+[F^+:\mathbb{Q}_p](f^+-1)/2)}$ when F is split.

PROOF. The split case in (i) follows from 4.4.8. Now consider (i) with F a field. By 4.5.10, there exists a κ-rational self-dual Lie type $\delta \in \mathrm{R}_{\overline{\kappa}}(\mathcal{O}_F)$ isogenous to $[\mathrm{Lie}(Y)]$, so by 4.2.8 there exists an $\mathcal{O}_F$-linear p-divisible group Y' over κ whose Lie type is δ and which admits an $\mathcal{O}_F$-linear κ-isogeny $Y' \to Y$ that can be chosen to have degree at most the desired bound. By 4.4.5, there exists a self-dual subset $\Psi \subset \mathrm{Hom}_{\mathbb{Q}_p\text{-alg}}(F, \overline{K}_0)$ whose reduction $\xi(\Psi)$ is equal to δ. By 4.4.6(ii), there exists a finite extension K of K_0 inside $\overline{K}_0$ and an $\mathcal{O}_F$-linear p-divisible group $\mathcal{Y}$ over $\mathcal{O}_K$ whose closed fiber is $\mathcal{O}_F$-linearly $\overline{\kappa}$-isomorphic to $Y'_{\overline{\kappa}}$. This proves (i).

Assume now $(F/F^+, \kappa)$ is bad, so F is a field, e is odd, $4|f$, and $[\kappa_F : (\kappa_F \cap \kappa)]$ is even. We know that Y is supersingular because F is a field. By 4.5.14, the finite field κ contains $\mathbb{F}_{p^2}$ and both striped Lie types in $\mathrm{R}_k(\mathcal{O}_F)$ are κ-rational. Let δ be one of the striped Lie types. By 4.2.8 there exists an $\mathcal{O}_F$-linear CM p-divisible group Y' over κ with δ as its Lie type and which admits an $\mathcal{O}_F$-linear κ-isogeny $Y' \to Y$ with degree at most the desired bound. Thus, (ii) follows via 4.5.11. The proof of (iii) is similar. □

Remark. As noted in 4.5.13, that Y'' in (ii) is a CM p-divisible group with a striped Lie type follows from the rest of 4.5.15(ii), and similarly for Y_{toy} in (iii).

4.5.16. Lemma. *Let $\overline{\mathbb{F}}_p$ be an algebraic closure of $\mathbb{F}_p$, and let $\kappa \subset \overline{\mathbb{F}}_p$ be the subfield of size p^4. Let Y_{toy} be a two-dimensional $W(\kappa)$-linear* CM *p-divisible group over $\overline{\mathbb{F}}_p$ whose Lie type is striped.*

(1) *There exists*
  - *a two-dimensional $W(\kappa)$-linear* CM *p-divisible group X_{toy} over $\overline{\mathbb{F}}_p$,*
  - *a $W(\kappa)$-linear isogeny $\rho_{\text{toy}} \colon X_{\text{toy}} \to Y_{\text{toy}}$ so that the $W(\kappa)$-stable finite subgroup scheme* $\ker(\rho_{\text{toy}})$ *is isomorphic to α_p,*
  - *a two-dimensional $W(\kappa)$-linear* CM *p-divisible group $\mathcal{X}_{\text{toy}}$ over $W(\overline{\mathbb{F}}_p)$ whose closed fiber is $W(\kappa)$-linearly isomorphic to X_{toy}*

  such that
  (i) $\ker(\rho_{\text{toy}})$ *is the unique subgroup scheme of X_{toy} of order p,*
  (ii) *the Lie type of X_{toy} is self-dual, and the p-adic* CM *type of $\mathcal{X}_{\text{toy}}$ is self-dual.*

(2) *Let $\mathcal{G}_{\text{toy}}$ be a finite flat subgroup scheme of $\mathcal{X}_{\text{toy}}[p]_{\mathcal{O}_K}$ of order p for a finite extension K of $K_0 = W(\overline{\mathbb{F}}_p)[1/p]$ contained in $\overline{K}_0$. The quotient $\mathcal{X}_{\text{toy}}[p]_{\mathcal{O}_K}/\mathcal{G}_{\text{toy}}$ with the induced action of $\mathbb{Z}_p + pW(\kappa)$ is a $W(\kappa)[1/p]$-linear* CM *lifting of Y_{toy} whose $\overline{K}_0$-valued p-adic* CM *type is self-dual.*

The justification for the notation Y_{toy} is explained in 4.5.12.

PROOF. The assertions in (1) are precisely the properties of toy model CM p-divisible groups that were discussed in 4.1.3 (resting on constructions in 4.1.2.5–4.1.2.6), and part (2) follows from (1). □

Remark. Up to unique isomorphism over Y_{toy}, there are exactly two $W(\kappa)$-linear isogenies ρ_{toy} over $\overline{\mathbb{F}}_p$ satisfying the requirements in 4.5.16(1). Indeed, passing to the Serre dual map, these correspond to the quotients of Y^t_{toy} by its two $W(\kappa)$-stable α_p-subgroups.

4.5.17. Proposition. *Suppose that F is a field and $4|f$. Let Y be an $\mathcal{O}_F$-linear* CM *p-divisible group of over $\overline{\mathbb{F}}_p$ whose Lie type is striped. There exists a finite extension field K of $K_0 = W(\overline{\mathbb{F}}_p)[1/p]$ with $[K : K_0]$ dividing $p^4 - 1$ and a self-dual F-linear* CM *p-divisible group $\mathcal{Y}$ over $\mathcal{O}_K$ whose closed fiber is F-linearly isomorphic to Y. Moreover $\mathcal{Y}$ can be chosen so that $\mathbb{Z}_p + p\mathcal{O}_F \subset$* $\mathrm{End}(\mathcal{Y})$ *inside* $\mathrm{End}^0(\mathcal{Y})$.

PROOF. Let $\kappa \subseteq \kappa_F$ be the subfield of order p^4. By 4.5.15(ii), the given $\mathcal{O}_F$-linear p-divisible group Y over $\overline{\mathbb{F}}_p$ with a *striped* Lie type can be written as a Serre-tensor product:

$$Y \simeq \mathcal{O}_F \otimes_{W(\kappa)} Y_{\text{toy}}$$

where Y_{toy} is is a 2-dimensional $W(\kappa)$-linear CM p-divisible group over $\overline{\mathbb{F}}_p$ whose Lie type is striped. Applying 4.5.16, we obtain

- a 2-dimensional $W(\kappa)$-linear CM p-divisible group $\mathcal{X}_{\text{toy}}$ over $W(\overline{\mathbb{F}}_p)$ whose p-adic CM type is self-dual,
- a $W(\kappa)$-linear isogeny $\rho_{\text{toy}} \colon X_{\text{toy}} \to Y_{\text{toy}}$, where X_{toy} is the closed fiber of $\mathcal{X}_{\text{toy}}$

such that the $W(\kappa)$-stable subgroup scheme $\ker(\rho_{\text{toy}})$ is isomorphic to α_p and is the unique finite subgroup scheme of X_{toy} of order p.

The representation space $\mathcal{X}_{\text{toy}}[p](\overline{K}_0)$ for $\text{Gal}(\overline{K}_0/K_0)$ is a character valued in the cyclic group $\kappa^\times$ of order p^4-1, so it is split by a finite extension K/K_0 with degree dividing p^4-1. Thus, we can construct a finite flat subgroup scheme $\mathcal{G}_{\text{toy}}$ of $\mathcal{X}_{\text{toy}}[p]_{O_K}$ with order p as in 4.5.16 via schematic closure of an $\mathbb{F}_p$-line in the Galois module. The closed fiber of $\mathcal{G}_{\text{toy}}$ is equal to $\ker(\rho_{\text{toy}})$ since X_{toy} has a unique finite subgroup scheme of order p.

Define the O_F-linear CM p-divisible groups $X_0 := O_F \otimes_{W(\kappa)} X_{\text{toy}}$ and $\mathcal{X} := O_F \otimes_{W(\kappa)} \mathcal{X}_{\text{toy}}$ over $\overline{\mathbb{F}}_p$ and $W(\overline{\mathbb{F}}_p)$ respectively. Clearly $\rho_0 := O_F \otimes_{W(\kappa)} \rho_{\text{toy}}$ is an O_F-linear isogeny from X_0 to Y. Note that the Lie type of the O_F-linear X_0 is self-dual and the p-adic CM type of the O_F-linear $\mathcal{X}$ is self-dual, because the respective statements hold for the $W(\kappa)$-linear X_{toy} and $\mathcal{X}_{\text{toy}}$ (as $[\text{Lie}(X_{\text{toy}})]$ is the reduction of the p-adic CM type of $\mathcal{X}_{\text{toy}}$).

Choose an isomorphism $O_F \simeq W(\kappa)^{\oplus r}$ as $W(\kappa)$-modules (so $r = [F:\mathbb{Q}_p]/4$), yielding compatible isomorphisms

$$\mathcal{X} \simeq (\mathcal{X}_{\text{toy}})^r, \quad X_0 \simeq (X_{\text{toy}})^r, \quad \ker(\rho_0) \simeq \ker(\rho_{\text{toy}})^r.$$

For the lift $\mathcal{G}_{\text{toy}}$ of $\ker(\rho_{\text{toy}})$ over O_K, $\mathcal{G} := (\mathcal{G}_{\text{toy}})^r$ is a finite flat subgroup scheme of $\mathcal{X}_{O_K}$ which lifts $\ker(\rho_0)$. Thus, the quotient

$$\mathcal{Y} := \mathcal{X}_{O_K}/\mathcal{G}$$

is a p-divisible group over O_K with an induced F-linear structure whose closed fiber is F-linearly isomorphic to Y and whose p-adic CM type is self-dual (because the p-adic CM type of $\mathcal{X}$ is self-dual). The containment $\mathbb{Z}_p + pO_F \subseteq \text{End}(\mathcal{Y})$ is clear because $\mathcal{G}$ is killed by p. □

4.5.18. Remark. In Appendices B.1 and B.2, we give two more proofs of the special case of 4.5.17 when F is unramified over $\mathbb{Q}_p$. The version in B.1 uses Raynaud's results on finite flat group schemes killed by p [**99**] (see B.1.16 and B.1.17), and the version in B.2 uses p-adic Hodge theory (see B.2.16). Originally 4.5.17 was proved by reducing to either B.1.17 or B.2.16 using 4.5.15(i). The original proof of the existence of a CM lifting up to isogeny also followed this route, using p-adic Hodge theory to produce self-dual CM liftings in the case of bad primes.

4.6. Complex conjugation and CM lifting

4.6.1. Conventions in the global case. Now we pass to the global setting. We fix a CM field L, with L^+ its maximal totally real subfield and ι the unique non-trivial automorphism of L over L^+.

Generally v denotes a p-adic place of L^+, w denotes a p-adic place of L (often with $w|v$), and $O_{L,v} := O_L \otimes_{O_{L^+}} O_{L^+,v} = \prod_{w|v} O_{L,w}$. The pair (L_v, L_v^+) satisfies the setup for the semi-local situation in 4.4.1. The local invariants for $L_v/\mathbb{Q}_p$ and $L_w^+/\mathbb{Q}_p$ are respectively denoted e_v, f_v, κ_v and e_w, f_w, κ_w. Also, as in our considerations in the semi-local setting, we let $L_{w,0}$ and $L_{v,0}^+$ denote the maximal unramified subextensions over $\mathbb{Q}_p$, and we define $L_{v,0} = \prod_{w|v} L_{w,0}$ (and $O_{L_{v,0}} = \prod_{w|v} O_{L_{w,0}}$ and $\kappa_{L_v} = \prod_{w|v} \kappa_{L_w}$).

For $\mathcal{O}_{L,p} := \mathcal{O}_L \otimes_{\mathbb{Z}} \mathbb{Z}_p = \prod_{w|p} \mathcal{O}_{L,w}$ and a field $\mathcal{K}$ of characteristic $p > 0$, the Grothendieck group $\mathrm{R}_{\mathcal{K}}(\mathcal{O}_L)$ of the category of finitely generated modules over $\mathcal{O}_L \otimes_{\mathbb{Z}} \mathcal{K} = \mathcal{O}_{L,p} \otimes_{\mathbb{Z}_p} \mathcal{K}$ naturally decomposes as a direct product

$$\mathrm{R}_{\mathcal{K}}(\mathcal{O}_{L,p}) \simeq \prod_{v|p} \mathrm{R}_{\mathcal{K}}(\mathcal{O}_{L,v}) \simeq \prod_{w|p} \mathrm{R}_{\mathcal{K}}(\mathcal{O}_{L,w})$$

in terms of the notation already introduced in our work in the local and semi-local cases. For an element $\delta \in \mathrm{R}_{\mathcal{K}}(\mathcal{O}_L)$ we write δ_v and δ_w to respectively denote the components of δ under these decompositions according to the p-adic places v of L^+ and w of L.

As usual, we fix an algebraically closed field k of characteristic p and let $K_0 = W(k)[1/p]$, and we generally work with finite extensions K of K_0 contained in a fixed algebraic closure $\overline{K}_0$ of K_0. For every place $w|p$ of L we define the index set

$$I_w := \mathrm{Hom}_{\mathbb{Z}_p\text{-alg}}(\mathcal{O}_{L_{w,0}}, W(k)) \xrightarrow{\simeq} \mathrm{Hom}_{\mathbb{Z}_p\text{-alg}}(\mathcal{O}_{L_{w,0}}, k) \xrightarrow{\simeq} \mathrm{Hom}_{\mathbb{F}_p\text{-alg}}(\kappa_w, k),$$

so this is naturally a torsor for $\mathrm{Gal}(\kappa_w/\mathbb{F}_p) \simeq \mathbb{Z}/f_w\mathbb{Z}$. Similarly, for every p-adic place v of L^+ we define

$$I_v := \mathrm{Hom}_{\mathbb{Z}_p\text{-alg}}(\mathcal{O}_{L_{v,0}}, W(k)) = \mathrm{Hom}_{\mathbb{F}_p\text{-alg}}(\kappa_v, k),$$

so there are canonical decompositions

$$\mathcal{O}_L \otimes_{\mathbb{Z}} W(k) \simeq \prod_{w|p} \mathcal{O}_{L,w} \otimes_{\mathbb{Z}_p} W(k) \simeq \prod_{w|p} \prod_{i \in I_w} \mathcal{O}_{L,w} \otimes_{\mathcal{O}_{L_{w,0}},i} W(k).$$

For each p-adic place w of L and each $i \in I_w$, we define

$$\epsilon_{w,i} \colon \mathrm{R}_k(\mathcal{O}_{L,p}) \longrightarrow \mathbb{Z}$$

to be the composition of the projection

$$\mathrm{R}_k(\mathcal{O}_{L,p}) \to \mathrm{R}_k(\mathcal{O}_{L,w}, i) := \mathrm{R}_k(\mathcal{O}_{L,w} \otimes_{\mathcal{O}_{L_{w,0}},i} W(k))$$

with the isomorphism

$$\mathrm{R}_k(\mathcal{O}_{L,w} \otimes_{\mathcal{O}_{L_{w,0}},i} W(k)) \xrightarrow{\simeq} R(k) \xrightarrow{\simeq} \mathbb{Z}$$

defined in 4.2.1. We also define

$$\epsilon_w := \sum_{i \in I_w} \epsilon_{w,i} \colon \mathrm{R}_k(\mathcal{O}_{L,p}) \longrightarrow \mathbb{Z};$$

this is the composition of the "dimension" as defined in the local theory at w and the projection of $\mathrm{R}_k(\mathcal{O}_{L,p})$ onto its w-factor.

An $\mathcal{O}_{L,p}$-linear p-divisible group X over a noetherian scheme S naturally decomposes as

$$X = \prod_{v|p} X[v^\infty] = \prod_{v|p} \prod_{w|v} X[w^\infty],$$

where each factor $X[v^\infty]$ (resp. $X[w^\infty]$) is an $\mathcal{O}_{L,v}$-linear (resp. $\mathcal{O}_{L,w}$-linear) p-divisible group over S. (The flatness property for the torsion-levels of these factors is proved by identifying each as the kernel of an endomorphism of X that is an isogeny on fibers over the noetherian S.) We say that X is a $\mathcal{O}_{L,p}$-linear CM *p-divisible group* if $\mathrm{height}(X[w^\infty]) = [L_w : \mathbb{Q}_p]$ for each p-adic place w of L.

4.6.2. Lie types and p-adic CM types in the global case. An element $\delta \in \mathrm{R}_\kappa(\mathcal{O}_L)$ is a *Lie type* if for every p-adic place w of L the w-component δ_w of δ is a Lie type; i.e., $0 \leqslant \epsilon_{w,i}(\delta_w) \leqslant e_w$. Likewise, δ is *κ-rational* for a finite subfield $\kappa \subset k$ when the w-component $\delta_w \in \mathrm{R}_k(\mathcal{O}_{L,w})$ is κ-rational for every w.

In terms of the semi-local theory applied to $L_v^+ \to L_v$ equipped with its involution ι_v for a p-adic place v of L^+, δ is *self-dual* if the v-component $\delta_v \in \mathrm{R}_k(\mathcal{O}_{L,v})$ is self-dual for all v (see 4.4.3). If v splits into two places w and w' of L then the self-duality condition at v means

$$\epsilon_{w,i}(\delta_w) + \epsilon_{w',i}(\delta_{w'}) = e_v$$

for all $i \in I_v = \mathrm{Hom}(\kappa_v, k) = \mathrm{Hom}(\kappa_w, k) = \mathrm{Hom}(\kappa_{w'}, k)$. If instead there is a unique place w of L over v, the self-duality condition at v means

$$\epsilon_{w,i}(\delta_w) + \epsilon_{w,i\circ\bar{\iota}}(\delta_w) = e_w$$

for all $i \in I_w = \mathrm{Hom}(\kappa_w, k)$, where $\bar{\iota}$ denotes the automorphism of κ_w induced by the complex conjugation ι on L.

Since the global complex conjugation ι on L induces the unique L_v^+-algebra involution ι_v of L_v for each p-adic place v of L^+, upon identifying $\mathrm{Hom}_{\mathbb{Q}_p\text{-alg}}(L_v, \overline{K}_0)$ with the subset of embeddings in $\mathrm{Hom}_{\mathbb{Q}\text{-alg}}(L, \overline{K}_0)$ inducing the p-adic place v on L^+ we see that a subset $\Phi_v \subset \mathrm{Hom}_{\mathbb{Q}_p\text{-alg}}\left(L_v, \overline{K}_0\right)$ is *self-dual* in the sense of the semi-local theory (see 4.4.3) precisely when $\mathrm{Hom}_{\mathbb{Q}_p\text{-alg}}\left(L_v, \overline{K}_0\right)$ is the disjoint union of Φ_v and $\Phi_v \circ \iota_v$.

The notions of *uniform* and *striped* in 4.5.3 apply to the v-component δ_v when L_v is a field (i.e., v has a unique place w over it on L). For example, when e_w is even (e.g., if L/L^+ is ramified at v) and the v-component $\delta_v = \delta_w \in \mathrm{R}_k(\mathcal{O}_{L,v})$ of a Lie type $\delta \in \mathrm{R}_k(\mathcal{O}_L)$ is supersingular then δ_v is *uniform* if

$$\epsilon_{w,i}(\delta_w) = e/2$$

for all $i \in I_w$. Likewise, if f_w is even then the v-component $\delta_v = \delta_w$ of a Lie type δ is *striped* if there exists a subset $J_w \subset I_w$ that is an orbit for the action of $\mathrm{Gal}(\kappa_w/\mathbb{F}_{p^2}) \subset \mathrm{Gal}(\kappa_w/\mathbb{F}_p)$ on I_w and satisfies

$$\epsilon_{w,i}(\delta_w) = \begin{cases} e_{w,i} & \text{if } i \in J_w \\ 0 & \text{if } i \in I_w - J_w. \end{cases}$$

If we trivialize the $\mathrm{Gal}(\kappa_w/\mathbb{F}_p)$-torsor I_w to identify I_w with $\mathbb{Z}/f_w\mathbb{Z}$ then J_w is either the subset of all even elements in I_w or the subset of all odd elements in I_w.

Remark. Let v be a p-adic place of L^+, and assume v is not split in L, so there is a unique place w in L above v and $L_v = L_w$ is a field. Whenever δ_v is self-dual or striped, necessarily δ_w is supersingular; i.e., $\epsilon_w(\delta_w) := \sum_{i\in I_w} \epsilon_{w,i}(\delta_w) = [L_w : \mathbb{Q}_p]/2$ and $0 \leqslant \epsilon_{w,i}(\delta_w) \leqslant e_w$ for all $i \in I_w$. Moreover, if δ_v is uniform and supersingular then it is self-dual and $\mathbb{F}_p$-rational, whereas if δ_v is striped then it is $\mathbb{F}_{p^2}$-rational but it is self-dual if and only if $f_w \equiv 2 \pmod 4$.

4.6.3. Self-duality and CM lifting revisited. We now explain how self-duality for both p-adic CM types and for Lie types are relevant in the study of CM lifting problems for abelian varieties. As a first step, it is convenient to record the following reformulation of the tangential algebraization criterion in 2.2.3 (for CM formal abelian schemes) in terms of our "self-duality" terminology for p-adic CM types:

4.6.3.1. Proposition. *Let $(R, \mathfrak{m})$ be a 1-dimensional $\mathbb{Z}_p$-flat complete noetherian local ring with residue characteristic p. Let A_0 be an abelian variety of dimension $g > 0$ over the residue field κ of R, and suppose there is given a* CM *field $L \subset \mathrm{End}^0(A_0)$ of degree $2g$, as well as an L_p-linear* CM *p-divisible group $\mathcal{X}$ over R whose closed fiber $\mathcal{X}_0$ is $A_0[p^\infty]$.*

Let $(\mathfrak{A}, \alpha\colon L \to \mathrm{End}^0(\mathfrak{A}))$ be the L-linear formal abelian scheme over R attached to $\mathcal{X}$ and the residual isomorphism $\mathcal{X}_0 \simeq A_0[p^\infty]$. Then $\mathfrak{A}$ is algebraizable if and only if for all places $v|p$ of L^+, the L_v-linear CM *p-divisible group $\mathcal{X}[v^\infty]$ has p-adic* CM *type at each geometric point of $R[1/p]$ that is self-dual in the sense of* 4.4.3.

Note that for every place w of L over v, the w-adic part $\mathcal{X}[w^\infty]$ automatically has height $[L_w : \mathbb{Q}_p]$ (so the L_v-linear p-divisible group $\mathcal{X}[v^\infty]$ is CM in the sense of 4.6.1). Indeed, this height condition even holds for the special fiber $A_0[w^\infty]$ (the w-adic factor of $A_0[p^\infty]$) because $\mathrm{M}^*(A_0[p^\infty])[1/p]$ is free of rank 1 as a module over $L_p \otimes_{\mathbb{Q}_p} W(\kappa)[1/p] = \prod_{w|p} L_w \otimes_{\mathbb{Q}_p} W(\kappa)[1/p]$ by 1.4.3.9.

The following necessary and sufficient lifting criterion in terms of self-duality for Lie types is proved in [**133**], but we give another proof resting on Proposition 4.6.3.1. A variant is given in 4.6.5.4.

4.6.3.2. Proposition. *Let L be a* CM *field and let A_0 be an $\mathcal{O}_L$-linear abelian variety over an algebraically closed field k of characteristic $p > 0$, with $2 \cdot \dim(A_0) = [L : \mathbb{Q}]$. There exists a finite extension K of $K_0 = W(k)[1/p]$ and an $\mathcal{O}_L$-linear abelian scheme $\mathcal{A}$ with special fiber $\mathcal{O}_L$-isomorphic to A_0 if and only if the Lie type $[\mathrm{Lie}(A_0[v^\infty])]$ is self-dual for each p-adic place v of L^+.*

PROOF. The necessity is immediate due to the fact that the p-adic CM type of such a lift is self-dual (due to CM theory in characteristic 0) and reduces to the Lie type of $A_0[p^\infty]$. For the sufficiency, we first note that (by 4.4.5) for every place v of L^+ above p there exists a self-dual subset

$$\Phi_v \subset \mathrm{Hom}_{\mathbb{Q}_p\text{-alg}}(\mathcal{O}_{L,v}, \overline{K}_0),$$

whose reduction $\xi_v(\Phi_v)$ is equal to the v-component of the Lie type of $A_0[p^\infty]$. Choose a finite extension K/K_0 inside $\overline{K}_0$ that contains the reflex field of (L_v, Φ_v) for all v. By 4.4.4(ii), for every v there exists an $\mathcal{O}_{L,v}$-linear lift $\mathcal{Y}_v$ of $A_0[v^\infty]$ over $\mathcal{O}_K$ with p-adic CM type Φ_v. We conclude by the Serre-Tate theorem and the algebraizability criterion in 4.6.3.1. ∎

Any L-linear CM abelian variety A in characteristic 0 is L-linearly isogenous to its dual A^t when using the L-action on A^t defined by composing the dual action with complex conjugation on L (as this ensures that A^t has the same CM type as A), so we are led to the following notion for p-divisible groups in characteristic p.

4.6.3.3. Definition. Let X_0 be an L_p-linear CM p-divisible group over a field of characteristic p. Equip the dual X_0^t with a structure of L_p-linear CM p-divisible group by letting $a \in L_p$ act (in the isogeny category) as the dual of the action of $\iota(a)$ on X_0. The L_p-linear X_0 is *self-dual up to isogeny* if X_0 and X_0^t are L_p-linearly isogenous.

The most basic example of this definition arises from CM abelian varieties in positive characteristic: if A_0 be an L-linear abelian variety of dimension $[L : \mathbb{Q}]/2$

over a field of characteristic $p > 0$ then the L_p-linear CM p-divisible group $A_0[p^\infty]$ is self-dual up to isogeny. In general, if an L_p-linear CM p-divisible group X_0 is self-dual up to isogeny then the following hold for p-adic places v of L^+:

(i) If v splits in L, say with $\{w, w'\}$ the places of L above v, there is an isogeny $X_0[w^\infty] \to X_0[w'^\infty]^t$ that is linear with respect to the isomorphism $L_w \simeq L_{w'}$ induced by ι. In particular, the sum of the respective slopes

$$(\dim X_0[w^\infty])/[L_w : \mathbb{Q}_p] \quad \text{and} \quad (\dim X_0[w'^\infty])/[L_{w'} : \mathbb{Q}_p]$$

of the isoclinic factors $X_0[w^\infty]$ and $X_0[w'^\infty]$ is 1 (see 4.2.6(iii)).

(ii) If there is a unique place $w|v$ of L then $X_0[w^\infty]$ is supersingular in the sense that it is isoclinic of slope $1/2$ (as we see by self-duality, since $\dim G + \dim G^t = \text{height}(G)$ for any p-divisible group G over a field of characteristic p).

We end our discussion of self-duality with a method for constructing self-dual Lie types arising from certain L_v-linear CM p-divisible groups.

4.6.3.4. Lemma. *Let X_0 be an $\mathcal{O}_{L,p}$-linear CM p-divisible group over a field $\mathcal{K}$ of characteristic $p > 0$, and assume that X_0 is self-dual up to isogeny. Let v be a place of L^+ above p, and assume either that v splits in L with $\mathcal{K} \cap \kappa_v = \mathbb{F}_p$ or that v does not split and $\kappa \cap \kappa_w = \mathbb{F}_p$ for the unique place w of L above v.*

The v-component $[\mathrm{Lie}(X_0)]_v = [\mathrm{Lie}(X_0[v^\infty])]$ of the Lie type of X_0 is self-dual.

Proof. Because the Lie type $[\mathrm{Lie}(X_0[w^\infty])]$ is rational over $\kappa \cap \kappa_w = \mathbb{F}_p$, the κ_w-rational $[\mathrm{Lie}(X_0[w^\infty])]$ is fixed under the action of $\mathrm{Gal}(\kappa_w/\mathbb{F}_p)$ for every place w of L above v. Thus, for each $w|v$ there exists an integer $c_w \geqslant 0$ such that $\epsilon_{w,i}(\delta_w) = c_w$ for all $i \in I_w$. It follows that $X_0[w^\infty]$ has slope $(c_w f_w)/[L_w : \mathbb{Q}_p] = c_w/e_w$. Since the $\mathcal{O}_{L,v}$-linear p-divisible group $X_0[v^\infty]$ is self-dual up to isogeny, we see that

- if v splits into two places w, w' in L, then $c_w + c_{w'} = e_v = e(L_v^+/\mathbb{Q}_p)$,
- if w is the unique place of L above v then e_w is even, $c_w = e_w/2$, and δ_w is supersingular and uniform.

In both cases, δ_v is self-dual. □

4.6.4. Proof of Theorem 4.1.1. Now we are finally in position to prove Theorem 4.1.1. As a preliminary reduction, we pass to the case $\mathcal{O}_L \subset \mathrm{End}(B)$ as follows. By 4.3.1(c), there exists a κ-isogeny $\gamma\colon B \to B'$ such that $\mathcal{O}_L \subseteq \mathrm{End}(B')$ via the equality $\mathrm{End}^0(B) = \mathrm{End}^0(B')$. More specifically, if we let $R = \mathcal{O}_L \cap \mathrm{End}(B)$ be the CM order in L then we may take γ to be the canonical isogeny

$$B \longrightarrow \mathcal{O}_L \otimes_R B =: B'$$

as in 4.3.1(c), using the Serre tensor construction over fields. We replace B with such a B' so that now we may and do assume $\mathcal{O}_L$ operates on B.

Brief outline of strategy for the proof of Theorem 4.1.1.

Step 1. Change B by an $\mathcal{O}_L$-linear κ-isogeny so that its Lie type becomes "as close to being self-dual as possible". To be precise, we find a B' that is $\mathcal{O}_L$-linearly κ-isogenous to B such that the v-component δ_v of $[\mathrm{Lie}(B')]$ is self-dual for every good p-adic place v of L^+. For any p-adic place v of L^+ that is bad for the $\mathcal{O}_L$-linear B, we arrange instead that δ_v is striped.

Step 2. Use 4.4.5, 4.4.6, and 4.6.3.1 to construct an L-linear abelian scheme $\mathcal{B}'$ over the ring of integers $\mathcal{O}_K$ of a finite extension field K of $W(\overline{\kappa})[1/p]$ such that the closed fiber of $\mathcal{B}'$ is L-linearly isomorphic to $B'_{\overline{\kappa}}$ (with B' as in Step 1).

Step 3. A deformation argument concludes the proof.

These steps are elaborated below.

4.6.4.1. Step 1. Let B be an $\mathcal{O}_L$-linear abelian variety of dimension $g = [L : \mathbb{Q}]/2$ over a finite field κ of characteristic p. Apply 4.5.15 to the $\mathcal{O}_{L,v}$-linear p-divisible group $B[v^\infty]$ over κ for each place v of L^+ above p. This provides $\mathcal{O}_{L,v}$-linear κ-isogenies $\psi'_v : Y'_v \longrightarrow B[v^\infty]$, so their direct product is an $\mathcal{O}_{L,p}$-linear κ-isogeny

$$\psi' : Y' := \prod_{v|p} Y'_v \longrightarrow \prod_{v|p} B[v^\infty] = B[p^\infty]$$

of $\mathcal{O}_{L,p}$-linear CM p-divisible groups over κ. This isogeny can be chosen so that its degree is at most

$$p^{\sum_{w|p} f_w \dim(B) + \sum_{w|p} [L_w:\mathbb{Q}_p](f_w-1)/2} \leqslant p^{[L:\mathbb{Q}]\dim(B)+[L:\mathbb{Q}]^2/2} = p^{4g^2},$$

and it gives an $\mathcal{O}_L$-linear abelian variety B' over κ and an $\mathcal{O}_L$-linear κ-isogeny $\psi : B' \to B$ of p-power degree such that the following conditions hold.

- There is an $\mathcal{O}_{L,p}$-linear κ-isomorphism
$$\psi_0 : B'[p^\infty] \simeq Y'$$
satisfying $\psi[p^\infty] = \psi' \circ \psi_0$ (so $\deg(\psi) = \deg(\psi_0)$).
- The Lie type δ_v of $B'[v^\infty]$ is self-dual for all p-adic places v of L^+ that are good relative to the $\mathcal{O}_{L,p}$-linear $B'[p^\infty]$.
- For the bad p-adic places v of L^+ (and w the unique place of L over v), the Lie type of $B'[v^\infty] = B'[w^\infty]$ is striped. Moreover, for an algebraic closure $\overline{\kappa}$ of κ, there exists an $\mathcal{O}_{L_{w,0}}$-linear CM p-divisible group $Y'_{0,w}$ over $\overline{\kappa}$ whose Lie type is striped and an $\mathcal{O}_{L,w}$-linear $\overline{\kappa}$-isomorphism
$$B'[w^\infty]_{\overline{\kappa}} \simeq \mathcal{O}_{L,w} \otimes_{\mathcal{O}_{L_{w,0}}} Y'_{0,w}.$$

4.6.4.2. Step 2. Let B' be the $\mathcal{O}_L$-linear abelian variety over κ as in 4.6.4.1. Fix an algebraic closure $\overline{\kappa}$ of κ, so the base change $B'_{\overline{\kappa}}$ makes sense. Let $K_0 = W(\overline{\kappa})[1/p]$. We claim that there exists a subfield $K \subset \overline{K}_0$ over K_0 of finite degree and an L-linear formal abelian scheme $\mathcal{B}'$ over $\mathcal{O}_K$ such that the closed fiber $\mathcal{B}'$ is L-linearly isomorphic to $B'_{\overline{\kappa}}$ and $\mathbb{Z} + p\mathcal{O}_L \subset \mathrm{End}(\mathcal{B}')$. (An inspection of the construction will show that the ramification of K over a suitable p-adic reflex field can be controlled; this is addressed in Step 3.)

For each p-adic place v of L^+ we will construct a subfield $K_v \subset \overline{K}_0$ over K_0 of finite degree and an L_v-linear CM p-divisible group $\mathcal{Y}_v$ over $\mathcal{O}_{K_v}$ such that: its $\overline{K}_0$-valued p-adic CM type is *self-dual*, its closed fiber is L_v-linearly isomorphic to the L_v-linear p-divisible group $B'[v^\infty]_{\overline{\kappa}}$, and its endomorphism ring contains $\mathbb{Z}_p + p\mathcal{O}_{L,v}$. Suppose we have constructed such K_v and $\mathcal{Y}_v$ for all $v|p$ on L^+. Let K be the compositum over K_0 inside $\overline{K}_0$ of all such fields K_v. The p-divisible group

$$\mathcal{Y} := \prod_{v|p} \mathcal{Y}_v$$

is a self-dual L_p-linear CM p-divisible group over $\mathcal{O}_K$ whose closed fiber is $B'[p^\infty]_{\overline{\kappa}}$, so the Serre–Tate deformation theorem yields the desired L-linear formal abelian scheme over $\mathcal{O}_K$; its algebraicity follows from 4.6.3.1.

CASE WHEN v IS A GOOD PLACE OF L^+ ABOVE p.

Since the Lie type δ_v of each component $B'[v^\infty]$ is self-dual, by 4.4.5 and 4.4.6 there exists

- a self-dual p-adic CM type $\Phi_v \subset \mathrm{Hom}(L_v, \overline{K}_0)$ such that $\xi_v(\Phi_v) = \delta_v$,
- a finite extension K_v of K_0 inside $\overline{K}_0$ that contains the reflex field $E(L_v, \Phi_v) \subset \overline{K}_0$,
- an $\mathcal{O}_{L,v}$-linear p-divisible group $\mathcal{Y}_v$ over $\mathcal{O}_{K_v}$ with p-adic CM type Φ_v.

The special fiber $\mathcal{Y}_{v,0}$ of $\mathcal{Y}_v$ is $\mathcal{O}_{L,v}$-linearly $\overline{\kappa}$-isomorphic to $B'[v^\infty]_{\overline{\kappa}}$ by 4.2.6(iv), and we can take K_v to be the reflex field $E(L_v, \Phi_v)$.

CASE WHEN v IS A BAD PLACE OF L^+ ABOVE p.

By 4.5.17, there exists a *self-dual* L_v-linear CM p-divisible group $\mathcal{Y}_v$ over $\mathcal{O}_{K_v}$, where $K_v \subset \overline{K}_0$ is a finite extension of K_0 with degree dividing $p^4 - 1$, such that $\mathbb{Z}_p + p\mathcal{O}_{L,v} \subset \mathrm{End}(\mathcal{Y}_v)$ and the closed fiber of $\mathcal{Y}_v$ is L_v-linearly isomorphic to the $\mathcal{O}_{L,v}$-linear p-divisible group $B'[v^\infty]_{\overline{\kappa}}$. (Note that in these cases $B'[v^\infty]$ is isoclinic of slope 1/2 due to Theorem 4.5.15(iii), since the definitions easily imply that striped Lie types have slope 1/2. Thus, bad v do not occur when 1/2 is not a slope of B'.) This finishes Step 2.

4.6.4.3. Step 3. Finally, we construct a 1-dimensional $\mathbb{Z}_p$-flat complete local noetherian domain D with residue field κ and an L-linear CM lifting of B' over D such that $\mathbb{Z} + p\mathcal{O}_L \subset \mathrm{End}(B')$. Let $\mathcal{B}'$ be the L-linear CM lifting of $B'_{\overline{\kappa}}$ built in Step 2 (over the $W(\overline{\kappa})$-finite $\mathcal{O}_K$), so $\mathcal{O}' := \mathbb{Z} + p\mathcal{O}_L \subset \mathrm{End}(\mathcal{B}')$. By construction, $\mathrm{Lie}(\mathcal{B}')_{\overline{K}_0}$ has p-adic CM type on L that is *self-dual*, so it is a genuine $\overline{\mathbb{Q}}_p$-valued CM type Φ on L. As for CM abelian varieties in general, necessarily the reflex field $E(L, \Phi) \subset \overline{K}_0$ is contained inside $K \subset \overline{K}_0$.

The induced p-adic place w on the reflex field $E(L, \Phi) \subset \overline{\mathbb{Q}}_p$ makes $E(L, \Phi)_w$ equal to the compositum over $\mathbb{Q}_p$ inside $\overline{\mathbb{Q}}_p$ of the reflex fields $E(L_v, \Phi_v)$ for the p-adic places v of L^+. Thus, due to the pro-cyclic structure of tame inertia and the construction of K, we see that K is tame over $E(L, \Phi)_w$ with ramification degree dividing $p^4 - 1$. If 1/2 is not a slope of B' then such bad $v|p$ on L^+ (with respect to L/L^+ and κ) do not occur, so in such cases K is unramified over $E(L, \Phi)_w$ (i.e., it is the completion of the maximal unramified extension of $E(L, \Phi)_w$).

A drawback of this construction is that $\mathcal{O}_K$ has residue field $\overline{\kappa}$ rather than κ. We will use deformation theory to improve the construction so that it avoids increasing the residue field without losing control of the ramification degree over the reflex field (at some p-adic place).

Pick an L-linear polarization of the generic fiber $\mathcal{B}'_K$ (as we may do by 2.2.2), and consider the L-linear $\mathcal{O}_K$-homomorphism $\phi : \mathcal{B}' \to \mathcal{B}'^t$ extending this. Its special fiber $\overline{\phi}$ is an L-linear polarization of $B'_{\overline{\kappa}}$, by 2.1.1.1. By 2.2.2, $\overline{\phi}$ descends to an L-linear polarization ϕ_0 of B'.

Consider the deformation functor of (B', ϕ_0, α_0) where $\alpha_0 : \mathcal{O}' \hookrightarrow \mathrm{End}(B')$ is the given inclusion. It has a formal deformation ring $\mathcal{R}$ by 1.4.4.5. By Lemma 2.1.1.2, every generic point of $\mathrm{Spec}(\mathcal{R})$ in characteristic 0 has $W(\kappa)$-finite closure. Thus, the canonical map $\mathcal{R} \to \mathcal{O}_K$ classifying the deformation we built over $\mathcal{O}_K$

factors through a $W(\kappa)$-finite quotient D of $\mathcal{R}$ that is a $W(\kappa)$-finite flat domain with local inclusion into $\mathcal{O}_K$, so it is an order in a finite extension of $W(\kappa)[1/p]$. Moreover, D has residue field κ since it is a quotient of $\mathcal{R}$.

Using pullback of the universal structure over $\mathcal{R}$ (which is algebraizable due to the polarization), B' admits an $\mathcal{O}'$-linear lift over D that descends $\mathcal{B}'$. The inclusion $D[1/p] \to K$ over $W(\kappa)[1/p]$ implies that the relative ramification degree of $D[1/p]$ over the reflex field of this L-linear CM lift (relative to the geometric generic point $D[1/p] \hookrightarrow K \hookrightarrow \overline{K}_0$ over $W(\kappa)[1/p]$) divides $p^4 - 1$ and is equal to 1 when $1/2$ is not a slope of B'. This completes the proof of Theorem 4.1.1.

4.6.5. Applications. We now record consequences of the proof of Theorem 4.1.1.

4.6.5.1. Proposition. *Let F be a finite extension of $\mathbb{Q}_p$, and Y_0 an $\mathcal{O}_F$-linear* CM *p-divisible over a finite field κ. Let $K_0 = W(\overline{\kappa})[1/p]$ and let Φ be a subset of* $\mathrm{Hom}(F, \overline{K}_0)$ *such that the Lie type of* $[\mathrm{Lie}(Y_0)]$ *in* $\mathrm{R}_\kappa(\mathcal{O}_F) \subset \mathrm{R}_{\overline{\kappa}}(\mathcal{O}_F)$ *equals* $\xi(\Phi)$.

There exists an $\mathcal{O}_F$-linear p-divisible group $\mathcal{Y}$ over a $\mathbb{Z}_p$-finite flat complete local noetherian domain $(R, \mathfrak{m}) \subset \overline{K}_0$ with residue field κ such that Φ is the p-adic CM *type of $\mathcal{Y}$ and the closed fiber of $\mathcal{Y}$ is $\mathcal{O}_F$-linearly isomorphic to Y_0.*

PROOF. Let $E \subset \overline{K}_0$ be the reflex field of the p-adic CM type (F, Φ), and let $K \subset \overline{K}_0$ be the compositum of E and K_0. There exists an $\mathcal{O}_F$-linear p-divisible group $\mathcal{X}$ over $\mathcal{O}_K$ with CM type (F, Φ), by 3.7.3(1). The closed fiber of $\mathcal{X}$ is an $\mathcal{O}_F$-linear p-divisible group over $\overline{\kappa}$ with Lie type $\xi(\Phi)$, so it is $\mathcal{O}_F$-linearly isomorphic to $(Y_0)_{\overline{\kappa}}$ by 4.2.6(iv). We conclude by a deformation theory argument (for p-divisible groups rather than for abelian schemes), using slicing as in 2.1.1. □

4.6.5.2. Corollary. *Let L be a* CM *field. Let B be an $\mathcal{O}_L$-linear abelian variety of dimension $g > 0$ over a finite field κ of characteristic p with $[L : \mathbb{Q}] = 2g$. There exists a $\mathbb{Z}_p$-finite flat local noetherian domain $(R, \mathfrak{m})$ with residue field κ and an $\mathcal{O}_L$-linear formal abelian scheme $\mathfrak{B}$ over* $\mathrm{Spf}(R)$ *whose closed fiber is $\mathcal{O}_L$-linearly κ-isogenous to B by an isogeny of p-power degree at most p^{4g^2}.*

PROOF. Construct a lifting of the $\mathcal{O}_{L,p}$-linear p-divisible group $B[p^\infty]$ by 4.6.5.1, and then apply the Serre-Tate deformation theorem. □

4.6.5.3. Proposition. *In the setting of Theorem* 4.1.1*, assume that every p-adic place of L^+ is good with respect to $(L/L^+, \kappa)$ in the sense of* 4.5.7*; i.e., for every p-adic place w of L that is inert over L^+, assume that at least one of the following conditions is satisfied:*

- *the absolute ramification index $e_w = e(L_w/\mathbb{Q}_p)$ is even,*
- *the residue field degree f_w is not divisible by* 4*,*
- $[\kappa_w : \kappa \cap \kappa_w]$ *is odd.*

Then the conclusion of Theorem 4.1.1 *is valid with an $\mathcal{O}_L$-linear lifting $\mathcal{B}'$.*

PROOF. The only step in 4.6.4 where we had to shrink the CM order to be smaller than $\mathcal{O}_L$ (or $\mathcal{O}_{L,v}$ in the semi-local arguments) was in the consideration of p-adic places on L^+ that are bad with respect to $(L/L^+, \kappa)$. □

The hypothesis in 4.6.5.3 that all p-adic places of L^+ are good with respect to $(L/L^+, \kappa)$ is automatically satisfied if $1/2$ is not a slope of B, or if $\kappa_w \cap \kappa = \mathbb{F}_p$ for every p-adic place w of L that is inert over L^+.

4.6.5.4. Proposition. *Let L be a* CM *field and let B be an $\mathcal{O}_L$-linear abelian variety over a finite field κ with $2\dim(B) = [L : \mathbb{Q}]$.*

(i) *If the Lie type $[\mathrm{Lie}(B)] \in \mathrm{R}_\kappa(\mathcal{O}_L)$ is self-dual then there exists an $\mathcal{O}_L$-linear abelian scheme $\mathcal{B}$ over a $\mathbb{Z}_p$-finite flat local domain with residue field κ and closed fiber that is $\mathcal{O}_L$-linearly κ-isomorphic to B.*

(ii) *The self-duality assumption on $[\mathrm{Lie}(B)]$ in* (i) *is satisfied if $\kappa \cap \kappa_w = \mathbb{F}_p$ for every place w of L above p, such as if $\kappa_w = \mathbb{F}_p$ for every place w of L above p.*

PROOF. Suppose first that the Lie type of B is self-dual. We know from 4.6.3.2 that the desired abelian scheme $\mathcal{B}$ exists except over a $W(\overline{\kappa})$-finite flat discrete valuation ring. We push down the base ring to be of the desired type (a local finite flat domain over $\mathbb{Z}_p$ with residue field κ) by the deformation argument as in 4.6.4.3. This proves (i), and (ii) is a restatement of 4.6.3.4. □

Remark. Proposition 4.6.5.4 is a strengthening of 4.6.3.2. The special case that $f_w = 1$ for every place w of L above p (at the end of 4.6.5.4(ii)) was proved directly in Proposition 2.2.1.

APPENDIX A

Some arithmetic results for abelian varieties

Although our discussion... is restricted to elliptic curves, the theory cannot be fully understood unless one considers them as special cases of abelian varieties.

— Goro Shimura

A.1. The p-part of Tate's work

The proof of Tate's isogeny theorem over finite fields of characteristic $p > 0$ (Theorem 1.6.1.1) consists of two separate parts: the case $\ell \neq p$ and the case $\ell = p$. The first of these is easier insofar as the ℓ-divisible groups are étale and hence can be treated within the more concrete framework of Galois modules. The case $\ell = p$ requires additional techniques from non-commutative algebra. An important application of this case is the complete characterization of endomorphism algebras of simple abelian varieties over finite fields as in Corollary 1.6.2.2 (more specifically, the determination of the local invariants of such division algebras at the p-adic places of the center).

Tate's proofs of these results were published by Milne and Waterhouse in [**79**]. The arguments use a lot of non-commutative algebra beyond what is generally familiar to algebraic number theorists. The exposition of the proof of Tate's isogeny theorem for $\ell = p$ that is given below reduces the input from non-commutative algebra to just basic properties of central simple algebras, and the application to computing local invariants of endomorphism algebras is also provided. This section is based on a set of notes by K. Eisenträger, and we are grateful to her for permission to present part of her notes in a more permanent form.

A.1.1. The isogeny theorem for $\ell = p$: first steps. Let k be a finite field of characteristic $p > 0$, $W = W(k)$, and $K_0 = W[1/p]$. We refer the reader to 1.4 for a review of notation and basic facts in the theory of p-divisible groups and their Dieudonné modules over a perfect field of characteristic p. In analogy with the ℓ-adic case we will now write T_pG to denote the (contravariant) Dieudonné module $\mathrm{M}^*(G)$ for a p-divisible group G over k, and $V_pG := \mathbb{Q}_p \otimes_{\mathbb{Z}_p} T_pG$. We also write T_pA and V_pA when $G = A[p^\infty]$ for an abelian variety A over k.

Let $D_k = W[\mathcal{F}, \mathcal{V}]$ denote the Dieudonné ring of k (non-commutative if $k \neq \mathbb{F}_p$). The left D_k-module T_pG is the replacement for the ℓ-adic Tate module in the "classical" case, even though it is contravariant in G; its D_k-action is the analogue of the Galois action on ℓ-adic Tate modules. In general V_pG is a K_0-module of rank equal to the height of G, and it also has a left module structure over the "Laurent polynomial ring" $D_k[1/p] = K_0[\mathcal{F}, 1/\mathcal{F}]$ that is non-commutative if $k \neq \mathbb{F}_p$.

Our aim is to prove the following result, which is Theorem 1.6.1.1 when $\ell = p$:

A.1.1.1. Theorem. *For any abelian varieties A and B over k, the $\mathbb{Z}_p$-linear map*

$$\mathbb{Z}_p \otimes_{\mathbb{Z}} \mathrm{Hom}(A, B) \to \mathrm{Hom}(A[p^\infty], B[p^\infty])$$

is an isomorphism.

The same methods as in the case $\ell \neq p$ show injectivity (Proposition 1.2.5.1) and reduce the proof of surjectivity to the case $B = A$ after inverting p on both sides. That is, we aim to prove that the injective map

$$\mathbb{Q}_p \otimes_{\mathbb{Q}} \mathrm{End}^0(A) \to \mathrm{End}^0(A[p^\infty]) \tag{A.1.1.1}$$

is an isomorphism for any abelian variety A over k. This will be proved by recasting the problem in terms of Dieudonné modules and using some module theory over non-commutative algebras.

Let $q = \#k = p^a$. The q-Frobenius endomorphism π of A acts $W(k)$-linearly on $\mathrm{M}^*(A[p^\infty])$ with the same characteristic polynomial (in $\mathbb{Z}[T]$) as it does as a $\mathbb{Z}_\ell$-linear endomorphism of the ℓ-adic Tate module of A for any $\ell \neq p$; this is true more generally for any element of $\mathrm{End}(A)$, by using Dieudonné modules in the role of Tate modules in the proof of the ℓ-adic cases ($\ell \neq p$) in [**82**, §19, Thm. 4]. Hence, we may use Tate's proof of surjectivity essentially verbatim (say, as in Appendix 1 of [**82**]) by replacing ℓ-adic Tate modules with Dieudonné modules throughout to deduce (via [**82**, App. 1, Lemma 6]) that the $\mathbb{Q}_p$-dimension of the left side of (A.1.1.1) is equal to the K_0-dimension of the centralizer of the semisimple K_0-linear endomorphism $\mathcal{F}^a = V_p(\pi)$ acting on the K_0-vector space $V_p(A)$ (i.e., $\dim_{K_0}\mathrm{End}_{K_0[\mathcal{F}^a]}(V_p(A))$). Thus, it is necessary and sufficient to prove that this K_0-dimension is equal to the $\mathbb{Q}_p$-dimension of $\mathrm{End}^0(A[p^\infty]) = \mathrm{End}_{D_k[1/p]}(V_p(A))^{\mathrm{opp}}$.

The central subalgebra $Z = \mathbb{Q}[\pi]$ in the finite-dimensional semisimple $\mathbb{Q}$-algebra $\mathrm{End}^0(A)$ must be semisimple and hence is a finite product of fields; we do not assume A to be k-simple, so in particular Z may not be a field. Consider the decomposition of A and $A[p^\infty]$ (in the isogeny sense) according to the idempotents of Z. The action by Z on $V_p(A)$ (through its action on A in the isogeny category of abelian varieties over k) commutes with the action by $D_k[1/p]$ on $V_p(A)$, and the element $\mathcal{F}^a \in D_k[1/p]$ acts as $V_p(\pi)$ on $V_p(A)$. Since there are no non-zero maps between pairs of factor abelian varieties or p-divisible groups over k on which the central subalgebra Z acts through distinct quotient fields, we may reduce the proof to the case when Z is a field.

Let $m \in \mathbb{Z}[T]$ be the common characteristic polynomial for $V_\ell(\pi)$ on the $\mathbb{Q}_\ell$-vector space $V_\ell(A)$ for all $\ell \neq p$ as well as for $V_p(\pi)$ on the K_0-vector space $V_p(A)$. By the faithfulness of $V_p(A)$ as a $\mathbb{Q}_p \otimes_{\mathbb{Q}} Z$-module, $\mathrm{rad}(m)$ is the minimal polynomial of $\pi \in Z$ over $\mathbb{Q}$. (In particular, when A is k-simple, so Z is a field, m is a power of this minimal polynomial over $\mathbb{Q}$.) Hence, $\mathbb{Q}_p \otimes_{\mathbb{Q}} Z \simeq \prod_{v|p} Z_v$ with $Z_v \simeq \mathbb{Q}_p[T]/(m_v)$ for the monic irreducible factors $m_v \in \mathbb{Z}_p[T]$ of m over $\mathbb{Q}_p$. Each $m_v(0)$ divides $m(0) = q^{\dim A}$, so all m_v have non-zero constant term. Since $\mathbb{Q}_p \otimes_{\mathbb{Q}} Z$ acts $D_k[1/p]$-linearly on $V_p(A)$, we get a decomposition of $D_k[1/p]$-modules $V_p(A) \simeq \prod_{v|p} V_p(G_v)$ where $\prod_{v|p} G_v$ is the isogeny decomposition of $A[p^\infty]$ with respect to the primitive idempotents of $\mathbb{Q}_p \otimes_{\mathbb{Q}} Z$. Since the central element $m_v(\mathcal{F}^a) \in D_k$ acts on G_v

through the element $m_v(\pi) = 0$ in Z_v, $V_p(G_v)$ is a left module over the quotient algebra $C_v = D_k[1/p]/D_k[1/p]m_v(\mathcal{F}^a)$. Using the compatible decompositions (as K_0-algebras and $\mathbb{Q}_p$-algebras respectively)

$$\mathrm{End}_{K_0[\mathcal{F}^a]}(V_p(A)) \simeq \prod_{v|p} \mathrm{End}_{K_0[\mathcal{F}^a]}(V_p(G_v))$$

and

$$\mathrm{End}_{D_k[1/p]}(V_p(A)) \simeq \prod_{v|p} \mathrm{End}_{C_v}(V_p(G_v)),$$

we are reduced to proving

$$\dim_{K_0} \mathrm{End}_{K_0[\mathcal{F}^a]}(M_v) \stackrel{?}{=} \dim_{\mathbb{Q}_p} \mathrm{End}_{C_v}(M_v)$$

for any $v|p$ and any left C_v-module M_v with finite $\mathbb{Q}_p$-dimension (for instance, $M_v = V_p(G_v)$). This general equality is valid when C_v is replaced with

$$D_k[1/p]/D_k[1/p]h(\mathcal{F}^a)$$

for any $h \in \mathbb{Q}_p[T]$ that is monic irreducible with non-zero constant term (i.e., $h \neq T$); a proof will be given in Corollary A.1.2.5 as a consequence of some general calculations in non-commutative algebra (having no logical dependence on abelian varieties or p-divisible groups).

A.1.2. Cyclic algebras. We now analyze the structure of the $\mathbb{Q}_p$-algebras C_v that arose above. The first key fact is that the C_v are always simple, with center Z_v. More generally:

A.1.2.1. Theorem. *Let $h \in \mathbb{Q}_p[T]$ be a monic irreducible polynomial distinct from T, and let $Z = \mathbb{Q}_p[T]/(h)$. The central Z-algebra $C = D_k[1/p]/D_k[1/p]h(\mathcal{F}^a)$ (with T acting as $\mathcal{F}^a$) is central simple.*

The proof of Theorem A.1.2.1 requires preparations. Recall that K_0 denotes the fraction field of $W(k)[1/p]$. We have $\dim_{K_0} C = \deg(h(T^a)) = a[Z : \mathbb{Q}_p]$ due to:

A.1.2.2. Lemma. *Let $\lambda \in K_0[T]$ be monic with degree $d > 0$ and non-vanishing constant term. Every $g \in D_k[1/p]$ can be uniquely written as*

$$g = h \cdot \lambda(\mathcal{F}) + (c_0 + c_1\mathcal{F} + \cdots + c_{d-1}\mathcal{F}^{d-1})$$

with $c_0, \ldots, c_{d-1} \in K_0$ and $h \in D_k[1/p]$. Thus, the left $D_k[1/p]$-module

$$D_k[1/p]/D_k[1/p] \cdot \lambda(\mathcal{F})$$

has dimension d as a left K_0-vector space.

PROOF. The uniqueness is proved by chasing "least-degree" coefficients (when non-zero). For existence it suffices to treat the cases $g = \mathcal{F}^r$ with $r \geqslant d$ and $r < 0$. These go as in the commutative case (for $r < 0$ we use the non-vanishing of the constant term of λ). $\square$

Since C is a $K_0 \otimes_{\mathbb{Q}_p} Z$-module with finite $\mathbb{Q}_p$-dimension, Lemma A.1.2.2 implies

$$\begin{aligned}\dim_Z C = [Z:\mathbb{Q}_p]^{-1}\dim_{\mathbb{Q}_p} C = &= \frac{[K_0:\mathbb{Q}_p]}{[Z:\mathbb{Q}_p]}\dim_{K_0} C\\ &= \frac{[K_0:\mathbb{Q}_p]}{[Z:\mathbb{Q}_p]}\cdot a[Z:\mathbb{Q}_p]\\ &= a[K_0:\mathbb{Q}_p].\end{aligned}$$

Choose any $\pi \in Z^\times$. Let $(K_0 \otimes_{\mathbb{Q}_p} Z)[\mathcal{F}]$ denote the Z-algebra defined by the relations $\mathcal{F}^a = 1 \otimes \pi$ and $\mathcal{F} \cdot x = (\sigma \otimes 1)(x) \cdot \mathcal{F}$ for $x \in K_0 \otimes_{\mathbb{Q}_p} Z$, with $\sigma \in \mathrm{Gal}(K_0/\mathbb{Q}_p)$ the absolute Frobenius. This is easily checked to be a central Z-algebra with dimension $a[K_0 : \mathbb{Q}_p] = a^2$. There is a unique map of central Z-algebras

$$(K_0 \otimes_{\mathbb{Q}_p} Z)[\mathcal{F}] \to C$$

sending $\mathcal{F}$ to $\mathcal{F}$ and K_0 to K_0 by the identity map, and this is clearly surjective. By consideration of Z-dimensions, it is an isomorphism. Theorem A.1.2.1 therefore will follow from:

A.1.2.3. Theorem. *The central Z-algebra $(K_0 \otimes_{\mathbb{Q}_p} Z)[\mathcal{F}]$ is Z-isomorphic to a matrix algebra over a cyclic Z-algebra Δ.*

Cyclic algebras are a special class of central simple algebras, studied by Dickson before the advent of the general theory.

PROOF. We first construct the cyclic Z-algebra Δ over which the Z-algebra $(K_0 \otimes_{\mathbb{Q}_p} Z)[\mathcal{F}]$ will be proved to be a matrix algebra. Let f be the absolute residue degree of Z, and let $g := \gcd(f, a)$. Let $Z_0' \subseteq Z$ denote the maximal unramified subextension of Z that embeds into K_0 (over $\mathbb{Q}_p$), so $g = [Z_0' : \mathbb{Q}_p]$. We arbitrarily choose one of the g embeddings $Z_0' \to K_0$ as cyclic extensions of $\mathbb{Q}_p$, and let K_0Z denote the resulting linearly disjoint compositum $K_0 \otimes_{Z_0'} Z$.

Clearly K_0Z/Z is unramified with degree a/g. Let

$$\theta : K_0 \otimes_{\mathbb{Q}_p} Z \to K_0Z$$

be the resulting projection map. Let Δ be the central Z-algebra $(K_0Z)[\mathcal{F}']$ defined with relations $(\mathcal{F}')^{a/g} = \pi \in Z^\times$ and $\mathcal{F}' \cdot x' = \sigma'(x') \cdot \mathcal{F}'$ where $\sigma' : K_0Z \simeq K_0Z$ is the Z_0'-automorphism of order a/g induced by σ^g on K_0 and the identity on Z. Thus, $\sigma'(\theta(x)) = \theta((\sigma^g \otimes 1)(x))$ for $x \in K_0 \otimes_{\mathbb{Q}_p} Z$, and we have a natural isomorphism of Z-algebras

$$K_0 \otimes_{\mathbb{Q}_p} Z \simeq \prod_{j=1}^{g} K_0Z \tag{A.1.2.1}$$

defined by

$$x \mapsto (\theta(x), \theta((\sigma \otimes 1)(x)), \dots, \theta((\sigma^{g-1} \otimes 1)(x))).$$

The central Z-algebra Δ is cyclic; in classical notation $\Delta = (K_0Z/Z, \sigma', \pi)$, where K_0Z is a cyclic (unramified) extension of Z and σ' is a chosen generator of its Galois group.

Now consider the natural map of central Z-algebras

$$\xi : (K_0 \otimes_{\mathbb{Q}_p} Z)[\mathcal{F}] \to \mathrm{Mat}_g(\Delta)$$

defined by

$$x \mapsto \begin{pmatrix} \theta(x) & 0 & \dots & 0 \\ 0 & \theta((\sigma \otimes 1)(x)) & 0 & 0 \\ \vdots & \vdots & \ddots & \vdots \\ 0 & 0 & \dots & \theta((\sigma^{g-1} \otimes 1)(x)) \end{pmatrix}$$

for $x \in K_0 \otimes_{\mathbb{Q}_p} Z$ and

$$\mathcal{F} \mapsto \begin{pmatrix} 0 & 1 & 0 & \dots & 0 \\ 0 & 0 & 1 & \dots & 0 \\ \vdots & 0 & \dots & \ddots & \vdots \\ \vdots & \vdots & 0 & \dots & 1 \\ \mathcal{F}' & 0 & 0 & \dots & 0 \end{pmatrix}$$

The map ξ is obviously a well-defined Z-algebra homomorphism, and it is easy to check that both sides have the same $\mathbb{Q}_p$-dimension. To prove that ξ is an isomorphism, it suffices to check surjectivity.

The decomposition (A.1.2.1) of $K_0 \otimes_{\mathbb{Q}_p} Z$ as a product of copies of K_0Z shows that the image of ξ contains the diagonal matrices whose entries are elements of $K_0Z \subseteq \Delta$. Moreover, $\xi(\mathcal{F}^g)$ is the diagonal matrix whose diagonal entries are all equal to $\mathcal{F}' \in \Delta$. Thus, by the definition of Δ we see that the image of ξ contains all diagonal matrices in $\mathrm{Mat}_g(\Delta)$. Left-multiplication by $\xi(\mathcal{F})$ carries the set of diagonal matrices onto the set of matrices of the form

$$\begin{pmatrix} 0 & \delta_2 & 0 & \dots & 0 \\ 0 & 0 & \delta_3 & \dots & 0 \\ \vdots & 0 & \dots & \ddots & \vdots \\ \vdots & \vdots & 0 & \dots & \delta_g \\ \mathcal{F}'\delta_1 & 0 & 0 & \dots & 0 \end{pmatrix}$$

with arbitrary $\delta_1, \dots, \delta_g \in \Delta$. The element $\mathcal{F}' \in \Delta$ has a multiplicative inverse (since $(\mathcal{F}')^{a/g} = \pi \in Z^\times$), so it follows that the image of ξ contains the matrix

$$\begin{pmatrix} 0 & 1 & 0 & \dots & 0 \\ 0 & 0 & 1 & \dots & 0 \\ \vdots & 0 & \dots & \ddots & \vdots \\ \vdots & \vdots & 0 & \dots & 1 \\ 1 & 0 & 0 & \dots & 0 \end{pmatrix}.$$

Powers of this matrix, together with arbitrary diagonal matrices in $\mathrm{Mat}_g(\Delta)$, generate $\mathrm{Mat}_g(\Delta)$ under the operations of multiplication and addition. This establishes surjectivity. □

A.1.2.4. Remark. For later purposes, it is convenient to describe the cyclic algebra Δ in Theorem A.1.2.3 in terms of an arithmetic Frobenius generator ϕ of the Galois group of the unramified extension K_0Z/Z. With notation as used above, we claim that $\Delta := (K_0Z/Z, \sigma', \pi)$ is Z-isomorphic to the cyclic Z-algebra $(K_0Z/Z, \phi, \pi^{f/g})$. This amounts to checking that ${\sigma'}^{f/g} = \phi$ in $\mathrm{Gal}(K_0Z/Z)$, so we just have to compare these on the residue field of K_0 (viewed inside that of K_0Z). Since σ' reduces to $x \mapsto x^{p^g}$ on the residue field of K_0, $\sigma'^{f/g}$ reduces to $x \mapsto x^{p^f}$ on this residue field. But f is the residual degree for Z over $\mathbb{Q}_p$, so we are done.

It is an elementary result in the theory of simple rings that (up to isomorphism) over a simple ring there is a unique simple left module and all finitely generated left modules are isomorphic to a direct sum of copies of the simple module. Thus, by Theorem A.1.2.1 there is (up to isomorphism) a unique simple left $D_k[1/p]$-module V killed by the central element $h(\mathcal{F}^a)$, and for any non-zero left module M over $C := D_k[1/p]/(h(\mathcal{F}^a))$ with finite $\mathbb{Q}_p$-dimension we have $M \simeq V^{\oplus r}$ for some $r \geqslant 1$. In particular, $\mathrm{End}_{D_k[1/p]}(M)$ is a matrix algebra over the division ring $\Delta := \mathrm{End}_C(V)$.

Since there is an isomorphism of left C-modules $C \simeq V^{\oplus r_0}$ for some $r_0 \geqslant 1$, so $\mathrm{Mat}_{r_0}(\Delta) = \mathrm{End}_C(C) = C^{\mathrm{opp}}$, Δ has center Z and in the Brauer group of Z the classes of Δ and C^{opp} coincide. Hence, the class of $\mathrm{End}_{D_k[1/p]}(M)^{\mathrm{opp}}$ in $\mathrm{Br}(Z)$ is the same as that of C for any non-zero finitely generated left C-module M. For any such M we have the following result that completes the final step in the proof of Theorem A.1.1.1:

A.1.2.5. Corollary. *Let $h \in \mathbb{Q}_p[T]$ be monic irreducible with $h \neq T$, and define $C := D_k[1/p]/D_k[1/p]h(\mathcal{F}^a)$. For any finitely generated left C-module M,*

$$\dim_{K_0} \mathrm{End}_{K_0[\mathcal{F}^a]}(M) = \dim_{\mathbb{Q}_p} \mathrm{End}_C(M). \tag{A.1.2.2}$$

Since $K_0[\mathcal{F}, 1/\mathcal{F}] = D_k[1/p]$ and $\mathrm{End}_C(M) = \mathrm{End}_{K_0[\mathcal{F}]}(M)$ (note that $\mathcal{F}$ acts invertibly on M), the equality $[K_0 : \mathbb{Q}_p] = a$ suggests the possibility that one may be able to deduce (A.1.2.2) by general nonsense with Galois descent. However, it seems that the proof requires fine structural information concerning C (such as Theorem A.1.2.1).

Proof. Let V be a simple left C-module and let $Z = \mathbb{Q}_p[T]/(h)$. We have $M \simeq V^{\oplus r}$ for some $r \geqslant 0$. The two sides of (A.1.2.2) for M are each r^2 times the two sides of (A.1.2.2) with V in the role of M. Hence, it suffices to verify this identity for one $M \neq 0$. We choose $M = C$. Since $\mathrm{End}_C(C) = C^{\mathrm{opp}}$ as $\mathbb{Q}_p$-algebras,

$$\dim_{\mathbb{Q}_p} \mathrm{End}_C(C) = [K_0 : \mathbb{Q}_p] \dim_{K_0}(C^{\mathrm{opp}}) = [K_0 : \mathbb{Q}_p] \deg h(T^a)$$

by Lemma A.1.2.2 (with $\lambda := h(T^a)$ satisfying $\lambda(0) \neq 0$ because the irreducible monic h is not T).

Since $\deg h = [Z : \mathbb{Q}_p]$ and $[K_0 : \mathbb{Q}_p] = a$, the dimension formula says

$$\dim_{\mathbb{Q}_p} \mathrm{End}_C(C) = \frac{[K_0 \otimes_{\mathbb{Q}_p} Z : \mathbb{Q}_p] a^2}{[K_0 : \mathbb{Q}_p]},$$

and so by expressing a K_0-dimension as $[K_0 : \mathbb{Q}_p]^{-1}$ times a $\mathbb{Q}_p$-dimension it is equivalent to prove

$$\dim_{\mathbb{Q}_p} \mathrm{End}_{K_0[\mathcal{F}^a]}(C) = [K_0 \otimes_{\mathbb{Q}_p} Z : \mathbb{Q}_p] a^2.$$

It therefore suffices to show that $\mathrm{End}_{K_0[\mathcal{F}^a]}(C)$ is isomorphic as a $\mathbb{Q}_p$-algebra to an $a \times a$ matrix algebra over $K_0 \otimes_{\mathbb{Q}_p} Z$.

Let $t \in Z = \mathbb{Q}_p[T]/(h)$ denote the non-zero residue class of T, so $Z = \mathbb{Q}_p[t]$. We saw above Theorem A.1.2.3 that $C \simeq (K_0 \otimes_{\mathbb{Q}_p} Z)[\mathcal{F}]$ as $K_0[\mathcal{F}^a]$-algebras, where the right side has relations $\mathcal{F}^a = 1 \otimes t$ and $\mathcal{F} \cdot x = (\sigma \otimes 1)(x)\mathcal{F}$ for all $x \in K_0 \otimes_{\mathbb{Q}_p} Z$. Hence, the $K_0[\mathcal{F}^a]$-linear endomorphism ring of C is the same as the $K_0 \otimes_{\mathbb{Q}_p} Z$-linear endomorphism ring of C. This is an $a \times a$ matrix algebra over $K_0 \otimes_{\mathbb{Q}_p} Z$ since the $K_0 \otimes_{\mathbb{Q}_p} Z$-module $C = (K_0 \otimes_{\mathbb{Q}_p} Z)[\mathcal{F}]$ is free of rank e. □

A.1.3. Local invariants. As an application of Theorem A.1.1.1, we now establish the formula for local invariants in Corollary 1.6.2.2(3). Let A be a simple abelian variety over a finite field k of size $q = p^a$, and let Δ denote its endomorphism algebra, so Δ is a central division algebra over the subfield $Z := \mathbb{Q}[\pi]$ (with π denoting the q-Frobenius endomorphism of A). Under the decomposition $\mathbb{Q}_p \otimes_{\mathbb{Q}} \Delta \simeq \prod_{v|p} \Delta_v$ with $\Delta_v := Z_v \otimes_Z \Delta$, we get a corresponding decomposition of $A[p^\infty]$ in the isogeny category of p-divisible groups over k, as a product of *non-zero* p-divisible groups G_v on which $\mathbb{Q}_p \otimes_{\mathbb{Q}} Z$ acts through its quotients Z_v. (We use the elementary injectivity aspect of Theorem A.1.1.1 to see that each G_v is non-zero.)

The Dieudonné functor is fully faithful and contravariant, and (A.1.1.1) is an isomorphism for finite k, so we may identify the central $\mathbb{Q}_p \otimes_{\mathbb{Q}} Z$-algebra $\mathbb{Q}_p \otimes_{\mathbb{Q}} \Delta$ with the opposite $\mathbb{Q}_p \otimes_{\mathbb{Q}} Z$-algebra $\mathrm{End}_{D_k[1/p]}(V_pA)^{\mathrm{opp}}$, where $Z = \mathbb{Q}[\pi]$ acts through functoriality via its action on A in the isogeny category over k. In particular, $V_p(\pi)$ is the action of the central element $\mathcal{F}^a \in D_k[1/p]$, so

$$\Delta_v \simeq \mathrm{End}^0(G_v) \simeq \mathrm{End}_{D_k[1/p]}(V_pG_v)^{\mathrm{opp}} \tag{A.1.3.1}$$

as finite-dimensional Z_v-algebras. Thus, the right side of (A.1.3.1) is a central simple Z_v-algebra. Our problem is to compute the image of its Brauer class under the isomorphism $\mathrm{Br}(Z_v) \simeq \mathbb{Q}/\mathbb{Z}$ defined according to the convention in 1.2.4.2.

Since π in the ring of integers of Z acts on A as the q-Frobenius over k, on the finite-dimensional $\mathbb{Q}_p$-vector space $V_p(G_v)$ the $\mathbb{Q}_p$-linear operator $\mathcal{F}^a = V_p(\pi)$ acts with a characteristic polynomial that divides the common characteristic polynomial $m \in \mathbb{Q}[T] \subset \mathbb{Q}_p[T]$ for the q-Frobenius on each $V_\ell(A)$ over $\mathbb{Q}_\ell$ (for $\ell \neq p$) and $V_p(A)$ over K_0. (For all $n \in \mathbb{Z}$, $m(n)$ is the degree of $n-\pi$ acting on A.) By the k-simplicity of A, m is a power of the minimal polynomial of $\pi \in Z$ over $\mathbb{Q}$ (as we noted in the discussion following Theorem A.1.1.1), so the decomposition $\mathbb{Q}_p \otimes_{\mathbb{Q}} Z \simeq \prod_{v|p} Z_v$ corresponds to the pairwise distinct monic irreducible factors m_v of m over $\mathbb{Q}_p$. That is, $Z_v = \mathbb{Q}_p(\pi)$ with $\pi \in Z_v$ having minimal polynomial m_v over $\mathbb{Q}_p$. In particular, $m_v(0)$ divides $m(0) = q^{\dim A}$, so each m_v has a non-zero constant term.

For each $v|p$, the Z_v-action on the non-zero $V_p(G_v)$ commutes with the $D_k[1/p]$-action (as it arises from an action of Z_v on the p-divisible group G_v in the isogeny category over k), and the central element $m_v(\mathcal{F}^a) \in D_k[1/p]$ acts as multiplication by the element $m_v(\pi) \in Z_v$ that is zero. In other words, $V_p(G_v)$ is a non-zero module over the ring

$$C_v := D_k[1/p]/D_k[1/p]m_v(\mathcal{F}^a)$$

that was described as a matrix algebra over a specific cyclic algebra in Theorem A.1.2.3 and Remark A.1.2.4 (taking $h = m_v$ there).

Now it remains to compute the local invariants for certain cyclic algebras over non-archimedean local fields. We begin by recording a general lemma.

A.1.3.1. Lemma. *Let F be a field and let F'/F be a cyclic extension. Fix a generator s' of $\Gamma = \mathrm{Gal}(F'/F)$. Let $\chi_{s'} : \Gamma \to \mathbb{Q}/\mathbb{Z}$ be the unique homomorphism sending s' to $1/[F' : F]$, and let $\theta_{s'} = \delta(\chi_{s'}) \in \mathrm{H}^2(\Gamma, \mathbb{Z})$ where $\delta : \mathrm{H}^1(\Gamma, \mathbb{Q}/\mathbb{Z}) \to \mathrm{H}^2(\Gamma, \mathbb{Z})$ is the connecting map arising from $0 \to \mathbb{Z} \to \mathbb{Q} \to \mathbb{Q}/\mathbb{Z} \to 0$.*

(1) *For* $c \in F^\times$, *the class of the cyclic algebra* $(F'/F, s', c)$ *in* $\mathrm{Br}(F)$ *is the image of* $c \bmod \mathrm{Nm}_{F'/F}(F'^\times)$ *under the Tate periodicity isomorphism*

$$F^\times/\mathrm{Nm}_{F'/F}(F'^\times) = \widehat{\mathrm{H}}^0(\Gamma, F'^\times) \overset{\cup\theta_s}{\simeq} \mathrm{H}^2(\Gamma, F'^\times) \subseteq \mathrm{Br}(F).$$

(2) *If* F'_0/F *is a subextension of* F'/F *and* $s'_0 = s'|_{F'_0}$ *then*

$$[(F'_0/F, s'_0, c)] = [(F'/F, s', c^{[F':F'_0]})]$$

in $\mathrm{Br}(F)$.

PROOF. The first part is an explicit cocycle calculation via unwinding definitions and using the cup-product compatibility of the δ-functorial map $\mathrm{H}^\bullet(\Gamma, \cdot) \to \widehat{\mathrm{H}}^\bullet(\Gamma, \cdot)$ (using Tate cohomology only in non-negative degrees). The second part follows from the first via the commutativity of the diagram

$$\begin{array}{ccc} F^\times/\mathrm{Nm}_{F'_0/F}(F_0'^\times) & \xrightarrow{\simeq} & \mathrm{H}^2(F'_0/F, F_0'^\times) \\ \downarrow & & \downarrow \mathrm{inf} \\ F^\times/\mathrm{Nm}_{F'/F}(F'^\times) & \xrightarrow[\simeq]{} & \mathrm{H}^2(F'/F, F'^\times) \end{array}$$

(which in turn follows from the definition of the horizontal isomorphisms and both the δ-functoriality and cup product compatibilities of inflation maps). □

Let F be a non-archimedean local field. By Lemma A.1.3.1 and the existence of unramified splitting fields for all Brauer classes of F, it follows that every element of $\mathrm{Br}(F)$ is represented by a cyclic algebra of the form $(F'/F, \phi, c)$ with F'/F a finite unramified extension, $\phi \in \mathrm{Gal}(F'/F)$ the *arithmetic* Frobenius generator, and $c \in F^\times$ an element. (We could also work with geometric Frobenius generators, but for the applications to abelian varieties it is the cyclic algebras resting on arithmetic Frobenius that naturally arise in the analysis of the Dieudonné modules; see Remark A.1.2.4.) Define

$$\mathrm{inv}_F : \mathrm{Br}(F) \simeq \mathbb{Q}/\mathbb{Z}$$

according to the convention in 1.2.4.2. The main result of A.1.3 is:

A.1.3.2. Theorem. *Let* F *be a non-archimedean local field and* F'/F *an unramified finite extension. Let* $\phi \in \mathrm{Gal}(F'/F)$ *be the arithmetic Frobenius element. For any* $c \in F^\times$, *the cyclic* F*-algebra* $(F'/F, \phi, c)$ *has local invariant in* $\mathbb{Q}/\mathbb{Z}$ *represented by* $\mathrm{ord}_F(c)/[F' : F]$.

PROOF. Using $\mathrm{ord}_F : F^\times/\mathrm{Nm}_{F'/F}(F'^\times) \simeq \mathbb{Z}/[F' : F]\mathbb{Z}$, by A.1.3.1 the element $[(F'/F, \phi, c)] \in \mathrm{Br}(F)$ only depends on the ratio $\mathrm{ord}_F(c)/[F' : F]$. In particular, we may change c by a unit multiple so as to reduce to the case when c is a power of a local uniformizer of F, and by passing to a suitable subextension over F we may reduce to the case when $\mathrm{ord}_F(c)$ is relatively prime to $[F' : F]$. In this case the class of c in $F^\times/\mathrm{Nm}_{F'/F}(F'^\times)$ has order $[F' : F]$, so the cyclic F-algebra $(F'/F, \phi, c)$ of degree $[F' : F]^2$ is necessarily a *division algebra* (due to period dividing the index).

A central division algebra Δ over F admits a unique valuation extending the normalized valuation on F [**107**, Ch. XII, §2], and there is a classical procedure that uses this valuation to compute $\mathrm{inv}_F([\Delta])$ as follows. Let

$$\mathrm{ord}_\Delta = \mathrm{ord}_F \circ \mathrm{Nrd}_{\Delta/F} : \Delta^\times \to \mathbb{Z}$$

(with $\mathrm{Nrd}_{\Delta/F}$ denoting the reduced norm), so if $[\Delta : F] = n^2$ then $(1/n)\mathrm{ord}_\Delta$ restricts to the normalized valuation on $F^\times$. By the Skolem-Noether theorem, there exists $\gamma \in \Delta^\times$ such that $\gamma x \gamma^{-1} = \phi(x)$ for all x in a copy of the degree-n unramified extension F_n/F inside Δ, with ϕ the arithmetic Frobenius generator of $\mathrm{Gal}(F_n/F)$. In particular, $\gamma^n \in F_n^\times$. Pick any such γ; in the case of cyclic algebras such a γ is provided as part of the construction (along with a generator of the cyclic Galois group).

Since we use the method of non-abelian cohomology (as in [**107**, Ch. X, §5]) to identify $\mathrm{H}^2(K/F, K^\times)$ with the subgroup of classes in $\mathrm{Br}(F)$ split by a Galois extension K/F and we use arithmetic Frobenius as the preferred topological generator for the Galois group of a finite field, after some explicit unwinding of definitions and calculation with 2-cocycles one finds

$$\mathrm{inv}_F([\Delta]) = \frac{\mathrm{ord}_\Delta(\gamma)}{n} = \frac{\mathrm{ord}_\Delta(\gamma^n)}{n^2}.$$

Now take Δ to be the cyclic algebra $(F'/F, \phi, c)$ equipped with its canonical choice of γ (so $n = [F' : F]$ and $\gamma^n = c$). We thereby get

$$\mathrm{inv}_F([\Delta]) = \frac{\mathrm{ord}_\Delta(c)}{n^2} \bmod \mathbb{Z} = \frac{\mathrm{ord}_F(c)}{n} \bmod \mathbb{Z}.$$

A.1.4. Applications to abelian varieties. Let A be a simple abelian variety over a finite field k of size $q = p^a$, and $\Delta := \mathrm{End}^0(A)$, so Δ is a central division algebra over the number field $Z = \mathbb{Q}(\pi)$ with q-Frobenius endomorphism $\pi \in \Delta$. We seek to compute local invariants $\mathrm{inv}_v(\Delta_v)$ for all place $v|p$ of Z. By (A.1.3.1) and A.1.2.4, the class of Δ_v in $\mathrm{Br}(Z_v)$ is the same as that of the cyclic Z_v-algebra

$$C_v = (K_0 Z_v/Z_v, \phi, \pi^{f_v/g_v})$$

that rests on the arithmetic Frobenius generator ϕ of $\mathrm{Gal}(K_0 Z_v/Z_v)$ and the element π^{f_v/g_v} in $Z_v^\times$ (where $f_v = f(v|p)$ and $g_v = \gcd(f_v, a)$).

Since $[K_0 Z_v : Z_v] = a/g_v$, the formula in Theorem A.1.3.2 gives

$$\mathrm{inv}_v(C_v) = \frac{1}{a/g_v} \cdot v(\pi^{f_v/g_v}) = \frac{f_v}{a} \cdot v(\pi)$$

in $\mathbb{Q}/\mathbb{Z}$. Let $e_v = e(v|p)$, so $e_v f_v = [Z_v : \mathbb{Q}_p]$. We have

$$v(q) = a \cdot v(p) = a \cdot e_v,$$

so $a = v(q)/e_v$. This implies that

$$\mathrm{inv}_v(C_v) = \frac{f_v}{a} \cdot v(\pi) = \frac{e_v f_v}{v(q)} \cdot v(\pi) = \frac{v(\pi)}{v(q)} \cdot [Z_v : \mathbb{Q}_p]$$

in $\mathbb{Q}/\mathbb{Z}$. This establishes the formula for local invariants in Corollary 1.6.2.2(3).

A.2. The Main Theorem of Complex Multiplication

In [**113**, Ch. IV, §18] the Main Theorem of Complex Multiplication is proved in terms of a style of algebraic geometry that has been largely abandoned. To help a wider audience appreciate many of the beautiful ideas in that proof, in this section we give a scheme-theoretic account of the argument. Our exposition centers on an algebraic formulation of the Main Theorem over $\overline{\mathbb{Q}}$ (Theorem A.2.4.4), without analytic uniformizations or the automorphism group of $\mathbb{C}$. In A.2.8 we use this version of the Main Theorem to recover a coordinate-free version of the traditional formulation of the Main Theorem in terms of analytic uniformizations and Riemann forms of certain polarizations.

Complex-analytic methods certainly have their place in the theory, but we prefer to minimize their appearance and keep proofs as algebraic as possible. The reason that we can achieve this is because of our systematic use of Serre's tensor construction (see 1.7.4), especially in its relative incarnation over base schemes with possibly mixed characteristic.

A.2.1. Notation. Here are some notational conventions to be used in A.2. We let $\overline{\mathbb{Q}}$ denote the algebraic closure of $\mathbb{Q}$ in $\mathbb{C}$, and the kernel of $\exp : \mathbb{C} \to \mathbb{C}^\times$ is denoted $\mathbb{Z}(1)$. We also write $M(1)$ to denote $\mathbb{Z}(1) \otimes_{\mathbb{Z}} M$ for any $\mathbb{Z}$-module M. For a scheme X over a field K and an automorphism $\gamma : K \to K$, X^γ denotes the K-scheme

$$\mathrm{Spec}(K) \times_{\mathrm{Spec}(\gamma),\mathrm{Spec}(K)} X$$

obtained by base change, so $(X^\gamma)^{\gamma'} = X^{\gamma\gamma'}$. The same notation is used for K-morphisms.

We indulge in one serious abuse of notation: if γ is an automorphism of K that restricts to an automorphism γ_0 of a subfield K_0 (the case of most interest being $K = \overline{\mathbb{Q}}$ and K_0 a number field) and if X is a K-scheme for which a K_0-descent X_0 is specified, we write X_0^γ to denote the K_0-scheme $X_0^{\gamma_0}$ descending the K-scheme X^γ. Hopefully this will not cause confusion.

For a CM field L, a CM abelian variety $(A, i : L \to \mathrm{End}^0(A))$ over a field is *principal* if the CM order $i^{-1}(\mathrm{End}(A))$ is equal to $\mathcal{O}_L$. If A is an abelian variety over a field K of characteristic 0 and $\overline{K}/K$ is an algebraic closure then $T_f(A)$ denotes the "adelic Tate module" $\varprojlim A[n](\overline{K}) \simeq \prod_\ell T_\ell(A)$ and $V_f(A)$ denotes the finite free $\mathbb{A}_f$-module $\mathbb{Q} \otimes_{\mathbb{Z}} T_f(A)$. The *adelic Weil pairing* for A is denoted

$$\langle \cdot, \cdot \rangle_A : V_f(A) \times V_f(A^t) \to \mathbb{A}_f(1).$$

A.2.2. Overview. Let (A, i) be an abelian variety over $\overline{\mathbb{Q}}$ with CM type (L, Φ). Let $\phi : A \to A^t$ be an L-linear polarization, using the L-action $i^t(c) := i(c^*)^t$ on A^t, with c^* denoting complex conjugation on the CM field L. (This is the dual CM structure in the sense of Definition 1.5.3.1, so (A^t, i^t) also has CM type (L, Φ).) Such a ϕ exists, by the complex-analytic theory (see Lemma 2.2.2). Let $E \subset \overline{\mathbb{Q}}$ be the reflex field (Definition 2.1.3), and pick $\sigma \in \mathrm{Gal}(\overline{\mathbb{Q}}/E)$ and a finite idele $s \in \mathbb{A}_{E,f}^\times$ that maps to $\sigma|_{E^{\mathrm{ab}}}$ under the global Artin map. The Main Theorem describes the $\mathbb{A}_{L,f}$-linear isomorphism $\sigma : V_f(A) \simeq V_f(A^\sigma)$ and the Weil pairing e_{ϕ^σ} on $V_f(A^\sigma)$ in terms of adelic operations on $V_f(A)$ and e_ϕ using s, and constructs (A^σ, i^σ) from (A, i) by an adelic procedure that does not use Galois automorphisms.

Specializing the arguments below to the 1-dimensional case gives a proof for elliptic curves that exhibits a different flavor from the traditional one (as in [**66**, Ch. 10]). Some technical issues in the general case simplify tremendously in the 1-dimensional case because any elliptic curve has a unique polarization of each positive square degree.

In A.2.3 we develop some concepts that are required in the proof of the Main Theorem, and in A.2.4 we give the setup for and statement of the Main Theorem. Before delving into the proof, in A.2.5 we work out a few consequences of the Main Theorem. The proof of the Main Theorem occupies A.2.6–A.2.7, Finally, in A.2.8 we use the "algebraic" version of the Main Theorem over $\overline{\mathbb{Q}}$ to deduce the traditional version in terms of analytic uniformizations. We also translate analytic refinements back into algebraic language (via a variant on the Serre tensor construction in the absence of CM by the maximal order) and give algebraic proofs of the latter.

A.2.3. $\mathbb{Q}$-polarizations. Let L be a CM field, with maximal totally real subfield L^+ of degree g over $\mathbb{Q}$. As above, $c \mapsto c^*$ denotes complex conjugation on L. Let $\Phi \subset \mathrm{Hom}(L, \overline{\mathbb{Q}})$ be a CM type; i.e., a set of representatives for the quotient of $\mathrm{Hom}(L, \overline{\mathbb{Q}})$ by the free action of the complex conjugation of L. Clearly Φ has size g and $[L : \mathbb{Q}] = 2g$. Finally, let $E \subset \overline{\mathbb{Q}}$ be the reflex field of (L, Φ) (as in Definition 2.1.3); this is the fixed field of the open subgroup of $\mathrm{Gal}(\overline{\mathbb{Q}}/\mathbb{Q})$ consisting of those elements σ such that $\sigma\Phi = \Phi$. (That is, composition with σ permutes the set Φ of embeddings.)

We shall be interested in studying pairs (A, i) where A is an abelian variety over $\overline{\mathbb{Q}}$ and $i : L \to \mathrm{End}^0(A)$ is a ring homomorphism making A an abelian variety of CM type (L, Φ). That is, the eigenspace decomposition for the $\overline{\mathbb{Q}}$-linear action of L on the g-dimensional tangent space of A consists of lines on which L acts through the mappings $\varphi : L \to \overline{\mathbb{Q}}$ given by the g elements $\varphi \in \Phi$. The dual A^t is endowed with the dual CM structure, so it also has CM type Φ.

Since $\overline{\mathbb{Q}}$ is algebraically closed, so $\mathrm{Hom}(X, Y) = \mathrm{Hom}(X_{\mathbb{C}}, Y_{\mathbb{C}})$ for any abelian varieties X and Y over $\overline{\mathbb{Q}}$ (Lemma 1.2.1.2), a polarization of A over $\overline{\mathbb{Q}}$ is the "same" as a polarization of $A_{\mathbb{C}}$ over $\mathbb{C}$. Thus, in terms of the analytic theory, such a polarization is encoded as a skew-symmetric bilinear pairing

$$\psi_{\mathbb{Z}} : \mathrm{H}_1(A(\mathbb{C}), \mathbb{Z}) \times \mathrm{H}_1(A(\mathbb{C}), \mathbb{Z}) \to \mathbb{Z}(1)$$

such that with respect to the complex structure induced by the $\mathbb{R}$-linear isomorphism

$$\mathrm{H}_1(A(\mathbb{C}), \mathbb{R}) \simeq \mathrm{Lie}(A(\mathbb{C}))$$

(defined by $\sigma \mapsto \int_\sigma \in \Omega^1(A(\mathbb{C}))^\vee = \mathrm{Cot}_0(A(\mathbb{C}))^\vee$), the $\mathbb{R}$-linear extension $\psi_{\mathbb{R}}$ satisfies

$$\psi_{\mathbb{R}}(cx, y) = \psi_{\mathbb{R}}(x, \overline{c}y)$$

for all $c \in \mathbb{C}$ and the resulting $\mathbb{R}$-valued symmetric bilinear form

$$(2\pi\sqrt{-1})^{-1}\psi_{\mathbb{R}}(\sqrt{-1} \cdot x, y) \tag{A.2.3.1}$$

(which is independent of the choice of $\sqrt{-1}$) is positive-definite.

In terms of the algebraic theory, $\psi_{\mathbb{Z}}$ may be encoded as a symmetric isogeny $\phi : A \to A^t$ such that $(1, \phi)^*(\mathscr{P})$ is ample on A, where $\mathscr{P}$ is the Poincaré bundle over $A \times A^t$. For our purposes it is much more convenient to work in the isogeny category, so the following notion is useful.

A.2.3.1. Definition. A $\mathbb{Q}$-*polarization* of an abelian variety over a field is a positive rational multiple of a polarization.

For an abelian variety A over $\overline{\mathbb{Q}}$, the positive rational multiple in this definition may be viewed within the $\mathbb{Q}$-vector space $\mathrm{Hom}^0(A, A^t) = \mathbb{Q} \otimes_{\mathbb{Z}} \mathrm{Hom}(A, A^t)$ or as a $\mathbb{Q}(1)$-valued skew-symmetric form on $\mathrm{H}_1(A(\mathbb{C}), \mathbb{Q})$ via the analytic theory. That is, a $\mathbb{Q}$-polarization of A is a positive rational multiple of a pairing $\psi_{\mathbb{Z}}$ as above or is a mapping $\phi : A \to A^t$ *in the isogeny category* of abelian varieties over $\overline{\mathbb{Q}}$ such that ϕ admits a positive integral multiple that is a polarization.

By Lemma 2.2.2(1), there exists a polarization $\phi : A \to A^t$ that is *L-linear*, or equivalently the associated Rosati involution on $\mathrm{End}^0(A)$ restricts to complex conjugation on L. The concept of *L-linear $\mathbb{Q}$-polarization* is defined in the evident manner. A useful formulation of the L-linearity condition on a $\mathbb{Q}$-polarization $\phi : A \to A^t$ is this: under the associated skew-symmetric Weil pairing

$$\text{(A.2.3.2)} \qquad \langle \cdot, \cdot \rangle_{\phi,\mathbb{Q}} : \mathrm{H}_1(A(\mathbb{C}), \mathbb{Q}) \times \mathrm{H}_1(A(\mathbb{C}), \mathbb{Q}) \to \mathbb{Q}(1),$$

for any $c \in L$ the adjoint to $\mathrm{H}_1(i(c))$ is $\mathrm{H}_1(i(c^*))$. In a special case, L-linearity is automatic:

A.2.3.2. Example. Suppose A is simple. In this case $i : L \to \mathrm{End}^0(A)$ is an isomorphism, by Proposition 1.3.6.4(1). Moreover, by Lemma 1.3.5.4, the Rosati involution for any polarization of A induces complex conjugation on L.

In the general (possibly non-simple) case, there is a useful refinement:

A.2.3.3. Lemma. *The set of elements in* $\mathrm{Hom}^0(A, A^t)$ *associated to L-linear $\mathbb{Q}$-polarizations is a principal homogeneous space for the action on A (in the isogeny category) by the subgroup of totally positive elements of $(L^+)^\times$.*

PROOF. Since L is its own centralizer in $\mathrm{End}^0(A)$, any two L-linear isogenies from A to A^t (in the isogeny category) are related through the L-action on A. Hence, given a single L-linear $\mathbb{Q}$-polarization $\phi : A \to A^t$, we just have to work out the condition on $c \in L^\times$ that $\phi \circ i(c)$ is a $\mathbb{Q}$-polarization.

Since ϕ is a symmetric L-linear isogeny, the symmetry condition on $\phi \circ i(c)$ says exactly that $i(c) = i(c^*)$ in $\mathrm{End}^0(A)$; i.e., $c \in L^+$. In [**82**, §22] there are given Hermitian forms that describe explicit L-linear polarizations on analytic models in Example 1.5.3 for a set of representatives of the L-linear isogeny classes of abelian varieties with each CM type (L, Φ) over $\mathbb{C}$. These explicit Hermitian forms show that the positivity condition on $\phi \circ i(c)$ says precisely that $\mathrm{H}_1(i(c))_{\mathbb{R}}$ preserves the positive-definiteness property of (A.2.3.1). This translates into the condition that for each $\varphi \in \Phi$ the totally real algebraic number $\varphi(c) \in \overline{\mathbb{Q}} \subset \mathbb{C}$ is *positive*. This is equivalent to the condition that the element $c \in L^+$ is totally positive since $\Phi|_{L^+}$ is the set of all embeddings of L^+ into $\mathbb{R}$ (as Φ is a CM type on L). □

A.2.3.4. Lemma. *Let (A_0, i_0) and (A'_0, i'_0) be abelian varieties with* CM *type (L, Φ) over a number field K. Assume K is so large that* $\mathrm{Hom}^0((A_0, i_0), (A'_0, i'_0))$ *is non-zero (see Proposition* 1.5.4.1*). If $\mathfrak{P}$ is a prime of good reduction for A_0 and A'_0, and $\overline{A}_0$ and $\overline{A}'_0$ denote the associated reductions over $\kappa(\mathfrak{P})$, then the injective reduction mapping*

$$\mathrm{Hom}^0((A_0, i_0), (A'_0, i'_0)) \to \mathrm{Hom}^0((\overline{A}_0, \overline{i}_0), (\overline{A}'_0, \overline{i}'_0))$$

is bijective.

PROOF. Since the abelian varieties all have dimension g with $[L : \mathbb{Q}] = 2g$, the source and target Hom^0's are each at most 1-dimensional over L due to the injectivity in Proposition 1.2.5.1. The reduction mapping is L-linear, so the assumption of non-vanishing in characteristic 0 gives the result. □

Consider triples (A, i, ϕ) where (A, i) is a CM abelian variety of type (L, Φ) over $\overline{\mathbb{Q}}$ and $\phi : A \to A^t$ is a $\mathbb{Q}$-polarization of A that is L-linear. Let E be the reflex field of (L, Φ). For $\sigma \in \mathrm{Gal}(\overline{\mathbb{Q}}/E)$, scalar extension along the isomorphism $\sigma : \overline{\mathbb{Q}} \simeq \overline{\mathbb{Q}}$ gives another triple $(A^\sigma, i^\sigma, \phi^\sigma)$ with $i^\sigma(c) := i(c)^\sigma \in \mathrm{End}^0(A^\sigma)$. The Lie algebras of A and A^σ satisfy $\mathrm{Lie}(A^\sigma) \simeq \sigma^*(\mathrm{Lie}(A))$ as $L \otimes_{\mathbb{Q}} \overline{\mathbb{Q}}$-modules, so (A^σ, i^σ) has CM type $\sigma\Phi = \Phi$. Hence, by Proposition 1.5.4.1 applied over $\overline{\mathbb{Q}}$, there exists an L-linear isogeny $A \to A^\sigma$. A fundamental fact is that such isogenies can be found that respect L-linear $\mathbb{Q}$-polarizations up to rational homothety:

A.2.3.5. Theorem. *The triples (A, i, ϕ) and $(A^\sigma, i^\sigma, \phi^\sigma)$ are isogenous up to $\mathbb{Q}^\times$-homothety. More precisely, there is an L-linear isomorphism $\xi : A \to A^\sigma$ in the isogeny category such that it is a "Frobenius lift" at a suitable place and for every L-linear $\phi : A \to A^t$ in the isogeny category we have $\xi^t \circ \phi^\sigma \circ \xi = q_{\xi,\sigma} \cdot \phi$ for some $q_{\xi,\sigma} \in \mathbb{Q}^\times_{>0}$; explicitly, $q_{\phi,\xi,\sigma} = \deg(\xi)^{1/g}$.*

If $\mathcal{O}_L \subset \mathrm{End}(A)$ then ξ can be chosen to be a morphism of abelian varieties.

The notion of degree for morphisms in the isogeny category is defined by $\mathbb{Z}$-homogeneity of degree for morphisms of abelian varieties. We do *not* claim that ξ is a morphism of abelian varieties (i.e., it might involve a non-trivial "denominator", so it may not be a genuine isogeny), but it will be important in the proof of the Main Theorem of Complex Multiplication that it can be chosen to be a morphism of abelian varieties when $\mathcal{O}_L \subset \mathrm{End}(A)$.

PROOF. Any two choices of ϕ are related by an $L^\times$-multiplier, so since ξ is required to be L-linear it is sufficient to work with a single ϕ. We fix a choice of ϕ that is an L-linear isogeny (as we have noted does exist, due to the equality of CM types). Once we find the desired ξ then computing the degree of both sides gives $q_{\xi,\sigma}^{2g} = \deg(\xi)\deg(\xi^t) = \deg(\xi)^2$, so $q_{\xi,\sigma} = \deg(\xi)^{1/g}$.

Let K/E be a finite Galois subextension of $\overline{\mathbb{Q}}$ that is a field of definition for (A, i, ϕ) in the sense that: (i) A descends to an abelian variety A_0 over K, (ii) the L-action on A (in the isogeny category over $\overline{\mathbb{Q}}$) descends to a ring homomorphism $i_0 : L \to \mathrm{End}^0(A_0)$, and (iii) there is an L-linear isogeny $\phi_0 : A_0 \to A_0^t$ whose $\overline{\mathbb{Q}}$-scalar extension is ϕ.

Choose a prime $\mathfrak{P}$ of K such that $\mathfrak{P}$ is unramified over its contraction $\mathfrak{p}$ in E, $\mathfrak{P}$ is a prime of good reduction for A_0, and $\sigma|_K = (\frac{K/E}{\mathfrak{P}})$. (There are infinitely many such $\mathfrak{P}$, by the Chebotarev density theorem.) In particular, since σ acting on $\mathcal{O}_K$ fixes $\mathfrak{P}$, it follows that $\mathfrak{P}$ is also a prime of good reduction for A_0^σ. Let $\mathscr{A}_0$ denote the Néron model for A_0 over $\mathcal{O}_{K,\mathfrak{P}}$, so this is an abelian scheme.

Viewing σ as an automorphism of $\mathcal{O}_{K,\mathfrak{P}}$, the abelian scheme $\mathscr{A}_0^\sigma$ is the Néron model of its generic fiber A_0^σ, and likewise the dual abelian scheme $\mathscr{A}_0^t$ is the Néron model of its generic fiber A_0^t. By the Néron mapping property for abelian schemes (i.e., the valuative criterion for properness and Weil's extension lemma), ϕ_0 extends uniquely to a mapping of abelian schemes $\mathscr{A}_0 \to \mathscr{A}_0^t$ which is moreover necessarily an isogeny on fibers by the argument in Remark 1.8.4.2.

Let q be the size of the residue field at the place of E below $\mathfrak{P}$, so σ induces the qth-power map on the residue field $\kappa(\mathfrak{P})$ at $\mathfrak{P}$. It follows from Lemma A.2.3.4 (whose non-vanishing hypothesis holds in our situation due to the existence of ϕ_0) that there exists a (unique) L-linear morphism $\xi : A_0 \to A_0^\sigma$ in the isogeny category over K (possibly not a morphism of abelian varieties; i.e., it may have a non-trivial "denominator") such that its reduction $\overline{\xi} : \overline{A}_0 \to \overline{A}_0^{(q)}$ over $\kappa(\mathfrak{P})$ is the relative q-Frobenius map $\mathrm{Fr}_{\overline{A}_0/\kappa(\mathfrak{P}),q}$, where $\overline{A}_0$ denotes the closed fiber of $\mathscr{A}_0$. Of course, ξ depends on the choice of $\mathfrak{P}$, and it is an isomorphism in the isogeny category over K by lifting an isogeny-inverse of its reduction $\mathrm{Fr}_{\overline{A}_0/\kappa(\mathfrak{P}),q}$. (Even though the q-Frobenius is a morphism of abelian varieties, it does not follow formally that ξ is necessarily a morphism of abelian varieties; see Remark 1.4.4.1.)

We claim that $\xi_{\overline{\mathbb{Q}}}$ works in the theorem, with the rational multiplier q. Equivalently, we claim that $\xi^t \circ \phi_0^\sigma \circ \xi = q\phi_0$. To check this equality as maps in the isogeny category of abelian varieties over K with good reduction at $\mathfrak{P}$, it is equivalent to check the corresponding equality for reductions over $\kappa(\mathfrak{P})$. The reduction of ϕ_0 is an L-linear isogeny $\overline{\phi}_0 : \overline{A}_0 \to \overline{A}_0^t$, and the desired identity among maps of special fibers is insensitive to replacing $\overline{\phi}_0$ with any $L^\times$-multiple. That is, we may work with any L-linear isogeny $\overline{A}_0 \to \overline{A}_0^t$ whatsoever, so we shall work with an L-linear *polarization* of $\overline{A}_0$ (which exists by Lemma 2.2.2).

It now suffices to prove more generally that if B is an abelian variety over a field κ of characteristic $p > 0$ and if q is a power of p then for any polarization $\phi : B \to B^t$ there is an equality

$$\mathrm{Fr}_{B/\kappa,q}^t \circ \phi^{(q)} \circ \mathrm{Fr}_{B/\kappa,q} = q\phi.$$

We may extend scalars to $\overline{\kappa}$ so that the ground field κ is algebraically closed. Hence, $\phi = \phi_{\mathcal{L}}$ for a line bundle $\mathcal{L}$ on A (even ample), where $\phi_{\mathcal{L}} : A \to A^t$ is $x \mapsto t_x^*(\mathcal{L}) \otimes \mathcal{L}^{-1}$. The formation of $\phi_{\mathcal{L}}$ is compatible with base change, so $\phi^{(q)} = \phi_{\mathcal{L}^{(q)}}$ with $\mathcal{L}^{(q)}$ denoting the pullback of $\mathcal{L}$ along the projection $A^{(q)} \to A$ (or equivalently, along the base change $\mathrm{Spec}(\kappa) \to \mathrm{Spec}(\kappa)$ given by the qth-power map on κ). By the functorial properties of the "Mumford construction" $\mathcal{L} \rightsquigarrow \phi_{\mathcal{L}}$,

$$\mathrm{Fr}_{B/\kappa,q}^t \circ \phi^{(q)} \circ \mathrm{Fr}_{B/\kappa,q} = \phi_{\mathrm{Fr}_{B/\kappa,q}^*(\mathcal{L}^{(q)})}.$$

Since $q\phi_{\mathcal{L}} = \phi_{\mathcal{L}^{\otimes q}}$, it now suffices to prove $\mathrm{Fr}_{B/\kappa,q}^*(\mathcal{L}^{(q)}) \simeq \mathcal{L}^{\otimes q}$ for any invertible sheaf $\mathcal{L}$ on B, where B is *any* κ-scheme. The composition of the relative q-Frobenius $\mathrm{Fr}_{B/\kappa,q} : B \to B^{(q)}$ and the projection $B^{(q)} \to B$ is the absolute q-Frobenius map $\mathrm{Fr}_{B,q} : B \to B$ that is the identity on topological spaces and the qth-power map on structure sheaves. Hence, $\mathrm{Fr}_{B/\kappa,q}^*(\mathcal{L}^{(q)}) \simeq \mathrm{Fr}_{B,q}^*(\mathcal{L})$, so it suffices to prove $\mathrm{Fr}_{B,q}^*(\mathcal{L}) \simeq \mathcal{L}^{\otimes q}$ for any invertible sheaf $\mathcal{L}$ on B. But this follows from $\mathcal{O}_B^\times$-valued Čech 1-cocycle calculations relative to a trivialization of $\mathcal{L}$ over B since $\mathrm{Fr}_{B,q}$ is the identity on the topological space B and the qth-power map on $\mathcal{O}_B$.

Finally, we check that ξ is a morphism of abelian varieties when $\mathcal{O}_L \subset \mathrm{End}(A)$. In such cases we have $\mathcal{O}_L \subset \mathrm{End}(A_0) = \mathrm{End}(\mathscr{A}_0)$. Since

$$M := \mathrm{Hom}((A_0, i_0), (A_0^\sigma, i_0^\sigma))$$

is non-zero, by Example 1.7.4.1 it is an invertible $\mathcal{O}_L$-module and the natural map $M \otimes_{\mathcal{O}_L} A_0 \to A_0^\sigma$ over K is an isomorphism. Thus, passing to Néron models over $\mathcal{O}_{K,\mathfrak{P}}$, the natural map $M \otimes_{\mathcal{O}_L} \mathscr{A}_0 \to \mathscr{A}_0^\sigma$ is an isomorphism. Hence, passing to special fibers, the natural map $M \otimes_{\mathcal{O}_L} \overline{A}_0 \to \overline{A}_0^{(q)}$ is an isomorphism. We may then conclude via Example 1.7.4.2 that every $\mathcal{O}_L$-linear homomorphism

$$\overline{A}_0 \to \overline{A}_0^{(q)} = M \otimes_{\mathcal{O}_L} \overline{A}_0$$

is induced by a unique element of M. Hence, the natural injective reduction map

$$\mathrm{Hom}(A_0, A_0^\sigma) \to \mathrm{Hom}(\overline{A}_0, \overline{A}_0^{(q)})$$

restricts to an *isomorphism* between subgroups of $\mathcal{O}_L$-linear homomorphisms (over K and $\kappa(\mathfrak{P})$ respectively), so an L-linear isogeny-category map $A_0 \to A_0^\sigma$ is a morphism of abelian varieties if and only if its reduction is so. Since ξ was constructed as such a map lifting the q-Frobenius morphism, we are done. □

In the language of the bilinear pairings (A.2.3.2), Theorem A.2.3.5 says that

$$\langle \mathrm{H}_1(\xi)(\cdot), \mathrm{H}_1(\xi)(\cdot)\rangle_{\phi^\sigma,\mathbb{Q}} = \deg(\xi)^{1/g} \cdot \langle \cdot, \cdot \rangle_{\phi,\mathbb{Q}}$$

and that we may choose an L-linear isogeny-category isomorphism $\varphi_\sigma : A^\sigma \to A$ over $\overline{\mathbb{Q}}$ such that φ_σ carries ϕ back to ϕ^σ up to $\mathbb{Q}^\times_{>0}$-multiple. The morphism φ_σ in the isogeny category is unique up to the action on A by elements $c \in L^\times$ such that simultaneous multiplication by c on both variables of the bilinear form $\langle \cdot, \cdot \rangle_{\phi,\mathbb{Q}}$ on $\mathrm{H}_1(A(\mathbb{C}), \mathbb{Q})$ preserves this bilinear form up to $\mathbb{Q}^\times_{>0}$-multiple. That is, for our L-linear $\mathbb{Q}$-polarization $\phi : A \to A^t$, the condition on c is that $i(c)^t \circ \phi \circ i(c)$ is a $\mathbb{Q}^\times_{>0}$-multiple of ϕ. But $i(c)^t = i^t(c^*)$ and ϕ is L-linear, so the condition on c is

$$\phi \circ i(cc^*) \overset{?}{=} q\phi = \phi \circ i(q)$$

for some $q \in \mathbb{Q}^\times_{>0}$. Equivalently, this says $\mathrm{Nm}_{L/L^+}(c) = q \in \mathbb{Q}^\times$; such a norm is necessarily positive since L is a CM field, so the positivity condition on q need not be explicitly required. (This is convenient when relating this situation to a purely algebraic one that can be formulated over any $\mathbb{Q}$-algebra.) We conclude:

A.2.3.6. Corollary. *The L-linear isogeny-category isomorphism φ_σ is unique up to the action of $T(\mathbb{Q}) \subseteq L^\times$ on A, where T is the $\mathbb{Q}$-group*

$$T := \ker(\mathrm{Res}_{L/\mathbb{Q}}(\mathbb{G}_\mathrm{m}) \overset{\mathrm{Nm}_{L/L^+}}{\twoheadrightarrow} \mathrm{Res}_{L^+/\mathbb{Q}}(\mathbb{G}_\mathrm{m})/\mathbb{G}_\mathrm{m}).$$

This T is the torus in Proposition 2.1.3.6.

A.2.3.7. Remark. The choice of ϕ is unique up to the action on A by totally positive elements of L^+ (by Lemma A.2.3.3), and the action by these elements commutes with the action by $T(\mathbb{Q}) \subseteq L^\times$, so the defining condition on φ_σ in Corollary A.2.3.6 is *independent* of the choice of ϕ. That is, such a φ_σ carries ϕ' to a positive rational multiple of ϕ'^σ for *every* L-linear $\mathbb{Q}$-polarization ϕ' of A.

A.2.3.8. Remark. The preceding argument also applies over ℓ-adic fields and adelic Tate modules to show that the elements of

$$\mathrm{Aut}_{\mathbb{A}_{L,f}}(V_f(A)) = \mathbb{A}^\times_{L,f}$$

preserving the $\mathbb{A}^\times_f$-homothety class of the $\mathbb{A}_f$-bilinear skew-symmetric Weil pairing

$$\langle \cdot, \cdot \rangle_\phi : V_f(A) \times V_f(A) \to \mathbb{A}_f(1)$$

are precisely the elements of $T(\mathbb{A}_f)$.

A.2.4. Algebraic formulation of the Main Theorem. We retain the same setup as above: we fix the triple (A, i, ϕ) over $\overline{\mathbb{Q}}$ with (A, i) of type (L, Φ) and ϕ an L-linear $\mathbb{Q}$-polarization of A. We also choose $\sigma \in \mathrm{Gal}(\overline{\mathbb{Q}}/E)$ and we pick an L-linear isogeny $\varphi_\sigma : A^\sigma \to A$ carrying ϕ back to ϕ^σ up to $\mathbb{Q}^\times_{>0}$-multiple (determined by the degree of φ_σ); this condition on φ_σ is *independent* of the choice of ϕ (by Remark A.2.3.7). Consider the $\mathbb{A}_{L,f}$-linear isomorphism

$$V_f(A) \overset{[\sigma]}{\simeq} V_f(A^\sigma) \overset{V_f(\varphi_\sigma)}{\simeq} V_f(A) \tag{A.2.4.1}$$

where $[\sigma]$ is the standard Galois action on $\overline{\mathbb{Q}}$-points (so in terms of skew-symmetric $\mathbb{A}_f(1)$-valued Weil pairings, $[\sigma]$ carries the bilinear form associated to ϕ^σ to the bilinear form associated to ϕ up to multiplication by the adelic cyclotomic character value $\chi(\sigma) \in \mathbb{A}_f^\times$ giving the action of σ on $\mathbb{A}_f(1)$).

The second step in (A.2.4.1) is equivariant for the bilinear forms defined by ϕ^σ and ϕ up to a positive rational multiple (namely $(\deg \varphi_\sigma)^{1/g}$), due to the defining condition on φ_σ. Hence, (A.2.4.1) is an $\mathbb{A}_{L,f}$-linear automorphism of the free $\mathbb{A}_{L,f}$-module $V_f(A)$ of rank 1 and it preserves the bilinear form associated to ϕ up to the $\mathbb{A}_f^\times$-multiplier $\chi(\sigma)(\deg \varphi_\sigma)^{1/g}$. The composite (A.2.4.1) is multiplication by an element $\mu_{\sigma,\varphi_\sigma} \in \mathbb{A}_{L,f}^\times = \mathrm{Res}_{L/\mathbb{Q}}(\mathbb{G}_\mathrm{m})(\mathbb{A}_f)$ that therefore preserves ϕ up to an $\mathbb{A}_f^\times$-multiple, so $\mu_{\sigma,\varphi_\sigma} \in T(\mathbb{A}_f)$ (by Remark A.2.3.8).

The $T(\mathbb{Q})$-ambiguity in the choice of φ_σ implies that the $T(\mathbb{Q})$-congruence class of the multiplier $\mu_{\sigma,\varphi_\sigma} \in T(\mathbb{A}_f)$ does not depend on the choice of φ_σ, and by Lemma A.2.3.3 it does not depend on the choice of ϕ (as φ_σ is L-linear). Since the pair (A, i) over $\overline{\mathbb{Q}}$ of type (L, Φ) is unique up to L-linear isogeny (Proposition 1.5.4.1), it follows that the choice of (A, i) does not affect the element we have just built in $T(\mathbb{Q})\backslash T(\mathbb{A}_f)$. Hence, we get a map of sets

$$\mathrm{Gal}(\overline{\mathbb{Q}}/E) \to T(\mathbb{Q})\backslash T(\mathbb{A}_f) \tag{A.2.4.2}$$

sending $\sigma \in \mathrm{Gal}(\overline{\mathbb{Q}}/E)$ to the common $T(\mathbb{Q})$-congruence class of elements $\mu_{\sigma,\varphi_\sigma} \in T(\mathbb{A}_f)$ for which the composite map in (A.2.4.1) is multiplication by $\mu_{\sigma,\varphi_\sigma}$, where φ_σ is as above. The map (A.2.4.2) only depends on (L, Φ) rather than on (A, i).

A.2.4.1. Lemma. *The map* (A.2.4.2) *is a continuous group homomorphism.*

PROOF. We first check that the map (A.2.4.2) is a group homomorphism. Choose $\sigma, \sigma' \in \mathrm{Gal}(\overline{\mathbb{Q}}/E)$, and pick φ_σ and $\varphi_{\sigma'}$. We need to express the composite isomorphism

$$V_f(A) \overset{[\sigma]}{\simeq} V_f(A^\sigma) \overset{V_f(\varphi_\sigma)}{\simeq} V_f(A) \overset{[\sigma']}{\simeq} V_f(A^{\sigma'}) \overset{V_f(\varphi_{\sigma'})}{\simeq} V_f(A)$$

as $\varphi_{\sigma'\sigma} \circ [\sigma'\sigma]$ for a suitable choice of $\varphi_{\sigma'\sigma}$. The isogeny-category isomorphism $A^{\sigma'\sigma} \simeq (A^\sigma)^{\sigma'} \overset{(\varphi_\sigma)^{\sigma'}}{\to} A^{\sigma'} \overset{\varphi_{\sigma'}}{\to} A$ carries ϕ back to $\phi^{\sigma'\sigma}$ up to a positive rational multiple and is L-linear, so we may take this composite morphism in the isogeny category as our *choice* of $\varphi_{\sigma'\sigma}$. Thus, it suffices to prove the commutativity of the outside edge of the diagram

$$\begin{array}{ccccc}
V_f(A) & \xrightarrow[\simeq]{[\sigma]} & V_f(A^\sigma) & \xrightarrow{\varphi_\sigma} & V_f(A) \\
& \searrow^{[\sigma'\sigma]} & \downarrow{\scriptstyle [\sigma']} & & \downarrow{\scriptstyle [\sigma']} \\
& & V_f(A^{\sigma'\sigma}) & \xrightarrow{(\varphi_\sigma)^{\sigma'}} & V_f(A^{\sigma'}) \\
& & & \searrow^{\varphi_{\sigma'}\circ(\varphi_\sigma)^{\sigma'}} & \downarrow{\scriptstyle \varphi_{\sigma'}} \\
& & & & V_f(A)
\end{array}$$

Since the left and bottom triangles clearly commute, it is therefore enough to prove that the inner square commutes. This amounts to the obvious identity $h^\tau(\tau(x')) = \tau(h(x'))$ for any $\overline{\mathbb{Q}}$-map $h : X' \to X$ between $\overline{\mathbb{Q}}$-schemes and any $x' \in X'(\overline{\mathbb{Q}})$.

We have verified the group homomorphism condition, so to check continuity it is enough to do so near the identity. Thus, we can restrict attention to $\sigma \in \mathrm{Gal}(\overline{\mathbb{Q}}/K)$ for a number field $K \subset \overline{\mathbb{Q}}$ over which (A, i, ϕ) is "defined" (any such K must contain E). Let (A_0, i_0, ϕ_0) be a descent of (A, i, ϕ) to such a K. This choice of descent naturally identifies A^σ with A carrying i^σ to i and ϕ^σ to ϕ. Taking φ_σ to be the resulting canonical identification $A^\sigma \simeq A$ associated to $\sigma \in \mathrm{Gal}(\overline{\mathbb{Q}}/K)$, the element $\mu_{\sigma,\varphi_\sigma} \in T(\mathbb{A}_f) \subseteq \mathbb{A}_{L,f}^\times$ is the multiplier for the $\mathbb{A}_{L,f}$-linear action by $\sigma \in \mathrm{Gal}(\overline{\mathbb{Q}}/K)$ on $V_f(A_0)$. Thus, the continuity of (A.2.4.1) on the open subgroup $\mathrm{Gal}(\overline{\mathbb{Q}}/K) \subseteq \mathrm{Gal}(\overline{\mathbb{Q}}/E)$ follows from the continuity of the Galois representation on the adelic Tate module of any abelian variety over a field of characteristic 0. ☐

A.2.4.2. Lemma. *The quotient space $T(\mathbb{Q})\backslash T(\mathbb{A}_f)$ is Hausdorff.*

This technical lemma is the key reason for working with the torus T rather than with the ambient torus $\mathrm{Res}_{L/\mathbb{Q}}(\mathbb{G}_{\mathrm{m}})$ (for which the analogue of the lemma is false whenever $\mathcal{O}_L^\times$ is infinite; i.e., whenever the CM field L is not imaginary quadratic).

PROOF. We shall prove that $T(\mathbb{Q})$ is a discrete subgroup of $T(\mathbb{A}_f)$. Since $T(\mathbb{A}_f)$ is a second countable Hausdorff space, we can test discreteness by studying sequences that converge to the identity. The canonical $\mathbb{Q}$-subgroup $\mathbb{G}_{\mathrm{m}} \subset \mathrm{Res}_{L/\mathbb{Q}}(\mathbb{G}_{\mathrm{m}})$ clearly is contained in T. Consider the resulting short exact sequence of $\mathbb{Q}$-groups

$$1 \to \mathbb{G}_{\mathrm{m}} \to T \to \overline{T} \to 1.$$

Granting for a moment that $\overline{T}(\mathbb{Q})$ is discrete in $\overline{T}(\mathbb{A}_f)$, it follows that a sequence in $T(\mathbb{Q})$ tending to 1 in $T(\mathbb{A}_f)$ has image in $\overline{T}(\mathbb{Q})$ that stabilizes at 1. Such a sequence is eventually a sequence in $\mathbb{G}_{\mathrm{m}}(\mathbb{Q}) = \mathbb{Q}^\times$ tending to 1 in $T(\mathbb{A}_f)$ and hence tending to 1 in $\mathbb{G}_{\mathrm{m}}(\mathbb{A}_f) = \mathbb{A}_f^\times$. But $\mathbb{Q}^\times$ is discrete in $\mathbb{A}_f^\times$ (as it meets the compact open subgroup $\widehat{\mathbb{Z}}^\times$ in $\{\pm 1\}$), so we would be done.

It remains to prove that $\overline{T}(\mathbb{Q})$ is discrete in $\overline{T}(\mathbb{A}_f)$. Since $\overline{T}(\mathbb{Q})$ is discrete in $\overline{T}(\mathbb{A}) = \overline{T}(\mathbb{R}) \times \overline{T}(\mathbb{A}_f)$ (due to the discreteness of $\mathbb{Q}$ in $\mathbb{A}$), it is enough to prove that $\overline{T}(\mathbb{R})$ is compact. (Note that $T(\mathbb{R})$ is not compact.) By Hilbert's Theorem 90 we have $\mathrm{H}^1(\mathbb{R}, \mathbb{G}_{\mathrm{m}}) = 1$, so the natural open embedding $T(\mathbb{R})/\mathbb{R}^\times \to \overline{T}(\mathbb{R})$ is a topological group isomorphism. We will prove that $T(\mathbb{R})/\mathbb{R}^\times$ is compact.

By definition of T, the group $T(\mathbb{R})$ is the closed subgroup of elements

$$(z_1, \ldots, z_g) \in \prod_{L^+ \hookrightarrow \mathbb{R}} (L \otimes_{L^+} \mathbb{R})^\times$$

whose image in $\prod_{L^+ \hookrightarrow \mathbb{R}} \mathbb{R}^\times$ under Nm_{L/L^+} lies in the diagonal $\mathbb{R}^\times$. This is identified with the subgroup of points $(z_1, \ldots, z_g) \in (\mathbb{C}^\times)^g$ such that $|z_j|$ is independent of j, and by working modulo the diagonally embedded subgroup $\mathbb{R}^\times$ we conclude that $T(\mathbb{R})/\mathbb{R}^\times$ is a quotient of $(S^1)^g$ and so is compact. □

By Lemma A.2.4.2, the continuous group homomorphism

$$\mathrm{Gal}(\overline{\mathbb{Q}}/E) \to T(\mathbb{Q})\backslash T(\mathbb{A}_f)$$

to an abelian target must factor through the *topological* abelianization of $\mathrm{Gal}(\overline{\mathbb{Q}}/E)$. The continuous Artin map $E^\times\backslash\mathbb{A}_E^\times \to \mathrm{Gal}(E^{\mathrm{ab}}/E)$ is a surjection and it identifies the abelianized Galois group with the topological group quotient of $E^\times\backslash\mathbb{A}_E^\times$ by the closure in $E^\times\backslash\mathbb{A}_E^\times$ of the image of the identity component of the archimedean part $(\mathbb{R} \otimes_{\mathbb{Q}} E)^\times$ [**4**, Ch. IX]. Since the reflex field E is a CM field and hence has no real places, the archimedean part $(\mathbb{R} \otimes_{\mathbb{Q}} E)^\times \simeq \prod_{w|\infty}(E_w)^\times$ is connected. The infinite divisibility of $(E_w)^\times$ for each $w|\infty$ on E therefore implies that the restricted Artin map $E^\times\backslash\mathbb{A}_{E,f}^\times \to \mathrm{Gal}(E^{\mathrm{ab}}/E)$ is surjective and identifies the abelianized Galois group with the maximal Hausdorff group quotient of $E^\times\backslash\mathbb{A}_{E,f}^\times$ (i.e., the quotient by the closure of the identity point).

We have built a natural continuous composite mapping of topological groups

$$E^\times\backslash\mathbb{A}_{E,f}^\times \twoheadrightarrow \mathrm{Gal}(E^{\mathrm{ab}}/E) \to T(\mathbb{Q})\backslash T(\mathbb{A}_f) \tag{A.2.4.3}$$

that only depends on the CM type (L, Φ) and *not* on the particular triple (A, i, ϕ) used to build it in the first place. It is therefore natural to demand a direct description of (A.2.4.3) in terms of (L, Φ).

A.2.4.3. Remark. Note that *any* continuous group homomorphism

$$h : E^\times\backslash\mathbb{A}_{E,f}^\times \to T(\mathbb{Q})\backslash T(\mathbb{A}_f)$$

factors continuously through the Artin quotient map $E^\times\backslash\mathbb{A}_{E,f}^\times \twoheadrightarrow \mathrm{Gal}(E^{\mathrm{ab}}/E)$. Indeed, h is a map to a Hausdorff target group $T(\mathbb{Q})\backslash T(\mathbb{A}_f)$, so it factors continuously through the maximal Hausdorff group quotient of the source.

We may rewrite the composite map (A.2.4.3) in the form

$$\mathrm{Res}_{E/\mathbb{Q}}(\mathbb{G}_{\mathrm{m}})(\mathbb{Q})\backslash\mathrm{Res}_{E/\mathbb{Q}}(\mathbb{G}_{\mathrm{m}})(\mathbb{A}_f) \to T(\mathbb{Q})\backslash T(\mathbb{A}_f).$$

The CM type (L, Φ) yields the reflex norm $\mathrm{N}_\Phi : \mathrm{Res}_{E/\mathbb{Q}}(\mathbb{G}_{\mathrm{m}}) \to T$ (Definition 2.1.3.3 and Proposition 2.1.3.6). Hence, it is reasonable to ask if the maps induced by the reflex norm on finite-adelic points and on rational points have any relation to the composite mapping (A.2.4.3). This is the content of the following formulation of the Main Theorem of Complex Multiplication:

A.2.4.4. Theorem (Main Theorem of Complex Multiplication; algebraic form)**.** *The mapping* (A.2.4.3) *is induced by* $s \mapsto \mathrm{N}_\Phi(s)^{-1} \in T(\mathbb{A}_f) \subseteq \mathbb{A}_{L,f}^\times$ *for* $s \in \mathbb{A}_{E,f}^\times$.

The proof of the Main Theorem is given in A.2.6–A.2.7. There is a further important aspect to the Main Theorem: giving an adelic description of $(A^\sigma, i^\sigma, \phi^\sigma)$

in terms of (A, i, ϕ). We shall address this aspect in Theorem A.2.7.2 in the case of CM order $\mathscr{O}_L$ and in Theorem A.2.9.1 in the general case.

A.2.4.5. Remark. The reason $\mathrm{N}_\Phi(s)^{-1}$ rather than $\mathrm{N}_\Phi(s)$ appears in the statement of the Main Theorem is due to our convention for defining the Artin map: it carries local uniformizers to arithmetic Frobenius elements. (See above Remark A.2.6.3.) If we were to use the alternative convention (common in algebraic geometry) that the Artin map carries local uniformizers to geometric Frobenius elements then the identification of $\mathrm{Gal}(E^{\mathrm{ab}}/E)$ with the maximal Hausdorff group quotient of $E^\times\backslash\mathbb{A}_{E,f}^\times$ would be modified by inversion and so the formula in the Main Theorem would lose the inversion.

A.2.5. Some applications. Before we take up the proof of Theorem A.2.4.4, we deduce several interesting arithmetic consequences. Pick $\sigma \in \mathrm{Gal}(\overline{\mathbb{Q}}/E)$ and choose $s \in \mathbb{A}_{E,f}^\times$ such that its image $r_E(s) \in \mathrm{Gal}(E^{\mathrm{ab}}/E)$ under the Artin map is $\sigma|_{E^{\mathrm{ab}}}$. By Remark A.2.4.3, the element $\mathrm{N}_\Phi(s)^{-1} \in T(\mathbb{A}_f)$ has class modulo $T(\mathbb{Q})$ that only depends on $r_E(s) = \sigma|_{E^{\mathrm{ab}}} \in \mathrm{Gal}(E^{\mathrm{ab}}/E)$ and not on the choice its lifting s through the Artin map.

By the Main Theorem, the element $\mathrm{N}_\Phi(s)^{-1} \in \mathbb{A}_{L,f}^\times = \mathrm{Aut}_{\mathbb{A}_{L,f}}(V_f(A))$ agrees modulo $T(\mathbb{Q})$ with the composite in (A.2.4.1) for any choice of L-linear isogeny $\varphi_\sigma : A^\sigma \to A$ carrying ϕ back to a (necessarily positive) rational multiple of ϕ^σ. Since this φ_σ is unique up to precisely the $T(\mathbb{Q})$-action on A, we may *uniquely* choose φ_σ to get the exact agreement: $\mathrm{N}_\Phi(s)^{-1} = V_f(\varphi_\sigma) \circ [\sigma]$. We have proved:

A.2.5.1. Proposition. *For all $\sigma \in \mathrm{Gal}(\overline{\mathbb{Q}}/E)$ and $s \in \mathbb{A}_{E,f}^\times$ such that $r_E(s) = \sigma|_{E^{\mathrm{ab}}}$, there is a unique L-linear isogeny $\lambda_{\sigma,s} : (A, i) \to (A^\sigma, i^\sigma)$ such that*

$$\mathrm{N}_\Phi(s^{-1}) \cdot V_f(\lambda_{\sigma,s}) = [\sigma],$$

where $[\sigma] : V_f(A) \simeq V_f(A^\sigma)$ is defined by σ-action on torsion points.

A.2.5.2. Remark. By construction, $\lambda_{\sigma,s}$ satisfies $\deg(\lambda_{\sigma,s})^{1/g} \in \mathbb{Q}^\times$ and carries ϕ^σ back to $\deg(\lambda_{\sigma,s})^{1/g}\phi$ for *all* L-linear polarizations ϕ of (A, i). Since $[\sigma]$ induces an isomorphism on underlying adelic Tate modules, the rational multiplier $\deg(\lambda_{\sigma,s})^{1/g}$ is easy to determine by computing the adelic degree on both sides of the identity $\mathrm{N}_\Phi(s^{-1})V_f(\lambda_{\sigma,s}) = [\sigma]$: it is the gth root of the generalized adelic lattice index

$$\begin{aligned}
[T_f(A) : \mathrm{N}_\Phi(s)T_f(A)] = \mathrm{Nm}_{L/\mathbb{Q}}(\mathrm{N}_\Phi(s)) &= \mathrm{Nm}_{L^+/\mathbb{Q}}(\mathrm{N}_\Phi(s)\mathrm{N}_{\Phi^*}(s)) \\
&= \mathrm{Nm}_{L^+/\mathbb{Q}}(\mathrm{N}_{E/\mathbb{Q}}(s)) \\
&= \mathrm{Nm}_{E/\mathbb{Q}}(s)^g
\end{aligned}$$

(the third equality due to N_Φ factoring through the $\mathbb{Q}$-subtorus $T \subset \mathrm{Res}_{L/\mathbb{Q}}(\mathbb{G}_{\mathrm{m}})$).

In other words, if we let $q_s \in \mathbb{Q}_{>0}^\times$ be the unique positive generator of the fractional $\mathbb{Q}$-ideal associated to the finite $\mathbb{Q}$-idele $\mathrm{N}_{E/\mathbb{Q}}(s)$ then $\lambda_{\sigma,s}$ carries ϕ^σ back to $q_s\phi$. We can describe q_s as a finite idele rather directly: $q_s = \chi(\sigma)\mathrm{N}_{E/\mathbb{Q}}(s)$ inside $\mathbb{A}_f^\times$, where $\chi : \mathrm{Gal}(\overline{\mathbb{Q}}/\mathbb{Q}) \to \widehat{\mathbb{Z}}^\times$ is the adelic cyclotomic character. Indeed, passing from $\mathbb{Q}$-ideles to fractional $\mathbb{Q}$-ideals gives the result, provided that $\chi(\sigma)\mathrm{N}_{E/\mathbb{Q}}(s) \in \mathbb{Q}_{>0}^\times$ inside $\mathbb{A}_f^\times$. By functoriality of the Artin map with respect to the extension $\mathbb{Q} \to E$, the image of $\mathrm{N}_{E/\mathbb{Q}}(s)$ in $\mathrm{Gal}(\mathbb{Q}^{\mathrm{ab}}/\mathbb{Q})$ is $\sigma|_{\mathbb{Q}^{\mathrm{ab}}}$. Since our local Artin map convention implies that the global Artin map for $\mathbb{Q}$ identifies $\mathrm{Gal}(\mathbb{Q}^{\mathrm{ab}}/\mathbb{Q})$ with $\widehat{\mathbb{Z}}^{\times} = \mathbb{Q}^{\times}\backslash\mathbb{A}_{\mathbb{Q}}^{\times}/\mathbb{R}_{>0}^{\times} = \mathbb{Q}_{>0}^{\times}\backslash\mathbb{A}_{\mathbb{Q},f}^{\times}$ via the *reciprocal* of the cyclotomic character, $\mathrm{Nm}_{E/\mathbb{Q}}(s)$ and $\chi(\sigma)^{-1}$ represent the same coset in $\mathbb{Q}_{>0}^{\times}\backslash\mathbb{A}_{\mathbb{Q},f}^{\times}$, as desired.

A.2.5.3. Remark. By replacing s with cs for $c \in E^{\times}$, we get via uniqueness that $\lambda_{\sigma,cs} = \mathrm{N}_{\Phi}(c)\lambda_{\sigma,s}$ since $\mathrm{N}_{\Phi}(cs) = \mathrm{N}_{\Phi}(c)\mathrm{N}_{\Phi}(s)$ with $\mathrm{N}_{\Phi}(c) \in T(\mathbb{Q}) \subseteq L^{\times}$.

Let $K \subset \overline{\mathbb{Q}}$ be a number field that is a field of definition for (A, i), so $E \subseteq K$. We let (A_0, i_0) be a descent of (A, i) to K. The ℓ-adic representations attached to A_0 (for all primes ℓ) define a continuous group homomorphism

$$\rho : \mathrm{Gal}(\overline{\mathbb{Q}}/K) \to \mathrm{Aut}_{\mathbb{A}_{L,f}}(V_f(A_0)) \simeq \mathbb{A}_{L,f}^{\times},$$

so this uniquely factors through a continuous homomorphism $\rho^{\mathrm{ab}} : \mathrm{Gal}(K^{\mathrm{ab}}/K) \to \mathbb{A}_{L,f}^{\times}$. Pick $s' \in \mathbb{A}_{K,f}^{\times}$, so the element $r_K(s') \in \mathrm{Gal}(K^{\mathrm{ab}}/K)$ acts on $V_f(A_0)$ via $\rho^{\mathrm{ab}}(r_K(s'))$. We want to describe this action:

A.2.5.4. Theorem. *For $s' \in \mathbb{A}_{K,f}^{\times}$ there is a unique $\lambda_{s'} \in L^{\times}$ such that*

$$\rho^{\mathrm{ab}}(r_K(s')) = \mathrm{N}_{\Phi}(\mathrm{Nm}_{K/E}(s'))^{-1} \cdot \lambda_{s'}$$

in $\mathbb{A}_{L,f}^{\times}$ Moreover, $s' \mapsto \lambda_{s'} \in L^{\times}$ is continuous for the discrete topology on $L^{\times}$.

PROOF. By functoriality of the Artin map, the restriction of $r_K(s')$ to $E^{\mathrm{ab}} \subseteq K^{\mathrm{ab}}$ is $r_E(\mathrm{Nm}_{K/E}(s'))$. Choose $\sigma' \in \mathrm{Gal}(\overline{\mathbb{Q}}/K) \subseteq \mathrm{Gal}(\overline{\mathbb{Q}}/E)$ lifting $r_K(s')$, so $\sigma'|_{E^{\mathrm{ab}}} = r_E(s)$ for $s = \mathrm{Nm}_{K/E}(s')$. Using the K-structure on (A, i) provided by the identification $A \simeq (A_0)_{\overline{\mathbb{Q}}}$, we get a canonical L-linear isomorphism $A^{\sigma'} \simeq A$ and hence by A.2.5.1 an L-linear isogeny $\lambda_{\sigma',s} : A \to A^{\sigma'} \simeq A$ such that $\rho(\sigma') = \mathrm{N}_{\Phi}(s)^{-1} \cdot V_f(\lambda_{\sigma',s'})$ as endomorphisms of $V_f(A)$. Since $\rho(\sigma') = \rho^{\mathrm{ab}}(r_K(s'))$, we see that $\lambda_{\sigma',s'}$ only depends on s' and not on σ'. But L is its own centralizer in $\mathrm{End}^0(A)$, so the L-linear $\lambda_{\sigma',s'}$ is multiplication by an element $\lambda_{s'} \in L^{\times}$. This completes the construction of $\lambda_{s'} \in L^{\times}$ such that $\rho^{\mathrm{ab}}(r_K(s')) = \mathrm{N}_{\Phi}(\mathrm{Nm}_{K/E}(s))^{-1} \cdot \lambda_{s'}$ in $\mathbb{A}_{L,f}^{\times}$.

It remains to prove continuity of $s' \mapsto \lambda_{s'}$ with respect to the discrete topology on $L^{\times}$. That is, we want $\lambda_{s'} = 1$ for $s' \in \mathbb{A}_{K,f}^{\times}$ sufficiently near 1. Since

$$\lambda_{s'} = \rho^{\mathrm{ab}}(r_K(s'))\mathrm{N}_{\Phi}(\mathrm{N}_{K/E}(s')) \in \mathbb{A}_{L,f}^{\times},$$

the mapping $s' \mapsto \lambda_{s'} \in L^{\times}$ is continuous for the topology on $L^{\times}$ induced by its inclusion into the group $\mathbb{A}_{L,f}^{\times}$ of finite L-ideles. In particular, for s' sufficiently near 1 in $\mathbb{A}_{K,f}^{\times}$ and a fixed choice of positive integer $M \geqslant 3$ we have several properties: $\lambda_{s'} \in \mathcal{O}_L^{\times}$, $\lambda_{s'} \equiv 1 \pmod{M}$, and $\lambda_{s'}$ lies in the CM order $\mathcal{O}_L \cap \mathrm{End}(A)$. Hence, for such s' the element $\lambda_{s'}$ is an automorphism of the abelian variety A acting trivially on the M-torsion.

If we can prove that $\lambda_{s'}$ has finite order then since $M \geqslant 3$ such an automorphism must be the identity and so we will be done. Pick an L-linear $\mathbb{Q}$-polarization ϕ for (A, i) (over $\overline{\mathbb{Q}}$). Taking $s' \in \mathbb{A}_{K,f}^{\times}$ sufficiently near 1, we can arrange that the element $r_K(s') \in \mathrm{Gal}(K^{\mathrm{ab}}/K)$ has a lift to $\mathrm{Gal}(\overline{\mathbb{Q}}/K)$ that acts trivially on a number field of definition for ϕ over K (after applying a base change to (A_0, i_0)). It therefore follows from the construction of $\lambda_{s'}$ that for such s' the element $\lambda_{s'} \in L^{\times}$ acts as an automorphism of A (not merely in the isogeny category) preserving ϕ up to a positive rational multiple. Degree considerations force this positive rational multiplier to be 1, and the automorphism group of a $\mathbb{Q}$-polarized abelian variety is *finite*. Hence, $\lambda_{s'}$ has finite order as desired. $\square$

A.2.5.5. Remark. Since the global Artin map $\mathbb{A}_K^\times \to \mathrm{Gal}(K^{\mathrm{ab}}/K)$ kills $K^\times$, by Theorem A.2.5.4 we get an algebraic Hecke character $\mathbb{A}_K^\times \to L^\times$ in the sense of Definition 2.4.3(b) via $s \mapsto \lambda_{s_f}$ (trivial on $K_\infty^\times$). Its algebraic part is $\mathrm{N}_\Phi \circ \mathrm{Nm}_{K/E}$.

We next use the Main Theorem (or rather, its consequence in Theorem A.2.5.4) to deduce some results concerning L-functions. First, we eliminate the intervention of $\overline{\mathbb{Q}}$ by working with an "abstract" number field as the base field, as follows.

Let A_0 be an abelian variety of dimension g over a number field K, and let $i_0 : L \to \mathrm{End}^0(A_0)$ be a CM structure on A_0. The extension $K((A_0)_{\mathrm{tor}})/K$ is abelian, so it uniquely embeds into K^{ab} over K. The field K has no real places because it contains a CM field. (Upon embedding K into $\overline{\mathbb{Q}}$ to define a $\overline{\mathbb{Q}}$-valued CM type Φ for (A_0, i_0), K contains the reflex field E for (L, Φ), and E is a CM field.) Hence, the group $K_\infty^\times = (\mathbb{R} \otimes_{\mathbb{Q}} K)^\times$ is connected and infinitely divisible.

The associated CM *character* is the map

$$\alpha_{(A_0,i_0)/K} : \mathbb{A}_K^\times \twoheadrightarrow \mathrm{Gal}(K^{\mathrm{ab}}/K) \to \mathbb{A}_{L,f}^\times$$

defined by the Galois action on torsion of A. The image of $K_\infty^\times$ in $\mathrm{Gal}(K^{\mathrm{ab}}/K)$ is trivial, so $\alpha_{(A_0,i_0)/K}(s)$ only depends on the finite component $s_f \in \mathbb{A}_{K,f}^\times$ for any $s \in \mathbb{A}_K^\times$, and it is invariant under $K^\times$-scaling of s.

The reflex norm

$$\mathrm{N}_\Phi : \mathrm{Res}_{E/\mathbb{Q}}(\mathbb{G}_{\mathrm{m}}) \to T \subseteq \mathrm{Res}_{L/\mathbb{Q}}(\mathbb{G}_{\mathrm{m}})$$

was defined in Definition 2.1.3.3 (and lands in T by Proposition 2.1.3.6). The specification of the CM type Φ and the reflex subfield $E \subseteq K$ depend on a choice of embedding of K into $\overline{\mathbb{Q}}$, but we have:

A.2.5.6. Lemma. *The composite $\mathbb{Q}$-homomorphism*

$$\mathrm{N}_{\Phi_K} : \mathrm{Res}_{K/\mathbb{Q}}(\mathbb{G}_{\mathrm{m}}) \overset{\mathrm{Nm}_{K/E}}{\to} \mathrm{Res}_{E/\mathbb{Q}}(\mathbb{G}_{\mathrm{m}}) \overset{\mathrm{N}_\Phi}{\to} \mathrm{Res}_{L/\mathbb{Q}}(\mathbb{G}_{\mathrm{m}})$$

is independent of the embedding of K into $\overline{\mathbb{Q}}$.

This lemma will be especially useful when applied to points valued in $\mathbb{A}_f$.

PROOF. By the Zariski-density of the $\mathbb{Q}$-points in any $\mathbb{Q}$-torus (due to the unirationality of tori over any field), it suffices to check that the composite map on $\mathbb{Q}$-points is independent of the $\overline{\mathbb{Q}}$-embedding of K. Thus, it suffices to prove that $\mathrm{N}_{\tau\Phi}(\tau x) = \mathrm{N}_\Phi(x)$ in $K^\times$ for any $x \in E^\times$ and $\tau \in \mathrm{Gal}(\overline{\mathbb{Q}}/E)$. By definition of the reflex norm in 2.1.3.3, it is equivalent that

$$\det{}_L(\tau(x) : t_{\tau\Phi} \simeq t_{\tau\Phi}) \overset{?}{=} \det{}_L(x : t_\Phi \simeq t_\Phi)$$

in L for all $x \in E$, where t_Φ is defined (as an $L \otimes_{\mathbb{Q}} E$-module) in the discussion preceding 2.1.3.3.

The natural map $\langle\tau\rangle : \prod_{\varphi\in\Phi} \overline{\mathbb{Q}} \simeq \prod_{\psi\in\tau\Phi} \overline{\mathbb{Q}}$ from $t_\Phi \otimes_E \overline{\mathbb{Q}}$ to $t_{\tau\Phi} \otimes_{\tau(E)} \overline{\mathbb{Q}}$ defined by $(x_\varphi) \mapsto (\tau(x_{\tau^{-1}\psi}))$ carries the "descent datum" action $(x_\varphi) \mapsto (g(x_{g^{-1}\varphi}))$ by

$g \in \mathrm{Gal}(\overline{\mathbb{Q}}/E)$ over to the action by $\tau g \tau^{-1} \in \mathrm{Gal}(\overline{\mathbb{Q}}/\tau(E))$ and respects the L-actions (but carries multiplication by $y \in E$ over to multiplication by $\tau(y) \in \tau(E)$). Thus, the isomorphism $\langle \tau \rangle$ descends to an isomorphism $t_\Phi \simeq t_{\tau\Phi}$ linear over the ring isomorphism $1 \otimes \tau : L \otimes_{\mathbb{Q}} E \simeq L \otimes_{\mathbb{Q}} \tau(E)$ (viewing E and $\tau(E)$ as subfields of $\overline{\mathbb{Q}}$). The equality of L-determinants for the x-action on t_Φ and the $\tau(x)$-action on $t_{\tau\Phi}$ therefore holds for any $x \in E$. □

A.2.5.7. Remark. By Theorem A.2.5.4, for each $s \in \mathbb{A}_K^\times$ there is a unique $\lambda_s \in L^\times$ depending only on s_f such that

$$\alpha_{(A_0,i_0)/K}(s) = \lambda_s \mathrm{N}_{\Phi_K}(s_f)^{-1},$$

and we have $\lambda_s = 1$ if s_f is sufficiently close to 1. Since $\alpha_{(A_0,i_0)/K}$ kills $K^\times$, it also follows that for $c \in K^\times$, $\lambda_c = \mathrm{N}_{\Phi_K}(c)$ in $L^\times$. Consideration of quadratic twists shows that $\alpha_{(A_0,i_0)/K}$ generally depends on (A_0, i_0) over K and not just on its $\overline{\mathbb{Q}}$-fiber (i.e., not just on (L, Φ)).

A.2.5.8. Remark. Two elementary properties of $s \mapsto \lambda_s$ are (i) for $s \in \mathbb{A}_K^\times$, the fractional L-ideal $[\mathrm{N}_{\Phi_K}(s_f)]_L$ associated to the idele $\mathrm{N}_{\Phi_K}(s_f) \in \mathbb{A}_{L,f}^\times$ is $\lambda_s \mathcal{O}_L$, and (ii) $\lambda_s \lambda_s^* = \mathrm{Nm}_{K/\mathbb{Q}}(s_f) \alpha_{(A,i)/K}(s) \alpha^*_{(A,i)/K}(s)$ in $\mathbb{A}_f^\times \subset \mathbb{A}_{L,f}^\times$. Identity (ii) follows from picking an embedding of K into $\overline{\mathbb{Q}}$ (so as to get a $\overline{\mathbb{Q}}$-valued CM type Φ on L from $A_0 \otimes_K \overline{\mathbb{Q}}$) and using the identity $\mathrm{N}_\Phi \cdot \mathrm{N}_{\Phi^*} = \mathrm{Nm}_{E/\mathbb{Q}}$ (see the proof of Proposition 2.1.3.6) and the transitivity of norms. To prove (i), first note that the automorphism $V_f(\lambda_s \cdot \mathrm{N}_{\Phi_K}(s_f^{-1})) = [\sigma]$ of $V_f(A_0)$ restricts to an automorphism of the adelic Tate module $T_f(A_0)$ and hence an automorphism of the free rank-1 module generated by $T_f(A_0)$ over the ring $\prod_{v \nmid \infty} \mathcal{O}_{L_v}$ of integral L-adeles. Hence, the fractional L-ideals $\lambda_s \mathcal{O}_L$ and $[\mathrm{N}_{\Phi_K}(s_f)]_L$ have the same order at all finite places of L, so these ideals coincide.

A.2.5.9. Theorem. *Let (A_0, i_0) be a* CM *abelian variety over a number field $K \subset \overline{\mathbb{Q}}$, with $i_0 : L \to \mathrm{End}^0(A_0)$ for a* CM *field L. Let $\lambda : \mathbb{A}_K^\times \to L^\times$ be the character $s \mapsto \lambda_{s_f}$ that is trivial on $K_\infty^\times$ and continuous for the discrete topology on $L^\times$. Pick a prime $\mathfrak{P}$ of K.*

(1) *The abelian variety A_0 has good reduction at $\mathfrak{P}$ if and only if $\lambda_{\mathfrak{P}} := \lambda|_{K_{\mathfrak{P}}^\times}$ is trivial on $\mathcal{O}_{K_{\mathfrak{P}}}^\times$.*
(2) *For a prime $\mathfrak{P}$ of good reduction and a uniformizer $\pi_{\mathfrak{P}}$ at $\mathfrak{P}$, the element $\lambda_{\mathfrak{P}}(\pi_{\mathfrak{P}}) \in L^\times$ lies in $\mathcal{O}_L$ and in the isogeny category over $\kappa(\mathfrak{P})$ it acts on the reduction of A_0 over $\kappa(\mathfrak{P})$ as the $q_{\mathfrak{P}}$-Frobenius endomorphism.*

PROOF. To check good reduction at $\mathfrak{P}$, we choose a rational prime ℓ distinct from the residue characteristic of $\mathfrak{P}$ and we need to determine if the action of an inertia group $I_{\mathfrak{P}}$ at $\mathfrak{P}$ is trivial on $V_\ell(A_0)$. The image of $I_{\mathfrak{P}}$ in $\mathrm{Gal}(K^{\mathrm{ab}}/K)$ is the image of $\mathcal{O}_{K_{\mathfrak{P}}}^\times$ under the Artin map, so it comes from ideles $s \in \mathbb{A}_K^\times$ with trivial ℓ-part. Hence, the formula

$$\alpha_{(A_0,i_0)/K}(s) = \lambda_s \mathrm{N}_{\Phi_K}(s_f)^{-1} \tag{A.2.5.1}$$

implies that $\alpha_{(A_0,i_0)/K}(s)$ has trivial ℓ-part for all $s \in \mathcal{O}_{K_{\mathfrak{P}}}^\times$ if and only if the value $\lambda_s \in L^\times$ viewed in $L_\ell^\times$ is trivial for all $s \in \mathcal{O}_{K_{\mathfrak{P}}}^\times$. That is, A_0 has good reduction at $\mathfrak{P}$ if and only if $\lambda_{\mathfrak{P}}$ is trivial on $\mathcal{O}_{K_{\mathfrak{P}}}^\times$.

Now choose $\mathfrak{P}$ at which there is good reduction, and pick ℓ as above. The preceding calculation shows that for a local uniformizer $\pi_{\mathfrak{P}}$, the arithmetic Frobenius element at $\mathfrak{P}$ acts on $V_\ell(A_0)$ via the action of $\lambda_{\mathfrak{P}}(\pi_{\mathfrak{P}}) \in L^\times$. Passing to the reduction $\overline{A}_0$ at $\mathfrak{P}$, it follows that the action by $\lambda_{\mathfrak{P}}(\pi_{\mathfrak{P}})$ on $V_\ell(\overline{A}_0)$ agrees with the action by the $q_{\mathfrak{P}}$-Frobenius endomorphism, so in $\mathrm{End}^0(\overline{A}_0)$ the element $\lambda_{\mathfrak{P}}(\pi_{\mathfrak{P}}) \in L^\times$ coincides with this Frobenius endomorphism. In particular, $\lambda_{\mathfrak{P}}(\pi_{\mathfrak{P}})$ acts as a genuine endomorphism of $\overline{A}_0$ (not just in the isogeny category) and so is integral over $\mathbb{Z}$. Thus, as an element of L it lies in $\mathcal{O}_L$. □

For any embedding $\tau : L \hookrightarrow \mathbb{C}^\times$, let $\lambda^\tau = \tau \circ \lambda$ with λ as in Theorem A.2.5.9. Define $\alpha_\infty = \mathrm{N}_{\Phi_K,\infty}^{-1} \cdot \lambda : \mathbb{A}_K^\times \to L_\infty^\times$, where $\mathrm{N}_{\Phi_K,\infty}$ is the composition of the map $\mathrm{N}_{\Phi_K} : \mathbb{A}_K^\times \to \mathbb{A}_L^\times$ from Lemma A.2.5.6 and the projection $\mathbb{A}_L^\times \twoheadrightarrow L_\infty^\times$. In particular, $\mathrm{N}_{\Phi_K,\infty}$ kills $\mathbb{A}_{K,f}^\times$. Clearly α_∞ is continuous, and it kills $K^\times$ due to Remark A.2.5.7. Hence, for each $\tau : L \hookrightarrow' \mathbb{C}$ and the corresponding archimedean place v of L, the composite map

$$\alpha^\tau : \mathbb{A}_K^\times \overset{\alpha_\infty}{\to} L_\infty^\times \twoheadrightarrow L_v^\times \overset{\tau}{\simeq} \mathbb{C}^\times$$

is a Hecke character and the restrictions $\alpha_{\mathfrak{P}}^\tau$ and $\lambda_{\mathfrak{P}}^\tau$ of α^τ and λ^τ to $K_{\mathfrak{P}}^\times \subset \mathbb{A}_K^\times$ coincide for all primes $\mathfrak{P}$ of K.

A.2.5.10. Remark. Since $\lambda|_{K^\times} = \mathrm{N}_{\Phi_K}$ (see Remark A.2.5.8), the restriction of the Hecke character α^τ to $K_\infty^\times$ is the continuous extension of the map

$$K^\times \overset{\mathrm{N}_{\Phi_K}}{\to} L^\times \overset{\tau}{\to} \mathbb{C}^\times$$

that is algebraic with respect to $\mathbb{Q}$ in the sense of Definition 2.4.1. More specifically, the $\mathbb{C}^\times$-valued α^τ (or, better, the $L^\times$-valued λ) is an algebraic Hecke character, as we noted in Remark A.2.5.5. Let us make this explicit by unraveling the proof of the equivalence of Definitions 2.4.3(a) and 2.4.3(b) in this case: we will compute $[\alpha^\tau](x\mathcal{O}_K)$ as an "algebraic" function of $x \in K^\times$ (viewed as the Zariski-dense locus of $\mathbb{Q}$-points in the $\mathbb{Q}$-variety $\underline{K^\times}$), where $[\alpha^\tau] : I_{\mathfrak{f}} \to \mathbb{C}^\times$ is the homomorphism (defined below) arising from α^τ on the group $I_{\mathfrak{f}}$ of fractional ideals relatively prime to a suitable non-zero ideal $\mathfrak{f}$ of $\mathcal{O}_K$.

Let $\mathfrak{f}$ be an non-zero ideal of $\mathcal{O}_K$ such that $\lambda : \mathbb{A}_K^\times \to L^\times$ is trivial on the open subgroup $U_{\mathfrak{f}} \subset \mathbb{A}_{K,f}^\times$ of integral finite ideles congruent to 1 modulo $\mathfrak{f}$, so the group $I_{\mathfrak{f}}$ of fractional ideals of K relatively prime to $\mathfrak{f}$ is naturally a quotient of $\mathbb{A}_{K,f}^\times/U_{\mathfrak{f}}$. Hence, α^τ induces a well-defined homomorphism $[\alpha^\tau] : I_{\mathfrak{f}} \to \tau(L)^\times \subset \mathbb{C}^\times$ that sends a prime $\mathfrak{P}$ to $\alpha_{\mathfrak{P}}^\tau(\pi_{\mathfrak{P}}) = \tau(\lambda_{\mathfrak{P}}(\pi_{\mathfrak{P}}))$ for any local uniformizer $\pi_{\mathfrak{P}}$ at $\mathfrak{P}$. On the subgroup $P_{\mathfrak{f}} \subseteq I_{\mathfrak{f}}$ of principal fractional ideals of the form $x\mathcal{O}_K$ with $x \in K^\times$ satisfying $x \equiv 1 \pmod{\mathfrak{f}}$, we have

$$[\alpha^\tau](x\mathcal{O}_K) = \alpha^\tau(x_f) = \alpha^\tau(x_\infty^{-1}) = \tau(\mathrm{N}_{\Phi_K,\infty}(x)) = \tau(\mathrm{N}_\Phi(\mathrm{Nm}_{K/E}(x)))$$

where $E \subset K$ be the reflex field of (L, Φ).

For each embedding $\sigma : E \hookrightarrow \overline{\mathbb{Q}}$, pick $\widetilde{\sigma} \in \mathrm{Gal}(\overline{\mathbb{Q}}/\mathbb{Q})$ lifting σ via the canonical inclusion of E into $\overline{\mathbb{Q}}$ given in the definition of the reflex field E. Let $\Sigma(\Phi, \tau)$ be the set of $\sigma \in \mathrm{Hom}(E, \overline{\mathbb{Q}})$ such that $\tau = \widetilde{\sigma} \circ \varphi$ for some (necessarily unique) $\varphi \in \Phi$.

For $x \equiv 1 \pmod{\mathfrak{f}}$ we have $x\mathcal{O}_K \in P_{\mathfrak{f}}$ and

$$[\alpha^\tau](x\mathcal{O}_K) = \tau(\mathrm{N}_{\Phi_K,\infty}(x)) = \prod_{\sigma \in \Sigma(\Phi,\tau)} \sigma(\mathrm{N}_{K/E}(x)) = \prod_{\psi: K \to \overline{\mathbb{Q}}, \psi|_E \in \Sigma(\Phi,\tau)} \psi(x),$$

where the second equality follows from the general identity $\tau \circ \mathrm{N}_\Phi = \prod_{\sigma \in \Sigma(\Phi,\tau)} \sigma$ as homomorphisms $E^\times \to \tau(L)^\times \subset \overline{\mathbb{Q}}^\times$ (an instance of (2.1.3.2) with τ serving in the role of j there). In particular, this shows that $\Sigma(\Phi,\tau)$ is independent of the choices of the lifts $\widetilde{\sigma}$, so if we define $\sigma^*(x) = \sigma(x^*)$ for all σ (with $x \in E$) then the choices of lifts can be made so that for every σ we have $\widetilde{\sigma^*} = c_\sigma \circ \widetilde{\sigma}$ for some $c_\sigma \in \mathrm{Gal}(\overline{\mathbb{Q}}/\mathbb{Q})$ of order 2. Hence, $\Sigma(\Phi,\tau)^* = \Sigma(\Phi^*,\tau)$, so the set Ψ_τ of ψ's in the final product formula for $[\alpha^\tau](x\mathcal{O}_K)$ is a $\overline{\mathbb{Q}}$-valued CM type on K.

For each τ, Theorem A.2.5.9(1) says that A_0 over K has good reduction at $\mathfrak{P}$ if and only if α^τ is unramified at $\mathfrak{P}$. By Theorem A.2.5.9(2) and the Riemann Hypothesis for abelian varieties over finite fields, the Hecke character $\|\cdot\|_K^{-1/2}\alpha^\tau$ takes values in the unit circle. This unitary character is non-trivial, for otherwise by working with degree-1 primes of K it would follow from the Riemann Hypothesis for abelian varieties that L contains square roots of infinitely many odd rational primes, an absurdity. (It can happen for CM elliptic curves that this non-trivial unitary character has finite order, and even order 2.) Hence, the Euler products that define each $\mathrm{L}(s,\alpha^\tau)$ are absolutely and uniformly convergent in half-planes $\mathrm{Re}(s) \geqslant 3/2 + \varepsilon$ for all $\varepsilon > 0$, and extend to holomorphic functions on $\mathbb{C}$. These Hecke L-functions compute the L-function of A_0:

A.2.5.11. Theorem. *For $s \in \mathbb{C}$ with* $\mathrm{Re}(s) > 3/2$,

$$\mathrm{L}(s, A_0) = \prod_{\tau: L \to \mathbb{C}} \mathrm{L}(s,\alpha^\tau).$$

In particular, the L*-function for A_0 has an analytic continuation to $\mathbb{C}$.*

Note that in the product in the theorem, we do repeat conjugate pairs of embeddings.

PROOF. We compare Euler factors at good and bad primes separately. Let $\mathfrak{P}$ be a prime of good reduction for A_0, and pick ℓ distinct from the residue characteristic of $\mathfrak{P}$. Let $\lambda(\mathfrak{P}) \in L^\times$ be the common value of λ on local ideles coming from uniformizers at $\mathfrak{P}$, and likewise for λ^τ. By Theorem A.2.5.9, we get

$$\begin{aligned}\det_{\mathbb{Q}_\ell}(1 - \lambda(\mathfrak{P})X) = \mathrm{N}_{L_\ell/\mathbb{Q}_\ell}(1 - \lambda(\mathfrak{P})X) &= \mathrm{N}_{L/\mathbb{Q}}(1 - \lambda(\mathfrak{P})X) \\ &= \mathrm{N}_{\mathbb{C}\otimes_{\mathbb{Q}} L/\mathbb{C}}(1 - (1 \otimes \lambda(\mathfrak{P}))X),\end{aligned}$$

and under the natural decomposition $\mathbb{C} \otimes_{\mathbb{Q}} L \simeq \prod_\tau \mathbb{C}$ we see that this norm is $\prod_\tau (1 - \lambda^\tau(\mathfrak{P})X)$. For $X = q_{\mathfrak{P}}^{-s}$ this yields the product of the Euler factors at $\mathfrak{P}$ for the right side of the proposed identity because α^τ and λ^τ have the same value on any idele with trivial archimedean component.

Now we turn to the bad primes for A_0. At such primes we know by Theorem A.2.5.9(1) that λ is non-trivial on $\mathcal{O}_{K_{\mathfrak{P}}}^\times$ and hence all Hecke characters α^τ are ramified at $\mathfrak{P}$. Thus, the Euler factor at $\mathfrak{P}$ for the L-function of each α^τ is trivial. We therefore have to prove that for ℓ as above, the abelian ℓ-adic representation $V_\ell(A_0)$ for $\mathrm{Gal}(\overline{\mathbb{Q}}/K)$ has vanishing subspace of inertial invariants at $\mathfrak{P}$. Since

$\lambda(\mathcal{O}_{K_{\mathfrak{P}}}^{\times}) \subset L^{\times}$ is non-trivial, it contains some element $x \in L^{\times} - \{1\}$. But $\lambda(\mathcal{O}_{K_{\mathfrak{P}}}^{\times})$ viewed in $L_{\ell}^{\times}$ is the image of the ℓ-part of $\alpha_{(A_0,i_0)/K}$ on $\mathcal{O}_{K_{\mathfrak{P}}}^{\times}$ (due to (A.2.5.1)), so the subspace of inertial invariants at $\mathfrak{P}$ is contained in the subspace of $V_{\ell}(A_0)$ killed by $x - 1 \in L^{\times}$. Hence, this subspace is 0. □

A.2.5.12. Remark. Theorem A.2.5.11 can be generalized to L-functions of abelian varieties that are merely potentially CM (i.e., acquire a CM structure over a finite extension of the base field). Note that if there is a CM structure over the base field then a CM reflex field lies in the base field, so a potentially CM abelian variety over a real number field (such as $\mathbb{Q}$) cannot have a CM structure over the base field.

Here is how $\lambda = \lambda_{(A_0,i_0)/K}$ behaves with respect to the action by $\mathrm{Aut}(L/\mathbb{Q})$.

A.2.5.13. Theorem. *For $\gamma \in \mathrm{Aut}(L/\mathbb{Q})$, $\lambda_{(A_0,i_0\circ\gamma^{-1})/K} = \gamma \circ \lambda_{(A_0,i_0)/K}$.*

PROOF. Pick an embedding of K into $\overline{\mathbb{Q}}$, so we get a $\overline{\mathbb{Q}}$-valued CM type Φ for (A_0, i_0) and a reflex field $E \subseteq K$. In this way, we canonically identify K^{ab} with a subfield of $\overline{\mathbb{Q}}$ over K. Pick $s \in \mathbb{A}_K^{\times}$ and $\sigma \in \mathrm{Gal}(\overline{\mathbb{Q}}/K)$ with $\sigma|_{K^{\mathrm{ab}}} = r_K(s)$. We need to prove

$$V_f(\lambda_{(A_0,i_0\circ\gamma^{-1})/K}(s)) \stackrel{?}{=} V_f(\gamma(\lambda_{(A_0,i_0)/K}(s))) = V_f(\gamma)(V_f(\lambda_{(A_0,i_0)/K}(s)))$$

in $\mathbb{A}_{L,f}^{\times}$ when the left and right sides are respectively viewed as endomorphisms of $V_f(A_0)$ with $\mathbb{A}_{L,f}$-module structures defined by $i_0 \circ \gamma^{-1}$ and i_0 respectively.

The $\overline{\mathbb{Q}}$-valued CM type for $(A_0, i_0 \circ \gamma^{-1})$ with respect to the chosen embedding $K \hookrightarrow \overline{\mathbb{Q}}$ defining Φ is $\Phi \circ \gamma^{-1}$. In other words (using the definition of λ via A.2.5.4) for $\mathrm{N}_{\Phi_K} := \mathrm{N}_{\Phi} \circ \mathrm{Nm}_{K/E}$ and $\mathrm{N}_{(\Phi\circ\gamma^{-1})_K} := \mathrm{N}_{\Phi\circ\gamma^{-1}} \circ \mathrm{Nm}_{K/E}$ we need to prove

$$\begin{aligned} \mathrm{N}_{(\Phi\circ\gamma^{-1})_K}(s) \cdot V_f(i_0 \circ \gamma^{-1})^{-1}([\sigma]) &\stackrel{?}{=} V_f(\gamma)(\mathrm{N}_{\Phi_K}(s) \cdot V_f(i_0)^{-1}([\sigma])) \\ &= V_f(\gamma)(\mathrm{N}_{\Phi_K}(s)) \cdot V_f(\gamma)(V_f(i_0)^{-1}([\sigma])) \end{aligned}$$

(where $V_f(i_0) : \mathbb{A}_{L,f} \simeq \mathrm{End}_{\mathbb{A}_{L,f}}(V_f(A_0))$ is the natural isomorphism). The second factors on the left and right sides clearly agree, so we just have to show

$$\gamma^{-1}(\mathrm{N}_{(\Phi\circ\gamma^{-1})_K}(s)) = \mathrm{N}_{\Phi}(s)$$

in $\mathbb{A}_{L,f}^{\times}$ for all $s \in \mathbb{A}_K^{\times}$. It suffices to prove the isomorphism $\gamma^{-1} : \mathbb{A}_L^{\times} \simeq \mathbb{A}_L^{\times}$ carries $\mathrm{N}_{\Phi\circ\gamma^{-1}}$ to N_{Φ} as maps from $\mathbb{A}_E^{\times}$.

In view of the algebraic definition of the reflex norm as an L-determinant, the problem reduces to constructing an $L \otimes_{\mathbb{Q}} E$-module isomorphism $L \otimes_{\gamma^{-1},L} t_{\Phi\circ\gamma^{-1}} \simeq t_{\Phi}$. We apply the extension of scalars $E \to \overline{\mathbb{Q}}$ and just have to compare dimensions of isotypic factors for the L-action. By definition via descent, the $L \otimes_{\mathbb{Q}} \overline{\mathbb{Q}}$-module

$$L \otimes_{\gamma^{-1},L} (t_{\Phi\circ\gamma^{-1}} \otimes_E \overline{\mathbb{Q}}) \simeq L \otimes_{\gamma^{-1},L} \prod_{\psi\in\Phi\circ\gamma^{-1}} \overline{\mathbb{Q}}$$

is $\overline{\mathbb{Q}}$-linearly isomorphic to a product of copies of $\overline{\mathbb{Q}}$ indexed by the set of elements $\psi \in \Phi \circ \gamma^{-1}$ such that L acts on the ψ-th copy via $\psi \circ \gamma$. Hence, the isotypic pieces are 1-dimensional over $\overline{\mathbb{Q}}$ and the eigencharacters are the elements of the set $(\Phi \circ \gamma^{-1}) \circ \gamma = \Phi$. □

Note that the definition of $\lambda_{(A_0,i_0)/K}$ involves the $\overline{\mathbb{Q}}$-valued CM type Φ on L, which depends on a choice of embedding $K \hookrightarrow \overline{\mathbb{Q}}$. In some references one finds a discussion of the behavior of $\lambda_{(A_0,i_0)/K}$ behavior with respect to the action by $\mathrm{Gal}(\overline{\mathbb{Q}}/\mathbb{Q})$, namely that $\lambda_{(A_0^\tau,i_0^\tau)/\tau(K)}(s^\tau) = \lambda_{(A_0,i_0)/K}(s)$. This says exactly that the character $\lambda_{(A_0,i_0)/K} : \mathbb{A}_K^\times \to L^\times$ is independent of the initial embedding of K into $\overline{\mathbb{Q}}$ used to define Φ, which in turn is a consequence of Lemma A.2.5.6 (due to Theorem A.2.5.4).

A.2.6. Hom-modules and fractional ideals. We now begin proving the algebraic form of the Main Theorem (i.e., Theorem A.2.4.4). By Remark A.2.3.7 the choice of L-linear $\mathbb{Q}$-polarization ϕ for (A,i) does not matter, and any two pairs (A,i) and (A',i') of type (L,Φ) over $\overline{\mathbb{Q}}$ are L-linearly isogenous, so in fact the choice of triple (A,i,ϕ) of type (L,Φ) does not matter. Thus, by the final part of Proposition 1.7.4.5, we may and do take (A,i) to be principal (i.e., $i^{-1}(\mathrm{End}(A)) = \mathcal{O}_L$); cf. Remark 2.5.1.1.

Unlike adelic multiplication operations in the isogeny category that we shall develop in A.2.8 when the CM order is not necessarily maximal, the Serre tensor construction from 1.7.4 is applicable at torsion levels in the principal case. That is, torsion-free modules over $\mathcal{O}_L$ are projective, whereas over proper orders in $\mathcal{O}_L$ it is much harder to control projectivity properties. This is why we have passed to the principal case.

Pick $\sigma \in \mathrm{Gal}(\overline{\mathbb{Q}}/E)$. Choose a finite Galois extension K/E inside $\overline{\mathbb{Q}}$ such that (A,i) descends to a pair (A_0,i_0) over K and such that the L-linear $\mathbb{Q}$-polarization ϕ of A descends to an L-linear $\mathbb{Q}$-polarization ϕ_0 of A_0. (We will have no need for ϕ_0 until after (A.2.7.3), so the reader may forget about it until then.) Make K big enough so that it splits L over $\mathbb{Q}$ (i.e., it contains the Galois closure of L in $\overline{\mathbb{Q}}$) and

$$\mathrm{Hom}((A_0,i_0),(A_0^\sigma,i_0^\sigma)) = \mathrm{Hom}((A,i),(A^\sigma,i^\sigma)). \tag{A.2.6.1}$$

In particular, this module of L-linear mappings over K is an invertible $\mathcal{O}_L$-module. Let $\mathfrak{a}_\sigma$ denote this abstract invertible $\mathcal{O}_L$-module; it will soon be naturally identified with a fractional L-ideal (once we choose a suitable auxiliary prime of K).

Let N be the product of all primes ℓ of $\mathbb{Q}$ that arise as the residue characteristic of a prime factor of $\mathrm{disc}(K/\mathbb{Q})$ or a prime of bad reduction for A_0. (Note that if ℓ is ramified in L then ℓ is ramified in K and hence divides N, as K is assumed to split L over $\mathbb{Q}$.) We will be interested in primes $\mathfrak{P}$ of $\mathcal{O}_K[1/N]$ such that $\sigma|_K = (\frac{K/E}{\mathfrak{P}})$. Before we pick such a $\mathfrak{P}$, we discuss an important feature of working over $\mathbb{Z}[1/N]$: the $L\otimes_{\mathbb{Q}}E$-module t_Φ that is implicit in the definition of the reflex norm (Definition 2.1.3.3) has a natural integral structure over $\mathbb{Z}[1/N]$.

To be precise, K contains all embeddings of L into $\overline{\mathbb{Q}}$ (due to the "largeness" of the choice of K), so t_Φ descends the $L\otimes_{\mathbb{Q}} K$-module $\prod_{\varphi\in\Phi} K$. Since the generically Galois extension $\mathcal{O}_E[1/N] \to \mathcal{O}_K[1/N]$ is finite étale, we may use finite étale descent in place of Galois descent to descend the $\mathcal{O}_L[1/N]\otimes_{\mathbb{Z}[1/N]}\mathcal{O}_K[1/N]$-module $\prod_{\varphi\in\Phi}\mathcal{O}_K[1/N]$ to an $\mathcal{O}_L[1/N]\otimes_{\mathbb{Z}[1/N]}\mathcal{O}_E[1/N]$-module $t_{\Phi,\mathbb{Z}[1/N]}$ such that $\mathbb{Q}\otimes_{\mathbb{Z}[1/N]} t_{\Phi,\mathbb{Z}[1/N]} \simeq t_\Phi$ as $L\otimes_{\mathbb{Q}}E$-modules. In particular, $t_{\Phi,\mathbb{Z}[1/N]}$ is a finite locally free module over each of $\mathcal{O}_L[1/N]$ and $\mathcal{O}_E[1/N]$. Thus, the reflex norm N_Φ between $\mathbb{Q}$-tori extends to a homomorphism between $\mathbb{Z}[1/N]$-tori

$$\mathrm{N}_{\Phi,\mathbb{Z}[1/N]} : \mathrm{Res}_{\mathcal{O}_E[1/N]/\mathbb{Z}[1/N]}(\mathbb{G}_m) \to \mathrm{Res}_{\mathcal{O}_L[1/N]/\mathbb{Z}[1/N]}(\mathbb{G}_m) \tag{A.2.6.2}$$

defined on R-points (for a $\mathbb{Z}[1/N]$-algebra R) by

$$\begin{aligned} x &\mapsto \det_{\mathcal{O}_L[1/N]\otimes_{\mathbb{Z}[1/N]}R}(x : t_{\Phi,\mathbb{Z}[1/N]} \otimes_{\mathbb{Z}[1/N]} R \simeq t_{\Phi,\mathbb{Z}[1/N]} \otimes_{\mathbb{Z}[1/N]} R) \\ &\in (\mathcal{O}_L[1/N] \otimes_{\mathbb{Z}[1/N]} R)^\times \end{aligned}$$

for $x \in (\mathcal{O}_E[1/N] \otimes_{\mathbb{Z}[1/N]} R)^\times$.

[The map $\mathrm{N}_{\Phi,\mathbb{Z}[1/N]}$ factors through the closed subtorus

$$T_{\mathbb{Z}[1/N]} = \ker\Big(\mathrm{Nm}_{\mathcal{O}_L[1/N]/\mathcal{O}_{L^+}[1/N]} : \mathrm{Res}_{\mathcal{O}_L[1/N]/\mathbb{Z}[1/N]}(\mathbb{G}_{\mathrm{m}}) \longrightarrow\!\!\!\!\!\rightarrow \mathrm{Res}_{\mathcal{O}_{L^+}[1/N]/\mathbb{Z}[1/N]}(\mathbb{G}_{\mathrm{m}})/\mathbb{G}_{\mathrm{m}}\Big)$$

because it does so on generic fibers over $\mathbb{Q}$, but we do not need to use $T_{\mathbb{Z}[1/N]}$ and so we will not mention it any further.]

It follows that $\mathrm{N}_\Phi : \mathbb{A}_{E,f}^\times \to \mathbb{A}_{L,f}^\times$ not only carries $\prod_{w\nmid N} \mathcal{O}_{E_w}^\times$ into $\prod_{v\nmid N} \mathcal{O}_{L_v}^\times$ but also respects *integrality* at the collection of places over any rational prime $\ell \nmid N$. Thus, for any place w of E over a prime $p \nmid N$, and any local uniformizer π_w of E_w, if we let $\iota_w : E_w^\times \to \mathbb{A}_{E,f}^\times$ be the natural inclusion map then the element $\mathrm{N}_\Phi(\iota_w(\pi_w)) \in \mathbb{A}_{L,f}^\times$ has local component 1 at all places v of L not over p and is *integral* at all places v of L over p. Moreover, the idele $\mathrm{N}_\Phi(\iota_w(\pi_w))$ changes by an integral unit multiple in all components at places of L over p when π_w is replaced with an $\mathcal{O}_{E_w}^\times$-multiple. Hence, the reflex norm induces a well-defined homomorphism on fractional ideal groups away from N: if $\mathfrak{p}$ is a prime ideal of E over $p \nmid N$ then we may define the fractional L-ideal

$$\mathrm{N}_\Phi(\mathfrak{p}) := \prod_{\mathfrak{q}|p\mathcal{O}_L} \mathfrak{q}^{\mathrm{ord}_\mathfrak{q}(\mathrm{N}_\Phi(\iota_\mathfrak{p}(\pi_\mathfrak{p})))} \tag{A.2.6.3}$$

for any uniformizer $\pi_\mathfrak{p}$ of $\mathcal{O}_{E_\mathfrak{p}}$; note that this ideal is *integral.*

Now pick a prime $\mathfrak{P}$ of K over a rational prime $p \nmid N$ such that $\sigma|_K = (\frac{K/E}{\mathfrak{P}})$. Let $\mathfrak{p}$ be the prime of E below $\mathfrak{P}$. Since $\mathcal{O}_L \subset \mathrm{End}(A_0)$, by the proof of Theorem A.2.3.5 there is a unique L-linear K-homomorphism of abelian varieties $\xi_{\sigma,\mathfrak{P}} : A_0 \to A_0^\sigma$ whose reduction is the relative q-Frobenius morphism

$$\mathrm{Fr}_{\overline{A}_0/\kappa(\mathfrak{P}),q} : \overline{A}_0 \to \overline{A}_0^{(q)},$$

where $q = \#\kappa(\mathfrak{p})$. (It is important for later purposes that $\xi_{\sigma,\mathfrak{P}}$ is not merely a map in the isogeny category.) The non-zero element

$$\xi_{\sigma,\mathfrak{P}} \in \mathfrak{a}_\sigma := \mathrm{Hom}((A_0, i_0), (A_0^\sigma, i_0^\sigma))$$

endows the 1-dimensional L-vector space $L\otimes_{\mathcal{O}_L} \mathfrak{a}_\sigma$ with a distinguished basis and so identifies $\mathfrak{a}_\sigma$ with a fractional L-ideal $\mathfrak{a}_{\sigma,\mathfrak{P}} \subset L$ that contains $\mathcal{O}_L$ and depends on $\mathfrak{P}$. Equivalently, composition with $\xi_{\sigma,\mathfrak{P}}$ defines an $\mathcal{O}_L$-linear embedding of invertible $\mathcal{O}_L$-modules

$$\mathcal{O}_L = \mathrm{Hom}((A_0, i_0), (A_0, i_0)) \overset{\xi_{\sigma,\mathfrak{P}}\circ(\cdot)}{\to} \mathrm{Hom}((A_0, i_0), (A_0^\sigma, i_0^\sigma)) = \mathfrak{a}_\sigma.$$

If we change the choice of prime $\mathfrak{P}$ of $\mathcal{O}_K[1/N]$ satisfying $(\frac{K/E}{\mathfrak{P}}) = \sigma|_K$ then the inverse-integral fractional L-ideal $\mathfrak{a}_{\sigma,\mathfrak{P}}$ is replaced with another such ideal.

The fractional ideal $\mathfrak{a}_{\sigma,\mathfrak{P}}$ has nothing to do with the L-linear $\mathbb{Q}$-polarization ϕ of A. It is important that $\mathfrak{a}_{\sigma,\mathfrak{P}}$ only depends on (L, Φ) and $\mathfrak{P}$:

A.2.6.1. Theorem. *With notation as above,* $\mathfrak{a}_{\sigma,\mathfrak{P}} = \mathrm{N}_\Phi(\mathfrak{p})^{-1}$ *with* $\mathfrak{p} := \mathfrak{P} \cap E$.

PROOF. Since $\xi_{\sigma,\mathfrak{P}}$ has p-power degree, $\mathcal{O}_L$ has p-power index in $\mathfrak{a}_{\sigma,\mathfrak{P}}$. Hence, the fractional ideal $\mathfrak{a}_{\sigma,\mathfrak{P}}$ is a unit away from p. By construction, the same holds for $\mathrm{N}_\Phi(\mathfrak{p})$. It is therefore enough to fix a place v of L over p and to compare ord_v's of $\mathfrak{a}_{\sigma,\mathfrak{P}}$ and $\mathrm{N}_\Phi(\mathfrak{p})^{-1}$. Rather than make such a comparison directly, it will be convenient to compare after raising both ideals to the $f(\mathfrak{P}|\mathfrak{p})$th-power. To make effective use of this, we need to describe $\mathfrak{a}^n_{\sigma,\mathfrak{P}}$ for positive integers n.

We claim that for any positive integer n, all L-linear maps $A \to A^{\sigma^n}$ descend to L-linear maps $A_0 \to A_0^{\sigma^n}$. This says that the $\mathrm{Gal}(\overline{\mathbb{Q}}/K)$-action on $\mathrm{Hom}(A, A^{\sigma^n})$ (via the K-structures A_0 and $A_0^{\sigma^n}$) is trivial on the subgroup of L-linear maps. It suffices to check such triviality in the isogeny category. The subspace of L-linear maps in the isogeny category is 1-dimensional as an L-vector space, and the L-action on A over $\overline{\mathbb{Q}}$ descends to A_0 over K, so it suffices to exhibit a single non-zero L-linear mapping $A_0 \to A_0^{\sigma^n}$. There is an L-linear isogeny $h_0 : A_0 \to A_0^\sigma$ (e.g., $\xi_{\sigma,\mathfrak{P}}$), so the composite map $h_0^{\sigma^{n-1}} \circ \cdots \circ h_0^\sigma \circ h_0$ does the job. We conclude that for all $n \geqslant 1$

$$\mathfrak{a}_{\sigma^n} := \mathrm{Hom}((A_0, i_0), (A_0^{\sigma^n}, i_0^{\sigma^n}))$$

is an invertible $\mathcal{O}_L$-module and it coincides with the analogous module of such mappings over $\overline{\mathbb{Q}}$.

By Example 1.7.4.1, the natural evaluation mapping

$$\mathfrak{a}_{\sigma^n} \otimes_{\mathcal{O}_L} A_0 \to A_0^{\sigma^n}$$

over K is an isomorphism and by the same argument with Néron models and Example 1.7.4.2 as in the proof of the final assertion in Theorem A.2.3.5 we see that the Hom-module $\mathfrak{a}_{\sigma^n}$ contains a unique element $\xi_{\sigma,n,\mathfrak{P}}$ that lifts the L-linear relative q^n-Frobenius morphism $\overline{A}_0 \to \overline{A}_0^{(q^n)}$ over $\kappa(\mathfrak{P})$. This identifies $\mathfrak{a}_{\sigma^n}$ with a fractional L-ideal $\mathfrak{a}_{\sigma,n,\mathfrak{P}}$. For example, $\mathfrak{a}_{\sigma,1,\mathfrak{P}} = \mathfrak{a}_{\sigma,\mathfrak{P}}$. What is the ideal $\mathfrak{a}_{\sigma,n,\mathfrak{P}}$ in general? Since $\mathrm{Fr}_{\overline{A}_0/\kappa(\mathfrak{P}),q^n}$ is roughly an n-fold composite of $\mathrm{Fr}_{\overline{A}_0/\kappa(\mathfrak{P}),q}$, the following answer is not a surprise:

A.2.6.2. Lemma. *For all $n \geqslant 1$, $\mathfrak{a}_{\sigma,n,\mathfrak{P}} = \mathfrak{a}^n_{\sigma,\mathfrak{P}}$.*

PROOF. Since all L-linear maps between A and A^σ are defined over K relative to the respective descents A_0 and A_0^σ, we have a canonical evaluation morphism

$$\mathfrak{a}_\sigma \otimes_{\mathcal{O}_L} A_0 \to A_0^\sigma$$

over K and it is an isomorphism. Hence, for any $\tau \in \mathrm{Gal}(\overline{\mathbb{Q}}/E)$, applying scalar extension along the isomorphism $\tau : K \simeq K$ (recall that K/E is Galois) and using the base-change compatibility of Serre's tensor construction yields a natural L-linear isomorphism

$$\mathfrak{a}_\sigma \otimes_{\mathcal{O}_L} A_0^\tau \simeq (\mathfrak{a}_\sigma \otimes_{\mathcal{O}_L} A_0)^\tau \simeq (A_0^\sigma)^\tau \simeq A_0^{\tau\sigma} \tag{A.2.6.4}$$

over K carrying $1 \otimes i_0^\tau$ to $i_0^{\tau\sigma}$.

By the definition of higher relative Frobenius maps, for any $\tau = \sigma^n$ with $n \in \mathbb{Z}^+$ the composite $\mathcal{O}_L$-linear module isomorphism

$$\begin{aligned} \mathfrak{a}_{\tau\sigma} : &= \mathrm{Hom}((A_0, i_0), (A_0^{\tau\sigma}, i_0^{\tau\sigma})) \\ &\overset{(\mathrm{A.2.6.4})}{\simeq} \mathrm{Hom}((A_0, i_0), (\mathfrak{a}_\sigma \otimes_{\mathcal{O}_L} A_0^\tau, 1 \otimes i_0^\tau)) \\ &\simeq \mathrm{Hom}((A_0, i_0), (\mathfrak{a}_\sigma \otimes_{\mathcal{O}_L} \mathfrak{a}_\tau \otimes_{\mathcal{O}_L} A_0, 1 \otimes 1 \otimes i_0)) \\ &\simeq \mathfrak{a}_\sigma \otimes_{\mathcal{O}_L} \mathfrak{a}_\tau \otimes_{\mathcal{O}_L} \mathrm{Hom}((A_0, i_0), (A_0, i_0)) \\ &\simeq \mathfrak{a}_\sigma \otimes_{\mathcal{O}_L} \mathfrak{a}_\tau \end{aligned}$$

carries $\xi_{\sigma,n+1,\mathfrak{P}}$ to $\xi_{\sigma,\mathfrak{P}} \otimes \xi_{\sigma,n,\mathfrak{P}}$. (The reader should check this assertion!) In terms of fractional ideals resting on the distinguished elements $\xi_{\sigma,m,\mathfrak{P}}$ for $m \in \mathbb{Z}^+$, this says $\mathfrak{a}_{\sigma,n+1,\mathfrak{P}} = \mathfrak{a}_{\sigma,\mathfrak{P}}\mathfrak{a}_{\sigma,n,\mathfrak{P}}$. Hence, by induction on $n \geqslant 1$ we conclude that $\mathfrak{a}_{\sigma,n,\mathfrak{P}} = \mathfrak{a}_{\sigma,\mathfrak{P}}^n$ for all $n \geqslant 1$. □

As a special case of Lemma A.2.6.2, since $\sigma|_K$ has order $f(\mathfrak{P}|\mathfrak{p})$, the fractional ideal $\mathfrak{a}_{\sigma,\mathfrak{P}}^{f(\mathfrak{P}|\mathfrak{p})}$ is associated to the invertible $\mathcal{O}_L$-module $\mathrm{End}_K(A_0, i_0) = \mathcal{O}_L$ endowed with the distinguished element $\pi_0 := \xi_{\sigma,f(\mathfrak{P}|\mathfrak{p}),\mathfrak{P}}$ that lifts the Frobenius $q_\mathfrak{P}$-endomorphism of the abelian variety $\overline{A}_0$ over the finite field $\kappa(\mathfrak{P})$ with size $q_\mathfrak{P} = q_\mathfrak{p}^{f(\mathfrak{P}|\mathfrak{p})}$. Hence, we conclude that $\mathfrak{a}_{\sigma,\mathfrak{P}}^{f(\mathfrak{P}|\mathfrak{p})} = \pi_0^{-1}\mathcal{O}_L$, so for any place v of L over p we have

$$\mathrm{ord}_v(\mathfrak{a}_{\sigma,\mathfrak{P}}^{f(\mathfrak{P}|\mathfrak{p})}) = -\mathrm{ord}_v(\pi_0).$$

Now recall the *Shimura–Taniyama formula* from (2.1.4.1):

$$\mathrm{ord}_v(\pi_0) = \mathrm{ord}_v(q_\mathfrak{P}) \cdot \frac{\#\Phi_v}{[L_v : \mathbb{Q}_p]}$$

with $\Phi_v \subseteq \Phi \subset \mathrm{Hom}_\mathbb{Q}(L, K)$ equal to the subset of elements $\varphi \in \Phi$ such that $\varphi : L \to K$ carries the $\mathfrak{P}$-adic place back to v. Since

$$\frac{\mathrm{ord}_v(q_\mathfrak{P})}{[L_v : \mathbb{Q}_p]} = \frac{f(\mathfrak{P}|p)\mathrm{ord}_v(p)}{[L_v : \mathbb{Q}_p]} = \frac{f(\mathfrak{P}|p)}{f(v|p)},$$

the problem $\mathrm{ord}_v(\mathfrak{a}_{\sigma,\mathfrak{P}}) \stackrel{?}{=} \mathrm{ord}_v(\mathrm{N}_\Phi(\mathfrak{p})^{-1})$ may be rephrased as

$$-\frac{f(\mathfrak{P}|p)}{f(v|p)} \cdot \#\Phi_v \stackrel{?}{=} \mathrm{ord}_v(\mathrm{N}_\Phi(\mathfrak{p})^{-f(\mathfrak{P}|\mathfrak{p})}),$$

or equivalently as

$$f(\mathfrak{p}|p)\#\Phi_v \stackrel{?}{=} f(v|p)\mathrm{ord}_v(\mathrm{N}_\Phi(\mathfrak{p})). \tag{A.2.6.5}$$

We will analyze the right side and eventually transform it into the left side.

Choose $\pi_\mathfrak{p} \in \mathcal{O}_E$ that has order 1 at $\mathfrak{p}$ and order 0 at all other places of E over p, so by (A.2.6.3) the fractional L-ideal $\mathrm{N}_\Phi(\mathfrak{p})$ is the "p-part" of the principal fractional ideal generated by $\det_L(\pi_\mathfrak{p} : t_\Phi \simeq t_\Phi) \in L^\times$. Hence, passing to the v-part of this L-determinant gives that

$$f(v|p)\mathrm{ord}_v(\mathrm{N}_\Phi(\mathfrak{p})) = f(v|p)\mathrm{ord}_v(\det_{L_v}(\pi_\mathfrak{p} : L_v \otimes_L t_\Phi \simeq L_v \otimes_L t_\Phi)).$$

For any $x \in L_v^\times$, $f(v|p)\mathrm{ord}_v(x) = \mathrm{ord}_p(\mathrm{N}_{L_v/\mathbb{Q}_p}(x))$ since p is unramified in L. Also, for any L_v-linear endomorphism $\mu : V \to V$ of a finite-dimensional L_v-vector space V we have $\mathrm{Nm}_{L_v/\mathbb{Q}_p}(\det_{L_v}(\mu)) = \det_{\mathbb{Q}_p}(\mu_{\mathbb{Q}_p})$ with $\mu_{\mathbb{Q}_p}$ denoting the underlying $\mathbb{Q}_p$-linear endomorphism. Taking μ to be multiplication by $\pi_\mathfrak{p}$ on $L_v \otimes_L t_\Phi$,

$$f(v|p)\mathrm{ord}_v(\mathrm{N}_\Phi(\mathfrak{p})) = \mathrm{ord}_p(\det_{\mathbb{Q}_p}(\pi_\mathfrak{p} : L_v \otimes_L t_\Phi \simeq L_v \otimes_L t_\Phi)).$$

The $L \otimes_{\mathbb{Q}} E$-module t_Φ is the "generic fiber" of an $\mathcal{O}_L[1/N] \otimes_{\mathbb{Z}[1/N]} \mathcal{O}_E[1/N]$-module $t_{\Phi,\mathbb{Z}[1/N]}$, extending scalars by $\mathbb{Z}[1/N] \to \mathbb{Z}_p$ and using the decomposition

$$L_v \otimes_L (L \otimes_{\mathbb{Q}} E) \simeq \prod_{w|p} L_v \otimes_{\mathbb{Q}_p} E_w$$

(and the analogous one with the Dedekind ring $\mathcal{O}_{L_v} \otimes_{\mathbb{Z}_p} \mathcal{O}_{E_w}$) gives a decomposition of $L_v \otimes_L t_\Phi$ into a product of w-parts over all $w|p$ such that the $\pi_{\mathfrak{p}}$-action respects the integral structure on this module decomposition. Thus, this action has integral unit $\mathbb{Z}_p$-determinant on w-factors for all $w \neq \mathfrak{p}$ (since $\pi_{\mathfrak{p}} \in \mathcal{O}_{E_w}^\times$ for all $w|p$ with $w \neq \mathfrak{p}$), so $f(v|p)\mathrm{ord}_v(\mathrm{N}_\Phi(\mathfrak{p}))$ is equal to ord_p of the $\mathbb{Q}_p$-determinant of the $\pi_{\mathfrak{p}}$-action on the $\mathfrak{p}$-part of $L_v \otimes_L t_\Phi$.

This latter action is $E_{\mathfrak{p}}$-linear, and for any $E_{\mathfrak{p}}$-linear endomorphism μ of a finite-dimensional $E_{\mathfrak{p}}$-vector space V we have

$$\mathrm{ord}_p(\det_{\mathbb{Q}_p}(\mu_{\mathbb{Q}_p})) = f(\mathfrak{p}|p)\mathrm{ord}_{\mathfrak{p}}(\det_{E_{\mathfrak{p}}}(\mu)),$$

where $\mu_{\mathbb{Q}_p}$ is the underlying $\mathbb{Q}_p$-linear endomorphism. This implies that

$$f(v|p)\mathrm{ord}_v(\mathrm{N}_\Phi(\mathfrak{p})) = f(\mathfrak{p}|p)\mathrm{ord}_{\mathfrak{p}}(\det_{E_{\mathfrak{p}}}(\mu))$$

with μ equal to scalar multiplication by $\pi_{\mathfrak{p}} \in E_{\mathfrak{p}}^\times$ on the $\mathfrak{p}$-part of $L_v \otimes_L t_\Phi$. The $E_{\mathfrak{p}}$-determinant of such a scalar action is just $\pi_{\mathfrak{p}}^{d_{\mathfrak{p}}}$, where $d_{\mathfrak{p}}$ is the $E_{\mathfrak{p}}$-dimension of the $\mathfrak{p}$-part of $L_v \otimes_L t_\Phi$. Hence,

$$f(v|p)\mathrm{ord}_v(\mathrm{N}_\Phi(\mathfrak{p})) = f(\mathfrak{p}|p)\mathrm{ord}_{\mathfrak{p}}(\pi_{\mathfrak{p}}^{d_{\mathfrak{p}}}) = f(\mathfrak{p}|p)d_{\mathfrak{p}}.$$

It follows that (A.2.6.5) is equivalent to the assertion $\#\Phi_v = d_{\mathfrak{p}}$. Under the extension of scalars $E \to K$ applied to t_Φ, the $\mathfrak{p}$-part of the $L_v \otimes_{\mathbb{Q}} E$-module $L_v \otimes_L t_\Phi$ is carried by the base change $E_{\mathfrak{p}} \to K_{\mathfrak{P}}$ over to the $\mathfrak{P}$-part of the $L_v \otimes_{\mathbb{Q}} K$-module $L_v \otimes_L (t_\Phi \otimes_E K)$. Thus, this $\mathfrak{P}$-part has $K_{\mathfrak{P}}$-dimension $d_{\mathfrak{p}}$. But by *construction* of the $L \otimes_{\mathbb{Q}} E$-module t_Φ via Galois descent, $t_\Phi \otimes_E K \simeq \prod_{\varphi \in \Phi} K$ as $L \otimes_{\mathbb{Q}} K$-modules. Hence, our problem is to prove that the $\mathfrak{P}$-part of the module $P_v = L_v \otimes_L (\prod_{\varphi \in \Phi} K)$ over the ring $L_v \otimes_{\mathbb{Q}} K \simeq \prod_{w'|p} L_v \otimes_{\mathbb{Q}_p} K_{w'}$ has $K_{\mathfrak{P}}$-dimension $\#\Phi_v$.

The module P_v decomposes as $\prod_{\varphi \in \Phi}(L_v \otimes_{L,\varphi} K)$, so it suffices to show that

$$P_{v,\varphi} := L_v \otimes_{L,\varphi} K$$

has vanishing $\mathfrak{P}$-part if $\varphi \notin \Phi_v$ and that it has $\mathfrak{P}$-part with $K_{\mathfrak{P}}$-dimension 1 for $\varphi \in \Phi_v$. By standard facts concerning completions of global fields, $L_v \otimes_{L,\varphi} K \simeq \prod_{w' \in \Sigma} K_{w'}$ where Σ is the set of places on K lifting v via φ. Thus, the $\mathfrak{P}$-part of $P_{v,\varphi}$ is at most 1-dimensional over $K_{\mathfrak{P}}$, and it is 1-dimensional precisely when φ pulls the $\mathfrak{P}$-adic place back to v. That is, the set of φ for which there is a non-zero $\mathfrak{P}$-part in $P_{v,\varphi}$ is precisely Φ_v. This completes the proof of Theorem A.2.6.1. □

By applying functoriality of the Serre tensor construction with respect to the natural $\mathcal{O}_L$-linear inclusion $\mathcal{O}_L \to \mathrm{N}_\Phi(\mathfrak{p})^{-1}$, we may restate Theorem A.2.6.1 in the following more convenient manner: for any prime $\mathfrak{P}$ of K over $p \nmid N$ such that $\sigma|_K = (\frac{K/E}{\mathfrak{P}})$, there is a unique $\mathcal{O}_L$-linear isomorphism

$$\theta_{\sigma,\mathfrak{P}} : \mathrm{N}_\Phi(\mathfrak{p})^{-1} \otimes_{\mathcal{O}_L} A_0 \simeq A_0^\sigma$$

of abelian varieties over K such that the diagram

(A.2.6.6)
$$\begin{array}{ccc} A_0 & \xrightarrow{\xi_{\sigma,\mathfrak{P}}} & A_0^{\sigma} \\ & \searrow & \simeq \uparrow \theta_{\sigma,\mathfrak{P}} \\ & & \mathrm{N}_{\Phi}(\mathfrak{p})^{-1} \otimes_{\mathcal{O}_L} A_0 \end{array}$$

commutes, where the top map is the unique lifting of the relative q-Frobenius for $\overline{A}_0$ over $\kappa(\mathfrak{P})$ and the diagonal map is induced by the inclusion of $\mathcal{O}_L$ into $\mathrm{N}_{\Phi}(\mathfrak{p})^{-1}$. The isomorphism $\theta_{\sigma,\mathfrak{P}}$ depends on the choice of $\mathfrak{P}$.

The isomorphism $\theta_{\sigma,\mathfrak{P}}$ uses a non-canonical K-descent (A_0, i_0) of (A, i) as well as a non-canonical choice of $\mathfrak{P}$. Our task in A.2.7 will be to improve this situation by constructing a *canonical* L-linear isomorphism

(A.2.6.7)
$$[\mathrm{N}_{\Phi}(s)^{-1}]_L \otimes_{\mathcal{O}_L} A \simeq A^{\sigma}$$

of abelian varieties over $\overline{\mathbb{Q}}$ for any $\sigma \in \mathrm{Gal}(\overline{\mathbb{Q}}/E)$ and $s \in \mathbb{A}_{E,f}^{\times}$ such that $r_E(s) = \sigma|_{E^{\mathrm{ab}}}$ (with (A, i) having CM order $\mathcal{O}_L$), where $[\cdot]_L$ denotes the fractional L-ideal associated to a finite L-idele.

We consider $\theta_{\sigma,\mathfrak{P}}$ to be an approximation to (A.2.6.7) "at level K" since the Artin map $\mathbb{A}_{E,f}^{\times} \twoheadrightarrow \mathrm{Gal}(K/E)^{\mathrm{ab}}$ carries the idele $s_{\mathfrak{p}} = \iota_{\mathfrak{p}}(\pi_{\mathfrak{p}})$ to $\sigma|_{E^{\mathrm{ab}} \cap K}$ (due to our convention on local uniformizers and arithmetic Frobenius elements under the Artin map!) and $[\mathrm{N}_{\Phi}(s_{\mathfrak{p}})]_L = \mathrm{N}_{\Phi}(\mathfrak{p})$. The map $\theta_{\sigma,\mathfrak{P}}$ generally does *not* descend the canonical isomorphism in (A.2.6.7) that will be constructed in A.2.7. Rather, $\theta_{\sigma,\mathfrak{P}}$ over $\overline{\mathbb{Q}}$ will be related to (A.2.6.7) via the action of an element of $\mathrm{N}_{\Phi}(E^{\times}) \subseteq T(\mathbb{Q}) \subseteq L^{\times}$.

A.2.6.3. Remark. The most important applications of the Main Theorem of Complex Multiplication are to abelian varieties over number fields, as in A.2.5. However, the situation is very much simplified by first proving the Main Theorem in the form given in A.2.4.4 (or A.2.7.2 below) that only involves abelian varieties over $\overline{\mathbb{Q}}$ (or more traditionally, over $\mathbb{C}$) and later deducing arithmetic consequences over number fields.

The main way in which this simplification manifests itself it is in the task of improving the construction of $\theta_{\sigma,\mathfrak{P}}$ over K to become canonical by passing up to $\overline{\mathbb{Q}}$. More specifically, to eliminate the intervention of $\mathfrak{P}$ we will have to apply certain procedures involving passage to a field that splits $A_0[M]$ for several relatively prime $M \geqslant 3$. By working over $\overline{\mathbb{Q}}$ from the outset we thereby avoid the unpleasant task of having to increase the base field at several places in the middle of an argument.

A.2.7. Polarizations and torsion. In A.2.6 the number field $K \subset \overline{\mathbb{Q}}$ was a rather general "sufficiently big" finite Galois extension of E, and we now need a few more "largeness" conditions on K. Choose a fixed auxiliary integer $M \geqslant 1$ (later we will take $M \geqslant 3$) and impose the additional "largeness" condition on K that the finite étale K-group $A_0[M]$ is constant. Thus, $A_0^{\sigma}[M]$ is constant, as is $(\mathfrak{a} \otimes_{\mathcal{O}_L} A_0)[M] \simeq \mathfrak{a} \otimes_{\mathcal{O}_L} (A_0[M])$ for any fractional ideal $\mathfrak{a}$ of L.

Recall that the choice of the prime $\mathfrak{P}$ of K was controlled by several ingredients: $\sigma|_K$ (via the condition $(\frac{K/E}{\mathfrak{P}}) = \sigma|_K$), the geometry of A_0, and the arithmetic of K and L (since we required the residue characteristic p of $\mathfrak{P}$ to not divide the product N of the residue characteristics of the primes of bad reduction for A_0 and the ramified primes for K and L over $\mathbb{Q}$). We impose the further condition on the residue characteristic p of $\mathfrak{P}$ that $p \nmid M$ (in addition to the condition $p \nmid N$).

The natural morphism $A_0 \to \mathrm{N}_\Phi(\mathfrak{p})^{-1} \otimes_{\mathcal{O}_L} A_0$ (see (A.2.6.3)) has p-power degree and so induces an isomorphism on M-torsion. Hence, we get a diagram on K-points

(A.2.7.1)
$$\begin{array}{ccc} A_0[M](K) & \xrightarrow{[\sigma]} & A_0^\sigma[M](K) \\ & \searrow{\scriptstyle \simeq} & \uparrow{\scriptstyle \simeq\ \theta_{\sigma,\mathfrak{P}}} \\ & & (\mathrm{N}_\Phi(\mathfrak{p})^{-1} \otimes_{\mathcal{O}_L} A_0)[M](K) \end{array}$$

that is the same as the associated diagram on $\overline{\mathbb{Q}}$-points (due to the constancy property of these M-torsion groups over $\operatorname{Spec} K$). The top map in this diagram is Galois-theoretic, and the other two sides arise from L-linear morphisms of abelian varieties over K. We claim that the diagram (A.2.7.1) commutes. Since $p \nmid M$ and the three abelian varieties involved all have good reduction at $\mathfrak{P}$, we may identify these K-point groups with the corresponding $\kappa(\mathfrak{P})$-point groups in the reductions of the abelian-scheme Néron models.

The diagram (A.2.7.1) viewed on $\kappa(\mathfrak{P})$-points is the same as the diagram induced on M-torsion by the reduction of (A.2.6.6) because the morphism $\xi_{\sigma,\mathfrak{P}}$ reduces to the morphism $\mathrm{Fr}_{\overline{A}_0/\kappa(\mathfrak{P}),q}$ whose effect on $\kappa(\mathfrak{P})$-points is *exactly* the action by the arithmetic Frobenius element in $\mathrm{Gal}(\kappa(\mathfrak{P})/\kappa(\mathfrak{p}))$ on such points. (Here, we use that the Galois-theoretic arithmetic Frobenius automorphism is given by the algebraic formula $t \mapsto t^{q_\mathfrak{p}}$.) This proves the commutativity of (A.2.7.1) as a consequence of the commutativity of (A.2.6.6).

To go further, we introduce a new "largeness" condition on K/E (that is to be imposed before choosing $\mathfrak{P}$): the extension K/E must contain a certain class field for E to now be described. Consider the reflex norm mapping $\mathrm{N}_\Phi : \mathbb{A}_{E,f}^\times \to \mathbb{A}_{L,f}^\times$ on $\mathbb{A}_f$-points. We have seen via (A.2.6.2) that for rational primes $\ell \nmid N$, the induced mapping on $\mathbb{Q}_\ell$-points

$$\mathrm{N}_{\Phi,\ell} : \prod_{w|\ell} E_w^\times = (E \otimes_\mathbb{Q} \mathbb{Q}_\ell)^\times \to (L \otimes_\mathbb{Q} \mathbb{Q}_\ell)^\times = \prod_{v|\ell} L_v^\times$$

carries $\mathcal{O}_{E_\ell}^\times$ into $\mathcal{O}_{L_\ell}^\times$ (where $E_\ell := \mathbb{Q}_\ell \otimes_\mathbb{Q} E$, $\mathcal{O}_{E_\ell} := \mathbb{Z}_\ell \otimes_\mathbb{Z} \mathcal{O}_E \simeq \prod_{w|\ell} \mathcal{O}_{E_w}$, and similarly for L). Define $U \subset \mathbb{A}_{E,f}^\times$ to be the open subgroup $\prod_\ell U_\ell$ where

$$U_\ell = \mathcal{O}_{E_\ell}^\times \cap \mathrm{N}_{\Phi,\ell}^{-1}(\{u \in \mathcal{O}_{L_\ell}^\times \mid u \equiv 1 \pmod{M}\})$$

for all rational primes ℓ (so $U_\ell = \mathcal{O}_{E_\ell}^\times$ for $\ell \nmid NM$). We require K to contain the class field for the open subgroup $E^\times E_\infty^\times U$, so the restricted Artin map $\mathbb{A}_{E,f}^\times \twoheadrightarrow \mathrm{Gal}(K/E)^{\mathrm{ab}}$ has kernel contained in $E^\times U$.

The conditions $\sigma|_K = (\frac{K/E}{\mathfrak{P}})$ and $r_E(s) = \sigma|_{E^{\mathrm{ab}}}$ force $s \cdot \iota_\mathfrak{p}(\pi_\mathfrak{p})^{-1}$ to be in the open kernel of the Artin map $E^\times \backslash \mathbb{A}_{E,f}^\times \twoheadrightarrow \mathrm{Gal}(K/E)^{\mathrm{ab}}$, so $s = \iota_\mathfrak{p}(\pi_\mathfrak{p}) uc$ in $\mathbb{A}_{E,f}^\times$ where $c \in E^\times$ and $u \in U$. Thus, the element $\mathrm{N}_\Phi(u) \in \mathbb{A}_{L,f}^\times$ is everywhere a local integral unit and satisfies $\mathrm{N}_\Phi(u) \equiv 1 \pmod{M}$ as an integral adele for L. (For later purposes, it is convenient to consider u as being determined by $s, \pi_\mathfrak{p}$, and c.) In particular, since $p \nmid M$, the finite L-idele $\mathrm{N}_\Phi(s^{-1}c) = \mathrm{N}_\Phi(\iota_\mathfrak{p}(\pi_\mathfrak{p})^{-1}u^{-1})$ that is possibly non-integral at p has component in $\prod_{\ell|M} L_\ell^\times$ that is an integral unit congruent to 1 modulo M.

Consider the commutative diagram of abelian varieties over K:

$$\begin{array}{ccc} \mathrm{N}_\Phi(\mathfrak{p})^{-1} \otimes_{\mathcal{O}_L} A_0 & \xrightarrow[\simeq]{\theta_{\sigma,\mathfrak{P}}} & A_0^\sigma \\ {\scriptstyle \mathrm{N}_\Phi(c)^{-1}}\downarrow\simeq & \nearrow\simeq & \\ \mathrm{N}_\Phi(c\mathfrak{p})^{-1} \otimes_{\mathcal{O}_L} A_0 & & \end{array} \tag{A.2.7.2}$$

where the diagonal mapping is defined to force commutativity; this diagonal mapping is an L-linear isomorphism. Letting $[x]_L$ denote the fraction L-ideal associated to a finite idele $x \in \mathbb{A}_{L,f}^\times$, we have

$$\begin{aligned} \mathrm{N}_\Phi(c\mathfrak{p}) = \mathrm{N}_\Phi(c) \cdot \mathrm{N}_\Phi(\mathfrak{p}) &= \mathrm{N}_\Phi(c) \cdot [\mathrm{N}_\Phi(\iota_\mathfrak{p}(\pi_\mathfrak{p}))]_L \\ &= [\mathrm{N}_\Phi(c \cdot \iota_\mathfrak{p}(\pi_\mathfrak{p}))]_L \\ &= [\mathrm{N}_\Phi(su^{-1})]_L \\ &= [\mathrm{N}_\Phi(s)]_L \end{aligned}$$

since $\mathrm{N}_\Phi(u) \in \prod_\ell \mathcal{O}_{L_\ell}^\times$ (by definition of U). Thus, the L-linear isomorphism of abelian varieties over K given by the diagonal map in (A.2.7.2) may be expressed as an abstract isomorphism

$$\theta_{\sigma,\mathfrak{P},\pi_\mathfrak{p},c,s,M} : [\mathrm{N}_\Phi(s^{-1})]_L \otimes_{\mathcal{O}_L} A_0 \simeq A_0^\sigma. \tag{A.2.7.3}$$

We shall prove that if $M \geqslant 3$ then the L-linear isomorphism

$$[\mathrm{N}_\Phi(s^{-1})]_L \otimes_{\mathcal{O}_L} A \simeq A^\sigma$$

obtained by applying the base change $K \to \overline{\mathbb{Q}}$ to (A.2.7.3) only depends on $\sigma \in \mathrm{Gal}(\overline{\mathbb{Q}}/E)$ and $s \in \mathbb{A}_{E,f}^\times$ (linked by the condition $r_E(s) = \sigma|_{E^{\mathrm{ab}}}$), and *not* on the choice of K/E, the choice of K-descent (A_0, i_0, ϕ_0) of (A, i, ϕ), or the choices of $\mathfrak{P}$, $\pi_\mathfrak{p}$, c, or M. We will also show that these canonical isomorphisms over $\overline{\mathbb{Q}}$ are well-behaved with respect to composition in $\mathrm{Gal}(\overline{\mathbb{Q}}/E)$ and multiplication in $\mathbb{A}_{E,f}^\times$. These matters will be settled in Theorem A.2.7.2 and Remark A.2.7.3.

Up to now, in A.2.6 and A.2.7 the L-linear $\mathbb{Q}$-polarization ϕ and its K-descent ϕ_0 on A_0 have played no role. We have arrived at the point where ϕ_0 will be used. We must endow $[\mathrm{N}_\Phi(s^{-1})]_L \otimes_{\mathcal{O}_L} A_0$ with a natural L-linear $\mathbb{Q}$-polarization $\phi_{0,s}$ over K such that $\theta_{\sigma,\mathfrak{P},\pi_\mathfrak{p},c,s,M}$ carries $\phi_{0,s}$ to a positive rational multiple of ϕ_0^σ (with the rational multiplier made explicit in terms of s). In view of how $\theta_{\sigma,\mathfrak{P},\pi_\mathfrak{p},c,s,M}$ is defined, and the fact that the mapping $\xi_{\sigma,\mathfrak{P}}$ in (A.2.6.6) that lifts $\mathrm{Fr}_{\overline{A}_0/\kappa(\mathfrak{P}),q_\mathfrak{p}}$ respects the $\mathbb{Q}$-polarizations ϕ_0 and ϕ_0^σ up to a positive rational multiple (as we showed in the proof of Theorem A.2.3.5), our task comes down to proving:

A.2.7.1. Lemma. *Let (A, i) be a* CM *abelian variety of type (L, Φ) over $\overline{\mathbb{Q}}$ with* CM *order $\mathcal{O}_L$, and let ϕ be an L-linear $\mathbb{Q}$-polarization on A. There is a unique way to assign to each $\alpha \in \mathbb{A}_{E,f}^\times$ an L-linear $\mathbb{Q}$-polarization ϕ_α of $[\mathrm{N}_\Phi(\alpha)]_L \otimes_{\mathcal{O}_L} A$ over $\overline{\mathbb{Q}}$ such that:*

1. *$\phi_1 = \phi$ via the canonical identification $\mathcal{O}_L \otimes_{\mathcal{O}_L} A = A$,*
2. *ϕ_α only depends on $[\mathrm{N}_\Phi(\alpha)]_L$,*
3. *if $h : [\mathrm{N}_\Phi(\alpha)]_L \to [\mathrm{N}_\Phi(\beta)]_L$ is an $\mathcal{O}_L$-linear map induced by multiplication by $c \in \mathrm{N}_\Phi(E^\times) \subseteq L^\times$ then the L-linear isogeny $h \otimes 1_A : [\mathrm{N}_\Phi(\alpha)]_L \otimes_{\mathcal{O}_L} A \to$*

$[\mathrm{N}_\Phi(\beta)]_L \otimes_{\mathcal{O}_L} A$ carries ϕ_α to $q_{c,\alpha,\beta}\phi_\beta$ with $q_{c,\alpha,\beta} \in \mathbb{Q}^\times_{>0}$ the unique positive generator of the fractional $\mathbb{Q}$-ideal $[\mathrm{Nm}_{E/\mathbb{Q}}(\beta/c\alpha)]_\mathbb{Q}$.

Moreover, $\deg \phi_\alpha = \deg \phi$ and if $K \subseteq \overline{\mathbb{Q}}$ is a subfield such that (A, i, ϕ) descends to a triple (A_0, i_0, ϕ_0) over K then each ϕ_α uniquely descends to an L-linear $\mathbb{Q}$-polarization $\phi_{0,\alpha}$ of A_0.

PROOF. Galois descent works for $\mathbb{Q}$-polarizations because descent theory is effective for polarizations. Hence, the compatibility of the Serre construction with respect to change of the base field (such as scalar extension along automorphisms of $\overline{\mathbb{Q}}$) and the uniqueness aspect in the lemma imply the final descent claim in the lemma via Galois descent. Thus, we may and do focus on the existence, uniqueness, and degree problems over $\overline{\mathbb{Q}}$. The uniqueness aspect is a formal consequence of (1) and (3), so it suffices to establish the existence aspect and that $\deg \phi_\alpha = \deg \phi$.

For the existence aspect, we first note that for fractional ideals $\mathfrak{a}$ and $\mathfrak{b}$ of L, the natural $\mathcal{O}_L$-linear map

$$\mathfrak{b}\mathfrak{a}^{-1} = \mathrm{Hom}_{\mathcal{O}_L}(\mathfrak{a}, \mathfrak{b}) \to \mathrm{Hom}(\mathfrak{a} \otimes_{\mathcal{O}_L} A, \mathfrak{b} \otimes A)$$

is an isomorphism (inspect homology lattices over $\mathbb{C}$ or use Example 1.7.4.2). Applying $\mathbb{Q} \otimes_\mathbb{Z} (\cdot)$ to both sides, we get a canonical isomorphism

$$L \simeq \mathrm{Hom}^0_L(\mathfrak{a} \otimes_{\mathcal{O}_L} A, \mathfrak{b} \otimes_{\mathcal{O}_L} A)$$

of L-vector spaces. The element $1 \in L$ therefore goes over to an L-linear morphism

$$j_{\mathfrak{a},\mathfrak{b}} : \mathfrak{a} \otimes_{\mathcal{O}_L} A \to \mathfrak{b} \otimes_{\mathcal{O}_L} A$$

in the isogeny category over $\overline{\mathbb{Q}}$ whose degree is the "generalized lattice index" $[\mathfrak{b} : \mathfrak{a}] = \mathrm{N}_{L/\mathbb{Q}}(\mathfrak{a}\mathfrak{b}^{-1})$. Indeed, if V/Λ is the canonical exponential uniformization of $A(\mathbb{C})$ then these assertions follow immediately from the identification of $\mathrm{H}_1(j_{\mathfrak{a},\mathfrak{b}}, \mathbb{Q})$ with the canonical L-linear isomorphism

$$\mathfrak{a} \otimes_{\mathcal{O}_L} \Lambda_\mathbb{Q} = \Lambda_\mathbb{Q} = \mathfrak{b} \otimes_{\mathcal{O}_L} \Lambda_\mathbb{Q}$$

(using the identifications $\mathfrak{a}_\mathbb{Q} = L = \mathfrak{b}_\mathbb{Q}$).

We define $j_\mathfrak{a} = j_{\mathfrak{a},\mathcal{O}_L}$ (with degree $\mathrm{N}_{L/\mathbb{Q}}(\mathfrak{a})$), so the L-linear map

$$j_\mathfrak{a}^*(\phi) := j_\mathfrak{a}^t \circ \phi \circ j_\mathfrak{a} : \mathfrak{a} \otimes_{\mathcal{O}_L} A \to (\mathfrak{a} \otimes_{\mathcal{O}_L} A)^t$$

in the isogeny category is a $\mathbb{Q}$-polarization with degree equal to $\mathrm{N}_{L/\mathbb{Q}}(\mathfrak{a})^2 \cdot \deg \phi$. Consider a fractional ideal $\mathfrak{a}$ of L such that the fractional ideal $\mathfrak{a}\overline{\mathfrak{a}}$ of L arises from a fractional ideal of $\mathbb{Q}$; equivalently, $\mathfrak{a}\overline{\mathfrak{a}} = q_\mathfrak{a}\mathcal{O}_L$ for a (necessarily unique) $q_\mathfrak{a} \in \mathbb{Q}^\times_{>0}$ (so $q_\mathfrak{a}^g = \mathrm{N}_{L/\mathbb{Q}}(\mathfrak{a})$). Roughly speaking, these ideals behave as if they are points of the reflex torus T. More precisely, a source of such fractional L-ideals is those of the form $\mathfrak{c} = [\mathrm{N}_\Phi(\gamma)]_L$ for $\gamma \in \mathbb{A}^\times_{E,f}$, in which case $q_\mathfrak{c} = \mathrm{Nm}_{E/\mathbb{Q}}(\gamma)$.

For any such $\mathfrak{a}$, we define the L-linear $\mathbb{Q}$-polarization

$$\phi_\mathfrak{a} = \frac{j_\mathfrak{a}^*(\phi)}{q_\mathfrak{a}}$$

whose degree is $(\mathrm{N}_{L/\mathbb{Q}}(\mathfrak{a})^2/q_\mathfrak{a}^{2g}) \deg \phi = \deg \phi$. It is easy to check that $\phi_\alpha := \phi_{[\mathrm{N}_\Phi(\alpha)]_L}$ has all of the desired properties. □

Lemma A.2.7.1 and the preceding discussion provide L-linear isomorphisms

$$\theta_{\sigma,\mathfrak{P},\pi_\mathfrak{p},c,s,M} : [\mathrm{N}_\Phi(s)]_L^{-1} \otimes_{\mathcal{O}_L} A_0 \simeq A_0^\sigma$$

of abelian varieties over K, carrying $\phi_{0,1/s}$ to a positive rational multiple of ϕ_0^σ; the rational multiplier must be 1 because $\deg(\phi^\sigma) = \deg\phi = \deg\phi_{1/s}$. We shall now prove that the isomorphism obtained from $\theta_{\sigma,\mathfrak{P},\pi_\mathfrak{p},c,s,M}$ after extension of scalars to $\overline{\mathbb{Q}}$ only depends on σ and s, so it can be called $\theta_{\sigma,s}$, and we will prove that $\theta_{\sigma,s}$ is well-behaved with respect to composition in $\mathrm{Gal}(\overline{\mathbb{Q}}/E)$ and multiplication in $\mathbb{A}_{E,f}^\times$.

Since $\mathrm{N}_\Phi(c)\cdot\mathrm{N}_\Phi(\mathfrak{p}) = [\mathrm{N}_\Phi(s)]_L$, we can define the L-linear isogeny

$$\theta_{\mathfrak{p},c} : A_0 \to [\mathrm{N}_\Phi(s)]_L^{-1}\otimes_{\mathcal{O}_L} A_0$$

over K to be the composite mapping

$$A_0 \longrightarrow \mathrm{N}_\Phi(\mathfrak{p})^{-1}\otimes_{\mathcal{O}_L} A_0 \xrightarrow[\simeq]{\mathrm{N}_\Phi(c)^{-1}\otimes 1} [\mathrm{N}_\Phi(s)]_L^{-1}\otimes_{\mathcal{O}_L} A_0.$$

This isogeny has p-power degree, so it induces an isomorphism on M-torsion. Since $\mathrm{N}_\Phi(sc^{-1}) = \mathrm{N}_\Phi(\iota_\mathfrak{p}(\pi_\mathfrak{p})u)$ with $p\nmid M$ and $\mathrm{N}_\Phi(u)\equiv 1 \pmod M$, it follows that on the level of geometric points the composite isomorphism

$$\begin{aligned} A_0[M] &\overset{\theta_{\mathfrak{p},c}}{\simeq} ([\mathrm{N}_\Phi(s)]_L^{-1}\otimes_{\mathcal{O}_L} A_0)[M] \\ &\simeq [\mathrm{N}_\Phi(s)]_L^{-1}\otimes_{\mathcal{O}_L} A_0[M] \\ &\simeq ([\mathrm{N}_\Phi(s)^{-1}]_L/M[\mathrm{N}_\Phi(s)^{-1}]_L)\otimes_{\mathcal{O}_L/M\mathcal{O}_L} A_0[M] \end{aligned}$$

of constant finite étale $\mathcal{O}_L/M\mathcal{O}_L$-modules over $\mathrm{Spec}\,K$ is induced by multiplication by the idele $\mathrm{N}_\Phi(s)^{-1}$ that is naturally a representative of a basis for the invertible $\mathcal{O}_L/M\mathcal{O}_L$-module

$$[\mathrm{N}_\Phi(s)^{-1}]_L/M[\mathrm{N}_\Phi(s)^{-1}]_L.$$

Consider induced maps on $\overline{\mathbb{Q}}$-points of M-torsion subgroups:

$$\begin{array}{ccc} A_0[M](\overline{\mathbb{Q}}) & \xrightarrow[\simeq]{[\sigma]} & A_0^\sigma[M](\overline{\mathbb{Q}}) \\ \theta_{\mathfrak{p},c}\big\downarrow\simeq & \nearrow_{\theta_{\sigma,\mathfrak{P},\pi_\mathfrak{p},c,s,M}}^{\simeq} & \\ ([\mathrm{N}_\Phi(s)]_L^{-1}\otimes_{\mathcal{O}_L} A_0)[M](\overline{\mathbb{Q}}) & & \end{array}$$

Since these $\overline{\mathbb{Q}}$-point groups coincide with K-point groups, the commutativity of this diagram follows from the commutativity of (A.2.7.1) and (A.2.7.2) (and the *definition* of $\theta_{\sigma,\mathfrak{P},\pi_\mathfrak{p},c,s,M}$). Via the isomorphism $(A_0)_{\overline{\mathbb{Q}}}\simeq A$ (carrying $(i_0)_{\overline{\mathbb{Q}}}$ and $(\phi_0)_{\overline{\mathbb{Q}}}$ to i and ϕ respectively, and likewise after base change by σ), the L-linear isomorphism $\theta_{\sigma,\mathfrak{P},\pi_\mathfrak{p},c,s,M}$ of abelian varieties over K must therefore induce the following composite of *canonical* maps on the $\overline{\mathbb{Q}}$-points of M-torsion subgroups:

$$\begin{aligned} ([\mathrm{N}_\Phi(s)]_L^{-1}\otimes_{\mathcal{O}_L} A)[M](\overline{\mathbb{Q}}) &\xrightarrow[\simeq]{} [\mathrm{N}_\Phi(s^{-1})]_L\otimes_{\mathcal{O}_L}(A[M](\overline{\mathbb{Q}})) \\ &\xrightarrow[\simeq]{\mathrm{N}_\Phi(s)} A[M](\overline{\mathbb{Q}}) \\ &\xrightarrow[\simeq]{\sigma} A^\sigma[M](\overline{\mathbb{Q}}). \end{aligned}$$

The steps in this composite are *independent* of all choices (including K/E and the descent (A_0, i_0, ϕ_0) over K) aside from the choices of σ, s, M, and for any $M'|M$ it is clear that on M'-torsion subgroups this intrinsic map on M-torsion restricts to the analogously described map arising from $\theta_{\sigma,\mathfrak{P},\pi_\mathfrak{p},c,s,M'}$. In other words, $\theta_{\sigma,\mathfrak{P},\pi_\mathfrak{p},c,s,M}$ and $\theta_{\sigma,\mathfrak{P},\pi_\mathfrak{p},c,s,M'}$ agree on M'-torsion subgroups if $M'|M$.

The automorphism $\theta_{\sigma,\mathfrak{P}',\pi'_{\mathfrak{p}'},c',s,M}\circ\theta^{-1}_{\sigma,\mathfrak{P},\pi_{\mathfrak{p}},c,s,M}$ lies in

$$\mathrm{Aut}(A_0^\sigma, i_0^\sigma, \phi_0^\sigma) = \mathrm{Aut}(A^\sigma, i^\sigma, \phi^\sigma),$$

where the preservation of the polarization is due to the fact that both θ's carry ϕ to ϕ^σ. We allow that $\mathfrak{P}'=\mathfrak{P}$ and $c'=c$ but $\pi'_{\mathfrak{p}}\neq\pi_{\mathfrak{p}}$. The preceding argument via (A.2.7.4) shows that this automorphism of A_0^σ induces the identity on M-torsion. Since the automorphism group of a $\mathbb{Q}$-polarized abelian variety is finite, this automorphism of A_0^σ has finite order. However, it acts as the identity on M-torsion, so as long as we restrict attention to $M\geqslant 3$ this finite-order automorphism is trivial. Thus, if $M\geqslant 3$ then the L-linear isomorphism

$$(\theta_{\sigma,\mathfrak{P},\pi_{\mathfrak{p}},c,s,M})_{\overline{\mathbb{Q}}} : [\mathrm{N}_\Phi(s)]_L^{-1}\otimes_{\mathcal{O}_L} A \simeq A^\sigma$$

of abelian varieties over $\overline{\mathbb{Q}}$ only depends on σ, s, and M, so we denote it $\theta_{\sigma,s,M}$.

For $M'|M$ with $M'\geqslant 3$, the same argument now shows that the L-linear automorphism $\theta_{\sigma,s,M}\circ\theta^{-1}_{\sigma,s,M'}$ of the abelian variety A^σ over $\overline{\mathbb{Q}}$ is trivial on M'-torsion and hence is trivial. Thus, for any $M_1, M_2\geqslant 3$ we have $\theta_{\sigma,s,M_1}=\theta_{\sigma,s,M_1M_2}=\theta_{\sigma,s,M_2}$, so $\theta_{\sigma,s,M}$ is independent of $M\geqslant 3$. (See Remark A.2.6.3.)

We summarize the conclusion of our efforts as follows:

A.2.7.2. Theorem. *Let (A,i) be a* CM *abelian variety of type (L,Φ) over $\overline{\mathbb{Q}}$, and assume (A,i) has* CM *order $\mathcal{O}_L$. For any $\sigma\in\mathrm{Gal}(\overline{\mathbb{Q}}/E)$ and $s\in\mathbb{A}_{E,f}^\times$ such that $r_E(s)=\sigma|_{E^{\mathrm{ab}}}$ there is a unique L-linear isomorphism*

$$\theta_{\sigma,s}=\theta_{\sigma,s,(A,i)} : [\mathrm{N}_\Phi(s)]_L^{-1}\otimes_{\mathcal{O}_L} A\simeq A^\sigma$$

such that for all $M\geqslant 1$ the isomorphism $[\sigma] : A[M](\overline{\mathbb{Q}})\simeq A^\sigma[M](\overline{\mathbb{Q}})$ is the composite

$$\text{(A.2.7.4)}\qquad \begin{array}{c} A[M](\overline{\mathbb{Q}}) \xrightarrow[\simeq]{\mathrm{N}_\Phi(s^{-1})} ([\mathrm{N}_\Phi(s^{-1})]_L/M[\mathrm{N}_\Phi(s^{-1})]_L)\otimes_{\mathcal{O}_L/M\mathcal{O}_L} A[M](\overline{\mathbb{Q}}) \\ \downarrow\simeq \\ ([\mathrm{N}_\Phi(s)]_L^{-1}\otimes_{\mathcal{O}_L} A)[M](\overline{\mathbb{Q}}) \\ \theta_{\sigma,s}\downarrow\simeq \\ A^\sigma[M](\overline{\mathbb{Q}}). \end{array}$$

For any L-linear $\mathbb{Q}$-polarization ϕ on A, the L-linear $\mathbb{Q}$-polarization $\phi_{1/s}$ of the $\mathcal{O}_L$-linear abelian variety $[\mathrm{N}_\Phi(s)]_L^{-1}\otimes_{\mathcal{O}_L} A$ and the L-linear $\mathbb{Q}$-polarization ϕ^σ of A^σ are intertwined by the isomorphism $\theta_{\sigma,s}$. Also, for any $c\in E^\times$ we have $\theta_{\sigma,cs}=\mathrm{N}_\Phi(c)\theta_{\sigma,s}$, and the formation of $\theta_{\sigma,s,(A,i)}$ is natural in the principal CM *abelian variety (A,i).*

PROOF. The uniqueness follows from the specification of the induced map on every torsion level (for all M). The existence is the content of the entire preceding analysis, and the proof of existence also gives the asserted behavior with respect to L-linear $\mathbb{Q}$-polarizations. The behavior with respect to replacing s with cs for $c\in E^\times$ comes from the uniqueness (or, less elegantly, from the construction). The naturality in (A,i) follows from the explicit canonical description (A.2.7.4) on every torsion level. □

A.2.7.3. Remark. We address the behavior of $\theta_{\sigma,s}$ with respect to composition in σ and multiplication in s. Pick $\sigma, \sigma' \in \mathrm{Gal}(\overline{\mathbb{Q}}/E)$ and $s, s' \in \mathbb{A}_{E,f}^{\times}$ such that $r_E(s) = \sigma|_{E^{\mathrm{ab}}}$ and $r_E(s') = \sigma'|_{E^{\mathrm{ab}}}$, so $r_E(s's) = (\sigma'\sigma)|_{E^{\mathrm{ab}}}$. We claim $\theta_{\sigma'\sigma,s's,(A,i)}$ is the composite isomorphism

$$\begin{aligned}[][\mathrm{N}_\Phi(s's)]_L^{-1} \otimes_{\mathcal{O}_L} A &\simeq [\mathrm{N}_\Phi(s')]_L^{-1} \otimes_{\mathcal{O}_L} ([\mathrm{N}_\Phi(s)]_L^{-1} \otimes_{\mathcal{O}_L} A)\\ &\overset{1\otimes\theta_{\sigma,s,(A,i)}}{\simeq} [\mathrm{N}_\Phi(s')]_L^{-1} \otimes_{\mathcal{O}_L} A^\sigma\\ &\overset{\theta_{\sigma',s',(A^\sigma,i^\sigma)}}{\simeq} (A^\sigma)^{\sigma'}\\ &\simeq A^{\sigma'\sigma}.\end{aligned}$$

This equality may be checked by comparing what happens on M-torsion for arbitrary $M \geqslant 3$. This is straightforward because $[\sigma] : A(\overline{\mathbb{Q}}) \simeq A^\sigma(\overline{\mathbb{Q}})$ is L-linear with respect to i and i^σ.

The descent to models over number fields has served its purpose, and for the rest of the proof of Theorem A.2.4.4 we will work exclusively over $\overline{\mathbb{Q}}$. Passing to the inverse limit on (A.2.7.4) and tensoring with $\mathbb{Q}$ gives a commutative diagram

(A.2.7.5)
$$\begin{array}{ccccc} V_f(A) & \xrightarrow{\simeq} & [\mathrm{N}_\Phi(s)]_L^{-1} \otimes_{\mathcal{O}_L} V_f(A) & \xrightarrow{\simeq} & V_f([\mathrm{N}_\Phi(s)]_L^{-1} \otimes_{\mathcal{O}_L} A) \\ & \searrow^{\mathrm{N}_\Phi(s)^{-1}} & \downarrow{\scriptstyle \mathrm{mult.}\ \simeq} & \nearrow^{\simeq} & \downarrow{\scriptstyle \simeq\ V_f(\theta_{\sigma,s})} \\ & & V_f(A) & \xrightarrow[\simeq]{} & V_f(A^\sigma) \end{array}$$

in which the composite across the top and right sides is $[\sigma]$ and the right diagonal and lower horizontal maps are defined by commutativity. Since $\mathrm{N}_\Phi(s) \in T(\mathbb{A}_f) \subseteq \mathbb{A}_{L,f}^\times$, the multiplication map by $\mathrm{N}_\Phi(s)^{-1}$ respects the $\mathbb{A}_f^\times$-homothety class of the skew-symmetric bilinear form induced by ϕ.

Suppose that the diagonal mapping $V_f(A) \to V_f([\mathrm{N}_\Phi(s)]_L^{-1} \otimes_{\mathcal{O}_L} A)$ in (A.2.7.5) is induced by an L-linear isogeny $\psi_s \in \mathrm{Hom}^0(A, [\mathrm{N}_\Phi(s)]_L^{-1} \otimes_{\mathcal{O}_L} A)$ *that respects* the L-linear $\mathbb{Q}$-polarizations ϕ and $\phi_{1/s}$ up to a (necessarily positive) rational multiple. The L-linear isogeny

$$\lambda_{\sigma,s} = \theta_{\sigma,s} \circ \psi_s \in \mathrm{Hom}^0(A, A^\sigma)$$

therefore carries ϕ to ϕ^σ up to a positive rational multiple and satisfies

$$V_f(\lambda_{\sigma,s}) \cdot \mathrm{N}_\Phi(s)^{-1} = [\sigma] \tag{A.2.7.6}$$

in $\mathrm{Hom}_{\mathbb{A}_{L,f}}(V_f(A), V_f(A^\sigma))$, so that would complete the proof of the Main Theorem! (By uniqueness, this map $\lambda_{\sigma,s}$ must coincide with the one introduced with the same notation in Proposition A.2.5.1, and in A.2.5.2 we saw that (A.2.7.6) forces $\lambda_{\sigma,s}$ to carry ϕ^σ back to $q_s\phi$ with q_s the unique positive generator of $[\mathrm{N}_{E/\mathbb{Q}}(s)]_{\mathbb{Q}}$.)

To construct ψ_s, first recall that for any $c \in E^\times$ the multiplier $\mathrm{N}_\Phi(c) \in T(\mathbb{Q}) \subseteq L^\times$ acting on A (in the isogeny category) preserves ϕ up to a rational multiple. Since $\theta_{\sigma,cs} = \mathrm{N}_\Phi(c)\theta_{\sigma,s}$, it follows that we only need to construct ψ_{cs} for some $c \in E^\times$ (and then $\mathrm{N}_\Phi(c) \cdot \psi_{cs}$ serves as ψ_s). Choosing $c \in \mathcal{O}_E - \{0\}$ that is sufficiently divisible (depending on s), since $\mathrm{N}_\Phi(cs) = \mathrm{N}_\Phi(c)\mathrm{N}_\Phi(s)$ we can replace s with cs to reduce to the case when the finite idele $\mathrm{N}_\Phi(s)$ in $\mathbb{A}_{L,f}^\times$ is everywhere integral as an adele. Hence, $\mathcal{O}_L \subseteq [\mathrm{N}_\Phi(s)]_L^{-1}$. In this case the canonical mapping

$$V_f(A) \to [\mathrm{N}_\Phi(s)]_L^{-1} \otimes_{\mathcal{O}_L} V_f(A)$$

is inverse to the vertical multiplication mapping in (A.2.7.5), so the map $A \to [\mathrm{N}_\Phi(s)]_L^{-1} \otimes_{\mathcal{O}_L} A$ induced by the canonical inclusion $\mathcal{O}_L \to [\mathrm{N}_\Phi(s)]_L^{-1}$ may be taken to be ψ_s. (The compatibility of this map with L-linear $\mathbb{Q}$-polarizations up to a $\mathbb{Q}_{>0}^\times$-multiple follows from property (3) in Lemma A.2.7.1, taking $\alpha = 1$ and $\beta = 1/s$.) The Main Theorem of Complex Multiplication is now proved.

A.2.7.4. Remark. Without recourse to replacing s with cs for a sufficiently divisible non-zero $c \in \mathcal{O}_L$ as in the preceding proof, the map ψ_s can always be described by the same recipe as at the end of the proof: it is the element in

$$\mathrm{Hom}_L^0(A, [\mathrm{N}_\Phi(s)]_L^{-1} \otimes_{\mathcal{O}_L} A) \simeq [\mathrm{N}_\Phi(s)]_L^{-1} \otimes_{\mathcal{O}_L} \mathrm{Hom}_L^0(A, A) \simeq L$$

corresponding to 1 (where the second isomorphism is induced by multiplication), due to Example 1.7.4.1.

A.2.8. Analytic version of Main Theorem and generalized Serre construction. To complete A.2, we explain how to express the Main Theorem in terms of adelic operations and analytic uniformizations. This recovers the traditional statement of the Main Theorem, except that we formulate things in more intrinsic terms. For example, our formulation of the analytic version of the Main Theorem does not involve bases of tangent spaces, in contrast with [**113**, 18.6]. We also explain how to translate such adelic operations into algebraic language by generalizing the Serre tensor construction from 1.7.4 within the framework of certain abelian varieties of CM type over a field; this provides an algebraic generalization of Theorem A.2.7.2 to the case of any CM order.

For CM abelian varieties over $\mathbb{C}$ with a fixed CM type (L, Φ), where $\Phi \subset \mathrm{Hom}(L, \mathbb{C}) = \mathrm{Hom}(L, \overline{\mathbb{Q}})$, morphisms among such objects uniquely and functorially descend to the subfield $\overline{\mathbb{Q}}$ (Lemma 1.2.1.2, Theorem 1.7.2.1). Thus, to avoid far-out things such as $\mathrm{Aut}(\mathbb{C})$ we shall consider our CM abelian varieties (A, i) with CM type (L, Φ) to be given over $\overline{\mathbb{Q}}$.

Consider the canonical analytic exponential uniformization $V/\Lambda \simeq A(\mathbb{C})$ of A. Clearly Λ is a module for the CM order $i^{-1}(\mathrm{End}(A)) \subseteq \mathcal{O}_L$, and $A(\mathbb{C})_{\mathrm{tor}} = \Lambda_\mathbb{Q}/\Lambda$. Hence, $\Lambda_\mathbb{Q}$ is a 1-dimensional L-vector space and the $\mathbb{R}$-vector space $\mathbb{R} \otimes_\mathbb{Q} \Lambda_\mathbb{Q} \simeq V$ with its complex structure is isomorphic to $(\mathbb{R} \otimes_\mathbb{Q} L)_\Phi \simeq \prod_{\varphi \in \Phi} \mathbb{C}_\varphi = \mathbb{C}^g$ as a $\mathbb{C} \otimes_\mathbb{Q} L$-module (where L acts on $\mathbb{C}_\varphi = \mathbb{C}$ through φ). Traditionally, an identification of V with $(\mathbb{R} \otimes_\mathbb{Q} L)_\Phi$ as $\mathbb{C} \otimes_\mathbb{Q} L$-modules is chosen in the analytic formulation of the Main Theorem because this identifies the "abstract" Λ with something more concrete, namely a $\mathbb{Z}$-lattice in $(\mathbb{R} \otimes_\mathbb{Q} L)_\Phi = \mathbb{C}^g$ that is stable under some order of $\mathcal{O}_L$. However, there is no essential benefit obtained, so we avoid such a choice.

Given $\sigma \in \mathrm{Aut}(\overline{\mathbb{Q}}/E)$, our goal is to describe an analytic uniformization of $A^\sigma(\mathbb{C})$ in terms of the canonical one for $A(\mathbb{C})$ such that the composite map on torsion subgroups

$$A(\mathbb{C})_{\mathrm{tor}} = A(\overline{\mathbb{Q}})_{\mathrm{tor}} \overset{\sigma}{\simeq} A^\sigma(\overline{\mathbb{Q}})_{\mathrm{tor}} = A^\sigma(\mathbb{C})_{\mathrm{tor}}$$

is easily described via an adelic operation on lattices arising from these analytic uniformizations.

We need to first discuss adelic operations on $\Lambda_\mathbb{Q}/\Lambda$. Somewhat more abstractly, let W be a finite-dimensional non-zero L-vector space and let $\Lambda \subset W$ be a $\mathbb{Z}$-lattice that is stable under some order $\mathcal{O}'$ of $\mathcal{O}_L$. We call such a lattice Λ an *order lattice* in W. There are many orders $\mathcal{O}'$ that preserve Λ, but there is a unique largest one, namely $\mathrm{End}_{\mathbb{Z}}(\Lambda)\cap L \subseteq \mathcal{O}_L$ inside $\mathrm{End}_{\mathbb{Q}}(W)$, and it is called the *endomorphism order* for Λ in L. The quotient W/Λ is a torsion $\mathbb{Z}$-module, so it is a torsion $\mathcal{O}'$-module.

Under contraction, each maximal ideal of $\mathcal{O}_L$ gives rise to one of $\mathcal{O}'$ (though several in $\mathcal{O}_L$ may give rise to the same one in $\mathcal{O}'$). For each maximal ideal v' of $\mathcal{O}'$ we write $\mathcal{O}'_{v'}$ to denote the corresponding completion of $\mathcal{O}'$, so this completion is a *local* order in the semi-local product $\prod_{v|v'} \mathcal{O}_v$ of the local rings for L at places v over v'. In particular, $\mathcal{O}'_{v'}$ has total ring of fractions $L_{v'} := \prod_{v|v'} L_v$.

All but finitely many maximal ideals v of $\mathcal{O}_L$ are unique over the contraction v' in $\mathcal{O}'$ and satisfy $\mathcal{O}'_{v'} = \mathcal{O}_v$ inside L_v. There is a canonical isomorphism

$$W/\Lambda \simeq \bigoplus_{v'} (W_{v'}/\Lambda_{v'}) \tag{A.2.8.1}$$

with $W_{v'} := \prod_{v|v'}(L_v \otimes_L W)$ a free module of rank 1 over $L_{v'} = \prod_{v|v'} L_v$ and $\Lambda_{v'} := \mathcal{O}'_{v'} \otimes_{\mathcal{O}'} \Lambda$. Concretely, $W_{v'}/\Lambda_{v'}$ is identified with the $\mathfrak{m}'_{v'}$-power torsion submodule of W/Λ. It is easy to check that the subset $\Lambda_{v'} \subset W_{v'}$ is the closure of Λ in $W_{v'}$, and that if we shrink the order $\mathcal{O}'$ to some $\mathcal{O}'' \subseteq \mathcal{O}'$ then for each maximal ideal v'' of $\mathcal{O}''$ we have $W_{v''} = \prod_{v'|v''} W_{v'}$ and $\Lambda_{v''} = \prod_{v'|v''} \Lambda_{v'}$. In this sense, the description (A.2.8.1) of the primary decomposition of W/Λ is well-behaved with respect to change in the choice of order $\mathcal{O}'$ in L preserving Λ (as there is a unique maximal such $\mathcal{O}'$ containing all others).

A.2.8.1. Lemma. *Let $\mathcal{O}'$ be the endomorphism order for Λ in L.*

(1) *For any $s \in \mathbb{A}_{L,f}^{\times}$, there is a unique order lattice $s\Lambda$ in W with endomorphism order $\mathcal{O}'$ such that $(s\Lambda)_{v'} = s_{v'}\Lambda_{v'}$ inside $W_{v'}$ for all v' on $\mathcal{O}'$, where $s_{v'} \in \prod_{v|v'} L_v^{\times}$ denotes the component of s along the set of places v of $\mathcal{O}_L$ over v'.*
(2) *There is a unique $\mathcal{O}'$-linear isomorphism $W/\Lambda \simeq W/(s\Lambda)$ such that on v'-factors it is the mapping $W_{v'}/\Lambda_{v'} \simeq W_{v'}/(s_{v'}\Lambda_{v'})$ induced by multiplication by $s_{v'}$ on $W_{v'}$.*
(3) *If $s, s' \in \mathbb{A}_{L,f}^{\times}$ then $s'(s\Lambda) = (s's)\Lambda$ inside W.*

A quick proof of Lemma A.2.8.1 can be given by working with $\mathbb{A}_f$-modules and $\widehat{\mathbb{Z}}$-algebras (such as $\prod_{v'} \mathcal{O}'_{v'}$), but we prefer to give a proof that is "intrinsic" to the order $\mathcal{O}'$ and avoids the crutch of the subring $\mathbb{Z}$ over which $\mathcal{O}'$ is finite and flat.

PROOF. For all but finitely many v', the endomorphism order $\mathcal{O}'$ for Λ is maximal at v' and $s_{v'} \in \mathcal{O}_{L_{v'}}^{\times}$. Hence, by multiplying Λ by a sufficiently divisible non-zero element of $\mathcal{O}_L$ we get an $\mathcal{O}_L$-lattice $\Lambda' \subseteq \Lambda$ such that $\prod_{v|v'} \Lambda'_v \subseteq s_{v'}\Lambda_{v'}$ for all v'. Moreover, for all but finitely many v' we have $\prod_{v|v'} \Lambda'_v = s_{v'}\Lambda_{v'} = \Lambda_{v'}$ inside $W_{v'}$. Our construction problem therefore takes place inside the $\mathcal{O}_L$-module W/Λ', and we need to look inside the I-torsion submodule for a sufficiently divisible non-zero ideal I of $\mathcal{O}_L$. Such a torsion submodule is of finite length over $\mathcal{O}_L$, and so decomposing along the finitely many primes in its support gives a solution to our $\mathcal{O}'$-module existence problem via finitely many local constructions. The uniqueness is seen in the same way.

The uniqueness implies $s'(s\Lambda) = (s's)\Lambda$. By working locally, we see that the endomorphism orders for Λ and $s\Lambda$ coincide. The existence and uniqueness of the desired isomorphism $W/\Lambda \simeq W/(s\Lambda)$ is proved by using the decomposition into v'-components for each v'. □

A.2.8.2. Example. If $\Lambda \subset W$ is a finitely generated $\mathcal{O}_L$-submodule then $s\Lambda = [s]_L\Lambda$, where $[s]_L$ is the fractional L-ideal associated to $s \in \mathbb{A}_{L,f}^{\times}$.

We can now prove the Main Theorem in its analytic guise as originally stated by Shimura and Taniyama in a coordinatized manner (but we avoid their "coordinates" on tangent spaces). There are two parts to this theorem: the first gives a description of the analytic uniformization of a Galois twist by an automorphism of $\overline{\mathbb{Q}}$ preserving the reflex field and the second describes the Riemann form of the corresponding Galois twist of an L-linear polarization (providing an analytic version of the observation in A.2.5.2 that $\lambda_{\sigma,s}$ carries ϕ^{σ} back to $q_s\phi$ with $q_s \in \mathbb{Q}_{>0}^{\times}$ the unique positive generator of $[\mathrm{Nm}_{E/\mathbb{Q}}(s)]_{\mathbb{Q}}$).

A.2.8.3. Theorem (Main Theorem of Complex Multiplication; analytic form). *Pick $\sigma \in \mathrm{Gal}(\overline{\mathbb{Q}}/E)$ and $s \in \mathbb{A}_{E,f}^{\times}$ such that $r_E(s) = \sigma|_{E^{\mathrm{ab}}}$ in $\mathrm{Gal}(E^{\mathrm{ab}}/E)$. Let ϕ be an L-linear $\mathbb{Q}$-polarization of A. There is a unique $\mathbb{C} \otimes_{\mathbb{Q}} L$-linear identification of $V = \mathrm{Lie}(A(\mathbb{C}))$ with $\mathrm{Lie}(A^{\sigma}(\mathbb{C}))$ such that the canonical analytic uniformization of $A^{\sigma}(\mathbb{C})$ is thereby identified with $V/\mathrm{N}_{\Phi}(1/s)\Lambda$ and the diagram*

$$\begin{array}{ccccc}
\Lambda_{\mathbb{Q}}/\Lambda & \xrightarrow{\simeq} & A(\mathbb{C})_{\mathrm{tor}} & \xleftarrow{\simeq} & A(\overline{\mathbb{Q}})_{\mathrm{tor}} \\
{\scriptstyle \mathrm{N}_{\Phi}(1/s)}\downarrow{\scriptstyle \simeq} & & & & {\scriptstyle \simeq}\downarrow{\scriptstyle [\sigma]} \\
\Lambda_{\mathbb{Q}}/\mathrm{N}_{\Phi}(1/s)\Lambda & \xrightarrow[\simeq]{} & A^{\sigma}(\mathbb{C})_{\mathrm{tor}} & \xleftarrow{\simeq} & A^{\sigma}(\overline{\mathbb{Q}})_{\mathrm{tor}}
\end{array}$$

commutes. Under the resulting identification of $\mathrm{H}_1(A^{\sigma}(\mathbb{C}), \mathbb{Q})$ with $\Lambda_{\mathbb{Q}}$, the $\mathbb{Q}(1)$-valued Riemann form $\Psi_{\phi^{\sigma}}$ is identified with $q_s\Psi_{\phi}$, where q_s is the unique positive generator of the fractional $\mathbb{Q}$-ideal $[\mathrm{Nm}_{E/\mathbb{Q}}(s)]_{\mathbb{Q}}$.

See Remark A.2.4.5 for the reason $1/s$ rather than s intervenes in this diagram.

PROOF. Let us first prove uniqueness. Any two isomorphisms $V \simeq \mathrm{Lie}(A^{\sigma}(\mathbb{C}))$ satisfying the requirements are related through the action of an automorphism of V that preserves Λ and induces on $V/\Lambda \simeq A(\mathbb{C})$ an analytic endomorphism that is the identity on all torsion points and so is the identity. This establishes uniqueness.

For existence, we use the algebraic form of the Main Theorem over $\overline{\mathbb{Q}}$ in Theorem A.2.4.4. By A.2.5.1, the Main Theorem provides a (necessarily unique) L-linear isomorphism in the isogeny category $\lambda_{\sigma,s} \in \mathrm{Hom}^0(A, A^{\sigma})$ (possibly not a genuine morphism of abelian varieties) such that the diagram

$$\begin{array}{ccc}
V_f(A) & \xrightarrow[\simeq]{[\sigma]} & V_f(A^{\sigma}) \\
{\scriptstyle \mathrm{N}_{\Phi}(1/s)}\downarrow{\scriptstyle \simeq} & \nearrow{\scriptstyle V_f(\lambda_{\sigma,s})} & \\
V_f(A) & &
\end{array} \tag{A.2.8.2}$$

commutes and $\lambda_{\sigma,s}$ carries ϕ^{σ} back to $q_s\phi$.

Letting $V_{\sigma}/\Lambda_{\sigma}$ be the canonical analytic uniformization of $A^{\sigma}(\mathbb{C})$, consider the induced L-linear map $\mathrm{H}_1(\lambda_{\sigma,s}) : \Lambda_{\mathbb{Q}} \simeq (\Lambda_{\sigma})_{\mathbb{Q}}$ on rational homologies. The $\mathbb{R}$-scalar extension of this $\mathbb{Q}$-vector space isomorphism is the $\mathbb{C}$-linear isomorphism $V \simeq V_{\sigma}$ induced by $\lambda_{\sigma,s}$ on tangent spaces at the identity. We claim that $\mathrm{H}_1(\lambda_{\sigma,s})$ carries $\mathrm{N}_{\Phi}(1/s)\Lambda$ isomorphically onto Λ_{σ}. It suffices to check this inside the $\mathbb{A}_f$-modules obtained through extension of scalars, but the commutative diagram (A.2.8.2) can be rewritten as a commutative diagram

$$\text{(A.2.8.3)} \qquad \begin{array}{ccc} \mathbb{A}_f \otimes_{\mathbb{Z}} \Lambda & \xrightarrow[\simeq]{[\sigma]_{\mathbb{A}_f}} & \mathbb{A}_f \otimes_{\mathbb{Z}} \Lambda_\sigma \\ {\scriptstyle \mathrm{N}_\Phi(1/s)}\downarrow & \nearrow{\scriptstyle \mathrm{H}_1(\lambda_{\sigma,s})} & \\ \mathbb{A}_f \otimes_{\mathbb{Z}} \Lambda & & \end{array}$$

where $[\sigma]_{\mathbb{A}_f}$ is obtained by applying the scalar extension $\widehat{\mathbb{Z}} \to \mathbb{A}_f$ to the isomorphism

$$\widehat{\mathbb{Z}} \otimes_{\mathbb{Z}} \Lambda = T_f(A) \overset{[\sigma]}{\simeq} T_f(A^\sigma) = \widehat{\mathbb{Z}} \otimes_{\mathbb{Z}} \Lambda_\sigma$$

of finite free $\widehat{\mathbb{Z}}$-modules. Hence, the image $\widehat{\mathbb{Z}} \otimes_{\mathbb{Z}} (\mathrm{N}_\Phi(1/s)\Lambda)$ of $\widehat{\mathbb{Z}} \otimes_{\mathbb{Z}} \Lambda$ along the left side of (A.2.8.3) (check this really is the image!) is carried by the $\mathbb{A}_f$-module map $\mathbb{A}_f \otimes_{\mathbb{Q}} \mathrm{H}_1(\lambda_{\sigma,s})$ isomorphically onto $\widehat{\mathbb{Z}} \otimes_{\mathbb{Z}} \Lambda_\sigma$. This confirms that $\mathrm{H}_1(\lambda_{\sigma,s})$ carries $\mathrm{N}_\Phi(1/s)\Lambda$ onto Λ_σ, since $\mathbb{Z} \to \widehat{\mathbb{Z}}$ is faithfully flat.

In this way, we get an analytic isomorphism $V/\mathrm{N}_\Phi(1/s)\Lambda \simeq V_\sigma/\Lambda_\sigma = A^\sigma(\mathbb{C})$. This identifies $A^\sigma(\mathbb{C})_{\mathrm{tor}}$ with $\Lambda_{\mathbb{Q}}/\mathrm{N}_\Phi(1/s)\Lambda$. Due to how this identification has been *constructed*, when it is used in conjunction with the identification $\Lambda_{\mathbb{Q}}/\Lambda \simeq A(\mathbb{C})_{\mathrm{tor}}$ we see that the map

$$A(\mathbb{C})_{\mathrm{tor}} = A(\overline{\mathbb{Q}})_{\mathrm{tor}} \overset{[\sigma]}{\simeq} A^\sigma(\overline{\mathbb{Q}})_{\mathrm{tor}} = A^\sigma(\mathbb{C})_{\mathrm{tor}}$$

is identified with the isomorphism $\Lambda_{\mathbb{Q}}/\Lambda \simeq \Lambda_{\mathbb{Q}}/\mathrm{N}_\Phi(1/s)\Lambda$ induced by $\mathrm{N}_\Phi(1/s)$-multiplication on primary components (as in Lemma A.2.8.1). $\square$

A.2.8.4. Remark. For a principal CM abelian variety (A, i) of type (L, Φ) over $\overline{\mathbb{Q}}$, Theorem A.2.8.3 and Example A.2.8.2 provide an L-linear analytic isomorphism

$$A^\sigma(\mathbb{C}) \simeq V/\mathrm{N}_\Phi(1/s)\Lambda \simeq [\mathrm{N}_\Phi(1/s)]_L \otimes_{\mathcal{O}_L} (V/\Lambda) \simeq ([\mathrm{N}_\Phi(1/s)]_L \otimes_{\mathcal{O}_L} A)(\mathbb{C}).$$

By GAGA and Lemma 1.2.1.2, this is induced by an L-linear isomorphism

$$A^\sigma \simeq [\mathrm{N}_\Phi(1/s)]_L \otimes_{\mathcal{O}_L} A$$

over $\overline{\mathbb{Q}}$. This latter isomorphism is precisely the isomorphism $\theta_{\sigma,s}$ that is uniquely characterized in Theorem A.2.7.2. In this sense, Theorem A.2.8.3 may be considered to be a generalization of Theorem A.2.7.2 to the non-principal case.

The interested reader may check as an exercise that the identification of Ψ_{ϕ^σ} with $q_s\Psi_\phi$ in Theorem A.2.8.3 is equivalent to the description of Ψ_{ϕ^σ} in terms of Ψ_ϕ given in [**113**, 18.6] using the classical description of Riemann forms of L-linear $\mathbb{Q}$-polarizations via 4-tuples $(L, \Phi, \mathfrak{a}, \zeta)$ (resting on choosing bases for tangent spaces as rank-1 free modules over $(\mathbb{R} \otimes_{\mathbb{Q}} L)_\Phi$).

We now finish the discussion of adelic operations by using a generalization the Serre tensor construction to give an algebraic formulation and proof of the adelic multiplication formalism in Theorem A.2.8.3.

For a CM field L, let (A, i) and (A', i') be L-linear CM abelian varieties over a field F of characteristic 0, so $\mathrm{Hom}((A', i'), (A, i))$ is a lattice in the L-vector space $\mathrm{Hom}^0((A', i'), (A, i))$ of dimension $\leqslant 1$. Hence, for any finite idele $s \in \mathbb{A}_{L,f}^\times$ another such lattice is $s\,\mathrm{Hom}((A',i'),(A,i))$. For a fixed (A,i), consider the following functor $T = T_{(A,i)}$ from the category of such pairs (A',i') (using L-linear maps over F) to the category of order lattices in finite-dimensional L-vector spaces:

$$T(A',i') = s\,\mathrm{Hom}((A',i'),(A,i)).$$

For $F = \mathbb{C}$, if we write $A(\mathbb{C}) = V/\Lambda$ and $A'(\mathbb{C}) = V'/\Lambda'$ then working locally over $\mathbb{Z}$ shows that $T(A',i') = \mathrm{Hom}((V'/\Lambda',i'),(V/s\Lambda,i_s))$ inside $\mathrm{Hom}_L(\Lambda'_{\mathbb{Q}},\Lambda_{\mathbb{Q}})$, with i_s denoting the action map on $s\Lambda$ by an order in L. Thus, an algebraic version of the adelic operation $\Lambda \rightsquigarrow s\Lambda$ on lattices is provided by the following lemma.

A.2.8.5. Lemma. *Let (A,i) over F be as above. The functor $T_{(A,i)}$ is represented by a pair (sA,i_s) and there is a canonical L-linear equality*

$$\mathrm{Hom}(A',sA) = \mathrm{Hom}^0(A',A) \cap \mathrm{Hom}_{\mathbb{A}_f[\mathrm{Gal}(\overline{F}/F)]}(T_f(A'),sT_f(A))$$

inside $\mathrm{Hom}_{\mathbb{A}_f[\mathrm{Gal}(\overline{F}/F)]}(V_f(A'),V_f(A))$.

In particular, there is a canonical isomorphism $(sA)^t \simeq s^{-1}A^t$ and the formation of the representing object sA commutes with extension of the base field F'/F in the sense that the canonical map $(sA)_{F'} \to s(A_{F'})$ is an isomorphism.

PROOF. It is harmless to multiply s by a sufficiently divisible non-zero integer so that it lies in the profinite completion $\mathcal{O}^\wedge$ of a common order $\mathcal{O} \subseteq \mathcal{O}_L$ acting on A and A'. Hence, suppressing explicit mention of the L-action maps, the subset $s\,\mathrm{Hom}(A',A) \subseteq \mathrm{Hom}(A',A)$ is given by local conditions: its ℓ-adic completion is $s_\ell(\mathbb{Z}_\ell \otimes_{\mathbb{Z}} \mathrm{Hom}(A',A))$ for all primes ℓ (with $s_\ell \in (\mathbb{Q}_\ell \otimes_{\mathbb{Q}} L)^\times$).

Under the injection of $\mathrm{Hom}(A',A)$ into the $\mathcal{O}_\ell$-module $\mathrm{Hom}(A'[\ell^\infty],A[\ell^\infty])$, the set $s\,\mathrm{Hom}(A,A')$ consists of those elements of $\mathrm{Hom}(A,A')$ that kill the kernel of the isogeny s_ℓ on $A'[\ell^\infty]$. By $\mathcal{O}_\ell$-linearity, it is equivalent to require that the dual map of ℓ-divisible groups kills the kernel of s_ℓ acting on $A^t[\ell^\infty]$ through duality (without the intervention of complex conjugation on L).

Hence, if we let $G_\ell \subset A^t[\ell^\infty]$ be the F-finite kernel of the isogeny given by the dual action of s_ℓ then $G_\ell = 0$ for all but finitely many ℓ and so there is a unique F-finite subgroup G in A^t whose ℓ-component is G_ℓ for all ℓ. Dualizing again, $T_{(A,i)}(A',i')$ consists of those $\mathcal{O}$-linear maps $A' \to A$ over F that lift (necessarily uniquely) through the isogeny $(A^t/G)^t \to A$. That is, $(A^t/G)^t$ represents the functor $T_{(A,i)}$. It is clear from this construction that the other properties hold. □

The proof of Lemma A.2.8.5 provides a canonical $\mathbb{A}_{L,f}$-linear and $\mathrm{Gal}(\overline{F}/F)$-equivariant isomorphism $V_f(sA) = V_f(A)$ identifying $T_f(sA)$ with $s(T_f(A))$, and this respects duality and extension of the base field. There is likewise a unique L-linear isomorphism $s'(sA) \simeq (s's)A$ over F compatible with the adelic Tate module description (so these isomorphisms are associative with respect to any three ideles).

By the construction, in the case $F = \mathbb{C}$ these Tate module descriptions are the $\widehat{\mathbb{Z}}$-scalar extensions arising from unique analytic isomorphisms $V/(s\Lambda) \simeq s(V/\Lambda)$. In this sense, the formation of sA is an algebraic substitute for adelic operations on uniformization lattices and so (via the Lefschetz principle) it follows that the CM type of sA coincides with that of A.

A.2.8.6. Example. Assume the CM order is $\mathscr{O}_L$. There are canonical L-linear isomorphisms $sA \simeq [s]_L \otimes_{\mathscr{O}_L} A$ over F that respect iteration of the idelic multiplication and Serre tensor operations, and for $s = \mathrm{N}_\Phi(1/s')$ with $s' \in \mathbb{A}_{E,f}^\times$, the resulting identification $V_f(A) = V_f(sA) \simeq V_f([s]_L \otimes_{\mathscr{O}_L} A)$ is $V_f(\psi_{s'})$ for $\psi_{s'}$ as in A.2.7 (due to Remark A.2.7.4).

The following lemma is obvious, and by the *proof* of Lemma A.2.7.1 it recovers the construction in Theorem A.2.7.2 in the principal case:

A.2.9. Lemma. *Let ϕ be an L-linear polarization of a pair (A, i) over $\overline{\mathbb{Q}}$ with* CM *type (L, Φ). For all $s \in \mathbb{A}_{E,f}^\times$ there is a unique L-linear polarization ϕ_s on $\mathrm{N}_\Phi(s)A$ such that the $\mathbb{A}_{L,f}$-linear identification $V_f(A) = V_f(\mathrm{N}_\Phi(s)A)$ carries e_{ϕ_s} to $q_s^{-1}e_\phi$ as adelic Weil pairings, where $q_s \in \mathbb{Q}_{>0}^\times$ is the unique positive generator of $[\mathrm{Nm}_{E/\mathbb{Q}}(s)]_\mathbb{Q}$.* □

The algebraic version of Theorem A.2.8.3 is:

A.2.9.1. Theorem. *Let (A, i) be a* CM *abelian variety over $\overline{\mathbb{Q}}$ with* CM *type (L, Φ). Choose $\sigma \in \mathrm{Gal}(\overline{\mathbb{Q}}/E)$ and $s \in \mathbb{A}_{E,f}^\times$ such that $r_E(s) = \sigma|_{E^{\mathrm{ab}}}$.*

There is a unique L-linear isomorphism $\theta_{\sigma,s} : \mathrm{N}_\Phi(1/s) \cdot A \simeq A^\sigma$ with respect to which the composite map

$$T_f(A) \overset{\sigma}{\simeq} T_f(A^\sigma) \overset{\theta_{\sigma,s}^{-1}}{\simeq} T_f(\mathrm{N}_\Phi(1/s) \cdot A) = \mathrm{N}_\Phi(1/s) T_f(A)$$

is multiplication by $\mathrm{N}_\Phi(1/s) \in \mathbb{A}_{L,f}^\times$. Moreover, for any L-linear polarization ϕ the isomorphism $\theta_{\sigma,s}$ carries ϕ^σ to $\phi_{1/s}$.

The analogue of Remark A.2.7.3 carries over by essentially the same argument (chasing actions on V_f's rather than on torsion subgroups).

PROOF. This follows from the analytic version in Theorem A.2.8.3, but we give an algebraic proof. In the special case that A has CM order $\mathscr{O}_L$, the theorem is exactly Theorem A.2.7.2. In the general case, uniqueness for $\theta_{\sigma,s}$ is clear and once existence is proved we obtain the final assertion from Lemma A.2.9.

To prove existence, first observe that if we construct $\theta_{\sigma,s}$ merely in the isogeny category subject to the condition that composing $V_f(\theta_{\sigma,s}^{-1})$ and $[\sigma]$ is multiplication by $\mathrm{N}_\Phi(1/s)$ on $V_f(A)$ then consideration with adelic Tate modules forces $\theta_{\sigma,s}$ to be a genuine isomorphism of abelian varieties. Thus, to prove existence it suffices to work in the isogeny category.

If $A' \to A$ is an L-linear isogeny-category isomorphism then it is equivalent to solve solve the existence problem for A or for A' (by functoriality of the idelic multiplication operation on abelian varieties with a fixed CM type). But every A of type (L, Φ) over $\overline{\mathbb{Q}}$ is L-linearly isogenous to one with CM order $\mathscr{O}_L$ (Proposition 1.7.4.5). Hence, we are done. □

Theorem A.2.9.1 accomplishes something remarkable: without Galois theory or complex analysis, it gives an adelic construction of $(A^\sigma, i^\sigma, \phi^\sigma)$ from (A, i, ϕ): this σ-twisted triple is $(\mathrm{N}_\Phi(1/s) \cdot A, i_{\mathrm{N}_\Phi(1/s)}, \phi_{1/s})$ with the canonical $\mathbb{A}_{L,f}$-linear identification $V_f(\mathrm{N}_\Phi(1/s)A) = V_f(A)$ carrying $e_{\phi_{1/s}}$ to $q_s e_\phi$, and the L-linear isomorphism $\theta_{\sigma,s} : \mathrm{N}_\Phi(1/s)A \simeq A^\sigma$ is uniquely characterized by the fact that it fits into the commutative diagram

$$\begin{array}{ccc} V_f(A) & \xrightarrow[\simeq]{[\sigma]} & V_f(A^\sigma) \\ {\scriptstyle \mathrm{N}_\Phi(1/s)}\downarrow\simeq & & \simeq\uparrow{\scriptstyle V_f(\theta_{\sigma,s})} \\ V_f(A) & \xrightarrow[\simeq]{} & V_f(\mathrm{N}_\Phi(1/s)A) \end{array}$$

with the bottom side given by the canonical identification (that in turn arises from a unique L-linear isogeny $\psi_s : A \to \mathrm{N}_\Phi(1/s)A$). In Example A.4.6.1 we will see how to use this analytic result to translate Shimura's notion of "field of moduli" for polarized CM abelian varieties into a notion that makes sense for a large class of algebraic Hecke characters (going beyond reflex norms N_Φ for CM types Φ on CM fields).

A.3. A converse to the Main Theorem of Complex Multiplication

In this section we prove Theorem 2.5.2, which is a converse to the Main Theorem. The argument will use constructions from the proof of the Main Theorem in A.2. The notation and hypotheses from the statement of Theorem 2.5.2 will be freely used in A.3.1–A.3.5 below without comment.

A.3.1. First we check the uniqueness of (A, α) up to L-linear isogenies. If (A, α) and (A', α') are two such pairs, then since they have the same CM type (L, Φ) there is (by Proposition 1.5.4.1) an L-linear isogeny $h : A_{\overline{\mathbb{Q}}} \to A'_{\overline{\mathbb{Q}}}$. The invertible L_ℓ-modules $V_\ell(A)$ and $V_\ell(A')$ have $\mathrm{Gal}(\overline{\mathbb{Q}}/K)$ acting through the same $L_\ell^\times$-valued character (namely the continuous ψ_ℓ that is uniquely attached to ϵ), so all L_ℓ-linear maps $V_\ell(A) \to V_\ell(A')$ are automatically $\mathrm{Gal}(\overline{\mathbb{Q}}/K)$-equivariant. Thus, for all $\sigma \in \mathrm{Gal}(\overline{\mathbb{Q}}/K)$ we have $V_\ell(\sigma^*(h)) = \rho_{A',\ell}(\sigma) \circ V_\ell(h) \circ \rho_{A,\ell}(\sigma)^{-1} = V_\ell(h)$, so $\sigma^*(h) = h$. That is, h is defined over K (and its K-descent is an L-linear isogeny). To prove the existence of (A, α), which is the hard part, we will use Galois descent.

A.3.2. By Proposition 1.5.4.1, there is a finite extension K'/K inside $\overline{\mathbb{Q}}$ and a CM abelian variety (A', α') over K' with CM type (L, Φ). Its associated algebraic Hecke character $\epsilon' : \mathbb{A}_{K'}^\times \to L^\times$ has algebraic part $\mathrm{N}_\Phi \circ \mathrm{Nm}_{K'/E} = \epsilon_{\mathrm{alg}} \circ \mathrm{Nm}_{K'/K}$. Thus, the algebraic Hecke characters ϵ' and $\epsilon \circ \mathrm{Nm}_{K'/K}$ have the same algebraic part. It follows from Remark 2.4.6(ii) that they are related through multiplication by a finite-order Hecke character of K', so upon replacing K' with a finite extension we can arrange that $\epsilon' = \epsilon \circ \mathrm{Nm}_{K'/K}$.

We may and do also assume that K'/K is Galois. By means of a suitable isogeny over K', we can arrange that $\mathcal{O}_L$ is the CM order (i.e., $\mathcal{O}_L$ acts on A' as an abelian variety, not only in the isogeny category); see Proposition 1.7.4.5. We shall construct an $\mathcal{O}_L$-linear descent datum on $(A', \alpha')_{\overline{\mathbb{Q}}}$ with respect to $\overline{\mathbb{Q}}/K$, and the resulting K-descent (A, α) of $(A', \alpha')_{\overline{\mathbb{Q}}}$ will then be checked to have associated algebraic Hecke character ϵ; it necessarily has CM type (L, Φ).

For any finite idele $\xi \in \mathbb{A}_{L,f}^\times$, let $[\xi]_L$ denote the associated fractional ideal of $\mathcal{O}_L$. In particular, if $s \in \mathbb{A}_{E,f}^\times$ then the reflex norm $\mathrm{N}_\Phi : \mathrm{Res}_{E/\mathbb{Q}}(\mathbb{G}_{\mathrm{m}}) \to \mathrm{Res}_{L/\mathbb{Q}}(\mathbb{G}_{\mathrm{m}})$ yields a fractional ideal $[\mathrm{N}_\Phi(s)]_L$. Recall from 1.7.4 that if X is any $\mathcal{O}_L$-module scheme over a scheme S and if M is any finite flat $\mathcal{O}_L$-module then there is an associated $\mathcal{O}_L$-module scheme $M \otimes_{\mathcal{O}_L} X$ over S that represents the functor $S' \rightsquigarrow M \otimes_{\mathcal{O}_L} X(S')$ on the category of S-schemes. In the special case that M is invertible and X is an abelian scheme with constant relative dimension g then $M \otimes_{\mathcal{O}_L} X$ is another such abelian scheme; if X with its $\mathcal{O}_L$-action is a CM abelian variety over a field of characteristic 0 then $M \otimes_{\mathcal{O}_L} X$ is another such (for invertible M) and it has the same CM type.

Choose $\sigma \in \mathrm{Gal}(\overline{\mathbb{Q}}/E)$ and $s \in \mathbb{A}_{E,f}^{\times}$ satisfying $r_E(s) = \sigma|_{E^{\mathrm{ab}}}$, and let I_s denote the fractional ideal $[\mathrm{N}_{\Phi}(1/s)]_L$, so for all $n \geqslant 1$ the finite idele $\mathrm{N}_{\Phi}(1/s) \in \mathbb{A}_{L,f}^{\times}$ defines a canonical generator of the $\mathcal{O}_L/(n)$-module I_s/nI_s. By the Main Theorem of Complex Multiplication in its adelic form for CM abelian varieties with maximal CM order (Theorem A.2.7.2), there is a unique L-linear isomorphism of abelian varieties

$$\theta_{\sigma,s} : [\mathrm{N}_{\Phi}(1/s)]_L \otimes_{\mathcal{O}_L} A'_{\overline{\mathbb{Q}}} = I_s \otimes_{\mathcal{O}_L} A'_{\overline{\mathbb{Q}}} \simeq (A'_{\overline{\mathbb{Q}}})^{\sigma}$$

over $\overline{\mathbb{Q}}$ such that for all $n \geqslant 1$ the natural isomorphism

$$[\sigma] : A'_{\overline{\mathbb{Q}}}[n](\overline{\mathbb{Q}}) \simeq (A'_{\overline{\mathbb{Q}}})^{\sigma}[n](\overline{\mathbb{Q}})$$

is given by the composite map

$$\begin{aligned} A'_{\overline{\mathbb{Q}}}[n](\overline{\mathbb{Q}}) \xrightarrow[\simeq]{\mathrm{N}_{\Phi}(1/s)} (I_s/nI_s) \otimes_{\mathcal{O}_L} A'_{\overline{\mathbb{Q}}}[n](\overline{\mathbb{Q}}) \xrightarrow[\simeq]{} & (I_s \otimes_{\mathcal{O}_L} A'_{\overline{\mathbb{Q}}})[n](\overline{\mathbb{Q}}) \\ \xrightarrow[\simeq]{\theta_{\sigma,s}} & (A'_{\overline{\mathbb{Q}}})^{\sigma}[n](\overline{\mathbb{Q}}) \end{aligned}$$

By uniqueness, for any $c \in E^{\times}$ we have that $\theta_{\sigma,cs}$ is the composition of $\theta_{\sigma,s}$ and the natural multiplication map by $\mathrm{N}_{\Phi}(c) \in L^{\times}$ carrying $I_{cs} = [\mathrm{N}_{\Phi}(1/(cs))]_L$ isomorphically to $I_s = [\mathrm{N}_{\Phi}(1/s)]_L$ as fractional ideals of $\mathcal{O}_L$.

The intervention of $\mathrm{N}_{\Phi}(1/s)$ rather than $\mathrm{N}_{\Phi}(s)$ in the definition of $\theta_{\sigma,s}$ is due to the use of the arithmetic normalization of the Artin map (implicit in the requirement $r_E(s) = \sigma|_{E^{\mathrm{ab}}}$). We are interested in applying this with $\sigma \in \mathrm{Gal}(\overline{\mathbb{Q}}/K) \subseteq \mathrm{Gal}(\overline{\mathbb{Q}}/E)$ and $s = \mathrm{Nm}_{K/E}(\xi)$ for any $\xi \in \mathbb{A}_{K,f}^{\times}$ such that $r_K(\xi) = \sigma|_{K^{\mathrm{ab}}}$ (so $r_E(s) = \sigma|_{E^{\mathrm{ab}}}$).

A.3.3. For $\xi \in \mathbb{A}_{K,f}^{\times}$ we claim that $[\mathrm{N}_{\Phi}(\mathrm{Nm}_{K/E}(\xi))]_L$ is the principal fractional ideal generated by $\epsilon(\xi) \in L^{\times}$. To prove this, first recall that $\mathrm{N}_{\Phi} \circ \mathrm{Nm}_{K/E} = \epsilon_{\mathrm{alg}}$. Thus, it suffices to prove more generally that if F is a number field and $\chi : \mathbb{A}_F^{\times} \to F'^{\times}$ is an algebraic Hecke character valued in the multiplicative group of some number field F' then for any $s \in \mathbb{A}_F^{\times}$, $\chi(s)$ is a generator of the fractional ideal $[(\chi_{\mathrm{alg}})_{\mathbb{A}}(s)]_{F'}$, where $(\chi_{\mathrm{alg}})_{\mathbb{A}} : \mathbb{A}_F^{\times} \to \mathbb{A}_{F'}^{\times}$ is the map induced by $\chi_{\mathrm{alg}} : \mathrm{Res}_{F/\mathbb{Q}}(\mathbb{G}_{\mathrm{m}}) \to \mathrm{Res}_{F'/\mathbb{Q}}(\mathbb{G}_{\mathrm{m}})$ on $\mathbb{A}$-points.

Note that χ annihilates the infinitely divisible identity component of the archimedean factor $F_{\infty}^{\times}$, and $(\chi_{\mathrm{alg}})_{\mathbb{A}} \cdot \chi^{-1}$ may be viewed as a continuous homomorphism $\mathbb{A}_F^{\times}/F^{\times} \to \mathbb{A}_{F'}^{\times}$. In particular, it carries the compact norm-1 idele class group of F into the maximal compact subgroup of $\mathbb{A}_{F'}^{\times}$. But composing this map with projection to $\mathbb{A}_{F',f}^{\times}$ kills the image of the identity component of the archimedean factor $F_{\infty}^{\times}$ and so carries $\mathbb{A}_{F,f}^{\times}$ into the maximal compact subgroup $\prod_{v \nmid \infty} \mathcal{O}_{F',v}^{\times}$ (since any finite idele of F can be realized as the finite part of a norm-1 idele of F whose archimedean factor is in the identity component). This says that $[(\chi_{\mathrm{alg}})_{\mathbb{A}}(s)]_{F'} = \chi(s)\mathcal{O}_{F'}$ as fractional ideals of $\mathcal{O}_{F'}$ for any $s \in \mathbb{A}_F^{\times}$, as desired.

For $\sigma \in \mathrm{Gal}(\overline{\mathbb{Q}}/K)$, choose $\xi_{\sigma} \in \mathbb{A}_{K,f}^{\times}$ such that $\sigma|_{K^{\mathrm{ab}}} = r_K(\xi_{\sigma})$ in $\mathrm{Gal}(K^{\mathrm{ab}}/K)$ and let $s_{\sigma} = \mathrm{Nm}_{K/E}(\xi_{\sigma}) \in \mathbb{A}_{E,f}^{\times}$. Thus, $[\mathrm{N}_{\Phi}(s_{\sigma})]_L = [(\epsilon_{\mathrm{alg}})_{\mathbb{A}}(\xi_{\sigma})]_L = \epsilon(\xi_{\sigma})\mathcal{O}_L$.

A.3.3.1. Lemma. *The composite L-linear isomorphism*

$$A'_{\overline{\mathbb{Q}}} \overset{\epsilon(\xi_\sigma)^{-1}}{\simeq} [\mathrm{N}_\Phi(1/s_\sigma)]_L \otimes_{\mathcal{O}_L} A'_{\overline{\mathbb{Q}}} \overset{\theta_{\sigma,s_\sigma}}{\simeq} (A'_{\overline{\mathbb{Q}}})^\sigma$$

is independent of the choice of ξ_σ.

In A.3.4 and A.3.5 we will prove that these isomorphisms for varying σ constitute a Galois descent datum on $(A', \alpha')_{\overline{\mathbb{Q}}}$ with respect to $\mathrm{Gal}(\overline{\mathbb{Q}}/K)$ such that the resulting descent (A, α) over K is the desired CM abelian variety.

Proof. Let $\xi'_\sigma \in \mathbb{A}_{K,f}^\times$ be another such choice, so $\xi'_\sigma\xi_\sigma^{-1} \in \ker(r_K) = \overline{K_\infty^\times K^\times}$ (closure in $\mathbb{A}_K^\times$). Thus, the element $\xi'_\sigma\xi_\sigma^{-1} \in \mathbb{A}_{K,f}^\times \simeq \mathbb{A}_K^\times/K_\infty^\times$ is in the closure of $K^\times$ in $\mathbb{A}_{K,f}^\times$, so for any open subset $U \subseteq \mathbb{A}_{K,f}^\times$ around the identity there exists $\lambda \in K^\times$ such that $\xi'_\sigma \in \lambda\xi_\sigma U$. Since $\ker(\epsilon)$ meets $\mathbb{A}_{K,f}^\times$ in an open subgroup and $[\mathrm{N}_\Phi(s_\sigma)]_L$ only depends on s_σ modulo the maximal compact open subgroup in $\mathbb{A}_{E,f}^\times$, it suffices to check two things: (a) if $\xi'_\sigma = \lambda\xi_\sigma$ for some $\lambda \in K^\times$ then the isomorphism in the statement of the Lemma and the analogous isomorphism for ξ'_σ define the same composite isomorphism $A'_{\overline{\mathbb{Q}}} \simeq (A'_{\overline{\mathbb{Q}}})^\sigma$, and (b) for any $\sigma \in \mathrm{Gal}(\overline{\mathbb{Q}}/E)$ and $s \in \mathbb{A}_{E,f}^\times$ such that $r_E(s) = \sigma|_{E^{\mathrm{ab}}}$, the isomorphism $\theta_{\sigma,s}$ is unaffected by replacing s with us for $u \in \mathbb{A}_{E,f}^\times \cap \ker(r_E)$ sufficiently near 1 (nearness possibly depending on s).

For (a), $s'_\sigma := \mathrm{Nm}_{K/E}(\xi'_\sigma)$ is equal to $\mathrm{Nm}_{K/E}(\lambda) \cdot s_\sigma$, so $\mathrm{N}_\Phi(s'_\sigma) = \epsilon(\lambda)\mathrm{N}_\Phi(s_\sigma)$. The required equality of $\overline{\mathbb{Q}}$-isomorphisms $A'_{\overline{\mathbb{Q}}} \simeq (A'_{\overline{\mathbb{Q}}})^\sigma$ therefore reduces to a general identity already noted in A.3.2: for any $c \in E^\times$ (such as $c = \mathrm{Nm}_{K/E}(\lambda)$), $\theta_{\sigma,cs}$ is the composition of $\theta_{\sigma,s}$ and the action of $\mathrm{N}_\Phi(c) \in L^\times$.

For (b), to study the general behavior of $\theta_{\sigma,s}$ under multiplication of s against any $u \in \ker(r_E) \cap \mathbb{A}_{E,f}^\times$ that is sufficiently near 1, we can restrict our attention to multiplication by $u \in \ker(r_E) \cap \prod_{v \nmid \infty} \mathcal{O}_{E,v}^\times$. If $s' = us$ for such a u then $[\mathrm{N}_\Phi(s')]_L = [\mathrm{N}_\Phi(s)]_L$, so $\theta_{\sigma,s}^{-1} \circ \theta_{\sigma,s'}$ is an automorphism of $[\mathrm{N}_\Phi(1/s)]_L \otimes_{\mathcal{O}_L} A'_{\overline{\mathbb{Q}}}$. Moreover, by choosing u sufficiently near 1 we can ensure that $\theta_{\sigma,s}$ and $\theta_{\sigma,s'}$ induce the same map on m-torsion for a fixed $m \geqslant 3$, so $\theta_{\sigma,s}^{-1} \circ \theta_{\sigma,s'}$ is trivial on the m-torsion.

By Lemma 2.2.2(1), there is an L-linear polarization ϕ of $A'_{\overline{\mathbb{Q}}}$. By Lemma A.2.7.1 and the subsequent construction of $\theta_{\sigma,s}$, the map $\theta_{\sigma,s}^{-1} \circ \theta_{\sigma,s'}$ commutes with a certain polarization $\phi_s = \phi_{s'}$ canonically associated to ϕ and the fractional ideal $[\mathrm{N}_\Phi(s)]_L = [\mathrm{N}_\Phi(s')]_L$. But an automorphism of a polarized abelian variety has finite order and hence is trivial when it is trivial on the m-torsion for some $m \geqslant 3$. □

Lemma A.3.3.1 provides an L-linear $\overline{\mathbb{Q}}$-isomorphism

$$c(\sigma) : A'_{\overline{\mathbb{Q}}} \simeq (A'_{\overline{\mathbb{Q}}})^\sigma$$

depending only on $\sigma \in \mathrm{Gal}(\overline{\mathbb{Q}}/K)$.

A.3.4. Since the isomorphisms $\theta_{\sigma,s}$ are L-linear and we can use $\xi_\sigma\xi_\tau$ as $\xi_{\sigma\tau}$, it is easy to check that the cocycle relation $c(\sigma\tau) \overset{?}{=} \sigma^*(c(\tau)) \circ c(\sigma)$ for $\sigma, \tau \in \mathrm{Gal}(\overline{\mathbb{Q}}/K)$ reduces to the general identity $\theta_{\sigma\tau,st} \overset{?}{=} \sigma^*(\theta_{\tau,t}) \circ (1 \otimes \theta_{\sigma,s})$ as isomorphisms

$$[\mathrm{N}_\Phi(1/st)]_L \otimes_{\mathcal{O}_L} A'_{\overline{\mathbb{Q}}} \simeq [\mathrm{N}_\Phi(1/s)]_L \otimes_{\mathcal{O}_L} ([\mathrm{N}_\Phi(1/t)]_L \otimes_{\mathcal{O}_L} A'_{\overline{\mathbb{Q}}}) \rightrightarrows ((A'_{\overline{\mathbb{Q}}})^\tau)^\sigma \simeq (A'_{\overline{\mathbb{Q}}})^{\sigma\tau}$$

for $\sigma, \tau \in \mathrm{Gal}(\overline{\mathbb{Q}}/E)$ and $s, t \in \mathbb{A}_{E,f}^\times$ such that $r_E(s) = \sigma|_{E^{\mathrm{ab}}}$ and $r_E(t) = \tau|_{E^{\mathrm{ab}}}$. This identity is immediately deduced from the unique characterization of $\theta_{\sigma\tau,st}$.

The 1-cocycle $\sigma \mapsto c(\sigma)$ on $\mathrm{Gal}(\overline{\mathbb{Q}}/K)$ valued in $\overline{\mathbb{Q}}$-isomorphisms is a descent datum if it satisfies a continuity condition, as follows. By the construction of

$$\theta_{\sigma,s} : [\mathrm{N}_\Phi(1/s)]_L \otimes_{\mathcal{O}_L} A'_{\overline{\mathbb{Q}}} \simeq (A'_{\overline{\mathbb{Q}}})^\sigma$$

in A.2.6–A.2.7, there is a finite extension K''/K' inside $\overline{\mathbb{Q}}$ that is Galois over K such that the L-linear $\theta_{\sigma,s}$ descends to an isomorphism $[\mathrm{N}_\Phi(1/s)]_L \otimes_{\mathcal{O}_L} A'_{K''} \simeq (A'_{K''})^{\sigma|_{K''}}$ for all $\sigma \in \mathrm{Gal}(\overline{\mathbb{Q}}/E)$ and $s \in \mathbb{A}^\times_{E,f}$ such that $r_E(s) = \sigma|_{E^{\mathrm{ab}}}$. (We do not expect that we can take $K'' = K'$.) Hence, for every $\sigma \in \mathrm{Gal}(\overline{\mathbb{Q}}/K)$ the isomorphism $c(\sigma)$ over $\overline{\mathbb{Q}}$ descends to an L-linear isomorphism $c_{K''}(\sigma) : A'_{K''} \simeq (A'_{K''})^{\sigma|_{K''}}$. Increase K'' to split $A'[m]$ for a fixed $m \geqslant 3$, so $c_{K''}(\sigma)$ is the identity on m-torsion when $\sigma|_{K''}$ is trivial.

Choose an L-linear polarization ϕ of A'; this exists by Lemma 2.2.2(1). By Lemma A.2.7.1, there is an associated L-linear polarization ϕ_α of $[\mathrm{N}_\Phi(\alpha)]_L \otimes_{\mathcal{O}_L} A'$ for all $\alpha \in \mathbb{A}^\times_{E,f}$. Using the properties of the ϕ_α as in Lemma A.2.7.1, the definition of $c(\sigma)$ via Lemma A.3.3.1 implies that $c_{K''}(\sigma)$ carries $\phi_{K''}$ to $(\phi_{K''})^{\sigma|_{K''}}$ for all $\sigma \in \mathrm{Gal}(\overline{\mathbb{Q}}/K)$. Hence, if $\sigma|_{K''}$ is the identity then $c_{K''}(\sigma)$ is an automorphism of a polarized abelian variety and is the identity on the m-torsion for some $m \geqslant 3$, so $c_{K''}(\sigma)$ is the identity for such σ.

It follows (using the cocycle relation) that $c_{K''}(\sigma)$ only depends on $\sigma|_{K''}$, so we can view $c_{K''}$ as a function on $\mathrm{Gal}(K''/K)$ that defines an L-linear Galois descent datum on $A'_{K''}$ with respect to K''/K. In other words, if we rename K'' as K' (as we may), then we get to the case where

$$\sigma \mapsto c_{K'}(\sigma) \in \mathrm{Isom}(A', A'^\sigma)$$

is a Galois descent datum on (A', α') with respect to K'/K. By descent theory, this uniquely determines an abelian variety A over K equipped with a ring homomorphism $\alpha : \mathcal{O}_K \to \mathrm{End}(A)$ and an L-linear isomorphism $A_{K'} \simeq A'$ respecting the Galois descent data on both sides.

A.3.5. Let $\epsilon_A : \mathbb{A}^\times_K \to L^\times$ be the algebraic Hecke character associated to (A, α). To prove that $\epsilon_A = \epsilon$, it suffices (by Lemma 2.4.9) to show that for a fixed choice of rational prime ℓ the continuous ℓ-adic characters $\psi_\ell, \psi'_\ell : \mathrm{Gal}(K^{\mathrm{ab}}/K) \rightrightarrows L^\times_\ell$ that are respectively uniquely associated to ϵ and ϵ_A coincide. By Theorem 2.5.1(ii), ψ'_ℓ computes the action of $\mathrm{Gal}(K^{\mathrm{ab}}/K)$ on $V_\ell(A)$. Also, by construction, $\psi_\ell(r_K(\xi)) = \epsilon(\xi)\mathrm{N}_\Phi(\mathrm{Nm}_{K/E}(1/\xi))_\ell$ for all $\xi \in \mathbb{A}^\times_{K,f}$. Thus, we want each $\sigma \in \mathrm{Gal}(\overline{\mathbb{Q}}/K)$ to act on $V_\ell(A)$ as multiplication by $\epsilon(\xi_\sigma)\mathrm{N}_\Phi(\mathrm{Nm}_{K/E}(1/\xi_\sigma))_\ell$ for any $\xi_\sigma \in \mathbb{A}^\times_{K,f}$ such that $r_K(\xi_\sigma) = \sigma|_{K^{\mathrm{ab}}}$.

Fix such a σ and ξ_σ, and let $s_\sigma = \mathrm{Nm}_{K/E}(\xi_\sigma)$. It is evident that σ acting on $V_\ell(A)$ is the composition of two maps:

(i) the canonical σ-pullback map $[\sigma] : V_\ell(A_{\overline{\mathbb{Q}}}) \simeq V_\ell((A_{\overline{\mathbb{Q}}})^\sigma)$,
(ii) the isomorphism on ℓ-adic Tate spaces induced by the descent-data isomorphism $(A_{\overline{\mathbb{Q}}})^\sigma \simeq A_{\overline{\mathbb{Q}}}$.

The definition of A by $\mathrm{Gal}(\overline{\mathbb{Q}}/K)$-descent of $A'_{\overline{\mathbb{Q}}}$ with respect to the 1-cocycle $\sigma \mapsto c(\sigma)$ provides a canonical isomorphism $A_{\overline{\mathbb{Q}}} \simeq A'_{\overline{\mathbb{Q}}}$ with respect to which the map $[\sigma]$ in (i) is identified with the composition of $V_\ell(\theta_{\sigma,s_\sigma})$ and the isomorphism

$$V_\ell(A_{\overline{\mathbb{Q}}}) \simeq [\mathrm{N}_\Phi(1/s_\sigma)]_L \otimes_{\mathcal{O}_L} V_\ell(A_{\overline{\mathbb{Q}}}) \simeq V_\ell([\mathrm{N}_\Phi(1/s_\sigma)]_L \otimes_{\mathcal{O}_L} A_{\overline{\mathbb{Q}}})$$

induced by multiplication by the ℓ-part $\mathrm{N}_\Phi(1/s_\sigma)_\ell$ of $\mathrm{N}_\Phi(1/s_\sigma) \in \mathbb{A}_{L,f}^\times$.

The descent data isomorphism $(A_{\overline{\mathbb{Q}}})^\sigma \simeq A_{\overline{\mathbb{Q}}}$ is (by definition in terms of $A'_{\overline{\mathbb{Q}}}$) inverse to the composition of θ_{σ,s_σ} and the multiplication map induced by the principal generator $\epsilon(1/\xi_\sigma) \in L^\times$ of $[\mathrm{N}_\Phi(1/s_\sigma)]_L$. By definition, the induced map $V_\ell((A_{\overline{\mathbb{Q}}})^\sigma) \simeq V_\ell(A_{\overline{\mathbb{Q}}})$ is the map in (ii) above.

Composing our descriptions of (i) and (ii), the two contributions from the mysterious θ_{σ,s_σ} cancel out and what remains is the product of multiplication by $\mathrm{N}_\Phi(1/s_\sigma)_\ell = \mathrm{N}_\Phi(\mathrm{Nm}_{K/E}(1/\xi_\sigma))_\ell \in L_\ell^\times$ and multiplication by $\epsilon(\xi_\sigma) \in L^\times$ on $V_\ell(A)$, exactly as desired. We have proved Theorem 2.5.2. □

A.4. Existence of algebraic Hecke characters

A.4.1. Let F and L be number fields, and let

$$\chi : \mathrm{Res}_{F/\mathbb{Q}}(\mathbb{G}_\mathrm{m}) \to \mathrm{Res}_{L/\mathbb{Q}}(\mathbb{G}_\mathrm{m})$$

be a $\mathbb{Q}$-homomorphism between the associated Weil restricted tori over $\mathbb{Q}$. In this section we explain a necessary and sufficient condition, due to Shimura, for the existence of an algebraic Hecke character $\epsilon : \mathbb{A}_F^\times \to L^\times$ with algebraic part $\epsilon_{\mathrm{alg}} = \chi$; see Theorem A.4.5.1. (We follow the notation in 2.4 for algebraic Hecke characters.) This is used in the proof of Propositions 3.7.3 and 3.7.11 (together with Casselman's converse 2.5.2 to the Main Theorem of Complex Multiplication) to construct CM p-divisible groups with a given p-adic CM type over the p-adic reflex field.

Shimura formulated his theorem in the situation that L is a CM field, F is a finite extension field of the reflex field E of a CM type (L, Φ), and χ is the composition

$$\mathrm{Res}_{F/\mathbb{Q}}(\mathbb{G}_\mathrm{m}) \xrightarrow{\mathrm{Nm}_{F/E}} \mathrm{Res}_{E/\mathbb{Q}}(\mathbb{G}_\mathrm{m}) \xrightarrow{\mathrm{N}_\Phi} \mathrm{Res}_{L/\mathbb{Q}}(\mathbb{G}_\mathrm{m})$$

of the field norm and the reflex norm. This special case is what we need for 3.7.3 and 3.7.11. Shimura's argument (in [**114**, §5, Thm. 11, Prop. 7]) works for general admissible $\mathbb{Q}$-homomorphisms between Weil restricted tori (as defined in A.4.4 and not necessarily coming from a CM type of a CM field), as we shall show.

We also prove two related results: a formula (in Corollary A.4.7.4(ii)) for the slopes of certain values of an algebraic Hecke character, generalizing the Shimura–Taniyama formula (2.1.4.1), and a result (in Theorem A.4.8.4) on the existence of Weil numbers with prescribed slopes in a given CM field.

A.4.2. We begin with generalities on tori over a field k and the associated geometric character groups (as Galois modules). In later sections we generally use $\mathsf{k} = \mathbb{Q}$.

Fix a separable closure k_s/k, and let $\Gamma = \mathrm{Gal}(\mathsf{k}_s/\mathsf{k})$. For any k-torus T, we write $\mathrm{X}^*(T)$ to denote the left Γ-module of characters $T_{\mathsf{k}_s} \to \mathbb{G}_\mathrm{m}$ over k_s. Explicitly, $\gamma \in \Gamma$ acts on a k_s-homomorphism $\chi : T_{\mathsf{k}_s} \to \mathbb{G}_\mathrm{m}$ by $\gamma.\chi = \gamma^*(\chi)$, using the k-structure on T and $\mathbb{G}_\mathrm{m}$ to identify $\gamma^*(\chi)$ as another such k_s-homomorphism. (Equivalently, the map $\gamma.\chi : T(\mathsf{k}_s) \to \mathsf{k}_s^\times$ carries $t \in T(\mathsf{k}_s)$ to $\gamma(\chi(\gamma^{-1}.t))$.)

Elements of $\mathrm{X}^*(T)$ are called *characters* of T, or *geometric characters* to emphasize that they are defined over k_s rather than over k. The functor $T \rightsquigarrow \mathrm{X}^*(T)$ is an anti-equivalence between the category of k-tori and the category of finite free $\mathbb{Z}$-modules equipped with a discrete Γ-action. We employ similar terminology for the $\mathbb{Z}$-dual lattice $\mathrm{X}_*(T)$ of (geometric) cocharacters.

Let F/k be a finite separable extension. The Weil restriction $\mathrm{Res}_{F/\mathsf{k}}(\mathbb{G}_{\mathrm{m}})$ is a k-torus, denoted as $\underline{F}^\times$ (with k understood from context), so there is a natural isomorphism $\underline{F}^\times(\mathsf{k}) \simeq F^\times$. If k is infinite then $F^\times$ is Zariski dense in $\underline{F}^\times$ (as tori are unirational over any field), so in many calculations with these tori over infinite fields it suffices to describe formulas on such elements rather than compute with functors on k-algebras.

There is a natural $\mathbb{Z}$-basis $(\xi^\tau)_{\tau \in \mathrm{Hom}_{\mathsf{k}}(F,\mathsf{k}_s)}$ of $\mathrm{X}^*(\underline{F}^\times)$ indexed by the k-embeddings $\tau\colon F \hookrightarrow \mathsf{k}_s$, as follows: the isomorphism of k_s-algebras

$$\mathsf{k}_s \otimes_{\mathsf{k}} F \simeq \prod_{\tau' : F \to \mathsf{k}_s} \mathsf{k}_s$$

defined by $a \otimes b \mapsto (a\tau'(b))$ yields a k_s-group isomorphism

$$\underline{F}^\times_{\mathsf{k}_s} \simeq \mathrm{Res}_{(\mathsf{k}_s \otimes_{\mathsf{k}} F)/\mathsf{k}_s}(\mathbb{G}_{\mathrm{m}}) = \mathrm{Res}_{(\prod_{\tau'} \mathsf{k}_s)/\mathsf{k}_s}(\mathbb{G}_{\mathrm{m}}) = \prod_{\tau' : F \to \mathsf{k}_s} \mathbb{G}_{\mathrm{m}},$$

under which the element $\xi^\tau \in \mathrm{X}^*(\underline{F}^\times)$ is defined to be the projection to the τ-factor. Thus, the effect of ξ^τ on the group $F^\times = \underline{F}^\times(\mathsf{k})$ of k-points is given by $\tau : F^\times \to \mathsf{k}^\times$, and for $\gamma \in \Gamma$ we see that $\gamma.\xi^\tau = \xi^{\gamma \circ \tau}$ (since the canonical k_s-algebra isomorphism $\mathsf{k}_s \otimes_{\gamma,\mathsf{k}_s} (\mathsf{k}_s \otimes_{\mathsf{k}} F) \simeq \mathsf{k}_s \otimes_{\mathsf{k}} F$ carries $1 \otimes \xi^\tau$ to $\xi^{\gamma\circ\tau}$). Also, for any k_s-algebra R, the map on R-points is projection to the τ-factor arising from the natural decomposition of R-algebras

$$R \otimes_{\mathsf{k}} F \simeq R \otimes_{\mathsf{k}_s} (\mathsf{k}_s \otimes_{\mathsf{k}} F) \simeq \prod_{\tau' : F \to \mathsf{k}_s} R$$

defined by $r \otimes x \mapsto (r\tau'(x))$.

Let $\mathbb{Z}^{\mathrm{Hom}_{\mathsf{k}}(F,\mathsf{k}_s)}$ be the set of all $\mathbb{Z}$-valued functions on the finite set $\mathrm{Hom}_{\mathsf{k}}(F, \mathsf{k}_s)$ of all k-embeddings of F into k_s. There is a natural left action of Γ on $\mathbb{Z}^{\mathrm{Hom}_{\mathsf{k}}(F,\mathsf{k}_s)}$ given by

$$(\gamma \cdot f)(\tau) = f(\gamma^{-1} \circ \tau)$$

for all $\gamma \in \Gamma$ and $\tau \in \mathrm{Hom}_{\mathsf{k}}(F, \mathsf{k}_s)$. The assignment $\tau \mapsto \xi^\tau$ extends to a natural Γ-equivariant isomorphism

$$\mathbb{Z}^{\mathrm{Hom}_{\mathsf{k}}(F,\mathsf{k}_s)} \simeq \mathrm{X}^*(\underline{F}^\times),$$

where each element $f \in \mathbb{Z}^{\mathrm{Hom}_{\mathsf{k}}(F,\mathsf{k}_s)}$ is identified with the character

$$\xi^f := \sum_{\tau \in \mathrm{Hom}_{\mathsf{k}}(F,\mathsf{k}_s)} f(\tau) \cdot \xi^\tau .$$

Strictly speaking, it is more accurate to write $\xi^\tau_{F/\mathsf{k}}$ rather than ξ^τ (especially when using several F/k at the same time), but the context will make the intended extension F/k clear.

For any finite separable extension field F'/F, we have two Γ-equivariant maps between the geometric character groups of $\underline{F'}^{\times}$ and $\underline{F}^{\times}$, in opposite directions:

- the pullback map $\mathrm{Nm}^*_{F'/F}\colon \mathbb{Z}^{\mathrm{Hom}_{\mathsf{k}}(F,\mathsf{k}_s)} \to \mathbb{Z}^{\mathrm{Hom}_{\mathsf{k}}(F',\mathsf{k}_s)}$ attached to the norm homomorphism $\mathrm{Nm}_{F'/F}\colon \underline{F'}^{\times} \to \underline{F}^{\times}$ is the pullback map of $\mathbb{Z}$-valued functions induced by the restriction map $\mathrm{Hom}_{\mathsf{k}}(F',\mathsf{k}_s) \to \mathrm{Hom}_{\mathsf{k}}(F,\mathsf{k}_s)$.
- the pullback map $j^*_{F\hookrightarrow F'}\colon \mathbb{Z}^{\mathrm{Hom}_{\mathsf{k}}(F',\mathsf{k}_s)} \to \mathbb{Z}^{\mathrm{Hom}_{\mathsf{k}}(F,\mathsf{k}_s)}$ attached to the natural injection $j_{F\hookrightarrow F'}\colon \underline{F}^{\times} \hookrightarrow \underline{F'}^{\times}$ is "integration along the fiber"; i.e.,
$$j^*_{F\hookrightarrow F'}(\widetilde{f})(\tau) = \sum_{\widetilde{\tau}\mapsto\tau} f(\widetilde{\tau})$$
for every $\widetilde{f} \in \mathbb{Z}^{\mathrm{Hom}_{\mathsf{k}}(F',\mathsf{k}_s)}$ and every $\tau \in \mathrm{Hom}_{\mathsf{k}}(F,\mathsf{k}_s)$, where $\widetilde{\tau}$ runs through all k-embeddings $F' \hookrightarrow \mathsf{k}_s$ extending τ.

These formulas will be used without comment in what follows.

By working with subfields in k_s of finite degree over k, we can express the above constructions in a form that is convenient for computations with varying F, as follows. Let $C(\Gamma;\mathbb{Z})$ be the $\mathbb{Z}$-module of locally constant $\mathbb{Z}$-valued functions on Γ. For any $\mathsf{k}' \subset \mathsf{k}_s$ of finite degree over k we let $\Gamma_{\mathsf{k}'} = \mathrm{Gal}(\mathsf{k}_s/\mathsf{k}')$, so we naturally identify $\mathrm{Hom}_{\mathsf{k}}(\mathsf{k}',\mathsf{k}_s)$ with $\Gamma/\Gamma_{\mathsf{k}'}$. For an abstract finite separable extension F of k, if we fix an k-embedding $\tau_0 : F \hookrightarrow \mathsf{k}_s$ then we get an open subgroup $\Gamma_F := \mathrm{Gal}(\mathsf{k}_s/\tau_0(F))$ in Γ (which depends on τ_0 when F/k is not Galois).

For $\gamma \in \Gamma$, we define the right translation operator $r(\gamma)$ on $C(\Gamma;\mathbb{Z})$ by
$$(r(\gamma)(f))(x) = f(x\gamma) \quad \text{for all} \quad x \in \Gamma;$$
this is a left action of Γ. Likewise, we define the left translation operator $\ell(\gamma)$ by $(\ell(\gamma)(f))(x) = f(\gamma^{-1}x)$; this is also a left action of Γ, and it commutes with the right translation action of Γ. Upon choosing an k-embedding $\tau_0 : F \to \mathsf{k}_s$, we obtain an isomorphism

$$\mathbf{i}_{\tau_0}\colon \mathrm{X}^*(\underline{F}^{\times}) \simeq \mathrm{Hom}_{\mathrm{Set}}(\Gamma/\Gamma_F,\mathbb{Z}) \simeq C(\Gamma;\mathbb{Z})^{r(\Gamma_F)} \tag{A.4.2.1}$$

onto the set of locally constant functions on Γ fixed under all *right* translations by elements of Γ_F. Under this isomorphism, for every $\sigma \in \Gamma$ the k-embedding $\sigma\circ\tau_0 : F \hookrightarrow \mathsf{k}_s$ corresponds to the characteristic function of $\sigma\Gamma_F$. The isomorphism (A.4.2.1) is equivariant for the *left* translation action of Γ on $C(\Gamma;\mathbb{Z})$ that sends $C(\Gamma;\mathbb{Z})^{r(\Gamma_F)}$ to itself.

The isomorphism $\mathbf{i}_{\tau_0}$ has the advantage of being compatible with the norm maps, in the following sense. Suppose that E is a subfield of F over k, $\tau_0' : E \hookrightarrow \mathsf{k}_s$ is the composition of the inclusion $E \hookrightarrow F$ with τ_0, and $\mathrm{Nm}_{F/E}\colon \underline{F}^{\times} \to \underline{E}^{\times}$ is the norm map between the induced k-tori. The diagram

$$\begin{array}{ccc}
\mathrm{X}^*(\underline{E}^{\times}) & \xrightarrow[\simeq]{\mathbf{i}_{\tau_0'}} & C(\Gamma;\mathbb{Z})^{r(\Gamma_E)} \\
{\scriptstyle \mathrm{Nm}^*}\downarrow & & \downarrow \\
\mathrm{X}^*(\underline{F}^{\times}) & \xrightarrow[\mathbf{i}_{\tau_0}]{\simeq} & C(\Gamma;\mathbb{Z})^{r(\Gamma_F)}
\end{array}$$

commutes, where the right vertical map is the inclusion as $\mathbb{Z}$-submodules of $C(\Gamma;\mathbb{Z})$.

As an application of these computations, consider subfields E and E' of F over k, and a k-homomorphism $\chi : \underline{F}^{\times} \to \underline{F'}^{\times}$ that factors through both $\mathrm{Nm}_{F/E}$ and

$\mathrm{Nm}_{F/E'}$. Any such χ factors through $\mathrm{Nm}_{F/(E\cap E')}$. Indeed, this is reduced to the equality

$$\mathrm{X}^*(\underline{E}^\times)\bigcap \mathrm{X}^*(\underline{E'}^\times) \stackrel{?}{=} \mathrm{X}^*(\underline{(E\cap E')}^\times)$$

inside $\mathrm{X}^*(\underline{F}^\times)$, which in turn is a consequence of the fact that $r(\Gamma_E)$ and $r(\Gamma_{E'})$ generate $r(\Gamma_{E\cap E'})$. We conclude that for any χ there is a unique minimal subfield $E(\chi) \subset F$ over k such that χ factors through $\mathrm{Nm}_{F/E(\chi)}$. The homomorphism χ is *primitive* if $E(\chi) = F$. This concept will be used only in the case $\mathsf{k} = \mathbb{Q}$.

A.4.3. The Serre torus. Let F be a number field. For any inclusion $U \subset U'$ of finite-index subgroup of $\mathcal{O}_F^\times$, the inclusion of Zariski closures $\overline{U} \subset \overline{U}'$ inside the $\mathbb{Q}$-torus $\underline{F}^\times$ is an equality on identity components since $\overline{U}'/\overline{U}$ is killed by the nonzero integer $[U' : U]$ and hence contains no nontrivial F-subtori. As a special case, the projection from $\mathcal{O}_F^\times$ onto the component group of its Zariski closure T in $\underline{F}^\times$ has kernel U_F with finite index, and the Zariski closure of U_F is equal to T.

To do computations with $\mathbb{Q}$-tori via character groups, we fix a choice of algebraic closure $\overline{\mathbb{Q}}$ of $\mathbb{Q}$ for the remainder of A.4, and for any subfield $L \subset \overline{\mathbb{Q}}$ we write Γ_L for $\mathrm{Gal}(\overline{\mathbb{Q}}/L)$. We also fix an embedding $\overline{\mathbb{Q}} \hookrightarrow \mathbb{C}$ and denote by ι the element of order 2 in $\Gamma_{\mathbb{Q}}$ that is induced by complex conjugation. It is a theorem of Artin [**67**, VI, Cor. 9.3; XI, 2.4, 2.9] that the involutions in $\Gamma_{\mathbb{Q}}$ constitute a single conjugacy class, so we get all such involutions as ι when we vary the choice of embedding $\overline{\mathbb{Q}} \hookrightarrow \mathbb{C}$. It is solely for convenience of exposition in what follows that we fix a choice of ι; Artin's theorem will be used to ensure that this choice does not matter.

A.4.3.1. Definition. The *Serre torus* (or *connected Serre group*) $\mathfrak{S}^F$ is the quotient of $\underline{F}^\times$ modulo the identity component of the Zariski closure of $\mathcal{O}_F^\times$.

An equivalent description is that $\mathfrak{S}^F$ is the quotient of $\underline{F}^\times$ modulo the Zariski closure of any finite-index subgroup $U \subset \mathcal{O}_F^\times$ with sufficiently divisible index in the finitely generated group $\mathcal{O}_F^\times$. The connectedness of the kernel of $\underline{F}^\times \twoheadrightarrow \mathfrak{S}^F$ implies that the Serre torus corresponds to a saturated $\mathbb{Z}$-submodule of $\mathrm{X}^*(\underline{F}^\times)$. Hence, this $\mathbb{Z}$-submodule is determined by its $\mathbb{Q}$-span inside $\mathrm{X}^*(\underline{F}^\times)_{\mathbb{Q}}$. Below we will describe this submodule explicitly. (We refer to [**105**, Ch. II], [**77**, §2], and [**78**, §1] for more information about the Serre torus.)

Among the finite-index subgroups of $\mathcal{O}_F^\times$ are the ones defined by congruence conditions: for a non-zero ideal ("modulus") $\mathfrak{m}$ of $\mathcal{O}_F$, we get the finite-index subgroup

$$\mathcal{O}_{F,\mathfrak{m}}^\times = \{u \in \mathcal{O}_F^\times \mid u \equiv 1 \pmod{\mathfrak{m}}\} = \ker(\mathcal{O}_F^\times \to (\mathcal{O}_F/\mathfrak{m})^\times).$$

Let $T_{\mathfrak{m}}$ be the torus quotient of $\underline{F}^\times$ modulo the Zariski closure of $\mathcal{O}_{F,\mathfrak{m}}^\times$, so each $T_{\mathfrak{m}}$ is an isogenous quotient of $\mathfrak{S}^F$. It is a theorem of Chevalley [**19**] that the $\mathcal{O}_{F,\mathfrak{m}}^\times$ are cofinal among all finite-index subgroups of $\mathcal{O}_F^\times$, so for sufficiently divisible $\mathfrak{m}$ the Zariski closure of $\mathcal{O}_{F,\mathfrak{m}}^\times$ in $\underline{F}^\times$ coincides with the identity component of the Zariski closure of $\mathcal{O}_F^\times$; i.e., we have $T_{\mathfrak{m}} = \mathfrak{S}^F$ as quotients of $\underline{F}^\times$ for such $\mathfrak{m}$. In particular, the subspaces $\mathrm{X}^*(T_{\mathfrak{m}})_{\mathbb{Q}} \subset \mathrm{X}^*(\underline{F}^\times)_{\mathbb{Q}}$ coincide for all $\mathfrak{m}$. This $\mathbb{Q}$-subspace meets $\mathrm{X}^*(\underline{F}^\times)$ in exactly the character group of the Serre torus.

A.4.3.2. Lemma. *The subgroup* $\mathrm{X}^*(\mathfrak{S}^F) \subset \mathrm{X}^*(\underline{F}^\times)$ *consists of those* $\xi = \sum_\tau a_\tau \xi^\tau$ *satisfying*

$$(\gamma - 1) \cdot (1 + \iota) \cdot \xi = 0, \;\; (1 + \iota) \cdot (\gamma - 1) \cdot \xi = 0$$

for all $\gamma \in \Gamma_{\mathbb{Q}}$. *Equivalently,* $\xi \in \mathrm{X}^*(\mathfrak{S}^F)$ *if and only if* $a_\tau + a_{\iota \cdot \tau}$ *is an integer* $w(\xi)$ *independent of* τ *and* $a_{\gamma\iota\cdot\tau} = a_{\iota\gamma\cdot\tau}$ *for all* $\gamma \in \Gamma_{\mathbb{Q}}$ *and all* τ.

Under the identification of $\mathrm{X}^*(\mathfrak{S}^F)$ *as a* $\Gamma_{\mathbb{Q}}$*-stable subgroup of* $\mathrm{X}^*(\underline{F}^\times)$, *we have*

$$\xi(x)(\iota \cdot \xi)(x) = \xi(x)\iota(\xi(x)) = \mathrm{Nm}_{F/\mathbb{Q}}(x)^{w(\xi)}$$

for all $\xi \in \mathrm{X}^*(\mathfrak{S}^F)$ *and all* $x \in F^\times$.

Proof. The equivalence of the two asserted descriptions is an elementary computation, as is the final displayed formula (granting the asserted description of $\mathrm{X}^*(\mathfrak{S}^F)$) since the subgroup $F^\times = \underline{F}^\times(\mathbb{Q}) \subset \underline{F}^\times(\overline{\mathbb{Q}})$ is fixed pointwise by $\Gamma_{\mathbb{Q}}$. Thus, it suffices to show that $\mathrm{X}^*(\mathfrak{S}^F)$ is defined by the conditions: $a_{\gamma\iota\cdot\tau} = a_{\iota\gamma\cdot\tau}$ for all γ and τ, and $a_\tau + a_{\iota\cdot\tau}$ is independent of τ.

These conditions clearly are insensitive to multiplying ξ by a non-zero integer, so by the saturatedness of $\mathrm{X}^*(\mathfrak{S}^F)$ in $\mathrm{X}^*(\underline{F}^\times)$ it is equivalent to show that these conditions define $\mathrm{X}^*(\mathfrak{S}^F)_{\mathbb{Q}}$ as a subspace of $\mathrm{X}^*(\underline{F}^\times)_{\mathbb{Q}} = \oplus \mathbb{Q}\xi^\tau$. Equivalently, we need to show that these conditions define the common subspace $\mathrm{X}^*(T_{\mathfrak{m}})_{\mathbb{Q}} \subset \mathrm{X}^*(\underline{F}^\times)_{\mathbb{Q}}$ for all non-zero ideals $\mathfrak{m}$ of $\mathcal{O}_F$.

The norm $\mathrm{Nm}_{F/\mathbb{Q}} : \underline{F}^\times \to \mathbb{G}_{\mathrm{m}}$ is $\sum_\tau \xi^\tau$, so any element $\xi \in \mathrm{X}^*(\underline{F}^\times)_{\mathbb{Q}}$ can be uniquely written as $a \cdot \mathrm{Nm}_{F/\mathbb{Q}} + \sum_\tau b_\tau \xi^\tau$ with $a, b_\tau \in \mathbb{Q}$ such that $\sum b_\tau = 0$. Explicitly, $a_\tau = a + b_\tau$ for all τ. Thus, the condition that $a_\tau + a_{\iota\cdot\tau}$ is independent of τ is exactly the condition that $b_{\iota\cdot\tau} = -b_\tau$ for all τ. The condition that $a_{\gamma\iota\cdot\tau} = a_{\iota\gamma\cdot\tau}$ for all $\gamma \in \Gamma_{\mathbb{Q}}$ and all τ is the same as the condition that $b_{\iota'\cdot\tau'} = b_{\iota\cdot\tau'}$ for all τ' and all $\Gamma_{\mathbb{Q}}$-conjugates ι' of ι.

Since the involutions in $\Gamma_{\mathbb{Q}}$ constitute a single conjugacy class (Artin's theorem), the proposed description of $\mathrm{X}^*(T_{\mathfrak{m}})_{\mathbb{Q}}$ for every $\mathfrak{m}$ is the system of equations $b_{c\cdot\tau} = -b_\tau$ for all τ and all involutions $c \in \Gamma_{\mathbb{Q}}$. This reformulation is exactly the description that is proved via general arguments with the $\mathbb{Q}$-linear representation theory of $\Gamma_{\mathbb{Q}}$ in [**105**, II, 3.1, A.2] (where ξ^τ is denoted as $[\tau]$). The key role of complex conjugations in these general arguments is the elementary observation that if T is an $\mathbb{R}$-torus then $T(\mathbb{R})$ is compact if and only if complex conjugation acts as negation on $\mathrm{X}^*(T)$, combined with the basic fact that if T is $\mathbb{Q}$-torus that is $\mathbb{Q}$-anisotropic then $T(\mathbb{A})/T(\mathbb{Q})$ is compact (so $T(\mathbb{R})/\Gamma$ is compact Hausdorff for any arithmetic subgroup $\Gamma \subset T(\mathbb{Q})$). □

The integer $w(\xi)$ in Lemma A.4.3.2 is called the *weight* of the character ξ of $\mathfrak{S}^F$. Explicitly, if we write $\xi = a \cdot \mathrm{Nm}_{F/\mathbb{Q}} + \sum_\tau b_\tau \xi^\tau$ in $\mathrm{X}^*(\underline{F}^\times)_{\mathbb{Q}}$ with $a \in \mathbb{Q}$ and $b_\tau \in \mathbb{Q}$ such that $\sum b_\tau = 0$ and $a + b_\tau \in \mathbb{Z}$ is independent of τ then $w(\xi) = 2a$. The map $\xi \mapsto w(\xi)$ is a $\Gamma_{\mathbb{Q}}$-invariant homomorphism $\mathrm{X}^*(\mathfrak{S}^F) \to \mathbb{Z}$, so it defines a $\mathbb{Q}$-rational cocharacter

$$w_F : \mathbb{G}_{\mathrm{m}} \to \mathfrak{S}^F;$$

this is called the *weight* cocharacter of $\mathfrak{S}^F$. (The composition of w_F with the isogeny from $\mathfrak{S}^F$ onto $T_{\mathfrak{m}}$ coincides with the morphism denoted $j^* : \mathbb{G}_{\mathrm{m}} \to T_{\mathfrak{m}}$ in [**105**, II, 3.2].) Our sign convention for the weight is opposite to that in [**77**, §2] and [**78**, §1] because we use the arithmetic normalization for local Artin maps (as Serre does in [**105**].)

In the theory of algebraic Hecke characters, CM fields are significant due to:

A.4.3.3. Lemma (Artin-Weil). *Let F_{cm} be the subfield of F generated by all CM fields contained in F.*[1] *The norm map $\mathrm{Nm}_{F/F_{\mathrm{cm}}}: \underline{F}^\times \to \underline{F_{\mathrm{cm}}}^\times$ induces an isomorphism*

$$\mathfrak{S}^F \simeq \mathfrak{S}^{F_{\mathrm{cm}}}.$$

PROOF. The norm map $\mathrm{Nm}_{F/F_{\mathrm{cm}}}: \underline{F}^\times \to \underline{F_{\mathrm{cm}}}^\times$ passes through the quotients by the Zariski closures of subgroups of units on both sides and induces a map $\mathfrak{S}^F \to \mathfrak{S}^{F_{\mathrm{cm}}}$ that corresponds to an inclusion $\mathrm{X}^*(\mathfrak{S}^{F_{\mathrm{cm}}}) \subset \mathrm{X}^*(\mathfrak{S}^F)$ of saturated lattices induced by the inclusion $\mathrm{X}^*(\underline{F_{\mathrm{cm}}}^\times) \hookrightarrow \mathrm{X}^*(\underline{F}^\times)$. Thus, it is equivalent to show that equality holds at the level of $\mathbb{Q}$-vector spaces; i.e., $\mathrm{X}^*(\mathfrak{S}^{F_{\mathrm{cm}}})_{\mathbb{Q}} = \mathrm{X}^*(\mathfrak{S}^F)_{\mathbb{Q}}$ inside $\mathrm{X}^*(\underline{F}^\times)_{\mathbb{Q}}$.

By Artin's theorem on the conjugacy of all involutions in $\Gamma_{\mathbb{Q}}$, the closed normal subgroup $H_c \subset \Gamma_{\mathbb{Q}}$ generated by the involutions has fixed field equal to the compositum $\overline{\mathbb{Q}}_{\mathrm{tr}}$ of all totally real number fields inside $\overline{\mathbb{Q}}$. The extension $\overline{\mathbb{Q}}_{\mathrm{cm}}/\overline{\mathbb{Q}}_{\mathrm{tr}}$ given by the compositum of all CM number fields inside $\overline{\mathbb{Q}}$ is a quadratic extension, and clearly the quotient map $\varepsilon : H_c \twoheadrightarrow \mathrm{Gal}(\overline{\mathbb{Q}}_{\mathrm{cm}}/\overline{\mathbb{Q}}_{\mathrm{tr}}) = \{1, -1\}$ carries all involutions to -1. Thus, the subgroup $H := \ker(\varepsilon) \subset \Gamma_{\mathbb{Q}}$ has $\overline{\mathbb{Q}}_{\mathrm{cm}}$ as its fixed field, so upon choosing an embedding $\tau_0: F \hookrightarrow \overline{\mathbb{Q}}$ to define the open subgroup $\Gamma_F \subset \Gamma_{\mathbb{Q}}$ we see that the subfield $F_1 \subset F$ corresponding to $\Gamma_F \cdot H$ is $F \bigcap \overline{\mathbb{Q}}_{\mathrm{cm}}$, which is the (intrinsic!) subfield F'_{cm} consisting of all elements of F on which the effect of complex conjugation under any embedding $j : F \hookrightarrow \mathbb{C}$ is independent of j.

By [**105**, II, 3.3, Cor.], $\mathrm{X}^*(\mathfrak{S}^F)_{\mathbb{Q}} = \mathrm{X}^*(\mathfrak{S}^{F_1})_{\mathbb{Q}}$ inside $\mathrm{X}^*(\underline{F}^\times)_{\mathbb{Q}}$. This gives the result with F'_{cm} in place of F_{cm}, so it settles the case when F contains a CM field and reduces the problem to the case of totally real fields when F does not contain a CM field. It therefore remains to show that if F is totally real then $\mathrm{Nm}_{F/\mathbb{Q}}$ induces an isomorphism $\mathfrak{S}^F \simeq \mathfrak{S}^{\mathbb{Q}} = \mathbb{G}_{\mathrm{m}}$. This isomorphism assertion follows from the description of the character group of the Serre torus $\mathfrak{S}^F$ in Lemma A.4.3.2, applied to totally real F. □

A.4.4. Admissible algebraic homomorphisms. For finite extensions F and L of $\mathbb{Q}$, a $\mathbb{Q}$-homomorphism $\chi: \underline{F}^\times \to \underline{L}^\times$ is *Hecke-admissible*, or *admissible*, if χ factors through the quotient homomorphism $\underline{F}^\times \twoheadrightarrow \mathfrak{S}^F$.[2] For any admissible χ, the associated primitive $\mathbb{Q}$-homomorphism $\chi' : \underline{E(\chi)}^\times \to \underline{L}^\times$ in the sense defined at the end of A.4.2 must also be admissible. Indeed, it suffices to check that for any subfield $E \subset F$, $\mathrm{X}^*(\underline{E}^\times) \bigcap \mathrm{X}^*(\mathfrak{S}^F) = \mathrm{X}^*(\mathfrak{S}^E)$, and this equality is immediate from the first characterization of the subgroup $\mathrm{X}^*(\mathfrak{S}^F) \subset \mathrm{X}^*(\underline{F}^\times)$ (and its analogue for E) in Lemma A.4.3.2. Since χ' is admissible and primitive, by Lemma A.4.3.3 it follows that $E(\chi) = E(\chi)_{\mathrm{cm}}$, so either $E(\chi) = \mathbb{Q}$ or $E(\chi)$ is a CM field.

We are interested in admissible homomorphisms because the algebraic part of any algebraic Hecke character χ is admissible. Indeed, let $\epsilon : \mathbb{A}_F^\times \to L^\times$ be an

[1] Often F_{cm} is called "the largest CM subfield of F". We take the subfield of F generated by the empty set to be $\mathbb{Q}$, so $F_{\mathrm{cm}} := \mathbb{Q}$ when F contains no CM subfield. This fits well with our later arguments. In other contexts, another uniform definition is more appropriate: the subfield F'_{cm} of elements of F on which the effect of complex conjugation under any embedding $j : F \hookrightarrow \mathbb{C}$ is independent of j. That is, $F'_{\mathrm{cm}} = F_{\mathrm{cm}}$ when F contains a CM field, and F'_{cm} is the maximal totally real subfield of F otherwise.

[2] In A.4.4–A.4.5 and A.4.7 the number field L does not have to be a CM field, contrary to the tacit convention in most of Chapter 2 and this appendix. However, by A.4.4.2 any admissible homomorphism factors through the natural injection $j: \underline{L_{\mathrm{cm}}}^\times \hookrightarrow \underline{L}^\times$. Thus, the reader may safely assume that L is either $\mathbb{Q}$ or a CM field.

algebraic Hecke character with $\chi = \epsilon_{\mathrm{alg}}$. For some modulus $\mathfrak{m}$, the character ϵ is trivial on the open subgroup

$$U_{\mathfrak{m}} = (F_\infty^\times)^0 \times \prod_{v|\mathfrak{m}} (1+\mathfrak{m}\mathcal{O}_{F,v})^\times \times \prod_{v\nmid\mathfrak{m}} \mathcal{O}_{F,v}^\times$$

of $\mathbb{A}_F^\times$, and $\epsilon|_{F^\times} = \chi|_{F^\times}$. Thus, χ is trivial on the finite-index subgroup $\mathcal{O}_{F,\mathfrak{m}}^\times = \mathcal{O}_F^\times \cap U_{\mathfrak{m}}$ in $\mathcal{O}_F^\times$, so χ kills the Zariski closure of $\mathcal{O}_{F,\mathfrak{m}}^\times$ and hence it factors through the quotient of $\underline{F}^\times$ modulo the identity component of this closure. That quotient is exactly the Serre torus. Here is another class of examples.

A.4.4.1. Example. Let Φ be a $\overline{\mathbb{Q}}$-valued CM type on a CM field L, and let $E \subset \overline{\mathbb{Q}}$ be its reflex field (so E is a CM field). We claim that the reflex norm $\mathrm{N}_\Phi : \underline{E}^\times \to \underline{L}^\times$ is admissible, or equivalently that the map $\mathrm{N}_\Phi : E^\times \to L^\times$ on $\mathbb{Q}$-points kills the congruence subgroup $\mathcal{O}_{E,\mathfrak{m}}^\times \subset \mathcal{O}_E^\times$ (of units congruent to 1 modulo $\mathfrak{m}$) for a sufficiently divisible non-zero ideal $\mathfrak{m}$ of $\mathcal{O}_E$. By a theorem of Chevalley [**19**] the congruence topology and profinite topology on $\mathcal{O}_E^\times$ *coincide*, so it is equivalent to show that $\mathrm{N}_\Phi(\mathcal{O}_E^\times)$ is finite. By continuity, the $\mathbb{Q}$-homomorphism N_Φ applied to adelic points carries the maximal compact subgroup of $\mathbb{A}_{E,f}^\times$ into the maximal compact subgroup of $\mathbb{A}_{L,f}^\times$. Since $\mathcal{O}_E^\times$ is the intersection of $E^\times$ with the maximal compact subgroup of $\mathbb{A}_{E,f}^\times$, and similarly for $\mathcal{O}_L^\times$, at the level of $\mathbb{Q}$-points it follows that $\mathrm{N}_\Phi(\mathcal{O}_E^\times) \subset \mathcal{O}_L^\times$.

The reflex norm N_Φ factors through the reflex torus $T \subset \underline{L}^\times$, so it suffices to show that $\mathcal{O}_L^\times \bigcap T(\mathbb{Q})$ inside $L^\times$ is finite. The subgroup $\mathcal{O}_{L^+}^\times \subset \mathcal{O}_L^\times$ has finite index (since L is a CM field), so it suffices to show that $\mathcal{O}_{L^+}^\times \bigcap T(\mathbb{Q})$ is finite. The norm Nm_{L/L^+} is squaring on $(L^+)^\times$, and this norm carries T into $\mathbb{G}_{\mathrm{m}} \subset \underline{L}^\times$, so elements of $\mathcal{O}_{L^+}^\times \bigcap T(\mathbb{Q})$ have square that lies in $\mathbb{Q}^\times$. Such squares are therefore in $\mathbb{Z}^\times = \{\pm 1\}$, so the desired finiteness is clear.

A.4.4.2. Lemma. *Let $\chi \colon \underline{F}^\times \to \underline{L}^\times$ be an admissible homomorphism over $\mathbb{Q}$. Let L_{cm} be the subfield of L generated by all* CM *subfields of L (so $L_{\mathrm{cm}} = \mathbb{Q}$ if L contains no* CM *subfield). Then χ factors through the natural injection $\underline{L_{\mathrm{cm}}}^\times \hookrightarrow \underline{L}^\times$.*

In particular, if L contains no CM *subfield then*

$$\chi = j_{\mathbb{Q}\hookrightarrow L} \circ (\mathrm{Nm}_{F/\mathbb{Q}})^n$$

for some integer n, where $\mathrm{Nm}_{F/\mathbb{Q}} : \underline{F}^\times \to \mathbb{G}_{\mathrm{m}}$ is the norm homomorphism and $j_{\mathbb{Q}\hookrightarrow L} : \mathbb{G}_{\mathrm{m}} \hookrightarrow \underline{L}^\times$ is the natural inclusion induced by $\mathbb{Q} \hookrightarrow L$.

PROOF. Since χ is admissible, by Lemma A.4.3.3 the induced map $\mathrm{X}^*(\underline{L}^\times) \to \mathrm{X}^*(\underline{F}^\times)$ factors through $\mathrm{X}^*(\underline{F_{\mathrm{cm}}}^\times)$. But the action of $\Gamma_{\mathbb{Q}}$ on this smaller target has all involutions acting through the same operation, so likewise the image T of χ is a $\mathbb{Q}$-torus inside $\underline{L}^\times$ such that all involutions in $\Gamma_{\mathbb{Q}}$ have the same effect on the $\mathbb{Z}[\Gamma_{\mathbb{Q}}]$-module quotient $\mathrm{X}^*(T)$ of $\mathrm{X}^*(\underline{L}^\times)$.

The maximal quotient V of $\mathrm{X}^*(\underline{L}^\times)_{\mathbb{Q}}$ on which all involutions in $\Gamma_{\mathbb{Q}}$ have the same effect is the same as the space of coinvariants under the closed normal subgroup H of $\Gamma_{\mathbb{Q}}$ topologically generated by products of two involutions. Clearly the field $\overline{\mathbb{Q}}^H = \overline{\mathbb{Q}}_{\mathrm{cm}}$ is the compositum all CM number fields inside $\overline{\mathbb{Q}}$, so V is a quotient of $\mathrm{X}^*(\underline{L'_{\mathrm{cm}}}^\times)_{\mathbb{Q}}$, where L'_{cm} is the subfield of elements of L on which the effect of complex conjugation under any embedding $j : L \hookrightarrow \mathbb{C}$ is independent of j. Hence, we can replace L with $L'_{\rm cm}$, so L is CM or totally real.

It remains to show that if L is totally real then χ factors through $\mathrm{Nm}_{F/\mathbb{Q}}$, or equivalently $\mathrm{X}^*(\chi) : \mathrm{X}^*(\underline{L}^\times) \to \mathrm{X}^*(\underline{F}^\times)$ is invariant under the effect of $\Gamma_{\mathbb{Q}}$ on the target. Since $\mathrm{X}^*(\chi)$ is $\Gamma_{\mathbb{Q}}$-equivariant, all involutions in $\Gamma_{\mathbb{Q}}$ act trivially on the image of $\mathrm{X}^*(\chi)$. By hypothesis, $\mathrm{X}^*(\chi)$ lands in $\mathrm{X}^*(\mathfrak{S}^F) \subset \mathrm{X}^*(\underline{F}^\times)$, so it suffices to check that an element of $\mathrm{X}^*(\mathfrak{S}^F)$ fixed by all involutions (equivalently, one involution) of $\Gamma_{\mathbb{Q}}$ is actually fixed by $\Gamma_{\mathbb{Q}}$. This is immediate from the first explicit description of $\mathrm{X}^*(\mathfrak{S}^F)$ in Lemma A.4.3.2. □

Let $\chi : \underline{F}^\times \to \underline{L}^\times$ be an admissible $\mathbb{Q}$-homomorphism, factoring uniquely as $h_\chi \circ \pi$ for a $\mathbb{Q}$-homomorphism $h_\chi \colon \mathfrak{S}^F \to \underline{L}^\times$ and the quotient map $\pi \colon \underline{F}^\times \to \mathfrak{S}^F$. The composition

$$h_\chi \circ w_F : \mathbb{G}_{\rm m} \to \underline{L}^\times$$

with the weight cocharacter of $\mathfrak{S}^F$ must be an integral power of $j_{\mathbb{Q}\hookrightarrow L}$, leading to:

A.4.4.3. Definition. Using notation as above, the *weight* $w(\chi)$ of the admissible homomorphism χ is the integer such that $h_\chi \circ w_F = (j_{\mathbb{Q}\hookrightarrow L})^{w(\chi)}$.

A.4.4.4. Proposition. *Let $h : \mathfrak{S}^F \to \underline{L}^\times$ be a $\mathbb{Q}$-homomorphism and let $\chi = h \circ \pi$ be the associated admissible $\mathbb{Q}$-homomorphism $\underline{F}^\times \to \underline{L}^\times$.*

(a) *For every embedding $\tau \in \mathrm{Hom}_{\rm ring}(L, \overline{\mathbb{Q}})$, when the pullback geometric character $h^*(\tau) := \xi^\tau \circ h$ on $\mathfrak{S}^F$ is identified with a geometric character of $\underline{F}^\times$ via composition with π we have*

$$h^*(\tau) + \iota \cdot h^*(\tau) = w(\chi) \cdot \mathrm{Nm}_{F/\mathbb{Q}}.$$

(b) *If L is a CM field with maximal totally real subfield L^+ then*

$$\mathrm{Nm}_{L/L^+} \circ \chi = j_{\mathbb{Q}\hookrightarrow L^+} \circ (\mathrm{Nm}_{F/\mathbb{Q}})^{w(\chi)}.$$

(c) *For all $y \in F^\times = \underline{F}^\times(\mathbb{Q})$ and $\tau : L \to \overline{\mathbb{Q}}$,*

$$\tau(\chi(y)) \cdot \iota(\tau(\chi(y))) = \mathrm{Nm}_{F/\mathbb{Q}}(y)^{w(\chi)} > 0.$$

PROOF. The identity in (a) is immediate from the final identity in Lemma A.4.3.2. The formula in (b) is a special case of (a) since $\mathrm{Gal}(L/L^+)$ is generated by complex conjugation when L is a CM field. The equality in (c) is a special case of (a). The inequality $\mathrm{Nm}_{F/\mathbb{Q}}(y)^{w(\chi)} > 0$ in (c) is obvious if F contains a CM field, and otherwise $F_{\rm cm} = \mathbb{Q}$ (so $\mathfrak{S}^F = \mathbb{G}_{\rm m}$), in which case the admissible χ is $\mathrm{Nm}^n_{F/\mathbb{Q}}$ for some $n \in \mathbb{Z}$ and $w(\chi) = 2n$ is even. □

Here are some elementary illustrations of Proposition A.4.4.4. Using (a), if χ factors through the $F/\mathbb{Q}$-norm then there exists an integer n such that $\chi = j_{\mathbb{Q}\hookrightarrow L} \circ (\mathrm{Nm}_{F/\mathbb{Q}})^n$ and $w(\chi) = 2n$. Also, in the special case that $F = \mathbb{Q}$ in (b), so $\chi(t) = t^n$ for some integer n, we have $w(\chi) = 2n$ and the formula in (b) becomes the obvious identity $(y^n)^2 = y^{2n}$. Finally, if L is a CM field then the displayed formula in (c) becomes

$$\mathrm{Nm}_{L/L^+}(\chi(y)) = \mathrm{Nm}_{F/\mathbb{Q}}(y)^{w(\chi)}$$

for all $y \in F^\times$. For instance, if Φ is a $\overline{\mathbb{Q}}$-valued CM type on a CM field L and if $E \subset \overline{\mathbb{Q}}$ is its reflex field then the reflex norm $\chi = \mathrm{N}_\Phi : \underline{E}^\times \to \underline{L}^\times$ (which is admissible, by Example A.4.4.1) has weight 1.

A.4.4.5. Corollary. *Let $\epsilon : \mathbb{A}_F^\times \to L^\times$ be an algebraic Hecke character with algebraic part $\chi : \underline{F}^\times \to \underline{L}^\times$. The character χ is admissible and $\epsilon(\mathbb{A}_F^\times) \subset L_{\mathrm{cm}}^\times$, where $L_{\mathrm{cm}}^\times$ is the subfield of L generated by all* CM *subfields of L.*

We emphasize that $L_{\mathrm{cm}} = \mathbb{Q}$ when L does not contain any CM subfield.

PROOF. The admissibility of χ was explained at the start of A.4.4, but we now recall the argument in order to introduce some notation that will be useful later in the proof. Let $\mathfrak{m}$ be a conductor of ϵ; i.e., ϵ is trivial on the open subgroup

$$U_{\mathfrak{m}} = (F_\infty^\times)^0 \times \prod_{v|\mathfrak{m}} (1 + \mathfrak{m}\mathcal{O}_{F,v})^\times \times \prod_{v\nmid\mathfrak{m}} \mathcal{O}_{F,v}^\times \subset \mathbb{A}_F^\times .$$

Thus, χ is trivial on $F^\times \bigcap U_{\mathfrak{m}}$, so it is admissible. In particular, if L does not contain a CM subfield then χ factors through a power of $\mathrm{Nm}_{F/\mathbb{Q}}$ and $w(\chi)$ is even.

Choose a finite place $\wp$ of F and a uniformizer $\pi_\wp$ of the valuation ring $\mathcal{O}_\wp$ of the completion $F_\wp$. For $q := \mathrm{N}(\wp)$, we shall show that the element $\epsilon(\pi_\wp) \in L^\times$ is a Weil q-number of weight $w(\chi)$ in the sense of Definition 1.6.1.2. There exists a positive integer $n > 0$, a global element $y \in F^\times$, and an element $u \in U_{\mathfrak{m}}$ such that $\pi_\wp^n = y \cdot u$. The rational number $\mathrm{Nm}_{F/\mathbb{Q}}(y)$ must be $\pm q^n$ for valuation-theoretic reasons. Clearly $\epsilon(\pi_\wp^n) = \chi(y)$, and by Proposition A.4.4.4(c) this is a Weil q^n-number in L of weight $w(\chi)$ in the sense of Definition 1.6.1.2(ii), so $\epsilon(\pi_\wp)$ is a Weil q-number of weight $w(\chi)$.

It is well-known that for every integer w and every positive integral power q_0 of a prime number p, every Weil q_0-number α of weight w lies in a CM field. Indeed, under any embedding $j : \mathbb{Q}(\alpha) \hookrightarrow \mathbb{C}$ the effect of complex conjugation on $j(\mathbb{Q}(\alpha))$ is determined by the formula $\alpha \mapsto q_0^w/\alpha$ that is independent of j. Thus, if $\mathbb{Q}(\alpha)$ is not totally real then it is a CM field, and in the totally real case certainly it is contained in a CM field. Note also that if α is totally real then it must be $\pm q_0^{w/2}$. Thus, if the Weil $q^{w(\chi)}$-number $\epsilon(\pi_\wp) \in L^\times$ is totally real (as must happen when L contains no CM subfield) then it coincides with $\pm q^{w(\chi)/2}$. But we have already noted that $w(\chi)$ is even when L contains no CM subfield, so in such cases $\epsilon(\pi_\wp) \in \mathbb{Q}^\times$.

To summarize, by treating the case of L not containing a CM field separately, we have shown that $\epsilon(\pi_\wp) \in L_{\mathrm{cm}}^\times$ for every finite place $\wp$ of F and every uniformizer $\pi_\wp$ of $\mathcal{O}_\wp$. Since $\mathbb{A}_F^\times/(F_\infty^\times)^0 U_{\mathfrak{m}}$ is generated by the images of $F^\times$ and such elements $\pi_\wp$ (the ratios $\pi'_\wp/\pi_\wp$ for two choices of a uniformizer at $\wp$ yield all elements of $\mathcal{O}_\wp^\times$), it follows that $\epsilon(\mathbb{A}_F^\times) \subset L_{\mathrm{cm}}^\times$. □

A.4.5. A theorem of Shimura. By Corollary A.4.4.5, for the question of existence of an algebraic Hecke character with a given algebraic part, we may restrict to the situation when the target number field is a CM field. Indeed, A.4.4.5 says that we may replace L by L_{cm} without changing the question. Moreover, if L_{cm} is not a CM field then $L_{\mathrm{cm}} = \mathbb{Q}$, in which case the question is answered in the affirmative: $\chi = \mathrm{Nm}_{F/\mathbb{Q}}^n$ for some $n \in \mathbb{Z}$, so $\epsilon_{\mathrm{cycl}} \circ \mathrm{Nm}_{F/\mathbb{Q}}$ is an algebraic Hecke character with χ as its algebraic part. Hence, there is no loss of generality in restricting attention to the case that L is a CM field, as we do in the next result.

A.4.5.1. Theorem (Shimura). *Let* $\chi\colon \underline{K}^\times \to \underline{L}^\times$ *be a* $\mathbb{Q}$*-homomorphism, with* L *a* CM *field having maximal totally real subfield* L^+. *There exists an algebraic Hecke character* $\epsilon\colon \mathbb{A}_K^\times \to L^\times$ *with* χ *as its algebraic part if and only if the following conditions are satisfied.*

(a) *The homomorphism* χ *is admissible; i.e. factors through the quotient map* $\underline{K}^\times \twoheadrightarrow \mathfrak{S}^K$.

(b) *For every finite idele* $s_f \in \mathbb{A}_{K,f}^\times$, *there exists* $x \in L^\times$ *such that*

$$\chi(s_f)\cdot \mathcal{O}_L = x \cdot \mathcal{O}_L \quad \text{and} \quad \mathrm{Nm}_{L/L^+}(x) = |s_f|_{\mathbb{A}_K}^{-w(\chi)} \in \mathbb{Q}_{>0}^\times,$$

where $w(\chi)$ *is the weight of* χ *defined in* A.4.4.3.

When the above conditions hold, for any integer $N > 0$ *there exists a rational prime* $p \nmid N$ *and an algebraic Hecke character* $\epsilon\colon \mathbb{A}_K^\times \to L^\times$ *with algebraic part* χ *such that* ϵ *is unramified at every finite place of* K *outside* p *and is at most tamely ramified at places of* K *above* p.

A.4.5.2. Remark. (1) The prime number p produced in the proof of A.4.5.1 is the norm of a finite place $\mathfrak{p}$ of the CM field L.

(2) When $N = 1$ in the last sentence of the statement of A.4.5.1, one can choose p such that

$$p \leqslant 2 \cdot |\mathrm{disc}(L(\mu_{mp_1\ldots p_s}))|^{A_1},$$

where A_1 is an absolute constant, m is the order of the group $\mu(L)$ of all roots of 1 in L, $p_1, \ldots, p_s$ are the distinct prime divisors of m, and $\mathrm{disc}(L(\mu_{m\cdot p_1\ldots p_s}))$ is the discriminant of the number field $L(\mu_{m\cdot p_1\ldots p_s})$. This refinement comes from the bound in [**63**, Theorem 1.1] for the least prime ideal in the Chebotarev density theorem, applied to the end of the proof of Lemma A.4.5.4 below (where the Chebotarev density theorem is invoked to produce a prime number p with the desired properties). Assuming the Generalized Riemann Hypothesis, the above bound on p can be improved to

$$p \leqslant 70 \cdot \log^2 |\mathrm{disc}(L(\mu_{m\cdot p_1\ldots p_s}))|$$

according to [**64**] and [**96**].

(3) The proof of A.4.5.1 shows that it can be strengthened as follows.

(a) Replace "for any integer $N > 0$" by

for any set S *of places of* L *with Dirichlet density strictly less than* $\prod_{i=1}^s (1 - 1/p_i)$, *where* $p_1, \ldots, p_i$ *are the distinct prime factors of the order of the group* $\mu(L)$ *of all roots of* 1 *in* L

(b) Replace "a rational prime $p \nmid N$" by

a finite place $\mathfrak{p}$ *of* L *outside* S *whose absolute norm* $q_\mathfrak{p}$ *is a rational prime number* p

The existence of such a place $\mathfrak{p}$ follows from the proof of Lemma A.4.5.4, where the Chebotarev density theorem is invoked to produce a place $\wp$ with the required properties.

A.4.5.3. Proof of necessity in A.4.5.1. We have seen in A.4.4 the necessity of condition (a). Also, we may and do assume that K contains a CM field. Indeed, otherwise $K_{\mathrm{cm}} = \mathbb{Q}$, so $\chi = (\mathrm{Nm}_{K/\mathbb{Q}})^n$ for some integer n, and hence the theorem holds because it holds for $K = \mathbb{Q}$ (by taking x in A.4.5.1(b) to be the positive rational generator of the fractional $\mathbb{Q}$-ideal associated to the finite $\mathbb{Q}$-idele s_f). Hence, now K is totally imaginary and so the restriction to $(\mathbb{R} \otimes_{\mathbb{Q}} K)^\times$ of every algebraic Hecke character of K is trivial.

Suppose that $\epsilon : \mathbb{A}_K^\times \to \underline{L}^\times$ is an algebraic Hecke character with algebraic part χ of conductor $\mathfrak{m}$. For every element $s \in \mathbb{A}_K^\times$, write s_f and s_∞ for its finite and archimedean parts respectively. Let $x := \epsilon(s) = \epsilon(s_f) \in L^\times$. We will show that $\chi(s_f) \cdot \mathcal{O}_L = x \cdot \mathcal{O}_L$ and $\mathrm{Nm}_{L/L^+}(x) = |s_f|_{\mathbb{A}_K}^{-w(\chi)}$. It suffices to show that the two equalities hold after both sides are raised to the nth power for some integer $n > 0$.

Choose a positive integer n such that $s^n \in K^\times U_{\mathfrak{m}}$ (as we may do by finiteness of generalized ideal class groups), and choose $y \in K^\times$ such that $y^{-1} \cdot s^n \in U_{\mathfrak{m}}$. Then $x^n = \epsilon(s)^n = \epsilon(s_f)^n = \chi(y)$. We have

$$x^n \cdot \mathcal{O}_L = \epsilon(s_f^n) \cdot \mathcal{O}_L = \chi(y) \cdot \mathcal{O}_L = \chi(s_f^n) \cdot \mathcal{O}_L\,;$$

the last equality holds because each finite component of $y^{-1} \cdot s_f^n$ is a local unit (and the continuous homomorphism induced by χ on adelic points carries the maximal compact subgroup into the maximal compact subgroup). Thus, $x \cdot \mathcal{O}_L = \chi(s_f) \cdot \mathcal{O}_L$.

We have seen in A.4.4.3 that $\mathrm{Nm}_{K/\mathbb{Q}}(y)^{w(\chi)} = \mathrm{Nm}_{L/L^+}(\chi(y)) > 0$, so the positive rational number $\mathrm{Nm}_{K/\mathbb{Q}}(y)^{w(\chi)}$ is the product of all normalized archimedean absolute values of $y^{w(\chi)}$. Let y_f be the finite adelic component of the global element $y \in K^\times$. The product formula applied to $y^{w(\chi)}$ tells us that $\mathrm{Nm}_{K/\mathbb{Q}}(y)^{w(\chi)} = |y_f|_{\mathbb{A}_K}^{-w(\chi)}$. Putting everything together, we get

$$\mathrm{Nm}_{L/L^+}(x^n) = \mathrm{Nm}_{L/L^+}(\chi(y)) = \mathrm{Nm}_{K/\mathbb{Q}}(y)^{w(\chi)} = |y_f|_{\mathbb{A}_K}^{-w(\chi)} = |s_f^n|_{\mathbb{A}_K}^{-w(\chi)}\,.$$

We have proved the necessity of A.4.5.1(b).

A.4.5.4. Lemma. *Let m be the order of the group $\mu(L)$ of all roots of 1 in L and let $p_1 = 2, p_2, \ldots, p_s$ be the distinct prime factors of m. For any finite set S of places of L, there exists a finite place $\mathfrak{p}$ outside S such that: the absolute norm $q_{\mathfrak{p}}$ of $\mathfrak{p}$ is an odd rational prime number, $q_{\mathfrak{p}} \equiv 1 \pmod{m}$, and $q_{\mathfrak{p}} \not\equiv 1 \pmod{mp_i}$ for $i = 1, \ldots, s$.*

PROOF. Consider the abelian extension $\mathbb{Q}(\mu_{m \cdot p_1 \cdots p_s})/\mathbb{Q}(\mu_m)$ whose Galois group Δ is naturally isomorphic to $\ker((\mathbb{Z}/m \cdot p_1 \cdots p_s\mathbb{Z})^\times \twoheadrightarrow (\mathbb{Z}/m\mathbb{Z})^\times)$, a cyclic group of order $\prod_{i=1}^s p_i$. The Galois group of $L(\mu_{m \cdot p_1 \cdots p_s})/L$ is canonically isomorphic to $\mathrm{Gal}(\mathbb{Q}(\mu_{m \cdot p_1 \cdots p_s})/(L \cap \mathbb{Q}(\mu_{m \cdot p_1 \cdots p_s})))$, a cyclic subgroup of the Galois group $\mathrm{Gal}(\mathbb{Q}(\mu_{m \cdot p_1 \cdots p_s})/\mathbb{Q}(\mu_m))$. If $L \cap \mathbb{Q}(\mu_{m \cdot p_1 \cdots p_s}) \supsetneqq \mathbb{Q}(\mu_m)$, then $L \cap \mathbb{Q}(\mu_{m \cdot p_1 \cdots p_s})$ must contain $\mathbb{Q}(\mu_{m \cdot p_i})$ for some $i = 1, \ldots, s$ since these are the minimal non-trivial subextensions of the field extension $\mathbb{Q}(\mu_{m \cdot p_1 \cdots p_s})/\mathbb{Q}(\mu_m)$. However L does not contain $\mu_{m \cdot p_i}$ for any i because there are only m distinct roots of 1 in L. Thus, $L \cap \mathbb{Q}(\mu_{m \cdot p_1 \cdots p_s}) = \mathbb{Q}(\mu_m)$ and the Galois group $\mathrm{Gal}(L(\mu_{m \cdot p_1 \cdots p_s})/L)$ is naturally isomorphic to Δ.

Every finite place $\mathfrak{p}$ of L relatively prime to m is unramified in $L(\mu_{m \cdot p_1 \cdots p_s})$, and the arithmetic Frobenius $\mathrm{Fr}_{\mathfrak{p}}$ in Δ is the class of $q_{\mathfrak{p}}$. Thus, $q_{\mathfrak{p}} \equiv 1 \pmod{m}$. The condition that $q_{\mathfrak{p}} \not\equiv 1 \pmod{mp_i}$ for all $i = 1, \ldots, s$ means that $\mathrm{Fr}_{\mathfrak{p}}$ is a generator of Δ. Consider the set of unramified primes $\mathfrak{p}$ of L not in S such that: $q_{\mathfrak{p}}$ is a rational prime number, $q_{\mathfrak{p}} \equiv 1 \pmod{m}$, and $q_{\mathfrak{p}} \not\equiv 1 \pmod{mp_i}$ for all $i = 1, \ldots, s$. By the Chebotarev density theorem, this set of $\mathfrak{p}$ has density $\prod_{i=1}^s (1 - 1/p_i) > 0$. □

A.4.5.5. PROOF OF SUFFICIENCY IN A.4.5.1.

If $K_{\mathrm{cm}} = \mathbb{Q}$, then χ is the composition of $\mathrm{Nm}_{K/\mathbb{Q}}^n$ with the injection $\mathbb{G}_{\mathrm{m}} \to \underline{L}^\times$ attached to $\mathbb{Q} \hookrightarrow L$. In this case condition (b) is automatically satisfied, and we can take ϵ to be the composition of $\mathrm{Nm}_{K/\mathbb{Q}} \colon \mathbb{A}_K^\times \to \mathbb{A}_\mathbb{Q}^\times$ with the n-th power of the algebraic Hecke character

$$\mathbb{A}^\times = \mathbb{Q}^\times \times \widehat{\mathbb{Z}} \times \mathbb{R}_{>0}^\times \xrightarrow{\mathrm{pr}} \mathbb{Q}_{>0}^\times$$

attached to the cyclotomic character for $\mathbb{Q}$; note that this latter algebraic Hecke character is unramified at every finite place of $\mathbb{Q}$ (but is ramified at ∞). In the rest of the proof we may and do assume that K contains a CM field.

For any idele $s \in \mathbb{A}_K^\times$, let s_f be the non-archimedean part of s. Condition (b) says that there exists an element $x \in L^\times$ satisfying

$$\chi(s_f) \cdot \mathcal{O}_L = x \cdot \mathcal{O}_L \qquad \text{and} \qquad \mathrm{Nm}_{L/L^+}(x) = |s_f|_{\mathbb{A}_K}^{-w(\chi)};$$

such an element x is uniquely determined by s_f up to multiplication by $u \in \mathcal{O}_L^\times$ such that $\mathrm{Nm}_{L/L^+}(u) = 1$. But $\mathcal{O}_{L^+}^\times$ has finite index in $\mathcal{O}_L^\times$ since L is a CM field, so there exists an integer $n > 0$ such that $u^n \in \mathcal{O}_{L^+}^\times$. It follows that $u^{2n} = \mathrm{Nm}_{L/L^+}(u^n) = 1$, so $u \in \mu(L)$. In other words, x is unique up to a $\mu(L)$-multiple.

According to Lemma A.4.5.4, there is a finite place $\mathfrak{p}$ of L such that $q_\mathfrak{p}$ is a rational prime number p relatively prime to N with $p \equiv 1 \pmod{m}$ and $(p-1)/m$ relatively prime to m. Thus, for the residue field $\kappa(\mathfrak{p})$ at $\mathfrak{p}$ the natural injection $j_\mathfrak{p} : \mu(L) \to \kappa(\mathfrak{p})^\times$ induced by reduction modulo $\mathfrak{p}$ admits a unique retraction $\bar{r}_\mathfrak{p}$; i.e., a homomorphism $\bar{r}_\mathfrak{p} \colon \kappa(\mathfrak{p})^\times \to \mu(L)$ such that $\bar{r}_\mathfrak{p} \circ j_\mathfrak{p} = \mathrm{id}_{\mu(L)}$.

Let $U_{L,f} := \prod_v \mathcal{O}_{L_v}^\times \subset \mathbb{A}_L^\times$ be the product of all local unit groups at the finite places of the number field L. Let $j \colon \mu(L) \to U_{L,f}$ be the diagonal embedding, and let $\overline{\mathrm{pr}}_\mathfrak{p} \colon U_{L,f} \to \kappa(\mathfrak{p})^\times$ be the composition of the $\mathfrak{p}$-th projection $U_{L,f} \twoheadrightarrow \mathcal{O}_\mathfrak{p}^\times$ and the reduction map $\mathcal{O}_\mathfrak{p}^\times \twoheadrightarrow \kappa(\mathfrak{p})^\times$. Clearly $\mathbf{r} := \bar{r}_\mathfrak{p} \circ \overline{\mathrm{pr}}_\mathfrak{p}$ is a retraction of j; i.e. $\mathbf{r} \circ j = \mathrm{id}_{\mu(L)}$.

Recall that we arranged for K to contain a CM field, so the archimedean part $(\mathbb{R} \otimes_\mathbb{Q} K)^\times$ of K is connected. The retraction $\mathbf{r}$ allows us to "cancel out" the indeterminacy up to $\mu(L)$ for the possible values of an algebraic Hecke character with algebraic part χ, as follows. For s and x as above, define $\epsilon(s) \in L^\times$ by

$$\epsilon(s) = x \cdot \mathbf{r}(\chi(s_f)/x_f).$$

Clearly $\epsilon(s)$ is well-defined, since $\mathbf{r} \circ j = \mathrm{id}_{\mu(L)}$. It is also clear that $s \mapsto \epsilon(s)$ is a homomorphism from $\mathbb{A}_K^\times$ to $L^\times$. If s is the image of a global element $y \in K^\times$ then we can take x to be $\chi(y)$ because

$$\mathrm{Nm}_{L/L^+}(\chi(y)) = \mathrm{Nm}_{K/\mathbb{Q}}(y)^{w(\chi)} = |s_f|_{\mathbb{A}_K}^{-w(\chi)}.$$

Indeed, the first equality is a general property for the admissible homomorphism χ discussed in A.4.4.3, and the second equality follows from the product formula for the global element $y \in K^\times$ (since $\mathrm{Nm}_{K/\mathbb{Q}}(y) > 0$, due to K containing a CM field). This proves that $\epsilon|_{K^\times} = \chi$.

Let $\mathfrak{m}$ be the product of all primes of $\mathcal{O}_K$ above p. We claim that $\epsilon(s) = 1$ for all s in the open subgroup $U_\mathfrak{m} \subset \mathbb{A}_K^\times$. For $\mathfrak{p}|\mathfrak{m}$, by consideration of topological nilpotence and pro-p groups we see via continuity that the $\mathfrak{p}$-th component $\chi(s_f)_\mathfrak{p}$ of $\chi(s_f)$ must lie in $1+\mathfrak{p}\mathcal{O}_{L_\mathfrak{p}}$, and we can take $x = 1$, so $\epsilon(s) = \mathbf{r}(\chi(s_f)) = \bar{r}(\bar{1}) = 1$.

Thus, ϵ is continuous, unramified outside p, and at most tamely ramified at places of K above p. This proves Theorem A.4.5.1. □

A.4.6. Field of moduli. We next wish to discuss the notion of "field of moduli" for admissible $\mathbb{Q}$-homomorphisms between certain $\mathbb{Q}$-tori. Before doing this, it is convenient to introduce some equivalent formulations of condition (b) in A.4.5.1. The statements involve basic terminology in the theory of Arakelov class groups, so we first digress to review those notions. (More information can be found in [**116**] and [**84**, Chapter III].)

A *replete* (or *Arakelov*, or *compactified*) divisor for a number field F is an element of

$$\bigoplus_{w|\infty} \mathbb{R}\cdot[w] \times \bigoplus_{v\nmid\infty} \mathbb{Z}\cdot[v],$$

where the sum is taken over all places of F. The *replete divisor* attached to an idele

$$s = \prod_{w|\infty} s_w \cdot \prod_{v\nmid\infty} s_v \ \in\ \mathbb{A}_F^\times$$

is

$$(s) = \sum_{w|\infty} -\log|s_w|_{F_w}\cdot[w] + \sum_{v\nmid\infty} \mathrm{ord}_v(s_v)\cdot[v],$$

where $|\cdot|_w$ for complex w denotes the square of the usual absolute value on $F_w \simeq \mathbb{C}$. In particular, the *replete principal divisor* (f) associated to any $f \in F^\times$ is

$$(f) = \sum_{w|\infty} -\log|f|_w\cdot[w] + \sum_{v\nmid\infty} \mathrm{ord}_v(f)\cdot[v]\,.$$

The *replete divisor class group* $\overline{\mathrm{Pic}}(\mathcal{O}_F)$ is the quotient of the group of all replete divisors of F modulo the group of principal ones. Equivalently, $\overline{\mathrm{Pic}}(\mathcal{O}_F) = \mathbb{A}_F^\times/F^\times\mathbb{K}_F$, where $\mathbb{K}_F \subset \mathbb{A}_F^\times$ denotes the maximal compact subgroup. This description equips the replete divisor class group with a natural structure of locally compact Hausdorff topological group having component group equal to the class group of F and identity component $(\oplus_{w|\infty}\mathbb{R}\cdot[w])/\ell(\mathcal{O}_F^\times)$, where $\ell(u) = \sum_{w|\infty}\log|u|_w\cdot[w]$ is the classical logarithm map on global units.

The *degree* of a replete divisor is defined to be

$$\deg\left(\sum_{w|\infty} a_w\cdot[w] + \sum_{v\nmid\infty} n_v\cdot[v]\right) = \sum_{w|\infty} a_w + \sum_{v\nmid\infty} n_v \log q_v\,,$$

where q_v is the cardinality of the residue field at $v\nmid\infty$. For a replete divisor (s) attached to an idele $s\in\mathbb{A}_F^\times$ we have $\deg((s)) = -\log|s|_{\mathbb{A}_F^\times}$. This definition passes to the quotient and defines a continuous homomorphism $\deg\colon \overline{\mathrm{Pic}}(\mathcal{O}_F)\to\mathbb{R}$ whose kernel is denoted by $\overline{\mathrm{Pic}}^0(\mathcal{O}_F)$.

Clearly $\overline{\mathrm{Pic}}^0(\mathcal{O}_F) = (\mathbb{A}_F^\times)^1/(F^\times\mathbb{K}_F)$, where $(\mathbb{A}_F^\times)^1$ is the group of ideles of norm 1, so this has component group equal to the usual class group and identity component equal to the torus of dimension r_1+r_2-1 that arises in the proof of Dirichlet's unit theorem. In particular, the degree-0 replete divisor class group is compact.

These constructions are functorial in the $\mathbb{Q}$-torus $\underline{F}^\times$. Indeed, if $\chi:\underline{F}^\times\to\underline{F'}^\times$ is any $\mathbb{Q}$-homomorphism (for number fields F and F') then the adelic description of replete divisor class groups via maximal compact subgroups of idele groups implies that the map induced by χ on adelic points defines a continuous homomorphism $\chi_{\overline{\mathrm{Pic}}} : \overline{\mathrm{Pic}}(\mathcal{O}_F) \to \overline{\mathrm{Pic}}(\mathcal{O}_{F'})$. The $\mathbb{Q}$-rational character group $\mathrm{Hom}_{\mathbb{Q}\text{-gp}}(\underline{F}^\times, \mathbb{G}_{\mathrm{m}}) = \mathrm{X}^*(\underline{F}^\times)^{\Gamma_{\mathbb{Q}}}$ is generated by $\mathrm{Nm}_{F/\mathbb{Q}}$, so $\mathrm{Nm}_{F'/\mathbb{Q}} \circ \chi = \mathrm{Nm}_{F/\mathbb{Q}}^n$ for some integer n and hence $|\chi(s)|_{\mathbb{A}_{F'}} = |s|_{\mathbb{A}_F}^n$. It follows that there is likewise an induced continuous homomorphism between degree-0 replete divisor class groups. This finishes our review of basic notions related to replete divisors.

Here are two equivalent versions of Theorem A.4.5.1(b) (using the notation in that result):

(i) For every idele $s \in \mathbb{A}_K^\times$ with $|s|_{\mathbb{A}_K} = 1$, the replete divisor $(\chi(s))$ for the number field L is principal.

(ii) The homomorphism

$$\chi_{\overline{\mathrm{Pic}}^0} : \overline{\mathrm{Pic}}^0(\mathcal{O}_K) \to \overline{\mathrm{Pic}}^0(\mathcal{O}_L)$$

between degree-0 replete divisor class groups is trivial.

The equivalence of (i) and (ii) is clear, and to relate (i) to Theorem A.4.5.1(b) we first note that if $s = (s_\infty, s_f) \in \mathbb{A}_K^\times$ with $|s|_{\mathbb{A}_K} = 1$ then

$$|s_f|_{\mathbb{A}_K}^{w(\chi)} = \mathrm{Nm}_{K/\mathbb{Q}}(s_\infty)^{w(\chi)} = \mathrm{Nm}_{L/L^+}(\chi(s_\infty)).$$

Thus, Theorem A.4.5.1(b) says that for each $s \in (\mathbb{A}_K^\times)^1$ there exists $x \in L^\times$ such that the replete divisor $(\chi(s)) - (x)$ on L is supported at ∞ and is killed by Nm_{L/L^+}. This is visibly equivalent to the principality of the replete divisor $(\chi(s))$.

Let $\chi : \underline{E}^\times \to \underline{L}^\times$ be an admissible $\mathbb{Q}$-homomorphism, with L a CM field and E *totally imaginary* (such as a CM field, though later it will be convenient to allow the generality of totally imaginary fields, such as number fields that contain a CM field). In particular, $(\mathbb{R} \otimes_{\mathbb{Q}} E)^\times$ is connected. Motivated by Theorem A.4.5.1, let $P = P(\chi)$ be the open subgroup of $\mathbb{A}_E^\times$ consisting of all ideles $(s_\infty, s_f) \in \mathbb{A}_E^\times$ such that there exists an element $x \in L^\times$ satisfying

$$\chi(s_f) \cdot \mathcal{O}_L = x \cdot \mathcal{O}_L \qquad \text{and} \qquad \mathrm{Nm}_{L/L^+}(x) = |s_f|_{\mathbb{A}_K}^{-w(\chi)}.$$

In terms of the above reformation of Theorem A.4.5.1(b) via replete divisors, P is generated by $(\mathbb{R} \otimes E)^\times$ and the subgroup of norm-1 ideles $s \in \mathbb{A}_E^\times$ such that the replete divisor of L attached to $\chi(s)$ is principal. Clearly P contains both $E^\times$ and the maximal compact subgroup $U_{E,f}$ of the non-archimedean part of $\mathbb{A}_E^\times$. By Theorem A.4.5.1, χ is the algebraic part of an algebraic Hecke character $\mathbb{A}_E^\times \to L^\times$ if and only if χ is admissible and $P = \mathbb{A}_E^\times$.

Let $F = F(\chi)$ be the finite abelian extension field of E corresponding to the open subgroup of finite index $P/E^\times \subset \mathbb{A}_E^\times/E^\times$ by class field theory. Thus, F is a subfield of the Hilbert class field of E, so $\mathrm{Gal}(F/E) \simeq \mathbb{A}_E^\times/(P \cdot E^\times)$ under the reciprocity law and $P = E^\times \cdot \mathrm{Nm}_{F/E}(\mathbb{A}_F^\times)$. Moreover if K/E is a finite extension of E (so K is totally imaginary since E is) then $\mathrm{Nm}_{K/E}(\mathbb{A}_K^\times) \subset P$ if and only if the natural homomorphism

$$\mathrm{Gal}(K^{\mathrm{ab}}/K) \to \mathrm{Gal}(F/E) \qquad \text{(A.4.6.1)}$$

is trivial (by the functoriality of the reciprocity law), or equivalently if and only if $F \subset K$. Thus, by Theorem A.4.5.1, $\chi \circ \mathrm{Nm}_{K/E}$ is the algebraic part of an algebraic

Hecke character of K (valued in $L^\times$) if and only if K contains $F(\chi)$. We call $F(\chi)$ the *field of moduli* of χ.

A.4.6.1. Example. Let $E \subset \overline{\mathbb{Q}}$ be the CM reflex field of a $\overline{\mathbb{Q}}$-valued CM type Φ of a CM field L. Consider a CM abelian variety $(A, i : L \to \mathrm{End}(A))$ over $\mathbb{C}$ and an L-linear $\mathbb{Q}$-polarization λ of (A, i). (Such a λ exists over any field of definition of A equipped with its L-action in the isogeny category, by Lemma 2.2.2.) This descends to a number field $K \subset \overline{\mathbb{Q}}$, by Theorem 1.7.2.1, so the group of elements $\sigma \in \mathrm{Gal}(\overline{\mathbb{Q}}/\mathbb{Q})$ such that $(A^\sigma, \lambda^\sigma)$ is L-linearly isomorphic to (A, λ) contains $\mathrm{Gal}(\overline{\mathbb{Q}}/K)$ and hence it is open. The number field $F \subset \overline{\mathbb{Q}}$ corresponding to this open subgroup is the "field of moduli" of (A, i, λ) in the sense defined by Shimura in [**115**, Chapter IV, §17]. We claim that this number field is an abelian extension of E and as such is exactly the field of moduli $F(\mathrm{N}_\Phi)$ attached to the admissible reflex norm N_Φ.

Passing to the induced $\mathbb{C} \otimes_{\mathbb{Q}} L$-linear isomorphism of tangent spaces $\mathrm{Lie}(A) \simeq \mathrm{Lie}(A^\sigma)$ shows that $\sigma\Phi = \Phi$ for any $\sigma \in \mathrm{Gal}(\overline{\mathbb{Q}}/F)$, so $E \subset F$ inside $\overline{\mathbb{Q}}$. Thus, the problem is to show that for $\sigma \in \mathrm{Gal}(\overline{\mathbb{Q}}/E)$, $\sigma \in \mathrm{Gal}(\overline{\mathbb{Q}}/F(\mathrm{N}_\Phi))$ if and only if $(A^\sigma, \lambda^\sigma)$ is L-linearly isomorphic to (A, λ). Note that the behavior of σ on $F(\mathrm{N}_\Phi)$ only depends on $\sigma|_{E^{\mathrm{ab}}}$, so to study this problem we pick a finite idele $s \in \mathbb{A}_{E,f}^\times$ that maps to $\sigma|_{E^{\mathrm{ab}}}$ under the reciprocity map. By Theorem A.2.9.1, $(A^\sigma, \lambda^\sigma)$ is L-linearly isomorphic to $(\mathrm{N}_\Phi(1/s) \cdot A, \lambda_{1/s})$ with $\lambda_{1/s}$ defined as in A.2.9. For the analytic uniformization V/Λ of $A(\mathbb{C})$ and any $s' \in \mathbb{A}_{L,f}^\times$ the analytic uniformization of $(s'A)(\mathbb{C})$ is $V/(s'\Lambda)$ for the lattice $s'\Lambda \subset \Lambda_{\mathbb{Q}}$, and the L-linear isogenies $V/\Lambda \to V/(s'\Lambda)$ correspond to precisely multiplication on V by $c \in L^\times$ such that $c\Lambda \subset s'\Lambda$. Hence, L-linear isomorphisms from A to A^σ (ignoring the polarizations) correspond exactly to $c \in L^\times$ such that $c\Lambda = \mathrm{N}_\Phi(1/s)\Lambda$ inside $\Lambda_{\mathbb{Q}}$, which is to say $\mathrm{N}_\Phi(s) \cdot \mathcal{O}_L = c^{-1} \cdot \mathcal{O}_L$.

The $\mathbb{Q}(1)$-valued analytic Weil pairing $e_{\lambda_{1/s}}$ on $\Lambda_{\mathbb{Q}}$ coincides with $q_s e_\lambda$ where $q_s \in \mathbb{Q}_{>0}^\times$ is the unique positive generator of the fractional $\mathbb{Q}$-ideal associated to the $\mathbb{Q}$-idele $\mathrm{Nm}_{E/\mathbb{Q}}(s)$ (see A.2.9), or equivalently $q_s = |s|_{\mathbb{A}_E}^{-1}$ (since $s_\infty = 1$). Thus, the compatibility with $\mathbb{Q}$-polarizations via c-multiplication means

$$q_s e_\lambda(cx, cy) = e_\lambda(x, y) \quad \text{for all} \quad x, y \in \Lambda_{\mathbb{Q}},$$

or in other words $q_s cc^* = 1$, which is to say $\mathrm{Nm}_{L/L^+}(1/c) = q_s = |s|_{\mathbb{A}_E}^{-1}$. To summarize, there is an L-linear isomorphism $(A^\sigma, \lambda^\sigma) \simeq (A, \lambda)$ if and only if the open subgroup $P(\mathrm{N}_\Phi) \subset \mathbb{A}_E^\times$ contains the idele $(1, s)$. That is, it is necessary and sufficient that $\sigma|_{E^{\mathrm{ab}}}$ is trivial on the finite abelian extension $F(\mathrm{N}_\Phi)$ of E. This completes the proof that $F(\mathrm{N}_\Phi)$ is the "field of moduli" in the sense of Shimura for every (A, i, λ) over $\mathbb{C}$.

Further justification for the terminology "field of moduli" will be given in Example A.4.6.3, Proposition A.4.6.4, and Corollary A.4.6.5.

A.4.6.2. Remark. The field of moduli of an admissible χ may decrease if one enlarges the target CM number field L. That is, if L' is a CM extension field of L and $j : \underline{L}^\times \to \underline{L'}^\times$ is the natural algebraic homomorphism induced by $L \hookrightarrow L'$, then we have $F(j \circ \chi) \subset F(\chi)$.

Similarly if K is a finite extension of E then the field of moduli $F(\chi \circ \mathrm{Nm}_{K/E})$ of $\chi \circ \mathrm{Nm}_{K/E} : \underline{K}^\times \to \underline{L}^\times$ is a subextension of the everywhere unramified abelian extension $KF(\chi)/K$. Explicitly, $F(\chi \circ \mathrm{Nm}_{K/E})$ is the class field for K associated to the preimage of P under $\mathrm{Nm}_{K/E} : \mathbb{A}_K^\times \to \mathbb{A}_E^\times$; equivalently, the natural map $\mathrm{Gal}(K^{\mathrm{ab}}/K) \to \mathrm{Gal}(F(\chi)/E)$ has kernel associated to $F(\chi \circ \mathrm{Nm}_{K/E})$. This kernel corresponds to the kernel of the natural map $\mathrm{Gal}(K \cdot F(\chi)/K) \to \mathrm{Gal}(F(\chi)/E)$.

A.4.6.3. Example. By Theorem A.4.5.1, the admissible $\chi \circ \mathrm{Nm}_{K/E}$ is the algebraic part of an algebraic Hecke character $\mathbb{A}_K^\times \to L^\times$ if and only if $F(\chi \circ \mathrm{Nm}_{K/E}) = K$, which is equivalent to $F(\chi)$ being contained in K (over E). That is, $F(\chi)$ is the unique minimal finite extension K/E such that $\chi \circ \mathrm{Nm}_{K/E}$ is the algebraic part of an $L^\times$-valued algebraic Hecke character of K.

As a special case, if Φ is a $\overline{\mathbb{Q}}$-valued CM type of a CM field L and if $E \subset \overline{\mathbb{Q}}$ is the CM reflex field then *every* field of definition $K \subset \overline{\mathbb{Q}}$ of an L-linear CM abelian variety over $\overline{\mathbb{Q}}$ with CM type Φ *must* contain $F(\mathrm{N}_\Phi)$ because $E \subset K$ (inside $\overline{\mathbb{Q}}$) and the Main Theorem of Complex Multiplication provides an algebraic Hecke character $\mathbb{A}_K^\times \to L^\times$ whose algebraic part is $\mathrm{N}_\Phi \circ \mathrm{Nm}_{K/E}$. The optimality of $F(\mathrm{N}_\Phi)$ in this respect is addressed in Corollary A.4.6.5.

The next result follows immediately from Theorem A.4.5.1 and the above discussion (especially (A.4.6.1)), together with the construction of $\epsilon : \mathbb{A}_{F(\chi)}^\times \to L^\times$ with algebraic part $\chi \circ \mathrm{Nm}_{K/E}$ via A.4.5.5.

A.4.6.4. Proposition. *Let $\chi : \underline{E}^\times \to \underline{L}^\times$ be an admissible $\mathbb{Q}$-homomorphism, with L a* CM *field and E totally imaginary. Let F/E be the field of moduli of χ.*

For any positive integer N, there exists an odd rational prime $p \nmid N$ and an algebraic Hecke character $\epsilon\colon \mathbb{A}_F^\times \to L^\times$ with algebraic part $\chi \circ \mathrm{Nm}_{F/E}$ such that ϵ is unramified outside p and at most tamely ramified at places of F above p. Moreover if K is a finite extension of E such that there exists an algebraic Hecke character $\mathbb{A}_F^\times \to L^\times$ with algebraic part $\chi \circ \mathrm{Nm}_{K/E}$, then $F \subset K$.

This has an important consequence in the setting of CM theory, in view of the admissibility of reflex norms (Example A.4.4.1):

A.4.6.5. Corollary. *Let L be a* CM *field and let Φ be a $\overline{\mathbb{Q}}$-valued* CM *type for L. Let $E \subset \overline{\mathbb{Q}}$ be the reflex field of (L, Φ) and let F/E be the field of moduli for the admissible reflex norm $\mathrm{N}_\Phi\colon \underline{E}^\times \to \underline{L}^\times$ (an everywhere unramified finite abelian extension of E). Let N be a positive integer.*

There exists an odd prime $p \nmid N$ and CM *abelian variety $(A, \alpha : L \to \mathrm{End}^0(A))$ over F with* CM *type (L, Φ) such that:*

(i) *A has good reduction outside p,*
(ii) *for all primes $\ell \neq p$, the ℓ-adic representation $\mathrm{Gal}(\overline{\mathbb{Q}}/F) \to \mathrm{Aut}_{L_\ell}(V_\ell(A)) = L_\ell^\times$ (which is unramified outside ℓp) is at most tamely ramified at places of F above p.*

For every such (A, α) over F there exists an L-linear polarization of A over F.

PROOF. The existence of a CM abelian variety $(A, \alpha : L \to \mathrm{End}^0(A))$ over F of CM type (L, Φ) with good reduction outside an odd prime number $p \nmid N$ and tame above p is an immediate consequence of Theorem 2.5.2 and Proposition A.4.6.4. Given such a CM abelian variety over F, Lemma 2.2.2 provides an L-linear polarization defined over F. □

A.4.7. Slopes. Fix an embedding $j_p : \overline{\mathbb{Q}} \hookrightarrow \overline{\mathbb{Q}}_p$, so we can speak of p-adic valuations for elements of $\overline{\mathbb{Q}}$. We aim to give a formula for the p-adic valuations of certain values of admissible algebraic Hecke characters of a number field K, with values in a CM field L, generalizing the Shimura–Taniyama formula in (2.1.4.1).

(A result on algebraic Hecke characters can generalize the Shimura–Taniyama formula because the Main Theorem of Complex Multiplication relates the Galois representation on a CM abelian variety over a number field to an algebraic Hecke character. Keep in mind, however, that the proof of the Main Theorem of Complex Multiplication *uses* the Shimura–Taniyama formula. Thus, the formula for p-adic valuations proved in this section does not lead to another proof of the Shimura–Taniyama formula. Instead, it puts that formula into a broader algebraic number theory context.)

In terms of notation from (A.4.2), the formula established for p-adic valuations will be expressed in terms of the basis $\{\xi^\sigma\}$ of $\mathrm{X}^*(\underline{F}^\times)$ indexed by embeddings $\sigma : F \to \overline{\mathbb{Q}} \subset \overline{\mathbb{Q}}_p$ (for $F = L$) and the basis $\{\nu_\tau\}$ of $\mathrm{X}_*(\underline{F}^\times)$ dual to $\{\xi^\tau\}$ (for $F = K$). The context will always make clear which F is intended, so we generally write ξ^σ rather than $\xi^\sigma_{F/\mathbb{Q}}$, and likewise for the dual basis.

To get started, we record an elementary result for general number fields.

A.4.7.1. Lemma. *Let K and L be number fields, and let $\chi : \underline{K}^\times \to \underline{L}^\times$ be a $\mathbb{Q}$-homomorphism. For every $y \in K^\times = \underline{K}^\times(\mathbb{Q})$ and every embedding $\sigma\colon L \to \overline{\mathbb{Q}}$ we have*

$$\sigma(\chi(y)) = \prod_{\tau \in \mathrm{Hom}(K,\overline{\mathbb{Q}})} \tau(y)^{\langle \xi^\sigma, \chi\circ\nu_\tau\rangle}.$$

PROOF. Consider the element

$$y' := \prod_{\tau \in \mathrm{Hom}(K,\overline{\mathbb{Q}})} \nu_\tau(\tau(y)) \in \underline{K}^\times(\overline{\mathbb{Q}}) = (\overline{\mathbb{Q}} \otimes_{\mathbb{Q}} K)^\times.$$

For $z \in \overline{\mathbb{Q}}^\times$ and the character $\xi^{\tau_1} = \xi^{\tau_1}_{K/\mathbb{Q}}$ attached to an embedding $\tau_1 \in \mathrm{Hom}(K,\overline{\mathbb{Q}})$, we have

$$\xi^{\tau_1}(\nu_\tau(z)) = \begin{cases} z & \text{if } \tau = \tau_1 \\ 1 & \text{if } \tau \neq \tau_1\,. \end{cases}$$

Thus, $\xi^{\tau_1}(y') = \tau_1(y) = \xi^{\tau_1}(y)$ for every $\tau_1 \in \mathrm{Hom}(K,\overline{\mathbb{Q}})$, so $y' = y$. We have shown that

$$y = \prod_{\tau \in \mathrm{Hom}(K,\overline{\mathbb{Q}})} \nu_\tau(\tau(y))$$

in $(\overline{\mathbb{Q}} \otimes_{\mathbb{Q}} K)^\times$ for all $y \in K^\times$.

Using this formula, in $\overline{\mathbb{Q}}^\times$ we compute

$$\begin{aligned}\sigma(\chi(y)) = \xi^\sigma(\chi(y)) &= \xi^\sigma\left(\chi\left(\prod_{\tau\in\mathrm{Hom}(K,\overline{\mathbb{Q}})} \nu_\tau(\tau(y))\right)\right)\\ &= \xi^\sigma\left(\prod_{\tau\in\mathrm{Hom}(K,\overline{\mathbb{Q}})} (\chi\circ\nu_\tau)(\tau(y))\right)\\ &= \prod_{\tau\in\mathrm{Hom}(K,\overline{\mathbb{Q}})} \tau(y)^{\langle\xi^\sigma,\,\chi\circ\nu_\tau\rangle}.\end{aligned}$$ □

Let ι be the complex conjugation of our chosen CM field L, and let $L^+ = L^{\iota=1}$ be the maximal totally real subfield of L. Let $\wp$ be a place above a rational prime p for our chosen number field K, and let v be a place of L above p, with v_0 be the place of L^+ induced by v. Let $\Psi_v = \mathrm{Hom}_{\mathbb{Q}_p}(L_v, \overline{\mathbb{Q}}_p) \subset \mathrm{Hom}(L, \overline{\mathbb{Q}})$ be the set of embeddings inducing v on L via $j_p : \overline{\mathbb{Q}} \hookrightarrow \overline{\mathbb{Q}}_p$. Similarly, let $\Psi_\wp = \mathrm{Hom}_{\mathbb{Q}_p}(K_\wp, \overline{\mathbb{Q}}_p) \subset \mathrm{Hom}(K, \overline{\mathbb{Q}})$ be the set of embeddings inducing $\wp$ on K via j_p.

A.4.7.2. Proposition. *Assume $\chi : \underline{K}^\times \to \underline{L}^\times$ as above is admissible with weight $w(\chi)$. Choose $y \in K^\times$ that is a local unit at all finite places of K different from $\wp$ and is a non-unit at $\wp$. Let $q = q_\wp^{\mathrm{ord}_\wp(y)} = |y|_{K_\wp}^{-1} \neq 1$, where $q_\wp = \#\kappa(\wp)$.*

(i) *The value $\chi(y) \in L^\times$ is a local unit at all finite places of L relatively prime to q and*

$$\mathrm{Nm}_{L/L^+}(\chi(y)) = q^{w(\chi)}\,.$$

(ii) *For a place $v|p$ of L, the v-slope of $\chi(y)$ normalized by q, defined to be*

$$\mathrm{slope}_{v,q}(\chi(y)) := \frac{\mathrm{ord}_v(\chi(y))}{\mathrm{ord}_v(q)},$$

is given by

$$\mathrm{slope}_{v,q}(\chi(y)) = \frac{\sum_{\sigma\in\Psi_v}\langle\xi^\sigma,\chi\circ\nu_\tau\rangle}{[L_v:\mathbb{Q}_p]}$$

for any $\mathbb{Q}_p$-embedding $\tau\colon K_\wp \to \overline{\mathbb{Q}}_p$.

(iii) *Let v_0 be a finite place of L^+ over p. If v_0 does not split in L then*

$$\mathrm{slope}_{v,q}(\chi(y)) = w(\chi)/2$$

for the place v of L above v_0. If v_0 splits in L then the slopes of $\chi(y)$ at the places v and $\overline{v}$ of L above v_0 satisfy

$$\mathrm{slope}_{v,q}(\chi(y)) + \mathrm{slope}_{\overline{v},q}(\chi(y)) = w(\chi)\,.$$

Part (i) says that $\chi(y)$ is a "Weil q-number of weight $w(\chi)$" in a sense that is more general than 1.6.1.2(ii) because q is an arbitrary non-trivial integral power of a prime number (presently restricted to the case of powers of $q_\wp$). That is, we allow the possibility $q < 1$, in which case $\mathrm{slope}_{v,q}(\chi(y)) \leqslant 0$ if $\chi(y) \in \mathcal{O}_{L_v}$. The proof of part (ii) does not use the hypothesis that χ is admissible. Note also that the dependence of q on y makes $\mathrm{slope}_{v,q}(\chi(y))$ invariant under replacing y with $1/y$.

PROOF. The statement (i) is immediate from Proposition A.4.4.4(b), since Nm_{L/L^+} takes totally positive values in $L^\times$ and for any prime ℓ the continuous map $K_\ell^\times \to L_\ell^\times$ induced by χ on $\mathbb{Q}_\ell$-points carries the maximal compact subgroup into the maximal compact subgroup (so $\chi(y)$ is a local unit at all ℓ-adic places of L when $\ell \neq p$). The statement (iii) follows from (i) and the definition of $\mathrm{slope}_{v,q}(\chi(y))$ in (ii). It remains to prove (ii).

We regard $\overline{\mathbb{Q}}$ as a subfield of $\overline{\mathbb{Q}}_p$ via the chosen embedding $j_p : \overline{\mathbb{Q}} \to \overline{\mathbb{Q}}_p$. Denote by ord_p the valuation on $\overline{\mathbb{Q}}_p$ normalized by $\mathrm{ord}_p(p) = 1$. Since y is assumed to be a unit at all p-adic places of K away from $\wp$, Lemma A.4.7.1 yields the formula

$$\mathrm{ord}_p(\sigma(\chi(y))) = \left(\sum_{\tau \in \Psi_\wp} \langle \xi^\sigma, \chi \circ \nu_\tau \rangle \right) \cdot \mathrm{ord}_p(\tau_0(y))$$

for any embedding $\sigma : L \to \overline{\mathbb{Q}} \subset \overline{\mathbb{Q}}_p$ and choice of embedding $\tau_0 : K \to \overline{\mathbb{Q}} \subset \overline{\mathbb{Q}}_p$ inducing the place $\wp$ on K (since $\mathrm{ord}_p(\tau_0(y)) = e(\wp|p)\mathrm{ord}_\wp(y)$ is the same integer for all τ_0). The action of $\mathrm{Gal}(\overline{\mathbb{Q}}/\mathbb{Q})$ on the basis $\{\xi^\sigma\}$ of $\mathrm{X}^*(\underline{L}^\times)$ is given by the permutation action $g.\xi^\sigma = \xi^{g \circ \sigma}$, and the same holds for the dual action on the basis $\{\nu^\tau\}$ of $\mathrm{X}_*(\underline{K}^\times)$, so by working over $\overline{\mathbb{Q}}_p$ rather than over $\overline{\mathbb{Q}}$ (and using decomposition groups inside $\mathrm{Gal}(\overline{\mathbb{Q}}/\mathbb{Q})$ at p-adic places) we conclude that the integer

$$r(\chi; \wp, \sigma) := \sum_{\tau \in \Psi_\wp} \langle \xi^\sigma, \chi \circ \nu_\tau \rangle$$

depends only on the p-adic place v of L induced by σ because (a) $\sum_{\tau \in \Psi_\wp} \chi \circ \nu_\tau$ is fixed under $\mathrm{Gal}(\overline{\mathbb{Q}}_p/\mathbb{Q}_p)$ and (b) the group $\mathrm{Gal}(\overline{\mathbb{Q}}_p/\mathbb{Q}_p)$ operates transitively on the set $\Psi_v = \mathrm{Hom}_{\mathbb{Q}_p\text{-alg}}(L_v, \overline{\mathbb{Q}}_p) \subset \mathrm{Hom}(L, \overline{\mathbb{Q}})$ of such σ. Denote this integer as $r(\chi; \wp, v)$.

Similarly, for any $\tau \in \Psi_\wp$ the sum $\sum_{\sigma \in \Psi_v} \langle \xi^\sigma, \chi \circ \nu_\tau \rangle$ depends only on the place $\wp$ of K induced by τ, so

$$\sum_{\sigma \in \Psi_v} \langle \xi^\sigma, \chi \circ \nu_\tau \rangle = \frac{[L_v : \mathbb{Q}_p]}{[K_\wp : \mathbb{Q}_p]} \cdot r(\chi; \wp, v)$$

for all $\tau \in \Psi_\wp$.

Thus, for any embeddings $\tau_0 \colon K \to \overline{\mathbb{Q}}_p$ $\sigma_0 : L \to \overline{\mathbb{Q}}_p$ respectively inducing $\wp$ on K and v on L, we have

$$\begin{aligned}
\frac{\mathrm{ord}_v(\chi(y))}{\mathrm{ord}_v(q)} = \frac{\mathrm{ord}_p(\sigma_0(\chi(y)))}{\mathrm{ord}_p(q)} &= \frac{r(\chi; \wp, v) \cdot \mathrm{ord}_p(\tau_0(y))}{\mathrm{ord}_p(q)} \\
&= \frac{\sum_{\sigma \in \Psi_v} \langle \xi^\sigma, \chi \circ \nu_\tau \rangle}{[L_v : \mathbb{Q}_p]} \cdot \frac{[K_\wp : \mathbb{Q}_p] \cdot \mathrm{ord}_p(\tau_0(y))}{\mathrm{ord}_p(q_\wp^{\mathrm{ord}_\wp(y)})} \\
&= \frac{\sum_{\sigma \in \Psi_v} \langle \xi^\sigma, \chi \circ \nu_\tau \rangle}{[L_v : \mathbb{Q}_p]}
\end{aligned}$$

since $\mathrm{ord}_p(q_\wp) = [\kappa(\wp) : \mathbb{F}_p] = [K_\wp : \mathbb{Q}_p]/e(\wp|p)$ and

$$\mathrm{ord}_p(\tau_0(y)) = \mathrm{ord}_\wp(y)/e(\wp|p) \neq 0.$$

This establishes (ii). □

A.4.7.3. Remark. During the proof of A.4.7.2(ii) we established the identity

$$\frac{\sum_{\sigma\in\Psi_v}\langle\xi^\sigma,\chi\circ\nu_{\tau_0}\rangle}{[L_v:\mathbb{Q}_p]}=\frac{\sum_{\tau\in\Psi_\wp}\langle\xi^{\sigma_0},\chi\circ\nu_\tau\rangle}{[K_\wp:\mathbb{Q}_p]}$$

for all $\tau_0\in\Psi_\wp$ and $\sigma_0\in\Psi_v$. Thus, for any $\sigma\in\Psi_v=\mathrm{Hom}_{\mathbb{Q}_p}(L_v,\overline{\mathbb{Q}}_p)$ we also have a "dual" formula

$$\mathrm{slope}_{v,q}(\chi(y))=\frac{\sum_{\tau\in\Psi_\wp}\langle\xi^\sigma\circ\chi,\nu_\tau\rangle}{[K_\wp:\mathbb{Q}_p]}$$

for the slope of $\chi(y)$, where the numerator is the sum of the coefficients (relative to the basis $\{\xi^\tau\}$ of $\mathrm{X}^*(\underline{K}^\times)$), at the locations corresponding to $\mathbb{Q}_p$-embeddings of $K_\wp$ into $\overline{\mathbb{Q}}_p$, of a single geometric character $\xi^\sigma\circ\chi$ of $\underline{K}^\times$. Recall that by hypothesis, y is a local unit at all p-adic places of K away from $\wp$ and a non-unit at $\wp$.

A.4.7.4. Corollary. *Let K and L be number fields, with L a* CM *field. Let $\epsilon\colon\mathbb{A}_K^\times\to L^\times$ be an algebraic Hecke character with (necessarily admissible) algebraic part $\chi\colon\underline{K}^\times\to\underline{L}^\times$. Let $\pi_\wp$ be a uniformizer of $\mathcal{O}_{K_\wp}$ for a prime $\wp|p$ of K, and let $q_\wp$ denote the size of the residue field at $\wp$.*

(i) *The value $\epsilon(\pi_\wp)\in L^\times$ is a Weil $q_\wp$-number of weight $w(\chi)$; i.e., $\epsilon(\pi_\wp)$ is a unit at every finite place of L relatively prime to $q_\wp$ and $\mathrm{Nm}_{L/L^+}(\epsilon(\pi_\wp))=q_\wp^{w(\chi)}$.*

(ii) *For any place $v|p$ of L, the Weil $q_\wp$-number $\epsilon(\pi_\wp)$ satisfies*

$$\mathrm{slope}_{v,q_\wp}(\epsilon(\pi_\wp)):=\frac{\mathrm{ord}_v(\epsilon(\pi_\wp))}{\mathrm{ord}_v(q_\wp)}=\frac{\sum_{\sigma\in\Psi_v}\langle\xi^\sigma,\chi\circ\nu_\tau\rangle}{[L_v\colon\mathbb{Q}_p]}$$

for any $\mathbb{Q}_p$-embedding $\tau:K_\wp\hookrightarrow\overline{\mathbb{Q}}_p$.

(iii) *If the place v_0 of L^+ below v does not split in L then $\mathrm{slope}_{v,q_\wp}(\epsilon(\pi_\wp))=w(\chi)/2$. If v_0 splits into two places v and $\overline{v}$ on L then*

$$\mathrm{slope}_{v,q_\wp}(\epsilon(\pi_\wp))+\mathrm{slope}_{\overline{v},q_\wp}(\epsilon(\pi_\wp))=w(\chi)\,.$$

PROOF. We have seen (i) in the proof of A.4.4.5. Here we repeat part of that argument to set up the notation for the proof of (ii) and (iii). Let $\mathfrak{m}$ be a conductor of ϵ, so ϵ is trivial on the open subgroup

$$U_\mathfrak{m}=(K_\infty^\times)^0\times\prod_{v|\mathfrak{m}}(1+\mathfrak{m}\mathcal{O}_{K,v})^\times\times\prod_{v\nmid\mathfrak{m}}\mathcal{O}_{K,v}^\times\subset\mathbb{A}_K^\times\,.$$

There exists a positive integer $n>0$, a global element $y\in K^\times$, and an element $u\in U_\mathfrak{m}$ such that $\pi_\wp^n=y\cdot u$. Then $\epsilon(\pi_\wp^n)=\chi(y)$ is a Weil $q_\wp^n$-number in L, and

$$\frac{\mathrm{ord}_v(\epsilon(\pi_\wp))}{\mathrm{ord}_v(q_\wp)}=\frac{\mathrm{ord}_v(\chi(y))}{\mathrm{ord}_v(q_\wp^n)}=\frac{\sum_{\sigma\in\Psi_v}\langle\xi^\sigma,\chi\circ\nu_\tau\rangle}{[L_v\colon\mathbb{Q}_p]}$$

by Proposition A.4.7.2(ii). This proves (ii), and (iii) follows from (i) and (ii). □

A.4.8. Construction of admissible homomorphisms. Let F be a number field and let $\mu\in\mathrm{X}_*(\underline{F}^\times)$ be a geometric cocharacter. Let $E=E(\mu)\subset\overline{\mathbb{Q}}$ denote the *field of definition* of μ over $\mathbb{Q}$; i.e., the unique minimal subfield of $\overline{\mathbb{Q}}$ such that μ descends to an E-homomorphism $\mu_0:\mathbb{G}_\mathrm{m}\to(\underline{F}^\times)_E$. Thus, the $\Gamma_\mathbb{Q}$-action on μ factors through the quotient set $\Gamma_\mathbb{Q}/\Gamma_E=\mathrm{Hom}(E,\overline{\mathbb{Q}})$ (carrying $1\bmod\Gamma_E$ to the given inclusion $\tau_0:E\hookrightarrow\overline{\mathbb{Q}}$).

We define the $\mathbb{Q}$-homomorphism

$$\chi(\mu)\colon \underline{E}^\times \to \underline{F}^\times$$

to be the composition of the Weil restriction $\mathrm{Res}_{E/\mathbb{Q}}(\mu_0) : \underline{E}^\times \to \mathrm{Res}_{E/\mathbb{Q}}(\underline{F}_E^\times)$ and the "norm" homomorphism $\mathrm{N}_{E/\mathbb{Q}} : \mathrm{Res}_{E/\mathbb{Q}}(\underline{F}_E^\times) \to \underline{F}^\times$ (as considered for rather general commutative linear algebraic $\mathbb{Q}$-groups in place of $\underline{F}^\times$ in Proposition 2.1.3.4 applied to $E/\mathbb{Q}$).[3]

There is a useful alternative characterization of $\chi(\mu)$ in terms of $\Gamma_{\mathbb{Q}}$-modules, as follows. This $\mathbb{Q}$-homomorphism is identified with a $\Gamma_{\mathbb{Q}}$-equivariant homomorphism $\mathrm{X}_*(\underline{E}^\times) \to \mathrm{X}_*(\underline{F}^\times)$, and $\mathrm{X}_*(\underline{E}^\times)$ has as a $\mathbb{Z}$-basis the $\Gamma_{\mathbb{Q}}$-orbit given by the dual basis $\{\nu_\tau\}$ of the basis $\{\xi^\tau\}$ of $\mathrm{X}^*(\underline{E}^\times)$. Hence, to determine $\chi(\mu)$ it suffices to compute the composition of $\chi(\mu)_{\overline{\mathbb{Q}}}$ with ν_τ for a single τ. By definition, there is a canonical inclusion $\tau_0 : E \to \overline{\mathbb{Q}}$, and we claim:

A.4.8.1. Proposition. *Inside* $\mathrm{X}_*(\underline{F}^\times)$, $\chi(\mu)_{\overline{\mathbb{Q}}} \circ \nu_\tau = \tau \cdot \mu$ *under the canonical identification of* $\mathrm{Hom}(E, \overline{\mathbb{Q}})$ *with* $\Gamma_{\mathbb{Q}}/\Gamma_E$. *In particular,* $\chi(\mu)_{\overline{\mathbb{Q}}} \circ \nu_{\tau_0} = \mu$.

PROOF. For $x \in \overline{\mathbb{Q}}^\times$, the element $\nu_\tau(x) \in \underline{E}^\times(\overline{\mathbb{Q}}) = \prod_{\tau'} \overline{\mathbb{Q}}^\times$ has τ'-component $x^{\langle \xi_{\tau'}, \nu_\tau \rangle}$ equal to x when $\tau' = \tau$ and equal to 1 otherwise. By definition, $\chi(\mu) = \mathrm{Nm}_{E/\mathbb{Q}} \circ \mathrm{Res}_{E/\mathbb{Q}}(\mu_0)$, and for any affine finite type $\mathbb{Q}$-group G (such as $\mathbb{G}_{\mathrm{m}}$ or $\underline{F}^\times$) we have canonically

$$(\mathrm{Res}_{E/\mathbb{Q}}(G_E))_{\overline{\mathbb{Q}}} = \mathrm{Res}_{(E \otimes_{\mathbb{Q}} \overline{\mathbb{Q}})/\overline{\mathbb{Q}}}(G_{E \otimes_{\mathbb{Q}} \overline{\mathbb{Q}}}) \simeq \prod_{\tau' : E \to \overline{\mathbb{Q}}} G_{\overline{\mathbb{Q}}}.$$

For any linear algebraic $\mathbb{Q}$-group G' and E-homomorphism $f : G'_E \to G_E$, this identifies $\mathrm{Res}_{E/\mathbb{Q}}(f)_{\overline{\mathbb{Q}}}$ with the direct product homomorphism $(\tau'^*(f))_{\tau'}$. Thus, the cocharacter

$$\mathrm{Res}_{E/\mathbb{Q}}(\mu_0)_{\overline{\mathbb{Q}}} \circ \nu_\tau$$

of $\prod_{\tau'} \underline{F}_{\overline{\mathbb{Q}}}^\times$ has τ-component $\tau^*(\mu_0)$ and all other components equal to 1.

For any commutative linear algebraic $\mathbb{Q}$-group C, applying scalar extension to $\overline{\mathbb{Q}}$ carries the norm $\mathrm{Res}_{E/\mathbb{Q}}(C_E) \to C$ to the homomorphism $\prod_\tau C_{\overline{\mathbb{Q}}} \to C_{\overline{\mathbb{Q}}}$ that is the product of all components. Hence, $\chi(\mu)_{\overline{\mathbb{Q}}} \circ \nu_\tau = \tau^*(\mu_0)$. The identification of $\mathrm{Hom}(E, \overline{\mathbb{Q}})$ with $\Gamma_{\mathbb{Q}}/\Gamma_E$ carries τ_0 to the trivial coset, and $\tau_0^*(\mu_0) = \mu$ (by definition of τ_0 and μ_0). Thus, $\tau^*(\mu_0) = \tau \cdot \mu$. □

Recall from A.4.3 that we fix an involution $\iota \in \Gamma_{\mathbb{Q}}$ (all choices of which are $\Gamma_{\mathbb{Q}}$-conjugate).

A.4.8.2. Corollary. *The* $\mathbb{Q}$*-homomorphism* $\chi(\mu)$ *is admissible if and only if*

$$(\gamma - 1)(\iota + 1) \cdot \mu = 0, \quad (\iota + 1)(\gamma - 1) \cdot \mu = 0$$

for all $\gamma \in \Gamma_{\mathbb{Q}}$. *When these conditions hold,* $(\iota + 1) \cdot \mu = wj$ *where* $j : \mathbb{G}_{\mathrm{m}} \to \underline{F}^\times$ *is the canonical cocharacter over* $\mathbb{Q}$ *and* w *is the admissible weight of* $\chi(\mu)$ *(in the sense of* A.4.4.3).

---

[3]In similar situation in 3.9.3 where the base field is $\mathbb{Q}_p$, the notation $\mathrm{N}\mu$ is used for the $\mathbb{Q}_p$-homomorphism between Weil restricted tori attached to a geometric cocharacter μ. We use the notation $\chi(\mu)$ here because the symbol χ is used to denote algebraic homomorphisms between Weil-restricted tori in this section.

PROOF. By Lemma A.4.3.2, $\chi(\mu)$ is admissible if and only if for all $\gamma \in \Gamma_{\mathbb{Q}}$ the elements $(\gamma - 1)(\iota + 1)$ and $(\iota + 1)(\gamma - 1)$ in $\mathbb{Z}[\Gamma_{\mathbb{Q}}]$ kill the $\Gamma_{\mathbb{Q}}$-equivariant map induced by $\chi(\mu)$ between geometric character groups. By duality, it is equivalent to check the same for the induced map between geometric cocharacter groups. But $\chi(\mu)_{\overline{\mathbb{Q}}} \circ \nu_\tau = \tau \cdot \mu$ (via the identification of $\mathrm{Hom}(E, \overline{\mathbb{Q}})$ with $\Gamma_{\mathbb{Q}}/\Gamma_E$ using τ_0), so by the $\Gamma_{\mathbb{Q}}$-equivariance of $\chi(\mu)_{\overline{\mathbb{Q}}}$ it is the same to say that all elements in the $\Gamma_{\mathbb{Q}}$-orbit of μ are killed by $(\gamma - 1)(\iota + 1)$ and $(\iota + 1)(\gamma - 1)$ for all $\gamma \in \Gamma_{\mathbb{Q}}$. Such annihilation for all γ on the single element μ easily implies the same on the entire $\Gamma_{\mathbb{Q}}$-orbit of μ, so the admissibility criterion is proved.

Now assume that $\chi(\mu)$ is admissible, say with w denoting its weight. We need to show that $(\iota + 1) \cdot \mu = wj$. The cocharacter j is $\sum_\sigma \nu_\sigma$ where σ varies through $\mathrm{Hom}(F, \overline{\mathbb{Q}})$ (since clearly $\xi^\sigma \circ j = \mathrm{id}$ for all σ), so it amounts to checking that $\langle \xi^\sigma, (\iota + 1) \cdot \mu \rangle = w$ for all σ. Fix a choice of σ. The defining property of the weight in Proposition A.4.4.4(a) gives that $(\iota + 1) \cdot (\xi^\sigma \circ \chi(\mu)) = w \,\mathrm{Nm}_{E/\mathbb{Q}}$, and for any $\tau \in \mathrm{Hom}(E, \overline{\mathbb{Q}})$ the pairing of both sides against any ν_τ yields that

$$\langle (\iota + 1)\xi^\sigma, \chi(\mu) \circ \nu_\tau \rangle = w.$$

But $\chi(\mu) \circ \nu_\tau = \tau \cdot \mu$ by A.4.8.1, and the pairing of geometric characters and cocharacters is $\Gamma_{\mathbb{Q}}$-equivariant, so $\langle \xi^\sigma, (\iota + 1)\tau \cdot \mu \rangle = w$ for all σ and τ. The admissibility of $\chi(\mu)$ implies that $(\iota+1)\tau \cdot \mu = (\iota+1) \cdot \mu$ (or simply restrict attention to the case that τ is the canonical inclusion $\tau_0 : E \hookrightarrow \overline{\mathbb{Q}}$), so $\langle \xi^\sigma, (\iota + 1) \cdot \mu \rangle = w$ for all σ, as desired. □

A.4.8.3. Example. Let $\Phi \subset \mathrm{Hom}(L, \overline{\mathbb{Q}})$ be a $\overline{\mathbb{Q}}$-valued CM type for a CM field L, and consider the geometric cocharacter

$$\mu_\Phi := \sum_{\sigma \in \Phi} \nu_\sigma ,$$

of $\underline{L}^\times$ characterized by the conditions

$$\langle \xi^\sigma, \mu_\Phi \rangle = \begin{cases} 1 & \text{if } \sigma \in \Phi \\ 0 & \text{if } \sigma \notin \Phi \end{cases}$$

for all $\sigma \in \mathrm{Hom}(L, \overline{\mathbb{Q}})$. The discussion immediately following Definition 2.1.3.3 shows that the field of definition $E(\mu_\Phi)$ of μ_Φ is the reflex field of (L, Φ). By Proposition 2.1.3.4, the $\mathbb{Q}$-homomorphism $\chi(\mu_\Phi)\colon \underline{E(\mu_\Phi)}^\times \to \underline{L}^\times$ is the reflex norm N_Φ for (L, Φ), so by the admissibility of reflex norms of CM types on CM fields (Example A.4.4.1) we see that $\chi(\mu_\Phi)$ is admissible. We now use the admissibility criterion in Corollary A.4.8.2 to reprove that $\chi(\mu_\Phi)$ is admissible.

Clearly $(\iota + 1) \cdot \mu_\Phi = \sum_\sigma \nu_\sigma$ is the canonical cocharacter $\mathbb{G}_{\mathrm{m}} \hookrightarrow \underline{L}^\times$ defined over $\mathbb{Q}$ (corresponding to the inclusion of $\mathbb{Q}$ into L), so $(\gamma - 1)(\iota + 1) \cdot \mu_\Phi = 0$ for all $\gamma \in \Gamma_{\mathbb{Q}}$. The vanishing of $(\iota + 1)(\gamma - 1) \cdot \mu_\Phi$ therefore amounts to the assertion that the elements $\iota\gamma$ and $\gamma\iota$ in $\Gamma_{\mathbb{Q}}$ have the same effect on μ_Φ for all $\gamma \in \Gamma_{\mathbb{Q}}$, or equivalently that all $\Gamma_{\mathbb{Q}}$-conjugates ι' of ι have the same effect on μ_Φ. Clearly $\iota' \cdot \mu_\Phi$ is uniquely characterized by the condition that $\langle \xi^\sigma, \iota' \cdot \mu_\Phi \rangle$ is equal to 1 when $\iota'\sigma \in \Phi$ and is equal to 0 otherwise. But $\iota'\sigma \in \Phi$ if and only if $\sigma \notin \Phi$ since Φ is a CM type on the CM field L, so the independence of ι' is proved. Hence, $\chi(\mu_\Phi)$ is admissible.

Example A.4.8.3 can be pushed further. Letting Φ be a $\overline{\mathbb{Q}}$-valued CM type for a CM field L, we have just seen that $\mu := \mu_\Phi$ has field of definition $E \subset \overline{\mathbb{Q}}$ equal to the reflex field $E(L, \Phi)$, and that $\chi := \chi(\mu)$ is the reflex norm : $\underline{E}^\times \to \underline{L}^\times$. Letting $\tau_0 : E \hookrightarrow \overline{\mathbb{Q}}$ be the canonical inclusion, Proposition A.4.8.1 implies that $\chi_{\overline{\mathbb{Q}}} \circ \nu_{\tau_0} = \mu = \mu_\Phi$. Thus, the displayed formula in A.4.7.4 (ii) becomes the formula

$$\frac{\mathrm{ord}_v(\epsilon(\pi_\wp))}{\mathrm{ord}_v(q_\wp)} = \frac{\#(\Psi_v \cap \Phi)}{[L_v : \mathbb{Q}_p]};$$

for any p-adic place $\wp$ of K and any algebraic Hecke character $\epsilon : \mathbb{A}_K^\times \to L^\times$ satisfying

$$\epsilon_{\mathrm{alg}} = \chi(\mu_\Phi) \circ \mathrm{Nm}_{K/E} = \mathrm{N}_\Phi \circ \mathrm{Nm}_{K/E}.$$

(To deduce this formula from A.4.7.4(ii), use that $\mathrm{Nm}_{K/E} : \underline{K}^\times \to \underline{E}^\times$ carries $\nu_{\tau_0'}$ to ν_{τ_0} for any $\tau_0' : K_\wp \hookrightarrow \overline{\mathbb{Q}}_p$ over τ_0, by the calculations at the end of A.4.3.)

The Main Theorem of Complex Multiplication (Theorem 2.5.1) implies that the algebraic Hecke character associated to any CM abelian variety over K with CM type (L, Φ) is such an ϵ, so we recover the Shimura–Taniyama formula (2.1.4.1) due to Theorem 2.5.1(v). (Recall, however, that our proof of the Main Theorem of Complex Multiplication used the Shimura–Taniyama formula at the step of arriving at the problem of verifying (A.2.6.5).)

As an application of criterion for admissibility of $\mathbb{Q}$-homomorphisms $\chi(\mu)$ in A.4.8.2, we obtain the following existence criterion for Weil numbers with prescribed slopes in a given CM field. (The proof is entirely algebraic number theory; it does not use abelian varieties.)

A.4.8.4. Proposition. *Let L be a CM field. Choose a prime p, and let $(\lambda_v)_{v|p}$ be a sequence of rational numbers indexed by the set of p-adic places v of L. For each v, let $\overline{v}$ be the composition of v with complex conjugation on L. Assume that the following properties hold:*

(a) *the denominator of λ_v divides $[L_v : \mathbb{Q}_p]$ for each p-adic place v of L,*
(b) *there exists $w \in \mathbb{Z}$ such that $\lambda_v + \lambda_{\overline{v}} = w$ for every p-adic place v of L. In particular, $\lambda_v = w/2$ if the place v_0 in L^+ below v does not split in L.*

There exists $q = p^n$ with $n \in \mathbb{Z}^+$ and a Weil q-number $x \in L^\times$ of weight w so that

$$\frac{\mathrm{ord}_v(x)}{\mathrm{ord}_v(q)} = \lambda_v$$

for every p-adic place v of L (so if $\lambda_v \geqslant 0$ for all $v|p$ then $x \in \mathcal{O}_L$).

PROOF. Recall that we fixed an embedding $j_p : \overline{\mathbb{Q}} \hookrightarrow \overline{\mathbb{Q}}_p$ and use it to identify each embedding $\sigma : L \to \overline{\mathbb{Q}}$ with the corresponding p-adic embedding $j_p \circ \sigma : L \to \overline{\mathbb{Q}}_p$. First we consider a p-adic place v_0 of L^+ which does not split in L, and we let v denote the unique place of L above v_0. For each $\sigma \in \Psi_v$ we have $\iota\sigma \in \Psi_v$ (since v is unique over v_0) and certainly $\sigma \neq \iota\sigma$ (since ι defines a non-trivial involution of L_v over $L^+_{v_0}$). We assign the value of one of n_σ or $n_{\iota\sigma}$ arbitrarily in $\mathbb{Z}$ and define the other so that $n_\sigma + n_{\iota\sigma} = w$. This implies that $\sum_{\sigma \in \Psi_v} n_\sigma = w[L_v : \mathbb{Q}_p]/2$ because $\sum_{\sigma \in \Psi_v} n_\sigma = \sum_{\sigma \in \Psi_v} n_{\iota \cdot \sigma}$. Finally, define

$$\mu_{v_0} = \sum_{\sigma \in \Psi_v} n_\sigma \cdot \nu_\sigma,$$

where ν_σ is the geometric cocharacter of $\underline{L}^\times$ attached to σ.

Next consider a p-adic place v_0 of L^+ which splits into two places of L, and arbitrarily choose a place v of L over v_0 (so the other place is $\overline{v}$). Using (a), we can choose a sequence of integers $(n_\sigma)_{\sigma\in\Psi_v}$ indexed by the set of $\mathbb{Q}_p$-embeddings $\sigma\colon L_v \to \overline{\mathbb{Q}}_p$ in such a way that

$$\sum_{\sigma\in\Psi_v} n_\sigma = [L_v : \mathbb{Q}_p]\cdot\lambda_v.$$

Also define

$$\mu_v = \sum_{\sigma\in\Psi_v} n_\sigma\cdot\nu_\sigma,\quad \mu_{\overline{v}} = \sum_{\sigma\in\Psi_{\overline{v}}} (w - n_{\iota\cdot\sigma})\cdot\nu_\sigma,\quad \mu_{v_0} = \mu_v + \mu_{\overline{v}}\,.$$

For our choice of $v|v_0$ on L and any $\sigma\in\Psi_{\overline{v}}$, we define $n_\sigma = w - n_{\iota\sigma}$, so

$$\sum_{\sigma\in\Psi_{\overline{v}}} n_\sigma = [L_{\overline{v}} : \mathbb{Q}_p]w - \sum_{\sigma'\in\Psi_v} n_{\sigma'} = [L_v : \mathbb{Q}_p]w - [L_v : \mathbb{Q}_p]\lambda_v = [L_v : \mathbb{Q}_p]\lambda_{\overline{v}}$$

(the final equality uses (b)).

To summarize, for *every* $v'|p$ on L we have $\sum_{\sigma'\in\Psi_{v'}} n_{\sigma'} = [L_{v'} : \mathbb{Q}_p]\lambda_{v'}$ (without regard to whether or not v' is the above "chosen" place over its restriction to L^+ in cases when there is more than one such place), and $n_\sigma + n_{\iota\sigma} = w$ for all $\sigma\in\mathrm{Hom}(L,\overline{\mathbb{Q}})$. Define a geometric cocharacter of $\underline{L}^\times$ by

$$\mu := \sum_{v_0} \mu_{v_0},$$

where v_0 runs through all places of L^+ above p, so this is equal to $\sum_\sigma n_\sigma\nu_\sigma$. Let $E = E(\mu)\subset\overline{\mathbb{Q}}$ be the minimal field of definition of μ, so by Proposition A.4.8.1 the $\mathbb{Q}$-homomorphism $\chi(\mu) : \underline{E}^\times \to \underline{L}^\times$ satisfies $\chi(\mu)\circ\nu_{\tau_0} = \mu$, where ν_{τ_0} is the geometric cocharacter of $\underline{E}^\times$ corresponding (in the dual of the character basis $\{\xi^\tau\}$) to the canonical inclusion $\tau_0 : E \hookrightarrow \overline{\mathbb{Q}}$.

Now we use Corollary A.4.8.2 to verify that $\chi(\mu)$ is admissible (i.e., it factors through the quotient map $\underline{E}^\times \twoheadrightarrow \mathfrak{S}^E$). Since L is a CM field, under the permutation action of $\Gamma_\mathbb{Q}$ on the basis $\{\xi^\sigma\}$ of $\mathrm{X}^*(\underline{L}^\times)$ all involutions in $\Gamma_\mathbb{Q}$ act in the same way (e.g., use that $\xi^\sigma(y) = \sigma(y)$ in $\overline{\mathbb{Q}}$ for $y\in\underline{L}^\times(\mathbb{Q}) = L^\times$). Thus, the same holds for the $\Gamma_\mathbb{Q}$-action on the dual basis $\{\nu_\sigma\}$ of $\mathrm{X}_*(\underline{L}^\times)$, and so likewise for their effect on every μ_{v_0} for p-adic places v_0 of L^+ (by separately treating the cases when v_0 splits or does not split in L). It follows that $\gamma^{-1}\iota\gamma\cdot\mu = \iota\cdot\mu$ for all $\gamma\in\Gamma_\mathbb{Q}$, so admissibility is reduced to checking that $(\iota+1)\cdot\mu$ is $\Gamma_\mathbb{Q}$-invariant. Since $\mu = \sum_\sigma n_\sigma\nu_\sigma$, and we arranged that $n_\sigma + n_{\iota\cdot\sigma} = w$ for all σ, it is clear that $(\iota+1)\cdot\mu = w\sum_\sigma \nu_\sigma$. Thus, the desired $\Gamma_\mathbb{Q}$-invariance (and hence admissibility of $\chi(\mu)$) holds, and the weight identity in A.4.8.2 implies that $\chi(\mu)$ has weight w.

Now we can finally construct the desired Weil number in L. Let $\wp$ be the prime of $\mathcal{O}_E$ over p induced by the composition of the canonical inclusion $E\hookrightarrow\overline{\mathbb{Q}}$ and the chosen embedding $j_p : \overline{\mathbb{Q}}\hookrightarrow\overline{\mathbb{Q}}_p$. Let a be the order of the class of $\wp$ in $\mathrm{Pic}(\mathcal{O}_E)$, and choose $y\in\mathcal{O}_E$ such that $y\mathcal{O}_E = \wp^a$. Let $q = \#(\mathcal{O}_E/\wp^a)$. Since $\chi(\mu)$ is admissible with weight w, Proposition A.4.7.2(i) (with $\tau = \tau_0$) ensures that $x := \chi(\mu)(y)$ is a Weil q-number of weight w such that

$$\mathrm{ord}_v(x) = \mathrm{ord}_v(q)\cdot\frac{\sum_{\sigma\in\Psi_v}\langle\xi^\sigma, \chi(\mu)\circ\nu_{\tau_0}\rangle}{[L_v : \mathbb{Q}_p]}$$

for every p-adic place v of L. But $\chi(\mu) \circ \nu_{\tau_0} = \mu$ by Proposition A.4.8.1, and $\langle \xi^\sigma, \mu \rangle = n_\sigma$ for all $\sigma \in \Psi_v$ (by definition of μ). Since $\sum_{\sigma \in \Psi_v} n_\sigma = [L_v : \mathbb{Q}_p]\lambda_v$, we are done. □

Remark. The proof of A.4.8.4 gives a trivial bound

$$\mathrm{ord}_p(q) \leqslant [E : \mathbb{Q}] \cdot h_E$$

for the exponent of the number q in A.4.8.4, where h_E is the class number of the minimal field of definition E of $\mu = \sum_{v_0} \mu_{v_0}$.

APPENDIX B

CM lifting via p-adic Hodge theory

... the theory of moduli spaces has the wonderful feature of having many doors, many techniques by which this theory can be developed. ... This is another part of its charm. ... one fact stands out: the story is not finished.

—David Mumford

B.1. A generalization of the toy model

We consider higher-dimensional generalizations of the p-divisible group over $\overline{\mathbb{F}}_p$ attached the toy model in 4.1.2 (see 4.1.3.2). These are CM p-divisible groups over $\overline{\mathbb{F}}_p$ of height $4m$ with action by $W(\mathbb{F}_{p^{4m}})$ whose Lie type is *striped* in the sense of 4.5.3 (b). We will show that such a CM p-divisible group admits a CM lift to characteristic 0, using Raynaud's method in [**99**]. The CM lift constructed this way is *different* from the one constructed in 4.5.17 using the toy model when $m > 1$.

B.1.1. In this section m is a fixed positive integer, $\mathcal{O} = W(\mathbb{F}_{p^{4m}})$, and (Y_0, β_0) is an $\mathcal{O}$-linear CM p-divisible group over $\overline{\mathbb{F}}_p$ whose Lie type $[\mathrm{Lie}(Y_0, \beta_0)]$ is *striped* in the sense of 4.5.3 (b). The $\mathbb{Q}_p$-algebra involution ι of $\mathcal{O}$ is induced by the unique element of order 2 in $\mathrm{Gal}(\mathbb{F}_{p^{4m}}/\mathbb{F}_p)$.

We recast the structure of the proof of 4.5.17 for (Y_0, β_0) under the above conditions in the following framework.

(a) Find a suitable $\mathcal{O}$-linear p-divisible group (X_0, γ_0) over $\overline{\mathbb{F}}_p$ whose Lie type $[\mathrm{Lie}(X_0, \gamma_0)]$ is self-dual, and an $\mathcal{O}$-linear isogeny

$$\rho_0 \colon (X_0, \gamma_0) \to (Y_0, \beta_0)\,.$$

(b) Denote by $(\mathcal{X}, \gamma)$ an $\mathcal{O}$-linear p-divisible group over $W(\overline{\mathbb{F}}_p)$ with closed fiber (X_0, γ). We know by 4.1.2.1 (2) that such an $\mathcal{O}$-linear p-divisible group $(\mathcal{X}, \gamma)$ exists and is unique up to unique isomorphism. Moreover the action of $\mathcal{O}$ on $\mathrm{Lie}(\mathcal{X})$ corresponds to a self-dual p-adic CM type for the fraction field of $\mathcal{O}$.

(c) Find a suitable finite extension field K of the fraction field K_0 of $W(\overline{\mathbb{F}}_p)$ and a suitable finite flat subgroup scheme $\mathcal{G} \subset \mathcal{X} \times_{\mathrm{Spec}(\mathcal{O})} \mathrm{Spec}(\mathcal{O}_K)$ such that the closed fiber of $\mathcal{G}$ is equal to $\ker(\rho_0)$.

The proof of 4.5.17 fits the above template, in which the $\mathcal{O}$-linear p-divisible groups (X_0, γ_0) and $(\mathcal{X}, \gamma)$ (over $\overline{\mathbb{F}}_p$ and $W(\overline{\mathbb{F}}_p)$ respectively) are both Serre tensor constructions built from $W(\mathbb{F}_{p^4})$-linear p-divisible groups X_{toy} and $\mathcal{X}_{\mathrm{toy}}$ over $\overline{\mathbb{F}}_p$ and $W(\overline{\mathbb{F}}_p)$ respectively. The finite flat group scheme $\mathcal{G}$ over $\mathcal{O}_K$ is isomorphic to the direct sum of m copies of a finite flat subgroup scheme of $\mathcal{X}_{\mathrm{toy}}$ of order p.

We will construct a CM lifting of the $\mathcal{O}$-linear p-divisible group (Y_0, β_0) with striped Lie type by following the same strategy as outlined in B.1.1 but with a different $\mathcal{O}$-linear isogeny ρ. In particular, for $m > 1$ the $\mathcal{O}$-linear p-divisible groups (X_0, γ_0) and (X, γ) considered in this section will *not* be Serre tensor constructions from a smaller discrete valuation ring R contained in $\mathcal{O}$ and an R-linear CM p-divisible group over $\overline{\mathbb{F}}_p$. The finite flat group schemes $\mathcal{X}[p]$ over $W(\overline{\mathbb{F}}_p)$, one for each $m \geqslant 1$, are "irreducible generalizations" of the finite flat group scheme $\mathcal{X}_{\rm toy}[p]$; in particular, $\mathcal{X}[p]$ is not a product of two non-trivial finite flat group schemes.

B.1.2. We will follow the notation scheme in 4.2 and 4.4 (reviewed below), and fix a positive integer m that serves as a stealth parameter.

(i) F is an unramified extension field of $\mathbb{Q}_p$ of degree $4m$. Let $\mathcal{O} = \mathcal{O}_F$ be the ring of integers in F, and let κ_F be the residue field of $\mathcal{O}$.
(ii) $\mathrm{Gal}(\kappa_F/\mathbb{F}_p)$ is identified with $\mathbb{Z}/4m\mathbb{Z}$ such that the generator "1 mod $4m$" of the cyclic group $\mathbb{Z}/4m\mathbb{Z}$ corresponds to the arithmetic Frobenius $x \mapsto x^p$ in $\mathrm{Gal}(\kappa_F/\mathbb{F}_p)$.
(iii) W denotes $W(\overline{\mathbb{F}}_p)$ and $\sigma : (x_0, x_1, x_2, \ldots) \mapsto (x_0^p, x_1^p, x_2^p, \ldots)$ is the Frobenius automorphism.
(iv) $I = \mathrm{Hom}_{\mathbb{Z}_p\text{-alg}}(\mathcal{O}, W) \xrightarrow{\simeq} \mathrm{Hom}_{\mathbb{F}_p\text{-alg}}(\kappa_F, \overline{\mathbb{F}}_p)$ is a torsor for the finite cyclic group $\mathrm{Gal}(\kappa_F/\mathbb{F}_p)$; we choose and fix an identification of I with $\mathbb{Z}/4m\mathbb{Z}$, compatible with the natural $\mathrm{Gal}(\kappa_F/\mathbb{F}_p)$-module structure for I and the isomorphism $\mathrm{Gal}(\kappa_F/\mathbb{F}_p) \xleftarrow{\simeq} \mathbb{Z}/4m\mathbb{Z}$ specified in (ii) above.

B.1.3. Construction of the $\mathcal{O}$-linear isogeny $\rho_0 \colon X_0 \to Y_0$.

- Let (Y_0, β_0) be an $\mathcal{O}$-linear p-divisible group over $\overline{\mathbb{F}}_p$ whose Lie type is striped in the sense of 4.5.3 (b). In particular, Y_0 is isoclinic of slope $1/2$. The Lie type $\mathrm{Lie}(Y_0, \beta_0)$ of (Y_0, β_0) corresponds to one of the two halves of I with a fixed parity. After changing the trivialization of the $(\mathbb{Z}/4m\mathbb{Z})$-torsor I if necessary, we may and do assume that the Lie type of (Y_0, β_0) is the subset $J := \{2, 4, 6, \ldots, 4m\} \subset I$ of even elements of I.
- $\rho_0 \colon X_0 \to Y_0$ is an $\mathcal{O}$-linear isogeny between $\mathcal{O}$-linear p-divisible groups over $\overline{\mathbb{F}}_p$ such that
  - the Lie type of X_0 is $J' := \{1, 3, 5, \ldots, 2m-1, 2m+2, 2m+4, \ldots, 4m\}$,
  - $G_0 := \ker(\rho_0)$ is a subgroup scheme of $X_0[p]$ of rank p^m that is stable under the action of $\mathcal{O}$,
  - the character of the $\mathcal{O}$-action on $\mathrm{Lie}(G_0)$ corresponds to the subset
  $$\{1, 3, \ldots, 2m-1\}$$
  of odd elements in J'.

The $\mathcal{O}$-linear isogeny ρ_0 will be made more explicit in B.1.4 and B.1.5 below. The notation in B.1.4 and B.1.5 is consistent with the notation in the proof of Propositions 4.2.6 and 4.2.7.

B.1.4. Explicit description of (Y_0, β_0). To describe the $\mathcal{O}$-linear p-divisible group Y_0 explicitly, we will write down its *covariant* crystalline Dieudonné module $N = \mathrm{E}_*(Y_0)$, naturally isomorphic to the Cartier module $\mathrm{C}_*(Y_0)$ of the smooth formal group attached to Y_0.[1] It is a free W-module of rank $4m$ equipped with

[1]Here we use the notation in B.3, which contains a summary of relations between different flavors of Dieudonné theories over a perfect field of characteristic $p > 0$.

a σ-linear endomorphism $\mathcal{F}$ and a σ^{-1}-linear endomorphism $\mathcal{V}$ such that $\mathcal{F}\mathcal{V} = p\,\mathrm{id}_N = \mathcal{V}\mathcal{F}$. In addition, the ring $\mathcal{O}$ operates on N.

We have a decomposition

$$W \otimes_{\mathbb{Z}_p} \mathcal{O} \simeq \prod_{i \in I} W_i$$

where each W_i is a copy of W such that the composition of the obvious embedding

$$\mathcal{O} \hookrightarrow W \otimes_{\mathbb{Z}_p} \mathcal{O} \simeq \prod_{i \in I} W_i$$

with the projection to W_i is the embedding $i\colon \mathcal{O} \hookrightarrow W$ for each $i \in I$. This decomposition of $W \otimes_{\mathbb{Z}_p} \mathcal{O}$ induces a decomposition

$$N \simeq \bigoplus_{i \in I} N_i\,,$$

where each N_i is a free W-module of rank one stable under the action of $\mathcal{O}$, and for $\eta \in \mathcal{O}$ the associated endomorphism $[\eta]$ of N is

$$[\eta]\left(\sum_{i \in I} n_i\right) = \sum_{i \in I} i(\eta)\, n_i$$

for $n_i \in N_i$.

We have

$$p\, N_{i+1} \subseteq \mathcal{F} \cdot N_i \subseteq N_{i+1}, \quad p\, N_{i-1} \subseteq \mathcal{V} \cdot N_i \subseteq N_{i-1}$$

for all $i \in I$, and

$$\mathcal{V} \cdot N_{i+1} = \begin{cases} p\, N_i & \text{if } i \in J \\ N_i & \text{if } i \notin J \end{cases},$$

where $J = \{2, 4, \ldots, 4m\}$ is the set of even elements in I. Consequently, we have

$$\mathcal{F} \cdot N_i = \begin{cases} N_{i+1} & \text{if } i \in J \\ p\, N_{i+1} & \text{if } i \notin J \end{cases}.$$

To summarize:

(a) the covariant crystalline Dieudonné module $\mathrm{E}_*(Y_0)$ of Y_0, naturally isomorphic to the Cartier module $\mathrm{C}_*(Y_0)$ of Y_0, is $\bigoplus_{i \in I} N_i$;

(b) the ring $\mathcal{O}$ operates on the factor N_i via the embedding $\mathcal{O} \overset{i}{\hookrightarrow} W$ for each $i \in I$,

(c) $\mathcal{F}$ and $\mathcal{V}$ operate according to the above displayed formulas.

Remark. The use of Dieudonné theory in B.1 is as follows.

(1) In B.1.4 above and B.1.5 below we write down the Dieudonné module of an $\mathcal{O}$-linear p-divisible group (Y_0, β_0) with a striped Lie type over $\overline{\mathbb{F}_p}$ and construct an $\mathcal{O}$-linear isogeny $\rho_0 : (X_0, \gamma) \to (Y_0, \beta_0)$; similar constructions appear in Chapter 4.

(2) In B.1.6 we use deformation theory for p-divisible groups to show that the $\mathcal{O}$-linear p-divisible group (X_0, γ_0) admits a $\mathcal{O}$-linear lift $\mathcal{X}$ over $W = W(\overline{\mathbb{F}_p})$ with a self-dual p-adic CM type for the action of $\mathcal{O}$ on $\mathrm{Lie}(\mathcal{X})$.

After this we apply Raynaud's theory to the $\mathcal{O}$-linear finite flat group scheme $\mathcal{X}[p]$; we don't use Dieudonné theory in B.1 after B.1.6.

B.1.5. Explicit description of (X_0, γ) and $\rho_0 \colon (X_0, \gamma_0) \to (Y_0, \beta_0)$. Let M be the submodule of N defined by

$$M = \sum_{1 \leqslant a \leqslant m} N_{2a-1} + \sum_{1 \leqslant a \leqslant m} p\, N_{2a} + \sum_{2m+1 \leqslant b \leqslant 4m} N_b \ \subset N .$$

It is easy to see that M is stable under $\mathcal{F}$, $\mathcal{V}$, and the action of $\mathcal{O}$, and that M has Lie type

$$J' := \{1, 3, 5, \ldots, 2m-1, 2m+2, 2m+4, \ldots, 4m\}$$

consisting of the odd integers between 1 and $2m$ and the even integers between $2m+1$ and $4m$. Clearly J' is self-dual in the sense that I is the disjoint union of J' and the translate $J' + 2m$ of J' by the unique element of order 2 in $\mathbb{Z}/4m\mathbb{Z}$. (See 4.4.3 for the notion of "self-dual" for elements of the Grothendieck group $\mathrm{R}_{\overline{\mathbb{F}}_p}(\mathcal{O})$.)

Let X_0 be the $\mathcal{O}$-linear p-divisible group over $\overline{\mathbb{F}}_p$ with M as its (covariant) Cartier–Dieudonné module, and let $\rho_0 \colon X_0 \to Y_0$ be the $\mathcal{O}$-linear isogeny corresponding to the inclusion map $M \hookrightarrow N$. The kernel $G_0 := \ker(\rho_0)$ is a finite group scheme over $\overline{\mathbb{F}}_p$ killed by p and equipped with an action by $\mathcal{O}$. The covariant crystalline Dieudonné module of G_0 is the submodule

$$\mathrm{E}_*(G_0) = (p^{-1}M_2 + p^{-1}M_4 + \cdots + p^{-1}M_{2m} + M)/M \subset p^{-1}M/M$$

of the covariant crystalline Dieudonné module $p^{-1}M/M$ of $X_0[p]$.

Remark. The class of the $(\mathcal{O} \otimes_{\mathbb{Z}_p} \overline{\mathbb{F}}_p)$-module $\mathrm{Lie}(G_0)$ in $\mathrm{R}_{\mathbb{F}}(\mathcal{O})$ corresponds to the subset

$$\{1, 3, \ldots, 2m-1\}$$

of I and the class of the $(\mathcal{O} \otimes_{\mathbb{Z}_p} \overline{\mathbb{F}}_p)$-module $\mathrm{E}_*(G_0)$ corresponds to the subset $\{2, 4, \ldots, 2m\}$. This reflects the general fact that for every finite group scheme H_0 over a perfect field k of characteristic p, we have a natural isomorphism

$$\mathrm{Lie}(H_0) = \ker\left(\mathrm{E}_*(H_0) \xrightarrow{\mathcal{V}} \mathrm{E}_*(H_0)^{(p)}\right) \otimes_{(k,\sigma^{-1})} k ,$$

where σ^{-1} is the automorphism $x \mapsto x^{1/p}$ of k and $\mathrm{E}_*(H_0)$ is the covariant crystalline Dieudonné module of H_0; see B.3.11.2.

B.1.6. Denote by $\mathrm{Def}(X_0, \gamma_0)$ the deformation functor which to every artinian local ring $(R, \mathfrak{m})$ with residue field $\overline{\mathbb{F}}_p$ associates the set of isomorphism classes of triples

$$\left(X, \mathcal{O} \longrightarrow \mathrm{End}_R(X), \delta \colon X_{\overline{\mathbb{F}}_p} \simeq X_0\right)$$

where $(X, \mathcal{O} \longrightarrow \mathrm{End}_R(X))$ is an $\mathcal{O}$-linear p-divisible group over R, and δ is an $\mathcal{O}$-linear isomorphism. By the crystalline deformation theory of Grothendieck–Messing ([**48**], [**75**]), the deformation functor $\mathrm{Def}(X_0, \gamma_0)$ is étale over W. Let $\mathcal{X}$ be the universal $\mathcal{O}$-linear p-divisible group over W lifting the $\mathcal{O}$-linear p-divisible group X_0.

Let K_0 be the field of fractions of W. Let K be the tame totally ramified extension field of K_0 of degree $p^{4m} - 1$. We will construct a finite flat subgroup scheme

$$\mathcal{G} \subset \mathcal{X}[p]_{\mathcal{O}_K}$$

over $\mathcal{O}_K$ whose closed fiber is the subgroup scheme $G_0 = \ker(\rho_0)$ in $X_0[p]$. The quotient

$$\mathcal{Y} := \mathcal{X}_{\mathcal{O}_K}/\mathcal{G}$$

is then a lift of Y_0 such that the action of $\mathbb{Z}_p + p\mathcal{O} \subset \mathcal{O}$ on Y_0 lifts to $\mathcal{Y}$.

B.1.7. In the rest of this section we will use the method of [**99**] to construct the desired subgroup $\mathcal{G}$ of $\mathcal{X}[p]_{\mathcal{O}_K}$. According to [**99**, 1.4.1, 1.5.1], there exist "structure constants" $(\gamma_i, \delta_i)_{i\in I}$ for the κ_F-linear finite flat commutative W-group scheme $\mathcal{X}[p]$ with

$$\gamma_i,\ \delta_i \in W,\ \ \gamma_i \cdot \delta_i = w \in p \cdot W^\times \qquad \text{for all } i \in I$$

such that the bi-algebra structure of the coordinate ring of $\mathcal{X}[p]$ can be explicitly described in terms of these constants. Here $w \in p \cdot W^\times$ is the universal constant defined in [**99**, §1.2, (11 bis)]. It can be expressed in terms of either Jacobi sums or Gauss sums, depends only on κ_F, and satisfies $w \equiv p! \pmod{p^2}$; see [**99**, 1.3.1].

The only other condition for the δ_i's is that

$$\begin{cases} \delta_i \in\ pW^\times \text{ if } i \in J' - 1 = \{4m, 2, 4, \dots, 2m-2, 2m+1, 2m+3, \dots, 4m-1\}, \\ \delta_i \in\ W^\times \text{ if } i \in \{1, 3, 5, \dots, 2m-3, 2m-1, 2m, 2m+2, \dots, 4m-2\}. \end{cases}$$

This condition is a consequence of [**99**, 1.5.4] and the fact that J' is the Lie type of X_0; see B.1.8. It is clear that for any other family of constants (γ_i', δ_i') satisfying the same conditions, one can find a family of units $(u_i \in W^\times)_{i\in I}$ such that

$$\delta_i' = u_i^p\, \delta_i\, u_{i+1}^{-1},\ \ \gamma_i = u_i^p\, \gamma_i'\, u_{i+1}^{-1} \qquad \text{for all } i \in I\,.$$

Thus, by [**99**, 1.5.1], (γ_i', δ_i') can also serve as a family of structure constants for $\mathcal{X}[p]$. The Hopf algebra structure of the coordinate ring R of $\mathcal{X}[p]$ and the action of κ_F on R are described in B.1.8 below.

B.1.8. Description of the κ_F-linear bi-algebra R.

B.1.8.1. The algebra structure for the coordinate ring R of the κ_F-linear W-group scheme $\mathcal{X}[p]$ is

$$R = W[X_1, \dots, X_{4m}]/(X_i^p - \delta_i X_{i+1} \,|\, i \in I)\,.$$

We write x_i for the image of X_i in R.

B.1.8.2. The co-algebra structure of R is determined by the formula

$$c(x_i) = x_i \otimes 1 + 1 \otimes x_i + \sum_{\chi'\,\chi'' = \chi_i} \frac{\gamma_{i-h} \cdots \gamma_{i-1}}{w_{\chi'}\, w_{\chi''}} \left(\prod\nolimits_{j\in I} x_j^{a_j'}\right) \otimes \left(\prod\nolimits_{j\in I} x_j^{a_j''}\right)$$

for each $i \in I$, to be read as follows.

(1) The summation runs through all pairs (χ', χ'') of $\mu_{p^{4m}-1}(W)$-valued non-trivial characters of $\kappa_E^\times$ such that $\chi'\,\chi'' = \chi_i$.

(2) $\chi_i : \kappa_E^\times \longrightarrow \mu_{p^{4m}-1}(W)$ is the fundamental character of κ_E indexed by i, namely the composition of the Teichmüller lifting $\kappa_E^\times \to W(\kappa_E)^\times$ with the restriction

$$i|_{W(\kappa_E)^\times} : W(\kappa_E)^\times \to W^\times$$

of i to $W(\kappa_E)^\times \subset W(\kappa_E)$. We have $\chi_i^p = \chi_{i+1}$ for all $i \in I$.

(3) $\underline{a}' = (a'_j)_{j\in I}$ and $\underline{a}'' = (a''_j)_{j\in I}$ are elements of $\{0,1,\dots,p-1\}^I$ corresponding to χ' and χ'' respectively:

$$\chi' = \prod_{j\in I} \chi_j^{a'_j}, \qquad \chi'' = \prod_{j\in I} \chi_j^{a''_j}, \qquad \underline{a}', \underline{a}'' \notin \{(0,\dots,0), (p-1,\dots,p-1)\}.$$

(4) $h = h(\chi', \chi'')$ is the unique integer such that $0 < h \leqslant 4m$ and

$$\begin{cases} a'_{i-h} + a''_{i-h} = p, & \\ a'_{i-k} + a''_{i-k} = p-1 & \text{for } 0 < k < h, \\ a'_j = a''_j = 0 & \text{otherwise.} \end{cases}$$

(5) $w_{\chi'}$ and $w_{\chi''}$ are units of W defined in [**99**, §1.2, (11)]. They can be written in terms of Jacobi sums or Gauss sums and depend only on κ_F, χ', and χ''. Moreover, we have

$$w_{\chi'} \equiv \prod_{j\in I} (a'_j\,!) \pmod p, \qquad w_{\chi''} \equiv \prod_{j\in I} (a''_j\,!) \pmod p.$$

See [**99**, 1.3.1].

The counit $\epsilon\colon R \longrightarrow W$ is given by $\epsilon(x_i) = 0$ for all $i \in I$.

B.1.8.3. The κ_F-action on R is the homomorphism $\kappa_E \longrightarrow \mathrm{End}_{\text{bi-alg}}(R)$ such that for every element $\eta \in \kappa_F$ the corresponding endomorphism $[\eta]^*$ of the bi-algebra R is

$$\begin{cases} [\eta]^*(x_i) = \chi_i(\eta)\, x_i & \text{for all } i \in I \text{ if } \eta \in \kappa_F^\times, \\ [0]^*(x_i) = 0 & \text{for all } i \in I \end{cases}$$

B.1.9. The Cartier dual of $\mathcal{X}[p]$.

B.1.9.1. Let $R' := \mathrm{Hom}_W(R, W)$ be the W-linear dual of R, endowed with the natural bi-algebra structure coming from the bi-algebra structure of R, so $\mathrm{Spec}(R')$ is the Cartier dual of $\mathcal{X}[p]$ over W. According to [**99**, 1.5.3], the structure constants for $\mathcal{X}[p]^t = \mathcal{X}[p]^D$ are $(\delta_i, \gamma_i)_{i\in I}$. Below is a quick review to establish the notation for later use.

B.1.9.2. Let the elements $y_{\underline{a}}$ for $\underline{a} \in \{0,1,\dots,p-1\}^I$ be the W-basis of R' dual to the W-basis

$$\left\{ \underline{x}^{\underline{a}} := \prod_{i\in I} x_i^{a_i} \;\middle|\; \underline{a} = (a_i)_{i\in I},\ 0 \leqslant a_i \leqslant p-1 \quad \text{for all } i \in I \right\}$$

of R, indexed by finite sequences $\underline{a} \in \{0,1,\dots,p-1\}^I$. Here x_i is the image of X_i in R as in B.1.8.1 for each $i \in I$. For each $i \in I$, let $\underline{e}_i$ be the element of $\{0,1,\dots,p-1\}^I$ whose i-th coordinate is 1 and all other coordinates are 0, and let $y_i := y_{\underline{e}_i}$. The action of κ_E on R dualizes to an action of κ_F on R' such that

$$[\eta]^*(y_i) = \chi_i(\eta)\, y_i, \quad [0]^*(y_i) = 0$$

for all $i \in I$ and $\eta \in \kappa_F^\times$.

B.1.9.3. It is easy to see that

$$\underline{y}^{\underline{b}} := \prod_{i\in I} y_i^{a_i} = w_{\chi_{\underline{a}}} \cdot y_{\underline{b}}$$

for each $\underline{b} \in \{0,1,\ldots,p-1\}$, where $\chi_{\underline{b}} := \prod_{i\in I} \chi_i^{b_i}$. Moreover, we have

$$y_i^p = \gamma_i \cdot y_{i+1}$$

for all $i \in I$. The co-algebra structure of R' is given by

$$c'(y_i) = y_i \otimes 1 + 1 \otimes y_i + \sum_{\chi'\,\chi''=\chi_i} \frac{\delta_{i-h}\cdots\delta_{i-1}}{w_{\chi'}\,w_{\chi''}} \left(\prod\nolimits_{j\in I} y_j^{a'_j}\right) \otimes \left(\prod\nolimits_{j\in I} y_j^{a''_j}\right),$$

read in the same way as B.1.8.2.

B.1.10. We will need to examine the closed fiber $X_0[p] = \mathcal{X}[p]_{\overline{\mathbb{F}}_p}$ of $\mathcal{X}[p]$ and its subgroup scheme G_0. By definition,

$$\mathcal{X}[p]_{\overline{\mathbb{F}}_p} = \operatorname{Spec}(R \otimes_W \overline{\mathbb{F}}_p),$$

with

$$R \otimes_W \overline{\mathbb{F}}_p = \overline{\mathbb{F}}_p[\bar{X}_1,\ldots,\bar{X}_{4m}]/\left(\bar{X}_i^p - \bar{\delta}_i\,\bar{X}_{i+1}\right)_{i\in I}$$

where $\bar{\delta}_i$ is the image of δ_i in $\overline{\mathbb{F}}_p$ and $\bar{X}_i$ is the image in $W[X_1,\ldots,X_{4m}]\otimes_W \overline{\mathbb{F}}_p$ of the element $X_i \in W[X_1,\ldots,X_{4m}]$. Recall that

$$\begin{aligned} &\bar{\delta}_i = 0 && \text{if} \quad i \in \{4m, 2, 4, \ldots, 2m-2,\ 2m+1, 2m+3, \ldots, 4m-1\},\\ &\bar{\delta}_i \in \overline{\mathbb{F}}_p^{\times} && \text{if} \quad i \in \{1, 3, \ldots, 4m-1,\ 2m, 2m+2, \ldots, 4m-4, 4m-2\}. \end{aligned}$$

B.1.10.1. The kernel of the Frobenius homomorphism

$$\operatorname{Fr}_{X_0[p]/\overline{\mathbb{F}}_p} : X_0[p] \longrightarrow X_0[p]^{(p)}$$

is the closed subscheme

$$\operatorname{Spec}\left(\overline{\mathbb{F}}_p[\bar{X}_1,\ldots,\bar{X}_{4m}] \middle/ \left(\begin{smallmatrix} \bar{X}_1^p,\bar{X}_3^p,\ldots,\bar{X}_{2m-1}^p,\ \bar{X}_{2m+2}^p,\bar{X}_{2m+4}^p,\ldots,X_{4m-2}^p,\bar{X}_{4m}^p \\ \bar{X}_2,\bar{X}_4,\ldots,\bar{X}_{2m-2},\bar{X}_{2m},\ \bar{X}_{2m+1},\bar{X}_{2m+3},\ldots,\bar{X}_{4m-1} \end{smallmatrix} \right)\right)$$

of $\overline{\mathbb{F}}_p[\bar{X}_1,\ldots,\bar{X}_{4m}]/(\bar{X}_i^p - \bar{\delta}_i\,\bar{X}_{i+1})_{i\in I}$.

B.1.10.2. Denote by $H_{J'-\{4m\}}$ the subscheme of $\ker\left(\operatorname{Fr}_{X_0[p]/\overline{\mathbb{F}}_p}\right)$ defined by the principal ideal (x_{4m}). It is easy to see that $H_{J'-\{4m\}}$ is a finite subgroup scheme of $X_0[p]$ of order p^{2m-1}, isomorphic to the product of $2m-1$ copies of α_p. More explicitly, the coordinate ring of $H_{J'-\{4m\}}$ is

$$\overline{\mathbb{F}}_p[\bar{X}_1,\ldots,\bar{X}_{4m}] \middle/ \left(\begin{smallmatrix} \bar{X}_1^p,\bar{X}_3^p,\ldots,\bar{X}_{2m-3}^p,\bar{X}_{2m-1}^p,\ \bar{X}_{2m+2}^p,\bar{X}_{2m+4}^p,\ldots,\bar{X}_{4m-2}^p \\ \bar{X}_2,\bar{X}_4,\ldots,\bar{X}_{2m-2},\bar{X}_{2m},\ \bar{X}_{2m+1},\bar{X}_{2m+3},\ldots,\bar{X}_{4m-1},\ \bar{X}_{4m} \end{smallmatrix} \right).$$

The co-multiplication on R induces a co-multiplication

$$\bar{x}_i \mapsto \bar{x}_i\otimes 1 + 1\otimes\bar{x}_i \qquad \text{for } i = 1,3,\ldots,2m-3,2m-1,\ 2m+2,2m+4,\ldots,4m-2$$

on the coordinate ring of $H_{J'-\{4m\}}$, where $\bar{x}_i$ denotes the image of $\bar{X}_i$ in the above quotient ring.

The character of the action of κ_E on the Lie algebra of $H_{J'-\{4m\}}$ corresponds to the subset $J' - \{4m\}$. As before, we say that $J' - \{4m\}$ is the *Lie type* of $H_{J'-\{4m\}}$.

B.1.10.3. For each subset

$$S \subset J' - \{4m\} = \{1, 3, \ldots, 2m-1,\ 2m+2, 2m+4, \ldots, 4m-2\},$$

there exists a unique subgroup scheme H_S of $H_{J'-\{4m\}}$ over $\overline{\mathbb{F}}_p$ stable under the action of κ_E whose Lie type is S. Namely, H_S is the subscheme of $H_{J'-\{4m\}}$ defined by the ideal generated by $\{\, \bar{x}_i \,|\, i \in J' - S\,\}$. Thus, $H_{\{1,3,\ldots,2m-1\}} = G_0$.

B.1.11. Lemma. *Let $\overline{K_0}$ be an algebraic closure of K_0. Suppose $\mathbf{z}$ is a non-zero point in $\mathcal{X}[p](\overline{K_0})$. Let $z_i = x_i(\mathbf{z})$ be the i-th coordinate of $\mathbf{z}$, $i \in I$. Let ord_p be the p-adic valuation on $\overline{K_0}$ normalized by $\mathrm{ord}_p(p) = 1$. Then*

$$\mathrm{ord}_p(z_i) = \frac{1}{p^{4m}-1} \sum_{a=0}^{4m-1} \mathrm{ord}_p(\delta_{i+a})\, p^{4m-1-a} = \frac{1}{p^m - 1} \sum_{b=0}^{4m-1} \mathrm{ord}_p(\delta_{i-1-b})\, p^b\,,$$

where each integer δ_j is either 0 or 1 as indicated in B.1.7. *Moreover, we have*

$$\begin{aligned} &\mathrm{ord}_p(z_{2m-1}) < \mathrm{ord}_p(z_{2m-3}) < \cdots < \mathrm{ord}_p(z_3) < \mathrm{ord}_p(z_1) \\ &\quad < \mathrm{ord}_p(z_{2m}) < \mathrm{ord}_p(z_{2m+2}) < \cdots < \mathrm{ord}_p(z_{4m-4}) < \mathrm{ord}_p(z_{4m-2}) \\ &\quad\quad < \mathrm{ord}_p(z_{2m-2}) < \mathrm{ord}_p(z_{2m-4}) < \cdots < \mathrm{ord}_p(z_2) < \mathrm{ord}_p(z_{4m}) \\ &\quad\quad\quad < \mathrm{ord}_p(z_{2m+1}) < \mathrm{ord}_p(z_{2m+3}) < \cdots < \mathrm{ord}_p(z_{4m-3}) < \mathrm{ord}_p(z_{4m-1}). \end{aligned}$$

PROOF. The first statement on the valuations $\mathrm{ord}_p(z_i)$ of the z_i's is [**99**, §3.4, (6)]. The inequality for these valuations is immediate from the fact that

$$\mathrm{ord}_p(\delta_i) = \begin{cases} 1 \text{ if } i \in \{4m, 2, 4, \ldots, 2m-2, 2m+1, 2m+3, \ldots, 4m-1\} = J' - 1, \\ 0 \text{ if } i \in \{1, 3, \ldots, 2m-3, 2m-1, 2m, 2m+2, \ldots, 4m-4, 4m-2\}; \end{cases}$$

B.1.12. Recall that K is the tame totally ramified extension of the fraction field K_0 of W with $[K : K_0] = p^{4m}-1$. Lemma B.1.11 tells us that $\mathcal{X}[p]_K := \mathcal{X}[p]_{O_K} \otimes_{O_K} K$ is isomorphic to the constant group scheme over K attached to $(\kappa_F, +)$. Choose and fix a non-trivial point $\mathbf{z}^{(0)}$ of $\mathcal{X}[p](K)$. Let $\mathcal{G}_{(0)}$ be the schematic closure in $\mathcal{X}[p]_K$ of the subgroup generated by $\mathbf{z}^{(0)}$.

It is easy to see from B.1.11 that the closed fiber of $\mathcal{G}_{(0)}$ is the subgroup scheme $H_{\{2m-1\}}$ of G_0 which is stable under the action of κ_F and has Lie type $\{2m-1\} \subset I$. We want to choose another $m-1$ non-trivial points $\mathbf{z}^{(1)}, \ldots, \mathbf{z}^{(m-1)}$ of $\mathcal{X}[p](K)$ such that the closed fiber of the schematic closure $\mathcal{G}_{(m-1)}$ in $\mathcal{X}[p]_K$ of the subgroup generated by $\mathbf{z}^{(0)}, \ldots, \mathbf{z}^{(m-1)}$ is equal to the kernel G_0 of the isogeny $\rho_0 \colon X_0 \to Y_0$.

Choose and fix a generator ζ of the finite cyclic group $\kappa_F^\times$. For $a = 1, \ldots, m-1$ let $\mathbf{z}^{(a)} = [\zeta^a](\mathbf{z}(0)) \in \mathcal{X}[p](K) - \{0\}$, where $[\zeta^a]$ denotes the action of the element ζ^a on $\mathcal{X}[p]$. Write $z_i^{(a)} = x_i(\mathbf{z}^{(a)})$ for the ith coordinate of $\mathbf{z}^{(a)}$, $i \in I$. We have $z_i^{(a)} = \chi_i(\zeta)^a$ for all $i \in I$ and all $a = 0, 1, 2, \ldots, m-1$.

B.1.13. Lemma. *The elements $\mathbf{z}^{(0)}, \ldots, \mathbf{z}^{(m-1)}$ in the $\mathbb{F}_p$-vector space $\mathcal{X}[p](K)$ are linearly independent over $\mathbb{F}_p$.*

PROOF. Consider $f_0, \ldots, f_{m-1} \in \mathbb{F}_p$ such that $[f_0](\mathbf{z}^{(0)}) + \ldots + [f_{m-1}](\mathbf{z}^{(m-1)}) = 0$. We have

$$0 = \sum_{a=0}^{m-1} [f_a](\mathbf{z}^{(a)}) = \left[\sum_{a=0}^{m-1} f_a\,\zeta^a\right](\mathbf{z}^{(0)}),$$

where $[f_a]$ denotes the action of $f_a \in \mathbb{F}_p \subset \kappa_F$ on the κ_F-linear group scheme $\mathcal{X}[p]$. Thus, $\sum_{a=0}^{m-1} f_a\,\zeta^a = 0$. But $\kappa_F = \mathbb{F}_p[\zeta]$ has $\mathbb{F}_p$-degree $4m > m$, so every f_a must vanish. □

B.1.14. Definition. Using notation as above, for each $a = 0, 1, \ldots, m-1$ denote by $\mathcal{G}_{(a)}$ the schematic closure in $\mathcal{X}[p]_{\mathcal{O}_K}$ of the subgroup of $\mathcal{X}[p](K)$ generated by $\mathbf{z}^{(0)}, \ldots, \mathbf{z}^{(a)}$.

By Lemma B.1.13, $\mathcal{G}_{(a)}$ is a finite flat subgroup scheme of $\mathcal{X}[p]_{\mathcal{O}_K}$ of rank p^{a+1} over $\mathcal{O}_K$. Moreover, the generic fiber $\mathcal{G}_{(a)} \times_{\mathrm{Spec}(\mathcal{O}_K)} \mathrm{Spec}(K)$ of $\mathcal{G}_{(a)}$ is isomorphic to the product of $a+1$ copies of the constant group scheme $\mathbb{Z}/p\mathbb{Z}$ over K. Note that $\mathcal{G}_{(a)}$ depends on the choice of the generator ζ of $\kappa_F^\times$.

B.1.15. Remark. It is easy to see that $\mathcal{G}_{(a)}$ descends uniquely to a closed subgroup scheme $\mathcal{G}'_{(a)}$ of $\mathcal{X}[p]_{\mathcal{O}_{K_1}}$, where K_1 is the tamely ramified subextension of K/K_0 such that $[K_1 : K_0] = (p^m - 1)/(p-1)$. The results B.1.16 and B.1.17 below hold with K replaced by K_1 and $\mathcal{G}_{(a)}$ replaced by $\mathcal{G}'_{(a)}$.

B.1.16. Theorem. *For $a = 0, 1, \ldots, m-1$, the closed fiber of $\mathcal{G}_{(a)}$ is equal to the subgroup scheme $H_{\{2m-2a-1, 2m-2a+1, \ldots, 2m-1\}}$ of G_0 (defined in* B.1.10*) that is stable under the action of κ_F and has $\{2m-2a-1,\ 2m-2a+1, \ldots, 2m-3,\ 2m-1\}$ as its Lie type. In particular the closed fiber of $\mathcal{G}_{(m-1)}$ is equal to the kernel G_0 of the $\mathcal{O}$-linear isogeny $\rho_0 \colon X_0 \to Y_0$.*

B.1.17. Corollary. *Let $\mathcal{Y}$ be the quotient p-divisible group*

$$\mathcal{Y} := \mathcal{X}_{\mathcal{O}_K}/\mathcal{G}_{(m-1)}$$

over $\mathcal{O}_K$.

(i) *This quotient $\mathcal{Y}$ is a p-divisible group over $\mathcal{O}_K$ whose closed fiber is Y_0 as a quotient of X_0 via ρ_0.*
(ii) *The action of the subring $\mathbb{Z}_p + p\mathcal{O} \subset \mathcal{O}$ on Y_0 lifts to the p-divisible group $\mathcal{Y}$ over $\mathcal{O}_K$.*
(iii) *The p-adic* CM *type of $\mathcal{Y}$ is the self-dual subset $J' \subset I$.*

Recall that I is identified with $\{1, 2, \ldots, 4m\}$, and

$$J' = \{1, 3, \ldots, 2m-1, 2m+2, 2m+4, \ldots, 4m-2, 4m\}.$$

PROOF OF THEOREM B.1.16.

The Cartier duals $\left(\mathcal{X}[p]/\mathcal{G}_{(a)}\right)^D$ and $\left(X_0[p]/H_{\{2m-2a-1, 2m-2a+1, \ldots, 2m-1\}}\right)^D$ are subgroups of the respective group schemes $\mathcal{X}[p]$ and $X_0[p]$. We will prove the following statement, equivalent to B.1.16, concerning the closed subgroup scheme $(\mathcal{X}[p]/\mathcal{G}_{(a)})^D$:

the closed fiber of $(\mathcal{X}[p]/\mathcal{G}_{(a)})^D$, *as a subscheme* of $X_0[p]^D$, *is equal to* $\left(X_0[p]/H_{\{2m-2a-1,2m-2a+1,\ldots,2m-1\}}\right)^D$.

Note that the closed subgroup $(\mathcal{X}[p]/\mathcal{G}_{(a)})^D \subset \mathcal{X}[p]^D$ is the schematic closure, computed inside $\mathcal{X}[p]^D$, of its own generic fiber, because $\mathcal{X}[p]/\mathcal{G}_{(a)}$ is finite flat over $\mathcal{O}_K$. In the rest of the proof we will compute the schematic closure of $(\mathcal{X}[p]/\mathcal{G}_{(a)})^D(K)$ in $\mathcal{X}[p]^D$.

STEP 1. Compute $\left(X_0[p]/H_{\{2m-2a-1,2m-2a+1,\ldots,2m-1\}}\right)^D$.

It is easy to see from B.1.10.1–B.1.10.3 that the inclusion

$$H_{\{2m-2a-1,2m-2a+1,\ldots,2m-1\}} \hookrightarrow X_0[p]$$

corresponds to the surjection

$$\begin{aligned} R\otimes_W\overline{\mathbb{F}}_p &\overset{h}{\twoheadrightarrow} (R\otimes_W\overline{\mathbb{F}}_p)/\left(\bar{x}_2,\bar{x}_4,\ldots,\bar{x}_{4m},\ \bar{x}_1,\bar{x}_3,\ldots,\bar{x}_{2m-2a-3}\right) \\ &\simeq \overline{\mathbb{F}}_p[\bar{Z}_{2m-2a-1},\bar{Z}_{2m-2a+1},\ldots,\bar{Z}_{2m-1}]/\left(\bar{Z}^p_{2m-2a-1},\bar{Z}^p_{2m-2a+1},\ldots,\bar{Z}^p_{2m-1}\right) \\ &=: \overline{\mathbb{F}}_p[\underline{\bar{Z}}]/(\underline{\bar{Z}}^p) \end{aligned}$$

of bi-algebras, where the co-multiplication for the last bi-algebra is determined by

$$\bar{Z}_i \mapsto \bar{Z}_i\otimes 1 + 1\otimes \bar{Z}_i \qquad \text{for } i = 2m-2a-1, 2m-2a+1,\ldots,2m-3,2m-1\,.$$

Since the bi-algebra $\overline{\mathbb{F}}_p[\underline{\bar{Z}}]/(\underline{\bar{Z}}^p)$ is self-dual, the $\overline{\mathbb{F}}_p$-linear transpose $h^\vee$ of the above surjection h is an inclusion

$$h^\vee\colon \overline{\mathbb{F}}_p[\underline{\bar{Z}}]/(\underline{\bar{Z}}^p) \longrightarrow R'\otimes_W\overline{\mathbb{F}}_p$$

such that $h^\vee(\bar{Z}_i) = \bar{y}_i$ for $i = 2m-2a-1, 2m-2a+1,\ldots,2m-1$, where $\bar{y}_i$ denotes the image in $R'\otimes_W\overline{\mathbb{F}}_p$ of the element $y_i \in R'$ defined in B.1.9.2.

Note that for each $i = 2m-2a-1, 2m-2a+1,\ldots,2m-1$ we have

$$\bar{y}_i^p = 0 \quad \text{in } R'\otimes_W\overline{\mathbb{F}}_p$$

because $\gamma_i \equiv 0 \pmod p$, and

$$c'(\bar{y}_i) = \bar{y}_i\otimes 1 + 1\otimes\bar{y}_i \quad \text{in } R'\otimes_W R'\otimes_W\overline{\mathbb{F}}_p$$

because $\delta_{i-1}\equiv 0 \pmod p$ (as expected since $h^\vee$ is a co-algebra homomorphism). The fact that $\left(X_0[p]/H_{\{2m-2a-1,2m-2a+1,\ldots,2m-1\}}\right)^D$ is the kernel of the surjective homomorphism

$$X_0[p]^D \longrightarrow \left(H_{\{2m-2a-1,2m-2a+1,\ldots,2m-1\}}\right)^D$$

means:

the subgroup scheme $(X_0[p]/H_{\{2m-2a-1,2m-2a+1,\ldots,2m-1\}})^D \hookrightarrow X_0[p]^D$ is defined by the ideal generated by the $a+1$ elements

$$\bar{y}_{2m-2a-1},\bar{y}_{2m-2a+1},\ldots,\bar{y}_{2m-1} \in R'\otimes_W\overline{\mathbb{F}}_p.$$

STEP 2. Equations for the generic fiber of $(\mathcal{X}[p]_{\mathcal{O}_K}/\mathcal{G}_{(a)})^D$.

There are three equivalent ways to think about the $(a+1)$ points $\mathbf{z}^{(0)},\dots,\mathbf{z}^{(a)}$:

- as $a+1$ homomorphisms $\xi_{(0)},\dots,\xi_{(a)}$ from the group scheme $\underline{\mathbb{Z}/p\mathbb{Z}}$ to
$$\mathcal{X}[p]_{\mathcal{O}_K},$$
- as $a+1$ homomorphisms $\xi^t_{(0)},\dots,\xi^t_{(a)}$ from $\mathcal{X}[p]^D_{\mathcal{O}_K}$ to the $\mathcal{O}_K$-group scheme
$$\mu_p = \mathrm{Spec}(\mathcal{O}_K[T]/(T^p-1))\,,$$
- as $a+1$ elements $\psi_{(0)},\dots,\psi_{(a)}$ in $R'\otimes_W \mathcal{O}_K$ such that
$$\psi^p_{(b)} = 1,\quad c'(\psi_{(b)}) = \psi_{(b)}\otimes\psi_{(b)}\,,$$
where $\psi_{(b)}$ is the function on $\mathcal{X}[p]^D_{\mathcal{O}_K}$ induced by $\xi^t_{(b)}$ for $b\in\{0,1,\dots,a\}$.

We know that $(\mathcal{X}[p]_{\mathcal{O}_K}/\mathcal{G}_{(a)})^D$ is the kernel of the natural homomorphism
$$h_{(a)}\colon (\mathcal{X}[p]_{\mathcal{O}_K})^D \longrightarrow \mathcal{G}^D_{(a)}\,.$$
Restricting $h_{(a)}$ to the generic fiber over $\mathrm{Spec}(K)$, we see that the generic fiber of
$$(\mathcal{X}[p]_{\mathcal{O}_K}/\mathcal{G}_{(a)})^D$$
is the closed subscheme of $\mathcal{X}[p]^D\times_{\mathrm{Spec}(W)}\mathrm{Spec}(K)$ defined by the ideal $\mathscr{I}^{(a)}_{\mathbb{Q}}\subset R'\otimes_W K$ generated by the $a+1$ elements
$$\psi_{(c)}-1,\quad c=0,1,\dots,a\,.$$

Explicitly, the generators $\psi_{(c)}-1$ of $\mathscr{I}^{(a)}_{\mathbb{Q}}$ are
$$\begin{aligned}\psi_{(c)}-1 &= \textstyle\sum_{\underline{b}\neq\underline{0}}\left(\prod_{i\in I}\chi_i(\zeta)^{cb_i}\,z_i^{b_i}\right)\cdot y_{\underline{b}}\\ &= \textstyle\sum_{\underline{b}\neq\underline{0}} w^{-1}_{\chi_{\underline{b}}}\cdot\left(\prod_{i\in I}\chi_i(\zeta)^{cb_i}\,z_i^{b_i}\right)\cdot\prod_{i\in I}y_i^{b_i},\end{aligned}$$
where $\underline{b}$ runs through the set of non-zero elements in $\{0,1,\dots,p-1\}^I$ and b_i denotes the ith component of $\underline{b}$. Here $z_i\in\mathcal{O}_K$ is the x_i-coordinate of the non-zero point $\mathbf{z}$ as in B.1.11, and the elements $y_i\in R'$ are defined in B.1.9.2.

As already noted, $(\mathcal{X}[p]/\mathcal{G}_{(a)})^D\subset\mathcal{X}[p]^D$ is the schematic closure of its own generic fiber. In other words, the closed subgroup $(\mathcal{X}[p]/\mathcal{G}_{(a)})^D$ is the subscheme of $\mathcal{X}[p]^D$ defined by the ideal
$$\mathscr{I}^{(a)} := \mathscr{I}^{(a)}_{\mathbb{Q}}\bigcap(R'\otimes_W\mathcal{O}_K)$$
in $R'\otimes_W\mathcal{O}_K$. We need to determine the ideal $\mathscr{I}^{(a)}$ in $R'\otimes_W\mathcal{O}_K$.

STEP 3. For each $c=0,1,\dots,a$, let $f_c(\underline{T}) = f_c(T_1,\dots,T_{4m})\in\mathcal{O}_K[T_1,\dots,T_{4m}]$ be the polynomial in $4m$ variables defined by
$$f_c(\underline{T}) := \sum_{\underline{b}\neq\underline{0}}\left(w^{-1}_{\chi_{\underline{b}}}\cdot\prod_{i\in I}\chi_i(\zeta)^{cb_i}\right)\cdot\prod_{i\in I}T_i^{b_i}\,.$$
We have $\psi_{(c)}-1 = f_c(z_1y_1,\dots,z_my_m)\in\mathscr{I}_{\mathbb{Q}}$ for $c=0,1,\dots,a$.

Consider the $(a+1)\times(a+1)$ minor
$$\mathrm{Jac}_{(a)}(\underline{f},\underline{T}) := \left(\frac{\partial f_c}{\partial T_i}\right)_{c=0,1,\dots,a;\; i=2m-2a-1,2m-2a+1,\dots,2m-1}$$

of the Jacobian matrix for the polynomials $f_0(\underline{T}), \dots, f_a(\underline{T})$. Since $w_{\chi_i} = 1$ for each $i \in I$, the value

$$\mathrm{Jac}_{(a)}(\underline{f}, \underline{T})\big|_{\underline{T}=\underline{0}}$$

of the Jacobian matrix $\mathrm{Jac}_{(a)}(\underline{f}, \underline{T})$ evaluated at $T_i \mapsto 0$ for all $i \in I$ is the $(a+1) \times (a+1)$ Vandermonde matrix attached to the $a+1$ roots of unity

$$\chi_{2m-2a-1}(\zeta), \chi_{2m-2a+1}(\zeta), \dots, \chi_{2m-1}(\zeta) \in \mu_{p^{4m}-1}(W).$$

Thus, $\mathrm{Jac}_{(a)}(\underline{f}, \underline{T}) \in \mathcal{O}_K^\times$ because the $\chi_i(\zeta)$'s have mutually distinct images in $\overline{\mathbb{F}}_p^\times$.

By the implicit function theorem, in the power series ring

$$\mathcal{O}_K[\![T_1, T_2, \dots, T_{2m-2a-2}\, T_{2m-2a}, T_{2m-2a+2}, \dots, T_{2m-2}, T_{2m},\ T_{2m+1}, \dots, T_{4m-1}, T_{4m}]\!]$$

there exist $a+1$ elements

$$\{\, g_j(\underline{T}) : j \in J_{(a)} \,\}, \qquad J_{(a)} := \{2m-2a-1, 2m-2a+1, \dots, 2m-1\}$$

with constant terms equal to 0 such that

$$z_j y_j - g_j(\underline{zy}) \in \mathscr{I}^{(a)}$$

for all $j \in J_{(a)}$, where $g_j(\underline{zy})$ denotes the "value" $g_j(\underline{T})|_{T_i = z_i y_i}$ (for all $i \in I - J_{(a)}$) of the power series $g_j(\underline{T})$ when the $4m-a-1$ variables T_i with $i \in I - J_{(a)}$ are replaced with $z_i y_i$. The resulting power series converges in $R' \otimes_W \mathcal{O}_K$ because z_i is an element of the maximal ideal $\mathfrak{m}_K$ of $\mathcal{O}_K$ for every $i \in I$. (Recall that $z_i \in \mathfrak{m}_K$ is the x_i-coordinate of the non-zero point $\mathbf{z}$ as in B.1.11 for each i, and the elements $y_i \in R'$ are defined in B.1.9.2.)

STEP 4. Because the $g_j(\underline{T})$'s have constant term 0, we see from B.1.11 that

$$z_j^{-1} \cdot g_j(\underline{zy}) \in \mathfrak{m}_K \cdot (R' \otimes_W \mathcal{O}_K)$$

for all $j \in J_{(a)}$. For each $j \in J_{(a)}$ we have

$$y_j - z_j^{-1}\, g_j(\underline{zy}) \in \mathscr{I}^{(a)}_{\mathbb{Q}} \cap R' \otimes_W \mathcal{O}_K = \mathscr{I}^{(a)},$$

an element of $\mathscr{I}^{(a)}$ congruent to y_j modulo $\mathfrak{m} \cdot (R' \otimes_W \mathcal{O}_K)$.

Consider the natural surjection

$$\pi\colon\ (R' \otimes_W \overline{\mathbb{F}}_p) / (\bar{y}_{2m-2a-1}, \bar{y}_{2m-2a+1}, \dots, \bar{y}_{2m-1}) \twoheadrightarrow \left(R' \otimes_W \mathcal{O}_K)/\mathscr{I}^{(a)}\right) \otimes_{\mathcal{O}_K} \overline{\mathbb{F}}_p .$$

We saw in Step 1 that the left side is the coordinate ring of $\left(X_0[p]/H_{J_{(a)}}\right)^D$, so it has dimension p^{4m-a-1} over $\overline{\mathbb{F}}_p$. The right side is the coordinate ring of the closed fiber of the Cartier dual of $\mathcal{X}/\mathcal{G}_{(a)}$, so it also has dimension p^{4m-a-1} over $\overline{\mathbb{F}}_p$. Therefore, the surjection π is an isomorphism. Consequently, by Nakayama's lemma the elements $y_j - z_j^{-1}\, g_j(\underline{zy})$ with $j \in J_{(a)}$ generate the ideal $\mathscr{I}^{(a)}$. We conclude by Step 1 that the closed fiber of the Cartier dual of $\mathcal{X}/\mathcal{G}_{(a)}$ is equal to the Cartier dual of $X_0[p]/H_{J_{(a)}}$. □

B.2. Construct CM lifting by p-adic Hodge theory

In this section we give another proof of the main result in B.1 by using integral p-adic Hodge theory. The strategy remains the same; we replace the theory in [**99**] with tools from recent progress in p-adic Hodge theory. Because we want to include the case $p = 2$, we will use Kisin's version in [**59**]. In B.4.2 and B.4.5 the definitions of some categories of p-linear algebra data are recalled, and in B.4.15 and B.4.17 we give formulas for closed fibers constructed from such p-linear algebra data.

B.2.1. We will follow the notation $(m, F, \sigma, I, \ldots)$ as in § B.1. In particular:

- W denotes $W(\overline{\mathbb{F}}_p)$, $K_0 = W[1/p]$, K is the tame totally ramified extension of K_0 of degree $p^{4m} - 1$, and $e = e(K/K_0) = p^{4m} - 1$.
- J is the subset of even elements in $I \simeq \mathbb{Z}/4m\mathbb{Z}$.
- $J' \subset I$ is the subset admitting as representatives exactly the odd integers between 1 and $2m - 1$ and the even integers between $2m + 2$ and $4m$.
- X_0 is an $\mathcal{O}$-linear p-divisible group of height $4m$ over $\overline{\mathbb{F}}_p$ whose Lie type $[\mathrm{Lie}(X_0)] \in \mathrm{R}_{\overline{\mathbb{F}}_p}(\mathcal{O})$ corresponds to the subset $J' \subset I$.
- Y_0 is an $\mathcal{O}$-linear p-divisible group of height $4m$ over $\overline{\mathbb{F}}_p$ whose Lie type $[\mathrm{Lie}(X_0)] \in \mathrm{R}_{\overline{\mathbb{F}}_p}(\mathcal{O})$ corresponds to the subset $J \subset I$.
- $\rho_0 \colon X_0 \to Y_0$ is an $\mathcal{O}$-linear isogeny over $\overline{\mathbb{F}}_p$ whose kernel $G_0 = \ker(\rho_0)$ is the finite subgroup scheme over $\overline{\mathbb{F}}_p$ of the $\mathcal{O}_E$-linear commutative group scheme $X_0[p]$ over $\overline{\mathbb{F}}_p$ characterized by the following properties.
  - G_0 has order p^m and $\dim_{\overline{\mathbb{F}}_p}(\mathrm{Lie}(G_0)) = m$. In particular the relative Frobenius $\mathrm{Fr}_{G_0/\overline{\mathbb{F}}_p} : G_0 \to G_0^{(p)}$ and the Verschiebung $\mathrm{Ver}_{G_0/\overline{\mathbb{F}}_p} : G_0^{(p)} \to G_0$ are both 0.
  - G_0 is stable under the action of $\mathcal{O}$ on X_0.
  - The Lie type $[\mathrm{Lie}(G_0)] \in \mathrm{R}_{\overline{\mathbb{F}}_p}(\mathcal{O})$ of G_0 corresponds to the subset of odd elements in J'.
  - $\mathcal{X}$ is the $\mathcal{O}_E$-linear p-divisible group over W whose closed fiber is the $\mathcal{O}_E$-linear p-divisible group X_0 over $\overline{\mathbb{F}}_p$. Equivalently, $\mathcal{X}$ is the $\mathcal{O}_E$-linear p-divisible group over W whose p-adic CM type is the subset $J' \subset I$.

B.2.2. In B.1.7–B.1.16 we used the theory in [**99**] to construct a finite flat subgroup scheme

$$\mathcal{G} \subset \mathcal{X}[p]_{\mathcal{O}_K}$$

over $\mathcal{O}_K$, denoted by $\mathcal{G}_{(m-1)}$ in B.1.16, such that the closed fiber of $\mathcal{G}$ is the kernel G_0 of the isogeny $\rho_0 \colon X_0 \to Y_0$. Thus, the quotient

$$\mathcal{X}_{\mathcal{O}_K}/\mathcal{G}$$

is a p-divisible group over $\mathcal{O}_K$ with action by the order $\mathbb{Z}_p + p\mathcal{O}_K \subset \mathcal{O}_K$ lifting $(Y_0, \mathbb{Z}_p + p\mathcal{O}_K \hookrightarrow \mathrm{End}(Y_0))$ and having self-dual p-adic CM type. In the rest of this section we will construct finite flat subgroup schemes

$$\mathcal{G}_{(0)}, \mathcal{G}_{(1)}, \ldots, \mathcal{G}_{(m-1)} \subset \mathcal{X}[p]_{\mathcal{O}_K}$$

as in B.1.16, but with Raynaud's theory in [**99**] replaced by integral p-adic Hodge theory as in [**59**]. See B.2.16 for the precise result (using notation defined below) and B.2.14 for the definition of the finite flat group schemes $\mathcal{G}_{(b)}$. For the problem at hand there is no obvious advantage to using p-adic Hodge theory except that the computation of the closure in $\mathcal{X}[p]_{\mathcal{O}_K}$ of a subgroup scheme of the generic fiber

$\mathcal{X}\times_{\mathrm{Spec}(W)}\mathrm{Spec}(K)$ is a little easier. However, it is likely that p-adic Hodge theory will be useful for the question **(sCML)** in 1.8.5 and 4.1.9.

At a technical level, an important step is to identify the Kisin module $(\mathfrak{M}, \phi_{\mathfrak{M}}, \psi_{\mathfrak{M}})$ corresponding to the $(\mathcal{O}/p\mathcal{O})$-linear finite flat group scheme $\mathcal{X}[p]_{\mathcal{O}_K}$. The rest of the computation is similar to but easier than the one in B.1.16. The steps for finding the desired Kisin module $(\mathfrak{M}, \phi_{\mathfrak{M}}, \psi_{\mathfrak{M}})$ are indicated in B.2.3–B.2.4.

B.2.3. We will apply p-adic Hodge theory to the two p-adic base fields K_0 and K.[2]

- Choose p as the generator of the maximal ideal of $\mathcal{O}_{K_0} = W$; this has Eisenstein minimal polynomial over W equal to $E_0(t) = t - p$.
- Let $\mathfrak{S}_0 = W[\![t]\!]$, and let $\epsilon_0 \colon \mathfrak{S}_0 \twoheadrightarrow W$ be the local W-algebra homomorphism such that $\epsilon_0(t) = p$.
- Let $\phi_{\mathfrak{S}_0} : \mathfrak{S}_0 \to \mathfrak{S}_0$ be the local ring homomorphism such that $\phi_{\mathfrak{S}_0}(t) = t^p$ and $\phi_{\mathfrak{S}_0}|_W = \sigma$, where σ is the canonical Frobenius on W.
- Choose a generator π of the maximal ideal of $\mathcal{O}_K$ such $\pi^e - p = 0$; recall that $e = p^{4m} - 1$. Let $E(u) = u^e - p$, the minimal polynomial of π with respect to K_0 in the variable u.
- Let $\mathfrak{S} = W[\![u]\!]$, and let $\epsilon \colon \mathfrak{S} \twoheadrightarrow \mathcal{O}_K$ be the local W-algebra homomorphism such that $\epsilon(u) = \pi$. Let $\phi_{\mathfrak{S}}$ the continuous ring endomorphism of $\mathfrak{S}$ such that $\phi_{\mathfrak{S}}(u) = u^p$ and $\phi_{\mathfrak{S}}|_W = \sigma$.
- Let $h \colon \mathfrak{S}_0 \to \mathfrak{S}$ be the injective local W-algebra homomorphism such that $h(t) = u^e$. It is clear that $\epsilon \circ h = \epsilon_0$ and $\phi_{\mathfrak{S}} \circ h = h \circ \phi_{\mathfrak{S}}$. We will abbreviate both $\phi_{\mathfrak{S}_0}$ and $\phi_{\mathfrak{S}}$ to ϕ unless this abuse leads to confusion.

B.2.4. How to find the Kisin module for $\mathcal{X}[p]$

1. We will first find an $\mathcal{O}$-linear object $(\mathfrak{N}, \phi_{\mathfrak{N}}, \psi_{\mathfrak{N}})$ in $\mathrm{BT}^{\phi,f}_{/\mathfrak{S}_0}$ which corresponds to the $\mathcal{O}$-linear p-divisible group $\mathcal{X}$ over $\mathcal{O}_{K_0}$.

2. Let $\mathfrak{M}_0 = \mathfrak{N}/p\mathfrak{N}$, let $\phi_{\mathfrak{M}_0} : \mathfrak{M}_0 \to \mathfrak{M}_0$ be the ϕ-linear map induced by $\phi_{\mathfrak{N}}$, and let $\psi_{\mathfrak{M}_0} : \mathfrak{M}_0 \to \phi^*\mathfrak{M}_0$ be the $\mathfrak{S}$-linear map induced by $\psi_{\mathfrak{N}}$. Then $(\mathfrak{M}_0, \phi_{\mathfrak{M}_0}, \psi_{\mathfrak{M}_0})$ is an $(\mathcal{O}/p\mathcal{O})$-linear object in the category $(\mathrm{Mod}/\mathfrak{S}_0)^c$ of connected Kisin modules, and the $(\mathcal{O}/p\mathcal{O})$-linear finite flat group scheme $\mathcal{G}_{\mathrm{Kis}}(\mathfrak{M}_0, \phi_{\mathfrak{M}_0}, \psi_{\mathfrak{M}_0})$ over $\mathcal{O}_{K_0}$ is $(\mathcal{O}/p\mathcal{O})$-linearly isomorphic to $\mathcal{X}[p]$ over $\mathcal{O}_{K_0}$.

3. Let $(\mathfrak{M}, \phi_{\mathfrak{M}}, \psi_{\mathfrak{M}})$ be the object of $(\mathrm{Mod}/\mathfrak{S})^c$ obtained from the object

$$(\mathfrak{M}_0, \phi_{\mathfrak{M}_0}, \psi_{\mathfrak{M}_0}) \in \mathrm{Ob}\left((\mathrm{Mod}/\mathfrak{S}_0)\right)$$

via the map $(\mathfrak{S}_0, \phi_{\mathfrak{S}_0}) \longrightarrow (\mathfrak{S}, \phi_{\mathfrak{S}})$, so $\mathcal{G}_{\mathrm{Kis}}(\mathfrak{M}, \phi_{\mathfrak{M}}, \psi_{\mathfrak{M}})$ is an $(\mathcal{O}/p\mathcal{O})$-linear finite flat group scheme over $\mathcal{O}_K$ that is $(\mathcal{O}/p\mathcal{O})$-linearly isomorphic to $\mathcal{X}[p]_{\mathcal{O}_K}$.

B.2.5. Definition. For a given a subset $\Lambda \subsetneq I$ of I, let $(\mathfrak{N}_\Lambda, \phi_\Lambda, \psi_\Lambda)$ be an $\mathcal{O}$-linear object in the category $\mathrm{BT}^{\phi,f}_{\mathfrak{S}_0}$ with the following properties.

(1) The $\mathfrak{S}_0$-module $\mathfrak{N}_\Lambda$ is a direct sum

$$\mathfrak{N}_\Lambda = \oplus_{i\in I}\, \mathfrak{N}_i$$

of invertible $\mathfrak{S}_0$-modules $\mathfrak{N}_i$ indexed by the set $I \simeq \mathbb{Z}/4m\mathbb{Z}$.

[2] The important question of the behavior of integral p-adic Hodge theory constructions with respect to change of the base field seems to be unresolved at present. For the tamely ramified extension K/K_0 we use Eisenstein polynomials $T - p$ for K_0 and $X^e - p$ for K because these are compatible with the map $h : \mathfrak{S}_0 \to \mathfrak{S}$ in B.2.3, so we do not encounter the general difficulty here.

(2) The ring $\mathcal{O}$ operates on each of the $\mathfrak{S}_0$-modules $\mathfrak{N}_i$ via the ring homomorphism

$$i\colon \mathcal{O} \to W \subset \mathfrak{S}_0\,.$$

In other words, the action $[\eta]\colon \mathfrak{N}_\Lambda \to \mathfrak{N}_\Lambda$ of any element $\eta \in \mathcal{O}$ is given by

$$[\eta]\left(\sum_{i\in I} x_i\right) = \sum_{i\in I} i(\eta)\cdot x_i \qquad \text{for all } x_i \in \mathfrak{N}_i, i \in I\,.$$

(3) $\mathfrak{S}_0 \cdot \phi_\Lambda(\mathfrak{N}_i) = \begin{cases} E_0(t)\cdot \mathfrak{N}_{i+1} & \text{if } i\in\Lambda, \\ \mathfrak{N}_{i+1} & \text{if } i\notin\Lambda. \end{cases}$

Remark. (a) It is clear that for any subset $\Lambda \subset I$ there exists an $\mathcal{O}$-linear $\mathfrak{S}_0$-module $\mathfrak{N}_\Lambda$ and a ϕ-linear map $\phi_\Lambda : \mathfrak{N}_\Lambda \to \mathfrak{N}_\Lambda$ satisfying the conditions (1)–(3) above.

(b) The condition that $\Lambda \neq I$ guarantees that the $4m$th iteration

$$\psi_\Lambda^{4m} := \phi^{4m-1\,*}\psi_\Lambda \circ \phi^{4m-2\,*}\psi_\Lambda \circ \cdots \phi^*\psi_\Lambda \circ \psi_\Lambda : \mathfrak{N}_\Lambda \longrightarrow \phi^{4m\,*}\mathfrak{N}_\Lambda$$

of the map

$$\psi_\Lambda : \mathfrak{N}_\Lambda \to \mathfrak{S}_0 \otimes_{\phi,\mathfrak{S}_0} \mathfrak{N}_\Lambda =: \phi^*\mathfrak{N}_\Lambda$$

defined by $(\mathfrak{N}_\Lambda, \phi_\Lambda)$ satisfies

$$\psi_\Lambda^{4m}(\mathfrak{N}_\Lambda) \subset E_0(t)\cdot \phi^{4m\,*}\mathfrak{N}_\Lambda\,;$$

hence the triple $(\mathfrak{N}_\Lambda, \phi_\Lambda, \psi_\Lambda)$ is an object in $\mathrm{BT}^{\phi,f}_{/\mathfrak{S}_0}$ having the properties in (a). Such an object in $\mathrm{BT}^{\phi,f}_{/\mathfrak{S}_0}$ is determined by the subset $\Lambda \subset I$ up to non-unique isomorphism.

(c) Any two objects in $\mathrm{BT}^{\phi,f}_{\mathfrak{S}_0}$ satisfying the properties (1)–(3) are $\mathcal{O}$-linearly isomorphic. The proof is easy and is omitted.

B.2.6. Lemma. *Using notation as in* B.2.5, *let $\mathcal{X}_\Lambda$ be the $\mathcal{O}$-linear connected p-divisible group over W attached to $(\mathfrak{N}_\Lambda, \phi_\Lambda, \psi_\Lambda)$. The p-adic* CM *type of $\mathcal{X}_\Lambda$ is the subset ${}^c\Lambda + 1 = {}^c(\Lambda+1)$ of I, which is to say the translate by* 1 *of the complement ${}^c\Lambda$ in I of the subset $\Lambda \subset I$. Equivalently, the Lie type of the closed fiber $X_{0,\Lambda}$ of $\mathcal{X}_\Lambda$ corresponds to the subset ${}^c\Lambda + 1 \subset I$.*

PROOF. According to B.4.15, the Lie algebra $\mathrm{Lie}(X_{0,\Lambda})$ is

$$\mathrm{Lie}(X_{0,\Lambda}) \simeq \mathrm{coker}\left(\mathfrak{N}_\Lambda/t\mathfrak{N}_\Lambda \xrightarrow{\psi_\Lambda \pmod t} \phi^*\mathfrak{N}_\Lambda/t\phi^*\mathfrak{N}_\Lambda\right).$$

It follows that for any $i \in I$, the $(i+1)$th component $\mathrm{Lie}(X_{0,\Lambda})_{i+1}$ of $\mathrm{Lie}(X_{0,\Lambda})$ is non-zero if and only if $i \notin \Lambda$, so the p-adic CM type of $\mathcal{X}_\Lambda$ is ${}^c\Lambda + 1 = {}^c(\Lambda+1)$. $\square$

Remark. One can compute the p-adic CM type of $\mathcal{X}_\Lambda$ directly using the formula in B.4.16, giving an alternative proof of B.2.6.

B.2.7. Definition. Let $\Lambda_0 := {}^cJ' - 1 = {}^c(J'-1)$ be the subset of $I \simeq \mathbb{Z}/4m\mathbb{Z}$ which has

$$\{1, 3, 5, \ldots, 2m-3, 2m-1, 2m, 2m+2, 2m+4, \ldots, 4m-2\}$$

as a set of representatives. Let $\mathcal{X} := \mathcal{X}_{\Lambda_0}$, and let X_0 be the closed fiber of $\mathcal{X}$, an $\mathcal{O}$-linear p-divisible group whose Lie type $[\mathrm{Lie}(X_0)]$ corresponds to the subset $J' \subset I$ (according to B.2.6).

Define the $(\mathcal{O}/p\mathcal{O})$-linear object $\left(\mathfrak{M}_0, \phi_{\mathfrak{M},0}, \phi_{\mathfrak{M},0}\right)$ in $(\mathrm{Mod}/\mathfrak{S}_0)^{\mathrm{c}}$ as follows.

- $\mathfrak{M}_0 := \mathfrak{N}_{\Lambda_0}$ as a $\mathfrak{S}_0$-module,
- $\phi_{\mathfrak{M}_0} : \mathfrak{M}_0 \to \mathfrak{M}_0$ is the ϕ-linear map induced by ϕ_{Λ_0},
- $\psi_{\mathfrak{M}_0} : \mathfrak{M}_0 \to \phi^*\mathfrak{M}_0$ is the $\mathfrak{S}_0$-linear map induced by ψ_{Λ_0},
- the action of $\kappa_F = \mathcal{O}/p\mathcal{O}$ on $\mathfrak{N}_{\Lambda_0}$ induces an action of κ_F on $\mathfrak{M}_0$ compatible with $\phi_{\mathfrak{M}_0}$ and $\psi_{\mathfrak{M}_0}$.

Define an object $(\mathfrak{M}, \phi_{\mathfrak{M}}, \phi_{\mathfrak{M}})$ in $(\mathrm{Mod}/\mathfrak{S})^{\mathrm{c}}$ as follows.[3]

- $\mathfrak{M} := \mathfrak{S} \otimes_{h,\mathfrak{S}_0} \mathfrak{M}_0$ as a $\mathfrak{S}$-module,
- $\phi_{\mathfrak{M}} : \mathfrak{M} \to \mathfrak{M}$ is the ϕ-linear map induced by $\phi_{\mathfrak{M}_0}$
- $\psi_{\mathfrak{M}} : \mathfrak{M} \to \phi^*\mathfrak{M}$ is the $\mathfrak{S}$-linear map induced by $\psi_{\mathfrak{M}_0}$.
- The action of κ_F on $(\mathfrak{M}_0, \phi_{\mathfrak{M}_0}, \psi_{\mathfrak{M}_0})$ induces an action of κ_F on $(\mathfrak{M}, \phi_{\mathfrak{M}}, \psi_{\mathfrak{M}})$.

In other words, $(\mathfrak{M}, \phi_{\mathfrak{M}}, \phi_{\mathfrak{M}})$ is obtained from $\left(\mathfrak{M}_0, \phi_{\mathfrak{M}_o}, \phi_{\mathfrak{M}_o}\right)$ by base change with respect to the map $h\colon (\mathfrak{S}_0, \phi_{\mathfrak{S}_0}) \longrightarrow (\mathfrak{S}, \phi_{\mathfrak{S}})$.

Remark. The finite flat group scheme over $\mathcal{O}_K$ associated to the Kisin module $(\mathfrak{M}, \phi_{\mathfrak{M}}, \phi_{\mathfrak{M}})$ is isomorphic to $\mathcal{X}[p]_{\mathcal{O}_K}$ by functoriality of the functor $\mathcal{G}_{\mathrm{Kis}}$ with respect to change in the base ring (a complete discrete valuation ring of mixed characteristic $(0,p)$).

B.2.8. Lemma. *The Kisin module $\mathfrak{M}$ has a direct sum decomposition*

$$\mathfrak{M} = \bigoplus_{i\in I} \mathfrak{M}_i\,,$$

where each component $\mathfrak{M}_i$ is a free $\overline{\mathbb{F}}_p[\![u]\!]$-module of rank 1, *and*

$$\overline{\mathbb{F}}_p[\![u]\!]\cdot\phi_{\mathfrak{M}}(\mathfrak{M}_i) = \begin{cases} E(u)\cdot\mathfrak{M}_{i+1} = u^e\cdot\mathfrak{M}_{i+1} & \text{if } i\in\Lambda_0,\\ \mathfrak{M}_{i+1} & \text{if } i\notin\Lambda_0,\end{cases}$$

with $\Lambda_0 \subset I$ defined as at the start of B.2.7.

Also, there exist $\overline{\mathbb{F}}_p[\![u]\!]$-bases v_i of the invertible $\overline{\mathbb{F}}_p[\![u]\!]$-modules $\mathfrak{M}_i$ such that

$$\phi_{\mathfrak{M}_i}(v_i) = \begin{cases} u^e\cdot v_{i+1} & \text{if } i\in\Lambda_0,\\ v_{i+1} & \text{if } i\notin\Lambda_0.\end{cases}$$

B.2.9. We would like to find all subgroup schemes of $\mathcal{X}[p]_{\mathcal{O}_K}$ with order p. Equivalently, we want to find all $\overline{\mathbb{F}}_p[\![u]\!]$-submodules of $\mathfrak{M}$ of the form $\overline{\mathbb{F}}_p[\![u]\!]\cdot y$ with the following properties:

(i) $\phi_{\mathfrak{M}}(y) \in \overline{\mathbb{F}}_p[\![u]\!]\cdot y$,
(ii) $\overline{\mathbb{F}}_p[\![u]\!]\cdot\phi_{\mathfrak{M}}(y) \supseteq u^e\,\overline{\mathbb{F}}_p[\![u]\!]\cdot y$,
(iii) $\overline{\mathbb{F}}_p[\![u]\!]\cdot y$ is an $\overline{\mathbb{F}}_p[\![u]\!]$-direct summand of $\mathfrak{M}$.

Write $y = \sum_{i\in I} f_i(u)\cdot v_i$ and $\phi_{\mathfrak{M}}(y) = g(u)\,y$; the above condition is translated into the following system of equations with $4m+1$ unknown elements

$$g(u), f_1(u), f_2(u), \ldots, f_{4m}(u)\; in\; \overline{\mathbb{F}}_p[\![u]\!]$$

[3] The $\mathfrak{S}_0$-module $\mathfrak{M}_0$ here should not be confused with the 0th component of the κ_E-linear $\mathfrak{S}$-module $\mathfrak{M}$.

subject to certain constraints on their orders:

There exist elements $f_i(u) \in \overline{\mathbb{F}}_p[\![u]\!]$ *(indexed by* $i \in I$*) and an element* $g(u) \in \overline{\mathbb{F}}_p[\![u]\!]$ *which satisfy the following properties.*

(a) $g(u) \cdot f_{i+1}(u) = \begin{cases} u^e \cdot \phi(f_i(u)) & = & u^e \cdot f_i(u)^p & \text{if } i \in \Lambda_0, \\ \phi(f_i(u)) & = & f_i(u)^p & \text{if } i \notin \Lambda_0, \end{cases}$

(b) $\mathrm{ord}_u(g(u)) \leqslant e$. *(In particular* $g(u) \neq 0$ *and* $f_i(u) \neq 0$ *for all* $i \in I$*),*

(c) *there exists* $i_1 \in I$ *such that* $f_{i_1}(u) \in \overline{\mathbb{F}}_p[\![u]\!]^\times$.

Write $\mu = \mathrm{ord}_u(g(u))$ and $d_i = \mathrm{ord}_u(f_i(u))$ for the "unknown" elements $g(u)$ and $f_i(u)$, $i \in I$; we arrive at the following system of *linear* equations for $4m + 1$ unknown *integers* μ and $d_1, d_2, \ldots, d_{4m}$, subject to certain linear inequalities.

(1) $d_{i+1} + \mu = \begin{cases} p\,d_i + e & \text{if } i \in \Lambda_0, \\ p\,d_i & \text{if } i \notin \Lambda_0, \end{cases}$

(2) $0 \leqslant \mu \leqslant e$ *and* $d_i \geqslant 0$ *for all* $i \in I$,

(3) *there exists* $i_1 \in I$ *such that* $d_i = 0$.

The system of equations (1)–(3) gives the u-adic orders of the components of any solution for the system of equations of (a)–(c), equivalent to the system of equations (i)–(iii). We will see that the system of equations (1)–(3) has a unique solution.

B.2.10. We proceed to solve the system of equations B.2.9 (1)–(3).

- Let U be the $\mathbb{Q}$-vector space with basis $\{\mathbf{u}_i \mid i \in I\}$.
- Consider the d_i's as "variables", and let $\mathbf{d} = \sum_{i\in I} d_i\, \mathbf{u}_i$.
- Let S be the $\mathbb{Q}$-linear endomorphism of U such that $S(\mathbf{u}_i) = \mathbf{u}_{i-1}$ for all $i \in I$.
- Let $\mathbf{1} := \sum_{i\in I} \mathbf{u}_i$, and let $\mathbf{u}_{\Lambda_0} := \sum_{j\in\Lambda_0} \mathbf{u}_j$.

The equations B.2.9 (1)–(3) can be written in the following vector form

$$(p \cdot \mathrm{id}_U - S) \cdot \mathbf{d} = \mu \cdot \mathbf{1} - e \cdot \mathbf{u}_{\Lambda_0}\,,$$

where μ appears as an integer-valued parameter between 0 and e, all components d_i of the solution $\mathbf{d}$ are natural numbers, and at least one of the components d_{i_0} is equal to 0.

Because $S^{4m} = \mathrm{id}_U$, we have

$$(p \cdot \mathrm{id}_U - S)^{-1} = (p^{4m} - 1)^{-1} \cdot (S^{4m-1} + p\,S^{4m-2} + \cdots + p^{4m-2}\,S + p^{4m-1})\,.$$

Using the fact that $S \cdot \mathbf{1} = \mathbf{1}$ and $e = p^{4m} - 1$, we get

$$\begin{aligned}
\mathbf{d} &= (p^{4m} - 1)^{-1} \cdot (p \cdot \mathrm{id}_U - S)^{-1} \cdot (\mu \cdot \mathbf{1} - e \cdot \mathbf{u}_{\Lambda_0}) \\
&= (p^{4m} - 1)^{-1} \cdot \left[\mu \cdot \left(\sum_{a=0}^{4m-1} p^a \right) \cdot \mathbf{1} - e \cdot \left(\sum_{a=0}^{4m-1} p^a \cdot S^{4m-1-a} \right) \cdot \mathbf{u}_{\Lambda_0} \right] \\
&= \frac{\mu}{p-1} \cdot \mathbf{1} - \frac{e}{p^{4m} - 1} \cdot \sum_{i\in I} \sum_{a=0}^{4m-1} p^{4m-1-a} \sum_{j\in\Lambda_0} \mathbf{u}_{j-a} \\
&= \frac{\mu}{p-1} \cdot \mathbf{1} - \sum_{i\in I} \left(\sum_{\substack{0\leqslant a\leqslant 4m-1 \\ a\in\Lambda_0 - i}} p^{4m-1-a} \right) \cdot \mathbf{u}_i \\
&= \frac{\mu}{p-1} \cdot \mathbf{1} - \sum_{i\in I} c_i \cdot \mathbf{u}_i,
\end{aligned}$$

where

$$c_i := \sum_{\substack{0\leqslant a\leqslant 4m-1\\ a\in\Lambda_0-i}} p^{4m-1-a}$$

for all $i\in I$. In other words,

$$d_i = \frac{\mu}{p-1} - c_i$$

for all $i\in I$. The p-adic expansions of the numbers c_i consist of 0's and 1's and are cyclic permutations of each other; c_{2m-1} is the largest and c_{4m-1} is the smallest among these $4m$ numbers. The p-adic expansions of c_{2m-1} and c_{4m-1} are depicted below.

$$c_{2m-1}:\ 11\underbrace{0101\ldots01}_{2m\text{-}2\text{ digits}}00\underbrace{1010\ldots10}_{2m\text{-}2\text{ digits}};\qquad c_{4m-1}:\ 00\underbrace{1010\ldots10}_{2m\text{-}2\text{ digits}}11\underbrace{0101\ldots01}_{2m\text{-}2\text{ digits}}$$

B.2.11. Lemma. *The equations* B.2.9 (1)–(3) *have a unique solution, given by*

$$\begin{cases} \mu &= (p-1)\cdot c_{2m-1} &= (p-1)\cdot \displaystyle\sum_{\substack{0\leqslant a\leqslant 4m-1\\ a\in\Lambda_0-2m+1}} p^{4m-1-a} \\ d_i &= \frac{\mu}{p-1}-c_i &= c_{2m-1}-c_i \qquad \textit{for all } i\in I \end{cases}$$

where

$$c_i := \sum_{\substack{0\leqslant a\leqslant 4m-1\\ a\in\Lambda_0-i}} p^{4m-1-a}$$

for all $i\in I$. The integers d_i satisfy the following inequalities:

$$\begin{aligned} 0 = d_{2m-1} < d_{2m-3} < \cdots < d_3 < d_1 \\ < d_{2m} < d_{2m+2} < \cdots < d_{4m-4} < d_{4m-2} \\ < d_{2m-2} < d_{2m-4} < \cdots < d_2 < d_{4m} \\ < d_{2m+1} < d_{2m+3} < \cdots < d_{4m-3} < d_{4m-1}\,. \end{aligned}$$

In particular,

$$d_1 < d_{2m} = e-\mu\,.$$

PROOF. We have seen the formulas for the c_i's and the d_i's in B.2.10. The value of the parameter μ is determined by the condition that $\min(\, d_i \mid i\in I\,)=0$. The ordering of the d_i's is easily determined from the p-adic expansions of the c_i's.

The equality $d_{2m} = e-\mu$ is the special case $i = 2m-1$ of equation B.2.9 (1), recorded here for future reference. One can also show by direct computation that $e-\mu-d_1 = p^{2m}-1$. □

B.2.12. With the solution of equations B.2.9 (1)–(3) available, we can solve the equations B.2.9 (a)–(c). Their solutions are far from unique, because one can scale the $f_i(u)$'s simultaneous by a unit $c(u)\in\overline{\mathbb{F}}_p[\![u]\!]^\times$ and multiply $g(u)$ by $\phi(c(u))$; the scaling corresponds to multiplying the corresponding solutions of equations B.2.9 (i)–(iii) by the unit $c(u)$. Thus, we *look for solutions of equations* B.2.9 (1)–(3) *with the normalization condition* $f_{2m-1}(u)=1$.

We solve the equations B.2.9 (i)–(iii) recursively, starting with $f_{2m-1}(u)=1$, and get

$$f_j(u) = \frac{u^{s_j}}{g(u)^{r_j}} \qquad j = 2m, 2m+1, \ldots, 4m, 1, 2, \ldots, 2m-2,$$

where

$$s_j = \begin{cases} e \cdot \sum\limits_{\substack{2m \leqslant i \leqslant j \\ i-1 \in \Lambda_0}} p^{j-i} & \text{for } 2m \leqslant j \leqslant 4m, \\ e \cdot \left[\sum\limits_{\substack{1 \leqslant i \leqslant j \\ i-1 \in \Lambda_0}} p^{j-i} + \sum\limits_{\substack{2m \leqslant i \leqslant 4m \\ i-1 \in \Lambda_0}} p^{j+4m-i} \right] & \text{for } 1 \leqslant j \leqslant 2m-2, \end{cases}$$

and

$$r_j = \begin{cases} \sum\limits_{0 \leqslant a \leqslant j-2m} p^{j-2m-a} & \text{for } 2m \leqslant j \leqslant 4m, \\ \sum\limits_{0 \leqslant a \leqslant j+2m} p^{j+2m-a} & \text{for } 1 \leqslant j \leqslant 2m-2. \end{cases}$$

For later reference, we set

$$s_{2m-1} = 0 = r_{2m-1}\,.$$

The above formulas depend on the unknown $g(u) \in \overline{\mathbb{F}}_p[\![u]\!]$ with $\mathrm{ord}_u(g(u)) = \mu$, where μ is given in B.2.11. The system of equations B.2.9 (i)–(iii) is now reduced to a single equation in the unknown element $g(u)$: set $i = 2m-2$ in equation (a) in B.2.9 and use the fact that $2m-2 \notin \Lambda_0$ to get

$$g(u) = f_{2m-2}(u)^p,$$

where $f_{2m-2}(u)$ is given above and depends on the unknown $g(u)$. Thus, the equation $g(u) = f^p_{2m-2}(u)$ is really an equation for $g(u)$. One can check that this equation for $g(u)$ is compatible with the requirement that

$$\mathrm{ord}_u(g(u)) = \mu = (p-1)\, c_{2m-1} = (p-1) \cdot \sum_{\substack{0 \leqslant a \leqslant 4m-1 \\ a \in \Lambda_0 - 2m+1}} p^{4m-1-a}\,.$$

Write the unknown $g(u) \in \overline{\mathbb{F}}_p[\![u]\!]$ as

$$g(u) = u^\mu \cdot h(u)\,, \qquad h(u) = \xi \cdot \left(1 + \sum_{j \in \mathbb{N}_{>0}} c_j\, u^j \right);$$

the above equation for $g(u)$ is transformed to the equation

$$h(u)^{\sum_{0 \leqslant a \leqslant 4m-1} p^a} = 1$$

in the unknown $h(u) \in \overline{\mathbb{F}}_p[\![u]\!]^\times$. It follows that $h(u) = \xi \in \mu_{e/(p-1)}(\overline{\mathbb{F}}_p)$; i.e., $h(u)$ is a constant in $\overline{\mathbb{F}}_p[\![u]\!]$, equal to a $((p^{4m}-1)/(p-1))$th root of unity in $\overline{\mathbb{F}}_p$. We have found all solutions of the system of equations B.2.9 (i)–(iii), summarized in B.2.13 below.

B.2.13. Lemma. *Let $N = e/(p-1) = (p^{4m}-1)/(p-1)$. Choose and fix a generator ζ of the group $\mu_N(\overline{\mathbb{F}}_p)$ of Nth roots of unity in $\overline{\mathbb{F}}_p$.*

There are exactly N admissible rank-1 subobjects of the object $(\mathfrak{M}, \phi_{\mathfrak{M}}, \psi_{\mathfrak{M}})$ (in the exact category $(\mathrm{Mod}/\mathfrak{S})^c$) that correspond to the N finite flat subgroups of

$\mathcal{X}[p]_{\mathcal{O}_K}$ of order p. Each of these N admissible subobjects is given by an invertible $\overline{\mathbb{F}}_p[\![u]\!]$-direct summand of the free $\overline{\mathbb{F}}_p[\![u]\!]$-module $\mathfrak{M}$ of the form

$$\overline{\mathbb{F}}_p[\![u]\!]\cdot y_a, \qquad a\in\mathbb{Z}/N\mathbb{Z}\,,$$

where

$$y_a := \sum_{j\in I}\xi^{-a\,r_j}\cdot u^{s_j-\mu\,r_j}\cdot v_j = \sum_{j\in I}\xi^{-a\,r_j}\cdot u^{d_j}\cdot v_j$$

and the non-negative integers μ, s_j and r_j are defined in B.2.11 *and* B.2.12.

Note that the action of $\mathcal{O}/p\mathcal{O}\simeq\kappa_F$ on $\mathfrak{M}$ induces a simply transitive action of $\kappa_F^\times/\overline{\mathbb{F}}_p^\times$ on the collection $\{\,\overline{\mathbb{F}}_p[\![u]\!]\cdot y_a \mid a\in\mathbb{Z}/N\mathbb{Z}\,\}$ of $\overline{\mathbb{F}}_p[\![u]\!]$-submodules of $\mathfrak{M}$.

B.2.14. Definition. Fix a generator ζ of the cyclic group $\kappa_F^\times$. For any element $\eta\in\kappa_F$, write $[\eta]\colon\mathfrak{M}\to\mathfrak{M}$ for the action of η on the κ_F-linear $\overline{\mathbb{F}}_p[\![u]\!]$-module $\mathfrak{M}$. (Recall that $[\eta](y):=i(\eta)\cdot y$ for all $\eta\in\kappa_F$, $i\in I$, and $y\in\mathfrak{M}_i$, and $(i+1)(\eta)=i(\eta)^p$ for all $i\in I$ and $\eta\in\kappa_F$.)

(1) Denote by $\mathfrak{M}_{(0)}$ the free $\overline{\mathbb{F}}_p[\![u]\!]$-submodule generated by the element

$$y_0 = \sum_{i\in I} u^{d_i}\cdot v_i$$

in B.2.13.

(2) For $b=1,2,\ldots,m-1$, let $\mathfrak{M}_{[b]}$ be the $\overline{\mathbb{F}}_p[\![u]\!]$-submodule of $\mathfrak{M}$ generated by the $b+1$ elements $y_0,[\zeta](y_0),\ldots,[\zeta^b](y_0)$ in $\mathfrak{M}$.

(3) For $b=1,2,\ldots,m-1$, let

$$\mathfrak{M}_{(b)} := \{x\in\mathfrak{M}\mid u^m\,x\in\mathfrak{M}_{[b]}\text{ for some }m\in\mathbb{N}\} = (\overline{\mathbb{F}}_p(\!(u)\!)\cdot\mathfrak{M}_{[b]})\bigcap\mathfrak{M}$$

be the saturation in $\mathfrak{M}$ of the $\overline{\mathbb{F}}_p[\![u]\!]$-submodule $\mathfrak{M}_{[b]}\subset\mathfrak{M}$.

B.2.15. Lemma. *Let b be an integer with $0\leqslant b\leqslant m-1$.*

(i) *The $\overline{\mathbb{F}}_p[\![u]\!]$-submodule $\mathfrak{M}_{[b]}$ of $\mathfrak{M}=\bigoplus_{i\in I}\overline{\mathbb{F}}_p[\![u]\!]\cdot v_i$ is a free $\overline{\mathbb{F}}_p[\![u]\!]$-module of rank $b+1$ with a basis of the form*

$$\begin{array}{lcllcl}
\mathbf{z}_0 & = & u^{d_{2m-1}}\cdot v_{2m-1} & + & \displaystyle\sum_{i\notin\{2m-2b-1,2m-2b+3,\ldots,2m-1\}} & \lambda_{0,i}\cdot v_i\\
\vdots & & \vdots & & \vdots & \\
\mathbf{z}_a & = & u^{d_{2m-2a-1}}\cdot v_{2m-2a-1} & + & \displaystyle\sum_{i\notin\{2m-2b-1,2m-2b+3,\ldots,2m-1\}} & \lambda_{a,i}\cdot v_i\\
\vdots & & \vdots & & \vdots & \\
\mathbf{z}_b & = & u^{d_{2m-2b-1}}\cdot v_{2m-2b-1} & + & \displaystyle\sum_{i\notin\{2m-2b-1,2m-2b+3,\ldots,2m-1\}} & \lambda_{b,i}\cdot v_i
\end{array}$$

where each $\lambda_{a,i}$ is an element of $\overline{\mathbb{F}}_p[\![u]\!]$ with

$$\mathrm{ord}_u(\lambda_{a,i})\geqslant d_i \qquad \text{for all } 0\leqslant a\leqslant b,\, i\notin\{2m-2b-1,2m-2b+1,\ldots,2m-1\}$$

and the d_i's are the natural numbers defined in B.2.11.

(ii) *The* $\overline{\mathbb{F}}_p[\![u]\!]$*-submodule* $\mathfrak{M}_{(b)}$ *of* $\mathfrak{M} = \bigoplus_{i\in I} \overline{\mathbb{F}}_p[\![u]\!]\cdot v_i$ *is a free* $\overline{\mathbb{F}}_p[\![u]\!]$*-module of rank* $b+1$ *with basis*

$$\left(u^{-d_{2m-2a-1}}\cdot \mathbf{z}_a\right)_{a=0,1,\ldots,b}.$$

(iii) $\mathfrak{M}_{(b)} \subset u^{-d_{2m-2b-1}}\cdot \mathfrak{M}_{[b]}$.

PROOF. Granting (i), the inequalities on the d_i's in B.2.11 yield

$$d_{2m-a-1} < \mathrm{ord}_u(\lambda_{a,i})$$

for all $0 \leqslant a \leqslant b$ and $i \notin \{2m-2b-1, 2m-2b+1, \ldots, 2m-1\}$, so (ii) and (iii) follow. It remains to prove (i).

We have

$$[\zeta^a](y_0) = \sum_{i\in I} i(\zeta^a)\cdot u^{d_i}\cdot v_i$$

for $a = 0, 1, \ldots b$, a system of $b+1$ linear equations with coefficients

$$C_{a,i} := i(\zeta)^a \qquad 0 \leqslant a \leqslant b,\ i \in I$$

indexed by $\{0, 1, \ldots, b\} \times I$. Look at the maximal minor $C_{(b)}$ of the $(b+1)\times(4m)$-matrix C where the second index i ranges through the subset

$$\{2m-2b-1, 2m-2b-3, \ldots, 2m-3, 2m-1\} \subset I.$$

This square matrix is a Vandermonde matrix given by the $b+1$ distinct roots of unity in $\overline{\mathbb{F}}_p$, so $C_{(b)}$ is an invertible matrix with entries in $\overline{\mathbb{F}}_p$. Multiplying the system of linear equations by the inverse of the matrix $C_{(b)}$, we get a system of generators of the $\overline{\mathbb{F}}_p[\![u]\!]$-module $\mathfrak{M}_{[b]}$ as described in (i). □

B.2.16. Proposition. *Let* b *be an integer with* $0 \leqslant b \leqslant m-1$.

(1) $\phi_{\mathfrak{M}}(\mathfrak{M}_{(b)}) \subset \mathfrak{M}_{(b)}$.

(2) $\psi_{\mathfrak{M}}(\mathfrak{M}_{(b)}) \subset \phi^*(\mathfrak{M}_{(b)}) \subseteq \phi^*\mathfrak{M} \cap \left(\overline{\mathbb{F}}_p(\!(u)\!)\cdot \phi^*(\mathfrak{M}_{(b)})\right)$, *where "*$\phi^*(\mathfrak{M}_{(b)})$*" in the final term is identified with its image in* $\phi^*\mathfrak{M}$ *via the natural injection* $\phi^*(\mathfrak{M}_{(b)}) \longrightarrow \phi^*\mathfrak{M}$.

(3) *Let* $\phi_{\mathfrak{M}_{(b)}} : \mathfrak{M}_{(b)} \to \mathfrak{M}_{(b)}$ *be the* ϕ*-linear map induced by* $\phi_{\mathfrak{M}}$, *and let*

$$\psi_{\mathfrak{M}_{(b)}} : \mathfrak{M}_{(b)} \to \phi^*(\mathfrak{M}_{(b)})$$

be the $\overline{\mathbb{F}}_p[\![u]\!]$*-linear map induced by* $\psi_{\mathfrak{M}}$. *The triple*

$$\left(\mathfrak{M}_{(b)}, \phi_{\mathfrak{M}_{(b)}} : \mathfrak{M}_{(b)} \to \mathfrak{M}_{(b)}, \psi_{\mathfrak{M}_{(b)}} : \mathfrak{M}_{(b)} \to \phi^*(\mathfrak{M}_{(b)})\right)$$

defines an admissible subobject of the object $(\mathfrak{M}, \phi_{\mathfrak{M}}, \psi_{\mathfrak{M}})$ *in the exact category* $(\mathrm{Mod}/\mathcal{O}_K)^c$; *it corresponds to a finite flat subgroup scheme* $\mathcal{G}_{(b)} \subset \mathcal{X}_{\mathcal{O}_K}$ *of order* p^{b+1}.

(4) *Identify the classical covariant Dieudonné module* $\mathrm{M}_*(G_{(b)})$ *of the closed fiber* $G_{(b)}$ *of* $\mathcal{G}_{(b)}$ *with a submodule of the Dieudonné module*

$$\mathrm{M}_*(X_0[p]) = \bigoplus_{i\in I} \mathrm{M}_*(X_0[p])_i \simeq \bigoplus_{i\in I} (\mathfrak{M}_i/u\mathfrak{M}_i)$$

of the p*-torsion in the* $\mathcal{O}$*-linear* p*-divisible group* X_0 *over* $\overline{\mathbb{F}}_p$, *a module scheme over* $\kappa_F = \mathcal{O}/p\mathcal{O}$. *We have*

$$\mathrm{M}_*(G_{(b)}) = \mathrm{M}_*(X_0[p])_{2m-2b-1} \oplus \mathrm{M}_*(X_0[p])_{2m-2b-3} \oplus \cdots \oplus \mathrm{M}_*(X_0[p])_{2m-1}.$$

In particular, the subgroup $G_{(b)}$ *of* X_0 *is stable under the action of* $\mathcal{O}$.

(5) *The closed fiber* $G_{(m-1)}$ *of the finite flat subgroup scheme*

$$\mathcal{G}_{(m-1)} \subset \mathcal{X}[p]_{\mathcal{O}_K}$$

is the kernel of the $\mathcal{O}$*-linear isogeny* $\rho_0 \colon X_0 \to Y_0$.

PROOF. Part (1) is clear. To prove (2), recall first from B.2.9 and B.2.11 that

$$\psi_{\mathfrak{M}}(\overline{\mathbb{F}}_p[\![u]\!]\cdot y_0) = u^{e-\mu}\cdot\phi^*(\overline{\mathbb{F}}_p[\![u]\!]\cdot y_0),$$

so

$$\psi_{\mathfrak{M}}(\mathfrak{M}_{[b]}) = u^{e-\mu}\cdot\phi^*(\mathfrak{M}_{[b]}).$$

We know from B.2.15 (iii) that $\mathfrak{M}_{(b)} \subset u^{-d_{2m-2b-1}}\cdot\mathfrak{M}_{[b]}$, and we have $d_{2m-2b-1} < e-\mu$ by the inequalities on the d_i's in B.2.11, so we get

$$\begin{aligned}\psi_{\mathfrak{M}}(\mathfrak{M}_{(b)}) \subseteq \psi_{\mathfrak{M}}(u^{-d_{2m-2b-1}}\cdot\mathfrak{M}_{[b]}) \subseteq u^{e-\mu-d_{2m-2b-1}}\cdot\phi^*(\mathfrak{M}_{[b]})\\ \subset \phi^*(\mathfrak{M}_{[b]}) \subseteq \phi^*(\mathfrak{M}_{(b)}).\end{aligned}$$

This proves (2).

The assertion in (3) is immediate from (1) and (2). Part (4) is a consequence of B.2.15 (ii) and the inequalities on the d_i's in B.2.11.

To prove (5), first note that we have $\mathcal{O}$-equivariant isomorphisms

$$\mathrm{M}_*(X_0[p]) \simeq \mathfrak{M}/u\mathfrak{M}$$

and

$$\mathrm{Lie}(\ker(\rho_0)) \simeq \ker\left(\mathfrak{M}/u\mathfrak{M} \xrightarrow{(1/c)\psi_{\mathfrak{M}}} \phi^*\mathfrak{M}/u\phi^*\mathfrak{M}\right)$$

from B.4.17. Since $\ker(\rho_0)$ is the unique subgroup of $X_0[p]$ of order p^m whose Lie type corresponds to the subset $\{1,3,5,\ldots,2m-1\}$ of I, $\mathrm{M}_*(\ker(\rho_0))$ is the sum of the eigensubspaces $\mathrm{M}_*(X_0[p])_i$ of $\mathrm{M}_*(X_0[p])$ for $i = 1,3,\ldots,2m-1$. Hence, (5) now follows from the special case $b = m-1$ in (4). □

B.2.17. Remark. Let $K_1 \subset K$ be the subextension of K over K_0 such that

$$[K_1 : K_0] = \frac{p^{4m}-1}{p-1}, \quad [K : K_1] = p-1.$$

It is clear from the formulas in B.2.11 that the integers d_i and μ are all divisible by $p-1$. From this it is not difficult to deduce that the admissible subobject $(\mathfrak{M}_{(b)}, \phi_{\mathfrak{M}_{(b)}}, \psi_{\mathfrak{M}_{(b)}})$ of $(\mathfrak{M}, \phi_{\mathfrak{M}}, \psi_{\mathfrak{M}})$ in the category $(\mathrm{Mod}/\mathcal{O}_K)^c$ is the base change, from $\mathcal{O}_{K_1}$ to $\mathcal{O}_K$, of an admissible subobject of $(\mathfrak{M}_1, \phi_{\mathfrak{M}_1}, \psi_{\mathfrak{M}_1})$ in $(\mathrm{Mod}/\mathcal{O}_{K_1})^c$, where $(\mathfrak{M}_1, \phi_{\mathfrak{M}_1}, \psi_{\mathfrak{M}_1})$ is the base extension from $\mathcal{O}_{K_0}$ to $\mathcal{O}_{K_1}$ of $(\mathfrak{M}_0, \phi_{\mathfrak{M}_0}, \psi_{\mathfrak{M}_0})$.

Correspondingly, the finite flat group scheme $\mathcal{G}_{(b)}$ in B.2.16 is the base change of a finite flat subgroup scheme of $\mathcal{X}[p]_{\mathcal{O}_{K_1}}$. Alternatively, one can replace K by K_1 and redo the constructions and proofs in this section. (See B.1.15 for a similar situation with Raynaud's method.)

B.3. Dieudonné theories over a perfect field of characteristic p

Throughout this section, k denotes a fixed perfect field of characteristic $p > 0$. We provide a synopsis and discussion of comparison for several Dieudonné theories. Each is a machine that translates a suitable category of commutative group schemes or p-divisible groups into a suitable category of p-linear algebra data.

B.3.1. Over k, we have the following Dieudonné theories.[4]

1. The classical contravariant Dieudonné modules for a finite commutative group scheme G of p-power order over k and for a p-divisible group X over k are respectively denoted $\mathrm{M}^*(G)$ and $\mathrm{M}^*(X)$.[5] The linear duals $\mathrm{M}_*(G)$ of $\mathrm{M}^*(G)$ and $\mathrm{M}_*(X)$ of $\mathrm{M}^*(X)$ are the classical covariant Dieudonné modules for G and X.
2. The crystalline theory ([**48**], [**75**], [**6**]) produces covariant crystals $\mathbb{E}_*(G)$ and $\mathbb{E}_*(X)$, as well as contravariant Dieudonné crystals $\mathbb{E}^*(G)$, $\mathbb{E}^*(X)$.[6] When evaluated at the natural PD extension $\mathrm{Spec}(k) \hookrightarrow \mathrm{Spec}(W(k))$ one obtains the covariant Dieudonné modules $\mathrm{E}_*(G)$ and $\mathrm{E}_*(X)$, as well as the contravariant Dieudonné modules $\mathrm{E}^*(G)$ and $\mathrm{E}^*(X)$. See B.3.5.
3. The covariant Cartier module $\mathrm{C}_*(X)$ of a formal p-divisible group X over k and its $W(k)$-linear dual $\mathrm{C}^*(X)$. Embedding a finite group scheme G over k into a p-divisible group X over k, one gets a covariant Dieudonné module $\mathrm{C}_*(G)$ and a contravariant Dieudonné module $\mathrm{C}^*(G)$ attached to G.

The relation between these Dieudonné theories is as follows. There are natural isomorphisms of functors

$$\mathrm{E}_*(\cdot) \simeq \mathrm{C}_*(\cdot) \quad \text{and} \quad \mathrm{E}^*(\cdot) \simeq \mathrm{C}^*(\cdot)\,,$$

so the crystalline Dieudonné modules coincide with the Cartier–Dieudonné modules, and the crystalline Dieudonné modules are the Frobenius twists of the corresponding classical Dieudonné modules:

$$\mathrm{E}_*(\cdot) \simeq \mathrm{C}_*(\cdot) \simeq \mathrm{M}_*(\cdot)^{(1)} \quad \text{and} \quad \mathrm{E}^*(\cdot) \simeq \mathrm{C}^*(\cdot) \simeq \mathrm{M}^*(\cdot)^{(1)};$$

in these natural isomorphisms, the functors are consider on the categories of finite commutative group scheme of p-power order or p-divisible groups. Each of the above functors, when evaluated at an object over k, has a natural structure as a module over the ring $W(k)$ of p-adic Witt vectors, and the superscript "$^{(1)}$" denotes the Frobenius twist $M \mapsto M^{(1)} := W(k) \otimes_{\sigma, W(k)} M$ for modules over $W(k)$; see B.3.2 below.

---

[4]The output of Zink's theory of displays, discussed in B.3.7 and B.4.7, when restricted to the category of all p-divisible groups over k is essentially the same as that of the covariant Cartier theory, so it is not listed below.

[5]The classical contravariant Dieudonné modules were initially defined by Dieudonné and Manin; a later uniform definition was given by Fontaine. The standard references are [**71**], [**29**], and [**41**]; see also [**86**].

[6]We decided to use the symbols $\mathbb{E}_*(G)$, $\mathbb{E}^*(G)$, $\mathrm{E}_*(G)$, $\mathrm{E}^*(G)$ for Dieudonné crystals and Dieudonné modules, to avoid any possible confusion with other Dieudonné theories. The notation $\mathrm{E}(G)$ appeared in the crystalline literature [**75**] for the universal vector extension of a p-divisible group G. Although EG is also the standard notation for the universal bundle over a classifying space BG for a group G, the danger of this conflict is minimal for this book.

B.3.2. Notation. Let $W = W(k)$ be the ring of p-adic Witt vectors with entries in the perfect base field k, and let $K_0 = W(k)[1/p]$ be its fraction field. Let σ be the Frobenius automorphism on $W(k)$ such that $\sigma(x_0, x_1, x_2, \ldots) = (x_0^p, x_1^p, x_2^p, \ldots)$.

B.3.2.1. For any of the Dieudonné theories reviewed below we will get, for any finite commutative group scheme of p-power order, p-divisible group, or commutative formal Lie group over k, a $W(k)$-module M of finite type equipped with $W(k)$-linear operators F and V,[7]

$$F\colon M^{(1)} := W(k)\otimes_{\sigma,W(k)} M \longrightarrow M, \quad V\colon M \longrightarrow M^{(1)}.$$

For any $n \in \mathbb{Z}$, define

$$M^{(n)} := W(k)\otimes_{\sigma^n,W(k)} M.$$

Note that the map $y \mapsto 1\otimes y$ from M to $M^{(n)}$ is a bijection. We can use this map to identify $M^{(n)}$ with M as an abelian group, but $M^{(n)}$ has a *new* $W(k)$-module structure such that

$$\xi \cdot_{\text{new}} y = \sigma^{-n}(\xi) \cdot_{\text{old}} y$$

for all $\xi \in W(k), y \in M$. We also have Frobenius twists of semi-linear maps; e.g.,

$$\ker\left({F_M}^{(-1)}\colon M \to M^{(-1)}\right) \simeq \left(\ker(F_M\colon M^{(1)} \to M)\right)^{(-1)}.$$

B.3.2.2. A general rule for Dieudonné theories is that for the contravariant theories, the operator F corresponds to the relative homomorphisms $\mathrm{Fr}_{G/k} : G \to G^{(p)}$ or $\mathrm{Fr}_{X/k} : X \to X^{(p)}$ and the operator V corresponds to the relative Verschiebung homomorphisms $\mathrm{Ver}_{G/k} : G^{(p)} \to G$ or $\mathrm{Ver}_{X/k} : X^{(p)} \to X$. Often $\mathrm{Fr}_{G/k}$ is shortened to $\mathrm{F}_{G/k}$ and $\mathrm{Ver}_{G/k}$ is shortened to $\mathrm{V}_{G/k}$.

In covariant theories, the operator F corresponds to the relative Verschiebung homomorphism Ver and the operator V corresponds to the relative Frobenius operator Fr. Recall that we have the identities

$$\mathrm{V}_{X/k} \circ \mathrm{F}_{X/k} = \mathrm{id}_X, \qquad \mathrm{F}_{X/k} \circ \mathrm{V}_{X/k} = \mathrm{id}_{X^{(p)}}$$

and similarly for finite commutative group schemes G over k.

B.3.2.3. Recall from 1.4.3.1 that the Dieudonné ring $D_k = W(k)[\mathcal{F}, \mathcal{V}]$ for the perfect field k has generators $\mathcal{F}$ and $\mathcal{V}$ and satisfies relations

$$\mathcal{F}\mathcal{V} = \mathcal{V}\mathcal{F} = p, \;\; \mathcal{F}c = \sigma(c)\mathcal{F}, \;\; c\mathcal{V} = \mathcal{V}\sigma(c)$$

for all $c \in W(k)$. For any of the Dieudonné theories (covariant or contravariant), the resulting semi-linear algebra data

$$(M, F : M^{(1)} \to M, V : M \to M^{(1)})$$

satisfies

$$V \circ F = p \cdot \mathrm{id}_{M^{(1)}} \quad \text{and} \quad F \circ V = p \cdot \mathrm{id}_M.$$

Suppose we have a $W(k)$-module M and $W(k)$-linear maps $F : M^{(1)} \to M$ and $V : M \to M^{(1)}$ such that $V \circ F = p \cdot \mathrm{id}_{M^{(1)}}$ and $F \circ V = p \cdot \mathrm{id}_M$. Define an operator $\mathcal{F}_M : M \to M$ on M by $\mathcal{F}_M(m) := F(1 \otimes m)$ for all $m \in M$. Similarly define an operator $\mathcal{V}_M : M \to M$ on M by requiring $1 \otimes \mathcal{V}_M(m) = V(m)$ for all $m \in M$.

[7]Traditionally one considers each of the Dieudonné modules M as a module over the Dieudonné ring D_k, in which case F and V are semi-linear operators. However to keep track of the Frobenius twists it is often more convenient to regard M as a module over the ring $W(k)$ of p-adic Witt vectors and "linearize" F and V as below.

Clearly $\mathcal{F}_M$ and $\mathcal{V}_M$ satisfy the defining relations for the Dieudonné ring D_k, so M acquires a D_k-module structure.

Conversely, suppose we have a module M over the Dieudonné ring D_k. Reversing the above procedure, we get $W(k)$-linear maps $F : M^{(1)} \to M$ and $V : M \to M^{(1)}$ such that $V \circ F = p \cdot \mathrm{id}_{M^{(1)}}$ and $F \circ V = p \cdot \mathrm{id}_M$.

B.3.2.4. Remark. Our general convention is that $\mathcal{F}$ and $\mathcal{V}$ are semi-linear operators on a $W(k)$-module M, giving M the structure of a D_k-module structure.

B.3.3. Witt covectors. We recall the definition and some basic properties of Witt covectors; more information can be found in [**41**, Ch. II].

B.3.3.1. For every commutative ring R, the set $CW(R)$ of all p-adic *Witt covectors* with entries in R is the set consisting of all sequences $\underline{a} = (\ldots, a_{-n}, \ldots, a_{-1}, a_0)$ such that

(i) $a_{-n} \in R$ for all $n \geqslant 0$,
(ii) there exists an $r \geqslant 0$ (depending on $\underline{a}$) such that the ideal generated by all a_{-n}'s with $n \geqslant r$ is *nilpotent* (i.e., some power of the ideal vanishes).

More generally, for any complete separated topological commutative ring R having a basis of open neighborhoods of 0 consisting of open ideals, we define $CW(R)$ by replacing condition (ii) with

(ii)′ for every open ideal I in R, there exist integers $r \geqslant 0$ and $N \geqslant 1$ such that the ideal $J_r = (a_{-n})_{n \geqslant r}$ generated by all a_{-n}'s with $n \geqslant r$ satisfies $J_r^N \subseteq I$.

B.3.3.2. The group law on $CW(R)$ is defined as follows: For any two elements

$$\underline{a} = (\ldots, a_{-n}, \ldots, a_{-1}, a_0) \quad \text{and} \quad \underline{b} = (\ldots, b_{-n}, \ldots, b_{-1}, b_0)$$

in $CW(R)$, their sum $\underline{c} = (\ldots, c_{-n}, \ldots, c_{-1}, c_0)$ is defined by

$$c_n = \lim_{m \to \infty} S_m(a_{-n-m}, \ldots, a_{-n}; b_{-n-m}, \ldots, b_{-n}),$$

where $S_m(a_{-n-m}, \ldots, a_{-n}; b_{-n-m}, \ldots, b_{-n})$ on the right side is constant for m sufficiently large if R has the discrete topology, and the polynomials

$$S_m(x_0, x_1, \ldots, x_m; y_0, y_1, \ldots, y_m) \in \mathbb{Z}[x_0, x_1, \ldots, x_m; y_0, y_1, \ldots, y_m]$$

for $m \geqslant 0$ are the universal polynomials which define the group laws on the p-adic Witt vectors; i.e.

$$\begin{aligned}(x_0, x_1, \ldots, x_n, \ldots) &+_{\mathrm{Witt}} (y_0, y_1, \ldots, y_n, \ldots) \\ &= (S_m(x_0, x_1, \ldots, x_m; y_0, y_1, \ldots, y_m))_{m \geqslant 0}\end{aligned}$$

characterized by

$$\sum_{i=0}^{m} p^a\, x_a^{p^{m-a}} + \sum_{i=0}^{m} p^a\, y_a^{p^{m-a}} = \sum_{i=0}^{m} p^a \cdot (S_a(x_0, x_1, \ldots, x_m; y_0, y_1, \ldots, y_m))^{p^{m-a}}$$

for all $m \geqslant 0$.

B.3.3.3. The group law for the functor CW is encoded in one formal power series

$$S(\underline{x};\underline{y}) = S(\ldots, x_{-2}, x_{-1}, x_0; \ldots, y_{-2}, y_{-1}, y_0) \in \mathbb{Z}[\![\underline{x}, \underline{y}]\!]$$

in infinitely many variables $\{x_{-n}, y_{-n}\}_{n\geqslant 0}$. This power series $S(\underline{x};\underline{y})$ is characterized by the equation

$$\sum_{n\in\mathbb{N}} p^{-n}\, x_{-n}^{p^n} + \sum_{n\in\mathbb{N}} p^{-n}\, y_{-n}^{p^n} = \sum_{n\in\mathbb{N}} p^{-n}\, S_{-n}(\underline{x};\underline{y}),$$

where for each $n \geqslant 0$ the power series $S_{-n}(\underline{x};\underline{y})$ is defined in terms of S by shifting the variables:

$$\begin{aligned} &S_{-n}(\ldots, x_{-2}, x_{-1}, x_0; \ldots, y_{-2}, y_{-1}, y_0) \\ &\qquad := S(\ldots, x_{-n-2}, x_{-n-1}, x_{-n}; \ldots, y_{-n-2}, y_{-n-1}, y_{-n}). \end{aligned}$$

The group law for the Witt covectors is

$$\begin{aligned} \underline{a} +_{CW} \underline{b} &= (\ldots, S_{-m}(\underline{a};\underline{b}), \ldots, S_{-1}(\underline{a};\underline{b}), S_0(\underline{a};\underline{b})) \\ &= (\ldots, S(\ldots, a_{-2}, a_{-1}; \ldots, b_{-2}, b_{-1}), S(\ldots, a_{-1}, a_0; \ldots, b_{-1}, b_0)) \end{aligned}$$

Alternatively, $S(\underline{x};\underline{y})$ is the limit

$$S(\underline{x};\underline{y}) = \lim_{m\to\infty} S_m(x_{-m}, x_{-m+1}, \ldots, x_{-1}, x_0; y_{-m}, y_{-m+1}, \ldots, y_{-1}, y_0);$$

this limit makes sense because

$$S_{m+1}(0, x_{-m}, \ldots, x_0; 0, y_{-m}, \ldots, y_0) = S_m(x_{-m}, \ldots, x_0; y_{-m}, \ldots, y_0)$$

for all $m \geqslant 0$. Note that each term of the power series $S(\underline{x};\underline{y})$ is homogeneous of weight 1 if x_{-n} and y_{-n} are given weight p^{-n} for every $n \geqslant 0$.

B.3.3.4. Restricting the group-valued functor CW to the category of commutative k-algebras defines the commutative group-valued functor CW_k over k.

For every k-algebra R, the group $CW_k(R)$ has a natural $W(k)$-module structure, as follows. For any

$$\underline{u} = (u_0, u_1, u_2, \ldots, u_n, \ldots) \in W(k) \quad\text{and}\quad \underline{a} = (\ldots, a_{-n}, \ldots, a_{-1}, a_0) \in W(R)$$

the product $\underline{b} = (\ldots, b_{-n}, \ldots, b_{-1}, b_0) = \underline{u}\cdot\underline{a}$ of $\underline{u}$ and $\underline{a}$ is given by

$$b_{-n} = P_m(\sigma^{-n-m}(u_0), \sigma^{-n-m}(u_1), \ldots, \sigma^{-n-m}(u_n); a_{-n-m}, a_{-n-m+1}, \ldots, a_{-1}, a_0)$$

for m sufficiently large, where the polynomials

$$P_m(x_0, x_1, \ldots, x_m; y_0, y_1, \ldots, y_m) \in \mathbb{Z}[x_0, x_1, \ldots, x_m; y_0, y_1, \ldots, y_m]$$

indexed by $m \geqslant 0$ are the universal polynomials which define the product law on the p-adic Witt vectors. In particular,

$$(\xi, 0, 0, \ldots)\cdot(\ldots, b_{-n}, \ldots, b_{-1}, b_0) = (\ldots, \sigma^{-n}(\xi)b_{-n}, \ldots, \sigma^{-1}(\xi)b_{-1}, \xi b_0)$$

for all $\xi \in k$ and all $(\ldots, b_{-n}, \ldots, b_{-1}, b_0) \in CW(R)$.

Define endomorphisms F and V of CW_k by

$$F(\ldots, a_{-m}, \ldots, a_{-2}, a_{-1}, a_0) = (\ldots, a_{-m}^p, \ldots, a_{-2}^p, a_{-1}^p, a_0^p)$$

and

$$V(\ldots, a_{-m}, \ldots, a_{-2}, a_{-1}, a_0) = (\ldots, a_{-m}, \ldots, a_{-2}, a_{-1})$$

for all k-algebras R and all $(\ldots, a_{-m}, \ldots, a_{-2}, a_{-1}, a_0) \in CW(R)$. We have

$$F\circ V = p\cdot \mathrm{id}_{CW_k} = V\circ F, \quad F\circ[a]_{CW_k} = [\sigma(a)]_{CW_k}\circ F, \quad [a]_{CW_k}\circ V = V\circ[\sigma(a)]_{CW_k}$$

for all $a \in W(k)$, where $[a]_{CW_k}$ denotes the endomorphism "multiplication by a" on the functor CW_k.

B.3.4. Let CW^u be the subfunctor of CW assigning to any commutative ring R the subset of $CW(R)$ consisting of all elements $(\ldots, a_{-n}, \ldots, a_{-1}, a_0)$ such that $a_{-n} = 0$ for all but finitely many n. In other words,

$$CW^u = \varinjlim \left(W_1 \xrightarrow{V} W_2 \xrightarrow{V} \cdots W_n \xrightarrow{V} W_{n+1} \cdots \right),$$

the direct limit of truncated Witt groups W_n with respect to the transition maps $V\colon W_n \to W_{n+1}$ that send $(x_0, x_1, \ldots, x_{n-1})$ to $(0, x_0, x_1, \ldots, x_{n-1})$ as usual. It is easy to see that CW^u is a subgroup of CW, and CW_k^u is stable under multiplication by elements of $W(k)$,

B.3.5. Classical Dieudonné theory.

B.3.5.1. For any finite commutative group scheme G over k of p-power order, the *classical contravariant Dieudonné module* $\mathrm{M}^*(G)$ is by definition the $W(k)$-module

$$\mathrm{M}^*(G) = \mathrm{Hom}_k(G, CW_k)$$

together with $W(k)$-linear maps

$$F_{\mathrm{M}^*(G)}\colon \mathrm{M}^*(G)^{(1)} \to \mathrm{M}^*(G), \quad V_{\mathrm{M}^*(G)}\colon \mathrm{M}^*(G) \to \mathrm{M}^*(G)^{(1)}$$

induced by the relative Frobenius homomorphism $\mathrm{Fr}_G\colon G \to G^{(p)}$ and the Verschiebung homomorphism $\mathrm{Ver}_G\colon G^{(p)} \to G$. The $W(k)$-module structure on $\mathrm{M}^*(G)$ comes from the $W(k)$-module structure on the functor CW_k. We have

$$V_{\mathrm{M}^*(G)} \circ F_{\mathrm{M}_*(G)} = p \cdot \mathrm{id}_{\mathrm{M}^*(G)^{(1)}}, \quad F_{\mathrm{M}^*(G)} \circ V_{\mathrm{M}^*(G)} = p \cdot \mathrm{id}_{\mathrm{M}^*(G)}.$$

B.3.5.2. If we identify $\mathrm{M}^*(G)^{(1)}$ with $\mathrm{M}^*(G)$ by the bijection $1 \otimes x \mapsto x$ for all $x \in \mathrm{M}^*(G)$ then $\mathrm{M}^*(G)$ becomes a left module over the Dieudonné ring D_k. Equivalently, the actions of $\mathcal{F}$ and $\mathcal{V}$ on $\mathrm{M}^*(G)$ are induced by the endomorphism F and V on CW_k defined in B.3.3.4.

B.3.5.3. The following facts are well-known and have been discussed in 1.4, especially in 1.4.3.2 and 1.4.3.7. The standard references are [**41**, Ch. III], [**29**], and [**71**]; see also [**86**].

(i) $\mathrm{M}^*(G)$ is a $W(k)$-module of finite length for every commutative finite group scheme G over k of p-power order.

(ii) The functor M^* induces an *anti*-equivalence between the abelian category of commutative finite groups schemes of p-power order over k and the abelian category of D_k-modules of finite length over $W(k)$. The inverse functor is

$$M \rightsquigarrow \mathrm{Hom}_{D_k}(M, CW),$$

assigning to any D_k-module M of finite $W(k)$-length the (representable) group functor

$$R \rightsquigarrow \mathrm{Hom}_{D_k}(M, CW(R))$$

on the category of k-algebras or the category of complete separated commutative topological k-algebras with a fundamental system of neighborhoods of 0 consisting of open ideals; the representing object is a finite k-algebra (or equivalently, a finite k-group scheme).

(iii) The cotangent space ω_G of G at the zero-section is naturally isomorphic to

$$\operatorname{coker}\left(F_{\mathrm{M}^*(G)}\colon \mathrm{M}^*(G)^{(1)} \to \mathrm{M}^*(G)\right).$$

When G is a BT_1-group over k (i.e., $G \simeq X[p]$ for some p-divisible group X over k), we have an isomorphism

$$\omega_G \xrightarrow{\simeq} \operatorname{image}\left(\mathrm{V}_{\mathrm{M}^*(\mathrm{G})}\colon \mathrm{M}^*(\mathrm{G}) \to \mathrm{M}^*(\mathrm{G})^{(1)}\right),$$

because the sequence

$$\mathrm{M}^*(G)^{(1)} \xrightarrow{F_{\mathrm{M}^*(G)}} \mathrm{M}^*(G) \xrightarrow{V_{\mathrm{M}^*(G)}} \mathrm{M}^*(G)^{(1)}$$

is exact at the middle term.

B.3.5.4. The example $G = \mu_{p^n}$ illustrates the difference between CW and CW^u. We have $\mathrm{Hom}_k(\mu_{p^m}, CW^u) = 0$ because $\mathrm{Hom}_k(\mu_{p^m}, W_n) = 0$ for all $n \geqslant 1$.[8] However, $\mathrm{M}^*(\mu_{p^n}) = \mathrm{Hom}_k(\mu_{p^m}, CW_k)$ is isomorphic to $W/p^m W$, and $V\colon \mathrm{M}^*(\mu_{p^n}) \to \mathrm{M}^*(\mu_{p^n})^{(1)}$ is an isomorphism.

We now construct an injection $\xi\colon \mu_{p^m} \to CW_k$ corresponding to a generator of the W-module $\mathrm{M}^*(\mu_{p^n})$ such that $V_{CW} \circ \xi = \xi$. Let

$$E(Y) = \prod_{\substack{n \in \mathbb{N}_{\geqslant 1} \\ \gcd(n,p)=1}} (1 - Y^n)^{\mu(n)/n} = \exp\left(-\sum_{m \in \mathbb{N}} \frac{Y^{p^m}}{p^m}\right) \in \mathbb{Z}_{(p)}[\![Y]\!]$$

be the classical Artin-Hasse exponential in the variable Y, and let

$$\psi(T) = T + \sum_{i \geqslant 2} b_i T^i \in \mathbb{Z}_{(p)}[\![T]\!]$$

be the inverse of $1 - E(Y)$ in the sense that

$$1 - E(\psi(T)) = T, \quad \psi(1 - E(Y)) = Y.$$

The map $\xi\colon \mu_{p^n} \longrightarrow CW_k$ is determined on R-valued points by

$$1 - t \mapsto (\ldots, \psi(t), \psi(t), \psi(t)) \in CW_k(R)$$

for all $t \in R$ with $t^{p^n} = 0$ (i.e., $(1-t)^{p^n} = 1$).

To see that ξ respects the group law, it suffices to check that the compatibility of the group law with inverse map

$$(\ldots, y, y, y, y) \mapsto E(y) \in \mathbb{G}_{\mathrm{m}}^{\wedge}(R)$$

that is defined for all topologically nilpotent elements y in a complete separated topological $\mathbb{Z}_{(p)}$-algebra R having a fundamental system of neighborhoods of 0 consisting of open ideals. This compatibility follows immediately from the description of the group law on CW in terms of the power series $S(\underline{x}; \underline{y})$.

[8] This fact caused problems in treatments of covariant Dieudonné theory using $\varinjlim W_n \simeq CW^u$ before [**41**], so the case of finite commutative k-groups of multiplicative type was treated separately.

B.3.5.5. We have a functorial isomorphism $\mathrm{M}^*(G^D) \simeq \mathrm{Hom}_W(\mathrm{M}^*(G), K_0/W)$, where G^D is the Cartier dual of G and K_0 is the fraction field of W.[9] Under this isomorphism, the map $F_{\mathrm{M}^*(G^D)}\colon \mathrm{M}^*(G^D)^{(1)} \to \mathrm{M}^*(G^D)$ corresponds to the transpose of $V_{\mathrm{M}^*(G)}\colon \mathrm{M}^*(G) \to \mathrm{M}^*(G)^{(1)}$. That is, there is a functorial pairing

$$\langle\ ,\ \rangle_{M^*(G)}\colon \mathrm{M}^*(G) \times \mathrm{M}^*(G^D) \longrightarrow K_0/W$$

such that

$$\langle Fx, y\rangle_{\mathrm{M}^*(G)} = \langle x, Vy\rangle_{\mathrm{M}^*(G)^{(1)}}$$

for all $x \in \mathrm{M}^*(G)^{(1)}$ and $y \in \mathrm{M}^*(G)$, where $\langle\ ,\ \rangle_{\mathrm{M}^*(G)^{(1)}}$ is the composition

$$(W \otimes_{\sigma,W} \mathrm{M}^*(G)) \times (W \otimes_{\sigma,W} \mathrm{M}^*(G^D)) \xrightarrow{W\otimes_{\sigma,W}\langle\ ,\ \rangle_{\mathrm{M}^*}} W \otimes_{\sigma,W} K_0/W \xrightarrow{w\otimes a \mapsto w\,\sigma(a)} K_0/W$$

Let $j_G\colon G \xrightarrow{\simeq} (G^D)^D$ be the canonical isomorphism from G to its double Cartier dual. We claim that the diagram

$$\begin{array}{ccc} \mathrm{M}^*(G) \times \mathrm{M}^*(G^D) & \xrightarrow{u\otimes v \mapsto v \otimes j_G(u)} & \mathrm{M}^*(G^D) \times \mathrm{M}^*((G^D)^D) \\ \downarrow{\scriptstyle \langle\ ,\ \rangle_{M^*(G)}} & & \downarrow{\scriptstyle \langle\ ,\ \rangle_{M^*(G^D)}} \\ K_0/W & \xrightarrow[=]{} & K_0/W \end{array}$$

commutes. By construction this is functorial in G, so if $G \hookrightarrow G'$ is an inclusion of finite commutative k-groups with p-power order then the surjectivity of $\mathrm{M}^*(G') \to \mathrm{M}^*(G)$ allows us to replace G with G'. Any G occurs as a k-subgroup of an abelian variety over k (see [**6**, 3.1.1]), so G is a k-subgroup of a p-divisible group. Thus, we may assume $G = X[p^n]$ for a p-divisible group X over k. By rewriting the pairing for $G = X[p^n]$ as valued in $(K_0/W)[p^n] = W/(p^n)$, we may pass to the inverse limit over n to get a W-valued perfect duality between $\mathrm{M}^*(X)$ and $\mathrm{M}^*(X^t)$. It is equivalent to show that this latter pairing is symmetric with respect to the isomorphism $X \simeq X^{tt}$, which is exactly [**6**, 5.3.5] (due to [**6**, (5.3.1.5), (5.3.1.6)]).[10]

B.3.5.6. Define the classical *covariant* Dieudonné module $\mathrm{M}_*(G)$ of a commutative finite group scheme G over k of p-power order by

$$\mathrm{M}_*(G) := \mathrm{M}^*(G^D)$$

with $F_{\mathrm{M}_*(G)} = F_{\mathrm{M}^*(G^D)}$ and $V_{\mathrm{M}_*(G)} = V_{\mathrm{M}^*(G^D)}$. Alternatively, one can define

$$\mathrm{M}_*(G) := \mathrm{Hom}_W(\mathrm{M}^*(G), K_0/W)\,,$$

[9] We *define* this isomorphism via the crystalline comparison isomorphism [**6**, 4.2.14, 4.2.17] and crystalline duality isomorphism [**6**, (5.2.7.1)] (applicable due to [**6**, 1.1.12–1.1.13]). An isomorphism is also given by [**41**, Ch. III, §5], but we do not use it. Other constructions of Dieudonné modules admit such an isomorphism: see [**86**, 3.19] and [**29**, (∗∗) on p. 67, Thm. (e) on p. 69].

[10] When using the alternative construction of Dieudonné modules in [**86**], this commutative diagram for finite G is part of [**86**, 3.19]. See [**6**, 5.2.13] for an analogous (and more general) double duality statement for the duality isomorphism $\Phi_G : \mathbf{\Delta}(G)^\vee[-1] \xrightarrow{\simeq} \mathbf{\Delta}(G^D)$, where $\mathbf{\Delta}(G)$ is the crystalline Dieudonné *complex* of G whose H^1 is the crystalline Dieudonné module $\mathrm{E}^*(G)$ of G discussed in B.3.8.2 and related to $\mathrm{M}^*(G)$ in B.3.10.

Note that [**6**, 5.2.13] asserts that the double duality diagram for the crystalline Dieudonné complex (in a derived category) *anti*-commutes. This anti-commutativity is closely related to the more familiar sign discrepancy in the double duality statement for abelian schemes in 1.4.2.5, 1.4.3.4 and [**86**, 1.8]; see also [**6**, 5.1.9] for a crystalline version.

define $F_{\mathrm{M}_*(G)}$ to be the W-linear dual of $V_{\mathrm{M}^*(G)}$, and define $V_{\mathrm{M}_*(G)}$ to be the W-linear dual of $F_{\mathrm{M}^*(G)}$.

B.3.5.7. The functor M_* gives an equivalence from the abelian category of finite commutative k-group schemes of p-power order to the category of left $W(k)[F,V]$-modules with finite $W(k)$-length. We have a natural isomorphism

$$\mathrm{Lie}(G) \simeq \ker\left(V_{\mathrm{M}_*(G)} \colon \mathrm{M}_*(G) \to \mathrm{M}_*(G)^{(1)}\right).$$

When G is a BT_1-group over k, we have an isomorphism

$$\mathrm{Lie}(G) \simeq \mathrm{image}\left(F_{\mathrm{M}_*(G)} \colon \mathrm{M}_*(G)^{(1)} \to \mathrm{M}_*(G)\right)$$

because the sequence

$$\mathrm{M}_*(G)^{(1)} \xrightarrow{F_{\mathrm{M}_*(G)}} \mathrm{M}_*(G) \xrightarrow{V_{\mathrm{M}_*(G)}} \mathrm{M}_*(G)^{(1)}$$

is exact at the middle term.

B.3.5.8. We can extend the classical contravariant Dieudonné theory M^* to p-divisible groups over k, as follows. For any p-divisible group X over k, define

$$\mathrm{M}^*(X) := \varprojlim \mathrm{M}^*(X[p^n]),$$

where the transition maps $\mathrm{M}^*(X[p^{n+1}]) \longrightarrow \mathrm{M}^*(X[p^n])$ are induced by the inclusions $X[p^n] \hookrightarrow X[p^{n+1}]$. The F and V operators for $\mathrm{M}^*(X[p^n])$ pass to the limit and define

$$F_{\mathrm{M}^*(X)} \colon \mathrm{M}^*(X)^{(1)} \to \mathrm{M}^*(X) \quad \text{and} \quad V_{\mathrm{M}^*(X)} \colon \mathrm{M}^*(X) \to \mathrm{M}^*(X)^{(1)}.$$

It is known that $\mathrm{M}^*(X)$ is a free $W(k)$-module of finite rank, equal to the height of X. Moreover the functor M^* induces an anti-equivalence of categories, from the category of p-divisible groups over k to the category of left $W(k)[F,V]$-modules free of finite rank over $W(k)$.

B.3.5.9. Similarly, define

$$\mathrm{M}_*(X) := \varprojlim \mathrm{M}_*(X[p^n]),$$

where the transition maps $\mathrm{M}_*(X[p^{n+1}]) \longrightarrow \mathrm{M}_*(X[p^n])$ are induced from the faithfully flat surjections $X[p^{n+1}] \twoheadrightarrow X[p^n]$. The functor M_* induces an equivalence from the category of p-divisible groups over k to the category of left $W(k)[F,V]$-modules free of finite rank over $W(k)$. We have a functorial isomorphism

$$\mathrm{M}_*(X) \simeq \mathrm{Hom}_{W(k)}(\mathrm{M}^*(X), W(k));$$

the operators $F_{\mathrm{M}_*(X)}$ and $V_{\mathrm{M}_*(X)}$ correspond to the transpose of $V_{\mathrm{M}^*(X)}$ and the transpose of $F_{\mathrm{M}^*(X)}$ respectively.

B.3.5.10. One can recover the Dieudonné theory for finite group schemes over k from the Dieudonné theory for p-divisible groups (or abelian varieties), as follows. A theorem of Raynaud asserts that every finite locally free commutative group scheme G over a local ring R can be embedded into the Jacobian of a smooth proper curve over R (with geometrically connected fibers); see [**6**, 3.1.1].

Start with a short exact sequence $0 \to G \to X \xrightarrow{\rho} Y \to 0$, where G is a given finite group scheme over k of p-power order, X and Y are p-divisible groups over k, and ρ is an isogeny over k. We have a natural isomorphism

$$\ker\Big((K/W)\otimes_W \mathrm{M}^*(Y) \xrightarrow{1\otimes \mathrm{M}^*(\rho)} (K/W)\otimes_W \mathrm{M}^*(X)\Big) \xrightarrow{\simeq} \operatorname{coker}\Big(\mathrm{M}^*(Y) \xrightarrow{\mathrm{M}^*(\rho)} \mathrm{M}^*(X)\Big)$$

and we take $\mathrm{M}^*(G)$ to be $\operatorname{coker}\Big(\mathrm{M}^*(Y) \xrightarrow{\mathrm{M}^*(\rho)} \mathrm{M}^*(X)\Big)$.

B.3.6. Cartier theory. The Cartier theory [**14**] for smooth commutative formal groups was systematically documented in [**69**]; we recommend [**136**] and refer to it for more information. Although Cartier theory is valid for commutative smooth formal groups, we will only discuss the special case when the base ring R is an algebra over $\mathbb{Z}_{(p)}$.

B.3.6.1. The following smooth commutative formal group $\widehat{W}_R$ plays a key role, a sort of "free generator" for the category of smooth commutative formal groups over R. For any commutative complete separated topological ring S over R having a fundamental system of open neighborhoods of 0 consisting of open ideals, $\widehat{W}_R(S)$ is the subgroup of $W(S)$, consisting of all p-adic Witt vectors $\underline{a} = (a_0, a_1, a_2, \ldots)$ with all $a_i \in S$ with the following property: for every open ideal $I \subset S$, there exists an integer $m \geqslant 0$ (depending on $\underline{a}$ and I) such that the ideal of S generated by all a_i's with $i \geqslant m$ is contained in I.

B.3.6.2. Define the Cartier ring $\mathrm{Cart}_p(R)$ to be the *opposite* ring of the ring $\mathrm{End}_R(\widehat{W})$ of endomorphisms of $\widehat{W}_R$.

(1) The ring $\mathrm{Cart}_p(R)$ contains two elements $\mathcal{V}$ and $\mathcal{F}$ and is complete with respect to the filtration by right ideals $\mathcal{V}^m\mathrm{Cart}_p(R)$. Moreover $\mathrm{Cart}_p(R)$ contains a family of elements $[a]$ indexed by the elements $a \in R$.
(2) The following identities hold.
   - $[0] = 0_{\mathrm{Cart}_p(R)}$, $[1] = 1_{\mathrm{Cart}_p(R)}$,
   - $[a]\cdot[b] = [ab]$ for all $a, b \in R$,
   - $\mathcal{F}[a] = [a^p]\mathcal{F}$ for all $a \in R$,
   - $[a]\mathcal{V} = \mathcal{V}[a^p]$ for all $a \in R$,
   - $\mathcal{F}\mathcal{V} = p$.
(3) There is an injective ring homomorphism $j\colon W(R) \to \mathrm{Cart}_p(R)$ such that

$$j(a_0, a_1, a_2, \ldots) = \sum_{n\geqslant 0} \mathcal{V}^n[a_n]\mathcal{F}^n \qquad \text{for all } (a_0, a_1, a_2, \ldots) \in W(R)\,.$$

(4) Every element of $\mathrm{Cart}_p(R)$ has a unique expression as a convergent sum

$$\sum_{m,n\geqslant 0} \mathcal{V}^m[a_{m,n}]\mathcal{F}^n\,, \qquad a_{m,n}\in R \ \text{ for all } \ m,n\,,$$

with the property that for every $m\geqslant 0$ there exists an integer $c(m)\geqslant 0$ such that $a_{m,n}=0$ for all $n\geqslant c(m)$.

(5) A left $\mathrm{Cart}_p(R)$-module M is *$\mathcal{V}$-reduced* if $\mathcal{V}\colon M\to M$ is injective and the natural map

$$M\to \varprojlim M/(\mathcal{V}^m\cdot M)$$

is an isomorphism. A $\mathcal{V}$-reduced left $\mathrm{Cart}_p(R)$-module M is *$\mathcal{V}$-flat* if $M/\mathcal{V}M$ is a flat R-module.

B.3.6.3. The Cartier module of a smooth commutative formal group X over R is defined to be

$$\mathrm{C}_*(X) := \mathrm{Hom}_R(\widehat{W}_R, X)$$

endowed with the left $\mathrm{Cart}_p(R)$-module structure from the right $\mathrm{Cart}_p(R)$-module structure on $\widehat{W}_R$. We have a natural isomorphism

$$\mathrm{C}_*(X)/\mathcal{V}\cdot\mathrm{C}_*(X) \xrightarrow{\simeq} \mathrm{Lie}(X)\,.$$

B.3.6.4. The main theorem of Cartier theory asserts that the functor $X\mapsto C_*(X)$ is an equivalence, from the category of smooth commutative formal groups over R to the category of $\mathcal{V}$-reduced $\mathcal{V}$-flat left $\mathrm{Cart}_p(R)$-modules. The inverse functor is

$$M \rightsquigarrow \left(S\rightsquigarrow \widehat{W}(S)\widehat{\otimes}_{\mathrm{Cart}_p(R)}M\right)\,,$$

where S denotes an arbitrary complete commutative topological R-algebra and $\widehat{\otimes}$ denotes a suitably defined completed tensor product; see [**136**, 4.23, 3.28] for details. The completed tensor product $\widehat{\otimes}$ can be replaced by the usual tensor product if the left $\mathrm{Cart}_p(R)$-module M is finitely generated, in which case the corresponding smooth commutative formal group is finite-dimensional.

B.3.6.5. In the case when $R=k$ is a perfect field of characteristic p, $\mathrm{Cart}_p(k)$ is the completion of the Dieudonné ring $W(k)[\mathcal{F},\mathcal{V}]$ with respect to the filtration by the right ideals $\mathcal{V}^n\cdot W(k)[\mathcal{F},\mathcal{V}]$. Note that for $R=k$ a perfect field of characteristic $p>0$, the description of the Cartier ring $\mathrm{Cart}_p(k)$ in B.3.6.2 (4) simplifies to

$$\mathrm{Cart}_p(k)=W(k)[\mathcal{F}][\![\mathcal{V}]\!]\,,$$

the ring of all power series in the element $\mathcal{V}$ with coefficients in $W(k)[\mathcal{F}]$. The $\mathcal{V}$-reduced $\mathcal{V}$-flat left $\mathrm{Cart}_p(k)$-modules corresponding to p-divisible smooth commutative formal groups over k are exactly those left $D_k = W(k)[\mathcal{F},\mathcal{V}]$-modules M free of finite rank over $W(k)$ such that there exists an integer $m\geqslant 0$ with $\mathcal{V}^m\cdot M\subseteq p\cdot M$.

B.3.6.6. We can use the procedure in B.3.5.10 to define a *covariant* Dieudonné theory for finite group schemes over k, as follows.

For a *connected* finite commutative k-group scheme G of p-power order, embed G in a p-divisible group X over k to get a short exact sequence $0 \to G \to X \xrightarrow{\rho} Y \to 0$. Then we have a natural isomorphism

$$\ker\left((K_0/W)\otimes_W \mathrm{C}_*(X) \xrightarrow{1\otimes \mathrm{C}_*(\rho)} (K_0/W)\otimes_W \mathrm{C}_*(Y)\right) \xrightarrow{\sim} \operatorname{coker}\left(\mathrm{C}_*(X) \xrightarrow{\mathrm{C}_*(\rho)} \mathrm{C}_*(Y)\right)$$

and we define $\mathrm{C}_*(G)$ to be

$$\ker\left((K_0/W)\otimes_W \mathrm{C}_*(X) \xrightarrow{1\otimes \mathrm{C}_*(\rho)} (K_0/W)\otimes_W \mathrm{C}_*(Y)\right) .$$

For a general finite commutative k-group scheme G of p-power order, decompose G as a product

$$G = G_{\mathrm{conn}} \times G^{\text{ét}} ,$$

and define

$$\mathrm{C}_*(G) := \mathrm{C}_*(G_{\mathrm{conn}}) \oplus \operatorname{Hom}_W\left(\mathrm{C}_*((G^{\text{ét}})^D), K_0/W\right) .$$

B.3.6.7. The functor $X \rightsquigarrow \mathrm{C}_*(X)$ gives an equivalence from the category p-divisible groups over k to the category of left D_k-modules which are free of finite rank over $W(k)$. The functor $G \rightsquigarrow \mathrm{C}_*(G)$ gives an equivalence from the category of finite commutative k-group schemes of p-power order to the category of left D_k-modules of finite length over $W(k)$. Similarly, the contravariant functors $X \rightsquigarrow \mathrm{C}^*(X)$ and $G \rightsquigarrow \mathrm{C}^*(G)$ give *anti*-equivalences of categories.

B.3.6.8. Similar to the case of classical Dieudonné theory, we have functorial isomorphisms

$$\mathrm{C}_*(G^D) \simeq \operatorname{Hom}_W(\mathrm{C}_*(G), K_0/W)$$

respecting the F and V operators. We define a contravariant theory over k by

$$\mathrm{C}^*(G) := \operatorname{Hom}_W(\mathrm{C}_*(G), K_0/W)$$

for any finite commutative k-group scheme G of p-power order, and define

$$\mathrm{C}^*(X) := \varprojlim \mathrm{C}^*(X[p^n])$$

for any p-divisible group X over k, where the transition maps $\mathrm{C}^*(X[p^{n+1}]) \to \mathrm{C}^*(X[p^n])$ in the projective system are induced by the inclusions $X[p^n] \hookrightarrow X[p^{n+1}]$.

One can verify various compatibilities; for instance $\mathrm{C}^*(X)$ defined above for a connected p-divisible group X over k is naturally isomorphic to the $W(k)$-linear dual of the Cartier module $\mathrm{C}^*(X^\wedge)$ of the smooth commutative formal group $X^\wedge$ attached to X.

B.3.7. Zink's theory of displays. We refer to [**139**] for the theory of displays; see also B.4.7 for the notation for displays. Let R be an excellent commutative ring, and assume either that p is nilpotent in R or that R is complete with respect to an ideal and p is topologically nilpotent.[11] What we need is the relation between

- a display
$$\mathscr{P} = \big(P,\, Q,\, F_P : P \to P,\, F_{1,Q} : Q \to P,\, V^\sharp : P \to W(R) \otimes_{F,W(R)} P\big)$$
over R, and
- the Cartier module $\mathrm{C}_*(\mathrm{BT}(\mathscr{P}))$ of the formal completion of the formal p-divisible group $\mathrm{BT}(\mathscr{P})$ over R attached to $\mathscr{P}$.

The answer provided in [**139**, Introduction] (see just above Theorem 5 there) is

$$\mathrm{C}_*(\mathrm{BT}(\mathscr{P})) \simeq \big(\mathrm{Cart}_p(R) \otimes_{W(R)} P\big) \Big/ \begin{pmatrix} F \otimes x - 1 \otimes F_P(x), \\ V \otimes F_{1,Q}(y) - 1 \otimes \;\; y \end{pmatrix}_{x \in P,\, y \in Q}$$

where

$$\big(F \otimes x - 1 \otimes F_P(x),\, V \otimes F_{1,Q}(y) - 1 \otimes y\big)_{x \in P,\, y \in Q}$$

is the $\mathrm{Cart}_p(R)$-submodule of $\mathrm{Cart}_p(R) \otimes_{W(R)} P$, generated by elements of the form $F \otimes x - 1 \otimes F_P(x)$ for $x \in P$ and $V \otimes F_{1,Q}(y) - 1 \otimes y$ for $y \in Q$.

When $R = k \supset \mathbb{F}_p$ is a perfect field of characteristic $p > 0$, for $X = \mathrm{BT}(\mathscr{P})$ the above formulas simplify to

- $\mathrm{C}_*(X) \simeq P$,
- the $W(k)$-linear operator $F_{\mathrm{C}_*(X)} : W(k) \otimes_{\sigma, W(k)} \mathrm{C}_*(X) \longrightarrow \mathrm{C}_*(X)$ is
$$F_{\mathrm{C}_*(X)}(w \otimes x) = w \cdot F_P(x)$$
for all $w \in W(k)$ and $x \in P$ after identifying $\mathrm{C}_*(X)$ with P,
- the $W(k)$-linear operator
$$V_{\mathrm{C}_*(X)} : \mathrm{C}_*(X) \longrightarrow W(k) \otimes_{\sigma, W(k)} \mathrm{C}_*(X)$$
is identified with $V^\sharp$.

B.3.8. Crystalline theory. We will restrict ourselves to the case of finite commutative group schemes G of p-power order or p-divisible groups X over a perfect field k of characteristic $p > 0$. Moreover, we will take an overly simplistic view about crystalline theory and look only at the "values" of the Dieudonné crystals when evaluated at surjections $W(k) \twoheadrightarrow k$ with the standard divided power structure on the kernel $pW(k)$. Some extra care is required when $p = 2$ (due to the possible failure of nilpotence of the divided power structure in such cases); these matters are left to the interested reader.

[11] The excellence hypothesis on R in [**139**] turned out to be superfluous, as was shown in [**68**].

B.3.8.1. For a p-divisible group X over k, a crystal $\mathbb{E}_*(X)$ attached to X is constructed in [**75**] via the crystalline nature of the universal vector extension of (deformations of) X. One gets a $W(k)$-module[12]

$$\mathrm{E}_*(X) := \mathbb{E}_*(X)_{\mathrm{Spec}(k)\hookrightarrow\mathrm{Spec}(W(k))}$$

and $W(k)$-linear maps

$$F_{\mathrm{E}_*(X)} : \mathrm{E}_*(X)^{(1)} \to \mathrm{E}_*(X) \quad \text{and} \quad V_{\mathrm{E}_*(X)} : \mathrm{E}_*(X) \to \mathrm{E}_*(X)^{(1)}$$

by functoriality applied to the relative Frobenius morphism $\mathrm{Fr}_{X/k}\colon X \to X^{(p)}$ for X and the Verschiebung homomorphism $\mathrm{Ver}_{X/k}\colon X^{(p)} \to X$.

B.3.8.2. For a finite commutative k-group scheme G of p-power order, one has a crystal

$$\mathbb{E}^*(G) := \mathscr{E}xt^1_{k/W(k)}(\underline{G}, \mathcal{O}_{k/W(k)})$$

and functorial morphisms

$$F\colon \sigma^*\mathbb{E}^*(G) \to \mathbb{E}^*(G) \qquad \text{and} \qquad V\colon \mathbb{E}^*(G) \to \sigma^*\mathbb{E}^*(G)$$

defined in [**6**, 3.1.5]. Let

$$\mathrm{E}_*(G) := \mathbb{E}^*(G)_{\mathrm{Spec}(k)\hookrightarrow\mathrm{Spec}(W(k))}$$

and let

$$F_{\mathrm{E}_*(G)}\colon \mathrm{E}_*(G)^{(1)} \to \mathrm{E}_*(G) \qquad \text{and} \qquad V_{\mathrm{E}_*(G)}\colon \mathrm{E}_*(G) \to \mathrm{E}_*(G)^{(1)}$$

be the $W(k)$-linear maps induced by the respective morphisms F and V for the crystal $\mathbb{E}^*(G)$.

B.3.8.3. For a p-divisible group X over k, define

$$\mathrm{E}^*(X) = \varprojlim \mathrm{E}^*(X[p^n])$$

as before. It is shown in [**6**, §3.2 ,§3.3] that the above definition coincides with the $W(k)$-linear dual of $\mathrm{E}_*(X)$, defined using the universal vector extension, so we will stick to the construction in [**6**].

As before, define a covariant theory for finite commutative k-group schemes of p-power order by

$$\mathrm{E}_*(G) := \mathrm{Hom}_{W(k)}(\mathrm{E}^*(G), K_0/W);$$

this is naturally isomorphic to $\mathrm{E}^*(G^D)$ and is compatible with the induced operators F and V. Similarly define

$$\mathrm{E}_*(X) := \varprojlim \mathrm{E}_*(X[p^n])\,,$$

for any p-divisible group X over k, with the induced operators F and V. We get covariant and a contravariant theories, for finite commutative k-group schemes of p-prime power and for p-divisible groups over k.

---

[12]Strictly speaking, we need to assume $p > 2$ so that the divided power structure on the ideal $pW(k) \subset W(k)$ is nilpotent.

B.3.9. Comparison of crystalline and Cartier theories. The left $W(k)[F,V]$-module $\mathrm{C}_*(X)$ for a connected p-divisible group X over k defined via Cartier theory is naturally isomorphic to the $W(k)[F,V]$-module underlying a display $\mathscr{P}$ over k which "produces" X as we saw in B.3.7. This coincides with the crystalline theory when both are applicable: there is a functorial isomorphism

$$\mathrm{C}_*(X) \simeq \mathbb{E}_*(X)_{\mathrm{Spec}(k)\hookrightarrow \mathrm{Spec}(W(k))}$$

for any connected p-divisible group X over k; see [**12**, (5.8)] and [**139**, Lemma 93]. Therefore, the covariant Dieudonné theory $\mathrm{C}_*(\cdot)$ obtained from Cartier theory coincides with the theory $\mathrm{E}_*(\cdot)$ obtained from the crystalline theory. Consequently, the contravariant theories $\mathrm{C}^*(\cdot)$ and $\mathrm{E}^*(\cdot)$ also coincide.

From now on we will suppress the notations $\mathrm{C}_*(\cdot)$ and $\mathrm{C}^*(\cdot)$, replacing them with $\mathrm{E}_*(\cdot)$ and $\mathrm{E}^*(\cdot)$. Note that we have "identified" connected p-divisible groups and their associated smooth commutative formal groups in the notation; we will do this whenever such abuse does not lead to confusion.

B.3.10. Comparison of the classical and crystalline theories. The basic comparison result between the classical and crystalline theories is found in [**6**, 4.2.14]: there exist functorial isomorphisms

$$\mathrm{E}^*(G) \simeq \mathrm{M}^*(G)^{(1)}, \quad \mathrm{E}_*(G) \simeq \mathrm{M}_*(G)^{(1)}$$

for every finite commutative k-group scheme G with p-power order. Taking the limit, we get functorial isomorphisms

$$\mathrm{E}^*(X) \simeq \mathrm{M}^*(X)^{(1)}, \quad \mathrm{E}_*(X) \simeq \mathrm{M}_*(X)^{(1)}$$

for every p-divisible group X over k.

Remark. The isomorphism $\mathrm{M}^*(X) \xrightarrow{\sim} \mathrm{E}^*(X)$ for any p-divisible group X over k given in [**74**, (15.3)] is σ-linear, for the $W(k)$-module structures explained previously. (The $W(k)$-module structure for either side of the isomorphism was not specified in [**74**, (15.3)].)

B.3.11. Basic differential invariants.

B.3.11.1. Let G be a finite commutative k-group scheme with p-power order.

- Write ℓ^G for the *co-Lie complex* of G over k, defined as the pull-back $\epsilon^*\mathrm{L}_\bullet^{G/k}$ of the cotangent complex $\mathrm{L}_\bullet^{G/k}$ of G along the zero section ϵ of G over k.
- The *Lie complex* $\ell^{G^\vee}$ of G is defined to be the k-linear (derived) dual of ℓ^G.
- The cohomologies of these two complexes are

$$\omega_G := \mathrm{H}^0(\ell^G), \quad \mathrm{n}_G := \mathrm{H}^{-1}(\ell^G); \quad \mathrm{Lie}(G) := \mathrm{H}^0(\ell^{G^\vee}), \quad \nu_G := \mathrm{H}^1(\ell^{G^\vee}).$$

Clearly we have natural isomorphisms $\mathrm{Lie}(G) \simeq \omega_G^\vee$ and $\nu_G \simeq \mathrm{n}_G^\vee$.

B.3.11.2. Formulas for the basic invariants. Let G be a finite commutative k-group scheme with p-power order. The invariants for G and its Cartier dual G^D

are given in terms of $\mathrm{E}_*(G)$ and $\mathrm{E}^*(G)$ as follows.

$$(1)\ \omega_G = \mathrm{Lie}(G)^\vee \simeq \left(\mathrm{coker}\left(\mathrm{E}^*(G)^{(1)} \xrightarrow{F_{\mathrm{E}^*(G)}} \mathrm{E}^*(G)\right)\right)^{(-1)}$$

$$(2)\ \nu_{G^D} \simeq \left(\mathrm{coker}\left(\mathrm{E}^*(G) \xrightarrow{V_{\mathrm{E}^*(G)}} \mathrm{E}^*(G)^{(1)}\right)\right)^{(-1)}$$

$$(3)\ \mathrm{Lie}(G) = \omega_G^\vee \simeq \left(\ker\left(\mathrm{E}_*(G) \xrightarrow{V_{\mathrm{E}_*(G)}} \mathrm{E}_*(G)^{(1)}\right)\right)^{(-1)}$$

$$(4)\ \mathrm{n}_{G^D} \simeq \left(\ker\left(\mathrm{E}_*(G)^{(1)} \xrightarrow{F_{\mathrm{E}_*(G)}} \mathrm{E}_*(G)\right)\right)^{(-1)}$$

$$(5)\ \mathrm{Lie}(G^D) \simeq \left(\ker\left(\mathrm{E}^*(G) \xrightarrow{V_{\mathrm{E}^*(G)}} \mathrm{E}^*(G)^{(1)}\right)\right)^{(-1)}$$

$$(6)\ \mathrm{n}_G \simeq \left(\ker\left(\mathrm{E}^*(G)^{(1)} \xrightarrow{F_{\mathrm{E}^*(G)}} \mathrm{E}^*(G)\right)\right)^{(-1)}$$

$$(7)\ \omega_{G^D} \simeq \left(\mathrm{coker}\left(\mathrm{E}_*(G)^{(1)} \xrightarrow{F_{\mathrm{E}_*(G)}} \mathrm{E}_*(G)\right)\right)^{(-1)}$$

$$(8)\ \nu_G \simeq \left(\mathrm{coker}\left(\mathrm{E}_*(G) \xrightarrow{V_{\mathrm{E}_*(G)}} \mathrm{E}_*(G)^{(1)}\right)\right)^{(-1)}$$

Note that (5)–(8) are dual to (1)–(4) and can be obtained from (1)–(4) with G replaced by its Cartier dual.

B.3.11.3. We have a 6-term exact sequence

$$0 \to \mathrm{n}_G \xrightarrow{\alpha} \mathrm{E}^*(G) \xrightarrow{F_{\mathrm{E}^*(G)}{}^{(-1)}} \mathrm{Lie}(G^D) \xrightarrow{\delta} \omega_G \xrightarrow{V_{\mathrm{E}^*(G)}{}^{(-1)}} \mathrm{E}^*(G) \xrightarrow{\beta} \nu_{G^D} \to 0$$

where α is the natural injection, β is the natural surjection, and δ is induced by $\mathrm{id}_{\mathrm{E}^*(G)^{(-1)}}$. Dually, we have a similar 6-term exact sequence

$$0 \to \mathrm{n}_{G^D} \to \mathrm{E}_*(G) \xrightarrow{F_{\mathrm{E}_*(G)}{}^{(-1)}} \mathrm{Lie}(G) \to \omega_{G^D} \xrightarrow{V_{\mathrm{E}_*(G)}{}^{(-1)}} \mathrm{E}_*(G) \to \nu_G \to 0$$

These two exact sequences are special cases of the 6-term exact sequences for finite commutative group schemes given in [**6**, 3.2.10] and part (i) in the subsequent Remarque there.

B.3.12. Remark. When G is a BT_1-group, or equivalently $G \simeq X[p]$ for some p-divisible group over k, the first exact sequence in B.3.11.3 splits into two short exact sequences

$$0 \longrightarrow \omega_G \xrightarrow{V_{\mathrm{E}^*(G)}{}^{(-1)}} \mathrm{E}^*(G) \xrightarrow{\beta} \nu_{G^D} \longrightarrow 0$$

and

$$0 \longrightarrow \mathrm{n}_G \xrightarrow{\alpha} \mathrm{E}^*(G) \xrightarrow{F_{\mathrm{E}^*(G)}{}^{(-1)}} \mathrm{Lie}(G^D) \longrightarrow 0$$

which are isomorphic to each other:

$$\begin{CD}
0 @>>> \omega_G @>{{V_{\mathrm{E}^*(G)}}^{(-1)}}>> \mathrm{E}^*(G) @>{\beta}>> \nu_{G^D} @>>> 0 \\
@. @V{{V_{\mathrm{E}^*(G)}}^{(-1)}}VV @V{=}VV @VV{{F_{\mathrm{E}^*(G)}}^{(-1)}}V @. \\
0 @>>> \mathrm{n}_G @>{\alpha}>> \mathrm{E}^*(G) @>{{F_{\mathrm{E}^*(G)}}^{(-1)}}>> \mathrm{Lie}(G^D) @>>> 0\,.
\end{CD}$$

Similarly, the second 6-term exact sequence in B.3.11.3 splits into two isomorphic short exact sequences

$$\begin{CD}
0 @>>> \omega_{G^D} @>{{V_{\mathrm{E}_*(G)}}^{(-1)}}>> \mathrm{E}_*(G) @>>> \nu_G @>>> 0 \\
@. @V{{V_{\mathrm{E}_*(G)}}^{(-1)}}VV @V{=}VV @VV{{F_{\mathrm{E}_*(G)}}^{(-1)}}V @. \\
0 @>>> \mathrm{n}_{G^D} @>>> \mathrm{E}_*(G) @>{{F_{\mathrm{E}_*(G)}}^{(-1)}}>> \mathrm{Lie}(G) @>>> 0
\end{CD}$$

when G is a BT_1-group over k.

B.3.13. Example. (Dieudonné modules of α-group schemes)
We have functorial isomorphisms $\mathrm{M}^*(G) \simeq \omega_G$ and $\mathrm{E}^*(G) \simeq \omega_G^{(1)} \simeq \mathrm{n}_G \simeq \nu_{G^D}$ for finite commutative k-group schemes G of α-type, explained in B.3.13.1 below.

When $G = \alpha_p$, both $\mathrm{M}^*(G)$ and $\mathrm{E}^*(G)$ are one-dimensional vector spaces over k, but the induced actions of $\mathrm{End}_k(\alpha_p) = k$ on these two Dieudonné modules are different. The endomorphism ring $\mathrm{End}_k(\alpha_p)$ of α_p is naturally identified with k, and the ring homomorphism

$$k = \mathrm{End}_k(\alpha_p) \longrightarrow \mathrm{End}_{W(k)}(\mathrm{M}^*(\alpha_p)) = k$$

is the identity map on k. The similarly-defined ring homomorphism

$$k = \mathrm{End}_k(\alpha_p) \longrightarrow \mathrm{End}_{W(k)}(\mathrm{E}^*(\alpha_p)) = k$$

is the Frobenius map $a \mapsto a^p$.

B.3.13.1. Suppose that G is a commutative finite group scheme over k of α-type, in the sense that both the relative Frobenius homomorphism $\mathrm{Fr}_G \colon G \to G^{(p)}$ and the Verschiebung homomorphism $\mathrm{Ver}_G \colon G^{(p)} \to G$ are zero; see also 3.1.9 and 3.1.10. The 6-term exact sequence for $\mathrm{E}^*(G)$ gives isomorphisms

$$\mathrm{n}_G \xrightarrow{\simeq} \mathrm{E}^*(G) \quad \text{and} \quad \mathrm{E}^*(G) \xrightarrow{\simeq} \nu_{G^D}\,.$$

On the other hand, we have a natural isomorphism

$$\mathrm{E}^*(G) \simeq k \otimes_{\sigma,k} \omega_G = \omega_G^{(1)}$$

because $\omega_G \simeq \mathrm{M}^*(G)$.

B.3.13.2. We can see the above natural isomorphisms "in action" in the context of one-dimensional abelian varieties. Let E be a supersingular elliptic curve over $k = \overline{\mathbb{F}}_p$, and let $G = \ker(\mathrm{Fr}_E : E \to E^{(p)})$ be the unique subgroup scheme of order p in E; it is isomorphic to α_p. Because the relative Frobenius Fr_E induces the zero map on the Lie algebras, the co-Lie complex ℓ^G of G is represented by the 2-term complex

$$\omega_{E^{(p)}} \xrightarrow{0} \omega_E$$

with $\omega_{E^{(p)}}$ in degree -1 and ω_E in degree 0. Thus,

$$\omega_G = \mathrm{H}^0(\ell^G) \simeq \omega_E \quad \text{and} \quad \mathrm{n}_G = \mathrm{H}^{-1}(\ell^G) \simeq \omega_{E^{(p)}} \simeq \omega_E^{(1)}.$$

Combined with the natural isomorphisms $\mathrm{M}^*(G) \simeq \omega_G$ and $\mathrm{E}^*(G) \simeq \mathrm{n}_G$ we saw in B.3.13.1, we get natural isomorphisms $\mathrm{M}^*(G) \simeq \omega_G \simeq \omega_E$ and $\mathrm{E}^*(G) \simeq \mathrm{n}_G \simeq \omega_{E^{(p)}} \simeq \omega_E^{(1)}$.

Again we see that the functorially defined actions of the endomorphism ring $\mathrm{End}(E)$ on $\mathrm{E}^*(G)$ and $\mathrm{M}^*(G)$ differ by a Frobenius twist in the following sense. We have

$$h_{\mathrm{E}^*(G)} = \mathrm{Fr}_p \circ h_{\mathrm{M}^*(G)},$$

where $\mathrm{Fr}_p : k \to k$ is the Frobenius automorphism $x \mapsto x^p$ of k and $h_{\mathrm{E}^*(G)}$ and $h_{\mathrm{M}^*(G)}$ are ring homomorphisms defined by the respective displayed formulas

$$h_{\mathrm{E}^*(G)}\colon \mathrm{End}(E) \to \mathrm{End}_{W(k)}(\mathrm{E}^*(G)) \simeq \mathrm{End}_k(\omega_E^{(1)}) = k$$

and

$$h_{\mathrm{M}^*(G)}\colon \mathrm{End}(E) \to \mathrm{End}_{W(k)}(\mathrm{M}^*(G)) \simeq \mathrm{End}_k(\omega_E) = k.$$

Remark. The diligent reader may have noticed an apparent puzzle when examining the dual version of the example in B.3.13.2. Forming the dual of the short exact sequence

$$0 \longrightarrow G \longrightarrow E \xrightarrow{\mathrm{F}_{E/k}} E^{(p)} \longrightarrow 0,$$

auto-duality via the unique principal polarization of E yields a short exact sequence

$$0 \longrightarrow G^D \longrightarrow E^{(p)} \xrightarrow{\mathrm{V}_{E/k}} E \longrightarrow 0.$$

The first exact sequence induces a natural isomorphism $\omega_G \simeq \omega_E$, while the second exact sequence induces a natural isomorphism $\omega_{G^D} \simeq \omega_{E^{(p)}}$. These isomorphisms suggest that the ring homomorphism $h_{G^D} : \mathrm{End}(E) \to k$ induced by the natural action of $\mathrm{End}(E)$ on ω_{G^D} is equal to the composition $\mathrm{Fr}_p \circ h_G$, where $h_G : \mathrm{End}(E) \to k$ is the ring homomorphism induced by the natural action of $\mathrm{End}(E)$ on ω_G and $\mathrm{Fr}_p : k \to k$ is the Frobenius automorphism of k. That would be in conflict with the Cartier duality between G and G^D.

This puzzle is resolved by noting that the auto-duality of E coming from the principal polarization induces a Rosati involution ι on $\mathrm{End}(E)$ (the same involution as obtained from the negative of the principal polarization), and $h_G \circ \iota = \mathrm{Fr}_p \circ h_G$. We also remark that $(\mathrm{Fr}_p)^2 \circ h_G = h_G$ because the ring $\mathrm{End}(E)$ is the maximal order of a quaternion division algebra over $\mathbb{Q}_p$ and the quotient of $\mathrm{End}(E)$ by its maximal ideal is isomorphic to $\mathbb{F}_{p^2}$.

B.4. p-adic Hodge theory and a formula for the closed fiber

The goal of this section is an *explicit formula* for the functor MCF in B.4.3, which sends an input linear algebra datum $(\mathfrak{M}, \phi_{\mathfrak{M}}, \psi_{\mathfrak{M}})$ in p-adic Hodge theory (whose definition is recalled in B.4.2) to the classical covariant Dieudonné module $\mathrm{M}_*(G_k)$ of the closed fiber G_k of the finite flat $\mathcal{O}_K$-group scheme G attached to the Kisin module $(\mathfrak{M}, \phi_{\mathfrak{M}}, \psi_{\mathfrak{M}})$ (for K as defined below).

B.4.1. Notation. Throughout this section k is a perfect field of characteristic $p > 0$, K is a finite totally ramified extension field of $K_0 = W(k)[1/p]$, $e = e(K/K_0)$ is the ramification index of K/K_0, $\pi \in \mathcal{O}_K$ is a fixed generator of the maximal ideal of $\mathcal{O}_K$, and $E(u) = u^e + a_{e-1}\,u^{e-1} + \ldots + a_1\,u + a_0$ is the Eisenstein minimal polynomial of π over K_0. In particular, $a_i \in p\,W(k)$ for all $i = 0, 1, \ldots, e-1$, and $c := a_0/p \in W(k)^\times$.

Let $\mathfrak{S} = W(k)[\![u]\!]$, and let $\phi\colon \mathfrak{S} \to \mathfrak{S}$ be the local endomorphism of $\mathfrak{S}$ such that $\phi(u) = u^p$ and $\phi|_{W(k)} = \sigma$ (the Frobenius automorphism of $W(k)$). Let S be the p-adic completion of the subring $W[u, E(u)^n/n!]_{n\geqslant 0} \subset K_0[u]$ generated by $W(k)[u]$ and all elements of the form $E(u)^n/n!$ for $n \geqslant 0$. It is easy to see that the natural inclusion $W(k)[u] \hookrightarrow S$ extends to a natural inclusion $\mathfrak{S} \hookrightarrow S$. The ring endomorphism ϕ of $\mathfrak{S}$ extends uniquely to a ring endomorphism of S, again denoted by ϕ.

Finally, we define $\mathrm{Fil}^1 S$ to be the kernel of the surjective $W(k)$-linear ring homomorphism $\epsilon\colon S \twoheadrightarrow \mathcal{O}_K$ such that $\epsilon(u) = \pi$. It is clear that $\mathrm{Fil}^1 S$ is the p-adic completion of the ideal of S generated by all elements of the form $E(u)^n/n!$ for some $n \geqslant 1$. Note that $\phi(E(u))/p \in S^\times$ because

$$(\phi(E(u))/p) - \sigma(c) \in \mathrm{Fil}^1 S + uW(k)[u] + pW(k)[u].$$

B.4.2. Recall from [**59**] that $(\mathrm{Mod}/\mathfrak{S})^c$ is the additive category whose objects are triples $(\mathfrak{M}, \phi_{\mathfrak{M}} : \mathfrak{M} \to \mathfrak{M}, \psi_{\mathfrak{M}} : \mathfrak{M} \to \phi^*\mathfrak{M})$ satisfying the following properties.

(1) $\mathfrak{M}$ is a finitely generated $\mathfrak{S}$-module killed by a power of p and having projective dimension $\leqslant 1$ over $\mathfrak{S}$.

(2) $\phi_{\mathfrak{M}} : \mathfrak{M} \to \mathfrak{M}$ is a ϕ-linear endomorphism of $\mathfrak{M}$; i.e., $\phi_{\mathfrak{M}}(am) = \phi(a)\phi_{\mathfrak{M}}(m)$ for every $a \in \mathfrak{S}$ and $m \in \mathfrak{M}$.[13]

(3) $\mathrm{coker}\,(1 \otimes \phi_{\mathfrak{M}} : \phi^*\mathfrak{M} \to \mathfrak{M})$ is killed by $E(u)$.

(4) The following diagram

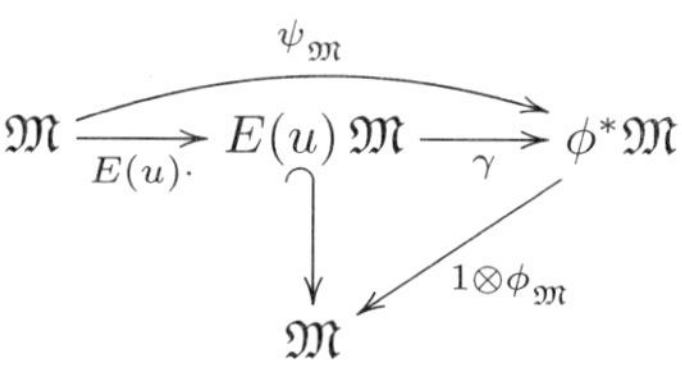

commutes. In other words, $\gamma\colon E(u)\mathfrak{M} \to \phi^*\mathfrak{M}$ is the restriction to $E(u)\mathfrak{M}$ of the inverse of the linearization $1 \otimes \phi_{\mathfrak{M}}$ of $\phi_{\mathfrak{M}}$, and ψ is the composition of γ with "multiplication with $E(u)$". Note that the $\mathfrak{S}$-linear map $\psi\colon \mathfrak{M} \to \phi^*\mathfrak{M}$ is uniquely defined for any pair $(\mathfrak{M}, \phi_{\mathfrak{M}})$ satisfying conditions (1)–(3) above.

(5) There exists a positive integer n such that the image of the composition

$$\psi^n_{\mathfrak{M}} := \phi^{n-1\,*}\psi_{\mathfrak{M}} \circ \phi^{n-2\,*}\psi_{\mathfrak{M}} \circ \cdots \circ \psi_{\mathfrak{M}} : \mathfrak{M} \longrightarrow \phi^{n\,*}\mathfrak{M} = \mathfrak{S}\otimes_{\mathfrak{S},\phi^n}\mathfrak{M}$$

is contained in $(p, u)\cdot\phi^{n\,*}\mathfrak{M} = (p, u)\cdot(\mathfrak{S}\otimes_{\mathfrak{S},\phi^n}\mathfrak{M})$.

(Loosely speaking, this says that $\psi_{\mathfrak{M}}$ is nilpotent modulo (p, u).)

A morphism in $(\mathrm{Mod}/\mathfrak{S})^c$ from an object $(\mathfrak{M}_1, \phi_1, \psi_1)$ to an object $(\mathfrak{M}_2, \phi_2, \psi_2)$ is a $\mathfrak{S}$-module homomorphism $h\colon \mathfrak{M}_1 \to \mathfrak{M}_2$ such that $\phi_{\mathfrak{M}_2} \circ h = h \circ \phi_{\mathfrak{M}_1}$; the last condition implies that $\psi_{\mathfrak{M}_2} \circ h = \phi^* h \circ \psi_{\mathfrak{M}_1}$. The category $(\mathrm{Mod}/\mathfrak{S})^c$ is an exact category.

[13] We often shorten $\phi_{\mathfrak{M}}$ to ϕ and $\psi_{\mathfrak{M}}$ to ψ if there is no danger of confusion.

It is proved in [**59**, 1.3.9] that there is an equivalence

$$\mathcal{G}_{\mathrm{Kis}}\colon\ (\mathrm{Mod}/\mathfrak{S})^c \xrightarrow{\simeq} (p\text{-}\mathrm{Gr}/\mathcal{O}_K)^c$$

of exact categories, where $(p\text{-}\mathrm{Gr}/\mathcal{O}_K)^c$ is the exact category of connected finite flat commutative group schemes over $\mathcal{O}_K$ of p-power order.

B.4.3. Let $(p\text{-}\mathrm{Gr}/k)^c$ be the abelian category of connected commutative finite group schemes of p-power order over k. Let $(\mathrm{Mod}/D_k)^c$ be the abelian category of left D_k-modules of finite $W(k)$-length on which V acts as a unipotent σ^{-1}-linear operator. Consider the functor

$$\mathrm{MCF} = \mathrm{M}_* \circ \mathrm{CF} \circ \mathcal{G}_{\mathrm{Kis}}$$

in the following commutative diagram of functors

$$\begin{array}{ccc}
(\mathrm{Mod}/\mathfrak{S})^c & \xrightarrow{\ \mathrm{MCF}\ } & (\mathrm{Mod}/D_k)^c \\
{\scriptstyle \mathcal{G}_{\mathrm{Kis}}}\downarrow & & \uparrow{\scriptstyle \mathrm{M}_*} \\
(p\text{-}\mathrm{Gr}/\mathcal{O}_K)^c & \xrightarrow[\mathrm{CF}]{} & (p\text{-}\mathrm{Gr}/k)^c
\end{array}$$

where CF is the functor which sends every connected finite flat commutative $\mathcal{O}_K$-group scheme G of p-power order to the closed fiber G_k of G. The group scheme G_k is a connected commutative finite group scheme over k, an object of the category $(p\text{-}\mathrm{Gr}/k)^c$. We need an explicit "formula" for the functor

$$\mathrm{MCF}\colon (\mathfrak{M}, \phi_{\mathfrak{M}}, \psi_{\mathfrak{M}}) \ \mapsto\ \mathrm{M}_*(\text{closed fiber of } G)\,,$$

where G is the finite flat group scheme attached to a connected Kisin module $(\mathfrak{M}, \phi_{\mathfrak{M}}, \psi_{\mathfrak{M}})$.

B.4.4. The functor $\mathcal{G}_{\mathrm{Kis}}$ is constructed from an equivalence of categories

$$p\text{-}\mathrm{Div}_{\mathrm{Kis}}\colon\ \mathrm{BT}^{\phi,f}_{/\mathfrak{S}} \xrightarrow{\simeq} (p\text{-}\mathrm{div}/\mathcal{O}_K)^f\ ,$$

similar to $\mathcal{G}_{\mathrm{Kis}}$, where $(p\text{-}\mathrm{div}/\mathcal{O}_K)^f$ is the category of connected p-divisible groups over $\mathcal{O}_K$ and $\mathrm{BT}^{\phi,f}_{/\mathfrak{S}}$ is a category of linear algebra data designed for describing p-divisible groups (and its definition is recalled in B.4.5 below).

The above functor $p\text{-}\mathrm{Div}_{\mathrm{Kis}}$, defined in [**59**, 1.2.8], is a composition of two functors

$$\begin{array}{ccc}
\mathrm{BT}^{\phi,f}_{/\mathfrak{S}} & \xrightarrow[\simeq]{\ p\text{-}\mathrm{Div}_{\mathrm{Kis}}\ } & (p\text{-}\mathrm{div}/\mathcal{O}_K)^f\ . \\
{\scriptstyle S\text{-}\mathrm{Win}}\ \searrow^{\simeq} & & \nearrow^{\simeq}\ {\scriptstyle \mathrm{BT}\circ\mathrm{Dsp}} \\
 & (\mathrm{Win}/S) &
\end{array}$$

We proceed to recall the definition of the categories $\mathrm{BT}^{\phi,f}_{/\mathfrak{S}}$ and (Win/S), and the functors S-Win, BT, and Dsp.

B.4.5. The definition of the exact category $\mathrm{BT}^{\phi,f}_{/\mathfrak{S}}$ is similar to that of $(\mathrm{Mod}/\mathcal{O}_K)^{\mathrm{c}}$. Its objects are quadruples $(\mathfrak{M}, \phi_{\mathfrak{M}} : \mathfrak{M} \to \mathfrak{M}, \psi_{\mathfrak{M}} : \mathfrak{M} \to \phi^*\mathfrak{M})$ satisfying the following properties.

(1) $\mathfrak{M}$ is a free $\mathfrak{S}$-module of finite rank.
(2) $\phi_{\mathfrak{M}} : \mathfrak{M} \to \mathfrak{M}$ is a ϕ-linear endomorphism of $\mathfrak{M}$; i.e., $\phi_{\mathfrak{M}}(am) = \phi(a)\phi_{\mathfrak{M}}(m)$ for all $a \in \mathfrak{S}$ and $m \in \mathfrak{M}$.
(3) $\mathrm{coker}\,(1 \otimes \phi_{\mathfrak{M}} : \phi^*\mathfrak{M} \to \mathfrak{M})$ is killed by $E(u)$.
(4) The diagram

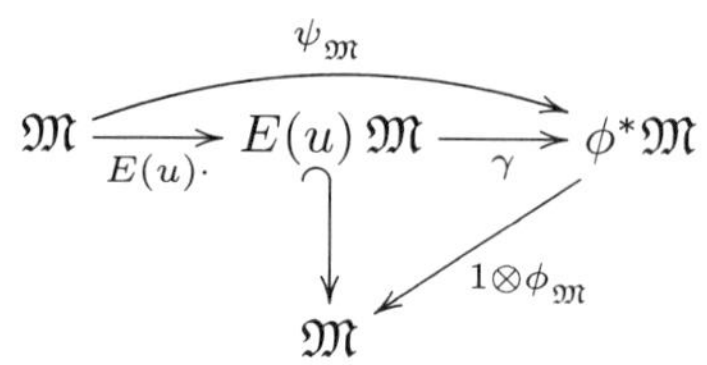

commutes.
(5) There exists a positive integer n such that the image of the composition

$$\psi^n_{\mathfrak{M}} := \phi^{n-1\,*}\psi_{\mathfrak{M}} \circ \phi^{n-2\,*}\psi_{\mathfrak{M}} \circ \cdots \circ \psi_{\mathfrak{M}} : \mathfrak{M} \longrightarrow \phi^{n\,*}\mathfrak{M}$$

is contained in $(p,u)\cdot\phi^{n\,*}\mathfrak{M}$.

A *morphism* in $\mathrm{BT}^{\phi,f}_{/\mathfrak{S}}$ from an object $(\mathfrak{M}_1, \phi_1, \psi_1)$ to an object $(\mathfrak{M}_1, \phi_2, \psi_2)$ is a $\mathfrak{S}$-module homomorphism $h\colon \mathfrak{M}_1 \to \mathfrak{M}_2$ such that $\phi_{\mathfrak{M}_2} \circ h = h \circ \phi_{\mathfrak{M}_1}$; the last condition implies that $\psi_{\mathfrak{M}_2} \circ h = \phi^* h \circ \psi_{\mathfrak{M}_1}$.

B.4.6. We recall the definition of the category (Win/S) of S-windows in [**137**]. An S-*window* is a quintuple $(\mathcal{M}, \mathcal{M}_1, \Phi, \Phi_1, \Psi)$ with the following properties.

(1) $\mathcal{M}$ is a finitely generated projective S-module.
(2) $\mathcal{M}_1$ is an S-submodule of M which contains $\mathrm{Fil}^1 S \cdot \mathcal{M}$ such that $\mathcal{M}/\mathcal{M}_1$ is a projective $\mathcal{O}_K$-module.
(3) $\Phi\colon \mathcal{M} \to \mathcal{M}$ and $\Phi_1\colon \mathcal{M}_1 \to \mathcal{M}$ are ϕ-linear maps of S-modules such that

$$\Phi(m_1) = p\,\Phi_1(m_1) \quad \text{and} \quad \Phi_1(am) = (\phi(a)/p)\cdot\Phi(m)$$

for all $m_1 \in \mathcal{M}$, $a \in \mathrm{Fil}^1 S$ and $m \in \mathcal{M}$. (Recall that $\phi(\mathrm{Fil}^1 S) \subset p\cdot S$.)
(4) $\mathcal{M}$ is generated by $\Phi_1(\mathcal{M}_1)$ and $\Phi(\mathcal{M})$ as an S-module.
(5) $\Psi\colon \mathcal{M} \to S\otimes_{\phi,S}\mathcal{M}$ is an S-linear map such that

$$\Psi(\Phi_1(m_1)) = 1 \otimes m_1 \quad \text{and} \quad \Psi(\Phi(m)) = p \otimes m$$

for all $m_1 \in \mathcal{M}_1$ and all $m \in \mathcal{M}$.[14]
(6) There exists $n \geqslant 0$ such that $\Psi^n(\mathcal{M}) \subset \mathrm{Fil}^1 S \cdot (S\otimes_{\phi^n,S}\mathcal{M})$, where

$$\Psi^n := \phi^{n-1\,*}\Psi \circ \phi^{n-2\,*}\Psi \circ \cdots \circ \Psi : \mathcal{M} \longrightarrow \phi^{n\,*}\mathcal{M} = S\otimes_{\phi^n,S}\mathcal{M}\,.$$

B.4.7. We recall the definition of the category (Dsp/R) of displays over a commutative ring R from [**139**, 1.1]. We will only be interested in the case where R is excellent, and either p is nilpotent or R is complete with respect to an ideal which contains p. Only the cases $R = \mathcal{O}_K$ and $R = k$ will be used later, where $\mathcal{O}_K$ and k are as in B.4.1.

We fix some notation for p-adic Witt vectors.

[14] It is shown in the discussion immediately following [**137**, Def. 1.2] that for any quadruple $(\mathcal{M}, \mathcal{M}_1, \Phi, \Phi_1)$ with properties (1)–(4), there exists a unique map Ψ which satisfies (5).

- Write $F\colon W(R) \to W(R)$ for the "Frobenius" ring endomorphism for $W(R)$, characterized by

$$\mathbf{w}_n(F(\underline{x})) = \mathbf{w}_{n+1}(\underline{x}) \qquad \text{for all } \underline{x} = (x_0, x_1, x_2, \ldots)\,,$$

  where

$$\mathbf{w}_n(\underline{x}) = x_0^{p^n} + p x_1^{n-1} + \cdots + p^{n-1} x_{n-1} + p^n x_n$$

  is the nth Witt polynomial.
- Denote by I_R the kernel of the surjective ring homomorphism $\mathbf{w}_0 : W(R) \to R$; i.e., I_R consists of all Witt vectors of the form $(0, x_1, x_2, \ldots)$ in $W(R)$.
- Let $V\colon W(R) \to W(R)$ be the additive endomorphism which sends each Witt vector $\underline{x} = (x_0, x_1, x_2, \ldots)$ to $(0, x_0, x_1, x_2, \ldots)$.

A *display* over a commutative ring R is a quintuple $\mathscr{P} = (P, Q, F_P, F_{1,Q}, V^\sharp)$ with the following properties.

(1) P is a finitely generated projective $W(R)$-module.
(2) Q is a submodule of P which contains $I_R \cdot P$.
(3) The quotient P/Q, with its natural R-module structure, is a projective R-module. (Consequently there exists a decomposition $P = L \oplus T$ of P as a direct sum of two projective $W(R)$-submodules L and P such that $Q = L \oplus I_R \cdot T$.)
(4) Both $F_P : P \to P$ and $F_{1,Q} : Q \to P$ are "F-linear" homomorphisms of $W(R)$-modules in the sense that

$$F_P(w \cdot x) = F(w) \cdot F_P(x) \quad \text{and} \quad F_{1,Q}(w \cdot y) = F(w) \cdot F_{1,Q}(y)$$

  for all $w \in W(R)$, $x \in P$ and $y \in Q$.
(5) $F_{1,Q}({}^V w \cdot x) = w \cdot F_P(x)$ for every $w \in W(R)$ and $x \in P$, where ${}^V w$ is the image of w under $V\colon W(R) \to W(R)$.
(6) P is generated by the subset $F_{1,Q}(Q)$ as a $W(R)$-module.
(7) $V^\sharp\colon P \to W(R) \otimes_{F, W(R)} P$ is a $W(R)$-linear homomorphism such that

$$V^\sharp(w \cdot F_P(x)) = pw \otimes x \quad \text{and} \quad V^\sharp(w \cdot F_{1,Q}(y)) = w \otimes y$$

  for all $w \in W(R)$, $x \in P$ and $y \in Q$.[15]
(8) There exists $n \geqslant 0$ such that the $W(R)$-module homomorphism

$$V^{\sharp\, n} := F^{n-1\,*} V^\sharp \circ F^{n-2\,*} V^\sharp \circ \cdots \circ V^\sharp \colon P \longrightarrow W(R) \otimes_{F^n, W(R)} P$$

  is congruent to 0 modulo the ideal $I_R + p\,W(R)$.

Note that the Lie algebra Lie(BT($\mathscr{P}$)) of the p-divisible group BT($\mathscr{P}$) attached to a display $\mathscr{P}$ as above is naturally isomorphic to P/Q; see [**139**, (157)] (in the proof of Theorem 81 there).

---

[15] It is shown in [**139**, Lemma 10] that for every quadruple $(P, Q, F_P, F_{1,Q})$ satisfying (1)–(6), there exists a *unique* $W(R)$-linear map $V^\sharp$ with the above properties. So the datum $V^\sharp$ here is redundant. We have inserted $V^\sharp$ into the definition because it is used in later definitions and constructions.

B.4.8. We have a commutative diagram of functors

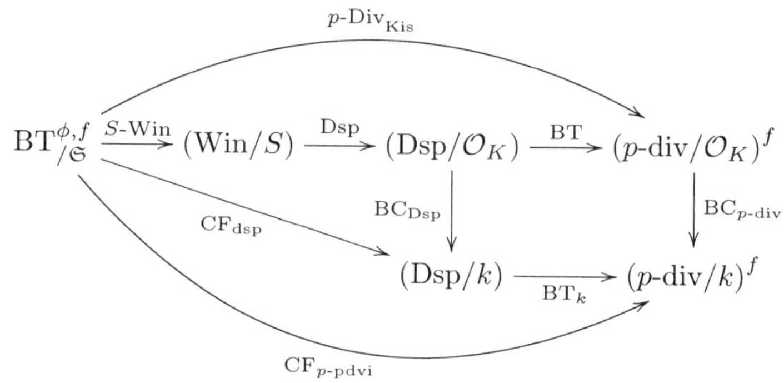

where

- BT and BT_k are the functors which associate to a display over $\mathcal{O}_K$ or k a connected p-divisible group over the same base ring; see [**139**, Thm. 81].
- $\mathrm{BC}_{\mathrm{Dsp}}$ is the base change functor, from displays over $\mathcal{O}_K$ to displays over k, defined in [**139**, Def. 20] and recalled in B.4.11 below.
- Dsp is the functor which to every S-window over $\mathcal{O}_K$ associates a display over $\mathcal{O}_K$, defined in the discussion immediately preceding [**137**, Prop. 1.3] and recalled in B.4.10 below.
- S-Win is the functor which to every object in $\mathrm{BT}^{\phi,f}_{/\mathfrak{S}}$ associates an S-window, defined in [**59**, 1.2.5] and recalled in B.4.9 below.
- $\mathrm{BC}_{p\text{-div}}$ is the base change functor from $\mathcal{O}_K$ to k, which associates to every connected p-divisible group $\mathcal{X}$ over $\mathcal{O}_K$ its closed fiber X_0 over k.

Our intermediate goal, before deriving a formula for the functor MCF, is to find an explicit formula for the functor

$$\mathrm{CF}_{\mathrm{dsp}} = \mathrm{BC}_{\mathrm{Dsp}} \circ \mathrm{Dsp} \circ S\text{-Win}\,.$$

B.4.9. The functor $S\text{-Win}\colon \mathrm{BT}^{\phi,f}_{/\mathfrak{S}} \longrightarrow (\mathrm{Win}/S)$ is defined in [**59**, 1.2.5] as follows. Given an object $(\mathfrak{M}, \phi_{\mathfrak{M}}, \psi_{\mathfrak{M}})$ in $\mathrm{BT}^{\phi,f}_{/\mathfrak{S}}$, the associated window $S\text{-Win}(\mathfrak{M}, \phi_{\mathfrak{M}}, \psi_{\mathfrak{M}})$ for the frame $S/\mathcal{O}_K$ is the quintuple $(\mathcal{M}, \mathcal{M}_1, \Phi, \Phi_1, \Psi)$ where:

- $\mathcal{M} := S \otimes_{\phi,\mathfrak{M}} \mathfrak{M}$
- $\mathcal{M}_1$ is defined by the Cartesian diagram[16]

$$\begin{array}{ccc} \mathcal{M}_1 & \hookrightarrow & \mathcal{M} = S \otimes_{\phi,\mathfrak{M}} \mathfrak{M} \\ \downarrow & \square & \downarrow {\scriptstyle 1\otimes\phi_{\mathfrak{M}}} \\ \mathrm{Fil}^1 S \cdot (S \otimes_{\mathfrak{S}} \mathfrak{M}) & \hookrightarrow & S \otimes_{\mathfrak{S}} \mathfrak{M} \end{array}$$

  (Note that $\mathrm{Fil}^1 S\cdot(S \otimes_{\mathfrak{S}} \mathfrak{M}) \xleftarrow{\simeq} \mathrm{Fil}^1 S \otimes_{\mathfrak{S}} \mathfrak{M}$ because

$$\mathcal{O}_K \otimes_{\mathfrak{S}} \mathfrak{M} \xrightarrow{E(u)\cdot} \mathcal{O}_K \otimes_{\mathfrak{S}} \mathfrak{M}$$

  is injective.)
- $\Phi\colon \mathcal{M} \to \mathcal{M}$ is the ϕ-linear map which sends $s \otimes m$ to $\phi(s) \otimes \phi_{\mathfrak{M}}(m)$ for any $s \in S$ and $m \in \mathfrak{M}$.

[16]The S-module $\mathcal{M}_1$ is denoted by $\mathrm{Fil}^1\mathcal{M}$ in [**59**, 1.2.5]

- $\Phi_1 = \frac{1}{p} \cdot (\Phi|_{\mathcal{M}_1})$, well-defined because $\phi(\mathrm{Fil}^1 S) \subset pS$. Note that $\Phi_1(\mathcal{M}_1)$ generates $\mathcal{M}$ over S because $\frac{1}{p}\phi(E(u)) \in S^{\times}$.
- $\Psi\colon \mathcal{M} \to S \otimes_{\phi,S} \mathcal{M}$ is the S-linear map such that $\Psi(\Phi_1(x_1)) = 1 \otimes x_1$ for all $x_1 \in \mathcal{M}_1$ and $\Psi(\Phi(x)) = p \otimes x$ for all $x \in \mathcal{M}$.

Note that $\mathcal{M}/\mathcal{M}_1 \xrightarrow{\simeq} \phi^*\mathfrak{M}/\mathrm{Fil}^1\phi^*\mathfrak{M}$, where $\phi^*\mathfrak{M} = \mathfrak{S} \otimes_{\phi,\mathfrak{S}} \mathfrak{M}$ and $\mathrm{Fil}^1\phi^*\mathfrak{M}$ is defined by the Cartesian diagram[17]

$$\begin{array}{ccc} \mathrm{Fil}^1\phi^*\mathfrak{M} & \hookrightarrow & \phi^*\mathfrak{M} = \mathfrak{S} \otimes_{\phi,\mathfrak{S}} \mathfrak{M} \\ \downarrow & \square & \downarrow {\scriptstyle 1\otimes\phi_{\mathfrak{M}}} \\ E(u)\cdot\mathfrak{M} & \hookrightarrow & \mathfrak{M} \end{array}$$

B.4.10. The functor $\mathrm{Dsp}\colon (\mathrm{Win}/S) \longrightarrow (\mathrm{Dsp}/\mathcal{O}_K)$ is defined in the discussion immediately preceding [**137**, Prop. 1.3]. It sends a quintuple $(\mathcal{M}, \mathcal{M}_1, \Phi, \Phi_1, \Psi)$ to the quintuple $\mathscr{P} = (P, Q, F_P, F_{1,Q}, V^{\sharp})$ with:

- $P := W(\mathcal{O}_K) \otimes_{\kappa,S} \mathcal{M}$, where
  - $\kappa\colon S \to W(\mathcal{O}_K)$ is the composition of two ring homomorphisms
  $$\kappa\colon S \xrightarrow{\delta} W(S) \xrightarrow{\beta} W(\mathcal{O}_K),$$
  - β is induced by the surjective continuous $W(k)$-linear ring homomorphism $\epsilon\colon S \to \mathcal{O}_K$ such that $\epsilon(u) = \pi$,
  - $\delta\colon S \to W(S)$ is the Cartier homomorphism, characterized by
  $$\mathbf{w}_n(\delta(a)) = \phi^n(a) \qquad \text{for all } a \in S, n \geqslant 0.$$
- $Q := \ker\left(W(\mathcal{O}_K) \otimes_{\kappa,S} \mathcal{M} \longrightarrow \mathcal{M}/\mathcal{M}_1\right)$, where the map is
  $$(a_0, a_1, a_2, \ldots) \otimes x \;\mapsto a_0 \cdot \bar{x}$$
  for all $(a_0, a_1, a_2, \ldots) \in W(\mathcal{O}_K)$ and $x \in \mathcal{M}$, with $\bar{x}$ denoting the image of x in $\mathcal{M}/\mathcal{M}_1$.
- $F_P : P \longrightarrow P$ is the map
  $$F_P : \xi \otimes x \;\mapsto\; F(\xi) \otimes \Phi(x) \qquad \text{for all } \xi \in W(\mathcal{O}_K), x \in \mathcal{M}.$$
- $F_{1,Q} : Q \longrightarrow P$ is the unique map such that
  $$F_{1,Q}(\xi \otimes x_1) = F(\xi) \otimes \Phi_1(x_1) \quad \text{and} \quad F_{1,Q}(V(\xi) \otimes x) = \xi \otimes \Phi(x)$$
  for all $\xi \in W(\mathcal{O}_K)$, $x_1 \in \mathcal{M}_1$ and $x \in \mathcal{M}$.
- $V^{\sharp}\colon W(\mathcal{O}_K) \otimes_{\kappa,S} \mathcal{M} \longrightarrow W(\mathcal{O}_K) \otimes_{F,W(\mathcal{O}_K)} W(\mathcal{O}_K) \otimes_{\kappa,S} \mathcal{M}$ is the unique $W(\mathcal{O}_K)$-linear map such that
  $$V^{\sharp}(w \otimes \Phi(x)) = p\,w \otimes 1 \otimes x \quad \text{and} \quad V^{\sharp}(w \otimes \Phi_1(x_1)) = w \otimes 1 \otimes x_1$$
  for all $x \in \mathcal{M}$ and $x_1 \in \mathcal{M}_1$.

---

[17]The formula $\mathcal{M}/\mathrm{Fil}^1\mathcal{M} \xrightarrow{\simeq} \phi^*\mathfrak{M}/(E(u)\cdot\mathfrak{M})$ in the third paragraph of the proof of [**59**, 1.2.5] should be interpreted as indicated in this Cartesian diagram; i.e. the term "$E(u)\cdot\mathfrak{M}$" in "$\phi^*\mathfrak{M}/(E(u)\cdot\mathfrak{M})$" should be understood to be the submodule $\mathrm{Fil}^1\phi^*\mathfrak{M}$ of $\phi^*\mathfrak{M}$ in our notation.

B.4.11. The base change functor $\mathrm{BC}_{\mathrm{Dsp}}$ from $\mathcal{O}_K$ to k is given in [**139**, Def. 20]. It sends a display $\mathscr{P} = (P, Q, F, F_1, V^\sharp)$ over $\mathcal{O}_K$ to a display $\widetilde{\mathscr{P}} = (\widetilde{P}, \widetilde{Q}, \widetilde{F}, \widetilde{F}_1, \widetilde{V}^\sharp)$ over k, where:

- $\widetilde{P} = W(k) \otimes_{W(\mathcal{O}_K)} P$,
- $\widetilde{Q} = \ker\Big(W(k) \otimes_{W(\mathcal{O}_K)} P \longrightarrow k \otimes_R (P/Q)\Big)$,
- $\widetilde{F}(a \otimes x) = F(a) \otimes F_P(x)$ for all $a \in W(k)$ and $x \in P$,
- $\widetilde{F}_1 \colon \widetilde{Q} \to \widetilde{P}$ is the unique F-linear homomorphism such that
$$\widetilde{F}_1(V(w) \otimes x) = w \otimes F(x) \quad \text{and} \quad \widetilde{F}_1(w \otimes y) = w \otimes F_1(y)$$
for all $w \in W(k)$, $x \in P$ and $y \in Q$,
- $\widetilde{V}^\sharp \colon W(k) \otimes_{W(\mathcal{O}_K)} P = \widetilde{P} \longrightarrow W(k) \otimes_{W(\mathcal{O}_K)} W(\mathcal{O}_K) \otimes_{F, W(\mathcal{O}_K)} P$ is defined by
$$\widetilde{V}^\sharp(w \otimes x) = w \otimes V^\sharp(x) \qquad \text{for all } w \in W(k), x \in P.$$

B.4.12. Recall from B.4.10 that $\delta \colon S \to W(S)$ is the Cartier homomorphism with respect to the Frobenius lifting $\phi \colon S \to S$. We have the following ring homomorphisms.

- Let $\kappa = W(\epsilon) \circ \delta : S \xrightarrow{\delta} W(S) \xrightarrow{W(\epsilon)} W(\mathcal{O}_K)$ be the composition of δ with the surjection $W(\epsilon) : W(S) \twoheadrightarrow W(\mathcal{O}_K)$ induced by the continuous $W(k)$-algebra surjection
$$\epsilon \colon S \twoheadrightarrow \mathcal{O}_K$$
with $\epsilon(u) = \pi$.
- Let $W(\mathrm{can})$ be the map from $W(\mathcal{O}_K)$ to $W(k)$ induced by the natural surjection
$$\mathrm{can} \colon \mathcal{O}_K \twoheadrightarrow k,$$
which identifies k with the $\mathcal{O}_K/\pi\mathcal{O}_K$.
- Let $j_{\mathfrak{S}\hookrightarrow S} : \mathfrak{S} \to S$ be the natural inclusion map from $\mathfrak{S}$ to S.
- Let $\rho := W(\mathrm{can}) \circ \kappa \circ \phi \circ j_{\mathfrak{S}\hookrightarrow S} : \mathfrak{S} \longrightarrow W(k)$ be the composition of the ring homomorphisms $\phi \circ j_{\mathfrak{S}\hookrightarrow S}$, κ, and $W(\mathrm{can})$.

B.4.13. Lemma. *Using the above notation, we have:*

(1) $\delta(u) = (u, 0, 0, 0 \ldots)$ *(the Teichmüller representative of u in $W(S)$),*
(2) $\rho(u) = 0$,
(3) $W(\mathrm{can}) \circ \kappa|_{W(k)} = \mathrm{id}_{W(k)}$,
(4) $\rho|_{W(k)} = \sigma$ *(the canonical lifting of Frobenius on $W(k)$).*

PROOF. The element $\delta(u) \in W(S)$ is characterized by the property that $\mathbf{w}_n(\delta(u)) = u^{p^n}$ for all $n \geqslant 0$, and the Teichmüller element $[u] := (u, 0, 0, \ldots) \in W(S)$ satisfies $\mathbf{w}_n([u]) = u^{p^n}$. This proves (1). It follows that
$$\rho(u) = W(\mathrm{can})(W(\epsilon)(u^p)) = W(\mathrm{can})(\pi^p)) = 0.$$
We have proved (2).

The restriction to $W(k) \subset S$ of the Cartier homomorphism δ is the Cartier homomorphism $\delta_{W(k)} \colon W(k) \longrightarrow W(W(k))$, characterized by the condition that $\mathbf{w}_n(\delta(\underline{a})) = F^n(\underline{a})$ for all $\underline{a} \in W(k)$. By the same argument as for (1), for any

Teichmüller element $[a] = (a, 0, 0, \ldots) \in W(k)$ we have $\delta([a]) = ([a], 0, 0, \ldots)$, the Teichmüller representative of $[a]$ in $W(W(k))$. Thus,

$$W(\mathrm{can})(\kappa([a])) = W(\mathrm{can})([a], 0, 0, 0, \ldots) = (a, 0, 0, 0, \ldots) \quad \text{for all} \quad a \in k.$$

Because the set of Teichmüller elements of the form $[a]$ $(a \in k)$ generate a p-adically dense subring of $W(k)$, we conclude that the restriction to $W(k)$ of the composition $W(\mathrm{can}) \circ \kappa$ is equal to $\mathrm{id}_{W(k)}$. We have proved (3). Now (4) follows immediately. □

B.4.14. We are ready to compute the functor $\mathrm{CF}_{\mathrm{dsp}}$:

$$\begin{tikzcd} \mathrm{BT}^{\phi,f}_{/\mathfrak{S}} \arrow[r, "\mathrm{SWin}"] \arrow[rrr, bend right, "\mathrm{CF}_{\mathrm{dsp}}"'] & (\mathrm{Win}/S) \arrow[r, "\mathrm{Dsp}"] & (\mathrm{Dsp}/\mathcal{O}_K) \arrow[r, "\mathrm{BC}_{\mathrm{Dsp}}"] & (\mathrm{Dsp}/k)\,. \end{tikzcd}$$

Let $(\mathfrak{M}, \phi_{\mathfrak{M}}, \psi_{\mathfrak{M}})$ be a given object in $\mathrm{BT}^{\phi,f}_{/\mathfrak{S}}$. Let $\widetilde{\mathscr{P}} = (\widetilde{P}, \widetilde{Q}, \widetilde{F}, \widetilde{F_1}, \widetilde{V}^{\sharp})$ be the display $\mathrm{CF}_{\mathrm{dsp}}(\mathfrak{M}, \phi_{\mathfrak{M}}, \psi_{\mathfrak{M}})$ over k of the p-divisible group $X := \mathrm{CF}_{p\text{-div}}(\mathfrak{M}, \phi_{\mathfrak{M}}, \psi_{\mathfrak{M}})$ over k. From the recipes in B.4.9–B.4.11, we see that

$$\widetilde{P} = W(k) \otimes_{W(\mathrm{can}), W(\mathcal{O}_K)} W(\mathcal{O}_K) \otimes_{\kappa, S} S \otimes_{\phi, \mathfrak{S}} \mathfrak{M} \simeq W(k) \otimes_{\rho, \mathfrak{S}} \mathfrak{M} \simeq W(k) \otimes_{\mathfrak{S}} \phi^*\mathfrak{M}\,.$$

The last isomorphism in the above displayed formula follows from B.4.13; the implicit ring homomorphism $\mathfrak{S} \to W(k)$ in the tensor product $W(k) \otimes_{\mathfrak{S}} \phi^*\mathfrak{M}$ is the $W(k)$-linear quotient map which sends u to 0.

Tracing through the recipes in B.4.9–B.4.11, we see that $\widetilde{F}\colon \widetilde{P} \to \widetilde{P}$ is induced by the ϕ-linear map $\phi_{\mathfrak{M}}: \mathfrak{M} \to \mathfrak{M}$. The rest of the data of the display $\widetilde{\mathscr{P}}$ is determined by $\widetilde{P}$ and $\widetilde{F}$. We summarize our conclusions in B.4.15 below.

B.4.15. Proposition. *For any object $(\mathfrak{M}, \phi_{\mathfrak{M}}, \psi_{\mathfrak{M}})$ in $\mathrm{BT}^{\phi,f}_{/\mathfrak{S}}$, the display*

$$\mathrm{CF}_{\mathrm{disp}}(\mathfrak{M}, \phi_{\mathfrak{M}}, \psi_{\mathfrak{M}}) = (\widetilde{P}, \widetilde{Q}, \widetilde{F}, \widetilde{F}_1, V^{\sharp})$$

over k of the closed fiber of the p-divisible group $p\text{-div}_{\mathrm{Kis}}(\mathfrak{M}, \phi_{\mathfrak{M}}, \psi_{\mathfrak{M}})$ attached to $(\mathfrak{M}, \phi_{\mathfrak{M}}, \psi_{\mathfrak{M}})$ is explicitly given as follows.

(1) $\widetilde{P} \simeq W(k) \otimes_{\rho, \mathfrak{S}} \mathfrak{M} \simeq W(k) \otimes_{\mathfrak{S}} \phi^*\mathfrak{M} \simeq \phi^*\mathfrak{M}/(u \cdot \phi^*\mathfrak{M})$,

(2) $\widetilde{F}\colon \widetilde{P} \to \widetilde{P}$ *is the map induced by $1 \otimes \phi$. In other words $\widetilde{F}$ sends the image in $\phi^*\mathfrak{M}/(u \cdot \phi^*\mathfrak{M})$ of the element $1 \otimes x \in \phi^*\mathfrak{M} = \mathfrak{S} \otimes_{\phi, \mathfrak{S}} \mathfrak{M}$ to the image in $\phi^*\mathfrak{M}/(u \cdot \phi^*\mathfrak{M})$ of the element $1 \otimes \phi(x) \in \mathfrak{S} \otimes_{\phi, \mathfrak{S}} \mathfrak{M}$, for all $x \in \mathfrak{M}$.*

(3) $\widetilde{Q} = \ker\big(\, \widetilde{P} \xrightarrow{\widetilde{F}} \widetilde{P} \twoheadrightarrow \widetilde{P}/(p \cdot \widetilde{P})\,\big)$, *where the second map is the quotient map from $\widetilde{P} = \phi^*\mathfrak{M}/(u \cdot \phi^*\mathfrak{M})$ to $\widetilde{P}/(p \cdot \widetilde{P}) = \phi^*\mathfrak{M}/((p, u) \cdot \phi^*\mathfrak{M})$.*

(4) $\widetilde{F}_1 = (1/p) \cdot \left(\widetilde{F}|_{\widetilde{Q}}\right) : \widetilde{Q} \longrightarrow \widetilde{P}$.

(5) $\widetilde{V}^{\sharp}\colon \widetilde{P} \to W(k) \otimes_{\sigma, W(k)} \widetilde{P}$ *is the unique $W(k)$-linear map such that*

$$\widetilde{V}^{\sharp}\left(w \cdot \widetilde{F}_1(y)\right) = w \otimes y \in W(k) \otimes_{\sigma, W(k)} \widetilde{P} \qquad \textit{for all } y \in \widetilde{Q}\,.$$

Moreover we have

$$\widetilde{V}^{\sharp}\left(w \cdot \widetilde{F}(x)\right) = pw \otimes x \in W(k) \otimes_{\sigma, W(k)} \widetilde{P} \qquad \textit{for all } x \in \widetilde{P}\,.$$

(6) *Alternatively, using the isomorphism* $\widetilde{P} \simeq W(k) \otimes_{\rho,\mathfrak{S}} \mathfrak{M}$, $\widetilde{V}^\sharp$ *is induced by* $\frac{1}{c}\psi_{\mathfrak{M}}$:

$$\widetilde{V}^\sharp = 1 \otimes ((1/c)\psi_{\mathfrak{M}}) : \ \widetilde{P} = W(k) \otimes_{\rho,\mathfrak{S}} \mathfrak{M} \xrightarrow{1\otimes(\frac{1}{c}\cdot\psi_{\mathfrak{M}})} W(k) \otimes_{\rho,\mathfrak{S}} \phi^*\mathfrak{M}.$$

(Recall that pc is the constant term of the Eisenstein polynomial

$$E(u) = u^e + a_{e-1}u^{e-1} + \ldots + a_1 u + pc$$

for the generator π of the maximal ideal of $\mathcal{O}_K$.)

In particular, the Lie algebra of the p-divisible group X over k is

$$\begin{aligned}\mathrm{Lie}(X) &\simeq W(k) \otimes_{\mathfrak{S}} \mathrm{coker}(\ \mathfrak{M} \xrightarrow{\psi_{\mathfrak{M}}} \phi^*\mathfrak{M}\ ) \\ &\simeq \mathrm{coker}(\ \mathfrak{M}/u\mathfrak{M} \xrightarrow{\psi \pmod{u}} \phi^*\mathfrak{M}/u\phi^*\mathfrak{M}\ ).\end{aligned}$$

PROOF. The assertions in (1) and (2) have been proved in B.4.14. Parts (3)–(5) follow from (1) and (2); for instance, the map $\widetilde{V}^\sharp \colon \widetilde{P} \to W(k) \otimes_{\sigma,W(k)} \widetilde{P}$ is uniquely determined by the linearization $1 \otimes \widetilde{F} \colon W(k) \otimes_{\sigma,W(k)} \widetilde{F} \to \widetilde{F}$ of $\widetilde{F}$ because

$$(1 \otimes \widetilde{F})(\widetilde{V}^\sharp(x)) = px \quad \text{and} \quad \widetilde{V}^\sharp(\widetilde{F}(x)) = p \otimes x \qquad \text{for all } x \in \widetilde{P}.$$

B.4.16. Remark. For any object $(\mathfrak{M}, \phi_{\mathfrak{M}}, \psi_{\mathfrak{M}})$ in $\mathrm{BT}^{\phi,f}_{/\mathfrak{S}}$, we have a natural isomorphism

$$\mathrm{Lie}\,(p\text{-div}_{\mathrm{Kis}}(\mathfrak{M}, \phi_{\mathfrak{M}}, \psi_{\mathfrak{M}})) \simeq \phi^*\mathfrak{M}/\mathrm{Fil}^1\phi^*\mathfrak{M},$$

where $\mathrm{Lie}\,(p\text{-div}_{\mathrm{Kis}}(\mathfrak{M}, \phi_{\mathfrak{M}}, \psi_{\mathfrak{M}}))$ denotes the Lie algebra of the p-divisible group $p\text{-div}_{\mathrm{Kis}}(\mathfrak{M}, \phi_{\mathfrak{M}}, \psi_{\mathfrak{M}})$ attached to $(\mathfrak{M}, \phi_{\mathfrak{M}}, \psi_{\mathfrak{M}})$, and

$$\mathrm{Fil}^1\phi^*\mathfrak{M} = \ker\left(\phi^*\mathfrak{M} \xrightarrow{1\otimes\phi} \mathfrak{M} \twoheadrightarrow \mathfrak{M}/(E(u)\cdot\mathfrak{M})\right)$$

as in B.4.9.

To see this, we proceed as in B.4.15, with the same notation:

- $(\mathcal{M}, \mathcal{M}_1, \Phi, \Phi_1, \Psi) = S\text{-Win}(\mathfrak{M}, \phi_{\mathfrak{M}}, \psi_{\mathfrak{M}})$,
- $(P, Q, F, F_1, V^\sharp) = \mathrm{Dsp}(\mathcal{M}, \mathcal{M}_1, \Phi, \Phi_1, \Psi)$,
- $X = \mathrm{BT}(P, Q, F, F_1, V^\sharp) = p\text{-div}_{\mathrm{Kis}}(\mathfrak{M}, \phi_{\mathfrak{M}}, \psi_{\mathfrak{M}})$.

It is shown in the proof of [**139**, Thm. 81] that $\mathrm{Lie}(X) \simeq P/Q$. We have seen that $P/Q \simeq \mathcal{M}/\mathcal{M}_1$ in B.4.10 and $\mathcal{M}/\mathcal{M}_1 \simeq \phi^*\mathfrak{M}/\mathrm{Fil}^1\phi^*\mathfrak{M}$ in B.4.9. The assertion follows.

B.4.17. Proposition. *For any object* $(\mathfrak{M}, \phi_{\mathfrak{M}}, \psi_{\mathfrak{M}})$ *in* $(\mathrm{Mod}/\mathfrak{S})^{\mathrm{c}}$, *the classical* covariant *Dieudonné module* $\mathrm{MCF}\,(\mathfrak{M}, \phi_{\mathfrak{M}}, \psi_{\mathfrak{M}}) = \left(\mathrm{M}_*(G_k), F_{\mathrm{M}_*(G_k)}, V_{\mathrm{M}_*(G_k)}\right)$ *of the closed fiber G_k of the connected finite flat commutative $\mathcal{O}_K$-group scheme* $G = \mathcal{G}_{\mathrm{Kis}}\,(\mathfrak{M}, \phi_{\mathfrak{M}}, \psi_{\mathfrak{M}})$ *attached to* $(\mathfrak{M}, \phi_{\mathfrak{M}}, \psi_{\mathfrak{M}})$ *can be explicitly described as follows.*

(1) *We have a functorial $W(k)$-module isomorphism*

$$\mathrm{M}_*(G_k) \simeq W(k) \otimes_{\mathfrak{S}} \mathfrak{M} \simeq \mathfrak{M}/u\cdot\mathfrak{M}.$$

(2) *The Frobenius map*

$$F_{\mathrm{M}_*(G_k)}: W(k) \otimes_{\sigma,W(k)} (\mathfrak{M}/u\mathfrak{M}) = W(k) \otimes_{\sigma,W(k)} \mathrm{M}_*(G_k) \longrightarrow \mathrm{M}_*(G_k) = \mathfrak{M}/u\mathfrak{M}$$

for $\mathrm{M}_*(G_k)$ *is given by*

$$F_{\mathrm{M}_*(G_k)}(w \otimes [x]) = [w \cdot \phi_{\mathfrak{M}}(x)] \qquad \text{for all } w \in W(k), x \in \mathfrak{M},$$

where $[x]$ *and* $[\phi_{\mathfrak{M}}(x)]$ *are the images in* $\mathfrak{M}/u\mathfrak{M}$ *of the elements* x *and* $\phi_{\mathfrak{M}}(x)$ *respectively.*

(3) $V_{\mathrm{M}_*(G_k)}: \mathfrak{M}/u\mathfrak{M} = \mathrm{M}_*(G_k) \longrightarrow W(k) \otimes_{\sigma,W(k)} \mathrm{M}_*(G_k) = \phi^*\mathfrak{M}/(u \cdot \phi^*\mathfrak{M})$ *is the map*

$$V_{\mathrm{M}_*(G_k)}([x]) = [(1/c) \cdot \psi_{\mathfrak{M}}(x)] \qquad \text{for all } x \in \mathfrak{M},$$

where $[x]$ *is the image of* x *in* $\mathfrak{M}/u\mathfrak{M}$ *and* $[(1/c) \cdot \psi_{\mathfrak{M}}(x)]$ *is the image of* $(1/c) \cdot \psi_{\mathfrak{M}}(x)$ *in* $\phi^*\mathfrak{M}/(u \cdot \phi^*\mathfrak{M})$.

In particular, the Lie algebra of the finite group scheme G_k *over* k *is*

$$\mathrm{Lie}(G_k) \simeq \ker\left(\mathfrak{M}/u\mathfrak{M} \xrightarrow{(1/c)\psi_{\mathfrak{M}}} \phi^*\mathfrak{M}/u\phi^*\mathfrak{M} \right).$$

PROOF. The equivalence $\mathcal{G}_{\mathrm{Kis}}$ between $(\mathrm{Mod}/\mathfrak{S})^{\mathrm{c}}$ and $(p\text{-}\mathrm{Gr}/\mathcal{O}_K)^{\mathrm{c}}$ in [**59**, 1.3.9] is deduced from the equivalence $p\text{-}\mathrm{Div}_{\mathrm{Kis}}$ between $\mathrm{BT}^{\phi,f}_{/\mathfrak{S}}$ and $(p\text{-div}/\mathcal{O}_K)^f$ as follows. Suppose that a connected finite flat commutative $\mathcal{O}_K$-group scheme G is the kernel of an isogeny $\gamma: \mathcal{X} \to \mathcal{X}'$ between connected p-divisible groups over $\mathcal{O}_K$, and this isogeny γ corresponds to a morphism f in the category $\mathrm{BT}^{\phi,f}_{/\mathfrak{S}}$, from an object $(\mathfrak{N}, \phi_{\mathfrak{N}}, \psi_{\mathfrak{N}})$ to an object $(\mathfrak{N}', \phi_{\mathfrak{N}'}, \psi_{\mathfrak{N}'})$. This G corresponds to the object $(\mathfrak{M}, \phi_{\mathfrak{M}}, \psi_{\mathfrak{M}})$, where $\mathfrak{M} = \mathrm{coker}(f)$ is the cokernel of the $\mathfrak{S}$-module homomorphism $f: \mathfrak{M} \to \mathfrak{N}$, $\phi_{\mathfrak{M}}$ is induced by $\phi_{\mathfrak{N}'}$, and $\psi_{\mathfrak{M}}$ is induced by $\psi_{\mathfrak{N}'}$.

For a p-divisible group X over k attached to a display $(\widetilde{P}, \widetilde{Q}, \widetilde{F}, \widetilde{F}_1, \widetilde{V}^\sharp)$ over k, recall that the crystalline Dieudonné module

$$\big(\mathrm{E}_*(X), F_{\mathrm{E}_*(X)}: W(k) \otimes_{\sigma,W(k)} \mathrm{E}_*(X) \to \mathrm{E}_*(X), V_{\mathrm{E}_*(X)}: \mathrm{E}_*(X) \to W(k) \otimes_{\sigma,W(k)} \mathrm{E}_*(X)\big)$$

is canonically isomorphic to

$$\left(\widetilde{P}, 1 \otimes \widetilde{F}: W(k) \otimes_{\sigma,W(k)} \widetilde{P}, \widetilde{V}^\sharp: \widetilde{P} \to W(k) \otimes_{\sigma,W(k)} \widetilde{P}\right).$$

Using this fact and tracing through the construction recalled in the previous paragraph, we see from B.4.15 that:

- the $W(k)$-module underlying the crystalline Dieudonné module $\mathrm{E}_*(G_k)$ of G_k is
$$\mathrm{E}_*(G_k) \simeq W(k) \otimes_{\rho,\mathfrak{S}} \mathfrak{M} \simeq \phi^*\mathfrak{M}/(u \cdot \phi^*\mathfrak{M}),$$
- the map
$$F_{\mathrm{E}_*(G_k)}: W(k) \otimes_{\sigma,W(k)} W(k) \otimes_{\rho,\mathfrak{S}} \mathfrak{M} \simeq W(k) \otimes_{\sigma,W(k)} \mathrm{E}_*(G_k) \longrightarrow \mathrm{E}_*(G_k) \simeq W(k) \otimes_{\rho,\mathfrak{S}} \mathfrak{M}$$
is given by
$$F_{\mathrm{E}_*(G_k)}(w_1 \otimes w_2 \otimes x) = (w_1 \cdot \sigma(w_2)) \otimes \phi_{\mathfrak{M}}(x) \qquad \text{for all } w_1, w_2 \in W(k), x \in \mathfrak{M},$$

- the map

$$D_{\mathrm{E}_*(G_k)} : W(k) \otimes_{\rho,\mathfrak{S}} \mathfrak{M} \simeq \mathrm{E}_*(G_k) \longrightarrow W(k) \otimes_{\sigma,W(k)} \mathrm{E}_*(G_k) \simeq W(k) \otimes_{\rho,\mathfrak{S}} \mathfrak{S} \otimes_{\phi,\mathfrak{S}} \mathfrak{M}$$

is given by

$$D_{\mathrm{E}_*(G_k)}(w \otimes x) = w \otimes ((1/c)\psi_{\mathfrak{M}}(x)) \qquad \text{for all } w \in W(k), x \in \mathfrak{M}.$$

The assertions in B.4.17 (1)–(3) follow from the above recipe and the fact that

$$\left(\mathrm{E}_*(G_k), F_{\mathrm{E}_*(G_k)}, V_{\mathrm{E}_*(G_k)}\right) \simeq W(k) \otimes_{\sigma,W(k)} \left(\mathrm{M}_*(G_k), F_{\mathrm{M}_*(G_k)}, V_{\mathrm{M}_*(G_k)}\right).$$

The final statement concerning $\mathrm{Lie}(G_k)$ follows from (3) and the general formula for the Lie algebra of a finite group scheme over k in terms of its classical covariant Dieudonné module $\big(\mathrm{M}_*(G_k), F_{\mathrm{M}_*(G_k)}, V_{\mathrm{M}_*(G_k)}\big)$. □

Notes on Quotes

Sources of quotes at the beginning of each chapter.

(0) The Zariski quote on page 1 is from the Preface of *Oscar Zariski: Collected Papers*, page xii of volumes I, II and page xiv of volumes III, IV.

(1) The Hilbert quote on page 13 is from page 182 of Hilbert's obituary by Olga Taussky in *Nature*, vol. **152**, pp. 182–183. Taussky, who heard Hilbert's comment, recorded it as follows.

> It is interesting to recall that, in connexion with a lecture by Prof. R. Fueter at the 1932 Zurich Congress, Hilbert asserted that the theory of complex multiplication (of elliptic modular functions) which forms a powerful link between number theory and analysis, is not only the most beautiful part of mathematics but also of all science.

(2) The Igusa quote on page 91 is from page 614 of Igusa's article *Arithmetic variety of moduli for genus two*, Annals of Math. vol. **72**, 1960, pp. 612–649.

(3) The Tate quote on page 137 is from page 158 of Tate's original article [**119**] on p-divisible groups.

(4) The Grothendieck quote on page 195 is in a letter from Grothendieck to Mumford dated September 4, 1968. The letter is included in *David Mumford Selected Papers* vol. II, Springer, 2010, pp. 735–737. The passage quoted is on page 736.

(5) The Shimura quote on page 249 is from a footnote on p. 96 of Shimura's book [**113**], the first page of Chapter 4.

(6) The Mumford quote on page 321 is from the Preface of *David Mumford Selected Papers, On the Classification of Varieties and Moduli Spaces*, Springer, 2004.

Glossary of Notations

$\mathrm{Hom}(A, B)$	K-rational homomorphisms for abelian varieties A, B over a field K, **10**[18]
$\mathrm{End}(A)$	ring of K-rational endomorphisms for an abelian variety A over a field K, **10**
$\mathrm{Hom}^0(A, B)$	$\mathbb{Q} \otimes_{\mathbb{Z}} \mathrm{Hom}(A, B)$, **10**
$\mathrm{End}^0(A)$	$\mathbb{Q} \otimes_{\mathbb{Z}} \mathrm{End}(A)$, the endomorphism algebra of an abelian variety A, **10**
A^t	the dual of an abelian variety A, **10**
$A \sim B$	abelian varieties A, B over K are isogenous over K, **10**
$\mathbb{A}_L$	the adele ring for a number field L, **10**
$\mathbb{A}$	the adele ring for $\mathbb{Q}$, **10**
$\mathbb{A}_{L,f}$	the ring of finite adeles for a number field L, **10**
$\mathbb{A}_f$	the ring of finite adeles for $\mathbb{Q}$, **10**
rec_L	the reciprocity law map for a number field L, **10**
rec_F	the Artin map for F, F either a number field or a local field, **10**
$\mathrm{Br}(K)$	the Brauer group of a field K, **18**
$\mathrm{Trd}_{D/Z}$	the reduced trace of a central simple algebra D over a field K, **18**
inv_K, inv_{L_v}	the invariant of an element of the Brauer group over a local field K or L_v, **19**
D_K	the Dieudonné ring for a perfect field $K \supset \mathbb{F}_p$, 38
N^D	the Cartier dual of a commutative finite locally free group scheme, **34**
N_Φ	the reflex norm of a CM type Φ, **94**
$\mathrm{Res}_{L/\mathbb{Q}}(\mathbb{G}_m)$	Weil's restriction of scalars for $L/\mathbb{Q}$ applied to $\mathbb{G}_{m/L}$, **95**
h_Φ	the cocharacter of $\mathrm{Res}_{L/\mathbb{Q}}(\mathbb{G}_m)$ attached to a CM type Φ for L, **95**
ϵ_{alg}	the algebraic part of an algebraic Hecke character $\epsilon\colon \mathbb{A}_F^\times \to K^\times$, 118
$S_\mathfrak{m}$	the Serre group with conductor $\mathfrak{m}$ for a number field F, **119**
$T_\mathfrak{m}$	the connected Serre group with conductor $\mathfrak{m}$ attached to a number field F, **119**
ψ_ℓ	the abelian ℓ-adic representation attached to an algebraic Hecke character ϵ, 121
$\mathfrak{S}^F$	the connected Serre group attached to a number field F, **119**, 299
S_F	the Serre group attached to a number field F, **120**
$\overline{\mathbb{F}}_p$	an algebraic closure of $\mathbb{F}_p$, 137–248
G_λ	a simple p-divisible group with slope λ over a perfect base field of characteristic p, **139**
M_λ	the contravariant Dieudonné module of G_λ, **139**
$X^{\text{ét}}$	the maximal étale quotient of a p-divisible group X, **140**
X^{mult}	the largest p-divisible subgroup of multiplicative type for a p-divisible group X, **140**

[18] A bold-faced page number indicates where a general notation is first defined or introduced.

$X_{(0,1)}$	the largest local-local part of a p-divisible group X, **140**
Fr_N	the relative Frobenius homomorphism of a group scheme N in characteristic p, **142**
Ver_N	the Verschiebung homomorphism of a group scheme N in characteristic p, **142**
$G[F,V]$, $X[F,V]$	the unique maximal α-subgroup of a finite commutative group scheme G (respectively a p-divisible group X) over a field K of characteristic $p>0$, **143**
$a(G)$	the a-number of a finite commutative group scheme G over a field K of characteristic $p>0$, **143**
X^t	the Serre dual of a p-divisible group X, **152**
$\mathrm{T}_p(X_\bullet)$	the essential projective limit attached to a p-divisible group, **148**
X	a p-divisible group with a-number $a(X)=2$, 156–166
Y_b	the α_p-quotient of $\mathcal{H}$ given by b, 159–166
(F,Φ)	p-adic CM type, decomposed as $(F_1,\Phi_1)\times\cdots\times(F_c,\Phi_c)$, **170**
$E(F_i,\Phi_i)$	reflex field of the p-adic CM type Φ_i, **170**
$E(F,\Phi)$	the compositum of the reflex fields $E(F_i,\Phi_i)$, **170**
$\underline{F}^\times$	$\mathrm{Res}_{F/\mathbb{Q}_p}(\mathbb{G}_m)$, the Weil restricted torus for a $\mathbb{Q}_p$-algebra with $[F:\mathbb{Q}_p]<\infty$, **171**
χ^σ	the geometric character of $\underline{F}^\times$ attached to $\sigma\in\mathrm{Hom}_{\mathbb{Q}_p,\mathrm{ring}}(F,\overline{\mathbb{Q}}_p)$, **171**
μ_Φ	the geometric cocharacter of $\underline{F}^\times$ attached to a p-adic CM type (F,Φ), **171**
$\mathrm{N}\mu_\Phi$	reflex norm attached to a p-adic CM type (F,Φ), **171**
ξ^σ	the geometric character of $\underline{F}^\times$ attached to a p-adic embedding $\sigma: F\to\overline{\mathbb{Q}}_p$, **187**
ν_σ	the geometric cocharacter of $\underline{F}^\times$ attached to a p-adic embedding $\sigma: F\to\overline{\mathbb{Q}}_p$, **187**
$E(\mu)$	the field of definition of a geometric p-adic cocharacter μ, **187**
$\mathrm{N}\mu$	the reflex norm of a geometric p-adic cocharacter μ, **187**
L	a CM field, **195**, 240–248
L^+	the maximal totally real subfield of L, **206**, 240–248
v	a p-adic place of L^+ above p, 206, 240–248
$\mathcal{O}_{L,v}$	$\mathcal{O}_L\otimes_{\mathcal{O}_{L^+}}\mathcal{O}_{L_v^+}$, **206**, 240–248
$[\mathrm{Lie}(Z_v)]$	the Lie type of an $\mathcal{O}_{L,v}$-linear CM p-divisible group Z_v, 206
$\Delta_{L^+},\Delta(L/L^+,\mathbb{F}_q))$	the set of all *bad* p-adic places of L_0 with respect to $\mathbb{F}_q$, **208**
Δ_L	the set of all p-adic places of L above Δ_{L^+}, **208**
$D_{p,\infty}$	the quaternion division algebra over $\mathbb{Q}$ ramified exactly at p and ∞, 221
E	a finite extension field of $\mathbb{Q}_p$, 211–220
	a commutative semisimple quadratic algebra over E^+, 228–240
E^+	a finite extension field of $\mathbb{Q}_p$, 228–240
E_0	the maximal unramified subextension of $E/\mathbb{Q}_p$, 211–220
	same as above when E is a field, equal to $E_0^+\times E_0^+$ if $E=E^+\times E^+$, 228–240
κ_E	residue field of E, 211–220
	either the quadratic extension field of κ_{E^+} or $\kappa_{E^+}\times\kappa_{E^+}$, 228–240
f, f_E	$\dim_{\mathbb{F}_p}(\kappa_E)$, 211–220, 228–240
e, $e(E/\mathbb{Q}_p)$	the ramification index of E, equals to $e(E_0/\mathbb{Q}_p)$ if $E\cong E_0\times E_0$, 211–220, 228–240
ι	the involution for E/E^+, 228–240, 240–248
k	an algebraically closed field of characteristic $p>0$, 211–248

	a perfect field of characteristic $p>0$, 343–370
I	$\mathrm{Hom}_{\mathbb{Z}_p\text{-alg}}(\mathcal{O}_{E_0}, W(k))$, 211–220, 228–240
κ	a finite field contained in k, 211–248
$\mathrm{R}_F(\mathcal{O}_{L_v})$	the Grothendieck group of $(\mathcal{O}_{L_v}\otimes_{\mathbb{Z}_p}F)$-modules of finite length, $F\supset\mathbb{F}_p$, 206
$\mathrm{R}_k(\mathcal{O}_E)$	the Grothendieck group of $(\mathcal{O}_E\otimes_{\mathbb{Z}_p}k)$-modules of finite length, 211–228, 229–240
$\mathrm{R}_\kappa(\mathcal{O}_E)$	κ-rational elements in $\mathrm{R}_k(\mathcal{O}_E)$, 211–228, 229–240
$\mathrm{R}_k^+(\mathcal{O}_E)$, $\mathrm{R}_\kappa^+(\mathcal{O}_E)$	effective elements in $\mathrm{R}_k(\mathcal{O}_E)$ and $\mathrm{R}_\kappa(\mathcal{O}_E)$ respectively, 211–228, 229–240
$\mathrm{R}_k(\mathcal{O}_E, i)$	the i-th component of $\mathrm{R}_k(\mathcal{O}_E)$ for any $i\in I$, 212–228, 229–240
ϵ	the degree map for $\mathrm{R}_k(\mathcal{O}_E)$, 212–228, 229–240,
ϵ_i	the i-th component of the degree map ϵ, 212–228, 229–240,
$[\mathrm{Lie}(X_0,\alpha_0)]$	the Lie type of an $\mathcal{O}_E$-linear CM p-divisible group (X_0,α_0), 212, 230
ξ	reduction map, from p-adic CM types to Lie types, 212, 230
δ_{unif}	the supersingular uniform Lie type, 234
$\delta_{\mathrm{str},J}$	the striped Lie type attached to a $\mathrm{Gal}(\kappa_E/\mathbb{F}_{p^2})$-orbit J, 234
$(Y_{\mathrm{toy}},\beta_{\mathrm{toy}})$	a CM p-divisible group with a striped Lie type, 236–239
w	a p-adic place of L above a place v of L_0, 240–248
$\mathcal{O}_w$	the ring of integers in the w-adic completion L_w of w, 240–248
$\mathcal{O}_{w,0}$	the maximal absolutely unramified subfield of L_w, 240–248
κ_w	the residue field of w, 240–248
I_w	$\mathrm{Hom}_{\mathbb{Z}_p\text{-alg}}(\mathcal{O}_{L_{w,0}}, W(k))$, 241–248
I_v	$\mathrm{Hom}_{\mathbb{Z}_p\text{-alg}}(\mathcal{O}_{L_{v,0}}, W(k))$, 241–248
$\mathrm{R}_k(\mathcal{O}_{L_v})$	the Grothendieck group of $(\mathcal{O}_{L_v}\otimes_{\mathbb{Z}_p}k)$-modules of finite length, 240–248
$\mathrm{R}_\kappa(\mathcal{O}_{L_v})$	the Grothendieck group of $(\mathcal{O}_{L_v}\otimes_{\mathbb{Z}_p}k)$-modules of finite length, equal to the subgroup of all κ-rational elements in $\mathrm{R}_k(\mathcal{O}_{L_v})$, 240–248
$\mathrm{R}_k(\mathcal{O}_{L,p})$	the Grothendieck group of $(\mathcal{O}_{L,p}\otimes_{\mathbb{Z}_p}k)$-modules of finite length, 240–248
ϵ_w	the degree map for $\mathrm{R}_k(\mathcal{O}_w)$, 241–248
$\epsilon_{w,i}$	the i-th component of ϵ_w, 241–248
$\underline{F}^\times$	$\mathrm{Res}_{F/\mathbb{Q}}(\mathbb{G}_m)$, the Weil restricted $\mathbb{Q}$-torus attached to a number field F, 296
ξ^τ	the character of $\underline{F}^\times$ attached to an embedding $\tau\in\mathrm{Hom}_{\mathrm{ring}}(F,\overline{\mathbb{Q}})$, 297
ξ^f	the character of $\underline{F}^\times$ attached to a $\mathbb{Z}$-valued function on $\mathrm{Hom}_{\mathrm{ring}}(F,\overline{\mathbb{Q}})$, 297
$E(\chi)$	the minimal subextension of F/L such that χ factors through $\mathrm{Nm}_{F/E(\chi)}$, 299
ι	a complex conjugation, 299–320
w_F	the weight cocharacter of the connected Serre group $\mathfrak{S}^F$, 300
F_{cm}	the subfield of F generated by all CM subfields of F, 301
$\mu(L)$	the group of all roots of 1 in a number field L, **305**
$\overline{\mathrm{Pic}}(\mathcal{O}_F)$	the replete divisor class group of $\mathcal{O}_F$, **308**
$F(\chi)$	the field of moduli of an admissible $\mathbb{Q}$-homomorphism χ, **310**
ν_τ	the geometric cocharacter of $\underline{F}^\times$ attached to an embedding $\tau: F\to\overline{\mathbb{Q}}$, 312
$E(\mu)$	the field of definition of a geometric cocharacter μ of a $\mathbb{Q}$-torus $\underline{K}^\times$, 315
$\chi(\mu)$	the homomorphism between Weil restricted tori attached to μ, 315

μ_Φ	the cocharacter of $\underline{L}^\times$ corresponding to a CM type Φ for a CM field L, 317
$\wp$	a finite place of a number field K above p, 313–318
v	a finite place of a CM field L above p, 313–318
$\underline{v}$	the place of the maximal totally real subfield L^+ in L below v, 313–318
Ψ_v	the set of all p-adic embeddings of L_v, 313–318
$\Psi_\wp$	the set of all p-adic embeddings of $K_\wp$, 313–318
I	identified with $\{1,2,3,\dots,4m\}$, 322–332, 333–342
J	the subset $\{2,4,6,\dots,4m\}$ of I, 322–342
J'	the subset $\{1,3,5,\dots,2m-1,2m+2,2m+4,\dots,4m\}$ of I, 322–342
$\mathcal{O}$	$W(\mathbb{F}_{p^{4m}})$, 321–342
(Y_0,β_0)	an $\mathcal{O}$-linear CM p-divisible group over $\overline{\mathbb{F}}_p$ with Lie type J, 322–342
ρ_0	an $\mathcal{O}$-linear isogeny from X_0 to Y_0 of degree p^m, 322–342
(X_0,γ_0)	an $\mathcal{O}$-linear CM p-divisible group over $\overline{\mathbb{F}}_p$ with Lie type J', 322–342
N, $\mathrm{D}_*(Y_0)$	the $\mathcal{O}$-liner Cartier module of (Y_0,β_0), 322–342
M, $\mathrm{D}_*(X_0)$	the $\mathcal{O}$-liner Cartier module of (X_0,γ_0), 324–342
G_0	$\ker(\rho_0)$, a finite group scheme over $\overline{\mathbb{F}}_p$ of order p^m 324–342
$H_{J'-\{4m\}}$	a subgroup scheme of X_0 isomorphic to α_p^{4m-1} with Lie type $J'-\{4m\}$, 327–342
H_S	the subgroup scheme of $H_{J'-\{4m\}}$ with Lie type $S\subset J'-\{4m\}$, 328–342
K_0	the fraction field of $W(\overline{\mathbb{F}}_p)$, 324–342
	the fraction field $W(k)[1/p]$ of $W(k)$, 360–370
K	the tame totally ramified extension field of K_0 of degree $p^{4m}-1$, 324–342
	a totally ramified finite extension field of $K_0=W(k)[1/p]$, 360–370
K_1/K_0	the subextension of K/K_0 such that $[K:K_1]=p-1$
$\mathcal{X}$, $\mathcal{X}_{\Lambda_0}$	the $\mathcal{O}$-linear lifting of X_0 over $W(\overline{\mathbb{F}}_p)$, 324–342, 336–342
$(\mathfrak{M}_0,\phi_{\mathfrak{M},0},\phi_{\mathfrak{M},0})$	the Kisin module for $\mathcal{X}[p]$ over $W(\overline{\mathbb{F}}_p)$, 336–342
$(\mathfrak{M},\phi_{\mathfrak{M}},\phi_{\mathfrak{M}})$	the Kisin module for the base change of $\mathcal{X}[p]$ to $\mathcal{O}_K$, 336–342
$\mathrm{E}_*(\cdot)$, $\mathrm{E}^*(\cdot)$	the covariant and contravariant crystal in crystalline Dieudonné theory, **354**–359
G	a finite commutative group scheme over the perfect base field k, 343–342
X	a p-divisible group over k, 343–342
$\mathrm{C}_*(X)$	the Cartier module of a smooth formal group X over k, **352**–359
$\mathrm{M}_*(G)$, $\mathrm{M}_*(X)$	covariant Dieudonné module for G and X, **354**–359
$\mathrm{M}^*(G)$, $\mathrm{M}^*(X)$	contravariant Dieudonné module for G and X, **354**–359
$\ell^G_\bullet$	the co-Lie complex of a finite group scheme G (over a perfect field k), **356**–359
n_G	$H^{-1}(\ell^G)$, **356**–359
ν_G	$H^1(\ell^{G^\vee})$, **356**–359
$E(u)$	the Eisenstein polynomial over K_0 for an local parameter π of K, 360–370
$\mathfrak{S}$	$W(k)[[u]]$, **334**–342, 360–370
$\mathfrak{S}_0$	$W(\overline{\mathbb{F}}_p)[[t]]$, **334**–342,
S	the p-adic completion of $W[u,E(u)^n/n!]_{n\in\mathbb{N}}$, **360**
$(\mathrm{Mod}/\mathfrak{S})^c$	the category of connected Kisin modules for finite flat group schemes over $\mathcal{O}_K$, **334**, 360
$(\mathrm{Mod}/\mathfrak{S}_0)^c$	the category of connected Kisin modules for finite flat group schemes over $W(\overline{\mathbb{F}}_p)$, **334**

$(p\text{-Gr}/k)^c$	the category of connected commutative finite group schemes of p-power order over k, **361**
$(\text{Mod}/D_k)^c$	Dieudonné modules for commutative finite groups schemes over k of local type, **361**
$(p\text{-div}/\mathcal{O}_K)^f$	the category of formal p-divisible groups over $\mathcal{O}_K$, **361**
$\text{BT}^{\phi,f}_{/\mathfrak{S}_0}$	the category of Kisin modules for formal p-divisible groups over $W(\overline{\mathbb{F}}_p)$, **334**
$\text{BT}^{\phi,f}_{/\mathfrak{S}}$	the category of Kisin modules for formal p-divisible groups over $\mathcal{O}_K$, **362**–370
(Win/S)	the category of S-windows, **362**–370
(Dsp/R)	the category of displays over a p-adic ring R, **362**–370

Bibliography

[1] A. A. Albert, *On the construction of Riemann matrices, I, II.* Ann. Math. **35** (1934), 1–28; **36** (1935), 376–394.

[2] A. A. Albert, *A solution of the principal problem in the theory of Riemann matrices.* Ann. Math. **35** (1934), 500–515.

[3] A. A. Albert, *Involutorial simple algebras and real Riemann matrices.* Ann. Math. **36** (1935), 886–964.

[4] E. Artin & J. Tate, *Class field theory.* AMS Chelsea Publishing, 2009.

[5] M. Artin, *Algebraization of formal moduli: I.* In *Global Analysis (Papers in honor of K. Kodaira)*, Univ. Tokyo Press, 1969.

[6] P. Berthelot, L. Breen, & W. Messing, *Théorie de Dieudonné cristalline II.* Lect. Notes Math. 930, Springer-Verlag, 1982.

[7] P. Berthelot, A. Grothendieck, & L. Illusie, *Théorie des intersections et théorème de Riemann-Roch.* Lect. Notes Math. 225, Springer-Verlag, 1971.

[8] P. Berthelot & A. Ogus, *F-isocrystals and de Rham cohomology I.* Inv. Math. **72** (1983), 159–199.

[9] S. Bosch, U. Güntzer, & R. Remmert, *Non-archimedean analysis.* Grundlehren Math. Wiss. 261, Springer-Verlag, 1983.

[10] S. Bosch, W. Lütkebohmert, & M. Raynaud, *Néron models.* Ergebnisse der Mathematik 21, Springer-Verlag, 1990.

[11] N. Bourbaki, *Algèbre, Chapitre 8, Modules et anneaux semi-simples.* Actualités Scientifiques et Industrielles 1261, Hermann, 1958.

[12] L. Breen, *Rapport sur la théorie de Dieudonné.* Astérisque **63** (1979), 39–66.

[13] C. Breuil, *Groupes p-divisibles, groupes finis et modules filtrés.* Ann. Math. **152** (2000), 489–549.

[14] P. Cartier, *Modules associés à un groupe formel commutatif. Courbes typiques.* C. R. Acad. Sci. **265** (1967), 129–132.

[15] J. Cassels, *Global fields.* In *Algebraic number theory* (J. Cassels & A. Fröhlich, ed.), Academic Press, 1967.

[16] C.-L. Chai & P. Norman, *Bad reduction of the Siegel moduli scheme of genus two with $\Gamma_0(p)$-level structure.* Amer. J. Math. **112** (1990), 1003–1071.

[17] C.-L. Chai & F. Oort, *Hypersymmetric abelian varieties.* Quat. J. Pure Appl. Math. **2** (Coates Special Issue) (2006), 1–27.

[18] C.-L. Chai & F. Oort, *Moduli of abelian varieties and p-divisible groups.* In *Arithmetic algebraic geometry*, (H. Darmon, D. A. Ellwood, B. Hassett & Y. Tschinkel, ed.), Clay Math. Proc. 8, 2009, 441–536.

[19] C. Chevalley, *Deux théorèmes d'arithmétique.* J. Math. Soc. Japan **3** (1951), 36–44.

[20] P. Colmez & J-M. Fontaine, *Construction des représentations p-adiques semi-stables.* Inv. Math. **140** (2000), 1–43.

[21] B. Conrad, *Finite-order automorphisms of a certain torus.* Michigan Math. J. **52** (2004), 423–434.

[22] B. Conrad, *Gross-Zagier revisited.* In *Heegner points and Rankin L-series*, MSRI Publ. 49, Cambridge Univ. Press, 2004.

[23] B. Conrad, *Chow's K/k-image and K/k-trace, and the Lang-Néron theorem.* Enseign. Math. (2) **52**, no. 1–2 (2006), 37–108.

[24] B. Conrad, *Deligne's notes on Nagata compactifications.* J. Ramanujan Math. Soc. **22** (2007), 205–257.

[25] B. Conrad, O. Gabber & G. Prasad, *Pseudo-reductive groups.* New Math. Monogr. 17, Cambridge Univ. Press, 2010.

[26] B. Conrad, *Lifting global representations with local properties.* Preprint, 2012.

[27] P. Deligne, *Application de la formule des traces aux sommes trigonométriques.* In *Cohomologie étales*, Séminaire de Géométrie Algébrique du Bois-Marie SGA4$\frac{1}{2}$, Lect. Notes Math. 569, Springer-Verlag 1977, 168–232.

[28] P. Deligne, *Motifs et groupes de Taniyama.* In *Hodge cycles, motives, and Shimura varieties*, Lect. Notes Math. 900, Springer-Verlag, 1982, 261–279.

[29] M. Demazure, *Lectures on p-divisible Groups.* Lect. Notes Math. 302, Springer-Verlag, 1972.

[30] M. Demazure & A. Grothendieck, *Schémas en groupes* I, II, III, Lect. Notes Math. 151, 152, 153, Springer-Verlag, 1970.

[31] M. Demazure & A. Grothendieck, *Schémas en groupes*, I, Documents mathématiques **7**, Soc. Math. France, 2011.

[32] M. Demazure & P. Gabriel, *Groupes algébriques.* Tome I: *Géométrie algébrique, généralités, groupes commutatifs.* Masson & Cie éditeur, North-Holland Publishing Company, 1970.

[33] M. Deuring, *Die Typen der Multiplikatorenringe elliptischer Funktionenkörper.* Abh. Math. Semin. Univ. Hambg. **14** (1941), 197–272.

[34] J. Dieudonné & A. Grothendieck, *Éléments de Géométrie Algébrique.* III. Publ. Math. IHES **11**, 1961.

[35] T. Ekedahl, *An effective version of Hilbert's irreducibility theorem.* Séminaire de Théorie des Nombres, Paris 1988–1989, Progr. Math. 91, Birkhäuser, 1990, 241–249.

[36] G. Faltings, *Finiteness theorems for abelian varieties over number fields.* In *Arithmetic Geometry* (G. Cornell & J. Silverman, ed.), Springer-Verlag, 1986, 9–27.

[37] G. Faltings, *Erratum Endlichkeitssätze für abelsche Varietäten über Zahlkörpern.* Inv. Math **75** (1984), 381.

[38] G. Faltings & C-L. Chai, *Degeneration of abelian varieties.* Ergeb. Math. Grenzgeb. 22, Springer-Verlag, 1990.

[39] B. Fantechi, L. Göttsche, L. Illusie, S. Kleiman, N. Nitsure, & A. Vistoli, *Fundamental algebraic geometry.* Math. Surveys Monogr. 123, Amer. Math. Soc., 2005.

[40] D. Ferrand, *Conducteur, descent, et pincement.* Bull. Soc. Math. France **131** (2003), 553–585.

[41] J-M. Fontaine, *Groupes p-divisible sur les corps locaux.* Astérisque **47–48** (1977).

[42] J-M. Fontaine, *Module galoisiens, modules filtrés et anneaux de Barsotti-Tate.* Astérisque **65** (1979), 3–80.

[43] J-M. Fontaine, *Sur certains types de représentations p-adiques du groupe de Galois d'un corps local; construction d'un anneau de Barsotti-Tate.* Ann. Math. **115**, 1982, 529–577.

[44] J-M. Fontaine, *Représentations p-adiques semi-stables.* Astérisque **223** (1994), 113–184.

[45] G. van der Geer & B. Moonen, *Abelian varieties.* In preparation.

[46] L. Gerritzen, *On multiplication algebras of Riemann matrices.* Math. Ann. **194** (1971), 109–122.

[47] W.-D. Geyer, *Galois groups of intersections of local fields.* Israel J. Math. **30** (1978), 382–396.

[48] A. Grothendieck, *Groupes de Barsotti-Tate et cristaux de Dieudonné.* Sém. Math. Supé. 45, Presses Univ. Montreal, 1970.

[49] M. Hazewinkel, *Formal groups and applications.* Academic Press, 1978.

[50] T. Honda, *Isogeny classes of abelian varieties over finite fields.* J. Math. Soc. Japan **20** (1968), 83–95.

[51] L. Illusie, *Déformation des groupes de Barsotti-Tate.* Astérisque **127** (1985), 151–198.

[52] A. Ivić, *The Riemann zeta-function: the theory of the Riemann zeta-function with applications.* John Wiley & Sons, 1985.

[53] N. Jacobson, *Basic algebra II.* W.H. Freeman & Co., 1989.

[54] A. J. de Jong, *Crystalline Dieudonné module theory via formal and rigid geometry.* Publ. Math. IHES **82**, 1995, 5–96.

[55] A. J. de Jong. *Homomorphisms of Barsotti-Tate groups and crystals in positive characteristic.* Inv. Math. **134** (1998), 301–333.

[56] A. J. de Jong & F. Oort, *Purity of the stratification by Newton polygons.* J. Amer. Math. Soc. **13** (2000), 209–241.

[57] N. Katz, *Serre–Tate local moduli.* Lect. Notes Math. 868, Springer-Verlag 1981, 138–202.

[58] M. Kisin, *Crystalline representations and F-crystals.* In *Algebraic geometry and number theory*, Progr. Math., 253, Birkhäuser, 2006, 459–496.

[59] M. Kisin, *Modularity of 2-adic Barsotti-Tate representations.* Inv. Math. **178** (2009), 587–634.

[60] D. Knutson, *Algebraic spaces*, Lect. Notes Math. 203, Springer-Verlag, 1971.

[61] A. Kresch, *Hodge-theoretic obstruction to the existence of quaternion algebras.* Bull. London Math. Soc. **35** (2003), 109–116.

[62] J. Kollár, *Quotients by finite equivalence relations.* Preprint, 2009.

[63] J. Lagarias, H. Montgomery & A. Odlyzko, *A bound for the least prime ideal in the Chebotarev density theorem.* Inv. Math. **54** (1979), 271–296.

[64] J. C. Lagarias & A. M. Odlyzko, *Effective versions of the Chebotarev density theorem*, in *Algebraic number fields* (A. Fröhlich, ed.), Academic Press, 1977, 409–464.

[65] S. Lang, *Complex multiplication.* Grundlehren Math. Wiss. 255, Springer-Verlag, 1983.

[66] S. Lang, *Elliptic functions* (2nd ed.). Graduate Texts in Math. 112, Springer-Verlag, 1987.

[67] S. Lang, *Algebra* (revised 3rd ed.). Graduate Texts in Math. 211, Springer-Verlag, 2002.

[68] E. Lau, *Displays and formal p-divisible groups*, Inv. Math. **171** (2008), 617–628.

[69] M. Lazard, *Commutative formal groups.* Lect. Notes Math. 443, Springer-Verlag, 1975.

[70] H. W. Lenstra Jr. & F. Oort, *Simple abelian varieties having a prescribed formal isogeny type.* J. Pure Appl. Algebra **4** (1974), 47–53.

[71] Y. Manin, *The theory of commutative formal groups over fields of finite characteristic.* Usp. Math. **18** (1963), 3–90; Russ. Math. Surveys **18** (1963), 1–80.

[72] H Matsumura, *Commutative algebra.* Benjamin, 1969; 2nd ed., 1980.

[73] H. Matsumura, *Commutative ring theory.* Cambridge Stud. Adv. Math. 8, Cambridge Univ. Press, 1989.

[74] B. Mazur & W. Messing, *Universal extensions and one-dimensional crystalline cohomology.* Lect. Notes Math. 370, Springer-Verlag, 1974.

[75] W. Messing, *The Crystals associated to Barsotti-Tate Groups: with applications to abelian schemes.* Lect. Notes Math. 264, Springer-Verlag, 1972.

[76] J. Milne, *Abelian varieties.* In *Arithmetic geometry* (G. Cornell & J. Silverman, ed.), Springer-Verlag, 1986.

[77] J. Milne, *Canonical models of (mixed) Shimura varieties and automorphic vector bundles.* In *Automorphic forms, Shimura varieties, and L-functions*, Vol. I, Perspect. Math. 10, Academic Press, 1990, 283–414.

[78] J. Milne & K.-Y. Shih, *Langlands' construction of the Taniyama group.* In *Hodge cycles, motives, and Shimura varieties.* Lect. Notes Math. 900, Springer-Verlag, 1982, 229–260.

[79] J. Milne & W. Waterhouse, *Abelian varieties over finite fields.* In *1969 number theory institute*, Proc. Sympos. Pure Math. 20, SUNY Stony Brook, NY, 1969), Amer. Math. Soc., 1971, 53–64.

[80] L. Moret-Bailly, *Pinceaux de variétés abéliennes.* Astérisque **129**, 1985.

[81] L. Moret-Bailly, *Construction de revêtements de courbes pointées.* J. Algebra **240** (2001), 505–534.

[82] D. Mumford, *Abelian varieties.* TIFR Stud. Math. 5, 1st ed. 1970; 2nd ed. TIFR and Oxford Univ. Press 1974; corrected reprint TIFR and Hindustan Book Agency 2008.

[83] D. Mumford, *Geometric invariant theory.* Ergeb. Math. Grensgeb. 34, 1st ed. Springer-Verlag, 1965; 3rd enlarged ed. by D. Mumford, J. Fogarthy & F. Kirwan, Springer-Verlag, 1994.

[84] J. Neukirch, *Algebraic number theory.* Grundlehren Math. Wiss. 322, Springer, 1999.

[85] P. Norman & F. Oort, *Moduli of abelian varieties.* Ann. Math., **112** (1980), 413–439.

[86] T. Oda *The first de Rham cohomology group and Dieudonné modules*, Ann. Sci. E.N.S. (4) **2** (1969), 63–135.

[87] F. Oort, *Commutative group schemes.* Lect. Notes in Math. 15, Springer-Verlag, 1966.

[88] F. Oort, *Finite group schemes, local moduli for abelian varieties, and lifting problems.* Compos. Math. **23** (1971), 265–296.

[89] F. Oort, *The isogeny class of a CM-type abelian variety is defined over a finite extension of the prime field.* J. Pure Appl. Algebra **3** (1973), 399–408.

[90] F. Oort, *Isogenies of formal groups.* Indag. Math., **37** (1975) 391–400.

[91] F. Oort, *Lifting algebraic curves, abelian varieties and their endomorphisms to characteristic zero.* In *Algebraic geometry, Bowdoin 1985* (S. J. Bloch, ed.), Proc. Sympos. Pure Math. 46, Part 2, Amer. Math. Soc. 1987, 165–195.

[92] F. Oort, *Endomorphism algebras of abelian varieties.* In *Algebraic geometry and commutative algebra: in honor of M. Nagata*, Vol. II, (H. Hijikata et al., ed.), Kinokuniya, 1988, 469–502.

[93] F. Oort, *CM-liftings of abelian varieties.* J. Alg. Geom. **1** (1992), 131–146.

[94] F. Oort, *Newton polygon strata in the moduli space of abelian varieties.* In *Moduli of abelian varieties* (C. Faber, G. van der Geer & F. Oort, ed.), Progr. Math. 195, Birkhäuser, 2001, 417–440.

[95] F. Oort, *Abelian varieties over finite fields.* Summer School on "Varieties over finite fields", Göttingen, 25-VI – 6-VII-2007. *Higher-dimensional geometry over finite fields*, Proc. NATO Adv. Stud. Inst. 2007 (Dmitry Kaledin & Yuri Tschinkel, ed.), IOS Press, 2008, 123–188.

[96] J. Oesterlé, *Versions effectives du théorème de Chetotarev sous l'hypothèse de Riemann généralisée*, Astérisque **61** (1979), 165–167.

[97] R. Pink, *The Mumford–Tate conjecture for Drinfeld modules.* Publ. RIMS, Kyoto Univ. **33** (1997), 393–425.

[98] M. Raynaud, *Passage au quotient par une relation d'équivalence plate.* 1967 Proc. Conf. Local Fields (Driebergen), Springer-Verlag, 78–85.

[99] M. Raynaud, *Schémas en groupes de type* $(p, \ldots, p)$. Bull. Soc. math. France **102** (1974), 241–280.

[100] D.S. Rim, *Formal deformation theory* In *Groupes de monodromie en géométrie algébrique. I.*, Lect. Notes Math. 288, Springer-Verlag, 1972, 32–132.

[101] D. Rydh, *Compactification of tame Deligne-Mumford stacks.* Preprint, 2012.

[102] M. Schlessinger, *Functors of Artin rings.* Trans. Amer. Math. Soc. **103** (1968), 208–222.

[103] J-P. Serre, *Local class field theory.* In *Algebraic number theory* (J. Cassels & A. Fröhlich, ed.). Academic Press, 1967, 129–161.

[104] J-P. Serre, *Complex multiplication.* In *Algebraic number theory* (J. Cassels & A. Fröhlich, ed.), Academic Press, 1967, 292–296.

[105] J-P. Serre, *Abelian* ℓ*-adic representations and elliptic curves.* W. A. Benjamin, 1968.

[106] J-P. Serre, *Linear representations of finite groups.* Grad. Texts in Math. 42, Springer-Verlag, 1977.

[107] J-P. Serre, *Local fields.* Grad. Texts in Math. 67, Springer-Verlag, 1979.

[108] J-P. Serre, *Groupes algébriques associés aux modules de Hodge-Tate.* Astérisque **65** (1979) 155–188.

[109] J-P. Serre & J. Tate, *Good reduction of abelian varieties.* Ann. Math. **88** (1965), 492–517.

[110] S. Shatz, *Group schemes, formal groups, and p-divisible groups.* In *Arithmetic geometry* (G. Cornell & J. Silverman, ed.), Springer-Verlag, 1986.

[111] G. Shimura & Y. Taniyama, *Complex multiplication of abelian varieties and its applications to number theory.* Publ. Math. Soc. Japan 6, 1961.

[112] G. Shimura, *On analytic families of polarized abelian varieties and automorphic functions.* Ann. Math. **78** (1963), 149–193.

[113] G. Shimura, *Introduction to the arithmetic theory of automorphic functions.* Princeton Univ. Press, 1971; reprinted 1994.

[114] G. Shimura, *On the zeta-function of an abelian variety with complex multiplication.* Ann. Math. **94** (1971), 504–533.

[115] G. Shimura, *Abelian varieties with complex multiplication and modular functions.* Princeton Univ. Press, 1998.

[116] L. Szpiro, *Degrés, intersections, hauteurs.* In Astérisque **127** (1985), 11–28.

[117] J. Tate, *Algebraic cycles and poles of zeta functions.* In *Arithmetical algebraic geometry* (Proc. Conf. Purdue Univ., 1963). Harper & Row, 1965, 93–110.

[118] J. Tate, *Endomorphisms of abelian varieties over finite fields.* Inv. Math. **2** (1966), 134–144.

[119] J. Tate, *p-divisible groups.* In *Proc. conference on local fields* (T. Springer, ed.), Springer-Verlag, 1967, 158–183.

[120] J. Tate, *Global class field theory.* In *Algebraic number theory* (J. Cassels & A. Fröhlich, ed.), Academic Press, 1967, 163–203.

[121] J. Tate, *Class d'isogenie des variétés abéliennes sur un corps fini* (d'après T. Honda). Séminaire Bourbaki, 1968/69, no. 352. Lect. Notes Math. 179, Springer-Verlag, 1971, 95–110.

[122] E. Titchmarsh & D. Heath-Brown, *The theory of the Riemann zeta-function*, 2nd ed., Clarendon Press, Oxford, 1986.

[123] J. Tate & F. Oort, *Group schemes of prime order.* Ann. Sc. Éc. Norm. Supé **3** (1970), 1–21.

[124] H. Umemura, *Formal moduli for p-divisible groups.* Nagoya Math. J. **42** (1971), 1–7.

[125] A. Vasiu & T. Zink, *Breuil's classification of p-divisible groups over regular local rings of arbitrary dimension.* In *Proceeding of Algebraic and Arithmetic Structures of Moduli Spaces, Hokkaido University, Sapporo, Japan, 2007*, Adv. Stud. Pure Math. 58 (2010), 461–479.

[126] W. Waterhouse, *Affine group schemes.* Grad. Texts in Math. 66 Springer-Verlag, 1979.

[127] A. Weil, *On a certain type of character of the idèle-class group of an algebraic number field.* In *Proc. intern. symp. on algebraic number theory*, Tokyo-Nikko, 1955, 1–7.

[128] A. Weil, *Basic number theory.* Grundlehren Math. 144, Springer-Verlag, 1967; 2nd ed. 1973.

[129] A. Wiles, *Modular elliptic curves and Fermat's Last Theorem.* Annals of Math. **141** (1995), 443–551.

[130] C.-F. Yu, *Lifting abelian varieties with additional structure.* Math. Z. **242** (2002), 427–441.

[131] C.-F. Yu, *On the supersingular locus in Hilbert-Blumenthal 4-folds.* J. Alg. Geom. **12** (2003), 653–698.

[132] C.-F. Yu, *On reduction of Hilbert-Blumenthal varieties.* Annales de l'Institut Fourier (Grenoble) **53** (2003), 2105–2154.

[133] C.-F. Yu, *The isomorphism classes of abelian varieties of CM-type.* J. Pure Appl. Algebra **187** (2004), 305–319.

[134] Y. Zarhin, *Abelian varieties in characteristic p.* Mat. Zametki **19** (1976), no. 3, 393–400.

[135] T. Zink, *Isogenieklassen von Punkten von Shimura-Mannigfaltigkeiten mit Werten in einem endlichen Körper.* Math. Nachr. **112** (1983), 103–124.

[136] T. Zink, *Cartiertheorie kommutative formaler Gruppen.* Teubner, 1984.

[137] T. Zink, *Windows for displays of p-divisible groups.* In *Moduli of Abelian Varieties*, Progr. Math. 195, Birkhäuser, 2001, 491–518.

[138] T. Zink, *On the slope filtration.* Duke Math. J. **109** (2001), 79–95.

[139] T. Zink, *The display of a formal p-divisible group.* *Astérisque* **278** (2002), 127–278.

Index

Selected Published Titles in This Series

195 **Ching-Li Chai, Brian Conrad, and Frans Oort,** Complex Multiplication and Lifting Problems, 2013
192 **Messoud Efendiev,** Attractors for Degenerate Parabolic Type Equations, 2013
191 **Grégory Berhuy and Frédérique Oggier,** An Introduction to Central Simple Algebras and Their Applications to Wireless Communication, 2013
190 **Aleksandr Pukhlikov,** Birationally Rigid Varieties, 2013
189 **Alberto Elduque and Mikhail Kochetov,** Gradings on Simple Lie Algebras, 2013
188 **David Lannes,** The Water Waves Problem, 2013
187 **Nassif Ghoussoub and Amir Moradifam,** Functional Inequalities: New Perspectives and New Applications, 2013
186 **Gregory Berkolaiko and Peter Kuchment,** Introduction to Quantum Graphs, 2013
185 **Patrick Iglesias-Zemmour,** Diffeology, 2013
184 **Frederick W. Gehring and Kari Hag,** The Ubiquitous Quasidisk, 2012
183 **Gershon Kresin and Vladimir Maz'ya,** Maximum Principles and Sharp Constants for Solutions of Elliptic and Parabolic Systems, 2012
182 **Neil A. Watson,** Introduction to Heat Potential Theory, 2012
181 **Graham J. Leuschke and Roger Wiegand,** Cohen-Macaulay Representations, 2012
180 **Martin W. Liebeck and Gary M. Seitz,** Unipotent and Nilpotent Classes in Simple Algebraic Groups and Lie Algebras, 2012
179 **Stephen D. Smith,** Subgroup complexes, 2011
178 **Helmut Brass and Knut Petras,** Quadrature Theory, 2011
177 **Alexei Myasnikov, Vladimir Shpilrain, and Alexander Ushakov,** Non-commutative Cryptography and Complexity of Group-theoretic Problems, 2011
176 **Peter E. Kloeden and Martin Rasmussen,** Nonautonomous Dynamical Systems, 2011
175 **Warwick de Launey and Dane Flannery,** Algebraic Design Theory, 2011
174 **Lawrence S. Levy and J. Chris Robson,** Hereditary Noetherian Prime Rings and Idealizers, 2011
173 **Sariel Har-Peled,** Geometric Approximation Algorithms, 2011
172 **Michael Aschbacher, Richard Lyons, Stephen D. Smith, and Ronald Solomon,** The Classification of Finite Simple Groups, 2011
171 **Leonid Pastur and Mariya Shcherbina,** Eigenvalue Distribution of Large Random Matrices, 2011
170 **Kevin Costello,** Renormalization and Effective Field Theory, 2011
169 **Robert R. Bruner and J. P. C. Greenlees,** Connective Real K-Theory of Finite Groups, 2010
168 **Michiel Hazewinkel, Nadiya Gubareni, and V. V. Kirichenko,** Algebras, Rings and Modules, 2010
167 **Michael Gekhtman, Michael Shapiro, and Alek Vainshtein,** Cluster Algebras and Poisson Geometry, 2010
166 **Kyung Bai Lee and Frank Raymond,** Seifert Fiberings, 2010
165 **Fuensanta Andreu-Vaillo, José M. Mazón, Julio D. Rossi, and J. Julián Toledo-Melero,** Nonlocal Diffusion Problems, 2010
164 **Vladimir I. Bogachev,** Differentiable Measures and the Malliavin Calculus, 2010
163 **Bennett Chow, Sun-Chin Chu, David Glickenstein, Christine Guenther, James Isenberg, Tom Ivey, Dan Knopf, Peng Lu, Feng Luo, and Lei Ni,** The Ricci Flow: Techniques and Applications: Part III: Geometric-Analytic Aspects, 2010

For a complete list of titles in this series, visit the
AMS Bookstore at **www.ams.org/bookstore/survseries/**.